Logic in Asia: Studia Logica Library

This book series promotes the advance of scientific research within the field of logic in Asian countries. It strengthens the collaboration between researchers based in Asia with researchers across the international scientific community and offers a platform for presenting the results of their collaborations. One of the most prominent features of contemporary logic is its interdisciplinary character, combining mathematics, philosophy, modern computer science, and even the cognitive and social sciences. The aim of this book series is to provide a forum for current logic research, reflecting this trend in the field's development.

The series accepts books on any topic concerning logic in the broadest sense, i.e., books on contemporary formal logic, its applications and its relations to other disciplines. It accepts monographs and thematically coherent volumes addressing important developments in logic and presenting significant contributions to logical research. In addition, research works on the history of logical ideas, especially on the traditions in China and India, are welcome contributions.

The scope of the book series includes but is not limited to the following:

- Monographs written by researchers in Asian countries.
- Proceedings of conferences held in Asia, or edited by Asian researchers.
- Anthologies edited by researchers in Asia.
- Research works by scholars from other regions of the world, which fit the goal of "Logic in Asia".

The series discourages the submission of manuscripts that contain reprints of previously published material and/or manuscripts that are less than 165 pages/ 90,000 words in length.

Please also visit our webpage: http://tsinghualogic.net/logic-in-asia/background/

Relation with Studia Logica Library

This series is part of the Studia Logica Library, and is also connected to the journal Studia Logica. This connection does not imply any dependence on the Editorial Office of Studia Logica in terms of editorial operations, though the series maintains cooperative ties to the journal.

This book series is also a sister series to Trends in Logic and Outstanding Contributions to Logic.

For inquiries and to submit proposals, authors can contact the editors-in-chief Fenrong Liu at fenrong@tsinghua.edu.cn or Hiroakira Ono at ono@jaist.ac.jp.

More information about this series at http://www.springer.com/series/13080

Syraya Chin-Mu Yang · Kok Yong Lee
Hiroakira Ono
Editors

Philosophical Logic: Current Trends in Asia

Proceedings of AWPL-TPLC 2016

Springer

Editors
Syraya Chin-Mu Yang
National Taiwan University
Taipei
Taiwan

Kok Yong Lee
National Chung Cheng University
Chiayi
Taiwan

Hiroakira Ono
Japan Advanced Institute of Science and Technology
Nomi
Japan

ISSN 2364-4613 ISSN 2364-4621 (electronic)
Logic in Asia: Studia Logica Library
ISBN 978-981-10-6354-1 ISBN 978-981-10-6355-8 (eBook)
DOI 10.1007/978-981-10-6355-8

Library of Congress Control Number: 2017950284

Printed on acid-free paper

This Springer imprint is published by Springer Nature
The registered company is Springer Nature Singapore Pte Ltd.
The registered company address is: 152 Beach Road, #21-01/04 Gateway East, Singapore 189721, Singapore

Preface

This new volume is a collection of selected papers presented at the Joint Conference of the Third Asian Workshop on Philosophical Logic and the Third Taiwan Philosophical Logic Colloquium (AWPL-TPLC 2016), held during 5–8 October 2016, at Department of Philosophy, National Taiwan University, Taiwan.

Ever since Gottlob Frege's *Begriffsschrift* (1879) and the pioneering works originating with the so-called new logic (to be compared with the then traditional Aristotelian syllogism), by George Boole, Georg Cantor, Giuseppe Peano, David Hilbert, Bertrand Russell and Alfred N. Whitehead, etc., logicians had concerned themselves with the establishment of the foundation of mathematics, which had led to various attempts to reduce arithmetic (and *a fortiori* mathematics) to logic. The new logic was thus entitled 'mathematical logic'. Russell during the 1910s and 1920s highlighted a more philosophical and methodological aspect of this new logic, suggesting that a logical theory not only builds up a solid foundation for mathematical reasoning, but also provides a comfortable framework for analysis of the nature of mathematical propositions and concepts. Basic concepts in mathematics and controversial issues regarding the foundation of mathematics can thus be explicitly articulated in a logical theory. Russell maintained that this methodology can be applied to philosophical propositions in general—the gist of his well-known logical atomism is that 'logic is fundamental in philosophy'.

It was soon realized that the framework of the new style of logical theory, taken as the orthodox or paradigm in the Fregean tradition, may not be comprehensible enough to deal with all kinds of philosophical propositions. It was well observed that certain types of propositions such as propositions concerning future events seemingly defy the basic logical laws of the orthodox logic, which suggests that the logic should be modified or revised. A variety of logical theories alternative to, or deviant from, the orthodox logic were proposed, such as Jan Łukasiewicz's three-valued propositional calculus (for future contingent propositions around 1917), Clarence Irving Lewis's strict implication calculus (for propositions with necessity/possibility, during the 1910s–1920s), Arend Heyting's intuitionistic logic (for a formal basis for Brouwer's intuitionism in the foundation of mathematics in 1930). In a more general sense, they were treated as the rivals of the so-called new logic, and gained the label 'non-classical logics', leaving the term 'classical logic' to the Fregean orthodox logic.

The flourishing of non-classical logics after late 1960s has proved that substantial analyses of certain philosophical propositions and the related concepts can be obtained based on the formulation of the correspondent non-classical logic and its semantical framework, such as modal logics for necessity/possibility and epistemic logic for knowledge/beliefs, to mention a few. This goal can be achieved typically by virtue of structural analyses of the models involved and aximatization, or theorization, of a group of related philosophical doctrines. The impact that non-classical logics have on philosophy is manifested by the fact that the term 'philosophical logic' has been widely used to cover the study of non-classical logics and the study of philosophical topics and concepts (typically, identity, existence, truth, predication, meaning, modality, etc.) that can be dealt with in the framework of non-classical logic. *The Journal of Philosophical Logic*, arguably the first journal mainly devoted to original works in philosophical logic, came into being in 1972 (the first volume of *The Journal of Symbolic Logic*, by comparison, was in 1936). And the first edition of *the Handbook of Philosophical Logic* (four volumes) was published during 1983–1989 (in comparison to the one volume *the Handbook of Mathematical Logic*, edited by Jon Barwise, North Holland, 1977). The second edition of *the Handbook of Philosophical Logic* has become an open-ended project, giving rise to an unlimited series of collection, and since 2001, 17 volumes have been published. This new edition of the Handbook has marked dramatic changes in the landscape of philosophical logic: New areas have been included by the establishment of a large number of new logics; while old areas were significantly enriched and expanded.

Since 2000, more and more logicians and philosophers in Asia area have paid great attention to philosophical logic. To promote mutual understanding and collaboration for future researchers in Asian on philosophical logic, a series of biennial conferences—*Asian Workshop on Philosophical Logic* (AWPLs), was initiated by Professor Hiroakira Ono (Japan Advanced Institute of Science and Technology, JAIST, Japan) and some others. *The First Asian Workshop on Philosophical Logic* (AWPL-2012) was held at JAIST in February 2012, followed by *The Second Asian Workshop on Philosophical Logic* (AWPL-2014) in April 2014 at Sun Yat-sen University, Guangzhou, China. A post-conference proceedings of AWPL-2014, *Modality, Semantics and Interpretations* (Ju Shier, Liu Hu and Hiroakira Ono, eds. 2015) was published as the first volume of the then newly established LIAA-book series ('Logic in Asia'), a subseries of Studia Logica Library, Springer. Almost at the same time, The Taiwan Philosophical Logic Colloquium (TPLCs), another series of biennial conferences in philosophical logic, hosted by the Department of Philosophy, National Taiwan University, Taiwan, with funding from the private sector, was established. A post-conference proceedings of TPLC-2014, *Structural Analysis of Non-classical Logics* (Syraya Chin-Mu Yang, Duen-Min Deng and Hanti Lin, eds.), was published as the second volume of LIAA-book series by Springer in 2015.

The aim of AWPLs and TPLCs is to provide a forum for dialogues amongst logic-minded philosophers and philosophically oriented logicians. The scope of AWPLs and TPLCs covers philosophical logic (in a broad sense), non-classical logics, algebraic logic, mereology, their applications to computer science, cognitive science, linguistics, game theory, and other social sciences, etc., and all kinds of semantics/logics relating to philosophical concepts and their applications in philosophical issues. It is dedicated to promoting both theoretical and empirical studies of logic (typically non-classical logics), with a close connection to disciplines that draw on diverse methods and approaches from philosophy, computer science, mathematics, psychology and linguistics.

The present volume, together with the two aforementioned volumes, *Modality, Semantics and Interpretations* and *Structural Analysis of Non-Classical Logics*, intends to flag a significant portion of the landscape of the development of philosophical logic in Asia at the early twenty-first century. We hope that these volumes can provide a useful and representative survey of the main fields to which distinguished logicians and philosophers in Asia, and perhaps in Australasian regions, have devoted their research.

In the opening chapter, "Representing and Completing Lattices by Propositions of Cover Systems," Robert Goldblatt continues his project of developing a theory of cover systems that encompasses non-distributive logics. In previous papers, he has used this to provide structural semantics for the logic of residuated ordered semigroups and quantales, intuitionistic modal logics, classical bilinear logic, relevant logic and the storage and consumption modalities of linear logic. The present article studies cover systems on the set of principal filters of a lattice and their role in lattice representations, obtaining presentations of the ideal completion and the MacNeille completion of the original lattice as lattices of propositions of a cover system. This is explored further for ortholattices, and for Heyting algebras in relation to Grothendieck topologies.

Hiroakira Ono's Chapter "A Uniform Algebraic Approach to Cut Elimination Via Semi-Completeness" introduces an algebraic condition 'semi-completeness' of sequent systems (due originally to Shoji Maehara, 1991). It is proved that the semi-completeness of a given system S without cut rule gives a sufficient condition of eliminating cut. Moreover, many of existing semantical proofs of cut elimination, using either Kripke semantics or algebraic one, can be naturally transformed into algebraic proofs of semi-completeness. The author concludes that semi-completeness offers a uniform algebraic way of understanding cut elimination, of both single- and multiple-succedent sequent systems, for a wide variety of non-classical propositional and predicate logics as well.

Chapter "Ancient Indian Logic, Pakṣa and Analogy" (by Jeffrey Paris and Alena Vencovská), provides a formalization which intends to capture the suggestion of B.K. Matilal, and earlier J.F. Staal, that the Indian Schema from Gotama's Nyāya–Sūtra should be understood in terms of an 'occurrence' relation linking events to their loci. It goes on to show that in consequence the Schema inherits a rational justification as analogical reasoning within Unary Pure Inductive Logic from the widely accepted principle of Atom Exchangeability, itself a property of the Carnap's Continuum of Inductive Methods.

Shih Ping Tung in Chapter "Provability and Decidability of Arithmetical Sentences", proves several results about the decidability, axiomatizability, recursive enumerability, etc., of certain rather basic families of arithmetic sentences. More specifically, it shows that the sets of all sentences of the form $\forall z\exists x\forall y\, f(x,y) - az \neq 0$, where $f(x,y) \in \mathbb{Z}[x, y]$ and $a \in \mathbb{Z}$, true in $\mathbb{N}$, $\mathbb{Z}$, and $\mathbb{Q}$ are axiomatizable, respectively. It follows that the sets of all sentences of the form $\exists z\forall x\exists y\, f(x,y) - az = 0$ true in $\mathbb{N}$, $\mathbb{Z}$ and $\mathbb{Q}$ are decidable, respectively. These results use earlier, seemingly more ground-breaking, work of the author's (from over 30 years ago) but nevertheless make a worthwhile contribution to the subject which may pave a new path to future work of researchers in this field.

In Chapter "On the Minimization Principle in the Boolean Approach to Causal Discovery", Jiji Zhang examines a Boolean approach to causal inference, which is rooted in John Mackie's celebrated INUS theory of causation and has been developed into several sophisticated data analysis methods for social scientists. The target of Zhang's criticism is the minimization steps in a most recent implementation of this approach, known as the method of Coincidence Analysis (CNA). Zhang presents prima facie counterexamples to the soundness of the minimization steps in CNA and discusses two possible responses to them. The author then argues that while one of the responses is viable, it renders the role of minimization much less substantial than it is usually intended to be.

Chapter "Contentual and Formal Aspects of Gentzen's Consistency Proofs", (by Ryota Akiyoshi and Yuta Takahashi) offers an analysis of Gentzen's second consistency proof for first-order arithmetic in 1936. Wilfried Sieg has recently highlighted two distinct notions of consistency proofs in Gentzen's series of work then: contentual correctness proofs (semantic manner in character, typically Gentzen's consistency proof in 1935); and formal correctness proofs (substantially a proof-theoretic approach, as shown in a 1938 paper). The authors show that Gentzen's 1936 proof is both contentual and formal. The connection between the contentual aspect of this proof and its formal aspect is specified and some consequences are noted.

Hao-Cheng Fu in Chapter "Saving Supervaluationism from the Challenge of Higher-Order Vagueness Argument" proposes a revised version of supervaluationism which could govern the puzzle of vagueness. Some plausible solutions to the problem of vagueness are reviewed, and shown to fail due to the phenomenon of higher order vagueness. Fu further argues that the proposed supervaluationism could be appealing if we can construct a dynamic model instead of the static model of supervaluationism.

Chapter "Cut Free Labelled Sequent Calculus for Dynamic Logic of Relation Changers", (by Ryo Hatano, Katsuhiko Sano, and Satoshi Tojo), provides a cut-free labelled sequent calculus **GDLRC** for van Benthem and Liu's dynamic logic of relation changers (**DLRC**, 2007), a variant of dynamic epistemic logic (**DEL**) that provides a general framework to capture many dynamic operators of **DEL** in terms of relation changing operation written by programs in propositional dynamic logic

(**PDL**). In contrast, proof theory for **DLRC** has not been well-studied except for a Hilbert-style axiomatization proposed in van Benthem and Liu's work. The authors further show that **GDLRC** is equipollent with the aforementioned Hilbert-style axiomatization.

Ryo Kashima in Chapter "On Second Order Propositional Intuitionistic Logics" studies two second-order propositional intuitionistic logics: the first one, with the full comprehension axiom; the other, including the constant domain axiom. The completeness theorems for these two logics with respect to corresponding Kripke models were proved by Sobolev (1977) and Gabbay (1974), respectively. Kashima here offers some slightly strong alternative proofs, using the technique of nested sequent calculi, and consider a closure condition on Kripke models, namely 'the domain of quantification is closed under any operation that is induced by a formula'. The author shows that this condition depends on the characterization of disjunction because at propositional level, a disjunction of the form '$A \vee B$', taking '$\vee$' as primitive, and the formula (for all x)$((A \rightarrow x) \rightarrow ((B \rightarrow x) \rightarrow x))$ induce different operations, whereas at second order level, $A \vee B$ can be defined by (for all x)$((A \rightarrow x) \rightarrow ((B \rightarrow x) \rightarrow x))$. Noticeably, this difference is critical in the argument on the constant domain condition in that if the language does not contain disjunction as primitive, the constant domain axiom is not required for the completeness with respect to constant domain models.

In Chapter "Classical Model Existence Theorem in Subclassical Predicate Logics. II", Jui-Lin Lee shows that there are some much weaker logics satisfying the classical model existence property (*CME*)—every consistent set has a classical model. By using weak deduction theorem, in propositional logics, Lee improves previous results and shows that some weak extensions of *BCI*/*BCIW* logic satisfy *CME*. Glivenko's Theorem for corresponding logics is also proved. In predicate logics, under such a weak propositional logic part, Lee uses the Herbrand-Henkin style approach (via prenex normal form theorem) and also the Hintikka style approach to construct weaker subclassical predicate logics which satisfy *CME*.

Ren–June Wang's Chapter "On Incorporating Reasoning Time into Epistemic Logic", introduces the notion of *reasoning-based knowledge*, a concept of knowledge taking reasoning time into account, in contrast with the *information-based knowledge*, which is normally formulated by the plain possible world semantics. Two formal systems, $\mathbf{tMEL}^{\mathbf{K}}$ and $\mathbf{tMEL}^{\infty}$, are proposed, with each of them having a device representing the information-based knowledge and reasoning-based knowledge, respectively. The author further applies $\mathbf{tMEL}^{\infty}$ to investigate the epistemic valid formula $\sim K(p \mathbin{\&} \sim Kp)$, with a detour to discuss the Moore's paradox from the reasoning-based knowledge perspective.

Sakiko Yamasaki and Katsuhiko Sano in Chapter "Proof-Theoretic Embedding from Visser's Basic Propositional Logic to Modal Logic K4 via Non-Labelled Sequent Calculi" employ **G3**-style *non-labelled* sequent calculi to establish a proof-theoretic embedding from Visser's Basic Propositional Logic **BPL** into modal logic **K4** via a variant of Gödel–McKinsey–Tarski translation sending an atom P to $P \mathbin{\&} \Box P$, where the logic **BPL** is obtained by dropping the requirement of reflexivity from Kripke semantics for intuitionistic logic. The authors first provide for **BPL** a **G3**-style non-labelled sequent calculus **G3BPL**, which enjoys cut elimination theorem and is proved to be sound and complete for intended Kripke semantics. Then they establish a proof-theoretic embedding where the above translation plays a key role in the proof for the direction of faithfulness, i.e. if the translation of a formula of **BPL** is provable in **K4** then the original formula is provable in **BPL**. By this proof-theoretic embedding, the authors also provide another syntactic proof of cut elimination theorem for **G3BPL** by reducing the admissibility of cut of **G3BPL** to that of a **G3**-style non-labelled sequent calculus for **K4**.

It is noteworthy that AWPL-TPLC 2016 organized two special workshops. The first one is a one-day plus workshop on Williamson's philosophy to honour Professor Timothy Williamson (Wykeham Professor of Logic, University of Oxford) for his outstanding contribution to philosophical logic (typically on identity, vagueness, knowledge first epistemology and metaphysics of modality). We regret that for the sake of the limitation of the space, papers based on talks at this workshop could not be included in this volume. A second workshop on mereology was organized to signify certain philosophical aspect of mereology. The remaining two chapters of this volume come out from presentations at this workshop.

In Chapter "Varieties of Parthood", Paul Hovda addresses formal patterns that emerge on the view that there are a variety of parthood relations, and that certain general principles govern these relations and connect them to one another. One such principle is shown to connect with the strong supplementation principle of classical mereology, and also to the classical notion of mereological fusion. The main line of development shows roughly that, for any given variety of parthood, relating objects in a given domain, for any larger domain that includes the given one, and where the new objects may have the old objects as parts in a new manner, then, on one general way of being a part, there is a 'unique' possible structure resulting, itself intimately related to classical mereology. The uniqueness is measured relative to a notion like the standard notion of isomorphism, but weaker, which we call 'quasi-isomorphism.'

In the final Chapter, "Infinite 'Atomic' Mereological Structures", Hsing–Chien Tsai shows that the strongest first-order atomic mereological theory which can be generated by axioms found in the literature, that is, General Extensional Mereology with the atomicity axiom, as well as a natural second-order extension of such a theory, that is, Classical Mereology with the atomicity axiom, still cannot secure an atomic domain, where a domain is atomic if it contains a collection of 'atoms' (an atom is something which has no proper parts) and each of its members is composed of some atoms and nothing else. These are further results of what have been done earlier by the author.

We owe thanks to all authors, the anonymous referees of the manuscripts, members of the organizing committee, participants in the joint conference, including all speakers, discussants, attendees and the staff of the Department of Philosophy, National Taiwan University. In particular, we would like to express our gratitude to Timothy Williamson, Jonardon Ganeri, Greg Restall, James Woodward, Nicholas J.J. Smith, Churn Jung Liau, Ma Minghui, Tatsuya Shimura,

Duen-Min Deng, Max Deutsch, Masashi Kasaki, Wai Chun Leong, Hanti Lin, Dan Marshall, Joshua Sack, Nicholas Silins, Satoru Suzuki, Yanjing Wang, Kai–Yee Wong, Yue Yang, Chi-Her Yang, for their contribution to AWPL-TPLC 2016 and to this volume. We are deeply indebted to Fenrong Liu, Hiroakira Ono and Rob Goldblatt for their long-term support for both AWPLs and TPLCs and the preparation of this volume. Thanks also go to Leana Li, Team Leader of Editor Human Sciences & Mathematics for her contribution to this volume. Finally and above all, we owe special thanks to Miss Wendy Huang. Without her exclusively financial support for the TPLCs and AWPL-TPLC 2016, this collection can only be a certain non-chunky object in this world.

Taipei, Taiwan — Syraya Chin-Mu Yang
Chiayi, Taiwan — Kok Yong Lee
Nomi, Japan — Hiroakira Ono
June 2017

References

Maehara, S. (1991). Lattice-valued representation of the cut elimination theorem. *Tsukuba Journal of Mathematics*. 15, 509–521.

Gabbay, Dov M. (1974). On 2nd order intuitionistic propositional calculus with full comprehension. *Archive for Mathematical Logic*. 16, 177–186.

Sobolev, S. K. (1977). The intuitionistic propositional calculus with quantifiers. *Mathematical Notes*. 22, 528–532. (English version of Mat. Zametki 22 (1977), 69–76).

Contents

Contributors

Ryota Akiyoshi Waseda Institute for Advanced Study, Tokyo, Japan

Hao-Cheng Fu Department of Philosophy, Chinese Culture University, Taipei, Taiwan

Robert Goldblatt School of Mathematics and Statistics/Centre for Logic, Language and Computation, Victoria University of Wellington, Wellington, New Zealand

Ryo Hatano Faculty of Science and Technology, Tokyo University of Science, Noda, Japan

Paul Hovda Department of Philosophy, Reed College, Portland, OR, USA

Ryo Kashima Department of Mathematical and Computing Science, Tokyo Institute of Technology, Tokyo, Japan

Jui-Lin Lee Center for General Education and Department of CSIE, National Formosa University, Yunlin, Taiwan

Hiroakira Ono Japan Advanced Institute of Science and Technology, Nomi, Japan

Jeffrey B. Paris School of Mathematics, The University of Manchester, Manchester, England

Katsuhiko Sano Graduate School of Letters, Hokkaido University, Hokkaido, Japan

Yuta Takahashi Research Fellow of Japan Society for the Promotion of Science, Nagoya University, Nagoya, Japan

Satoshi Tojo School of Information Science, Japan Advanced Institute of Science and Technology, Nomi, Japan

Hsing-Chien Tsai Department of Philosophy, National Chung-Cheng University, Min-Hsiung, Chia-Yi, Taiwan

Shih Ping Tung Department of Applied Mathematics, Chung Yuan Christian University, Taoyuan, Taiwan

Alena Vencovská School of Mathematics, The University of Manchester, Manchester, England

Ren-June Wang Department of Philosophy, National Chung-Cheng University, Chiayi County, Taiwan

Sakiko Yamasaki Graduate School of Humanities, Tokyo Metropolitan University, Tokyo, Japan

Jiji Zhang Department of Philosophy, Lingnan University, Tuen Mun, Hong Kong

Representing and Completing Lattices by Propositions of Cover Systems

Robert Goldblatt

Abstract Cover systems abstract from the properties of open covers in topology, and have been used to construct lattices of propositions for various modal and non-modal substructural logics. Here we explore cover systems on the set of principal filters of a lattice and their role in lattice representations. A particular system with finite covers gives a lattice of propositions isomorphic to the ideal completion of the original lattice. Relaxing the finiteness condition yields another cover system whose lattice of propositions gives a presentation of the MacNeille completion. This is analysed further for ortholattices. For Heyting algebras a stricter cover relation is shown to have the properties of a Grothendieck topology while its lattice of propositions is still the MacNeille completion.

1 Introduction

In several previous articles the author has used the notion of a *cover system* to develop structural semantics for propositional and quantifed versions of various logics. These include the logic of residuated ordered semigroups and quantales (Goldblatt 2006), intuitionistic modal logics (Goldblatt 2011), classical bilinear logic (Goldblatt 2011), relevant logic (Goldblatt 2011, Chap. 6), and the modalities ! and ? of linear logic (Goldblatt 2016).

Cover systems abstract from the properties of open covers in topology, and in a categorical setting have led to the development of the Kripke-Joyal semantics for intuitionistic logic in topoi (Mac Lane and Moerdijk 1992). This diverges from Kripke's own semantics by giving a non-classical interpretation to disjunction while still validating the distribution of conjunction over disjunction. In some of the above-mentioned articles we adapted cover systems to model logics that lack this distribution.

R. Goldblatt (✉)
School of Mathematics and Statistics/Centre for Logic, Language and Computation, Victoria University of Wellington, Wellington, New Zealand
e-mail: rob.goldblatt@msor.vuw.ac.nz

S.C.-M. Yang et al. (eds.), *Philosophical Logic: Current Trends in Asia*,
Logic in Asia: Studia Logica Library, DOI 10.1007/978-981-10-6355-8_1

The present article explores the use of cover systems to construct representations of lattices in general, including non-distributive ones. In the early days of lattice theory, Birkhoff and Frink (1948) defined a *meet-representation* of a lattice L as a map $a \mapsto |a|$ from elements a of L to certain sets $|a|$ which "sends lattice meets into set-intersections", i.e. $|a \sqcap b| = |a| \cap |b|$ where $\sqcap$ denotes meet and $\cap$ denotes intersection. They observed that any lattice can be given a faithful meet-representation by taking $|a|$ to be set of all filters of L containing a, or just the set of all principal filters containing a. Here we focus on the set S_L of principal filters. We construct a cover system that defines a closure operator j on subset of S_L such that a meet-representation results in which lattice joins are sent to *closures* of set-unions, i.e. $|a \sqcup b| = j(|a| \cup |b|)$.

So in this context the basic role of cover systems is to provide a way of defining closure operators. The lattice L is isomorphically embedded into a *complete* lattice of fixed points of the closure operator j. We call these fixed points *propositions* because of their role in modelling formulas of logical systems. The lattice *Prop*($\mathscr{S}_L$) of propositions is a *completion* of L: a complete lattice having a sublattice isomorphic to L. The first cover system we define has finite covers and makes *Prop*($\mathscr{S}_L$) isomorphic to the ideal completion of L. Dropping the finiteness condition produces another cover system whose lattice of propositions is a presentation of the MacNeille completion of L. We then analyse this situation further when L is an ortholattice or a Heyting algebra.

Here is an outline of the paper. After setting out the basic theory of closure operators and cover systems in Sect. 2, we proceed in Sect. 3 to define a cover system on S_L whose covers are finite and show that L embeds isomorphically into the resulting *Prop*($\mathscr{S}_L$). In Sect. 4 we show that if L is distributive then $\mathscr{S}_L$ has a property that forces *Prop*($\mathscr{S}_L$) to be a complete Heyting algebra. Then in Sect. 5 we show that if L itself is a Heyting algebra, then the embedding into *Prop*($\mathscr{S}_L$) preserves the Heyting implication operation. Section 6 proves that *Prop*($\mathscr{S}_L$) is isomorphic to the ideal completion of L, i.e. the complete lattice of all ideals of L. In Sect. 7 a different cover system $\mathscr{S}^m_L$ is defined by dropping the finiteness restriction on covers. It is shown that the resulting lattice of propositions *Prop*($\mathscr{S}^m_L$) is isomorphic to the MacNeille completion of L. Section 8 considers the case that L is an ortholattice and reviews the embedding of L into a complete lattice *Orth*$(S_L, \perp_L)$ of "bi-orthogonally closed" sets defined using an orthogonality relation $\perp_L$ on S_L. We show that *Orth*$(S_L, \perp_L) =$ *Prop*($\mathscr{S}^m_L$). Finally, in Sect. 9 we return to the Heyting algebra case to define a more restricted cover system $\mathscr{S}^G_L$ that has additional properties making it a Grothendieck topology, but defines the same lattice of propositions as $\mathscr{S}^m_L$.

2 Closure Operators and Cover Systems

Let $(L, \sqsubseteq)$ be a *poset*, comprising a partial ordering $\sqsubseteq$ on a non-empty set L. We write $\bigsqcup X$ for the *join* (= least upper bound), and $\bigsqcap X$ for the *meet* (= greatest lower bound) of a set $X \subseteq L$, when these bounds exist. The smaller symbols $\sqcup$ and $\sqcap$ are used for the binary join and meet operations. The poset is a *lattice* if $a \sqcap b$ and $a \sqcup b$

exist for every $a, b \in L$, and so $\bigsqcup X$ and $\bigsqcap X$ exist for every finite non-empty X. A poset is *order-complete*, or just *complete*, if every subset has a meet, or equivalently if every subset has a join.

A function $j : L \to L$ is defined to be

- *monotone* if it is order-preserving, i.e. $a \sqsubseteq b$ implies $ja \sqsubseteq jb$;
- *inflationary* if $a \sqsubseteq ja$, for all $a \in L$;
- *idempotent* if $j(ja) = ja$, for all $a \in L$;
- a *closure operator* if it is monotone, inflationary and idempotent.

An element $a \in L$ is called *j-closed* if $ja = a$, i.e. if a is a *fixed point* under j. If $(L, \sqsubseteq)$ is a *complete* poset and j is a closure operator, the set L^j of j-closed elements is closed under meets $\bigsqcap X$, and so is order-complete under the same partial ordering and with same meet operation. The join operation $\bigsqcup^j$ in $(L^j, \sqsubseteq)$ is given by $\bigsqcup^j X = j(\bigsqcup X)$ for all $X \subseteq L^j$.

Our aim is to explore ways in which a given lattice can be represented as a lattice of j-closed elements for a closure operator j on some complete lattice $(L, \sqsubseteq)$ whose elements are *sets* and in which $\sqsubseteq$ is the set inclusion relation $\subseteq$, the meet $\bigsqcap \Theta$ is the set-intersection $\bigcap \Theta$, and the join $\bigsqcup \Theta$ is set-union $\bigcup \Theta$; hence in $(L^j, \subseteq)$ the meet will be $\bigcap \Theta$ and the join $j(\bigcup \Theta)$. In pursuit of that aim we consider structures of the form

$$\mathscr{S} = (S, \preccurlyeq, \lhd, \dots)$$

that fulfil the following description:

- $\preccurlyeq$ is a binary relation on set S that is a *preorder*, i.e. reflexive and transitive. We may write $y \succcurlyeq x$ when $x \preccurlyeq y$, and say that y *refines* x.
- $\lhd$ is a binary relation from S to its powerset $\mathscr{P}S$. When $x \lhd C$, where $x \in S$ and $C \subseteq S$, we say that x *is covered by* C, and write this also as $C \rhd x$, saying that C *covers* x or that C is an *x-cover*.

In such a structure, an *up-set* is a subset X of S that is closed upwardly under refinement: $y \succcurlyeq x \in X$ implies $y \in X$. For each $X \subseteq S$,

$$\uparrow X = \{y \in S : (\exists x \in X)\, x \preccurlyeq y\}$$

is the smallest up-set including X. For $x \in S$, we write $\uparrow x$ for $\uparrow\{x\} = \{y : x \preccurlyeq y\}$, the smallest up-set containing x.

The collection $Up(\mathscr{S})$ of all up-sets of $\mathscr{S}$ is a complete poset under the partial order $\subseteq$ of set inclusion, with the meet and join of any collection Θ of up-sets being its set intersection $\bigcap \Theta$ and union $\bigcup \Theta$, respectively. $Up(\mathscr{S})$ has least element $\emptyset$ and greatest element S.

Now for each subset X of S, define

$$j_{\lhd} X = \{x \in S : \exists C\, (x \lhd C \subseteq X)\}. \tag{1}$$

A property can be thought of as being *locally true* of x if x is covered by a set whose members that have this property, i.e. if there is some C such that $x \lhd C$ and each member of C has the property. In this sense, x belongs to $j_{\lhd}X$ just when the property of *being a member of* X is locally true of x. So $j_{\lhd}X$ can be thought of as the collection of "local members" of X. X is called *localised* if $j_{\lhd}X \subseteq X$, i.e. if every local member of X is an actual member of X.

The operation $j_{\lhd}$ is $\subseteq$-monotonic on subsets of S: $X \subseteq Y$ implies $j_{\lhd}X \subseteq j_{\lhd}Y$. To discuss further properties of $j_{\lhd}$ we will say that a subset Y of S *refines* a subset X, or X *is refined by* Y, if $Y \subseteq \uparrow X$, i.e. if every member of Y refines some member of X. Consider the following possible axioms, for an arbitrary $x \in S$:

- *Refinement*: if $x \preccurlyeq y$, then every x-cover can be refined to a y-cover, i.e. if $C \rhd x$, then there exists a C' such that $y \lhd C' \subseteq \uparrow C$.
- *Existence*: there exists an x-cover $C \subseteq \uparrow x$;
- *Refined Transitivity*: if $x \lhd C$ and for all $y \in C$, $y \lhd C_y$, then $\bigcup_{y\in C} C_y$ can be refined to an x-cover, i.e. there exists a C' such that $x \lhd C'$ and $C' \subseteq \uparrow \bigcup_{y\in C} C_y$.

Theorem 1 *(1) $\mathscr{S}$ satisfies the Refinement axiom iff $j_{\lhd}X$ is an up-set whenever X is an up-set.*
(2) $\mathscr{S}$ satisfies the Existence axiom iff $j_{\lhd}$ is inflationary on $Up(\mathscr{S})$.
(3) $\mathscr{S}$ satisfies Refined Transitivity iff $j_{\lhd}j_{\lhd}X \subseteq j_{\lhd}X$ for all up-sets X.

Proof (1) Let $\mathscr{S}$ satisfy Refinement. If X is an up-set and $y \succcurlyeq x \in j_{\lhd}X$, then there is an x-cover $C \subseteq X$, and this can be refined to y-cover $C' \subseteq \uparrow C \subseteq \uparrow X = X$. Hence $y \in j_{\lhd}X$. This shows that $j_{\lhd}X$ is an up-set.
Conversely, suppose $j_{\lhd}$ maps $Up(\mathscr{S})$ into itself. Let $C \rhd x$ and $x \preccurlyeq y$. Since $C \subseteq \uparrow C$ it follows that $x \in j_{\lhd}\uparrow C$. But $\uparrow C$ is an up-set, hence so is $j_{\lhd}\uparrow C$ by hypothesis, and therefore $y \in j_{\lhd}\uparrow C$. This means that there exists a y-cover $C' \subseteq \uparrow C$, confirming that the Refinement axiom holds.

(2) Let $\mathscr{S}$ satisfy Existence. Then if X is an up-set and $x \in X$, there is a C with $x \lhd C \subseteq \uparrow x \subseteq X$. Hence $x \in j_{\lhd}X$. This shows $X \subseteq j_{\lhd}X$ as required.
Conversely, if $j_{\lhd}$ is inflationary on $Up(\mathscr{S})$, then for any x we have $x \in \uparrow x \subseteq j_{\lhd}\uparrow x$. This implies that there is some C such that $x \lhd C \subseteq \uparrow x$, confirming Existence.

(3) Assume Refined Transitivity. If X is an up-set and $x \in j_{\lhd}j_{\lhd}X$, then there is an x-cover $C \subseteq j_{\lhd}X$, so for all $y \in C$ there is a y-cover $C_y \subseteq X$. Hence by assumption there exists a C' such that

$$x \lhd C' \subseteq \uparrow \bigcup_{y\in C} C_y \subseteq \uparrow X = X.$$

This implies that $x \in j_{\lhd}X$, confirming that $j_{\lhd}j_{\lhd}X \subseteq j_{\lhd}X$.
For the converse, let $x \lhd C$ and for all $y \in C$, $y \lhd C_y$. Define X to be the up-set $\uparrow(\bigcup_{y\in C} C_y)$. Then $y \lhd C_y \subseteq X$ and hence $y \in j_{\lhd}X$, for all $y \in C$. Thus $x \lhd C \subseteq j_{\lhd}X$, showing $x \in j_{\lhd}j_{\lhd}X$. So if $j_{\lhd}j_{\lhd}X \subseteq j_{\lhd}X$ we get $x \in j_{\lhd}X$, hence there is an x-cover $C' \subseteq X$, as required to prove Refined Transitivity. □

We now define a structure $\mathscr{S}$ to be a *cover system* if it satisfies the Refinement, Existence and Refined Transitivity axioms. From Theorem 1 we see that this is equivalent to requiring that $j_\lhd$ be a closure operator on the complete poset $(Up(\mathscr{S}), \subseteq)$ of up-sets of $\mathscr{S}$.

Drágalin (1979, p. 72) defined an operation $\mathbf{D}$ on *down*-sets of a preorder by taking a function Q assigning to each $x \in S$ a collection $Q(x)$ of subsets of S, and putting $\mathbf{D}Y = \{x \in S : \forall C \in Q(x), C \cap Y \neq \emptyset\}$. He gave conditions on Q ensuring that $\mathbf{D}$ is a closure operator, and interpreted $C \in Q(x)$ by saying that 'C is a path starting from the moment x'. Now if we define $x \lhd C$ to mean $C \in Q(x)$, then for any up-set X it follows that $S \setminus X$ is a down-set and $j_\lhd X = S \setminus (\mathbf{D}(S \setminus X)$, so in this sense Drágalin's approach is dual to that of cover systems. See (Bezhanishvili and Holliday 2016) for a detailed discussion of this relationship.

Typically we will be dealing with cover systems that satisfy a strong version of Refined Transitivity, namely the property

- *Transitivity*: if $x \lhd C$ and for all $y \in C$, $y \lhd C_y$, then $\bigcup_{y \in C} C_y$ itself is an x-cover.

A *proposition* in a cover system is an up-set X that is localised, i.e. $j_\lhd X \subseteq X$, hence $j_\lhd X = X$. In general, a set X is a proposition iff $X = \uparrow X = j_\lhd X$. $j_\lhd \uparrow X$ is the smallest proposition that includes an arbitrary X, and $j_\lhd \uparrow x$ is the smallest proposition containing the element x. The smallest proposition including an up-set X is just $j_\lhd X$, so in fact $j_\lhd$ maps $Up(\mathscr{S})$ onto the set $Prop(\mathscr{S})$ of all localised up-sets of a cover system $\mathscr{S}$. Indeed, $Prop(\mathscr{S})$ is precisely the earlier described set L^j of fixed points of j when $L = Up(\mathscr{S})$ and $j = j_\lhd$. Thus $Prop(\mathscr{S}) = Up(\mathscr{S})^{j_\lhd}$ and forms a complete lattice under the inclusion order, in which any collection Θ of propositions has meet $\bigsqcap \Theta = \bigcap \Theta$ and join $\bigsqcup \Theta = j_\lhd \bigcup \Theta$.

3 Finite Covers for Representing a Lattice

Let $(L, \sqsubseteq, \sqcup, \sqcap)$ be an arbitrary lattice, which we typically refer to just as L. We will construct a cover system $\mathscr{S}_L$ on the set of principal filters of L, and show that L can be isomorphically embedded into its lattice of propositions $Prop(\mathscr{S}_L)$. For each $a \in L$, let

$$[a) = \{c \in L : a \sqsubseteq c\}$$

be the *principal filter* of L *generated by* a. This generator is unique, as $[a) = [b)$ iff $a = b$.

From the definition of joins it is immediate that $[a) \cap [b) = [a \sqcup b)$, and more generally that

$$\bigcap_{a \in X} [a) = [\bigsqcup X) \tag{2}$$

whenever the join of a subset X exists in L.

Let S_L be the set of all principal filters of L. For each $a \in L$, let $|a| = \{x \in S_L : a \in x\}$ be the set of all principal filters containing a. In general, if $x \in S_L$ then $a \in x$ iff $[a) \subseteq x$, so $|a|$ is the up-set $\uparrow[a)$ in the poset $(S_L, \subseteq)$. In other words, if we define g_x to be the unique generator of principal filter x, i.e. $x = [g_x)$, then in $(S_L, \subseteq)$ we have $\uparrow x = |g_x|$.

The function $a \mapsto |a|$ maps L *injectively* into the set of up-sets of $(S_L, \subseteq)$, since if $|a| = |b|$ then $\uparrow[a) = \uparrow[b)$, hence $[a) = [b)$ and so $a = b$. This function from L into $Up(S_L, \subseteq)$ also preserves lattice meets, since $|a \sqcap b| = |a| \cap |b|$ and more generally

$$|\bigsqcap X| = \bigcap_{a \in X} |a| \tag{3}$$

whenever the meet of a subset X exists in L. However the function need not preserve lattice joins, for instance the principal filter $[a \sqcup b)$ will contain $a \sqcup b$ but may not contain either a or b, in which case $|a \sqcup b| \neq |a| \cup |b|$.

To overcome this we introduce a cover relation $\lhd$ on S_L. For $x \in S_L$ and $C \subseteq S_L$, define

$$x \lhd C \iff C \text{ is finite and } \bigcap C \subseteq x. \tag{4}$$

Then the structure $\mathscr{S}_L := (S_L, \subseteq, \lhd)$ satisfies the axioms of a cover system, as we now show:

- *Refinement*: let $C \rhd x$ and $x \subseteq y$. Then $\bigcap C \subseteq x \subseteq y$. So in this case we get a y-cover C' refining C just by putting $C' = C$.
- *Existence*: for $x \in S_L$, let $C = \{x\}$. Then $C \subseteq \uparrow x$ as $\uparrow x = \{y \in S_L : x \subseteq y\}$, and $\bigcap C = x$ so C is an x-cover included in $\uparrow x$ as required.
- *Transitivity*: let $x \lhd C$ and for all $y \in C$, $y \lhd C_y$. Put $C' = \bigcup_{y \in C} C_y$. Then for each $y \in C$, $C_y \subseteq C'$ and so $\bigcap C' \subseteq \bigcap C_y \subseteq y$, the last inclusion holding because $y \lhd C_y$. Hence $\bigcap C' \subseteq \bigcap C \subseteq x$. But C' is finite, being the union of finitely many finite sets, so we have $x \lhd C'$, as required to establish the (stronger than Refined) Transitivity axiom.

In the cover system $\mathscr{S}_L$, each up-set of the form $|a|$ is localised, i.e. $j_\lhd |a| \subseteq |a|$, and hence is a proposition. For if $x \lhd C \subseteq |a|$, then $a \in \bigcap |a| \subseteq \bigcap C \subseteq x$, and so $x \in |a|$. Thus the map $a \mapsto |a|$ embeds L into the complete lattice *Prop*($\mathscr{S}_L$) of propositions of the cover system $\mathscr{S}_L$. To show that this map preserves binary joins we observe first that if X is any subset of L that has a join $\bigsqcup X$ in L, then using (2), for all $x \in S_L$,

$$\bigcap\{[a) : a \in X\} \subseteq x \text{ iff } [\bigsqcup X) \subseteq x \text{ iff } \bigsqcup X \in x. \tag{5}$$

Consequently, if X is any *finite non-empty* subset of L, then

$$x \lhd \{[a) : a \in X\} \text{ iff } \bigsqcup X \in x. \tag{6}$$

From this it follows that for all $a, b \in L$,

$$j_{\lhd}(|a| \cup |b|) = |a \sqcup b|. \tag{7}$$

Proof Since $|a| \cup |b| \subseteq |a \sqcup b|$, we have $j_{\lhd}(|a| \cup |b|) \subseteq j_{\lhd}|a \sqcup b| \subseteq |a \sqcup b|$. Conversely, if $x \in |a \sqcup b|$, then $a \sqcup b \in x$, so by (6) it follows that $x \lhd \{[a), [b)\} \subseteq |a| \cup |b|$, hence $x \in j_{\lhd}(|a| \cup |b|)$.

Equation (7) confirms that the map $a \mapsto |a|$ preserves binary joins between L and $Prop(\mathscr{S}_L)$. Altogether then we have established the following.

Theorem 2 *Every lattice L has an embedding into the complete lattice $Prop(\mathscr{S}_L)$ of all localised up-sets of the cover system $\mathscr{S}_L$, by a lattice monomorphism that also preserves any existing infinite meets of L.* □

Note that as $j_{\lhd}$ is inflationary we have $S_L = j_{\lhd} S_L$ and this is the greatest element of $Prop(\mathscr{S}_L)$. But if L has a greatest element 1, then it belongs to all principal filters, so $|1| = S_L$ and the map $a \mapsto |a|$ preserves greatest elements.

$Prop(\mathscr{S}_L)$ always has least element $j_{\lhd}\emptyset = \{x \in S_L : x \lhd \emptyset\}$. By (4) $x \lhd \emptyset$ iff $L = \bigcap \emptyset \subseteq x$ iff the generator of the principal filter x is a least element of L. Thus $j_{\lhd}\emptyset = \emptyset$ if L has no least element, while if it does have least element 0, then $j_{\lhd}\emptyset = \{[0)\} = L = |0|$, so the map $a \mapsto |a|$ preserves least elements.

4 Distributive Lattices

The lattice representation just given will now be adapted to the case that the lattice L is *distributive*, meaning that it satisfies the equation

$$a \sqcap (b \sqcup c) = (a \sqcap b) \sqcup (a \sqcap c),$$

and hence satisfies

$$a \sqcap (\bigsqcup X) = \bigsqcup \{a \sqcap b : b \in X\} \tag{8}$$

for any non-empty finite $X \subseteq L$.

A cover system $\mathscr{S}$ will be called *localic* if, for each point x of $\mathscr{S}$,

$$\text{every x-cover can be refined to an x-cover that is included in} \uparrow\text{x}, \tag{9}$$

i.e. if $x \lhd C$, then there an x-cover C' with $C' \subseteq \uparrow C$ and $C' \subseteq \uparrow x$.

Theorem 3 *If L is a distributive lattice, then its associated cover system $\mathscr{S}_L$ is localic.*

Proof Suppose $x \lhd C$ in S_L. If $C = \emptyset$, then C itself is an x-cover refining C and included in $\uparrow x$. Otherwise, when $C \neq \emptyset$, the set $X = \{a \in L : [a) \in C\}$ is finite and non-empty, and $\{[a) : a \in X\} = C \rhd x$, so $\bigsqcup X \in x$ by (6). Hence $g_x \sqsubseteq \bigsqcup X$, where g_x is the generator of x. Applying the distributive law (8), we see that

$$\bigsqcup\{g_x \sqcap a : a \in X\} = g_x \sqcap (\bigsqcup X) = g_x \in x.$$

Defining $C' = \{[g_x \sqcap a) : a \in X\}$ then gives $x \lhd C'$ by (6). But if $[a) \in C$, then since $[a) \subseteq [g_x \sqcap a)$ it follows that $[g_x \sqcap a) \in \uparrow C$. Hence $C' \subseteq \uparrow C$, i.e. C' refines C in $\mathscr{S}_L$. Also, as $x = [g_x) \subseteq [g_x \sqcap a)$ we get that $C' \subseteq \uparrow x$, completing the proof that $\mathscr{S}_L$ satisfies the localic condition. □

Combining this result with Theorem 5 gives

Corollary 1 *Every distributive lattice L can be isomorphically embedded into the lattice of all propositions of the localic cover system* $\mathscr{S}_L$. □

It turns out that when $\mathscr{S}$ is localic, the lattice $Prop(\mathscr{S}) = Up(\mathscr{S})^{j_\lhd}$ is not just distributive, but satisfies the law (8) for *every* subset X, including $X = \emptyset$. In general a complete lattice in which finite meets distribute over arbitrary joins as in (8) is called a *locale*. For example, the complete lattice $Up(\mathscr{S})$ of up-sets of any cover system is a locale, since in its case (8) amounts to the set-theoretic fact that

$$X \cap \bigcup \Theta = \bigcup\{X \cap Y : Y \in \Theta\}$$

for any up-set X and any collection Θ of up-sets.

Note that a complete distributive lattice need not be a locale: it may be that (8) holds for finite X but not in general:

Example 1 Let L be the set of closed subsets of the real line $\mathbb{R}$ under its standard topology. L is a complete lattice under the inclusion order, with $\bigsqcap \Theta = \bigcap \Theta$ and $\bigsqcup \Theta = \mathrm{cl} \bigcup \Theta$, where cl denotes topological closure. In fact $\bigsqcup \Theta = \bigcup \Theta$ when Θ is a finite set of closed subsets, so this lattice is distributive. Each singleton subset $\{p\}$ belongs to L, but $\{p\}$ is not open and so $p \in \mathrm{cl}(\mathbb{R}\backslash\{p\}) = \mathrm{cl} \bigcup_{r \neq p}\{r\}$. Thus $\{p\} \sqcap (\bigsqcup_{r \neq p}\{r\}) = \{p\}$. However $\bigsqcup_{r \neq p}(\{p\} \sqcap \{r\}) = \emptyset$, so this is a failure of (8). □

The proof that $Prop(\mathscr{S})$ is a locale when $\mathscr{S}$ is localic involves further theory of closure operators which we now outline. A *nucleus* on a lattice L (Simmons 1978) is a closure operator j that satisfies $ja \sqcap jb \sqsubseteq j(a \sqcap b)$, and hence preserves meets because the reverse inequality holds by monotonicity of j. In fact a closure operator j is a nucleus iff for $a, b \in L$,

$$a \sqcap jb \sqsubseteq j(a \sqcap b). \tag{10}$$

Certainly (10) holds if j is $\sqcap$-preserving, by the inflationary property of j. Conversely, by using (10) twice (and commutativity of $\sqcap$) we reason that

$$ja \sqcap jb \sqsubseteq j(ja \sqcap b) \sqsubseteq jj(a \sqcap b) \sqsubseteq j(a \sqcap b).$$

Theorem 4 *A cover system $\mathscr{S}$ is localic iff $j_{\lhd}$ is a nucleus on $Up(\mathscr{S})$.*

Proof Take $X, Y \in Up(\mathscr{S})$. If $x \in X \cap j_{\lhd}Y$, then $x \in X$ and there is an x-cover $C \subseteq Y$, hence $\uparrow x \subseteq X$ and $\uparrow C \subseteq Y$. But if $\mathscr{S}$ is localic, there is a C' with

$$x \lhd C' \subseteq \uparrow x \cap \uparrow C \subseteq X \cap Y,$$

hence $x \in j_{\lhd}(X \cap Y)$. This shows that $X \cap j_{\lhd}Y \subseteq j_{\lhd}(X \cap Y)$, so $j_{\lhd}$ satisfies (10) on $Up(\mathscr{S})$.

Conversely, assume $j_{\lhd}$ is a nucleus on $Up(\mathscr{S})$. Let C be an x-cover. Since $C \subseteq \uparrow C$, we have $x \in j_{\lhd}\uparrow C$ and therefore $x \in \uparrow x \cap j_{\lhd}\uparrow C$. (10) on $Up(\mathscr{S})$ then implies $x \in j_{\lhd}(\uparrow x \cap \uparrow C)$, which implies that there is an x-cover included in both $\uparrow x$ and $\uparrow C$, showing that $\mathscr{S}$ is localic. □

Now if j is a nucleus on a locale L, then the complete lattice L^j of fixed points of j is also a locale (Macnab 1976, 1981). For, if $a \in L$ and $X \subseteq L$, then in L,

$$a \sqcap j(\bigsqcup X) \sqsubseteq j(a \sqcap \bigsqcup X) = j(\bigsqcup\{a \sqcap b : b \in X\}).$$

But $\bigsqcup\{a \sqcap b : b \in X\} \sqsubseteq a \sqcap (\bigsqcup X)$ in any lattice when the two joins involved exist, so $j \bigsqcup\{a \sqcap b : b \in X\} \sqsubseteq ja \sqcap j(\bigsqcup X)$ by monotonicity of j. It follows that if $a \in L^j$, then

$$a \sqcap j(\bigsqcup X) = j(\bigsqcup\{a \sqcap b : b \in X\}).$$

But when also $X \subseteq L^j$, this last equation expresses that finite meets distribute over arbitrary joins in L^j, i.e. L^j is a locale.

Consequently, if $\mathscr{S}$ is localic then the lattice $Prop(\mathscr{S})$ of fixed points of the nucleus $j_{\lhd}$ is a locale, and in particular if L is a distributive lattice then $Prop(\mathscr{S}_L)$ is a locale. Corollary 1 thus implies the well-known fact that any distributive lattice can be isomorphically embedded into a locale. We elaborate on this in Sect. 6.

5 Heyting Algebras

A *Heyting algebra* is a lattice with a least element and a binary "implication" operation $a \Rightarrow b$ satisfying

$$c \sqsubseteq a \Rightarrow b \text{ iff } c \sqcap a \sqsubseteq b. \tag{11}$$

This operation is thereby uniquely determined, since (11) implies that

$$a \Rightarrow b = \bigsqcup\{c \in L : c \sqcap a \sqsubseteq b\}. \tag{12}$$

Heyting algebras provide the standard algebraic semantics for intuitionistic propositional logic. They are always distributive.

Any locale becomes a Heyting algebra when $a \Rightarrow b$ is *defined* by equation (12). Conversely, a complete Heyting algebra satisfies distributivity of finite meets over arbitrary joins (8). So the notions of locale and complete Heyting algebra are equivalent. The complete distributive lattice of Example 1 is not a locale.

For any cover system $\mathcal{S}$ it is readily seen that the operation $\Rightarrow$ in the locale $Up(\mathcal{S})$, as given according to (12) by the equation

$$X \Rightarrow Y = \bigcup\{Z \in Up(\mathcal{S}) : Z \cap X \subseteq Y\},$$

in fact has the simpler characterisation

$$X \Rightarrow Y = \{x \in S : \uparrow x \cap X \subseteq Y\}. \tag{13}$$

This means that $x \in X \Rightarrow Y$ iff $x \preccurlyeq y \in X$ implies $y \in Y$, reflecting the Kripke semantics for intuitionistic implication.

For the case of a locale of the form $Prop(\mathcal{S})$, i.e. when $\mathcal{S}$ is localic, we turn to more theory of nuclei. Any nucleus j on a Heyting algebra L satisfies $a \Rightarrow jb = j(a \Rightarrow jb)$. Hence if $b \in L^j$, then $a \Rightarrow b \in L^j$ for all $a \in L$. In particular, L^j is closed under the operation $\Rightarrow$, and is a Heyting algebra under this same operation (Macnab 1976). Also, as (11) holds for all $c \in L$, it holds for all $c \in L^j$. Therefore if $a, b \in L^j$ we get that $a \Rightarrow b \in L^j$ and

$$a \Rightarrow b = \bigsqcup\{c \in L^j : c \sqcap a \sqsubseteq b\} = j \bigsqcup\{c \in L^j : c \sqcap a \sqsubseteq b\},$$

where $\bigsqcup$ is the join operation of L. In other words, when $a, b \in L^j$, the set $\{c \in L^j : c \sqcap a \sqsubseteq b\}$ has the same join in L as it does in L^j, and this join is the Heyting implication of a and b in both L and L^j.

This theory tells us that if $\mathcal{S}$ is localic, then if X and Y belong to the locale $Prop(\mathcal{S})$, and $X \Rightarrow Y$ is the set given by (13), then $X \Rightarrow Y \in Prop(\mathcal{S})$ and

$$X \Rightarrow Y = \bigcup\{Z \in Prop(\mathcal{S}) : Z \cap X \subseteq Y\} = j_{\lhd} \bigcup\{Z \in Prop(\mathcal{S}) : Z \cap X \subseteq Y\}.$$

Theorem 5 *Every Heyting algebra L can be embedded into the locale of all propositions of the localic cover system $\mathcal{S}_L$, by a Heyting algebra monomorphism.*

Proof Since every Heying algebra is distributive, we know from the previous two Sections that $\mathcal{S}_L$ is localic and the map $a \mapsto |a|$ is a lattice homomorphism embedding L injectively into the the locale $Prop(\mathcal{S}_L)$. It remains to show this map preserves the Heyting implication operations of L and $Prop(\mathcal{S}_L)$, i.e. that $|a| \Rightarrow |b| = |a \Rightarrow b|$, or equivalently by (13) that

$$\{x \in S_L : \uparrow x \cap |a| \subseteq |b|\} = |a \Rightarrow b|. \tag{14}$$

But for $x \in S_L$ we have $\uparrow x = |g_x|$ where g_x is the generator of the principal filter x, so $\uparrow x \cap |a| = |g_x| \cap |a| = |g_x \sqcap a|$. Hence $\uparrow x \cap |a| \subseteq |b|$ iff $|g_x \sqcap a| \subseteq |b|$ iff $g_x \sqcap a \sqsubseteq b$ iff (by (11)) $g_x \sqsubseteq a \Rightarrow b$ iff $a \Rightarrow b \in x$ iff $x \in |a \Rightarrow b|$. That proves (14) as required. □

6 Ideal Completion

The lattice of propositions $Prop(\mathscr{S}_L)$ can be seen as an alternative presentation of the *ideal completion* of the lattice L. We now explain this.

Recall that an *ideal* of a lattice L is any non-empty subset $I \subseteq L$ such that $a \sqcup b \in I$ iff $a, b \in I$. Equivalently, I is closed under the binary operation $\sqcup$ and closed downwards under the partial order of L: $b \sqsubseteq a \in I$ implies $b \in I$. The intersection of any collection of ideals is an ideal, so each subset J has a smallest ideal $(J]$ including it. This has

$$(J] = \{b \in L : b \sqsubseteq \bigsqcup X \text{ for some finite } X \subseteq J\}.$$

In particular, for each $a \in L$, the set $(a] = \{b \in L : b \sqsubseteq a\}$ is the *principal ideal generated by* a, the smallest ideal containing a.

The set $\hat{L}$ of all ideals of L forms a *complete* lattice when ordered by set inclusion $\subseteq$, with the meet and join of any collection Θ of ideals being its set intersection $\bigcap \Theta$ and the ideal $(\bigcup \Theta]$ generated by the union $\bigcup \Theta$, respectively.

Theorem 6 *The lattices $(\hat{L}, \subseteq)$ and $(Prop(\mathscr{S}_L), \subseteq)$ are isomorphic.*

Proof For each ideal I of L, define

$$X_I := \{x \in S_L : g_x \in I\} = \{[a) : a \in I\}.$$

Then the isomorphism is provided by the map $I \mapsto X_I$. To prove this we first have to show that X_I is a proposition of $\mathscr{S}_l$. That it is a $\subseteq$-up-set follows readily as if $x \in X_I$ and $x \subseteq y$, then $g_y \sqsubseteq g_x \in I$, hence $g_y \in I$ and so $y \in X_I$. To show that $j_\triangleleft X_I \subseteq X_I$, take any $x \in j_\triangleleft X_I$. Then there is a C with $x \triangleleft C \subseteq X_I$. Define $X_C := \{g_y : y \in C\}$. Since C is finite, so too is X_C. Since $C \subseteq X_I$ we get that $X_C \subseteq I$, and so $\bigsqcup X_C \in I$. But $C = \{[a) : a \in X_C\}$, so by (6), $\bigsqcup X_C \in x$. Hence $g_x \sqsubseteq \bigsqcup X_C$, forcing $g_x \in I$ and so $x \in X_I$.

This confirms that X_I is a localised up-set, and so $I \mapsto X_I$ is a map from $\hat{L}$ into $Prop(\mathscr{S}_L)$. In the reverse direction, take any $X \in Prop(\mathscr{S}_L)$ and define $I_X := \{a \in L : [a) \in X\}$. Then I_X is an ideal, since if $b \sqsubseteq a \in I_X$, then $[a) \in X$ and $[a) \subseteq [b)$, hence $[b) \in X$ so $b \in I_X$; and if $a, b \in I_X$ then as $[a) \cap [b) = [a \sqcup b)$, from (4) we get $[a \sqcup b) \triangleleft \{[a), [b)\} \subseteq X$, making $[a \sqcup b) \in j_\triangleleft X \subseteq X$ and hence $a \sqcup b \in I_X$. Thus indeed $I_X \in \hat{L}$. Since

$$I_{X_I} = \{a \in L : [a) \in X_I\} = \{a \in L : a \in I\} = I,$$

and

$$X_{I_X} = \{[a) : a \in I_X\} = \{[a) : [a) \in X\} = X,$$

we see that the maps $I \mapsto X_I$ and $X \mapsto I_X$ are mutually inverse, so each is a bijection.

Finally, it is straightforward that $I \subseteq I'$ iff $X_I \subseteq X'_I$, or equivalently that $X \subseteq X'$ iff $I_X \subseteq I'_X$, so these maps are mutually inverse order-isomorphisms between $(\hat{L}, \subseteq)$ and $(Prop(\mathscr{S}_L), \subseteq)$. □

Theorem 6 and the results of the Sect. 4 reproduce the well-known result of Stone (1937) that the ideal completion $\hat{L}$ of a distributive lattice satisfies the distribution law (8) for arbitrary subsets $X \subseteq \hat{L}$, so is a locale.

In the proof of Theorem 6, if $X = |a|$, then

$$I_X = \{b \in L : [b) \in |a|\} = \{b : b \sqsubseteq a\} = (a].$$

So the embedding $a \mapsto |a|$ of L into $Prop(\mathscr{S}_L)$ composes with the isomorphism $X \mapsto I_X$ from $Prop(\mathscr{S}_L)$ onto $\hat{L}$ to give the map $a \mapsto (a]$, which is the standard embedding of L into its ideal completion $\hat{L}$.

We can identify L with its isomorphic image under this embedding, by identifying a with $(a]$, and thereby view $\hat{L}$ as a complete extension of L. The image is *join dense* in $\hat{L}$ in the sense that every member of $\hat{L}$ is a join of members of the image. This is because each ideal I is the join of $\{(a] : a \in I\}$ in $\hat{L}$.

7 Covers for the MacNeille Completion

The MacNeille completion (MacNeille 1937) of a lattice (or poset) L is a complete lattice $\bar{L}$ containing an isomorphic image of L that is join dense and also *meet dense* in $\bar{L}$. Meet density means that each member of the completion is also a meet of members of the image. Possession of these two density properties characterises $\bar{L}$ uniquely up to isomorphism.

$\bar{L}$ can be constructed as follows. For each subset J of L, let J^u be the set of upper bounds of J and J^l the set of lower bounds. Composing these two operations, put $mJ = J^{ul}$, the set of lower bounds of the set of upper bounds of J. Then m is a closure operator on the complete lattice of all subsets of L under the inclusion ordering. $\bar{L}$ is defined to be the set of all m-closed (i.e. $J = J^{ul}$) subsets of L. It is a complete lattice under the ordering $\subseteq$, with $\bigsqcap \Theta = \bigcap \Theta$ and $\bigsqcup \Theta = m(\bigcup \Theta)$ in $\bar{L}$, where Θ is any collection of m-closed sets. Each m-closed set is an ideal of L, so $\bar{L}$ is a subset of $\hat{L}$. (See (Davey and Priestley 1990, pp. 40–44) for a comprehensive discussion of this construction.)

Any set of the form J^l is m-closed, including the set $\{a\}^l = \{b \in L : b \sqsubseteq a\} = (a]$ for each $a \in L$. The function $a \mapsto (a]$ is an order-invariant injection of $(L, \sqsubseteq)$ into $(\bar{L}, \subseteq)$ having the property that it preserves all joins and meets that exist in L.

Our purpose now is to define a new cover system $\mathscr{S}_L^m = (S_L, \subseteq, \blacktriangleleft)$, on the set of principal filters of L, whose lattice of proposition gives an alternative presentation of the MacNeille completion of L. The relation $\blacktriangleleft$ is obtained by dropping the finiteness condition on covers from (4), simply putting

$$x \blacktriangleleft C \text{ iff } \bigcap C \subseteq x. \tag{15}$$

The proof that the resulting structure $\mathscr{S}_L^m$ is a cover system is essentially the same as that for $\mathscr{S}_L$, without the requirement of finiteness in showing that $\mathscr{S}_L^m$ satisfies Transitivity. Note that since $\lhd$ is a subrelation of $\blacktriangleleft$ we have $j_\lhd X \subseteq j_\blacktriangleleft X$ in general, so any $j_\blacktriangleleft$-closed up-set is $j_\lhd$-closed. Hence $Prop(\mathscr{S}_L^m)$ is a subset of $Prop(\mathscr{S}_L)$. For each $a \in L$ the set $|a|$ is $j_\blacktriangleleft$-closed and so belongs to $Prop(\mathscr{S}_L^m)$. The map $a \mapsto |a|$ is an embedding of L into $Prop(\mathscr{S}_L^m)$ that preserves any meets of L by (3), and also preserves any join $\bigsqcup X$ that exists in L, i.e.

$$|\bigsqcup X| = j_\blacktriangleleft \bigcup\{|a| : a \in X\}. \tag{16}$$

To see this, observe that if $x \in |\bigsqcup X|$, then from (5), $x \blacktriangleleft \{[a) : a \in X\}$, and since $[a) \in |a|$, this implies $x \in j_\blacktriangleleft \bigcup\{|a| : a \in X\}$. The converse holds as $|\bigsqcup X|$ is a $j_\blacktriangleleft$-closed superset of $\bigcup\{|a| : a \in X\}$.

Lemma 1 *For any $X \subseteq S_L$ and $x \in S_L$, $x \in j_\blacktriangleleft X$ iff $\bigcap X \subseteq x$.*

Proof If $\bigcap X \subseteq x$, then $x \blacktriangleleft X \subseteq X$, hence $x \in j_\blacktriangleleft X$. For the converse, if $x \blacktriangleleft C \subseteq X$ for some C, then $\bigcap X \subseteq \bigcap C \subseteq x$. □

Theorem 7 *The lattices $(\bar{L}, \subseteq)$ and $(Prop(\mathscr{S}_L^m), \subseteq)$ are isomorphic.*

Proof The isomorphism is given by the maps $I \mapsto X_I$ and $X \mapsto I_X$ from Theorem 6 when restricted to $\bar{L}$ and $Prop(\mathscr{S}_L^m)$.

First we take an $I \in \bar{L}$ and show that $X_I \in Prop(\mathscr{S}_L^m)$. Since I is an ideal, it follows as before that X_I is a $\subseteq$-up-set. But whereas Theorem 6 showed that X_I is $j_\lhd$-closed, here we have to show the stronger property that it is $j_\blacktriangleleft$-closed. For this, let $x \in j_\blacktriangleleft X_I$. Then $\bigcap X_I \subseteq x$ by Lemma 1. Next we show that $g_x \in I^{ul}$: if $a \in I^u$, then for all $y \in X_I$ we have $g_y \in I$, hence $g_y \sqsubseteq a$ and so $a \in y$. This shows $a \in \bigcap X_I$, implying that $a \in x$ and so $g_x \sqsubseteq a$ as required to confirm $g_x \in I^{ul}$. But $I^{ul} = I$ as I is m-closed, giving $g_x \in I$ and thus $x \in X_I$. This proves that $j_\blacktriangleleft X_I \subseteq X_I$, making X_I $j_\blacktriangleleft$-closed and hence a member of $Prop(\mathscr{S}_L^m)$.

In the reverse direction, let $X \in Prop(\mathscr{S}_L^m)$. We show $I_X \in \bar{L}$. Let $a \in {I_X}^{ul}$. Then $\bigcap X \subseteq [a)$. For if $b \in \bigcap X$, then for all $c \in I_X$ we have $[c) \in X$, so $b \in [c)$, hence $c \sqsubseteq b$. This shows $b \in {I_X}^u$, hence $a \sqsubseteq b$ as $a \in {I_X}^{ul}$. So $b \in [a)$, as required to prove $\bigcap X \subseteq [a)$. From this Lemma 1 gives $[a) \in j_\blacktriangleleft X$. But X is $j_\blacktriangleleft$-closed, so $[a) \in X$ and thus $a \in I_X$. That proves that ${I_X}^{ul} \subseteq I_X$, making I_X m-closed and hence a member of $\bar{L}$.

The upshot is that the bijections $I \mapsto X_I$ and $X \mapsto I_X$ map $\bar{L}$ into $Prop(\mathscr{S}_L^m)$ and vice versa, respectively. Since they both preserve set inclusion, they provide the asserted order-isomorphism. □

A more abstract proof of this result can be obtained by showing that the image of the embedding $a \mapsto |a|$ is join and meet dense in $Prop(\mathcal{S}_L^m)$. For join density, if $X \in Prop(\mathcal{S}_L^m)$, then as X is an up-set it is equal to $\bigcup\{\uparrow x : x \in X\} = \bigcup\{|g_x| : x \in X\}$, hence as X is $j_\blacktriangleleft$-closed it is the join of $\{|g_x| : x \in X\}$ in $Prop(\mathcal{S}_L^m)$. For meet density we have $X = \bigcap\{|a| : X \subseteq |a|\}$. To show this, let x belong to the intersection on the right of the equation. Then $\bigcap X \subseteq x$, for if $a \in \bigcap X$, then $X \subseteq |a|$, hence by assumption $x \in |a|$, so $a \in X$. Since $\bigcap X \subseteq x$ we have $x \in j_\blacktriangleleft X$ by Lemma 1. But X is $j_\blacktriangleleft$-closed, so then $x \in X$ as required.

Thus $\{|a| : a \in L\}$ is join and meet dense in $Prop(\mathcal{S}_L^m)$, making the latter isomorphic to the MacNeille completion of L.

8 Ortholattices

Ortholattices have a unary *orthocomplementation* operation $a \mapsto a'$ and can be represented as a lattice of closed sets of a closure operator that is constructed from a binary *orthogonality* relation $\perp$. We now review how this works and then show that this closure operator is characterisable as the j-operator of the cover relation $\blacktriangleleft$ of (15).

An ortholattice can be defined as a lattice $(L, \sqcap, \sqcup, 0, ')$ with least element 0 and unary operation $'$ that is

- *antitone*: $a \sqsubseteq b$ implies $b' \sqsubseteq a'$;
- *involutary*: $a'' = a$; and
- *orthocomplemented*: $a \sqcap a' = 0$.

Then $0'$ is a greatest element, usually denoted 1, and has $a \sqcup a' = 1$. Ortholattices obey De Morgan's laws for distribution of complements over meets and joins. In particular $\bigsqcup X = (\bigsqcap_{a \in X} a')'$ whenever $\bigsqcup X$ exists.

Consider a structure $(S, \perp)$, where $\perp$ is a symmetric relation on S. For $X \subseteq S$, let $X^\perp = \{x \in S : x \perp y \text{ for all } y \in X\}$. This operation is antitone for inclusion: $X \subseteq Y$ implies $Y^\perp \subseteq X^\perp$. The symmetry of $\perp$ ensures that $X \subseteq X^{\perp\perp}$, and together with antitonicity this implies $X^{\perp\perp\perp} = X^\perp$.

X will be called *orthoclosed* if $X = X^{\perp\perp}$. Thus any set of the form $X^\perp$ is orthoclosed. The map $X \mapsto X^{\perp\perp}$ is a closure operator on the lattice of all subsets of S. Thus the set $Orth(S, \perp)$ of all orthoclosed sets is a complete lattice under the ordering $\subseteq$, with $\bigsqcap \Theta = \bigcap \Theta$ and $\bigsqcup \Theta = (\bigcup \Theta)^{\perp\perp}$, where Θ is any collection of orthoclosed sets. $Orth(S, \perp)$ has least element $S^\perp$ and is closed under the antitone operation $X \mapsto X^\perp$, which is involutory on orthoclosed sets. This operation is orthocomplemented, which makes $Orth(S, \perp)$ an ortholattice, precisely when $(S, \perp)$ has the following property (Hedlíková and Pulmannová 1991, Lemma 1.1):

$$\text{if } x \perp x \text{ then } x \perp y \text{ for all } y \in S. \tag{17}$$

Now define a relation $\perp_L$ on the set S_L of principal filters of an ortholattice L by putting

$$x \perp_L y \iff g_x \sqsubseteq g_y{}' \quad (\text{iff} \quad g_y{}' \in x). \tag{18}$$

If $g_x \sqsubseteq g_y{}'$ then $g_y = g_y{}'' \sqsubseteq g_x{}'$, so $\perp_L$ is symmetric. If $g_x \sqsubseteq g_x{}'$ then $g_x = g_x \sqcap g_x{}' = 0$, so $x = [0) = L$ and $g_x \sqsubseteq g_y{}'$ for all $y \in S_L$. Thus $\perp_L$ satisfies (17) and $Orth(S_L, \perp_L)$ is a complete ortholattice as above, with $S^{\perp_L} = \{[0)\}$.

Next, observe that for all $a \in L$,

$$|a'| = |a|^{\perp_L}. \tag{19}$$

For, if $x \in |a'|$, then for all $y \in |a|$ we have $g_y \sqsubseteq a$, so $g_x \sqsubseteq a' \sqsubseteq g_y{}'$, hence $x \perp_L y$, showing $x \in |a|^{\perp_L}$. Conversely, if $x \in |a|^{\perp_L}$ then $x \perp_L [a)$ as $[a) \in |a|$, so $g_x \sqsubseteq a'$ and $x \in |a'|$.

It follows that $|a| = |a''| = |a|^{\perp_L\perp_L}$ so $|a|$ belongs to $Orth(S_L, \perp_L)$. The injection $a \mapsto |a|$ embeds L into $Orth(S_L, \perp_L)$, preserving any existing meets, and preserving orthocomplements by (19). Hence it preserves any existing joins as $\bigsqcup X = (\bigsqcap_{a \in X} a')'$.[1]

Lemma 2 *For any $X \subseteq S_L$ and $x \in S_L$, $x \in X^{\perp_L\perp_L}$ iff $\bigcap X \subseteq x$.*

Proof Let $x \in X^{\perp_L\perp_L}$. To show $\bigcap X \subseteq x$, let $a \in \bigcap X$. Put $y = [a')$. Then $y \in X^{\perp_L}$, for if $z \in X$ then $a \in z$ so $g_z \sqsubseteq a$, hence $a' \sqsubseteq g_z{}'$ so $g_z{}' \in y$, showing $y \perp_L z$. Since $y \in X^{\perp_L}$ and $x \in X^{\perp_L\perp_L}$, this now gives $x \perp_L y$, hence $a = g_y{}' \in x$ as required.

Conversely, assume $\bigcap X \subseteq x$. We show that if $y \in X^{\perp_L}$ then $x \perp_L y$, as follows. If $y \in X^{\perp_L}$, then for all $z \in X$ we have $y \perp_L z$, hence $z \perp_L y$ and so $g_y{}' \in z$. This shows that $g_y{}' \in \bigcap X$. Therefore $g_y{}' \in x$, giving $x \perp_L y$, as required for $x \in X^{\perp_L\perp_L}$. □

A fact that we will use below is that any subset of S_L of the form $Z^{\perp_L}$ is a $\subseteq$-up-set. For if $x \in Z^{\perp_L}$ and $x \subseteq y$, then for all $z \in z$ we have $x \perp_L z$, hence $g_z{}' \in x$, so $g_z{}' \in y$ and therefore $y \perp_L z$. Thus $y \in Z^{\perp_L}$.

Theorem 8 *For any $X \subseteq S_L$, $X^{\perp_L\perp_L} = j_\blacktriangleleft X$. Hence $Orth(S_L, \perp_L) = Prop(\mathscr{S}_L^m)$.*

Proof $X^{\perp_L\perp_L} = j_\blacktriangleleft X$ by Lemmas 1 and 2.

Now if $X \in Prop(\mathscr{S}_L^m)$ then $X = j_\blacktriangleleft X$, hence $X = X^{\perp_L\perp_L}$ so $X \in Orth(S_L, \perp_L)$. Conversely, if $X \in Orth(S_L, \perp_L)$, then $X = X^{\perp_L\perp_L}$ and so $X = j_\blacktriangleleft X$. Also X is an up-set, since every set of the form $Z^{\perp_L}$ is an up-set. So X is a localised up-set of $\mathscr{S}_L^m$ and belongs to $Prop(\mathscr{S}_L^m)$. □

This shows that $Orth(S_L, \perp_L)$ is (isomorphic to) the MacNeille completion of the ortholattice L.

---

[1]In (Goldblatt 1975) the relation $\perp_L$ was defined on the set of all *proper filters* of L by putting $x \perp_L y$ iff $\exists a \in L$ such that $a' \in x$, $a \in y$. That is equivalent to (18) when restricted to principal filters, and is also *irreflexive*, and hence satisifies (17) vacuously, when restricted to proper filters.

9 Stricter Cover Systems for Heyting Algebras

The localic condition (9) that causes propositions to form a locale can be strengthened to the requirement that every x-cover *itself* be included in the up-set $\uparrow x = \{y : x \preccurlyeq y\}$, i.e. $C \rhd x$ implies $C \subseteq \uparrow x$. This condition, which we shall call *strictly localic*, is intrinsically built into generalisations of topological open-cover systems to 'Grothendieck (pre)topologies' on categories, and other abstract settings (Johnstone 1982; Mac Lane and Moerdijk 1992; Johnstone 2002; Bell 2005).[2] These often involve further conditions on the cover relation. For instance, in (Bell 2005) a cover system on a pre-ordered set is called a *Grothendieck topology* if it is strictly localic, every cover is an up-set, and whenever $x \lhd C$ and C' is an up-set such that $C \subseteq C' \subseteq \uparrow x$, then $x \lhd C'$. This implies further properties, including that the covers of any element form a filter of up-sets in $\uparrow x$ (i.e. if $x \lhd C$ and $x \lhd C'$ then $x \lhd C \cap C'$).

In this final section we construct another cover relation $\trianglelefteq$ on S_L that forms such a Grothendieck topology when L is a Heyting algebra, and defines the same closure operator as $\blacktriangleleft$, thereby defining the same lattice of propositions representing L.

We continue to work in the poset $(S_L, \subseteq)$ in which $\uparrow x = \{y : x \subseteq y\}$, and define

$$x \trianglelefteq C \text{ iff } C = \uparrow C \text{ and } \bigcap C = x. \tag{20}$$

Theorem 9 *If L is a Heyting algebra, then the structure $\mathscr{S}_L^G = (S_L, \subseteq, \trianglelefteq)$ is a strictly localic cover system in which all covers are up-sets and whenever $x \trianglelefteq C$ and $C \subseteq C' = \uparrow C' \subseteq \uparrow x$, then $x \trianglelefteq C'$.*

Proof In the definition of $x \trianglelefteq C$, the necessary condition $C = \uparrow C$ just means that the cover C is an up-set. The condition $\bigcap C = x$ implies that if $y \in C$ then $x \subseteq y$, hence C is a subset of $\uparrow x$ in $(S_L, \subseteq)$. Thus $\mathscr{S}_L^G$ is strictly localic.

Suppose now that $x \trianglelefteq C \subseteq C' = \uparrow C' \subseteq \uparrow x$. Then $\bigcap C' \subseteq \bigcap C = x$, and also $x \subseteq \bigcap C'$ as $C' \subseteq \uparrow x$, so $\bigcap C' = x$. Hence $x \trianglelefteq C'$ by (20).

It remains then to verify that $\trianglelefteq$ satisfies the axioms for a cover system:

- *Refinement*: this part requires the Heyting implication operation $\Rightarrow$ of L. Suppose $x \trianglelefteq C$ and $x \subseteq y$. We have to refine C to a y-cover C'. We show this using only that $\bigcap X \subseteq x \subseteq y$. Define

  $$C' = \uparrow\{[g_z \sqcap g_y) : z \in C\}.$$

  C' is an up-set by definition. Since $z = [g_z) \subseteq [g_z \sqcap g_y)$ we have $C' \subseteq \uparrow C$, so C' refines C. Since $y \subseteq [g_z \sqcap g_y)$ we have $C' \subseteq \uparrow y$, so $y \subseteq \bigcap C'$. Thus to prove that $y \trianglelefteq C'$ it remains to show that $\bigcap C' \subseteq y$.

[2]Those settings typically use down-sets rather than up-sets. Here we are following the order-convention most commonly used in the relational semantics of intuitionistic, relevant and other substructural logics.

Let $a \in \bigcap C'$. Then for all $z \in C$, $g_z \sqcap g_y \sqsubseteq a$ and so $g_z \sqsubseteq g_y \Rightarrow a$ (11), hence $g_y \Rightarrow a \in z$. This shows that $g_y \Rightarrow a \in \bigcap C$. *But* $\bigcap C \subseteq x$ *because* $x \trianglelefteq C$, so $g_y \Rightarrow a \in x$ and hence $g_x \sqsubseteq g_y \Rightarrow a$. Thus $g_x \sqcap g_y \sqsubseteq a$. But $g_y \sqsubseteq g_x$ as $x \subseteq y$, so $g_y = g_x \sqcap g_y \sqsubseteq a$ and therefore $a \in y$ as required to prove that $\bigcap C' \subseteq y$.

- *Existence*: for $x \in S_L$, let $C = \uparrow x$. Then $\bigcap C = \bigcap \uparrow x = x$ and C is an up-set, so $x \trianglelefteq C$. This shows that X has a $\trianglelefteq$-cover $C \subseteq \uparrow x$, as required.
- *Transitivity*: let $x \trianglelefteq C$ and for all $y \in C$, $y \trianglelefteq C_y$. Put $C' = \bigcup_{y \in C} C_y$. Then $\bigcap C' \subseteq x$ by the same argument given earlier for the relation $\lhd$. But for each $y \in C$, by the strictly localic condition we have $x \subseteq y$ and $C_y \subseteq \uparrow y \subseteq \uparrow x$. Hence $C' \subseteq \uparrow x$ and so $x \subseteq \bigcap C'$. Thus $\bigcap C' = x$. But C' is an up-set, being a union of up-sets, so we get $x \trianglelefteq C'$ as required to prove Transitivity. □

It is clear that $\trianglelefteq$ need not be the same relation as $\blacktriangleleft$ in general, in which case $\mathscr{S}_L^G \neq \mathscr{S}_L^m$. For we always have $x = \bigcap\{x\}$, hence $x \blacktriangleleft \{x\}$ (and indeed $x \lhd \{x\}$), but $\{x\}$ may not be an up-set, in which case $x \trianglelefteq \{x\}$ fails. Nonetheless we have

Theorem 10 *If L is a Heyting algebra, then $j_\trianglelefteq = j_\blacktriangleleft$ and hence $Prop(\mathscr{S}_L^G) = Prop(\mathscr{S}_L^m)$.*

Proof $x \trianglelefteq C$ implies $\bigcap C \subseteq x$ hence $x \blacktriangleleft C$, so for any X we have $j_\trianglelefteq X \subseteq j_\blacktriangleleft X$.

For the converse inclusion, let $x \in j_\blacktriangleleft X$, so that $x \blacktriangleleft C \subseteq X$ for some set C. Now in the proof of the case of the Refinement axiom in Theorem 9, we showed that if $x \subseteq y \in S_L$ and

$$C' = \uparrow\{[g_z \sqcap g_y) : z \in C\},$$

then C' is an up-set included in $\uparrow C$ and in $\uparrow y$, and $\bigcap C' = y$. This proof depended on having $x \trianglelefteq C$, but that assumption was only used to infer that $\bigcap C \subseteq x$, which itself still holds here because now $x \blacktriangleleft C$.

Thus these properties of C' still hold here for any y with $x \subseteq y$. In particular they hold when $y = x$, in which case we conclude that C' is an up-set with $C' \subseteq \uparrow C$ and $\bigcap C' = x$. Hence $x \trianglelefteq C'$. But $C \subseteq X$ and X is an up-set, so $\uparrow C \subseteq X$. Altogether this gives $x \trianglelefteq C' \subseteq X$, hence $x \in j_\trianglelefteq X$.

Thus $j_\trianglelefteq X = j_\blacktriangleleft X$ for all X, so every $j_\trianglelefteq$-closed up-set is a $j_\blacktriangleleft$-closed up-set and vice versa. □

References

Bell, J. L. (2005). Cover schemes, frame-valued sets and their potential uses in spacetime physics. In A. Reimer (Ed.), *Spacetime physics research trends, horizons in World physics* (Vol. 248). New York: Nova Science Publishers. http://publish.uwo.ca/~jbell.

Bezhanishvili, G., & Holliday, W. H. (2016). Locales, nuclei, and Dragalin frames. In L. Beklemishev, S. Demri, & A. Máté (Eds.), *Advances in modal logic* (Vol. 11, pp. 177–196). London: College Publications.

Birkhoff, G., & Frink, O. (1948). Representations of lattices by sets. *Transactions of the American Mathematical Society*, *64*(2), 299–316.
Davey, B. A., & Priestley, H. A. (1990). *Introduction to lattices and order*. Cambridge: Cambridge University Press.
Drágalin, A. G. (1988). *Mathematical intuitionism: Introduction to proof theory*. Providence: American Mathematical Society. Translation of the 1979 Russian original by Elliott Mendelson.
Goldblatt, R. I. (1975). The Stone space of an ortholattice. *Bulletin of the London Mathematical Society*, *7*, 45–48.
Goldblatt, R. (2006). A Kripke-Joyal semantics for noncommutative logic in quantales. In G. Governatori, I. Hodkinson, & Y. Venema (Eds.), *Advances in modal logic* (Vol. 6, pp. 209–225). London: College Publications. www.aiml.net/volumes/volume6/.
Goldblatt, R. (2011). Cover semantics for quantified lax logic. *Journal of Logic and Computation*, *21*(6), 1035–1063. http://doi.org/10.1093/logcom/exq029.
Goldblatt, R. (2011). Grishin algebras and cover systems for classical bilinear logics. *Studia Logica*, *99*(1–3), 203–227.
Goldblatt, R. (2011). *Quantifiers, propositions and identity: Admissible semantics for quantified modal and substructural logics* (Vol. 38)., Lecture notes in logic No. 38, Cambridge: Cambridge University Press and the Association for Symbolic Logic.
Goldblatt, R. (2016). Cover systems for the modalities of linear logic. arXiv:1610.09117.
Hedlíková, J., & Pulmannová, S. (1991). Orthogonality spaces and atomistic orthocomplemented lattices. *Czechoslovak Mathematical Journal*, *41*(1), 8–23.
Johnstone, P. T. (1982). *Stone spaces*. Cambridge: Cambridge University Press.
Johnstone, P. T. (2002). *Sketches of an elephant: A topos theory compendium*. Oxford: Oxford University Press.
Mac Lane, S., & Moerdijk, I. (1992). *Sheaves in geometry and logic: A first introduction to topos theory*. New York: Springer.
Macnab, D. S. (1976). *An Algebraic Study of Modal Operators on Heyting Algebras, with Applications to Topology and Sheafification*. Ph.D thesis, University of Aberdeen.
Macnab, D. S. (1981). Modal operators on Heyting algebras. *Algebra Universalis*, *12*, 5–29.
MacNeille, H. M. (1937). Partially ordered sets. *Transactions of the American Mathematical Society*, *42*, 416–460.
Simmons, H. (1978). A framework for topology. In A. Macintyre, L. Pacholski, & J. Paris (Eds.), *Logic Colloqium '77* (pp. 239–251)., Studies in logic. North-Holland.
Stone, M. H. (1937–8). Topological representations of distributive lattices and Brouwerian logics. *Casopis pro Pestovani Matematiky a Fysiky*, *67*, 1–25.

A Uniform Algebraic Approach to Cut Elimination via Semi-completeness

Hiroakira Ono

Abstract This is an attempt to present a uniform algebraic framework for semantical approach to cut elimination. The basic idea came from a paper by S. Maehara of 1991. Using the notion of *quasi-homomorphisms* essentially due to Maehara, an algebraic condition called *semi-completeness* of a given sequent system is introduced. It is shown that for a given sequent system **S**, semi-completeness for the system $\mathbf{S}^-$ implies cut elimination for **S**, where $\mathbf{S}^-$ is obtained from **S** by deleting cut rule. In the present paper it is confirmed that many of existing semantical proofs of cut elimination using either Kripke semantics or algebraic one will fall into our algebraic framework. In fact, semi-completeness is considered to be an intelligible algebraic criterion of cut elimination which is applicable to both single- and multiple-succedent sequent systems for wide variety of nonclassical logics, including modal logics and substructural logics. For modal logics and intuitionistic logic, connections of quasi-homomorphisms with *downward saturations*, the conditions which are used in the constructions of *canonical* Kripke models, will be clarified. On the other hand, for substructural logics, quasi-homomorphisms will be discussed in relation to *quasi-embeddings* which are crucial in algebraic approaches to cut elimination. In the last three sections, semi-completeness arguments are extended so as to cover semantical proofs of cut elimination for nonclassical *predicate logics*. This can be carried out by generalizing quasi-homomorphisms on *expanded algebraic structures*. In this way, semi-completeness will provide a unified view of comprehending various semantical approaches to cut elimination.

1 Maehara's Approach Revisited

In his paper Maehara (1991), S. Maehara gave a *semi-algebraic* proof of cut elimination for both classical and intuitionistic sequent systems for simple type theory.[1]

H. Ono (✉)
Japan Advanced Institute of Science and Technology, Nomi, Japan
e-mail: ono@jaist.ac.jp

[1]For further exposition of Maehara's semantical proof of cut elimination for second order calculi, see a recent paper by Arai (2017).

S.C.-M. Yang et al. (eds.), *Philosophical Logic: Current Trends in Asia*,
Logic in Asia: Studia Logica Library, DOI 10.1007/978-981-10-6355-8_2

Based on his idea, we present here a uniform algebraic framework for semantical approaches to cut elimination. To clarify the essence of his work and to show its generality, we introduce the notion of *semi-completeness* of a given sequent system, which means the completeness with respect to *quasi-homomorphisms* instead of homomorphisms (or valuations).[2]

In the next section, the notion of semi-completeness is introduced and it is shown that semi-completeness for a cut-free system implies cut elimination. To see how our idea will work, we will discuss semi-completeness of sequent systems for both modal logic **S4** and intuitionistic logic in Sects. 3 and 4, respectively, comparing it with standard semantical proof using Kripke semantics of cut elimination. It will be pointed out that required properties of quasi-homomorphisms follow exactly from conditions of *downward saturation* in *canonical Kripke models*. Then, by taking the *complex algebras* of Kripke frames, standard semantical methods for cut elimination using Kripke frames can be naturally incorporated into our algebraic framework. On the other hand, semi-completeness of sequent systems for basic substructural logics will be discussed in relation to *quasi-embeddings* in their algebraic proof of cut elimination in our joint paper Belardinelli et al. (2004) with F. Belardinelli and P. Jipsen. As a matter of fact, quasi-embeddings which came also from Maehara (1991) can be regarded as an alternative way of presenting quasi-homomorphisms.

The last three sections will be devoted to show how we can extend these arguments about semi-completeness to predicate logics. For substructural predicate logics, this was essentially done in Belardinelli et al. (2004) in terms of quasi-embeddings. On the other hand, there may be some difficulties in transforming semantical proofs of cut elimination using Kripke semantics into algebraic ones, when a predicate logic under consideration is complete with respect to *Kripke frames with varying domains*, like intuitionistic predicate logic and modal predicate logic for **S4**. To overcome this we will introduce *expanded algebraic structures* and *general quasi-homomorphisms* on them in Sect. 8. Then, semi-completeness with respect to expanded algebraic structures is shown to work well in these cases.

In this way, semi-completeness arguments using (general) quasi-homomorphisms are shown to cover many of existing standard semantical methods of cut elimination for various sequent systems, and thus we can conclude that they will provide a uniform algebraic framework for semantical proofs of cut elimination. This will answer a question which we posed in my previous paper Ono (2015). We note here that in Lahav and Avron (2014), authors introduced a semantical framework using Kripke-type semantics, in order to discuss related problems. It is noticed in Sect. 6 in Lahav and Avron (2014) that some obstacles may happen in treating single-succedent sequent systems like systems for substructural logics.

[2]See Footnote 3.

2 Semi-completeness

We will explain Maehara's idea in Maehara (1991) firstly in the context of modal propositional logics. We take $\wedge$, $\vee$, $\neg$ and $\Box$ for logical connectives of our language, and use $\cap$, $\cup$, $'$ and $\Box$ for expressing corresponding operations of modal algebras. In the following, Ω_M denotes the set of all formulas of this language. We say a nonempty subset Z of Ω_M is *subformula-closed* when for any subformula β of α if $\alpha \in Z$ then $\beta \in Z$.

Definition 1 (*partial homomorphisms on modal algebras*) Let $\mathbf{A} = \langle A, \cap, \cup, ', \Box \rangle$ be any modal algebra. A mapping f from a subformula-closed subset Z of Ω_M to A is a *partial homomorphism* on $\mathbf{A}$ if it satisfies the following conditions.

1. $f(\alpha \wedge \beta) = f(\alpha) \cap f(\beta)$ for $\alpha \wedge \beta \in Z$,
2. $f(\alpha \vee \beta) = f(\alpha) \cup f(\beta)$ for $\alpha \vee \beta \in Z$,
3. $f(\neg\alpha) = f(\alpha)'$ for $\neg\alpha \in Z$,
4. $f(\Box\alpha) = \Box f(\alpha)$ for $\Box\alpha \in Z$.

When $Z = \Omega_M$, the mapping f is simply called a *homomorphism* (on $\mathbf{A}$).[3]

The set Z in Definition 1 is called the *domain* of f. The following will be almost trivial. For, it is enough to take an arbitrary element in A for each propositional variable which does not belong to the domain of a given partial homomorphism.

Lemma 1 *Every partial homomorphism to an algebra* $\mathbf{A}$ *can be extended to a homomorphism to* $\mathbf{A}$.

The notion of partial homomorphisms can be generalized in the following way.

Definition 2 (*quasi-homomorphisms on modal algebras*) Let $\mathbf{A} = \langle A, \cap, \cup, ', \Box \rangle$ be any modal algebra. A *quasi-homomorphism* on $\mathbf{A}$ is a pair of mappings k and K from a subformula-closed subset Z of Ω_M to A, which satisfies the following conditions.

1. $k(\alpha) \le K(\alpha)$ for $\alpha \in Z$,
2. $k(\alpha \wedge \beta) \le k(\alpha) \cap k(\beta)$ and $K(\alpha) \cap K(\beta) \le K(\alpha \wedge \beta)$ for $\alpha \wedge \beta \in Z$,
3. $k(\alpha \vee \beta) \le k(\alpha) \cup k(\beta)$ and $K(\alpha) \cup K(\beta) \le K(\alpha \vee \beta)$ for $\alpha \vee \beta \in Z$,
4. $k(\neg\alpha) \le K(\alpha)'$ and $k(\alpha)' \le K(\neg\alpha)$ for $\neg\alpha \in Z$,
5. $k(\Box\alpha) \le \Box k(\alpha)$ and $\Box K(\alpha) \le K(\Box\alpha)$ for $\Box\alpha \in Z$.

Mappings k and K are sometimes called the *lower* and the *upper* mappings of this quasi-homomorphism.

---

[3]The word *valuation* in place of *homomorphism* might be suitable. But, for further generalization, here we call it homomorphism, considering Ω_M as a freely generated modal algebra. By the same reason, we use the word *quasi-homomorphism* instead of *quasi-valuation* in Definition 2 given below.

The set Z in Definition 2 is called also the *domain* of (k, K). It is easily shown that if $k(\alpha) = K(\alpha)$ for each $\alpha \in Z$ for a quasi-homomorphism (k, K) with the domain Z, then k (and hence K) is a partial homomorphism from Z to $\mathbf{A}$. Because of this, sometimes we identify a partial homomorphism f with a quasi-homomorphism (f, f). We note that we cannot expect the result corresponding to Lemma 1 for quasi-homomorphisms. By using induction on the complexity of a formula α, the following can be shown easily.

Lemma 2 *Suppose that both a partial homomorphism f and a quasi-homomorphism (k, K) are mappings from Z to an algebra $\mathbf{A}$ such that $k(p) \leq f(p) \leq K(p)$ for each propositional variable $p \in Z$. Then $k(\alpha) \leq f(\alpha) \leq K(\alpha)$ for every formula $\alpha \in Z$.*

The following result plays a crucial role in the proof of the main theorem of Maehara (1991).

Theorem 3 *For formulas $\alpha_1, \ldots, \alpha_m, \beta_1, \ldots, \beta_n$, if*

$$g(\alpha_1) \cap \ldots \cap g(\alpha_m) \leq g(\beta_1) \cup \ldots \cup g(\beta_n) \tag{1}$$

holds for every homomorphism g on an algebra $\mathbf{A}$, then

$$k(\alpha_1) \cap \ldots \cap k(\alpha_m) \leq K(\beta_1) \cup \ldots \cup K(\beta_n) \tag{2}$$

holds for every quasi-homomorphism (k, K) from any domain Z to $\mathbf{A}$ such that all formulas $\alpha_1, \ldots, \alpha_m, \beta_1, \ldots, \beta_n$ belong to Z.

Proof For any given quasi-homomorphism (k, K), take a partial homomorphism f from Z to $\mathbf{A}$ satisfying $k(p) \leq f(p) \leq K(p)$ for each variable $p \in Z$. (For instance, let $f(p) = k(p)$ for each $p \in Z$). By Lemma 2, the inequality $k(\gamma) \leq f(\gamma) \leq K(\gamma)$ holds for every formula $\gamma \in Z$, and thus in particular for each α_i and β_j. Take an arbitrary homomorphism f^* to $\mathbf{A}$ which is an extension of f, whose existence is assured by Lemma 1. By our assumption,

$$f^*(\alpha_1) \cap \ldots \cap f^*(\alpha_m) \leq f^*(\beta_1) \cup \ldots \cup f^*(\beta_n)$$

holds. Thus $k(\alpha_1) \cap \ldots \cap k(\alpha_m) \leq f^*(\alpha_1) \cap \ldots \cap f^*(\alpha_m) \leq f^*(\beta_1) \cup \ldots \cup f^*(\beta_n) \leq K(\beta_1) \cup \ldots \cup K(\beta_n)$ holds. □

Note that in (2) we understand that $1 \leq K(\beta_1) \cup \ldots \cup K(\beta_n)$ when $m = 0$ and that $k(\alpha_1) \cap \ldots \cap k(\alpha_m) \leq 0$ when $n = 0$. We follow the same conventions for g in (1).

For a given sequent system $\mathbf{S}$ for a normal modal logic $\mathbf{M}$, by an $\mathbf{M}$-algebra we mean any modal algebra such that any formula α which is provable in $\mathbf{S}$ is valid in it. By standard argument using Lindenbaum algebras, we have that a formula α is provable in $\mathbf{M}$ if and only if it is valid in any $\mathbf{M}$-algebra. In particular, soundness of a modal logic $\mathbf{M}$ can be expressed as follows.

Lemma 4 *Suppose that* **S** *is a sequent system for a given modal logic* **M**. *If a sequent* $\alpha_1, \ldots, \alpha_m \Rightarrow \beta_1, \ldots, \beta_n$ *is provable in* **S**, *then* $g(\alpha_1) \cap \ldots \cap g(\alpha_m) \leq g(\beta_1) \cup \ldots \cup g(\beta_n)$ *holds for any* **M**-*algebra* **A** *and any homomorphism* g *on* **A**.

Combining this with Theorem 3, we have the following.

Corollary 5 *Suppose that* **S** *is a sequent system for a given modal logic* **M**. *If a sequent* $\alpha_1, \ldots, \alpha_m \Rightarrow \beta_1, \ldots, \beta_n$ *is provable in* **S**, *then*

$$k(\alpha_1) \cap \ldots \cap k(\alpha_m) \leq K(\beta_1) \cup \ldots \cup K(\beta_n)$$

holds for every quasi-homomorphism (k, K) *from any domain* Z *to* **A** *such that all formulas* $\alpha_1, \ldots, \alpha_m, \beta_1, \ldots, \beta_n$ *belong to* Z.

Suppose that sequents $\alpha \Rightarrow \beta$ and $\beta \Rightarrow \gamma$ are both provable in **S**. But it may happen that $\alpha \Rightarrow \gamma$ is not provable in **S** if cut rule is not admissible in **S**. Corresponding to this, though $k(\alpha) \leq K(\beta)$ and $k(\beta) \leq K(\gamma)$ hold for a given (k, K) to an algebra **A** by the above corollary, we are not sure whether $k(\alpha) \leq K(\gamma)$ holds or not. When the converse of Corollary 5 holds for a sequent system **S**, we say that **S** is *semi-complete*.

Definition 3 (*semi-completeness for modal logics*) A sequent system **S** is *semi-complete* with respect to a class $\mathscr{C}$ of modal algebras, when for all formulas $\alpha_1, \ldots, \alpha_m, \beta_1, \ldots, \beta_n$, if the inequality

$$k(\alpha_1) \cap \ldots \cap k(\alpha_m) \leq K(\beta_1) \cup \ldots \cup K(\beta_n) \tag{3}$$

holds for each **M**-algebra $\mathbf{A} \in \mathscr{C}$ and each quasi-homomorphism (k, K) on **A** with a domain Z such that $\alpha_1, \ldots, \alpha_m, \beta_1, \ldots, \beta_n \in Z$, then the sequent $\alpha_1, \ldots, \alpha_m \Rightarrow \beta_1, \ldots, \beta_n$ is provable in **S**.

When **S** is semi-complete with respect to a singleton set $\{\mathbf{A}\}$, we say simply that **S** is semi-complete with respect to the algebra **A**. By Corollary 5 with Definition 3, we have immediately the following.

Lemma 6 *Let* **S** *be a sequent system for a modal logic* **M** *and* $\mathbf{S}^-$ *be the sequent system obtained from* **S** *by deleting the cut rule. If* $\mathbf{S}^-$ *is semi-complete with respect to a class* $\mathscr{C}$ *of* **M**-*algebras, then cut elimination holds for* **S**.

Sometimes the following contrapositive form of semi-completeness may be more convenient. That is, a sequent system **S** is *semi-complete* with respect to $\mathscr{C}$, whenever

> *if a sequent* $\alpha_1, \ldots, \alpha_m \Rightarrow \beta_1, \ldots, \beta_n$ *is not provable in* **S**, *then there exists* (k, K) *on some modal algebra* $\mathbf{A} \in \mathscr{C}$ *such that*
>
> $$k(\alpha_1) \cap \ldots \cap k(\alpha_m) \not\leq K(\beta_1) \cup \ldots \cup K(\beta_n). \tag{4}$$

If such a homomorphism g is found as (k, K) in (4), or more precisely, if (g, g) satisfies (4) for a mapping g, the condition implies completeness of **S** with respect to $\mathscr{C}$. As a matter of fact, semi-completeness with respect to $\mathscr{C}$ means completeness of S with respect to quasi-homomorphisms on algebras in $\mathscr{C}$. The main reason why we focus on quasi-homomorphims in the context of cut elimination comes from the fact that in many cases it is much easier to find a necessary quasi-homomorphim than to give a required homomorphism in a definite form.

3 Quasi-homomorphisms and Downward Saturations

To explain how Lemma 6 on semi-completeness can be applied, we take here a sequent system **GS4** for the modal logic **S4** as an example. The following proof is obtained essentially by *algebraizing* a semantical proof of cut elimination for **GS4** using Kripke frames, which is shown e.g. in Ono (2015). Our sequent system **GS4** is obtained from the system **LK** for classical logic by adding the following two rules $(\Box \Rightarrow)$ and $(\Rightarrow \Box)$ for $\Box$;

$$\frac{\alpha, \Gamma \Rightarrow \Delta}{\Box\alpha, \Gamma \Rightarrow \Delta}\ (\Box \Rightarrow) \qquad\qquad \frac{\Box\Gamma \Rightarrow \alpha}{\Box\Gamma \Rightarrow \Box\alpha}\ (\Rightarrow \Box)$$

To simplify our arguments, each sequent is supposed to be of the form $\Sigma \Rightarrow \Theta$ where Σ and Θ are finite (possibly empty) sets of formulas. Thus, our system has neither exchange rules nor contraction rules. Let **GS4**$^-$ be the system **GS4** without cut rule. In the following, we will show the semi-completeness of **GS4**$^-$ in a stronger form. That is, we will give *uniformly* an **S4**-algebra **B** and a quasi-homomorphism (k, K) from the set Ω_M of all modal formulas to **B** (hence both k and K are total mappings) such that for every sequent $\alpha_1, \ldots, \alpha_m \Rightarrow \beta_1, \ldots, \beta_n$, if it is not provable in **GS4**$^-$ then the corresponding inequality (3) does not hold for the quasi-homomorphism (k, K) on this algebra **B**. For this purpose, we refer our semantical proof of cut elimination for **GS4** in Ono (2015), which is briefly sketched below.

We say that a pair (Σ, Θ) of (possibly infinite) subsets Σ and Θ of Ω_M is **GS4**$^-$-*consistent* (in Ω_M) if the sequent $\alpha_1, \ldots, \alpha_m \Rightarrow \beta_1, \ldots, \beta_n$ is not provable in **GS4**$^-$ for any $\alpha_1, \ldots, \alpha_m \in \Sigma$ and any $\beta_1, \ldots, \beta_n \in \Theta$. A **GS4**$^-$-consistent pair (Σ, Θ) is *maximal* **GS4**$^-$*-consistent* (in Ω_M), if it is **GS4**$^-$-consistent but neither $(\Sigma \cup \{\gamma\}, \Theta)$ nor $(\Sigma, \Theta \cup \{\gamma\})$ is **GS4**$^-$-consistent for any $\gamma \in \Omega_M \backslash (\Sigma \cup \Theta)$. It can be shown e.g. by using Zorn's lemma that each **GS4**$^-$-consistent pair (Σ_0, Θ_0) can be extended to a maximal **GS4**$^-$-consistent pair (Σ, Θ). As **GS4**$^-$ does not have cut rule, the union of $\Sigma \cup \Theta$ is not always equal to Ω_M for a maximal **GS4**$^-$-consistent pair (Σ, Θ).

Let V be the set of all maximal **GS4**$^-$-consistent pairs. Clearly, the power set $\wp(V)$ of V forms a Boolean algebra with respect to set theoretical operations $\cap, \cup$ and $'$. Further, we introduce a unary operation $\Box^\circ$ on $\wp(V)$, by defining $\Box^\circ S =$

$\{(\Sigma, \Theta) \in V : \textit{for every } (\Gamma, \Delta) \in V, \textit{ if } \Sigma_{\Box} \subseteq \Gamma_{\Box} \textit{ then } (\Gamma, \Delta) \in S\}$ for each subset S of V. Here $\Lambda_{\Box}$ denotes the set $\{\alpha : \Box\alpha \in \Lambda\}$ for every subset Λ of Ω_M. It is easily verified that the algebra $\mathbf{B} = \langle \wp(V), \Box^{\circ} \rangle$ forms an **S4**-algebra. We can show the following basic result. See e.g. Ono (2015) for the details.

Lemma 7 (downward saturation in modal logics) *The following holds for every maximal* **GS4**$^{-}$*-consistent pair* (Σ, Θ).

(1a) *if* $\beta \wedge \gamma \in \Sigma$ *then both* β *and* γ *are in* Σ,
(1b) *if* $\beta \wedge \gamma \in \Theta$ *then either* β *or* γ *are in* Θ,
(2a) *if* $\beta \vee \gamma \in \Sigma$ *then either* β *or* γ *are in* Σ,
(2b) *if* $\beta \wedge \gamma \in \Theta$ *then both* β *and* γ *are in* Θ,
(3a) *if* $\neg\beta \in \Sigma$ *then* β *is in* Θ,
(3b) *if* $\neg\beta \in \Theta$ *then* β *is in* Σ,
(4a) *if* $\Box\beta \in \Sigma$ *then* $\beta \in \Gamma$ *for each* $(\Gamma, \Delta) \in V$ *such that* $\Sigma_{\Box} \subseteq \Gamma_{\Box}$,
(4b) *if* $\Box\beta \in \Theta$ *then* $\beta \in \Delta$ *for some* $(\Gamma, \Delta) \in V$ *such that* $\Sigma_{\Box} \subseteq \Gamma_{\Box}$.

Next, we introduce mappings k and K on $\mathbf{B}$ by $k(\alpha) = \{(\Sigma, \Theta) \in V : \alpha \in \Sigma\}$ and $K(\alpha) = \{(\Sigma, \Theta) \in V : \alpha \notin \Theta\}$ for each $\alpha \in \Omega_M$. As our proof of the following lemma shows, every condition for downward saturation is reflected exactly by a condition for quasi-homomorphisms through mappings k and K.

Lemma 8 *The pair* (k, K) *is a quasi-homomorphism from* Ω_M *to* $\mathbf{B}$.

Proof First, we show that $k(\alpha) \subseteq K(\alpha)$. If $(\Sigma, \Theta) \in k(\alpha)$, then $\alpha \in \Sigma$ and hence α cannot be a member of Θ because of the **GS4**$^{-}$-consistency of (Σ, Θ). Thus, $(\Sigma, \Theta) \in K(\alpha)$. The fact that k and K satisfy remaining conditions of quasi-homomorphisms can be derived by Lemma 7. To see this, let us give only proofs of the following two conditions $k(\Box\alpha) \subseteq \Box^{\circ}k(\alpha)$ and $\Box^{\circ}K(\alpha) \subseteq K(\Box\alpha)$. Suppose that $(\Sigma, \Theta) \in k(\Box\alpha)$. This means that $\Box\alpha \in \Sigma$. To show that $(\Sigma, \Theta) \in \Box^{\circ}k(\alpha)$, take any $(\Gamma, \Delta) \in V$ such that $\Sigma_{\Box} \subseteq \Gamma_{\Box}$. By Lemma 7 (4a), $\alpha \in \Gamma$. This means that $(\Gamma, \Delta) \in k(\alpha)$. Thus, $(\Sigma, \Theta) \in \Box^{\circ}k(\alpha)$. Next suppose that $(\Sigma, \Theta) \notin K(\Box\alpha)$. Then $\Box\alpha \in \Theta$. By Lemma 7 (4b), $\alpha \in \Delta$ holds for some $(\Gamma, \Delta) \in V$ such that $\Sigma_{\Box} \subseteq \Gamma_{\Box}$. That is, $(\Gamma, \Delta) \notin K(\alpha)$ for some (Γ, Δ) with $\Sigma_{\Box} \subseteq \Gamma_{\Box}$. Hence, $(\Sigma, \Theta) \notin \Box^{\circ}K(\alpha)$. □

Lemma 9 *The sequent system* **GS4**$^{-}$ *is semi-complete with respect to the* **S4**-*algebra* $\mathbf{B}$.

Proof Let $\alpha_1, \ldots, \alpha_m \Rightarrow \beta_1, \ldots, \beta_n$ be any sequent which is not provable in **GS4**$^{-}$, i.e., the pair $(\{\alpha_1, \ldots, \alpha_m\}, \{\beta_1, \ldots, \beta_n\})$ is **GS4**$^{-}$-consistent. Then there exists a maximal **GS4**$^{-}$-consistent pair (Σ, Θ) such that $\alpha_i \in \Sigma$ for each $i \leq m$ and $\beta_j \in \Theta$ for each $j \leq n$. This implies that $(\Sigma, \Theta) \in k(\alpha_1) \cap \ldots \cap k(\alpha_m)$, while $(\Sigma, \Theta) \notin K(\beta_1) \cup \ldots \cup K(\beta_n)$. Thus, the inequality (3) does not hold. □

Then we have the following from Lemma 6 together with Lemma 9.

Corollary 10 *Cut elimination holds for* **GS4**.

Semi-completeness of sequent systems for other modal logics, including **GK** and **GKT** discussed in Ono (2015), can be shown in the same way. The above argument looks quite similar to standard proof of completeness of modal logics using canonical frames. In fact, if we replace $\mathbf{GS4}^-$ by **GS4** in the above proof, the algebra **B** is exactly the complex algebra of the canonical frame for **S4** and the mapping K, which is equal to k, is the *canonical valuation* on it.

To derive cut elimination, it is not necessary to show the semi-completeness with respect to a fixed algebra in a uniform way. That is, it is enough to give such an algebra and a quasi-homomorphism, *depending on* each unprovable sequent. This can be done by modifying the above proofs slightly, as shown below.

For a given sequent $\alpha_1, \ldots, \alpha_m \Rightarrow \beta_1, \ldots, \beta_n$, let $\Omega^\sharp$ be the set of all subformulas of formulas in this sequent. We will restrict our attention only to formulas in $\Omega^\sharp$. Let $V^\sharp$ be the set of all maximal $\mathbf{GS4}^-$-consistent pairs in $\Omega^\sharp$. As $\Omega^\sharp$ is finite, so is $V^\sharp$. We can carry on the rest of arguments in the same as before, only by replacing Ω by $\Omega^\sharp$ and V by $V^\sharp$. (See Sect. 1.5 of Ono (2015) for the details.) In this way, we can get a *finite* **S4**-algebra $\mathbf{B}^\sharp$ and a quasi-homomorphism $(k^\sharp, K^\sharp)$ such that the inequality (3) does not hold for these formulas $\alpha_1, \ldots, \alpha_m, \beta_1, \ldots, \beta_n$. Hence, we have the following in place of Lemma 9.

Lemma 11 *The sequent system* $\mathbf{GS4}^-$ *is semi-complete with respect to the set of all finite* **S4**-*algebras.*

This implies the finite model property of the modal logic **S4**. In fact, if a formula α is not provable in **S4**, then the sequent $\Rightarrow \alpha$ is not provable in $\mathbf{GS4}^-$. By Lemma 11 together with Theorem 3, there exist a finite **S4**-algebra **A** and a homomorphism f on **A** such that $1 \not\leq f(\alpha)$. This means the finite model property of **S4**.

4 A Semi-complete Sequent System for Intuitionistic Logic

The above proof of cut elimination via semi-completeness can be applied to sequent systems for some other logics which are Kripke complete. To see this, in this section we will outline a proof of cut elimination for a sequent system **LJ**′ for intuitionistic logic, which is sometimes called **G3im**. The system **LJ**′ is a multiple-succedent system obtained from the sequent system **LK** for classical logic by restricting the right implication rule to the following form;

$$\frac{\alpha, \Gamma \Rightarrow \beta}{\Gamma \Rightarrow \alpha \to \beta}\ (\Rightarrow\to)$$

We assume that our language consists of primitive logical connectives $\wedge, \vee, \to$ and a constant symbol 0. As usual, the negation $\neg\alpha$ of α is defined by $\alpha \to 0$. For convenience sake, the constant 0 is regarded as a subformula of any formula. Let Ω_I be the set of all formulas in this language. As before, we suppose that each sequent is of the form $\Sigma \Rightarrow \Theta$ with finite sets Σ and Θ of formulas. For this language, we need to modify the definition of quasi-homomorphisms slightly, as shown below.

Definition 4 (*quasi-homomorphisms on Heyting algebras*) Let $\mathbf{A} = \langle A, \cap, \cup, \rightarrow, 0_A \rangle$ be any Heyting algebra.[4] Here, 0_A denotes the smallest element of A. A *quasi-homomorphism* on $\mathbf{A}$ is a pair of mappings k and K from a subformula-closed subset Z of Ω_I to A which satisfy the following conditions.

1. $k(\alpha) \leq K(\alpha)$ for $\alpha \in Z$,
2. $k(\alpha \wedge \beta) \leq k(\alpha) \cap k(\beta)$ and $K(\alpha) \cap K(\beta) \leq K(\alpha \wedge \beta)$ for $\alpha \wedge \beta \in Z$,
3. $k(\alpha \vee \beta) \leq k(\alpha) \cup k(\beta)$ and $K(\alpha) \cup K(\beta) \leq K(\alpha \vee \beta)$ for $\alpha \vee \beta \in Z$,
4. $k(0) = 0_A$,
5. $k(\alpha \rightarrow \beta) \leq K(\alpha) \rightarrow k(\beta)$ and $k(\alpha) \rightarrow K(\beta) \leq K(\alpha \rightarrow \beta)$ for $\alpha \rightarrow \beta \in Z$.

Now, it is easy to see that statements which correspond to Theorem 3 and Corollary 5 hold also for the sequent system $\mathbf{LJ}'$ and any Heyting algebra $\mathbf{A}$. The semi-completeness of $\mathbf{LJ}'^-$, i.e. $\mathbf{LJ}'$ without cut rule, can be shown similarly, again with the help of Kripke frames as follows. Let W be the set of all maximal $\mathbf{LJ}'^-$-consistent pairs. We introduce a binary relation $\leq$ of W by the condition that $(\Sigma, \Theta) \leq (\Gamma, \Delta)$ if and only if $\Sigma \subseteq \Gamma$. Clearly, $\leq$ is a preorder on W. Now, let $\mathscr{U}(W)$ is the set of all upward closed subsets of W with respect to $\leq$. By standard arguments, we can show that the algebra $\mathbf{C}$ defined by $\langle \mathscr{U}(W), \cap, \cup, \rightarrow, \emptyset \rangle$ forms a Heyting algebra. Here, for $S, T \in \mathscr{U}(W)$, the set $S \rightarrow T$ is defined by $\{(\Sigma, \Theta) : \textit{if } (\Gamma, \Delta) \in S \textit{ then } (\Gamma, \Delta) \in T \textit{ for every } (\Gamma, \Delta) \textit{ such that } (\Sigma, \Theta) \leq (\Gamma, \Delta)\}$. The set $S \rightarrow T$ can be shown to be a member of $\mathscr{U}(W)$. Corresponding to Lemma 7, we have the following.

Lemma 12 (downward saturation in intuitionistic logic) *The following holds for every* (Σ, Θ) *of* W.

(1a) *if* $\beta \wedge \gamma \in \Sigma$ *then both* β *and* γ *are in* Σ,
(1b) *if* $\beta \wedge \gamma \in \Theta$ *then either* β *or* γ *are in* Θ,
(2a) *if* $\beta \vee \gamma \in \Sigma$ *then either* β *or* γ *are in* Σ,
(2b) *if* $\beta \wedge \gamma \in \Theta$ *then both* β *and* γ *are in* Θ,
(3) $0 \notin \Sigma$,
(4a) *if* $\beta \rightarrow \gamma \in \Sigma$, *then either* $\beta \in \Delta$ *or* $\gamma \in \Gamma$ *for each* $(\Gamma, \Delta) \in W$ *such that* $\Sigma \subseteq \Gamma$,
(4b) *if* $\beta \rightarrow \gamma \in \Theta$, *then* $\beta \in \Gamma$ *and* $\gamma \in \Delta$ *for some* $(\Gamma, \Delta) \in W$ *such that* $\Sigma \subseteq \Gamma$.

Proof We will give here a proof of (4a) and (4b). Suppose that $\beta \rightarrow \gamma \in \Sigma$. Moreover, suppose that there exists $(\Gamma, \Delta) \in W$ such that $\Sigma \subseteq \Gamma$ such that $\beta \notin \Delta$ or $\gamma \notin \Gamma$. Then, both $(\Gamma, \Delta \cup \{\beta\})$ and $(\{\gamma\} \cup \Gamma, \Delta)$ are $\mathbf{LJ}'^-$-inconsistent, which means that both $\Gamma \Rightarrow \Delta, \beta$ and $\gamma, \Gamma \Rightarrow \Delta$ are provable in $\mathbf{LJ}'^-$. Hence, $\beta \rightarrow \gamma, \Gamma \Rightarrow \Delta$ must be provable in it. But as (Γ, Δ) is $\mathbf{LJ}'^-$-consistent, $\beta \rightarrow \gamma$ cannot be a member of Γ. On the other hand, $\beta \rightarrow \gamma \in \Sigma \subseteq \Gamma$. This is a contradiction.

[4]By abuse of symbols, we use $\rightarrow$ for both logical connective and algebraic operation.

Suppose next that $\beta \to \gamma \in \Theta$. Then, $(\Sigma \cup \{\beta\}, \{\gamma\})$ must be $\mathbf{LJ}'^{-}$-consistent. For, otherwise $\Sigma \Rightarrow \beta \to \gamma$ is provable by using the right implication rule of $\mathbf{LJ}'$ and hence (Σ, Θ) is $\mathbf{LJ}'^{-}$-inconsistent, which contradicts our assumption. Now, take a maximal $\mathbf{LJ}'^{-}$-consistent pair (Γ, Δ) which is an extension of $(\Sigma \cup \{\beta\}, \{\gamma\})$. Then $\Sigma \subseteq \Gamma$, $\beta \in \Gamma$ and $\gamma \in \Delta$. □

As before, we define k and K by $k(\alpha) = \{(\Sigma, \Theta) \in W : \alpha \in \Sigma\}$ and $K(\alpha) = \{(\Sigma, \Theta) \in W : \alpha \notin \Theta\}$. Then, by using Lemma 12, it is easily seen that the following holds.

Lemma 13 *The pair (k, K) is a quasi-homomorphism on the Heyting algebra* **C**.

In addition, we can show a result similar to Lemma 11. Thus, we have the following.

Theorem 14 *The sequent system* $\mathbf{LJ}'^{-}$ *is semi-complete with respect not only to the Heyting algebra* **C**, *but also to set of all finite Heyting algebras. Thus, cut elimination holds for* $\mathbf{LJ}'$.

5 Quasi-homomorphisms and Quasi-embeddings

We have shown that semantical proofs of cut elimination in Ono (2015) using Kripke frames can be transformed into algebraic proofs of cut elimination through semi-completeness. This was carried out quite similarly to the proof of algebraic completeness of a given modal logic **M** by using the complex algebra of canonical frame of **M** when **M** is canonical.

In this section, we will explain that algebraic proofs of cut elimination developed in Belardinelli et al. (2004) can be treated also within the present framework, and in fact they can be regarded as algebraic proofs of cut elimination via semi-completeness of sequent systems. As shown below, required quasi-homomorphisms in our proofs can be derived directly from *quasi-embedding* in Theorem 5.3 of Belardinelli et al. (2004). This will make an interesting contrast with proofs in previous sections, since arguments developed in Belardinelli et al. (2004) has a close resemblance to *MacNeille completions* while arguments in the previous sections to *canonical extensions* (see e.g. Galatos et al. (2007) for the details).

We will show semi-completeness of the sequent system $\mathbf{FL_{ew}}$ for substructural logic with exchange rule and weakening rule, as an example, by following an algebraic proof of cut elimination for $\mathbf{FL_{ew}}$ described in details in Belardinelli et al. (2004).[5] We are going to outline the proof of semi-completeness of $\mathbf{FL_{ew}}$, by giving a brief sketch of the corresponding proof in Belardinelli et al. (2004), with a slight modification of expressions and statements, when necessary, in order to adjust

[5]The very basic idea of algebraic proofs of cut elimination in Belardinelli et al. (2004) came partly from Maehara (1991), as mentioned in it.

them to the present paper. Semi-completeness for other sequent systems discussed in Belardinelli et al. (2004) can be also obtained in the same way. Thus, our algebraic approach can cover both single- and multiple-succedent sequent systems for various nonclassical logics, including modal logics and substructural logics. (See Galatos et al. (2007) for general information on substructural logics.)

The language for the sequent system $\mathbf{FL_{ew}}$ is obtained from the language for $\mathbf{LJ}'$ in the previous section by adding a logical connective *fusion* $\cdot$ and a constant symbol 1. Each sequent of $\mathbf{FL_{ew}}$ is of the form $\Sigma \Rightarrow \theta$ with a *multiset* Σ of formulas and a formula θ. Both Σ and θ may possibly be empty.

The sequent system $\mathbf{FL_{ew}}$ is obtained from the single succedent sequent system $\mathbf{LJ}$ for intuitionistic logic, by deleting contraction rule but having the following rules for fusion:

$$\frac{\alpha, \beta, \Gamma \Rightarrow \delta}{\alpha \cdot \beta, \Gamma \Rightarrow \delta}\ (\cdot \Rightarrow) \qquad \frac{\Gamma \Rightarrow \alpha \quad \Sigma \Rightarrow \beta}{\Gamma, \Sigma \Rightarrow \alpha \cdot \beta}\ (\Rightarrow \cdot)$$

For constant symbols *zero* 0 and *unit* 1, we assume the following initial sequents and rules: (1) $0 \Rightarrow$, (2) $\Rightarrow 1$,

$$\frac{\Gamma \Rightarrow}{\Gamma \Rightarrow 0}\ (\Rightarrow 0) \qquad \frac{\Gamma \Rightarrow \delta}{1, \Gamma \Rightarrow \delta}\ (1 \Rightarrow)$$

In the following, Ω_S denotes the set of all formulas in this language. Again for convenience sake, both constant symbols 0 and 1 are regarded as subformulas of every formula.

Definition 5 ($\mathbf{FL_{ew}}$-*algebras*) An algebra $\mathbf{A} = \langle A, \cap, \cup, \cdot, \rightarrow, 1\rangle$ is a *commutative residuated lattice* if it satisfies the following[6]:

1. $\langle A, \cap, \cup\rangle$ is a lattice,
2. $\langle A, \cdot, 1\rangle$ is a commutative monoid with the unit element 1,
3. $x \cdot y \leq z$ iff $x \leq (y \rightarrow z)$, for any $x, y, z \in A$.

An algebra $\mathbf{A} = \langle A, \cap, \cup, \cdot, \rightarrow, 0, 1\rangle$ is an $\mathbf{FL_{ew}}$-*algebra*, if $\langle A, \cap, \cup, \cdot, \rightarrow, 1\rangle$ is a commutative residuated lattice such that $0 \leq x \leq 1$ holds for every $x \in A$.

It is easy to see that for each sequent, it is provable in the sequent system $\mathbf{FL_{ew}}$ if and only if it is valid in every $\mathbf{FL_{ew}}$-algebra. Similarly to Definition 4, we define quasi-homomorphisms to an arbitrary $\mathbf{FL_{ew}}$-algebra as follows.

Definition 6 (*quasi-homomorphisms on residuated lattices*)Let $\mathbf{A} = \langle A, \cap, \cup, \cdot, \rightarrow, 0_A, 1_A\rangle$ be any $\mathbf{FL_{ew}}$-algebra. A *quasi-homomorphism* from Ω_S to $\mathbf{A}$ is a pair of mappings k and K from a nonempty, subformula-closed subset Z of Ω_S to A which satisfy the following conditions.

[6] Again, we use the same symbol $\cdot$ for both fusion and monoid operation.

1. $k(\alpha) \leq K(\alpha)$ for $\alpha \in Z$,
2. $k(\alpha \wedge \beta) \leq k(\alpha) \cap k(\beta)$ and $K(\alpha) \cap K(\beta) \leq K(\alpha \wedge \beta)$ for $\alpha \wedge \beta \in Z$,
3. $k(\alpha \vee \beta) \leq k(\alpha) \cup k(\beta)$ and $K(\alpha) \cup K(\beta) \leq K(\alpha \vee \beta)$ for $\alpha \vee \beta \in Z$,
4. $k(\alpha \cdot \beta) \leq k(\alpha) \cdot k(\beta)$ and $K(\alpha) \cdot K(\beta) \leq K(\alpha \cdot \beta)$ for $\alpha \cdot \beta \in Z$,
5. $k(0) = 0_A$ and $K(1) = 1_A$,
6. $k(\alpha \to \beta) \leq K(\alpha) \to k(\beta)$ and $k(\alpha) \to K(\beta) \leq K(\alpha \to \beta)$ for $\alpha \to \beta \in Z$.

Because of our language Ω_S and of the interpretation of commas in antecedents of sequents by the monoid operation $\cdot$, it is necessary to modify the definition of semi-completeness in the following way (when the system is a single-succedent one).

Definition 7 (*semi-completeness for substructural logics*) A sequent system **S** for a substructural logic is *semi-complete* with respect to a class $\mathscr{C}$ of residuated lattices, when for all formulas $\alpha_1, \ldots, \alpha_m, \beta$, if the inequality

$$k(\alpha_1) \cdot \ldots \cdot k(\alpha_m) \leq K(\beta) \tag{5}$$

holds for each residuated lattice $\mathbf{A} \in \mathscr{C}$ and each quasi-homomorphism (k, K) on $\mathbf{A}$ with a domain Z such that $\alpha_1, \ldots, \alpha_m, \beta \in Z$, then the sequent $\alpha_1, \ldots, \alpha_m \Rightarrow \beta$ is provable in **S**.

In the following, U denotes the set of all (finite, possibly empty) multisets of elements of Ω_S. The empty multiset is denoted by ε, and the multiset-union of Σ and Γ is denoted by $\Sigma \cdot \Gamma$, or also by Σ, Γ, when no confusions may occur for every Σ and $\Gamma \in U$. Clearly, $\mathbf{U} = \langle U, \cdot, \varepsilon \rangle$ is a commutative monoid with the unit element ε. From this **U**, we can construct a special $\mathbf{FL_{ew}}$-algebra $\mathbf{C_U}$ as shown in Belardinelli et al. (2004). We describe it briefly in the following. Let us take any *closure operator* C on $\wp(U)$, i.e. a unary function C which satisfies that for all $X, Y \in \wp(U)$,

1. $X \subseteq C(X)$,
2. $CC(X) \subseteq C(X)$,
3. $X \subseteq Y$ implies $C(X) \subseteq C(Y)$,
4. $C(X) * C(Y) \subseteq C(X * Y)$.

Here, $*$ is defined by $S * T = \{\Sigma \cdot \Gamma : \Sigma \in S \text{ and } \Gamma \in T\}$ for $S, T \in \wp(U)$. A subset X of U is *C-closed* if $C(X) = X$. Let $C(\wp(U))$ denote the set of all C-closed subsets. Define operations $\sqcup$, $\otimes$ and $\mapsto$ on $C(\wp(U))$ as follows. For all C-closed sets X and Y:

- $X \sqcup Y = C(X \cup Y)$,
- $X \otimes Y = C(X * Y)$,
- $X \mapsto Y = \{\Gamma \in U : \{\Gamma\} * X \subseteq Y\}$.

Lemma 3.2 in Belardinelli et al. (2004) says that the algebra $\mathbf{C_U} = \langle C(\wp(U)), \cap, \sqcup, \otimes, \mapsto, O, C(\{\varepsilon\})\rangle$ is a commutative residuated lattice with the greatest element U and the least element $C(\emptyset)$. Moreover, it is a *complete* algebra, i.e. both infinite meet

$\bigcap_{i \in I} X_i$ and infinite join $\bigsqcup_{i \in I} X_i$ always exist for an arbitrary subset $\{X_i : i \in I\}$ of $C(\wp(U))$.

Now, we will introduce a special closure operator C^* on $\wp(U)$. For each Σ and θ, where Σ is a member of U and θ is either a formula or empty, $\Sigma \lhd \theta$ means that the sequent $\Sigma \Rightarrow \theta$ is provable in $\mathbf{FL}^-_{\mathbf{ew}}$, i.e. $\mathbf{FL}_{\mathbf{ew}}$ without cut rule, and $[\Sigma; \theta]$ denotes the set $\{\Gamma \in U : \Sigma, \Gamma \lhd \theta\}$. For each $X \subseteq U$, a subset $C^*(X)$ of U is defined to be the intersection of all subsets Y of U of the form $[\Sigma; \theta]$ such that $X \subseteq Y$. Lemma 5.2 in Belardinelli et al. (2004) says that this C^* is in fact a closure operator such that $C(\{0\}) = C^*(\emptyset)$ and $C^*(\{\varepsilon\}) = U$. Now let G be the set $C^*(\wp(U))$.

Lemma 15 *The algebra* $\mathbf{G} = \langle G, \cap, \sqcup, \otimes, \mapsto, C^*(\emptyset), U\rangle$ *is an* $\mathbf{FL}_{\mathbf{ew}}$*-algebra.*

We define k and K by $k(\alpha) = C^*(\{\alpha\}) = \{\Sigma \in U : \textit{if } \Gamma, \alpha \lhd \theta \textit{ then } \Gamma, \Sigma \lhd \theta$ *for all* $\Gamma, \theta\}$, and $K(\alpha) = [\varepsilon; \alpha] = \{\Sigma \in U : \Sigma \lhd \alpha\}$. Obviously, both k and K are mappings from Ω_S to G. The following lemma can be shown in parallel with *quasi-embedding theorem*, i.e. Theorem 5.3 in Belardinelli et al. (2004).

Lemma 16 *The pair* (k, K) *is a quasi-homomorphism from* Ω_S *to the* $\mathbf{FL}_{\mathbf{ew}}$*-algebra* $\mathbf{G}$.

Proof To avoid confusions, it should be remarked first that *the quasi-embedding k in* Belardinelli et al. (2004) *is expressed here by K which will be shown to be the upper quasi-homomorphism.* To confirm that (k, K) satisfies each of conditions for quasi-homo-morphisms, in the proof of Theorem 5.3 of Belardinelli et al. (2004) we replace first a, b by formulas α, β, respectively, and next we take $k(\alpha), k(\beta)$ (with k in the sense of the present paper) for U, V, respectively. Then, by Theorem 5.3 of Belardinelli et al. (2004), we have that $\alpha \wedge \beta \in k(\alpha) \cap k(\beta)$. Since $k(\alpha) \cap k(\beta)$ is a C^*-closed set and $k(\gamma)$ $(= C^*(\{\gamma\}))$ is the smallest C^*-closed set containing $\{\gamma\}$ for each γ, it follows that $k(\alpha \wedge \beta) = C^*(\{\alpha \wedge \beta\}) \subseteq k(\alpha) \cap k(\beta)$. Thus, the second conditions of quasi-homomorphisms are satisfied. Similarly for the third and fourth. To see that the sixth condition is satisfied, it is enough to take $K(\alpha), k(\beta)$ for U, V for the left inequality and to take $k(\alpha), K(\beta)$ for U, V for the right inequality. □

Theorem 17 *The sequent system* $\mathbf{FL}^-_{\mathbf{ew}}$ *is semi-complete with respect to the* $\mathbf{FL}_{\mathbf{ew}}$*-algebra* $\mathbf{G}$*. Hence, cut elimination holds for* $\mathbf{FL}_{\mathbf{ew}}$.

Proof Suppose that $\alpha_1, \ldots, \alpha_m \Rightarrow \beta$ be any sequent which is not provable in $\mathbf{FL}^-_{\mathbf{ew}}$, and moreover that $k(\alpha_1) \otimes \ldots \otimes k(\alpha_m) \subseteq K(\beta)$ holds. Since the singleton multiset $\langle\alpha_i\rangle$ belongs to $k(\alpha_i)$ for each i by the definition of k, the multiset union $\langle\alpha_1\rangle \cdots \langle\alpha_m\rangle$ of all $\langle\alpha_i\rangle$ $(1 \leq i \leq m)$, which is equal to the multiset $\langle\alpha_1, \ldots, \alpha_m\rangle$, belongs to $k(\alpha_1) \otimes \ldots \otimes k(\alpha_m)$. Hence, $\langle\alpha_1, \ldots, \alpha_m\rangle \in K(\beta)$. This means that $\alpha_1, \ldots, \alpha_m \lhd \beta$. But, then $\alpha_1, \ldots, \alpha_m \Rightarrow \beta$ must be provable in $\mathbf{FL}^-_{\mathbf{ew}}$, which is a contradiction. Thus, $k(\alpha_1) \otimes \ldots \otimes k(\alpha_m) \not\subseteq K(\beta)$. □

Similarly, we can prove semi-completeness and cut elimination for the sequent system $\mathbf{FL}_{\mathbf{e}}$, the system obtained from $\mathbf{FL}_{\mathbf{ew}}$ by deleting weakening rule, which is for intuitionistic linear logic, and also for the single succedent sequent system **LJ** for intuitionistic logic.

Semi-completeness of sequent systems for substructural logics with respect to finite algebras also holds, though to these logics we cannot apply the method given in the previous sections. The original idea is due to Lafont (1997); Okada and Terui (1999), and a proof in the present context was described in details in Sect. 7 of Belardinelli et al. (2004), which we will sketch briefly below.

We will modify the proof in the above in the following way. Suppose that a sequent $\alpha_1, \ldots, \alpha_m \Rightarrow \beta$ is given. We introduce a new relation $\Sigma \lhd^{\natural} \theta$ for each $\Sigma \in U$ and each θ which is either a formula or empty, as follows. Recall that $\Sigma \lhd \theta$ denotes the provability of the sequent $\Sigma \Rightarrow \theta$ in $\mathbf{FL}^{-}_{\mathbf{ew}}$.

- For any sequent $\Sigma \Rightarrow \theta$ which *appears in the proof-search tree of the sequent* $\alpha_1, \ldots, \alpha_m \Rightarrow \beta$, define that $\Sigma \lhd^{\natural} \theta$ holds if and only if $\Sigma \lhd \theta$ holds. When $\Sigma \Rightarrow \theta$ does not appear in it, we define that $\Sigma \lhd^{\natural} \theta$ holds always.

Using this relation $\lhd^{\natural}$, we define a closure operator $C^{\natural}$ in the same way as before. Then, it can be shown that $C^{\natural}(X) = U$ for all $X \subseteq U$ except finitely many X. Because of this, the algebra $\mathbf{G}^{\natural}$ which is constructed similarly to $\mathbf{G}$ becomes finite, and the inequality (5) fails in $\mathbf{G}^{\natural}$. Note that $\mathbf{G}^{\natural}$ is obtained dependently on a given sequent $\alpha_1, \ldots, \alpha_m \Rightarrow \beta$. Thus, we have the following. The corresponding statement can be shown also for $\mathbf{FL}^{-}_{\mathbf{e}}$ and $\mathbf{LJ}^{-}$.

Lemma 18 *The sequent system* $\mathbf{FL}^{-}_{\mathbf{ew}}$ *is semi-complete with respect to the set of all finite* $\mathbf{FL}_{\mathbf{ew}}$*-algebras.*

Remark 19 Algebras and quasi-homomorphisms in the proof of semi-completeness of this section look totally different from those in the previous two sections. As we mentioned already at the beginning of this section, The construction of the algebra $\mathbf{G}$ in this section has a strong resemblance to MacNeille completions, while the constructions of algebras in the previous ones are to canonical extensions. Also, quasi-homomorphisms in this section are defined by using the notion of *provability*, while those in the previous sections are defined by using *consistency*, i.e. the negation of provability.

But these differences are on the surface, and as a matter of fact, there are close connections between them, as we will explain below. For comparison, we consider the case for *multiple-succeedent sequent systems*. Then, quasi-homomorphisms (k, K) defined just above Lemma 16 will be modified into the following form:

- $k(\alpha) = \{(\Sigma, \Theta) \in U^2 : \textit{if } \Gamma, \alpha \lhd \Theta, \Delta \textit{ then } \Gamma, \Sigma \lhd \Theta, \Delta \textit{ for all } (\Gamma, \Delta) \in U^2\}$,
- $K(\alpha) = \{(\Sigma, \Theta) \in U^2 : \Sigma \lhd \alpha, \Theta\}$

In the following, we assume moreover that the following two conditions hold in a given sequent system $\mathbf{S}$.

1. Structural rules, i.e. weakening, contraction and exchange rules, are admissible.
2. Every consistent pair can be extended to a maximal consistent pair.

Obviously, each sequent system for intuitionistic logic and for classical modal logics satisfies these conditions. With the help of weakening rules, we can see that every inconsistent pair (Σ, Θ), i.e., each pair (Σ, Θ) for which $\Sigma \lhd \Theta$ holds, belongs always to both $k(\alpha)$ and $K(\alpha)$ for every α. Thus, it will be enough to pay attention only to *consistent pairs* in U^2.

We will focus attention further on the set M of all maximal consistent pairs, which is a subset of U^2. Now, for each formula α, let $k_M(\alpha)$ and $K_M(\alpha)$ be subsets of M obtained from $k(\alpha)$ and $K(\alpha)$, respectively, defined in the above by replacing U^2 by M. Then the following relations can be proved, which say that $k_M(\alpha)$ and $K_M(\alpha)$ thus obtained are equal exactly to those k and K, respectively, introduced in Sect. 3 and also in Sect. 4.

- $(\Sigma, \Theta) \in k_M(\alpha)$ if and only if $\alpha \in \Sigma$,
- $(\Sigma, \Theta) \in K_M(\alpha)$ if and only if $\alpha \notin \Theta$.

To show this, first suppose that $(\Sigma, \Theta) \in k_M(\alpha)$. If $\alpha \notin \Sigma$ then $\Sigma, \alpha \lhd \Theta$ holds as (Σ, Θ) is maximal consistent. Then, $\Sigma, \alpha \lhd \Theta, \Theta$ and hence $\Sigma, \Sigma \lhd \Theta, \Theta$ follows by the definition of $k_M(\alpha)$. But, the last relation implies $\Sigma \lhd \Theta$ by exchange and contraction rules, which contradicts the consistency of (Σ, Θ). Thus, α must be a member of Σ. Conversely, suppose that $\alpha \in \Sigma$. Then $\Gamma, \alpha \lhd \Theta, \Delta$ implies $\Gamma, \Sigma \lhd \Theta, \Delta$ by weakening rules. Thus, $(\Sigma, \Theta) \in k_M(\alpha)$. The second relation follows immediately from the fact that $\Sigma \lhd \alpha, \Theta$ if and only if $\alpha \notin \Theta$, when (Σ, Θ) is maximal consistent. □

6 Semi-completeness of Sequent Systems for Substructural Predicate Logics

We will discuss now how to extend semi-completeness results obtained so far to the case of predicate logics. First, we will discuss cut elimination for substructural predicate logics which are predicate extensions of sequent systems for substructural propositional logics in the previous section. Here, by the *predicate extension* of a given sequent system **S** for a propositional logic, we mean the sequent system obtained from **S** by adding the following rules for quantifiers $\forall$ and $\exists$.[7]

$$\frac{\alpha[t/x], \Gamma \Rightarrow \Delta}{\forall x\alpha, \Gamma \Rightarrow \Delta}\ (\forall \Rightarrow) \qquad \frac{\Gamma \Rightarrow \alpha[z/x], \Sigma}{\Gamma \Rightarrow \forall x\alpha, \Sigma}\ (\Rightarrow \forall)$$

$$\frac{\alpha[z/x], \Gamma \Rightarrow \Delta}{\exists x\alpha, \Gamma \Rightarrow \Delta}\ (\exists \Rightarrow) \qquad \frac{\Gamma \Rightarrow \alpha[t/x], \Sigma}{\Gamma \Rightarrow \exists x\alpha, \Sigma}\ (\Rightarrow \exists)$$

[7]When **S** is a single-succedent system, Δ contains at most one formula and Σ is empty.

Here, t is a term, x and z are individual variables, and $\alpha[z/x]$ ($\alpha[t/x]$) are the formula obtained from α by replacing all free occurrences of x in α by z (by t, respectively). Moreover, in applications of ($\Rightarrow \forall$) and ($\exists \Rightarrow$), the variable z should not occur as a free variable in the lower sequent.

In his paper Maehara (1991), the author has dealt essentially with the case for classical predicate logic. In the following, we will show how to extend our arguments about sequent systems for substructural propositional logics in Sect. 5 to those for their predicate extensions. As a matter of fact, this was done already in Belardinelli et al. (2004) for the predicate extension $\mathbf{QFL_{ew}}$ of the sequent system $\mathbf{FL_{ew}}$. So, it suffices for us to translate the arguments using the notion of quasi-embeddings in Belardinelli et al. (2004) into our framework.

For the simplicity's sake, we suppose that our language $\mathscr{L}$ has neither function symbols nor constant symbols. In our algebraic semantics for predicate logics, quantifiers $\forall$ and $\exists$ are interpreted by infinite meets and infinite joins, respectively, in a given algebra $\mathbf{A}$. To guarantee the existence of these infinite meets and infinite joins, we should take a *complete* $\mathbf{FL_{ew}}$*-algebra* $\mathbf{A}$, i.e. an $\mathbf{FL_{ew}}$-algebra in which both $\bigcap_{i\in I} a_i$ and $\bigcup_{i\in I} a_i$ exist for an arbitrary (possibly infinite) subset $\{a_i : i \in I\}$ of A.

Definition 8 (*algebraic structures*) A pair $\langle \mathbf{A}, D\rangle$ is an *algebraic structure* for the commutative, integral substructural predicate logic, if $\mathbf{A}$ is a complete $\mathbf{FL_{ew}}$-algebra and D is an nonempty set.

The above set D is called an *individual domain*. To define our interpretation, we introduce first a new constant symbol $\hat{d}$ for each $d \in D$ and take the language $\mathscr{L}[D]$ obtained from $\mathscr{L}$ by adding every $\hat{d}$ for $d \in D$. Let $\Omega_{QS[D]}$ be the set of all first-order sentences in the language $\mathscr{L}[D]$. Here a *first-order sentence* means a first-order formula containing no free variables. For convenience's sake, every formula of the form $\alpha[\hat{d}/x]$ for $d \in D$ is regarded as a subformula of both $\forall x\alpha$ and $\exists x\alpha$.

Definition 9 (*partial homomorphisms on algebraic structures*) For a given algebraic structure $\langle \mathbf{A}, D\rangle$ for the commutative, integral substructural predicate logic, a mapping f from a subformula-closed subset Z of $\Omega_{QS[D]}$ to A (the underlying set of $\mathbf{A}$) is a *partial homomorphism* on $\langle \mathbf{A}, D\rangle$ if it satisfies the following conditions.

1. $f(\alpha \wedge \beta) = f(\alpha) \cap f(\beta)$ for $\alpha \wedge \beta \in Z$,
2. $f(\alpha \vee \beta) = f(\alpha) \cup f(\beta)$ for $\alpha \vee \beta \in Z$,
3. $f(\alpha \cdot \beta) = f(\alpha) \cdot f(\beta)$ for $\alpha \cdot \beta \in Z$,
4. $f(\alpha \to \beta) = f(\alpha) \to f(\beta)$ for $\alpha \to \beta \in Z$,
5. $f(0) = 0_{\mathbf{A}}$ and $f(1) = 1_{\mathbf{A}}$,
6. $f(\forall x\alpha) = \bigcap\{f(\alpha[\hat{d}/x]) : d \in D\}$ for $\forall x\alpha \in Z$,
7. $f(\exists x\alpha) = \bigcup\{f(\alpha[\hat{d}/x]) : d \in D\}$ for $\exists x\alpha \in Z$.

A triple $\langle \mathbf{A}, D; f\rangle$ with an algebraic structure $\langle \mathbf{A}, D\rangle$ and a partial homomorphism f on it is called an *algebraic model*.

Like Lemma 1, we can show that every partial homomorphism to an algebraic structure $\langle \mathbf{A}, D \rangle$ can be extended to a homomorphism to $\langle \mathbf{A}, D \rangle$. When free variables $x_1, \ldots, x_n$ are all free variables in a given formula β, we say that β is *true* in an algebraic model $\langle \mathbf{A}, D; f \rangle$ whenever $f(\beta[\hat{d}_1/x_1, \ldots, \hat{d}_n/x_n]) = 1$ for all $d_1, \ldots, d_n \in D$. A formula β is *valid* in an algebraic structure $\langle \mathbf{A}, D \rangle$ if it is true in every algebraic model $\langle \mathbf{A}, D; f \rangle$ for any partial homomorphism f on it. Similarly, the validity of a given sequent containing free variables can be defined. It can be shown e.g. in Ono (1993) that the completeness of $\mathbf{QFL_{ew}}$ holds with respect to the class of all algebraic structures for the commutative, integral substructural predicate logic.

Similarly we can define a *quasi-homomorphism (k,K) on an algebraic structure* $\langle \mathbf{A}, D \rangle$ as a pair of mappings k and K from a nonempty, subformula-closed subset Z of $\Omega_{QS[D]}$ to A by extending Definition 6 as follows.

7. $k(\forall x\alpha) \subseteq \bigcap\{k(\alpha[\hat{d}/x]) : d \in D\}$ and $\bigcap\{K(\alpha[\hat{d}/x]) : d \in D\} \subseteq K(\forall x\alpha)$ for $\forall x\alpha \in Z$,
8. $k(\exists x\alpha) \subseteq \bigcup\{k(\alpha[\hat{d}/x]) : d \in D\}$ and $\bigcup\{K(\alpha[\hat{d}/x]) : d \in D\} \subseteq K(\exists x\alpha)$ for $\exists x\alpha \in Z$.

Semi-completeness of a sequent system for a substructural predicate logic is defined in the same way as Definition 7, simply by replacing "residuated lattices" by "algebraic structures (which are pairs of complete residuated lattices and individual domains)". Now, if we define mappings k and K in the same way as in Sect. 5, we can show that the pair (k, K) satisfies also above conditions 7 and 8, as shown essentially in Lemma 6. 1 of Belardinelli et al. (2004) which extends quasi-embedding theorem to predicate logics. Thus. we have the following.

Theorem 20 *The sequent system* $\mathbf{QFL^{-}_{ew}}$ *is semi-complete. Hence, cut elimination holds for* $\mathbf{QFL_{ew}}$.

The method can be applied easily for other predicate extensions of sequent systems which were discussed in Belardinelli et al. (2004).

7 A Semantical Proof of Cut Elimination for Intuitionistic Predicate Logic

In this section we show first that proofs developed in our Sects. 2 and 3, i.e. proofs of cut elimination based on Kripke frames, can be also extended to predicate extensions. On the other hand, since Kripke frames used in these proofs are those with *varying domains*, constructing algebraic structures using complex algebras will not work. Thus, there seem to be a certain obstacle of getting an alternative algebraic proof from our semantical proof obtained in this section. Nevertheless, as we show in the next section, the proof in the present section can be naturally transformed into semi-completeness arguments if we generalize notions of both algebraic structures and quasi-homomorphisms.

We will consider here the predicate extension **QLJ**$'$ of **LJ**$'$ for intuitionistic logic. **QLJ**$'$ is a multiple-succedent sequent system, *but in which* ($\Rightarrow \forall$) of **QLJ**$'$ *is restricted to a single-succedent form.*[8] Let **QLJ**$'^-$ be the system obtained from **QLJ**$'$ by deleting cut rule. Again, just for the simplicity's sake, we assume that our language $\mathscr{L}'$ has neither function symbols nor constant symbols. Let us take an increasing sequence of sets $\{E_m\}$ $(m \geq 0)$ of *new constant symbols* such that (1) $E_0 \subseteq E_1 \subseteq E_2 \subseteq \cdots$ and (2) both the set E_0 and the difference $E_{n+1} \backslash E_n$ are countably infinite for every $n \geq 0$. The union $\bigcup_m E_m$ is denoted by E. The set of all first-order sentences in the language $\mathscr{L}'[E_m]$ ($\mathscr{L}'[E]$), i.e. $\mathscr{L}'$ with individuals in E_m (in E) as new constant symbols, is denoted by Ω_m (and $\Omega_{QI[E]}$, respectively).

We will give first a semantical proof of cut elimination for **QLJ**$'$, and then show how to incorporate it into our algebraic framework. Our semantical proof of cut elimination will proceed similarly to those in Sect. 4. As before, we can define the notions of **QLJ**$'^-$*-consistency in* Ω_m and *maximal* **QLJ**$'^-$*-consistency in* Ω_m for each $m \geq 0$, a pair (Σ, Θ) of sets Σ and Θ of first-order sentences of Ω_m. Let Y be the set of all triples of the form $\langle \Sigma, \Theta, m \rangle$ such that (Σ, Θ) is maximal **QLJ**$'^-$-consistent in Ω_m for some $m \geq 0$. Next we introduce a binary relation $\leq$ on Y by the condition that $\langle \Sigma, \Theta, m \rangle \leq \langle \Gamma, \Delta, n \rangle$ if and only if $m \leq n$ and $\Sigma \subseteq \Gamma$. Clearly, the relation $\leq$ is a preorder. Similarly to Lemma 12, we have the following.

Lemma 21 (downward saturation in intuitionistic predicate logic) *The following holds for any triple* $\langle \Sigma, \Theta, m \rangle$ *in* Y.

(1a) *if* $\beta \wedge \gamma \in \Sigma$ *then both* β *and* γ *are in* Σ,
(1b) *if* $\beta \wedge \gamma \in \Theta$ *then either* β *or* γ *are in* Θ,
(2a) *if* $\beta \vee \gamma \in \Sigma$ *then either* β *or* γ *are in* Σ,
(2b) *if* $\beta \wedge \gamma \in \Theta$ *then both* β *and* γ *are in* Θ,
(3) $0 \notin \Sigma$,
(4a) *if* $\beta \to \gamma \in \Sigma$, *then either* $\beta \in \Delta$ *or* $\gamma \in \Gamma$ *for each* $(\Gamma, \Delta) \in V$ *such that* $\Sigma \subseteq \Gamma$,
(4b) *if* $\beta \to \gamma \in \Theta$, *then* $\beta \in \Gamma$ *and* $\gamma \in \Delta$ *for some* $(\Gamma, \Delta) \in V$ *such that* $\Sigma \subseteq \Gamma$.
(5a) *if* $\forall x\beta \in \Sigma$, *then* $\beta[\hat{d}/x] \in \Gamma$ *for each* $d \in E_n$ *and each* $\langle \Gamma, \Delta, n \rangle \in Y$ *such that* $\langle \Sigma, \Theta, m \rangle \leq \langle \Gamma, \Delta, n \rangle$,
(6b) *if* $\exists x\beta \in \Theta$, *then* $\beta[\hat{d}/x] \in \Theta$ *for each* $d \in E_m$.

But, the above properties are not enough to derive cut elimination result. We need further requirements in the following.

Definition 10 (*downward Henkin saturation*) A triple $\langle \Sigma, \Theta, m \rangle \in Y$ is *downward Henkin saturated* if it satisfies that

(5b) if $\forall x\beta \in \Theta$, then $\beta[\hat{d}/x] \in \Delta$ for some $d \in E_n$ and some $\langle \Gamma, \Delta, n \rangle \in Y$ such that $\langle \Sigma, \Theta, m \rangle \leq \langle \Gamma, \Delta, n \rangle$,

[8] Otherwise, the *axiom of constant domain* $\forall x(\alpha \vee \beta) \to (\forall x\alpha \vee \beta)$, which is not provable in intuitionistic predicate logic, becomes provable in the system, where x does not occur in β as a free variable.

(6a) if $\exists x\beta \in \Sigma$, then $\beta[\hat{d}/x] \in \Sigma$ for some $d \in E_m$.

We can show the following lemma by applying standard arguments but to the sequent system without cut rule **QLJ**$'^{-}$.

Lemma 22 *If a pair* (Σ, Θ) *of sets of first-order sentences of* Ω_m *is* **QLJ**$'^{-}$*-consistent in* Ω_m *then there exists a downward Henkin saturated triple* $\langle \Gamma, \Delta, m+1\rangle$ *such that* $\Sigma \subseteq \Gamma$ *and* $\Theta \subseteq \Delta$.

Let Y^+ be the set of all downward Henkin saturated triples. The pair $\langle Y^+, \leq\rangle$ is a preordered set. Define a mapping D from Y^+ to the set E by $D(\langle \Sigma, \Theta, m\rangle) = E_m$, which says that the set of individual domain for the state $\langle \Sigma, \Theta, m\rangle$ is E_m. Obviously, the triple $\langle Y^+, \leq, D\rangle$ determines a *canonical Kripke frame for intuitionistic logic with varying domains*. Note here that $\langle \Sigma, \Theta, m\rangle \leq \langle \Gamma, \Delta, n\rangle$ implies that $D(\langle \Sigma, \Theta, m\rangle) = E_m \subseteq E_n = D(\langle \Gamma, \Delta, n\rangle)$ as $m \leq n$.

We introduce a valuation v on this Kripke frame as follows: For each n-ary predicate symbol P and each member $\langle \Sigma, \Theta, m\rangle$ of Y^+, $v(P(\hat{d}_1, \ldots, \hat{d}_n), \langle \Sigma, \Theta, m\rangle)$ is *true* if and only if $P(\hat{d}_1, \ldots, \hat{d}_n) \in \Sigma$ for all $d_1, \ldots, d_n \in E_m$. As usual, v can be extended inductively to a relation $\models$ which expresses the *truth at a state of* Y^+ in a model $\langle Y^+, \leq, D, W\rangle$. Then the following lemma holds.

Lemma 23 (partial truth lemma) *For each* $\langle \Sigma, \Theta, m\rangle \in Y^+$ *and each sentence* $\alpha \in \Omega_m$,

1. *if* $\alpha \in \Sigma$ *then* $\langle \Sigma, \Theta, m\rangle \models \alpha$,
2. *if* $\alpha \in \Theta$ *then* $\langle \Sigma, \Theta, m\rangle \not\models \alpha$.

Proof This is proved by induction on the complexity of the sentence α (cf. e.g. proof of Lemma 11 in Ono (2015)). We will show this when α is of the form $\forall x\beta$. First, suppose that $\forall x\beta \in \Sigma$. By Lemma 21 (5a), $\beta[\hat{d}/x] \in \Gamma$ for each $d \in E_n$ and each $\langle \Gamma, \Delta, n\rangle \in Y^+$ such that $\langle \Sigma, \Theta, m\rangle \leq \langle \Gamma, \Delta, n\rangle$. By using the hypothesis of induction, we have that for any $\langle \Gamma, \Delta, n\rangle$, if $\langle \Sigma, \Theta, m\rangle \leq \langle \Gamma, \Delta, n\rangle$ then $\langle \Gamma, \Delta, n\rangle \models \beta[\hat{d}/x]$ for each $d \in D(\langle \Gamma, \Delta, n\rangle)$. Thus, $\langle \Sigma, \Theta, m\rangle \models \forall x\beta$ by the definition of $\models$ of Kripke models for intuitionistic logic.

Next suppose that $\forall x\beta \in \Theta$. Since $\langle \Sigma, \Theta, m\rangle$ is downward Henkin saturated, $\beta[\hat{d}/x] \in \Delta$ for some $d \in E_n$ and some $\langle \Gamma, \Delta, n\rangle \in Y$ such that $\langle \Sigma, \Theta, m\rangle \leq \langle \Gamma, \Delta, n\rangle$. By using the hypothesis of induction, $\langle \Gamma, \Delta, n\rangle \not\models \beta[\hat{d}/x]$ for some $d \in D(\langle \Gamma, \Delta, n\rangle)$ and some $\langle \Gamma, \Delta, n\rangle \in Y^+$ such that $\langle \Sigma, \Theta, m\rangle \leq \langle \Gamma, \Delta, n\rangle$. This implies that $\langle \Sigma, \Theta, m\rangle \not\models \forall x\beta$. □

Suppose that an arbitrary sequent $\Pi \Rightarrow \Lambda$ is given, where both Π and Λ are finite sets of first-order formulas in the language $\mathscr{L}'$. If they contain some free variables, then we replace distinct free variables among them by distinct constant symbols of E in all formulas in Π and Λ. Let Π' and Λ' are finite sets of first-order sentences in $\Omega_{QI[E]}$ thus obtained from Π and Λ. As they are finite, we can assume that all formulas in them belong to Ω_m for some m. It is clear that $\Pi \Rightarrow \Lambda$ is provable in **QLJ**$'^{-}$ if and only if $\Pi' \Rightarrow \Lambda'$ is provable in **QLJ**$'^{-}$.

Now suppose that $\Pi' \Rightarrow \Lambda'$ is not provable in $\mathbf{QLJ}'^-$, i.e. the pair (Π', Λ') is $\mathbf{QLJ}'^-$-consistent in Ω_m. By Lemma 22, there exists a downward Henkin saturated triple $\langle \Gamma, \Delta, m+1 \rangle$ in Y^+ such that $\Pi' \subseteq \Gamma$ and $\Lambda' \subseteq \Delta$. By Lemma 23, $\langle \Gamma, \Delta, m+1 \rangle \models \alpha$ holds for each formula $\alpha \in \Pi'$ while $\langle \Gamma, \Delta, m+1 \rangle \not\models \beta$ holds for each $\beta \in \Lambda'$. Therefore, $\Pi' \Rightarrow \Lambda'$ is false in this model. By the soundness of intuitionistic predicate logic, $\Pi' \Rightarrow \Lambda'$ is not provable in $\mathbf{QLJ}'$. By taking a contraposition, we have the following.

Theorem 24 (cut elimination) *If a sequent $\Gamma \Rightarrow \Delta$ is provable in* $\mathbf{QLJ}'$*, it is provable in* $\mathbf{QLJ}'$ *without using cut rule.*

8 General Quasi-homomorphisms on Expanded Algebraic Structures

While a semantical proof of cut elimination for the sequent system $\mathbf{QLJ}'$ for intuitionistic predicate logic is given in the previous section, there seems to be a certain difficulty in transforming the present proof into algebraic one in such a way as we have done in Sect. 4. To overcome the difficulty, we will expand the notion of algebraic structures and interpretations on them. This idea was inspired by Part 3 Sect. 5 of the work by Dragalin (1988). (See the chapter on algebraic semantics in Gabbay et al. (2009) for the details. Also see Chap. 1 of Goldblatt (2011) in which closely related topics are discussed but in a slightly different context.) As before, an *algebraic structure for intuitionistic predicate logic* is defined to be a pair $\langle \mathbf{A}, D \rangle$ of a complete Heyting algebra $\mathbf{A}$ and a nonempty set D.

Definition 11 (*definiteness functions and expanded algebraic structures*) For a given algebraic structure $\langle \mathbf{A}, D \rangle$ for intuitionistic predicate logic, a mapping ϕ from D to A is called a *definiteness function* (or, simply a *d-function*) on it, if it satisfies $\bigcup\{\phi(d) : d \in D\} = 1_{\mathbf{A}}$. Any triple $\langle \mathbf{A}, D, \phi \rangle$ with an algebraic structure $\langle \mathbf{A}, D \rangle$ with a d-function ϕ on it is called an *expanded algebraic structure* for intuitionistic predicate logic.

In Dragalin (1988), $\phi(d)$ is denoted by $\|d\|$ and is read as *measure of definiteness* of an individual object d. Each expanded algebraic structure can be regarded as a simplified form of *Heyting-valued structures* in Gabbay et al. (2009), where ϕ is read as *measure of existence*.

Definition 12 (*partial general homomorphisms*) For a given expanded algebraic structure $\langle \mathbf{A}, D, \phi \rangle$, a mapping f from a subformula-closed subset Z of Ω_{QI} to A is a *partial general homomorphism* on $\langle \mathbf{A}, D, \phi \rangle$ if it satisfies the following conditions.

1. $f(\alpha \wedge \beta) = f(\alpha) \cap f(\beta)$ for $\alpha \wedge \beta \in Z$,
2. $f(\alpha \vee \beta) = f(\alpha) \cup f(\beta)$ for $\alpha \vee \beta \in Z$,
3. $f(\alpha \to \beta) = f(\alpha) \to f(\beta)$ for $\alpha \to \beta \in Z$,

4. $f(0) = 0_{\mathbf{A}}$ and $f(1) = 1_{\mathbf{A}}$,
5. $f(\forall x\alpha) = \bigcap\{\phi(d) \to f(\alpha[\hat{d}/x]) : d \in D\}$ for $\forall x\alpha \in Z$,
6. $f(\exists x\alpha) = \bigcup\{\phi(d) \wedge f(\alpha[\hat{d}/x]) : d \in D\}$ for $\exists x\alpha \in Z$.

The pair (ϕ, f) of a d-function ϕ and a partial general homomorphism f can be considered also as an interpretation on an algebraic structure $\langle \mathbf{A}, D\rangle$. Because of this, the quadruple $\langle \mathbf{A}, D, \phi; f\rangle$ is sometimes called an *general algebraic model* over an algebraic structure $\langle \mathbf{A}, D\rangle$ for intuitionistic predicate logic. When $\phi(d) = 1_{\mathbf{A}}$ holds for any $d \in D$ for a given d-function in the above, any partial general homomorphism is a partial homomorphism and hence a general algebraic model can be identified with an algebraic model of the previous section.

For a formula β such that $x_1, \ldots, x_n$ are all free variables in it, we say that β is *true in a general algebraic model* $\langle \mathbf{A}, D, \phi; f\rangle$ whenever $(\phi(d_1) \cap \ldots \cap \phi(d_n)) \to f(\beta[\hat{d}_1/x_1, \ldots, \hat{d}_n/x_n]) = 1$, or equivalently $(\phi(d_1) \cap \ldots \cap \phi(d_n)) \le f(\beta[\hat{d}_1/x_1, \ldots, \hat{d}_n/x_n])$, holds for all $d_1, \ldots, d_n \in D$. A formula β is *g-valid* if it is true in every general algebraic model, which is equivalent to say that β is true for any partial general homomorphism on every expanded algebraic structure. The *g-validity* can be defined similarly for an arbitrary sequent.

Theorem 25 (Completeness of $\mathbf{QLJ}'$ with respect to general algebraic models) *A sequent is g-valid if and only if it is provable in* $\mathbf{QLJ}'$.

Proof The only-if part follows from the completeness with respect to algebraic models, as every algebraic structure can be regarded as a particular expanded algebraic structure whose d-function ϕ takes always the value 1_A for all $d \in D$. To see that the converse direction holds, it suffices to check that the g-validity of the lower sequent follows from the g-validity of the upper sequent for each rule of $\mathbf{QLJ}'$ for quantifiers. This is almost obvious for $(\Rightarrow \forall)$ and $(\exists \Rightarrow)$.

We consider the rule $(\forall \Rightarrow)$. Suppose that $\forall x\alpha, \Gamma \Rightarrow \Delta$ is inferred from $\alpha[y/x], \Gamma \Rightarrow \Delta$. Let γ and δ be the conjunction of formulas in Γ and the disjunction of formulas in Δ, respectively. Take an arbitrary element $d \in D$. For simplicity's sake, we assume that no free variables appear in $\alpha[\hat{d}/x]$, γ and δ. From our assumption that $\alpha[y/x], \Gamma \Rightarrow \Delta$ is g-valid, $\phi(d) \to ((f(\alpha[\hat{d}/x]) \cap f(\gamma)) \to f(\delta)) = 1$, or equivalently, $\phi(d) \cap f(\alpha[\hat{d}/x]) \le f(\gamma) \to f(\delta)$ holds for any general algebraic model $\langle \mathbf{A}, D, \phi; f\rangle$. On the other hand, it holds that $\phi(d) \cap f(\forall x\alpha) = \phi(d) \cap \bigcap\{\phi(e) \to f(\alpha[\hat{e}/x]) : e \in D\} \le \phi(d) \cap f(\alpha[\hat{d}/x])$. Combining these two inequalities, we have that $\phi(d) \cap f(\forall x\alpha) \le f(\gamma) \to f(\delta)$, and hence that $\phi(d) \le (f(\forall x\alpha) \cap f(\gamma)) \to f(\delta)$ for every $d \in D$. Therefore, $1 = \bigcup\{\phi(d) : d \in D\} \le (f(\forall x\alpha) \cap f(\gamma)) \to f(\delta)$. Thus, the lower sequent is true in the general algebraic model $\langle \mathbf{A}, D, \phi; f\rangle$.

Next, consider the rule $(\Rightarrow \exists)$, and suppose that $\Gamma \Rightarrow \Delta, \exists x\alpha$ is inferred from $\Gamma \Rightarrow \Delta, \alpha[y/x]$. Suppose that $\Gamma \Rightarrow \Delta, \alpha[y/x]$ is g-valid. Thus, $\phi(d) \cap f(\gamma) \le f(\delta) \cup f(\alpha[\hat{d}/x])$ holds for any general algebraic model $\langle \mathbf{A}, D, \phi; f\rangle$, by using the same abbreviation and assumption in the above. Then, $\phi(d) \cap f(\gamma) \le f(\delta) \cup (\phi(d) \cap f(\alpha[\hat{d}/x])) \le f(\delta) \cup \bigcup\{\phi(e) \cap f(\alpha[\hat{e}/x]) : e \in D\}$. Hence, $1 = \bigcup\{\phi(d) : d \in D\} \le f(\gamma) \to [f(\delta) \cup \bigcup\{\phi(e) \cap f(\alpha[\hat{e}/x]) : e \in D\}] = f(\gamma) \to (f(\delta) \cup f(\exists x\alpha))$.

Thus, the lower sequent is true in this general algebraic model. □

Definition 12 will suggest us how to extend the notion of quasi-homomorphisms on expanded algebraic structures.

Definition 13 (*general quasi-homomorphisms*) For a given expanded algebraic structure $\langle \mathbf{A}, D, \phi \rangle$, a pair (k, K) of mappings k and K from a nonempty, subformula-closed subset Z of $\Omega_{QI[E]}$ to A is a *general quasi-homomorphism* on $\langle \mathbf{A}, D, \phi \rangle$ if they satisfy the following conditions.

1. $k(\alpha) \leq K(\alpha)$ for $\alpha \in Z$,
2. $k(\alpha \wedge \beta) \leq k(\alpha) \cap k(\beta)$ and $K(\alpha) \cap K(\beta) \leq K(\alpha \wedge \beta)$ for $\alpha \wedge \beta \in Z$,
3. $k(\alpha \vee \beta) \leq k(\alpha) \cup k(\beta)$ and $K(\alpha) \cup K(\beta) \leq K(\alpha \vee \beta)$ for $\alpha \vee \beta \in Z$,
4. $k(0) = 0_A$,
5. $k(\alpha \to \beta) \leq K(\alpha) \to k(\beta)$ and $k(\alpha) \to K(\beta) \leq K(\alpha \to \beta)$ for $\alpha \to \beta \in Z$.
6. $k(\forall x\alpha) \leq \bigcap\{\phi(d) \to k(\alpha[\hat{d}/x]) : d \in D\}$ and $\bigcap\{\phi(d) \to K(\alpha[\hat{d}/x]) : d \in D\} \leq K(\forall x\alpha)$ for $\forall x\alpha \in Z$,
7. $k(\exists x\alpha) \leq \bigcup\{\phi(d) \cap k(\alpha[\hat{d}/x]) : d \in D\}$ and $\bigcup\{\phi(d) \cap K(\alpha[\hat{d}/x]) : d \in D\} \leq K(\exists x\alpha)$ for $\exists x\alpha \in Z$.

These inequalities in the above were discussed also in Part 3 Sect. 5 of Dragalin (1988), in which these triples of ϕ, k and K satisfying the above conditions are called *semivaluations*. Using these conditions, the author of Dragalin (1988) has derived cut elimination for $\mathbf{QLJ}'$ semantically.

It is obvious that Theorem 3 holds also between general homomorphisms and general quasi-homomorphisms. Thus the next lemma follows from Theorem 25, which corresponds to Lemma 4. Suppose that any sequent $\alpha_1, \ldots, \alpha_m \Rightarrow \beta_1, \ldots, \beta_n$ of the original language $\mathcal{L}'$ is given. Let $\mathbf{x} = x_1, \ldots, x_k$ be a fixed enumeration of distinct variables in this sequent. Suppose also that an expanded algebraic structure $\langle \mathbf{A}, D, \phi \rangle$ is given. Expressions $\alpha_i[\hat{\mathbf{d}}]$ and $\beta_i[\hat{\mathbf{d}}]$ denote formulas obtained from α_i and β_j, respectively, by replacing each free occurrence of x_h by a constant symbol $\hat{d}_h$ for $1 \leq h \leq k$, for any k-tuple $\mathbf{d}$ $(= d_1, \ldots, d_k)$ of elements of D.

Lemma 26 *For any sequent $\alpha_1, \ldots, \alpha_m \Rightarrow \beta_1, \ldots, \beta_n$ of the language $\mathcal{L}'$, if it is provable in $\mathbf{QLJ}'$ then*

$$(\phi(d_1) \cap \ldots \cap \phi(d_k)) \cap k(\alpha_1[\hat{\mathbf{d}}]) \cap \ldots \cap k(\alpha_m[\hat{\mathbf{d}}]) \leq K(\beta_1[\hat{\mathbf{d}}]) \cup \ldots \cup K(\beta_n[\hat{\mathbf{d}}])$$

holds for any expanded algebraic structure $\langle \mathbf{A}, D, \phi \rangle$ for intuitionistic predicate logic, any general quasi-homomorphism (k, K) on it and any k-tuple $\mathbf{d}$ $(= d_1, \ldots, d_k)$ of elements of D.

The semi-completeness with respect to a class of expanded algebraic structures can be defined as before, The following proof of semi-completeness of $\mathbf{QLJ}'^{-}$, i.e. $\mathbf{QLJ}'$ without cut rule, goes essentially in the same way as that in Dragalin (1988). But here we will clarify exact correspondences between downward Henkin saturated triples and general quasi-homomorphisms.

Theorem 27 (semi-completeness of $\mathbf{QLJ}'^{-}$ with respect to expanded algebraic structures) *For all formulas $\alpha_1, \ldots, \alpha_m, \beta_1, \ldots, \beta_n$, if the inequality*

$$(\phi(d_1) \cap \ldots \cap \phi(d_k)) \cap k(\alpha_1[\hat{\mathbf{d}}]) \cap \ldots \cap k(\alpha_m[\hat{\mathbf{d}}]) \leq K(\beta_1[\hat{\mathbf{d}}]) \cup \ldots \cup K(\beta_n[\hat{\mathbf{d}}])$$

holds for any expanded algebraic structure $\langle \mathbf{A}, D, \phi \rangle$ for intuitionistic predicate logic, any k-tuple $\mathbf{d}$ $(= d_1, \ldots, d_k)$ of elements of D and any general quasi-homomorphism (k, K) on $\langle \mathbf{A}, D, \phi \rangle$ with a domain Z such that $\alpha_1, \ldots, \alpha_m, \beta_1, \ldots, \beta_n \in Z$, then the sequent $\alpha_1, \ldots, \alpha_m \Rightarrow \beta_1, \ldots, \beta_n$ is provable in $\mathbf{QLJ}'^{-}$.

As we have already had a semantical proof of cut elimination for $\mathbf{QLJ}'$ in the previous section, our proof of semi-completeness of $\mathbf{QLJ}'^{-}$ will go in the same way as the proof of semi-completeness of $\mathbf{LJ}'^{-}$ given in Sect. 4. Recall that we have introduced a preorder $\leq$ on the set Y^+ of all downward Henkin saturated triples. Now, let $\mathscr{U}(Y^+)$ be the set of all upward closed subsets of Y^+ with respect to $\leq$. Then the algebra $\mathbf{H}$ defined by $\langle \mathscr{U}(Y^+), \cap, \cup, \rightarrow, \emptyset \rangle$ is a complete Heyting algebra ordered by the set inclusion with the greatest element Y^+. Define a mapping ϕ from the set E of all new constant symbols to $\mathscr{U}(Y^+)$ by $\phi(d) = \{\langle \Sigma, \Theta, m \rangle : d \in E_m\}$ for each $d \in E$. It is easy to see that each $\phi(d)$ is upward closed. For each $\langle \Gamma, \Delta, n \rangle \in Y^+$, we can take an element $e \in E_n$, as E_n is always non-empty. Then, $\langle \Gamma, \Delta, n \rangle \in \phi(e)$ and hence $\bigcup \{\phi(d) : d \in E\} = Y^+ (= 1_{\mathbf{H}})$. Therefore, $\langle \mathbf{H}, E, \phi \rangle$ forms an expanded algebraic structure. Just as arguments in Sect. 4, define k and K by $k(\alpha) = \{\langle \Sigma, \Theta, m \rangle \in Y^+ : \alpha \in \Sigma\}$ and $K(\alpha) = \{\langle \Sigma, \Theta, m \rangle \in Y^+ : \alpha \notin \Theta\}$ for any $\alpha \in \Omega_{QI}$.

Lemma 28 *The pair (k, K) is a general quasi-homomorphism on the expanded algebraic structure $\langle \mathbf{H}, E, \phi \rangle$.*

Proof Conditions from 1 to 5 in Definition 13 can be shown similarly to Lemma 13 using downward saturation. For, the condition 6, suppose that $\langle \Sigma, \Theta, m \rangle \in k(\forall x \alpha)$. This means that $\forall x \alpha \in \Sigma$. Then by Lemma 21 (5a), we have that $\alpha[\hat{d}/x] \in \Gamma$ for any $d \in E_n$ and any $\langle \Gamma, \Delta, n \rangle \in Y^+$ such that $\langle \Sigma, \Theta, m \rangle \leq \langle \Gamma, \Delta, n \rangle$. In other words, for any $\langle \Gamma, \Delta, n \rangle \in Y^+$ such that $\langle \Sigma, \Theta, m \rangle \leq \langle \Gamma, \Delta, n \rangle$, if $\langle \Gamma, \Delta, n \rangle \in \phi(d)$ then $\alpha[\hat{d}/x] \in \Gamma$ for each $d \in E$. Thus, $\langle \Sigma, \Theta, m \rangle \in \bigcap \{\phi(d) \rightarrow k(\alpha[\hat{d}/x]) : d \in E\}$ by using also the hypothesis of induction. Similarly, $\bigcap \{\phi(d) \rightarrow K(\alpha[\hat{d}/x]) : d \in E\} \subseteq K(\forall x \alpha)$ follows from the first condition (5b) of downward Henkin saturation. Consider next the second inclusion in the condition 7. Taking the contraposition, assume that $\langle \Sigma, \Theta, m \rangle \notin K(\exists x \alpha)$. From the definition of K, it follows that $\exists x \alpha \in \Theta$. By Lemma 21 (6b), $\alpha[\hat{d}/x] \in \Theta$ for each $d \in E_m$. That is, if $\langle \Sigma, \Theta, m \rangle \in \phi(d)$ then $\langle \Sigma, \Theta, m \rangle \notin K(\alpha[\hat{d}/x]$ for any d. Thus $\langle \Sigma, \Theta, m \rangle \notin \bigcup \{\phi(d) \cap K(\alpha[\hat{d}/x]) : d \in E\}$ □

In this way, we can translate a semantical proof of cut elimination of $\mathbf{QLJ}'$ given in the previous section into an algebraic proof via semi-completeness. Thus cut elimination theorem for $\mathbf{QLJ}'$ follows Lemma 26 and Theorem 27.

Cut elimination for predicate extensions of sequent systems for modal logics can be shown similarly. We will explain it briefly below. An *algebraic structures for modal predicate logics* is a pair $\langle \mathbf{A}, D \rangle$ of a complete Boolean algebra with a modality $\mathbf{A}$ and a non-empty set D.

Definition 14 (*expanded algebraic structures for modal predicate logics*) A triple $\langle \mathbf{A}, D, \phi \rangle$ is an *expanded algebraic structures for modal predicate logics*, if $\langle \mathbf{A}, D \rangle$ is an algebraic structures for modal predicate logics and ϕ is a mapping from D to A satisfying (1) $\bigcup\{\phi(d) : d \in D\} = 1_{\mathbf{A}}$ and (2) $\phi(d) \leq \Box\phi(d)$ for all $d \in D$. The mapping ϕ is called a definiteness function on an algebraic structure $\langle \mathbf{A}, D \rangle$ for modal predicate logics.

We can introduce *general quasi-homomorphisms on expanded algebraic structures for modal predicate logics* similarly to Definition 13. In this case, each expanded algebraic structure $\langle \mathbf{A}, D, \phi \rangle$ for modal predicate logics is a simplified form of *modal valued structures* in Gabbay et al. (2009). The additional condition that $\phi(d) \leq \Box\phi(d)$ for all $d \in D$ is necessary to show that the g-validity is preserved by such rules as ($\Rightarrow \Box$) in which all formulas in the antecedent of the lower sequent must be *boxed*. Conversely, this condition is verified in our *canonical Kripke frame for modal predicate logics with varying domains*, which can be defined similarly as the canonical Kripke frame for intuitionistic predicate logics with varying domains discussed above. In fact, this condition is shown to correspond to the statement that aRb implies $D(a) \subseteq D(b)$ for all a, b, which is of course true. In this way, we can give an alternative algebraic proof of cut elimination for sequent systems of various modal predicate logics, though we omit the details here.

Theorem 29 *The predicate extension of each of sequent systems* **GK**, **GKT** *and* **GS4** *is semi-complete with respect to corresponding expanded algebraic structures for modal predicate logics. Hence, cut elimination holds for each of them.*

Acknowledgements The author would like to express many thanks to C.-M. Yang and K. Sano for their helpful comments and suggestions, and also to T. Kowalski for valuable discussions.

References

Arai, T. (2017). Cut-eliminability in second order logic calculi, from https://arxiv.org/abs/1701.00929, submitted on 2017.

Belardinelli, F., Jipsen, P., & Ono, H. (2004). Algebraic aspects of cut elimination. *Studia Logica, 77*, 209–240.

Dragalin, A. G. (1988). *Mathematical intuitionism: introduction to proof theory*: American Mathematical Society.

Gabbay, D. M., Shehtman, V., & Shehtman, D. (2009). *Quantification in nonclassical logic I* (Vol. 153)., Studies in Logic and the Foundations of Mathematics: Elsevier.

Galatos, N., Jipsen, P., Kowalski, T., & Ono, H. (2007). *Residuated lattices: an algebraic glimpse at substructural logics* (Vol. 151)., Studies in Logic and the Foundations of Mathematics: Elsevier.

Goldblatt, R. (2011). *Quantifiers, propositions and identity: admissible semantics for quantified modal and substructural logics.*, Lecture Notes in Logic, Association for Symbolic Logic: Cambridge University Press.

Lafont, Y. (1997). The finite model property for various fragments of linear logic. *Journal of Symbolic Logic*, *62*, 1202–1208.

Lahav, O., & Avron, A. (2014). A unified semantic framework for fully-structural propositional sequent systems. *ACM Transactions on Computational Logic*, *165*, 26–48.

Maehara, S. (1991). Lattice-valued representation of the cut elimination theorem. *Tsukuba Journal of Mathematics*, *15*, 509–521.

Okada, M., & Terui, K. (1999). The finite model property for various fragments of intuitionistic linear logic. *Journal of Symbolic Logic*, *64*, 790–802.

Ono, H. (1993). Semantics for substructural logics. In K. Dosen & P. Schroeder-Heister (Eds.), *Substructural logics* (Vol. 2, pp. 259–291)., Studies in Logic and Computation: Oxford University Press.

Ono, H. (2015). Semantical approach to cut elimination and subformula property in modal logic. In C.-M. Yang, D.-M. Deng, & H. Lin (Eds.), *Structural Analysis of Non-Classical Logics, The Proceedings of the Second Taiwan Philosophical Logic Colloquium* (pp. 1–15)., Logic in Asia: Springer.

Ancient Indian Logic, Pakṣa and Analogy

Jeffrey B. Paris and Alena Vencovská

Abstract The authors B.K. Matilal, and earlier J.F. Staal, have suggested that the Indian Schema (erstwhile Hindu Syllogism) from Gotama's Nyāya-Sūtra should be understood in terms of an occurrence relation of an *event* happening at a *locus* (pakṣa). Building on a reading of the Indian Schema as analogical reasoning, that is, smoke on the hill and the prior example of smoke in the kitchen being associated with fire in the kitchen leading to the conclusion that there is fire on the hill, we have previously suggested and investigated a formalisation of the Schema within a binary predicate language. In this paper we consider a version derived therefrom using just unary predicates. We go on to show that within the context of Pure Inductive Logic this formal version has a rational justification, in the sense that from the Principle of Atom Exchangeability (akin to Carnap's Attribute Symmetry and a property of Carnap's Continuum of Inductive Methods) it follows that under the assumptions of the Schema fire on the hill must be at least as probable as absence of fire on the hill.

Keywords Indian schema · Pakṣa · Nyāya-Sūtra · Analogy · Pure inductive logic · Rationality

1 Introduction

In Sūtras 32–36 of the Nyāya-Sūtra ($\sim$200BCE-150CE), Gotama presents a schema of reasoning consisting of:

⋆ Statement of the thesis.
⋆ Statement of a reason.

Jeffrey B. Paris and Alena Vencovská—Supported by a UK Engineering and Physical Sciences Research Council Research Grant [Number EP/L023989/1].

J.B. Paris (✉) · A. Vencovská
School of Mathematics, The University of Manchester, Manchester M13 9PL, England
e-mail: jeff.paris@manchester.ac.uk

A. Vencovská
e-mail: alena.vencovska@manchester.ac.uk

S.C.-M. Yang et al. (eds.), *Philosophical Logic: Current Trends in Asia*,
Logic in Asia: Studia Logica Library, DOI 10.1007/978-981-10-6355-8_3

★ An example supporting the reason on the grounds of similarity to the present case.
★ Application of the above to the present case.
★ Conclusion.

The ubiquitous illustration of the schema runs as follows:

★ There is fire on the hill.
★ For there is smoke.
★ (Wherever there is smoke, there is fire), as in the kitchen.
★ This is such a case (smoke on the hill).
★ Therefore it is so, (fire on the hill).

It is generally acknowledged that the five limb schema as above was intended to convince someone else, hence the apparent repetition is to be mitigated by the listener's growing comprehension as the schema unfolds. As an argument to convince oneself it is commonly rephrased in a three line form as, for example,[1]

★ (Wherever there is smoke, there is fire), as in the kitchen.
★ There is smoke on the hill.
★ Therefore there is fire on the hill.

In this form the *Indian Schema* as we shall call it, bears an evident resemblance to the 'Socrates is Mortal' example of Aristotle's BARBARA syllogism, at least from the perspective of 'Western deductive logic'. From that viewpoint the key difference is the inclusion of the seemingly redundant 'as in the kitchen'.

However, as we have briefly documented in Paris and Vencovská (2016a, b), historical and circumstantial evidence give grounds for arguing that it is the universal generalization 'Wherever there is smoke, there is fire' that is out of place and that the example itself alone is being taken to sum up or crystalize the available evidence. When the example is so taken to encapsulate the relevant evidence,[2] the schema may be rephrased as

★ When there was smoke in the kitchen there was fire.
★ There is smoke on the hill. $\underline{\text{A}}$
★ Therefore there is fire on the hill.

The plan now is to propose some possible formalizations of $\underline{\text{A}}$ within Pure Inductive Logic[3] (PIL for short) and show that they can be justified as rational on the grounds of following from principles usually accepted in that subject as rational.

Before doing so we should point out, as B.K. Matilal does in Matilal (2001), p. 197, that Westerners have frequently misunderstood, and in consequence criticized, the

---

[1]There is a fine edge here as to correct ordering of the lines which for the purposes of this paper we shall gloss over.

[2]Which we have termed in Paris and Vencovská (2016a) a *good example*.

[3]See Paris and Vencovská (2015), Chap. 1 for a full, or below for a short, introduction to Pure Inductive Logic. A key feature therein is that statements are assigned (subjective) probabilities rather than just 0,1 truth values.

Indian Schema as rendering a *valid* conclusion. He goes on to suggest the reasoning may be more correctly understood as *inductive*, and for practical purposes providing knowledge of the real world. In accord with this sentiment then we will in what follows be interpreting the 'Therefore' in the final line of A as the less demanding and more realistic 'provides grounds for accepting'.

2 The Pakṣa Formalisation

In our previous attempts (Paris and Vencovská 2016a, b) at formalising A we worked within Unary Predicate Logic, so using S, F, h and k in the obvious sense we employed $S(k)$ to express *There is smoke in the kitchen*, $F(h)$ to express *There is fire on the hill* etc. Within PIL, A then becomes the assertion that given $S(h)$ and in the absence of any other pertinent information[4] adding $S(k) \rightarrow F(k)$ (or $S(k) \leftrightarrow F(k)$ or $S(k) \wedge F(k)$[5]) should provide grounds for accepting $F(h)$. In Paris and Vencovská (2016a) we showed that this does indeed follow from certain well accepted rational symmetry principles and in consequence is itself rational.

Some authors however, notably B.K. Matilal (2001) and earlier J.F. Staal (2001), have suggested that it is much closer to the Indian way of thinking to formalise A by employing a binary relation standing for 'occurring at'. According to Staal in Indian logic an entity is never regarded in isolation but always considered as occurring at a locus, and the fundamental relation which underlies all expressions is that between an entity and its locus (pakṣa). Using R for this relation and f, s, h, k for $fire, smoke, hill$ and $kitchen$ respectively, A becomes the claim that in the absence of any other pertinent information $R(s, h) \wedge (R(s, k) \rightarrow R(f, k))$ (or possibly $R(s, h) \wedge (R(s, k) \leftrightarrow R(f, k))$ or $R(s, h) \wedge (R(s, k) \wedge R(f, k))$ should provide grounds for accepting $R(f, h)$. In Paris and Vencovská (2017) we investigated to what extent PIL supports this version as a rational inference. A feature of that approach was the duality it conferred on the pakṣa relation in the sense that we could equally consider the hill to be the location of the smoke or the smoke to be the location of the hill. While this may seem a little strange the sometimes cited example of the pot being on the table or the table being under the pot more plainly exhibits the inherent symmetry here.

Whether or not this is a feature which we arguably should, or should not, incorporate in our formalisation in this current paper we shall make the distinction and simply consider the hill and kitchen as the locations. In that case it seems that, for example, $R(s, k)$ can without loss be simplified to $K(s)$ where K is now a unary relation (i.e. predicate) symbol standing for 'at the kitchen' and s is as before a constant symbol standing for 'smoke'. Within this formulation then A might become the claim that:

---

[4]I.e. in the sense of Paris and Vencovská (2016a) that this is a *good* example.

[5]As we argue in Paris and Vencovská (2016a) it is not obvious which of these versions correspond to what was originally being envisaged here.

In the absence of any other pertinent information $H(s) \wedge (K(s) \rightarrow K(f))$ *(or possibly* $H(s) \wedge (K(s) \leftrightarrow K(f))$ *or* $H(s) \wedge (K(s) \wedge K(f)))$ *provides grounds for accepting* $H(f)$.

(†)

To facilitate investigating the justification for this claim within the context of PIL we first need to summarise some necessary background from PIL. Further details can be found for example in Paris and Vencovská (2015).

3 Unary Pure Inductive Logic

As far as this paper is concerned we may take our framework for Unary Pure Inductive Logic (UPIL) as Predicate Logic employing a language L_q with unary relation symbols $R_1, R_2, \ldots, R_q$, countably many constants $a_1, a_2, a_3, \ldots$ and no function symbols nor equality. Let SL_q denotes the set of sentences of L_q.

A probability function on L_q is a function $w : SL_q \rightarrow [0, 1]$ such that for any $\theta, \phi, \exists x\, \psi(x)$ from SL_q,

(i) If $\models \theta$ then $w(\theta) = 1$.

(ii) If $\theta \models \neg\phi$ then $w(\theta \vee \phi) = w(\theta) + w(\phi)$.

(iii) $w(\exists x\, \psi(x)) = \lim_{n\to\infty} w\left(\bigvee_{i=1}^{n} \psi(a_i)\right)$.

Any function w satisfying the above conditions has the properties we usually expect of probability (see Paris and Vencovská 2015, Proposition 3.1), in particular if ψ logically implies θ then $w(\psi) \leq w(\theta)$.

Given a probability function w and $\theta, \phi \in SL_q$ with $w(\phi) > 0$ we define the conditional probability of θ given ϕ as usual by

$$w(\theta \mid \phi) = \frac{w(\theta \wedge \phi)}{w(\phi)}\,. \tag{1}$$

With a fixed $\phi \in SL_q$, $w(\phi) > 0$, the function defined by (1) is also a probability function.[6]

Apart from the above conditions defining probability, within PIL some further principles are usually imposed on any 'rational' probability function proposed to be adopted in a situation where nothing is known about any intended interpretation of the constant and relation symbols, i.e. a zero knowledge or blank slate situation. Within UPIL (i.e. where all the relation symbols are unary) one of the most widely assumed rational principles that a probability function w on L_q should satisfy is *Atom Exchangeability*, which we will now explain.

The *atoms* of L_q are the 2^q formulae $\alpha_1(x), \alpha_2(x), \ldots, \alpha_{2^q}(x)$ of the form

[6]To avoid any problems with conditioning on sentences with zero probability we shall identify expressions like $w(\phi \mid \psi) \geq w(\eta)$ with $w(\phi \wedge \psi) \geq w(\eta) \cdot w(\psi)$.

$$\bigwedge_{j=1}^{q} R_j^{\epsilon_j}(x)$$

where the $\epsilon_j \in \{0, 1\}$ and $R_j^{\epsilon_j}$ is R_j if $\epsilon_j = 1$ and $\neg R_j$ if $\epsilon_j = 0$.

For (distinct) constants $b_1, b_2, \ldots, b_n$ (coming from the constants $a_1, a_2, a_3, \ldots$ of L_q) a *state description* for $b_1, b_2, \ldots, b_n$ is a sentence of the form[7]

$$\bigwedge_{i=1}^{n} \alpha_{h_i}(b_i) \tag{2}$$

for some $1 \leq h_1, h_2, \ldots, h_n \leq 2^q$. By a theorem of H. Gaifman, see Gaifman (1964) (or in the notation of this paper see Paris and Vencovská 2015, Chap. 7) a probability function on L_q is uniquely determined by its values on the state descriptions.

Given a state description as in (2) we define its *spectrum* to be the multiset

$$\{m_1, m_2, \ldots, m_{2^q}\}$$

where $m_r = |\{j \mid h_j = r\}|$.[8]

We are now ready to state, for w a probability function on L_q,

3.1 *The Principle of Atom Exchangeability, Ax*

For Θ a state description, $w(\Theta)$ depends only on the spectrum of Θ.

So this principle asserts that all that matters when assigning a probability to a state description is how many of the instantiating constants are indistinguishable, and how many not so. Expressed in this way then the principle appeals to the *irrelevance* of all other factors in determining the probability of a state description, for example the irrelevance of the extent to which the instantiating constants might be distinguishable (i.e. the number of predicates on which they disagree).[9] On this point it is worth mentioning that the founding fathers of Inductive Logic, R. Carnap and earlier W.E. Johnson, independently proposed another principle of irrelevance, which subsequently became known as Johnson's Sufficientness Postulate, from which Atom Exchangeability is a consequence, see for example Paris and Vencovská (2015), Lemma 17.1, and this principle (which characterises Carnap's Continuum of Inductive Methods when the unary language has at least two predicate symbols) has been

---

[7]Up to logical equivalence.

[8]It is customary to omit the zero entries from this multiset since their number is determined by 2^q and the number of non-zero m_r.

[9]An entertaining argument in support of this view is given by D. Miller in (1974) (and also Paris and Vencovská 2016d).

widely promoted as furnishing a constraint on the logical or rational assignment of probabilities in a zero knowledge, blank slate, situation. In this sense then Atom Exchangeability inherits (at least) all of the perceived rationality of Johnson's Sufficientness Postulate.

By an easy consequence of Gaifman's Theorem mentioned earlier (see for example Paris and Vencovská 2015, Lemma 7.3) if w satisfies Ax,[10] then it satisfies:

3.2 The Principle of Constant Exchangeability, Ex

For $\Theta(a_1, a_2, \ldots, a_m) \in SL_q$ *and* σ *a permutation of* $\mathbb{N}^+ = \{1, 2, 3, \ldots\}$,

$$w(\theta(a_1, a_2, \ldots, a_m)) = w(\theta(a_{\sigma(1)}, a_{\sigma(2)}, \ldots, a_{\sigma(m)})).$$

Constant Exchangeability, Ex, is widely advocated in Inductive Logic as a requirement for a zero-knowledge rational probability function w (much more so than Ax in fact) since in that situation there is no reason to treat any one constant differently from any other.

We remark that Ax can also be construed as a symmetry principle since it can be shown to be equivalent to Ex plus the assertion that for any permutation τ of $\{1, 2, \ldots, 2^q\}$ and state description as in (2),

$$w\Big(\bigwedge_{i=1}^{n} \alpha_{h_i}(b_i)\Big) = w\Big(\bigwedge_{i=1}^{n} \alpha_{\tau(h_i)}(b_i)\Big).$$

In what follows it will be useful to note that if w is a probability function on L_q satisfying Atom Exchangeability and $1 \leq n < q$ then w restricted to SL_n ($\subset SL_q$) is a probability function on L_n satisfying Atom Exchangeability.

Furthermore, we will need to explicitly mention two particular probability functions on L_q, namely $c_\infty^{L_q}$ and $c_0^{L_q}$ from Carnap's Continuum of Inductive Methods. The probability function $c_\infty^{L_q}$ can be specified by

$$c_\infty^{L_q}\Big(\bigwedge_{i=1}^{n} \alpha_{h_i}(b_i)\Big) = 2^{-nq}$$

where the b_i are distinct constants from among $a_1, a_2, a_3, \ldots$. Another way of saying this is that it treats all the (distinct) $R_j(a_i)$ as stochastically independent with probability $1/2$.

$c_\infty^{L_q}$ satisfies Ax (as do all the functions in Carnap's Continuum of Inductive Methods). However it is generally considered as an unsatisfactory choice of rational probability function in the situation of zero knowledge since it denies any form of

[10] As formulated here.

inductive influence. For example it gives

$$c_{\infty}^{L_q}(R_1(a_{n+1}) \mid R_1(a_1) \wedge R_1(a_2) \wedge \ldots, \wedge R_1(a_n)) = 1/2 \tag{3}$$

no matter how large n may be.

The probability function $c_0^{L_q}$ can be specified by

$$c_0^{L_q}\Big(\bigwedge_{i=1}^{m} \alpha_{h_i}(b_i)\Big) = \begin{cases} 2^{-q} & \text{if } h_1 = h_2 = \ldots = h_m, \\ 0 & \text{otherwise.} \end{cases}$$

Again this function is generally considered as an unsatisfactory choice of rational probability function in the situation of zero knowledge since it is too ready to jump to categorical conclusions, for example giving

$$c_0^{L_q}(R_1(a_2) \mid R_1(a_1)) = 1. \tag{4}$$

4 The Main Theorems

Recall that L_2 is the unary language for PIL as above with $q = 2$. Without loss of generality let $H, K, s, f,$ stand for R_1, R_2, a_1, a_2 (etc.). In the Appendix we shall give a proof of the following (and subsequent) results.

Theorem 1 *Let w be a probability function on L_2 satisfying Ax. Then*
(a) $w(H(f) \mid H(s) \wedge (K(s) \rightarrow K(f)) \geq 1/2$,
(b) $w(H(f) \mid H(s) \wedge (K(s) \leftrightarrow K(f)) \geq 1/2$,
(c) $w(H(f) \mid H(s) \wedge (K(s) \wedge K(f)) \geq 1/2$,
with equality in any of these just if $w = c_{\infty}^{L_2}$.[11]

This theorem shows that *if* we assume that the Indian Schema can be formalised in one of the ways given in (†), and so in particular that all the relevant evidence is captured by the example and the direct observation of smoke on the hill, then any assignment of probabilities which is rational in the sense of satisfying Ax and not altogether rejecting inductive influence (i.e. not $c_{\infty}^{L_2}$) will necessarily have to give 'fire on the hill' a probability strictly greater than a half. In other words strictly greater than 'no fire on the hill' and in this sense 'fire on the hill' will be judged to be the best bet, or in the words of (†), the evidence 'provides grounds for accepting'.

At this point a natural question to ask is whether adding the experience of more than just one kitchen might not increase the probability of fire on the hill still further. This is indeed essentially the case:

---

[11]This same result holds if we take w a probability function on L_q satisfying Ax since then w restricted to SL_2 ($\subset SL_q$) will still satisfy Ax on this smaller language. The only change would be that in the last line it is w restricted to the SL_2 ($\subset SL_q$) that equals $c_{\infty}^{L_2}$.

Theorem 2 *Let w be a probability function on L_{n+2} satisfying Ax. Then*

(a) $w(H(f) \mid H(s) \wedge \bigwedge_{j=1}^{n+1}(K_j(s) \to K_j(f))) \geq w(H(f) \mid H(s) \wedge \bigwedge_{j=1}^{n}(K_j(s) \to K_j(f)))$,

(b) $w(H(f) \mid H(s) \wedge \bigwedge_{j=1}^{n+1}(K_j(s) \leftrightarrow K_j(f))) \geq w(H(f) \mid H(s) \wedge \bigwedge_{j=1}^{n}(K_j(s) \leftrightarrow K_j(f)))$,

(c) $w(H(f) \mid H(s) \wedge \bigwedge_{j=1}^{n+1}(K_j(s) \wedge K_j(f))) \geq w(H(f) \mid H(s) \wedge \bigwedge_{j=1}^{n}(K_j(s) \wedge K_j(f)))$,

and in each case strictly greater than unless $w = c_\infty^{L_{n+2}}$, in which case they all take value $1/2$, or $n > 0$ and $w = c_0^{L_{n+2}}$, in which case they all take the value 1.

The version of the Indian Schema which we have considered here is based on Sūtra 36 of the Nyāya-Sūtra which is commonly referred to as a 'homogeneous example'. We have variously formalised this as

$$K(s) \to K(f), \quad K(s) \leftrightarrow K(f) \tag{5}$$

or

$$K(s) \wedge K(f). \tag{6}$$

However in Sūtra 37 Gotama describes another sort of example, a heterogeneous example, for which the time honoured example is a lake which, being devoid of fire is devoid also of smoke (see Paris and Vencovská 2016a for a fuller discussion of this case). Proceeding as in the homogeneous example it seems we may variously formalize this evidence as

$$\neg L(f) \to \neg L(s), \quad \neg L(f) \leftrightarrow \neg L(s), \quad \neg L(f) \wedge \neg L(s).$$

The first two of these are logically equivalent to (5), respectively, so the same Theorem 2 parts (a),(b) still hold if we add in some such heterogeneous examples. For the remaining part (c) of that theorem we can now extend it to

Theorem 3 *Let w be a probability function on L_{n+2} satisfying Ax. Then for $\epsilon_1, \epsilon_2, \ldots, \epsilon_{n+1} \in \{0, 1\}$,*

$$w(H(f) \mid H(s) \wedge \bigwedge_{j=1}^{n+1}(K_j^{\epsilon_j}(s) \wedge K_j^{\epsilon_j}(f)) \geq w(H(f) \mid H(s) \wedge \bigwedge_{j=1}^{n}(K_j^{\epsilon_j}(s) \wedge K_j^{\epsilon_j}(f)),$$

and strictly greater than unless $w = c_\infty^{L_{n+2}}$, in which case they both take value $1/2$, or $w = c_0^{L_{n+2}}$, in which case they both take the value 1.

From Theorems 1(c) and 3 one might have hoped that we would have had that

$$w(H(f) \mid H(s) \wedge \bigwedge_{j=1}^{n+1} (K_j^{\epsilon_{1_j}}(s) \wedge K_j^{\epsilon_{2_j}}(f)) > 1/2$$

whenever $\epsilon_{1_j} = \epsilon_{2_j}$ for a majority of the $1 \leq j \leq n+1$. However with $w \neq c_0^{L_{n+2}}$ satisfying Ax all these left had sides are exactly 1/2 unless *all* the $\epsilon_{1_j} = \epsilon_{2_j}$. (For more on this see the *Constant Analogy Principle* from Howarth et al. 2016 and Paris and Vencovská 2016c, Appendix.)

5 Weakening Atom Exchangeability

Given Atom Exchangeability (Ax), each of the formalisations in terms of implication, bi-implication and conjunction appears equally favoured and yield, in Theorems 1, 2 and 3, all that one might hope for in terms of analogical reasoning. Atom Exchangeability is however a rather strong assumption so it would be natural to ask if we could not get away with less, and at the same time possibly find grounds for distinguishing the three formalisations.

The natural step down from Ax is the combination of Ex and Predicate Exchangeability (Px) plus Strong Negation (SN) since both Px and SN are consequences of Ax, are directly justified by symmetry considerations, and together they are still strictly weaker than Ax.

5.1 The Principle of Predicate Exchangeability, Px

For $\theta \in SL$ and relation symbols R_i, R_j, if θ' is the result of transposing R_i, R_j throughout θ then $w(\theta) = w(\theta')$.

5.2 The Strong Negation Principle, SN

For $\theta \in SL$, $w(\theta) = w(\theta')$ where θ' is the result of replacing each occurrence of the relation symbol R in θ by $\neg R$.

With Ex + Px + SN in place of Ax Theorem 1 still holds. However Theorems 2 and 3 do not, in each case we can find probability functions satisfying Ex + Px + SN for which these fail. Indeed for each of

$$w(H(f) \mid H(s) \wedge (K(s) \rightarrow K(f)) \geq w(H(f) \mid H(s)), \tag{7}$$

$$w(H(f) \mid H(s) \wedge (K(s) \leftrightarrow K(f)) \geq w(H(f) \mid H(s)), \tag{8}$$

$$w(H(f) \mid H(s) \wedge (K(s) \wedge K(f)) \geq w(H(f) \mid H(s)), \tag{9}$$

we can find probability functions satisfying Ex + Px + SN for which they fail. (More details are given in the appendix.) From the standpoint then that more examples should strengthen the analogical influence Ex + Px + SN alone seems not to confirm this intuition.

Nevertheless we do at least have the many-kitchen generalizations of Theorem 1(b)&(c) for w satisfying Ex + Px + SN, namely:

$$w(H(f) \mid H(s) \wedge \bigwedge_{j=1}^{n} (K_j(s) \leftrightarrow K_j(f)) \geq 1/2, \tag{10}$$

$$w(H(f) \mid H(s) \wedge \bigwedge_{j=1}^{n} (K_j(s) \wedge K_j(f)) \geq 1/2. \tag{11}$$

Whether or not case (a) of the many kitchens version of Theorem 1 holds (for all n), i.e.

$$w(H(f) \mid H(s) \wedge \bigwedge_{j=1}^{n} (K_j(s) \rightarrow K_j(f)) \geq 1/2, \tag{12}$$

is currently open.

It is interesting to note however that in a later volume five addition to the Nyāya-Sūtra the author, Vātsyāyana (~375CE-450CE), specifically points out the incorrectness of a variant of the Indian Schema involving two examples, the so called 'saying too much' rejoinder, see Vidyabhusana 1974, p. 172.

6 Conclusion

We have shown that various logical formulations of the Indian Schema which attempt to capture the notion of pakṣa, as suggested by J.F. Staal and B.K. Matilal, within *Unary* Pure Inductive Logic are actually consequences of the arguably rational principle of Atom Exchangeability. By this we are certainly do not wish to argue that the ancients were somehow aware of this principle. Rather we simply intend to point out that the everyday common-senseness of the the Indian Schema does in fact have a *formal* justification as rational within the context of UPIL.

Appendix

We begin with a proof of Theorem 3 (which eclipses Theorem 2(c)). Let $\alpha_i(x)$ for $i = 1, 2, \ldots, 2^{n+2}$ denote the atoms of L_{n+2}, for simplicity of notation let $\alpha_1(x), \alpha_2(x), \alpha_3(x), \alpha_4(x)$ be the atoms

$$\alpha_1(x) = H(x) \wedge \bigwedge_{j=1}^{n+1} K_j^{\epsilon_j}(x),$$

$$\alpha_2(x) = \neg H(x) \wedge \bigwedge_{j=1}^{n+1} K_j^{\epsilon_j}(x),$$

$$\alpha_3(x) = H(x) \wedge \bigwedge_{j=1}^{n} K_j^{\epsilon_j}(x) \wedge K_{n+1}^{1-\epsilon_{n+1}}(x).$$

$$\alpha_4(x) = \neg H(x) \wedge \bigwedge_{j=1}^{n} K_j^{\epsilon_j}(x) \wedge K_{n+1}^{1-\epsilon_{n+1}}(x).$$

Then using (P2) the inequality in Theorem 3 becomes[12]

$$\frac{X}{Z} \geq \frac{T}{Y} \tag{13}$$

where

$$\begin{aligned}
X &= w(\alpha_1(f) \wedge \alpha_1(s)),\\
Z &= w(\alpha_1(f) \wedge \alpha_1(s)) + w(\alpha_2(f) \wedge \alpha_1(s)),\\
T &= w(\alpha_1(f) \wedge \alpha_1(s)) + w(\alpha_3(f) \wedge \alpha_3(s)) + w(\alpha_1(f) \wedge \alpha_3(s)) + w(\alpha_3(f) \wedge \alpha_1(s)),\\
Y &= w(\alpha_1(f) \wedge \alpha_1(s)) + w(\alpha_3(f) \wedge \alpha_3(s)) + w(\alpha_1(f) \wedge \alpha_3(s)) + w(\alpha_3(f) \wedge \alpha_1(s))\\
&\quad + w(\alpha_2(f) \wedge \alpha_1(s)) + w(\alpha_4(f) \wedge \alpha_1(s)) + w(\alpha_2(f) \wedge \alpha_3(s)) + w(\alpha_4(f) \wedge \alpha_3(s)).
\end{aligned}$$

Noticing that by Ax $w(\alpha_i(f) \wedge \alpha_i(s))$ has the same value, x say, for each i and similarly $w(\alpha_i(f) \wedge \alpha_j(s))$ has the same value, y say, whenever $i \neq j$, (13) simplifies to

$$\frac{x}{x+y} \geq \frac{2x+2y}{2x+6y}, \tag{14}$$

equivalently $xy \geq y^2$. So it is enough to show that $x \geq y$, $x > 0$ with $x = y$ just if $w = c_\infty^{L_{n+2}}$ and $y = 0$ just if $w = c_0^{L_{n+2}}$.

To this end we need de Finetti's Representation Theorem, (de Finetti 1974) (or in the notation of this paper see Paris and Vencovská 2015, p. 90) which tells us that there is a normalized σ-additive measure μ on the Borel subsets of

[12]As we shall shortly see there can be no problems with zero divisors here.

$$\mathbb{D} = \{\langle x_1, x_2, \ldots, x_{2^{n+2}}\rangle \in \mathbb{R}^{2^{n+2}} \mid x_i \geq 0, \sum_i x_i = 1\}$$

which is invariant under permutations of the coordinates and such that for $1 \leq i, j \leq 2^{n+2}$,

$$w(\alpha_i(f) \wedge \alpha_j(s)) = \int_{\mathbb{D}} x_i x_j \, d\mu(\vec{x})$$

Hence,

$$\begin{aligned} x \geq y &\iff \int_{\mathbb{D}} x_1^2 - x_1 x_2 \, d\mu(\vec{x}) \geq 0 \\ &\iff \int_{\mathbb{D}} x_1^2 + x_2^2 - 2x_1 x_2 \, d\mu(\vec{x}) \geq 0 \text{ (by invariance)} \\ &\iff \int_{\mathbb{D}} (x_1 - x_2)^2 \, d\mu(\vec{x}) \geq 0. \end{aligned} \tag{15}$$

So $x \geq y$. Since

$$1 = w\big(\bigvee_{i,j=1}^{2^{n+2}} (\alpha_i(f) \wedge \alpha_j(s))\big) = 2^{n+2}x + 2^{n+2}(2^{n+2} - 1)y$$

$x > 0$ now also follows.

If $x = y$ then from (15) and invariance we must have that for $\vec{x}$ in the support of μ all the x_i must be equal (to $2^{-(n+2)}$) which forces w to be the probability function $c_\infty^{L_{n+2}}$. Similarly if $y = 0$ then

$$w\big(\bigwedge_{i=1}^{m} \alpha_{h_i}(a_i)\big) = \begin{cases} 2^{-(n+2)} & \text{if } h_1 = h_2 = \ldots = h_m, \\ 0 & \text{otherwise,} \end{cases}$$

since by Ax the $w(\bigwedge_{i=1}^m \alpha_h(a_i))$ are all equal for $h = 1, 2, \ldots, 2^{n+2}$, and this uniquely determines w to be $c_0^{L_{n+2}}$.

Turning to Theorem 2 for (a), with the above notation this becomes

$$\frac{2^{n+1}x + (3^{n+1} - 2^{n+1})y}{2^{n+1}x + (2.3^{n+1} - 2^{n+1})y} \geq \frac{2^{n+1}x + (4.3^{n+1} - 2^{n+1})y}{2^{n+1}x + (8.3^{n+1} - 2^{n+1})y} \tag{16}$$

which simplifies to $xy \geq y^2$. As above this always holds, and with equality just if $x = y$, in which case $w = c_\infty^{L_{n+2}}$ or $y = 0$ in which case $w = c_0^{L_{n+2}}$.

For Theorem 2 for (b) we similarly obtain the inequality to be equivalent to

$$\frac{2^{n+1}x}{2^{n+1}x + 2^{n+1}y} \geq \frac{2^{n+1}x + (4.2^n - 2^{n+1})y}{2^{n+1}x + (8.2^n - 2^{n+1})y} \tag{17}$$

which again reduces to $xy \geq y^2$ and yields the same conclusion.

Theorem 1 now follows since it is easy to check that since $x \geq y$ in each case the right hand sides of (14), (16), (17) must be at least $1/2$ (for any $n \geq 0$).

Turning now to the assumption that w merely satisfies Ex+Px+SN, let $\alpha_1(x), \alpha_2(x), \ldots, \alpha_{2^{n+2}}(x)$ be as usual the atoms of L_{n+2} and notice that in this case $w(\alpha_i(s) \wedge \alpha_j(f))$ takes the same value, d_r say, for all pairs i, j such that $\alpha_i(x)$ and $\alpha_j(x)$ differ on exactly r predicate symbols.

Notice too that (10)–(12) are respectively equivalent to

$$w(H(s) \wedge H(f) \wedge \bigwedge_{i=1}^{n}((K_j(s) \wedge K_j(f)) \vee (\neg K_j(s) \wedge \neg K_j(f)) \vee (\neg K_j(s) \wedge K_j(f)))$$
$$\geq w(H(s) \wedge \neg H(f) \wedge \bigwedge_{i=1}^{n}((K_j(s) \wedge K_j(f)) \vee (\neg K_j(s) \wedge \neg K_j(f)) \vee (\neg K_j(s) \wedge K_j(f))), \tag{18}$$

$$w(H(s) \wedge H(f) \wedge \bigwedge_{i=1}^{n}((K_j(s) \wedge K_j(f)) \vee (\neg K_j(s) \wedge - K_j(f))))$$
$$\geq w(H(s) \wedge \neg H(f) \wedge \bigwedge_{i=1}^{n}((K_j(s) \wedge K_j(f)) \vee (\neg K_j(s) \wedge \neg K_j(f))), \tag{19}$$

$$w(H(s) \wedge H(f) \wedge \bigwedge_{i=1}^{n}((K_j(s) \wedge K_j(f)))$$
$$\geq w(H(s) \wedge \neg H(f) \wedge \bigwedge_{i=1}^{n}((K_j(s) \wedge K_j(f))). \tag{20}$$

Both (19) and (20) now simplify to $d_0 \geq d_1$. To show this inequality $d_0 \geq d_1$ we note that as with Ax we have a version of de Finetti's Theorem, the only difference is that the measure μ is only required to be invariant under permutations of coordinates $1, 2, \ldots, 2^{n+1}$ when the corresponding permutation of atoms $\alpha_1(x), \alpha_2(x), \ldots, \alpha_{2^{n+1}}(x)$ follows from Px+SN. Still, this enables us, just as in the proof for Ax as above, to show that for any $1 \leq i, j \leq 2^{n+1}$,

$$d_0 = w(\alpha_i(s) \wedge \alpha_i(f)) \geq w(\alpha_i(s) \wedge \alpha_j(f)), \tag{21}$$

so in particular $d_0 \geq d_1$ and $d_0 = d_1$ just if $w = c_\infty^{L_{n+1}}$. This establishes (19) and (20).

Unfortunately (18) does not simplify so sweetly and whether or not it holds is currently open.

In order to describe probability functions satisfying Ex + Px + SN which are counter-examples to (7)–(9) we need to introduce some notation.

Let $\alpha_i(x)$ for $i = 1, 2, \ldots, 2^q$ be the atoms of L_q. Given $\langle c_1, c_2, \ldots, c_{2^q}\rangle \in \mathbb{R}^{2^q}$ with $c_1, c_2, \ldots, c_{2^q} \geq 0$ and $\sum_{i=1}^{2^q} c_i = 1$ define the probability function $w_{\vec{c}}$ on L_q by

$$w_{\vec{c}}\Big(\bigwedge_{i=1}^{m} \alpha_{h_i}(b_i)\Big) = \prod_{i=1}^{m} c_{h_i}.$$

Clearly $w_{\vec{c}}$ satisfies Ex.

In the case of L_2 let $\alpha_1(x), \alpha_2(x), \alpha_3(x), \alpha_4(x)$ be respectively

$$H(x) \wedge K(x), \quad H(x) \wedge \neg K(x), \quad \neg H(x) \wedge K(x), \quad \neg H(x) \wedge \neg K(x)$$

and

$$w = 4^{-1}(w_{\langle 1/2,1/2,0,0\rangle} + w_{\langle 1/2,0,1/2,0\rangle} + w_{\langle 0,1/2,0,1/2\rangle} + w_{\langle 0,0,1/2,1/2\rangle}).$$

Then w satisfies Ex + Px + SN. However it is straightforward to check that (7)–(9) all fail for w.

References

de Finetti, B. (1974). *Theory of probability* (Vol. 1). New York: Wiley.

Gaifman, H. (1964). Concerning measures on first order calculi. *Israel Journal of Mathematics*, *2*, 1–18.

Howarth, E., Paris, J. B., & Vencovská, A. (2016). An examination of the sep candidate analogical inference rule within pure inductive logic. *Journal of Applied Logic*, *14*, 22–45.

Matilal, B. K. (2001). Introducing Indian logic. In J. Ganeri (Ed.), *Indian logic, a reader* (pp. 183–215). London: Routledge.

Miller, D. (1974). Popper's qualitative theory of versimilitude. *British Journal for the Philosophy of Science*, *25*, 166–177.

Paris, J. B., & Vencovsk, A. (2015). *Pure inductive logic*. Cambridge: in the Association Cambridge University Press.

Paris, J. B. & Vencovská, A. (2016a). The Indian schema as analogical reasoning. MIMS eprints. Accessed April 2016, http://eprints.ma.man.ac.uk/2436/01/covered/MIMS_ep2016_10.pdf

Paris, J. B., & Vencovská, A. (2016b). The Indian schema analogy principles. *IfCoLog Journal of Logics and their Applications*, *3*(5), 903–927.

Paris, J. B. & Vencovská, A. (2016c). A note on 'A note on Mohist logic'. Accessed October 2016, http://www.maths.manchester.ac.uk/~jeff/

Paris, J.B. & Vencovská, A. (2016d). Translation invariance and Miller's weather example. MIMS eprints. Accessed October 2016, http://eprints.ma.man.ac.uk

Paris, J.B. & A.Vencovská. (2017). Ancient Indian Logic and Analogy. In S. Ghosh & S. Prasad (Eds.), *Proceedings of the Seventh Indian Conference on Logic and its Applications, ICLA, Kanpur* (pp. 198–210). Berlin: Springer. LNCS 10119.

Staal, J. F. (2001). The concept of Pakṣa in Indian logic. In J. Ganeri (Ed.), *Indian logic: A reader* (pp. 151–161). London: Routledge.

Vidyabhusana, S. C. (1974). In Basu, B. D. (Ed.), *The Nyāya Sūtras of Gotama* (Vol. 8)., Sacred books of the Hindus New York: AMS Press, INC.

Provability and Decidability of Arithmetical Sentences

Shih Ping Tung

Abstract Gödel's incompleteness theorem tells that the elementary theories of **N** and **Z** are not axiomatizable, respectively. J. Robinson proved that the elementary theory of **Q** is undecidable, hence not axiomatizable. We may ask what subsets of these elementary theories are decidable or axiomatizable. We call an arithmetical sentence ϕ an $\exists^n\forall\exists$ sentence if it is logically equivalent to a sentence of the form $\exists x_1 \cdots \exists x_n \forall y \exists z\, \psi(x_1, \ldots, x_n, y, z)$, where $\psi(x_1, x_2, \ldots, x_n, y, z)$ is a quantifier-free formula. In this talk we show that the set of all true $\exists^n\forall\exists$ sentences in **N**, **Z**, and **Q** are axiomatizable, respectively. Also, let $f(x, y) \in \mathbf{Z}[x, y]$ and $a \in \mathbf{Z}$. The sets of all sentences of the form $\forall z \exists x \forall y\, f(x, y) - az \neq 0$ true in **N**, **Z**, and **Q** are also axiomatizable, respectively. Therefore, the sets of sentences of the form $\exists z \forall x \exists y\, f(x, y) - az = 0$ true in **N**, **Z**, and **Q** are decidable, respectively. Over **N**, we have shown that this decision problem is NP-hard, co-NP-hard, and in PSPACE. Therefore, if NP $\neq$ co-NP, then this decision problem is in PSPACE $\setminus$ (NP $\cup$ co-NP).

1 Introduction

Gödel's incompleteness theorem tells that the elementary theories of the set of natural numbers **N**, hence the set of integers **Z**, are not axiomatizable. Thus, these two theories are undecidable. J. Robinson proved that the theory of rational number field **Q** is undecidable (Robinson 1949), hence not axiomatizable either. On the other hand, the elementary theories of **R** and **C** are axiomatizable by the axioms of real closed fields and the axioms of algebraically closed fields, respectively. Thus, the elementary theories of **R** and **C** are both decidable. We may ask what subsets of those unaxiomatizable theories are axiomatizable or decidable. One of the most well-known problems of this kind is Hilbert's Tenth Problem. Given a polynomial $f(x_1, \ldots, x_n)$ with n variables and coefficients in a ring R, we often ask whether $f(x_1, \ldots, x_n) = 0$ solvable in R or not? This question is equivalent to ask whether

S.P. Tung (✉)
Department of Applied Mathematics, Chung Yuan Christian University,
Taoyuan 32023, Taiwan
e-mail: sptung@cycu.edu.tw

S.C.-M. Yang et al. (eds.), *Philosophical Logic: Current Trends in Asia*,
Logic in Asia: Studia Logica Library, DOI 10.1007/978-981-10-6355-8_4

$$\exists x_1 \cdots \exists x_n\, f(x_1, \ldots, x_n) = 0$$

true in R or not. Is there an algorithm which decides the solvability over $\mathbf{Z}$ of the equation $f(x_1, \ldots, x_n) = 0$ for an arbitrarily given polynomial $f(x_1, \ldots, x_n) \in \mathbf{Z}[x_1, \ldots, x_n]$? This is the well known Hilbert's Tenth Problem over $\mathbf{Z}$. Hilbert's Tenth Problem over $\mathbf{N}$ and over $\mathbf{Z}$ are undecidable if the number of variables n is sufficiently large (Davis 1973). We still do not know the answer of the Hilbert's Tenth Problem over $\mathbf{Q}$. Solving equations in one unknown over $\mathbf{Z}$ or $\mathbf{Q}$ are taught in high schools. It may surprise the non-specialists that the case of two unknowns already constitutes some of the most challenging problems in number theory. Right now, we only have algorithms for solving certain specially structured diophantine equations in more than one variable (Baker 1990; Wüstholz 2002).

The solvability of equations is asking the truth of the sentences involving only existential quantifiers. Most mathematical sentences contain both universal and existential quantifiers. For example, *every element has an additive inverse*; $\forall x \exists y\; x + y = 0$, which is true in $\mathbf{Z}$ but not in $\mathbf{N}$. Also, *every non-zero element has a multiplicative inverse*; $\forall x \exists y\; x \neq 0 \rightarrow xy = 1$, which is true in $\mathbf{Q}$ but not in $\mathbf{Z}$.

From logical point of view, the reason why we do not have an algorithm to decide the solvability of diophantine equations, say equations in two unknowns over $\mathbf{Z}$, is because the set of true sentences in $\mathbf{Z}$ of the form $\forall x \forall y\, f(x, y) \neq 0$ is not known to be recursively enumerable. Every axiomatizable set is a recursively enumerable set (Enderton 2001). This means we need some breakthroughs in number theory to show this set of sentences is axiomatizable. On the other hand, Matijasevich and Robinson proved that the set of the sentences of the form

$$\exists x \exists y \forall z \exists w\, f(x, y, z) - w = 0,$$

where $f(x, y, z) \in \mathbf{Z}[x, y, z]$, true in $\mathbf{N}$ is undecidable (Matijasevich and Robinson 1974). Many other cases of the equations with four quantifiers over $\mathbf{N}$ are undecidable (Jones 1981), hence not axiomatizable either. In this paper we try to find the borderline between the axiomatizable and the unaxiomatizable of the arithmetical sentences according to the numbers or the alternations of the quantifiers of the sentences. In some cases, adding one more quantifier similar sets of sentences will move from axiomatizable to unaxiomatizable.

Gödel's completeness theorem and Hilbert's irreducibility theorem are two main tools for us to prove the results of this paper. In order to study what is called "inverse Galois problem" Hilbert proved this celebrated theorem. In its simplest form it say that if $f(t, x) \in \mathbf{Q}[t, x]$ is an irreducible polynomial over $\mathbf{Q}$, then there is a rational number a such that $f(a, x)$ is still irreducible over $\mathbf{Q}$. This theorem has various generalizations and many applications in number theory and algebra (Lang 1997; Schinzel 1982). This theorem and its applications are also studied with logical methods (Fried and Jarden 2005). In this paper, we give another application of this important theorem.

2 Notations and Preliminaries

An arithmetical sentence is a sentence constructed by the language L, which contains the usual logical symbols $\wedge$ (and), $\vee$ (or), $\neg$ (not), $\Rightarrow$ (implies), $\Leftrightarrow$ (if and only if); the quantifiers $\forall$ (for every), $\exists$ (there exists); the mathematical symbols $+$ (addition), $\cdot$ (multiplication), $=$ (equals); constants 0, 1; and variables $x, y, z, \ldots$. We use $\bar{x}$ to denote a finite sequence of variables $(x_1, \ldots, x_m)$ for $m > 1$. If $m = 1$, then we simply write it as x. For simplicity, we write that $\underbrace{1 + 1 + \cdots + 1}_{n}$ as n, and similarly, $\underbrace{x + x + \cdots + x}_{n}$ as nx. The atomic formulas in the language L should be of the form $f(\bar{x}) = g(\bar{y})$, where $\bar{x} = (x_1, \ldots x_m)$, $\bar{y} = (y_1, \ldots, y_n)$, and $f(\bar{x}), g(\bar{y})$ can be written as polynomials with coefficients in $\mathbf{N}$. For convenience, we write the above atomic formula as $h(\bar{z}) = 0$, where $\bar{z}$ contains variables $x_1, \ldots x_m$ and $y_1, \ldots, y_n$, is a polynomial with coefficients in $\mathbf{Z}$. Let Q denote the quantifier $\forall$ or $\exists$. We call ψ a $Q_1^m Q_2^n Q_1^r$ sentence if and only if ψ is logically equivalent to a sentence of the form

$$Q_1 x_1 \cdots Q_1 x_m Q_2 y_1 \cdots Q_2 y_n Q_1 z_1 \cdots Q_1 z_r \ \phi(\bar{x}, \bar{y}, \bar{z}),$$

where $\phi(\bar{x}, \bar{y}, \bar{z})$ is a quantifier-free formula with variables $\bar{x} = (x_1, \ldots, x_m)$, $\bar{y} = (y_1, \ldots, y_n)$, $\bar{z} = (z_1, \ldots, z_r)$. We also write ψ as $Q_1^m \bar{x} Q_2^n \bar{y} Q_1^r \bar{z} \, \phi(\bar{x}, \bar{y}, \bar{z})$. If m, n or r equal 1, we then omit the superscript, respectively. Let S be a set of arithmetical sentences. An arithmetical sentence ϕ is provable by S, denoted by $S \vdash \phi$, if ϕ is deducible from S. S is axiomatizable by a set T of arithmetical sentences if T is a recursive set, and for every sentence ϕ in S, $T \vdash \phi$. Let M be an algebraic structure. M is a model of S, denoted by $M \models S$, if every sentence in S is true in M.

Define $x \le y$ if $\exists z\, x + z = y$. PA^- is the following set of sentences:

1. $\forall x\, x + 1 \neq 0$
2. $\forall x\, x \neq 0 \Rightarrow 1 \le x$
3. $\forall x\, x + 0 = x$
4. $\forall x\, x \cdot 1 = x$
5. $\forall x\, x \cdot 0 = 0$
6. $\forall x \forall y\, (x + y = y + x) \wedge (x \cdot y = y \cdot x)$
7. $\forall x \forall y\, x \le y \vee y \le x$
8. $\forall x \forall y \forall z\, ((x + y) + z = x + (y + z)) \wedge ((x \cdot y) \cdot z = x \cdot (y \cdot z))$
9. $\forall x \forall y \forall z\, x \cdot (y + z) = x \cdot y + x \cdot z$
10. $\forall x \forall y \forall z\, (x + y = x + z \Rightarrow y = z) \wedge ((x \cdot y = x \cdot z \wedge x \neq 0) \Rightarrow y = z)$.

If M is a model of PA^-, then $\mathbf{N}$ can be embedded as an initial segment of M. For convenience, we simply take that $\mathbf{N} \subseteq M$.

For every positive integer $n \ge 1$, ψ_n is the sentence

$$\forall x \exists y\, (x = ny + 0) \vee (x = ny + 1) \vee \cdots \vee (x = ny + i) \vee \cdots \vee (x = ny + (n-1)),$$

where $0 \leq i \leq n-1$. Sentence ψ_n means that we have the Division Algorithm for this positive integer n. Let $D = \{\psi_n \mid n \in \mathbf{N} \setminus \{0\}\}$. P_D is the set of sentences PA^- combining with D.

With the predicate $\leq$ we can define discretely ordered rings. A *discretely ordered ring* is a linearly ordered ring in which there is no element between 0 and 1.

In this paper, Z_D is the set of the axioms of discretely ordered commutative rings combining with D. If R is a discretely ordered ring, then $\mathbf{Z}$ can be embedded in R. As above, we take $\mathbf{Z} \subseteq R$ as a subring. For every element $x \in R$, let $|x| = x$ if $x \geq 0$, otherwise, $|x| = -x$. Then $|x| < |y|$ if $x \in \mathbf{Z}$ and $y \in R \setminus \mathbf{Z}$.

It is easy to obtain the following lemmas.

Lemma 2.1 *If $M \models Z_D$, then M is an integral domain with characteristic 0.*

Lemma 2.2 *Let $a, b \in \mathbf{Z}$, b is not divisible by a, and M is a discretely ordered ring. Then there is no $c \in M$ such that $ac = b$.*

Let $M \models P_D$. Then M can be extended to be a ring R by adding the additive inverse for each element of M, and we call R be the corresponding ring of M. From the sentences

$$\forall x \forall y\, x \leq y \vee y \leq x$$

and

$$\forall x\, x \neq 0 \Rightarrow 1 \leq x$$

in P_T, we obtain that R is a discretely ordered ring. Also, for each positive integer n, $M \models \psi_n$, hence $R \models \psi_n$. We obtain that $R \models Z_D$.

Let $M \models Z_D$, $x, y \in M$, and n a positive integer, we define that

$$x \equiv y \pmod{n}$$

if there is a z in M such that $x = y + nz$. Also, let $\bar{a} = (a_1, \ldots, a_r)$, $\bar{b} = (b_1, \ldots, b_r)$ $\in M^r$. We write that

$$\bar{a} \equiv \bar{b} \pmod{n}$$

if for every i, $1 \leq i \leq r$, $a_i \equiv b_i \pmod{n}$. For every positive integer n, ψ_n implies that for every $a \in M$ there is an i, $0 \leq i < n$, such that

$$a \equiv i \pmod{n}.$$

Similarly, for every $\bar{a} \in M^r$ there is a $\bar{b} = (b_1, \ldots, b_r) \in \mathbf{N}^r$ with $0 \leq b_i < n$, such that

$$\bar{a} \equiv \bar{b} \pmod{n}.$$

It is easy to see the following lemmas.

Lemma 2.3 *Let $M \models Z_D$, $f(\bar{x}) \in \mathbf{Z}[\bar{x}]$, n a positive integer, and $\bar{a}, \bar{b} \in M^r$. If $\bar{a} \equiv \bar{b} \pmod{n}$, then*

$$f(\bar{a}) \equiv f(\bar{b}) \pmod{n}.$$

The axioms for the fields with characteristic 0 can be written with the language L. A field F is a *purely transcendental* extension field of $\mathbf{Q}$ if every element in $F \setminus \mathbf{Q}$ is transcendental over $\mathbf{Q}$, or equivalently, for every $f(x) \in \mathbf{Z}[x]$ with $\deg(f(x)) > 1$ and irreducible over $\mathbf{Q}$, $F \models \forall x\ f(x) \neq 0$.

Let R be an integral domain, and K be its quotient field. Let $f(x) = \sum_{i=0}^{n} a_i x^i$ be a polynomial over $\mathbf{Z}$. Then there are $p \neq 0, q$ in R such that $f(q/p) = 0$ if and only if

$$f(q/p) = \sum_{i=0}^{n} a_i (q/p)^i = \sum_{i=0}^{n} a_i q^i p^{n-i} = 0.$$

Thus, $K \models \exists x\ \sum_{i=0}^{n} a_i x^i = 0$ if and only if

$$R \models \exists x \exists y \sum_{i=0}^{n} a_i x^i y^{n-i} = 0 \wedge y \neq 0.$$

Now, let M be a model of P_D and let R be a model of Z_D containing M as the set of non-negative elements of R. Let $f(x) \sum_{i=0}^{n} a_i x^i$ be a polynomial over $\mathbf{Z}$. Then, $R \models \exists x \exists y\ f(x, y) = 0$ if and only if

$$M \models \exists x \exists y\ f(x, y) f(-x, y) f(x, -y) f(-x, -y) = 0.$$

Now, let T be the set of sentences of the form $\forall x\ f(x) \neq 0$ for every $f(x) \in \mathbf{Z}[x]$ with $\deg(f(x)) > 1$ and irreducible over $\mathbf{Q}$. T_Z be the set of sentences of the form

$$\forall x \forall y \sum_{i=0}^{n} a_i x^i y^{n-i} \neq 0 \vee y = 0$$

for every $f(x) = \sum_{i=0}^{n} a_i x^i \in \mathbf{Z}[x]$ with $\deg(f(x)) > 1$ and irreducible over $\mathbf{Q}$. T_N be the set of sentences of the form

$$\forall x \forall y \left(\sum_{i=0}^{n} a_i x^i y^{n-i}\right)\left(\sum_{i=0}^{n} a_i (-x)^i y^{n-i}\right) \neq 0 \vee y = 0$$

for every $f(x) = \sum_{i=0}^{n} a_i x^i \in \mathbf{Z}[x]$ with $\deg(f(x)) > 1$ and irreducible over $\mathbf{Q}$.

In this paper, Q_T denotes the set of the axioms of fields with characteristic 0 combining with the set T. Similarly, P_T and Z_T denote the set of the axioms of P_D combining with the set T_N and the set of the axioms of Z_D with the set T_Z, respectively.

Q can be embedded in every field with characteristic 0. Again, for convenience, if $K \models Q_T$, we take **Q** as a subfield of K. Then, K is a purely transcendental extension field of **Q**. Let M be a model of P_T, R the corresponding ring of M. From the observation above, we know that $R \models Z_T$. Similarly, if $R \models Z_T$ and K is the quotient field of R, then $K \models Q_T$.

3 Hilbertian Properties

A field K is a *Hilbertian field* or Hilbert's irreducibility theorem holds for K if the following statement is true (Fried and Jarden 2005; Schinzel 1982).

For every polynomial $F_0 \in K[\bar{X}] \setminus \{0\}$, $F_j \in K[\bar{X}, \bar{Y}] \setminus K[\bar{X}]$ $(1 \leq j \leq h)$ irreducible over K there exists a $\bar{x} \in K^r$ such that $F_0(\bar{x}) \neq 0$ and all $F_j(\bar{x}, \bar{Y})$ are irreducible over K.

The set of all elements $\bar{x} \in K^r$ satisfying the conditions of the above statement, denoted by $H_K(F_1, \ldots, F_h; F_0)$, is called a *Hilbert set* of K^r for some positive integer r. Every Hilbert set is infinite. In particular, a Hilbertian field must be infinite (Fried and Jarden 2005). An integral domain R with its quotient field K is a *Hilbertian domain* if every Hilbert subset of K^r contains some elements in R^r (Fried and Jarden 2005). (In this paper, we concern only the integral domains with characteristic 0. Thus we won't be bothered by the separability of the extension.) As above, a Hilbertian domain R is infinite, and for every Hilbert set H of K^r, $H \cap R^r$ is infinite. Every purely transcendental proper extension of a field is Hilbertian (Schinzel 1982, Theorem 47). Thus, if $M \models Q_T$ then M is Hilbertian. In particular, **Q** is a Hilbertian field and **Z** is a Hilbertian domain (Schinzel 1982, Theorem 46). Following the proof of Theorem 47 in Schinzel (1982) we also have the result below.

Proposition 3.1 *If $M \models Z_T$, then M is a Hilbertian domain.*

Proof Let $M \models Z_T$ and K be its quotient field. From Lemma 2.2, K is a purely transcendental extension field of **Q**. Let $F_0 \in K[\bar{X}] \setminus 0$, $F_j \in K[\bar{X}, \bar{Y}] \setminus K[\bar{X}]$ $(1 \leq j \leq h)$ be irreducible over K. There exists a purely transcendental extension field K_1 of **Q** with generators $s_1, \ldots, s_m$ in M, and K a purely transcendental extension of K_1, such that $F_0 \in K_1[\bar{X}]$, $F_j \in K_1[\bar{X}, \bar{Y}]$ $(1 \leq j \leq h)$. Then, for every $\bar{x} \in K_1^r$, and every $1 \leq j \leq h$, $F_j(\bar{x}, \bar{Y})$ is irreducible over K_1 implies that $F_j(\bar{x}, \bar{Y})$ is irreducible over K. Let $R = \mathbf{Z}[s_1, \ldots, s_m]$, then K_1 is a quotient field of R. R is a Hilbertian domain. (Fried and Jarden 2005, Proposition 13.4.1). This means the Hilbert set

$$H_{K_1}(F_1, \ldots, F_h; F_0) \subseteq H_K(F_1, \ldots, F_h; F_0)$$

contains some elements in R^r for some integer r. Since $R \subseteq M$, every Hilbert subset of K^r must contain some elements in M^r. Therefore, M is a Hilbertian domain. □

Following fact is given as an exercise in Fried and Jarden (2005, Chap. 12, Exercise 1, p. 228).

Proposition 3.2 *Let K be a Hilbertian field and let $f \in K[\bar{X}, y]$, $\deg_y(f) > 0$, be a polynomial such that for every $\bar{x} \in K^r$ there exists an $y \in K$ such that $f(\bar{x}, y) = 0$. Then $f(\bar{X}, Y)$ has a factor of degree 1 in Y.*

Before we show a similar result for the models of Z_T, we give a needed lemma.

Lemma 3.3 *Let $M \models Z_T$, and $f(X) = a(X)/b(X)$, where $a(X), b(X) \in \mathbf{Z}[X]$, $a(X)$ and $b(X)$ are relatively prime, and $\deg(b(X)) > 0$. Then there are only finitely many elements $t \in M$ such that $f(t) \in M$.*

Proof Let $f(X) = a(X)/b(X)$ as above. There exist a $q(X)$ and an $r(X)$ in $\mathbf{Q}[X]$ such that

$$a(X) = q(X)b(X) + r(X),$$

where $r(X) \neq 0$ and $\deg(r(X)) < \deg(b(X))$. Since $b(X) \in \mathbf{Z}[X]$, the denominators of the coefficients of $q(X)$ are in $\mathbf{Z}$. Let D be the least common multiple of the denominators of the coefficients of $q(X)$.

This is an $s \in \mathbf{N}$ sufficiently large such that for every $t \in \mathbf{Z}$ with $|t| > s$, $0 < |r(t)/b(t)| < 1/D$. Then,

$$f(t) = a(t)/b(t) = q(t) + r(t)/b(t)$$

is not an element of M if $|t| > s$. Therefore, there are only finitely many elements $t \in M$ such that $f(t) \in M$. □

Proposition 3.4 *Let $M \models Z_T$, K be its quotient field, and $f \in \mathbf{Z}[X, Y]$ with $\deg_Y(f) > 0$. If there is a $c \in M$ such that $M \models \forall x \exists y\ f(x, y) = c$, then there is a $\gamma \in \mathbf{Z}$ and an $g(X) \in \mathbf{Q}[X]$ such that $f(X, g(X)) \equiv \gamma$.*

Proof Let $M \models Z_T$, then K be the quotient field of M. $K[Y]$ is a unique factorization domain (Hungerford 1974). We factor $f(X, Y) - c$ over $K[Y]$, and with Gauss Lemma we may write that

$$f(X, Y) - c = f_0(X, Y) \prod_{i=1}^{n} (q_i(X)Y - p_i(X)),$$

where $p_i(X), q_i(X) \in M[X]$ are relatively prime over K, and $f_0(X, Y) \in K[X, Y]$ has no linear factor in Y. From Lemma 2.2, K is a purely transcendental extension field of $\mathbf{Q}$. Let $H(X)$ be the leading term of $f(X, Y)$ with respect to X, then $H(X) \in \mathbf{Z}[X]$. Since $q_i(X)$ is a factor of $H(X)$ over K, we obtain that $q_i(X) \in \mathbf{Q}[X]$. Then, $q_i(X) \in (\mathbf{Q}[X] \cap M[X]) = \mathbf{Z}[X]$. Suppose that $\deg(q_i(X)) > 0$ for each $1 \leq i \leq n$. By Lemma 3.3, there are only finitely many elements $x \in M$ such that $p_i(x)/q_i(x) \in \mathbf{M}$ for some i. This means that there are only finitely many elements t in M such that equation $\prod_{i=1}^{n}(q_i(X)Y - p_i(X)) = 0$ is solvable in M. Since M is a Hilbertian domain, there are infinitely many elements $r \in M$ such that $f_0(r, Y)$ is irreducible (Fried and Jarden 2005). This means that $f_0(r, Y) = 0$ is not solvable in M for infinitely many r in M. Hence there are elements $x \in M$ such that $f(x, Y) - c = 0$ is not solvable in M. This contradicts that

$$M \models \forall x \exists y\, f(x, y) - c = 0.$$

Thus there exists an i such that $\deg(q_i(X)) = 0$. Let $g(X) = p_i(X)/q_i(X) \in \mathbf{K}[X]$, we have $f(X, g(X)) \equiv c$. □

With a stronger result on Hilbert sets we also obtain a similar result for the model of P_T.

Corollary 3.5 *Let $M \models P_T$, R be the ring by adding the additive inverse for each element of M, and K the quotient field of R. Let $f \in \mathbf{Z}[X, Y]$ with* $\deg_Y(f) > 0$. *If there is a $c \in R$ such that $M \models \forall x \exists y\, f(x, y) = c$, then there is a $g(X) \in K[X]$ such that $f(X, g(X)) \equiv c$.*

Proof From the proof of Proposition 3.1, we know that for every Hilbert set H in K there is a finitely generated purely transcendental extension field K_1 over $\mathbf{Q}$ such that H contains a Hilbert set H_1 of K_1. Every Hilbert set in K_1 contains an arithmetic progression in $\mathbf{Z}$ (Schinzel 2000, Theorem 46). Hence, every Hilbert set of K contains an arithmetic progression in $\mathbf{Z}$. In particular, every Hilbert set of K includes infinitely many elements of $\mathbf{N}$. This then implies that every Hilbert set of K includes infinitely many elements of M. With the same argument of Proposition 3.4 we may obtain the conclusion. □

4 Arithmetical Sentences

We first extend the results proved in Tung (1985) to sentences with one more quantifier alternations. The difficult parts of the proofs have been done there. Although the proofs for the axiomatizability of the $\exists^n\forall\exists$ sentences over $\mathbf{N}$, $\mathbf{Z}$ or $\mathbf{Q}$ are similar, the computational complexities of the related decision problems may be different if P $\neq$ NP. For example, the decision problems for $\exists\forall$ sentences true in $\mathbf{N}$ and $\mathbf{Z}$ are NP-complete, respectively. However, the decision problem for $\exists\forall$ sentences true in $\mathbf{Q}$ is in P (Tung 1995).

Theorem 4.1 *Let $n \in \mathbf{N}$ and ϕ be an $\exists^n\forall\exists$ sentence.*
(a) $\mathbf{N} \models \phi$ *if and only if* $P_T \vdash \phi$.
(b) $\mathbf{Z} \models \phi$ *if and only if* $Z_T \vdash \phi$.
(c) $\mathbf{Q} \models \phi$ *if and only if* $Q_T \vdash \phi$.

Proof (a) Clearly, $\mathbf{N} \models P_T$. If $P_T \vdash \phi$, then $\mathbf{N} \models \phi$. Similar arguments are valid for (b) and (c). Hence, we will only prove the "if" parts of this theorem (also the theorems stated in the rest of the paper.)

Now, let ϕ be an $\exists^n\forall\exists$ sentence logically equivalent to $\exists\bar{x}\forall y\exists z\, \psi(\bar{x}, y, z)$, where $\psi(\bar{x}, y, z)$ is a quantifier-free formula, true in $\mathbf{N}$. Then there exists $\bar{a} \in \mathbf{N}^n$ such that $\mathbf{N} \models \forall y\exists z\, \psi(\bar{a}, y, z)$. Let M be an arbitrary model of P_T. Then M preserves all the true $\forall\exists$ sentences in $\mathbf{N}$ (Tung 1986, Theorem 1.1). Hence, $M \models \forall y\exists z\, \psi(\bar{a}, y, z)$. Since $\bar{a} \in \mathbf{N}^n \subseteq M^n$,

$$M \models \exists\bar{x}\forall y\exists z\, \psi(\bar{x}, y, z).$$

By Completeness Theorem, $P_T \vdash \phi$.

With the same proof of (a), and the Theorem 2.1 of Tung (1986), we may prove (b).

(c) We first note following fact. Let $f(\bar{x}) \in \mathbf{Q}[\bar{x}]$, and D be the least common multiple of the denominators of the coefficients of $f(\bar{x})$. Then $Df(\bar{x}) \in \mathbf{Z}[\bar{x}]$. Clearly, if M is a field with characteristic 0, then for any $\bar{a} \in M^r$, $f(\bar{a}) = 0$ if and only if $Df(\bar{a}) = 0$. In this paper, we study only the fields with characteristic 0. We may extend the language L to L' by containing all the rational numbers as our constants. We have that every arithmetical sentence written in L' can be transformed to a sentence in L with the same truth value.

Now, let ϕ be an $\exists^n\forall\exists$ sentence logically equivalent to $\exists\bar{x}\forall y\exists z\, \psi(\bar{x}, y, z)$, where $\psi(\bar{x}, y, z)$ is a quantifier-free formula, true in $\mathbf{Q}$. Then there exists $\bar{a} \in \mathbf{Q}^n$ such that $\mathbf{Q} \models \forall y\exists z\, \psi(\bar{a}, y, z)$. We may transform the sentence $\forall y\exists z\, \psi(\bar{a}, y, z)$ to an $\forall\exists$ sentence $\forall y\exists z\, \phi(y, z)$ with the language L such that $\mathbf{Q} \models \forall y\exists z\, \psi(\bar{a}, y, z)$ if and only if $\mathbf{Q} \models \forall y\exists z\, \phi(y, z)$. Now, with the same argument as the proof of (a) and Theorem 3.1 of Tung (1986) we may prove (c). □

Remarks

1. Axiomatizable sets are recursively enumerable (Enderton 2001). Form Theorem 4.1, the set of true $\exists^n\forall\exists$ sentences in $\mathbf{N}$ is recursively enumerable. However, from the negative answer to the Hilbert's Tenth Problem over $\mathbf{N}$, this set is not decidable. This is also the case for $\mathbf{Z}$.

2. We still do not have an algorithm for solving diophantine equations with two unknowns in general. This is caused by the fact that we do not know whether the sets of the sentences of the form

$$\forall x\forall y\, f(x, y) \neq 0$$

true in $\mathbf{N}$, $\mathbf{Z}$, and $\mathbf{Q}$, respectively, are recursive enumerable or not. Bounds of the solutions for some specially structured diophantine equations in two variables are known. These bounds usually are very high (Baker 1990; Wüstholz 2002). For these equations we then may use exhaust-search to decide the solvability in $\mathbf{Z}$. Breakthroughs in number theory may be needed to show the axiomatizability of the sets of the sentences of the form

$$\forall x\forall y\, f(x, y) \neq 0$$

true in $\mathbf{N}$, $\mathbf{Z}$, and $\mathbf{Q}$, respectively.

However, following special sets of arithmetical sentences containing two universal quantifiers are axiomatizable.

Kneser Theorem Let K be a field and n an integer ≥ 2. Let $a \in K$. The binomial $x^n - a$ is reducible over K if and only if either $a + b^p$ for some prime divisor p of n and $b \in K$ or $4|n$ and $a = -4b^4,\ b \in K$.

Theorem 4.2 *Let ϕ be the sentence $\forall z \exists x \forall y\ f(x, y) - az \neq 0$, where $f(x, y) \in \mathbf{Z}[x, y]$ and $a \in \mathbf{Z}$.*
(a) $\mathbf{N} \models \phi$ if and only if $P_T \vdash \phi$.
(b) $\mathbf{Z} \models \phi$ if and only if $Z_T \vdash \phi$.
(c) $\mathbf{Q} \models \phi$ if and only if $Q_T \vdash \phi$.

Proof (a) Let M be an arbitrary model of P_T. We want to show that if $\mathbf{N} \models \phi$, then $M \models \phi$. We first prove some degenerated cases. Suppose that $a = 0$. Then ϕ is reduced to $\exists x \forall y\ f(x, y) \neq 0$. This is an $\exists\forall$ sentence. If $\mathbf{N} \models \phi$, then $M \models \phi$ (Tung 1985). Thus, we assume that $a \neq 0$. If $a < 0$, then ϕ is equivalent to

$$\forall z \exists x \forall y \ - f(x, y) + az \neq 0.$$

Thus, we also assume that $a > 0$. Suppose that $\deg_x(f(x, y)) = 0$. Then ϕ is reduced to

$$\forall z \forall y\ f(y) - az \neq 0.$$

Assume that this sentence is true in $\mathbf{N}$ but false in M. This is equivalent to that the congruence equation

$$f(y) \equiv 0 \pmod{a}$$

is not solvable in $\mathbf{N}$, but solvable in M. Hence, there is an $r \in M, r > 0$, such that $f(r) \equiv 0 \pmod{a}$. Since $M \models \psi_a$, there is an i, $0 \leq i < a$, $r \equiv i \pmod{a}$. By Lemma 2.1,

$$f(i) \equiv 0 \pmod{a}.$$

This contradiction shows that if

$$\mathbf{N} \models \forall z \forall y\ f(y) - az \neq 0,$$

then

$$M \models \forall z \forall y\ f(y) - az \neq 0.$$

Now, we suppose that $\deg_y(f(x, y)) = 0$. Then ϕ is reduced to

$$\forall z \exists x\ f(x) - az \neq 0.$$

Assume that this sentence is true in $\mathbf{N}$ but false in M. This means that

$$M \models \exists z \forall x\ f(x) - az = 0.$$

This is possible only if $f(x) \equiv ar$ for an $r \in \mathbf{N}$. If so, then

$$\mathbf{N} \models \exists z \forall x\, f(x) - az = 0$$

too. This is a contradiction. Thus, we may assume that $\deg_x(f(x, y)) > 0$ and $\deg_x(f(x, y)) > 0$.

Assume that ϕ is false in M. This means that

$$M \models \exists z \forall x \exists y\, f(x, y) - az = 0.$$

Let $c \in M$ such that $M \models \forall x \exists y\, f(x, y) - ac = 0$. By adding the additive inverses of the elements of M, we extend M to a ring R. $R \models Z_T$ and R is a Hilbertian domain by Proposition 3.1. Also, let K be the quotient field of R. From Corollary 3.5, there is a $g(x) \in \mathbf{K}[x]$ such that $y - g(x)$ is a factor of $f(x, y) - ac$ over $\mathbf{K}$. This implies $f(x, g(x)) \equiv ac$.

We claim that if $f(x, g(x)) \equiv ac$ for a $g(x) \in K[x]$, then $g(x) \in \mathbf{Q}[x]$. We first prove the case that $\deg(g(x)) = 0$. Let $g(x) \equiv d \in K \setminus \mathbf{Q}$. Then

$$f(x, y) - ac = \sum_{i=0}^{s} F_i(y)x^i = (y - d)(\sum_{i=0}^{s} f_i(y)x^i),$$

where $F_i(y) \in \mathbf{Z}[y]$ for $0 < i \leq s$, $F_0(y) \in R[y]$ and $f_i(y) \in R[y]$ for $0 \leq i \leq s$. Since $\deg_x(f(x, y)) > 0$, $s \neq 0$. Then the leading term of $f(x, y) - ac$ with respect to x is $F_s(y) = (y - d)f_s(y) \in \mathbf{Q}[y]$. Clearly, $F_s(d) = 0$, but this contradicts that $d \in K \setminus \mathbf{Q}$ is transcendental over $\mathbf{Q}$. Therefore, if $f(x, y) - ac$ has a factor of the form $y - d$ with $d \in K$, then $d \in \mathbf{Q}$.

Now suppose that $f(x, g(x)) \equiv ac$ for a $g(x) \in K[x]$ with $\deg(g(x)) > 0$. Assume that $g(x) \in K[x] \setminus \mathbf{Q}[x]$. Let $g(x) = \sum_{j=0}^{t} a_j x^j$, where $t > 0$ and $a_j \in K$. Let s, $0 \leq s \leq t$, be the index such that if $j > s$ then $a_j \in \mathbf{Q}$ but $a_s \in K \setminus \mathbf{Q}$. Let

$$F(x, y) = f(x, y + \sum_{j=s+1}^{t} a_j x^j).$$

Since $f(x, y) \in \mathbf{Q}[x, y]$ and $\sum_{j=s+1}^{t} a_j x^j \in \mathbf{Q}[x]$, $F(x, y) \in \mathbf{Q}[x, y]$. Note that $\deg_y(F(x, y)) = \deg_y(f(x, y)) > 0$. We also have that

$$F(x, \sum_{j=0}^{s} a_j x^j) = f(x, \sum_{j=0}^{s} a_j x^j + \sum_{j=s+1}^{t} a_j x^j) = f(x, g(x)) \equiv ac.$$

If $s = 0$, this means that all the coefficients of $g(x)$, except the constant term a_0, are in $\mathbf{Q}$. We also have that $F(x, a_0) = f(x, g(x)) \equiv ac$. With the same argument as the case of $y - d$ with $d \in K$ is a factor of $f(x, y) - ac$, we will obtain that $a_0 \in \mathbf{Q}$. This contradicts that $g(x)$ is not in $\mathbf{Q}[x]$. Hence, we obtain that $s > 0$.

Now, write that $F(x, y) = \sum_{j=0}^{m} f_j(x)y^j$, where $f_j(x) \in \mathbf{Q}[x]$. Then,

$$F(x, \sum_{i=0}^{s} a_i x^i) = \sum_{j=0}^{m} f_j(x)(\sum_{i=0}^{s} a_i x^i)^j.$$

For each j, let D_j be the degree of $f_j(x)(\sum_{i=0}^{s} a_i x^i)^j$ and C_j be the leading coefficient of $f_j(x)(\sum_{i=0}^{s} a_i x^i)^j$. Since $\deg_y(F(x, y)) > 0$ and $\deg(\sum_{i=0}^{s} a_i x^i) > 0$, $D_j > 0$ for some j. For each $j, 0 \le j \le m$, there is an $h_j(z) \in \mathbf{Q}[z], \deg(h_j(z)) = j$, such that $C_j = h_j(a_s)$. Let $D > 0$ be the maximum of $D_j, 0 \le j \le m$, and S be the set of indices such that $j \in S$ if and only if $D_j = D$. Let $G(z) = \sum_{j \in S} h_j(z) \in \mathbf{Q}[z]$. Since the degrees of $h_j(z)$ are all different, $G(z) \not\equiv 0$. $G(a_s)$ is the leading coefficient of the polynomial $F(x, \sum_{i=0}^{s} a_i x^i)$. Since $F(x, \sum_{i=0}^{s} a_i x^i) - ac = 0$, $G(a_s) = 0$. This contradicts that a_s is transcendental over $\mathbf{Q}$. Thus, $g(x)$ must be in $\mathbf{Q}[x]$. This finishes the proof of our claim.

Then $f(x, g(x)) \equiv ac \in \mathbf{Q} \cap M$. We obtain that $ac \in \mathbf{N}$, hence, $c \in \mathbf{N}$. Thus,

$$M \models \forall x \exists y\, f(x, y) - ac = 0,$$

and $\forall x \exists y\, f(x, y) - ac = 0$ is an $\forall\exists$ sentence in L. An $\forall\exists$ sentence is true in $\mathbf{N}$ if and only if it is true in M, since $M \models P_T$ (Tung 1985). Thus, $\mathbf{N} \models \forall x \exists y\, f(x, y) - ac = 0$, where $c \in \mathbf{N}$. We obtain that

$$\mathbf{N} \models \exists z \forall x \exists y\, f(x, y) - az = 0.$$

This contradicts that $\mathbf{N} \models \phi$, hence, $M \models \phi$. We then prove that if $\mathbf{N} \models \phi$, then $M \models \phi$ for every model M of P_T. By Completeness Theorem, $P_T \vdash \phi$.

(b) Its proof is similar with the proof of (a), hence we omit the proof.
(c) Let M be an arbitrary model of Q_T. We show that if $\mathbf{Q} \models \phi$, then $M \models \phi$. We first prove some degenerated cases. Suppose that $\deg_y(f(x, y)) = 0$. Then ϕ is reduced to $\forall z \exists x\, f(x) - az \neq 0$. Let M be a model of Q_T. Assume that

$$M \models \exists z \forall x\, f(x) - az = 0.$$

This is possible only if $f(x) \equiv c$. If $a = 0$, then $c = 0$. Thus, $\exists z \forall x\, f(x) - az = 0$ is further reduced to $\exists z \forall x\, 0 = 0$ which is true in $\mathbf{Q}$. If $a \neq 0$, then ϕ is reduced to $\exists z\, c - az = 0$. Clearly, $\exists z\, c - az = 0$ is also true in $\mathbf{Q}$. Hence if $\mathbf{Q} \models \forall z \exists x\, f(x) - az \neq 0$ then $M \models \forall z \exists x\, f(x) - az \neq 0$.

Now, assume that $\deg_y(f(x, y)) > 0$, and $\deg_x(f(x, y)) = 0$. Then ϕ is reduced to

$$\forall z \forall y\, f(y) - az \neq 0.$$

Hence, $\mathbf{Q} \models \forall z \forall y\, f(y) - az \neq 0$. Suppose that $a \neq 0$. Take a $y \in \mathbf{Q}$, and let $z = f(y)/a$, then $f(y) - az = 0$. We have that $\mathbf{Q} \models \exists z \exists y\, f(y) - az = 0$. This contradiction shows that $a = 0$. Then, ϕ is further reduced to $\forall y\, f(y) \neq 0$, where $f(y) \in \mathbf{Q}[y]$. This means $f(y) = 0$ is not solvable in $\mathbf{Q}$. For any model M of Q_T, $f(y) = 0$ is solvable in $\mathbf{Q}$ if and only if $f(y) = 0$ is solvable in M. Hence if $\mathbf{Q} \models \forall z \forall y\, f(y) - az \neq 0$, then $M \models \forall z \forall y\, f(y) - az \neq 0$.

We next prove the case that $\deg_y(f(x, y)) = 1$. Then,

$$f(x, y) = F_1(x)y - F_0(x),$$

where $F_1(x), F_0(x) \in \mathbf{Q}[x]$. Let $c \in M$ such that

$$M \models \forall x \exists y\, F_1(x)y - F_0(x) - ac = 0.$$

Then for every $\alpha \in M$, $F(\alpha) = 0$ implies that $F_0(\alpha) - ac = 0$. First assume that there is an $\alpha \in \mathbf{Q}$ and $F_1(\alpha) = 0$. Then from $F_0(\alpha) - ac = 0$ we obtain that $c \in \mathbf{Q}$. As explained in the proof of (c) of Theorem 4.1,

$$\forall x \exists y\, F_1(x)y - F_0(x) - ac = 0$$

can be transformed to an $\forall\exists$ sentence in L with the same truth value. Since $M \models Q_T$, every $\forall\exists$ sentence is true in M if and only if it is true in $\mathbf{Q}$ (Tung 1986). Thus,

$$\mathbf{Q} \models \forall x \exists y\, F_1(x)y - F_0(x) - ac = 0,$$

and

$$\mathbf{Q} \models \exists z \forall x \exists y\, F_1(x)y - F_0(x) - az = 0.$$

Now, assume that for every $\alpha \in \mathbf{Q}$, $F_1(\alpha) \neq 0$. Let c be an arbitrary element in $\mathbf{Q}$. For every $\alpha \in \mathbf{Q}$, let $\beta = (F_0(\alpha) + ac)/F_1(\alpha)$, $F_1(\alpha)(\beta) - F_0(\alpha) - ac = 0$. Again,

$$\mathbf{Q} \models \exists z \forall x \exists y\, F_1(x)y - F_0(x) - az = 0.$$

Both cases contradict $\mathbf{Q} \models \forall z \exists x \forall y\, f(x, y) - az \neq 0$. Hence,

$$M \models \forall z \exists x \forall y\, f(x, y) - az \neq 0.$$

Now, we assume that $\deg_x(f(x, y)) > 0$ and $\deg_y(f(x, y)) > 1$. Suppose that $\mathbf{Q} \models \phi$ but ϕ is false in M. This means that

$$M \models \exists z \forall x \exists y\, f(x, y) - az = 0.$$

Let $c \in M$ such that

$$M \models \forall x \exists y\, f(x, y) - ac = 0.$$

If $c \in \mathbf{Q}$, then

$$\forall x \exists y\, f(x, y) - ac = 0$$

can be transformed to an ∀∃ sentence with the same truth value. Thus,

$$\mathbf{Q} \models \forall x \exists y\, f(x, y) - ac = 0$$

Tung (1986), and $\mathbf{Q} \models \exists z \forall x \exists y\, f(x, y) - az = 0$. This is a contradiction. Thus, $c \in M \setminus \mathbf{Q}$. From Proposition 3.2 and

$$M \models \forall x \exists y\, f(x, y) - ac = 0,$$

there are $P(x)$, $Q(x) \in M[x]$ relatively prime, and $G(x, y) \in M[x, y]$ such that

$$f(x, y) - ac = (Q(x)y - P(x))G(x, y).$$

Let S be a transcendence base of M over $\mathbf{Q}$ with $T \in S$ such that $ac \in \mathbf{Q}(T)$. Note that M is also a purely transcendental extension over $\mathbf{Q}(T)$. Then, $P(x)$, $Q(x)$ and $G(x, y)$ are polynomials with coefficients in $\mathbf{Q}(T)$. Let

$$ac = r_1(T)/r_2(T)$$

where $r_1(T), r_2(T) \in \mathbf{Q}[T]$. We then have that

$$f(x, y) - r_1(T)/r_2(T) = (Q(x)y - P(x))G(x, y),$$

where $Q(x)$, $P(x)$ and $G(x, y)$ are polynomials with coefficients in $\mathbf{Q}(T)$. Moreover, $Q(x)$ must divide the leading term of $f(x, y)$ with respect to y, which is a polynomial of x over $\mathbf{Q}$, $Q(x) \in \mathbf{Q}[x]$. We may view T as an independent variable. Thus, the coefficients of $P(x)$ and $G(x, y)$ are the rational functions with variable T. Write that $P(x) \equiv p(x, T)$, where $p(x, T)$ is a polynomial with respect to x, but a rational function with respect to T over $\mathbf{Q}$. We similarly define $g(x, y, T)$ for $G(x, y)$. Let E be the finite set of rational numbers which contains the zeros of $r_2(T)$ and the zeros of the denominators of $p(x, T)$ and $g(x, y, T)$. Hence, for every $\gamma \in \mathbf{Q} \setminus E$,

$$f(x, y) - r_1(\gamma)/r_2(\gamma) = (Q(x)y - p(x, \gamma))g(x, y, \gamma).$$

For every $\alpha \in \mathbf{Q}$ and $\gamma \in \mathbf{Q} \setminus E$, $Q(\alpha) \neq 0$ implies

$$f(\alpha, p(\alpha, \gamma)/Q(\alpha)) - r_1(\gamma)/r_2(\gamma) = 0.$$

Suppose that $Q(\alpha) = 0$. Since

$$M \models \forall x \exists y\, (Q(x)y - p(x, T))g(x, y, T) = 0,$$

$$M \models \exists y\, [-p(\alpha, T)g(\alpha, y, T)] = 0.$$

This means that $-p(\alpha, T) = 0$ or there is a $\gamma \in M$ such that $g(\alpha, \gamma, T) = 0$. If $-p(\alpha, T) = 0$, then $-p(\alpha, T)$ is a rational function with variable T, which is identically equal to 0. Hence $-p(\alpha, \delta) = 0$ for every $\delta \in \mathbf{Q} \setminus E$.

Now, if $g(\alpha, \gamma, T) = 0$ for a $\gamma \in M$, then γ must be algebraic over $\mathbf{Q}(T)$. Since $T \in S$, which is a base of M over $\mathbf{Q}$, every element of $M \setminus \mathbf{Q}(T)$ is transcendental over $\mathbf{Q}(T)$. We obtain that $\gamma \in \mathbf{Q}(T)$, and $\gamma = h_1(T)/h_2(T)$, where $h_1(z), h_2(z)$ are polynomials in $\mathbf{Q}[z]$. Then, $g(\alpha, h_1(T)/h_2(T), T) = 0$. Equation $h_2(z) = 0$ has at most finitely many solutions in $\mathbf{Q}$. Choose a $\delta \in \mathbf{Q} \setminus E$ such that $h_2(\delta) \neq 0$. Let $\beta = h_1(\delta)/h_2(\delta) \in \mathbf{Q}$, then

$$g(\alpha, \beta, \delta) = g(\alpha, h_1(\delta)/h_2(\delta), \delta) = 0.$$

In summary, let $\delta \in \mathbf{Q} \setminus E$ and $h_2(\delta) \neq 0$. Then, for every $\alpha \in \mathbf{Q}$, if $Q(\alpha) \neq 0$, take $\beta = p(\alpha, \delta)/Q(\alpha)$ we have

$$f(\alpha, \beta) - r_1(\delta)/r_2(\delta) = 0.$$

If $Q(\alpha) = 0$, then take $\beta = h_1(\delta)/h_2(\delta)$ we have

$$f(\alpha, \beta) - r_1(\delta)/r_2(\delta) = -p(\alpha, \delta)g(\alpha, \beta, \delta) = 0$$

too. This implies that

$$\mathbf{Q} \models \exists z \forall x \exists y\, f(x, y) - az = 0,$$

where $z = r_1(\delta)/(a \cdot r_2(\delta))$. This contradicts $\mathbf{Q} \models \phi$. Hence, if $\mathbf{Q} \models \phi$, then $M \models \phi$ for every model M of Q_T. Again, by Completeness Theorem, $Q_T \vdash \phi$. □

Corollary 4.3 *The sets of sentences of the form* $\exists z \forall x \exists y\, f(x, y) - az = 0$, *where* $f(x, y) \in \mathbf{Z}[x, y]$ *and* $a \in \mathbf{Z}$, *true in* $\mathbf{N}, \mathbf{Z}$ *or* $\mathbf{Q}$ *are decidable, respectively.*

Proof The set of sentences of the form $\exists z \forall x \exists y\, f(x, y) - az = 0$, where $f(x, y) \in \mathbf{Z}[x, y]$ and $a \in \mathbf{Z}$, true in $\mathbf{N}$ is recursively enumerable since it is provable by P_T by Theorem 4.1. Similarly, the set of sentences of the form

$$\forall z \exists x \forall y\, f(x, y) - az \neq 0,$$

the negation of

$$\exists z \forall x \exists y\, f(x, y) - az = 0,$$

true in $\mathbf{N}$ is recursively enumerable too by Theorem 4.2. One of these two sentences must be true in $\mathbf{N}$. Hence, the set of sentences of the form

$$\exists z \forall x \exists y\, f(x, y) - az = 0,$$

where $f(x, y) \in \mathbf{Z}[x, y]$ and $a \in \mathbf{Z}$, true in $\mathbf{N}$ is decidable. With the same proofs, we have the similar decidable results over $\mathbf{Z}$ or $\mathbf{Q}$ too. □

Remarks

1. Instead of giving the algorithms we use logical method to prove the decidability of those decision problems given in Corollary 4.3. There are some benefits to use this method. For example, the proof is much more elegant, and we know the axiomatizability of the sentences. However, we do not know the computational complexities of those decision problems. Solving quadratic diophantine equations with two unknowns is NP-complete. More specifically, the decision problem of determining whether the sentences of the form

$$\exists x \exists y\, ax^2 + by - c = 0,$$

where $a, b, c \in \mathbf{N}$, true in $\mathbf{N}$ or not is NP-complete (Manders and Adleman 1978). Also, the decision problem of determining the sentences of the form

$$\forall x \exists\, f(x, y) = 0,$$

where $f(x, y) \in \mathbf{Z}[x, y]$, true in $\mathbf{N}$ or not is co-NP-complete (Tung 1987). These sentences are only the special cases of the sentences of the form

$$\exists z \forall x \exists y\, f(x, y) - az = 0,$$

where $f(x, y) \in \mathbf{Z}[x, y]$ and $a \in \mathbf{Z}$. This means that the decision problem given in the Corollary 4.3 for the case over $\mathbf{N}$ is both NP-hard and co-NP-hard. We have an algorithm in PSPACE to decide this decision problem (Tung 2017). If NP $\neq$ co-NP, then NPC $\cap$ co-NP= $\emptyset$, and NP $\cap$ co-NPC= $\emptyset$ (Garey and Johnson 1979). This means this decision is in PSAPCE \ (NP $\cup$ co-NP) if NP $\neq$ co-NP. In certain sense, we have known the sharp bounds for the computational complexity of this decision problem. Theorems 4.1 and 4.2 tell us that given an arbitrary polynomial $f(x, y) \in \mathbf{Z}[x, y]$ and $a \in \mathbf{Z}$,

$$\exists z \forall x \exists y\, f(x, y) - az = 0$$

can be proved or disproved by the axioms of P_T. However, there is no theorem-prover can prove or disprove these sentences by P_T, or any other axiom systems, in polynomial time if NP $\neq$ co-NP.

2. Matijasevich and Robinson (1974) proved that the set of sentences true in $\mathbf{N}$ of the form

$$\exists x \exists y \forall z \exists w\, f(x, y, z) - w = 0$$

is undecidable. Over $\mathbf{N}$, quantifier prefixes over diophantine equations classified by their decidability were studied by Jones (1982). We may see that many cases over $\mathbf{N}$ with only four quantifiers are undecidable. Thus, with just one more quantifier we have attained undecidability.

Finally, we give a set of sentences true in **Z**, which is axiomatizable by Z_T, but the similar set of sentences true in **N** is not axiomatizable. We first demonstrate the case over **N**.

Proposition 4.4 *The set of sentences of the form*

$$\forall y_1 \forall y_2 \cdots \forall y_9 \exists z\, f(y_1, y_2, \ldots, y_9, z) = 0,$$

where $f(y_1, y_2, \ldots, y_9, z) \in \mathbf{Z}[y_1, y_2, \ldots, y_9, z]$, *true in* **N** *is not axiomatizable.*

Proof It is known that there is no algorithm which decides the solvability of diophantine equations over **N** with 9 unknowns (Jones 1981). By checking all the possible combinations of natural numbers, the set of sentences of the form

$$\exists y_1 \exists y_2 \cdots \exists y_9\, f(y_1, y_2, \ldots, y_9) = 0,$$

where $f(y_1, y_2, \ldots, y_9) \in \mathbf{Z}[y_1, y_2, \ldots, y_9]$, true in **N** is a recursively enumerable set. Hence the set of sentences of the form

$$\forall y_1 \forall y_2 \cdots \forall y_9\, f(y_1, y_2, \ldots, y_9) \neq 0,$$

where $f(y_1, y_2, \ldots, y_9) \in \mathbf{Z}[y_1, y_2, \ldots, y_9]$ true in **N**, is not recursively enumerable. For any integer y, $y \neq 0$ if and only if $\mathbf{N} \models \exists z\, y^2 - z - 1 = 0$. Thus,

$$\mathbf{N} \models \forall y_1 \forall y_2 \cdots \forall y_9\, f(y_1, y_2, \ldots, y_9) \neq 0$$

if and only if

$$\mathbf{N} \models \forall y_1 \forall y_2 \cdots \forall y_9 \exists z\, (f(y_1, y_2, \ldots, y_9))^2 - z - 1 = 0.$$

Hence, both sets of sentences are not recursively enumerable. Every axiomatizable set is a recursively enumerable set (Enderton 2001). Thus, the set of sentences of the form

$$\forall y_1 \forall y_2 \cdots \forall y_9 \exists z\, f(y_1, y_2, \ldots, y_9, z) = 0,$$

where $f(y_1, y_2, \ldots, y_9, z) \in \mathbf{Z}[y_1, y_2, \ldots, y_9, z]$, true in **N** is not axiomatizable. □

For any integer x, $x \neq 0$ if and only if

$$\mathbf{Z} \models \exists y \exists z\, x - (2y + 1)(3z + 1) = 0 \text{ (Tung 1985)}.$$

With the same argument like Proposition 4.4 we also have following proposition.

Proposition 4.5 *The set of sentences of the form*

$$\forall y_1 \forall y_2 \cdots \forall y_m \exists z_1 \exists z_2\, f(y_1, y_2, \ldots, y_m, z_1, z_2) = 0,$$

where $f(y_1, y_2, \ldots, y_m, z_1, z_2) \in \mathbf{Z}[y_1, y_2, \ldots, y_m, z_1, z_2]$, *true in* $\mathbf{Z}$ *is not axiomatizable for sufficiently large* m.

However, we have following axiomatizable result over $\mathbf{Z}$.

Theorem 4.6 *Let* $f(\bar{x}, \bar{y}, z) \in \mathbf{Z}[\bar{x}, \bar{y}, z]$. $\mathbf{Z} \models \exists\bar{x}\forall\bar{y}\exists z\, f(\bar{x}, \bar{y}, z) = 0$ *if and only if* $Z_T \vdash \exists\bar{x}\forall\bar{y}\exists z\, f(\bar{x}, \bar{y}, z) = 0$.

Proof Let $f(\bar{x}, \bar{y}, z) \in \mathbf{Z}[\bar{x}, \bar{y}, z]$, and

$$\mathbf{Z} \models \exists\bar{x}\forall\bar{y}\exists z\, f(\bar{x}, \bar{y}, z) = 0.$$

There is an $\bar{a} \in \mathbf{Z}^n$ such that

$$\mathbf{Z} \models \forall\bar{y}\exists z\, f(\bar{a}, \bar{y}, z) = 0.$$

Then there is a finite set

$$S = \{g_1(\bar{y}), g_2(\bar{y}), \ldots, g_s(\bar{y})\}$$

of polynomials over $\mathbf{Q}$ such that $f(\bar{a}, \bar{y}, g(\bar{y})) \equiv 0$ for each $g(\bar{y}) \in S$, and for every $\bar{b} \in \mathbf{Z}^m$ there is $g(\bar{y}) \in S$ such that $g(\bar{b}) \in \mathbf{Z}$ (Tung 1985). Notice that for every $\bar{b} \in \mathbf{Z}^m$ and every $g(\bar{y}) \in S$, $f(\bar{a}, \bar{b}, g(\bar{b})) = 0$. Let D_i be the least common multiple of the denominators of the coefficients of $g_i(\bar{y}) \in S$, and $D = \prod_{i=1}^{s} D_i$. Then $Dg_i(\bar{y}) \in \mathbf{Z}[\bar{y}]$ for every i, $1 \leq i \leq s$. Let M be an arbitrary model of Z_T. For every $\bar{b} \in M^m$, $g_i(\bar{b}) \in M$ if and only if

$$Dg_i(\bar{b}) \equiv 0 \pmod{D}$$

in M. Let $\bar{\alpha}$ be an arbitrary element in M^m. From ψ_T there is a $\bar{b} \in \mathbf{Z}^m$ such that

$$\bar{\alpha} \equiv \bar{b} \pmod{D}.$$

Also, there is a $g_i(\bar{y}) \in S$ such that $g_i(\bar{b})$ is an integer, hence

$$Dg_i(\bar{b}) \equiv 0 \pmod{D}.$$

By Lemma 2.1,

$$Dg_i(\bar{\alpha}) \equiv Dg_i(\bar{b}) \equiv 0 \pmod{D}.$$

This means that $g_i(\bar{\alpha}) \in M$. Thus, for every $\bar{\alpha}$ in M^m there is a $g_i(\bar{y}) \in S$ such that $g_i(\bar{\alpha}) \in M$. We have that $f(\bar{a}, \bar{\alpha}, g_i(\bar{\alpha})) = 0$ for each $g_i(\bar{y}) \in S$. Hence, for for every $\bar{\alpha}$ in M^m there is a $\beta = g_i(\bar{\alpha}) \in M$ for some i such that $f(\bar{a}, \bar{\alpha}, \beta) = 0$. Since $\bar{a} \in \mathbf{Z}^n \subseteq M^n$, we have that

$$M \models \exists\bar{x}\forall\bar{y}\exists z f(\bar{x}, \bar{y}, z) = 0.$$

By Completeness Theorem, $Z_T \vdash \exists\bar{x}\forall\bar{y}\exists z f(\bar{x}, \bar{y}, z) = 0$. □

Final Remark

For any $x \in \mathbf{Q}$, $x \neq 0$ if and only if

$$\mathbf{Q} \models \exists y\, xy = 1.$$

With the same argument as Proposition 4.4, if the set of the sentences of the form

$$\forall \bar{y} \exists z\, f(\bar{y}, z) = 0,$$

where $\bar{y} = (y_1, y_2, \ldots, y_m)$ for arbitrary m and $f(\bar{y}, z) \in \mathbf{Z}[\bar{y}, z]$, true in $\mathbf{Q}$ is axiomatizable, then it gives the positive answer to the Hilbert's Tenth Problem over $\mathbf{Q}$. We still don't know the answer of it. On the other hand, there is a polynomial time algorithm to decide whether the sentences of the form

$$\forall y \exists z\, f(y, z) = 0,$$

where $f(y, z) \in \mathbf{Z}[y, z]$, true in $\mathbf{Q}$ or not (Tung 1987).

References

Baker, A. (1990). *Transcendental number theory* (2nd ed.). Cambridge: Cambridge University Press.

Davis, M. (1973). Hilbert's tenth problem is unsolvable. *The American Mathematical Monthly, 80*, 233–269.

Enderton, H. B. (2001). *A mathematical introduction to logic* (2nd ed.). San Diego: Harcourt/Academic Press.

Fried, M. D., & Jarden, M. (2005). *Field arithmetic* (2nd ed.). Berlin: Springer.

Garey, M. R., & Johnson, D. S. (1979). *Computers and intractability: A Guide to theory of NP-completeness*. San Francisco: W. H. Freeman.

Hungerford, T. W. (1974). *Algebra*. New York: Springer.

Jones, J. P. (1981). Classification of quantifier prefixes over diophantine equations. *Z. Math. Logik Grundlag. Math., 27*, 403–410.

Jones, J. P. (1982). Universal diophantine equation. *Journal of Symbolic Logic, 47*, 549–571.

Lang, S. (1997). *Survey of diophantine geometry*. Berlin: Springer.

Manders, K., & Adleman, L. (1978). NP-complete decision problems for binary quadratics. *Journal of Computer and System Sciences, 16*, 168–184.

Matijasevich, Y. V., & Robinson, J. (1974). Two universal 3-quantifier representations of recursively enumerable sets. *Teoriya Algorifmov i Matematicheskaya Logika*, (volume dedicated to A. A. Markov), 112–123, Vychislitel'nyĭ Tsentr, Akademiya Nauk SSSR, Moscow, (Russian).

Robinson, J. (1949). Definability and decision problems in arithmetic. *Journal of Symbol Logic, 14*, 98–114.

Schinzel, A. (1982). Diophantine equations with parameters. In J. V. Arimitage (Ed.), *Journées arithmetiques 1980* (pp. 211–217)., London mathematical society lecture note series Cambridge: Cambridge University Press.

Schinzel, A. (2000). *Polynomials with special regard to reducibility*. Cambridge: Cambridge University Press.

Tung, S. P. (1985). On weak number theories. *Japanese Journal of Mathematics, 11*, 203–232.

Tung, S. P. (1986). Provability and decidability of arithmetical universal-existential sentences. *Bulletin of the London Mathematical Society*, *18*, 241–247.
Tung, S. P. (1987). Computational complexities of diophantine equations with parameters. *Journal of Algorithms*, *8*, 324–336.
Tung, S. P. (1995). Computational complexity of arithmetical sentences. *Information and Computation*, *120*, 315–325.
Tung, S. P. (2017). *Diophantine equations with three quantifiers*, manuscript.
Wüstholz, G. (Ed.). (2002). *A Panorama of number theory or the view from Baker's garden*. Cambridge: Cambridge University Press.

On the Minimization Principle in the Boolean Approach to Causal Discovery

Jiji Zhang

Abstract I examine the role of minimization in the Boolean approach to causal discovery, focusing on a recent development of the approach, implemented in a method called *Coincidence Analysis (CNA)*. I present some *prima facie* counterexamples to the soundness of the minimization steps in CNA, or to the validity of a minimization principle that is naturally suggested by the minimization steps. I discuss two possible responses to the challenge, and argue that while one (but not the other) of them is viable, it renders the role of minimization steps inessential in an important sense. I end by suggesting that the trouble with the minimization principle arises out of *locally* uninstantiated regularities that regularity theorists have no reason to dismiss.

1 Introduction

One of the best known accounts of causation is John Mackie's (1965, 1974) INUS-condition account. Roughly, when we say that (the presence of) a factor C is a cause of (the presence of) a factor E, we typically mean, among other things, that C is (at least) an Insufficient but Non-redundant part of an Unnecessary but Sufficient condition for E.[1] A key insight underlying this account is that factors of interest that enter causal analysis are usually not related as simple sufficient and necessary conditions for one another; instead, the causal law or regularity (in a given background "causal field") for an effect E is typically a complex statement involving multiple factors that describes a condition that is sufficient and necessary for E(given the background).[2] Moreover, the complex sufficient and necessary condition can be

[1]The parenthetical qualifier "at least" is there to accommodate extreme cases where a single factor is already sufficient or is part of a condition that is both sufficient and necessary.
[2]As I will make explicit later, the notions of causal structure, sufficiency, necessity, etc. that concern us in this paper are all "local", in the sense that they are applied or judged in a limited frame and relative to certain background conditions, or, in Mackie's term, relative to a "causal field".

J. Zhang (✉)
Department of Philosophy, Lingnan University, Tuen Mun, Hong Kong
e-mail: jijizhang@ln.edu.hk

S.C.-M. Yang et al. (eds.), *Philosophical Logic: Current Trends in Asia*,
Logic in Asia: Studia Logica Library, DOI 10.1007/978-981-10-6355-8_5

formulated in disjunctive normal form – a disjunction of conjunctions in which each conjunct either expresses the presence of a factor or expresses the absence of a factor – such that each disjunct/conjunction describes a (possibly complex) sufficient condition for the effect that is *minimal* in the sense that no conjunct can be taken away without affecting the said sufficiency. Then every (positive or negative) factor that appears in the complex statement is (at least) an INUS condition for the effect.

Based on this account, Mackie reconstructed and generalized J.S. Mill's (1843) "canons of induction". For example, the celebrated Method of Difference can be viewed as a method to identify an antecedent factor that is (at least) an INUS condition for the effect of interest. Viewed this way, there is no need to make (as Mill apparently did) the implausibly strong assumption that the sufficient and necessary condition for the effect of interest involves one and only one factor among the given antecedent factors (or in other words, the otherwise complex disjunctive normal form is assumed to be degenerate and involves one and only one factor from the given set of factors). It is sufficient to assume that it is some complex condition involving possibly multiple members of the given set of factors that is sufficient and necessary for the effect. Under this assumption, it is natural to seek to identify the complex condition or part thereof from observations of cases that feature different combinations of the factors. When all the logically possible combinations of the antecedent factors are observed and the presence or absence of the effect recorded in each case, the task becomes one of identifying a Boolean function of some sort that is compatible with a truth table.

This Boolean approach to causal discovery was developed into a sophisticated data analysis method for social scientists, known as *Qualitative Comparative Analysis* (*QCA*), by Charles Ragin (1987, 2000), and was more recently generalized into a method for discovering complex causal structures, called *Coincidence Analysis* (*CNA*), by Michael Baumgartner (2009a, b). Both have found interesting applications in social science (e.g., Ragin and Alexandrovna Sedziaka 2013; Baumgartner and Epple 2014; Baumgartner and Thiem 2015). In these methods, in addition to Mackie's requirement that each sufficient condition in the disjunctive normal form should be minimal, the whole disjunction as a necessary condition is also required to be minimal, in the sense that no disjunct can be taken away without affecting the necessity. The main purpose of this paper is to examine in some detail the rationale of these minimization steps as they are used in CNA. My main contention is that the rationale is problematic if one understands it as an ontological minimization principle constraining what causal regularities are, as Mackie apparently did (with regard to the minimization of sufficient conditions).

I will proceed as follows. In Sect. 2, I review the general setup of the Boolean approach and the method of CNA, and formulate a minimization principle corresponding to the minimization steps in CNA. Then, in Sect. 3, I present prima facie counterexamples to the soundness of the minimization steps of CNA and hence to the validity of the minimization principle. I argue that it is not plausible to maintain the minimization principle in view of the examples. However, it is possible to defend the minimization steps without sticking to the minimization principle, as I will discuss in Sect. 4. But the defense relies on a weak interpretation of the output of CNA, given which the point of minimization becomes much less substantial than it first appears. Under the weak interpretation, the minimization steps are not intended to implement substantial or defining constraints on what causal regularities are, and do not serve to narrow down the space of possible causal structures. Instead, minimization serves to *reveal* common features shared by some possible causal structures, which is useful but not essential, for the common features may be detected in other ways. I conclude in Sect. 5 with some remarks on the connection to the so-called causal Minimality condition in the literature on probabilistic causal models.

2 The Boolean Approach to Causal Discovery

To prepare the ground for subsequent discussions, I will review the core of the Boolean approach as it is presented by Baumgartner (2009b), with some simplifications when the more complicated details do not matter for the present purpose. We will be concerned with type-level causation or causal generalization, as opposed to toke-level or singular causation. The causal relata are taken to be event types (or types of whatever relata for token-level causation), which are also referred to as binary *factors* (or binary *variables*) which may or may not be instantiated in particular cases. When a factor X is instantiated in a particular case, we also say that it is present in that case, or that the value of X as a variable is 1 in that case. Throughout the paper, I assume we are interested in a finite set of factors **F**, called a *factor frame*, which is supposed to satisfy certain conditions to be specified below. All factors in **F** are assumed to be observable in that the data – called *coincidence data* or *coincidence list* – consist of a list of cases and in each case every factor $X \in \mathbf{F}$ is either observed to be present (denoted, with abuse of notation, simply by "X", or by "$X = 1$"), or observed to be absent (represented by "$\sim X$", or by "$X = 0$"). The causal relations among factors in **F** obtain in relation to some background or environment, often called a "causal field", which is left implicit. All subsequent talk of sufficiency or necessity is also understood to be relative to the background.

If a factor Y is featured (either positively or negatively) in a causal generalization regulating factor X, Y is said to be *causally relevant* to X, or simply a possible causal factor of X. Some factors in **F** are *endogenous* in the sense that **F** contains some of their possible causal factors. If a factor is not endogenous, then it is said to be *exogenous*. The causes of an exogenous factor are hence outside of **F** and not explicitly modeled. We assume that the structure of causal relevance is *acyclic*: there are no $X_1, \ldots, X_n \in \mathbf{F}$, such that X_i is causally relevant to X_{i+1} for all $1 \leq i \leq n-1$, and X_n also causally relevant to X_1. This assumption of acyclicity (together with the assumption that **F** is finite) entails that **F** must contain some exogenous factors.

As Baumgartner (2009b, p. 82) noted, the identification of the subset of endogenous factors (what he called "potential effects") in **F** is not a proper part of CNA. Although the data may also indicate which factors are exogenous and which are endogenous under some assumptions, it is better to regard the correct partition of **F** into a set of exogenous factors and a set of endogenous factors as an input, which may be inferred from temporal information, background knowledge of causal order, and

Table 1 A sample coincidence list on five factors

	A	*B*	*C*	*D*	*E*
Case 1	1	1	1	1	1
Case 2	1	1	0	1	1
Case 3	1	0	1	1	1
Case 4	1	0	0	1	1
Case 5	0	1	1	1	1
Case 6	0	1	0	1	1
Case 7	0	0	1	0	1
Case 8	0	0	0	0	0

so on, in addition to the data. In any case, how to identify the subset of endogenous variables is not relevant to the purpose of this paper, and I will simply assume that it is known that $\mathbf{F} = \mathbf{U} \cup \mathbf{W}$, where $\mathbf{U}$ is a (non-empty) set of exogenous factors and $\mathbf{W}$ is a (non-empty) set of endogenous factors.

We now turn to the more important assumptions. First, the causal relations to be discovered are assumed to be deterministic. More precisely, for every endogenous factor $X \in \mathbf{W}$, there is a set of factors, $\mathbf{PA}_X \subseteq \mathbf{F}\backslash\{X\}$, that are *directly causally relevant* to X (relative to the frame $\mathbf{F}$), and the presence or absence of X is fully determined by the configuration of $\mathbf{PA}_X$. In other words, X (as a variable) is a function of (the variables in) $\mathbf{PA}_X$: $X = f_X(\mathbf{PA}_X)$, or using Baumgartner's notation, $f_X(\mathbf{PA}_X) \Rightarrow X$. I will refer to f_X, or $f_X(\mathbf{PA}_X) \Rightarrow X$, as expressing the causal regularity for X, relative to the frame and the background.[3]

The task of causal discovery is to infer, as much as possible, $\mathbf{PA}_X$ and f_X for every $X \in \mathbf{W}$, from data represented by a coincidence list for the factors in $\mathbf{F}$. Here is a simple example adapted from one in (Baumgartner 2009b), which I will use for illustration in this section. Suppose $\mathbf{F} = \{A, B, C\} \cup \{D, E\}$, where $\{A, B, C\}$ is the set of exogenous factors and $\{D, E\}$ endogenous. A possible coincidence list is described in Table 1 (where "1" denotes presence and "0" denotes absence).

A second important assumption is that all cases in the given coincidence list are generated without confounding by factors outside of the frame. There are stronger and weaker ways to make this assumption precise. Baumgartner (2009b, p. 80) carefully formulated an assumption of *homogeneity* that is just enough to prevent unsound causal inference due to confounding. For the present purpose, we can adopt a slightly stronger formulation that is easier to state, and nothing important in this paper hinges on this difference. Given an endogenous factor X, call a factor outside the frame that

[3]One may also think of $f_X(\mathbf{PA}_X) \Rightarrow X$ as a result of restricting a "global" regularity for X, $f_X(\mathbf{PA}_X, \mathbf{B}) \Rightarrow X$ (where $\mathbf{B}$ denotes factors outside of the frame that are directly causally relevant to X but are not caused by any factor in the frame), to the given frame and the given background. By the assumption of no confounding (which is introduced next), the configuration of $\mathbf{B}$ remains constant in the background, in which case it seems most natural to derive $f_X(\mathbf{PA}_X)$ by plugging the configuration of $\mathbf{B}$ into $f_X(\mathbf{PA}_X, \mathbf{B})$.

is directly causally relevant to X (relative to the frame) and is not caused by any factor in the frame a *potential confounder* for X (cf. note 3). We assume that for every endogenous factor $X \in \mathbf{W}$, every potential confounder for X remains constant (i.e., constantly present or constantly absent) across all cases in the given coincidence list. This assumption ensures that variations of an endogenous factor X across cases must be due to variations of $\mathbf{PA}_X$ across cases.

Third, a notable feature of the example in Table 1 is that all logically possible configurations of the exogenous factors (A, B and C in this case) are contained in the list. Given the previous assumptions of determinism and no confounding (and the assumption of acyclicity), this is equivalent to saying that all coincidences of the factors in $\mathbf{F}$ that are compatible with the causal regularities (for the endogenous factors) are observed. We assume that the given coincidence list has this feature, which Baumgartner called the *Principle of Empirical Exhaustiveness*.

Under these assumptions, the Boolean method CNA aims to uncover, for each endogenous factor X, the causal regularity for X. Since all factors are assumed to be binary, f_X is a Boolean function. Without loss of generality, we can take f_X to be expressed in disjunctive normal form in terms of factors in $\mathbf{PA}_X$, such that each disjunct/conjunction expresses a sufficient condition for the presence of X (relative to the background), where "sufficiency" is understood in the regularity-theoretic sense that given the background, X is present in all cases that satisfy the condition (and there are cases that satisfy the condition), and the whole disjunction expresses a necessary condition for the presence of X, also in the regularity-theoretic sense.

In this sense of sufficiency, given an coincidence list that satisfies the principle of empirical exhaustiveness, a conjunction φ is sufficient for (the presence of) X if and only if in the coincidence list, there is a case in which φ is satisfied and X is present, and there is no case in which φ is satisfied but X is absent. In the first step of CNA, for each $X \in \mathbf{W}$, conjunctions involving factors in $\mathbf{F}\backslash\{X\}$ that are sufficient for X are identified. For example, given the coincidence list in Table 1, the sufficient conditions for D include $A\&B\&C\&E$, $A\&B\&{\sim}C\&E$, $A\&{\sim}B\&C\&E$, $A\&{\sim}B\&{\sim}C\&E$, ${\sim}A\&B\&C\&E$, ${\sim}A\&B\&{\sim}C\&E$, $A\&B\&C$, $A\&B\&E$, $A\&C\&E$, $B\&C\&E$, ..., $A\&B$, $A\&C$, $A\&E$, $B\&C$, $B\&E$, ..., A, B. Similarly, the sufficient conditions for E include $A\&B\&C\&D$, $A\&B\&{\sim}C\&D$, $A\&{\sim}B\&C\&D$, $A\&{\sim}B\&{\sim}C\&D$, ${\sim}A\&B\&C\&D$, ${\sim}A\&B\&{\sim}C\&D$, ${\sim}A\&{\sim}B\&C\&{\sim}D$, $A\&B\&C$, $A\&B\&D$, $A\&C\&D$, $B\&C\&D$, ..., $A\&B$, $A\&C$, $A\&D$, $B\&C$, $B\&D$, $C\&D$, ..., A, B, C, D.

Now comes the main focus of this paper, the minimization steps. The first minimization step is to pick out the *minimally sufficient conditions*, in accordance with Mackie's requirement of non-redundancy in his INUS-condition account. A conjunction φ that is sufficient for X is minimally sufficient just in case if any conjunct in φ is taken away, the remaining is not sufficient for X. Applying this definition to our running example, there are only two minimally sufficient conditions for D: A and B, and there are four minimally sufficient conditions for E: A, B, C, and D. Call this the *step of minimizing sufficiency*.

Then, for each $X \in \mathbf{W}$, the disjunction of all minimally sufficient conditions for X is formed. That disjunction will constitute a necessary condition for X, and according to Mackie's theory, that disjunction already qualifies as a candidate for expressing the causal regularity for X. In our running example, for instance, the said disjunction for D is $A \vee B$, and the said disjunction for E is $A \vee B \vee C \vee D$. However, in Ragin's and Baumgartner's development, a further minimization step is required. The necessary condition for X needs also to be minimized. A disjunction $\varphi_1 \vee \ldots \vee \varphi_n$ that is necessary for X is *minimally necessary* just in case if any disjunct φ_i is taken away, the remaining is not necessary for X. Applying this criterion in the running example, $A \vee B$ is already a minimally necessary condition for D. However, $A \vee B \vee C \vee D$ is not a minimally necessary condition for E, for it contains proper parts that are already necessary for E, namely, $A \vee B \vee C$, and also $C \vee D$.

In general, the *step of minimizing necessity* is to obtain, for each $X \in \mathbf{W}$, minimally necessary conditions (as disjunctions of minimally sufficient conditions) for X. The results, also called *minimal theories*, are then taken as proper candidates for the causal regularity for X. In the running example, the method will output $A \vee B \Rightarrow D$ as the only candidate for expressing the causal regularity for D; for E, however, the method will output two candidates: $A \vee B \vee C \Rightarrow E$, and $C \vee D \Rightarrow E$. Putting them together, two possible causal structures emerge:

$$\begin{array}{lll} & A \vee B \Rightarrow D, \text{ and } A \vee B \vee C \Rightarrow E & \text{(common cause structure)} \\ \text{or} & A \vee B \Rightarrow D, \text{ and } C \vee D \Rightarrow E & \text{(causal chain structure)} \end{array} \quad (1)$$

The method, without additional background knowledge, will not further eliminate one structure in favor of the other, and rightly so, for the data truly underdetermine the causal structure here, as is often the case. What can be reasonably demanded of the method is that the true causal structure remains at least one of the candidates in the output, in the ideal situation where the assumptions of determinism, no confounding, and empirical exhaustiveness are all met. The method is sound if and only if under these assumptions, its output necessarily contains the true causal structure.

The following principle is an obvious statement of the rationale behind the minimization steps:

Minimization Principle (MP): Given any frame and any background that satisfy the assumptions of determinism, no confounding, and empirical exhaustiveness, for every endogenous variable X, the true causal regularity for X (relative to the frame and the background) can be expressed in disjunctive normal form, as $\varphi_1 \vee \ldots \vee \varphi_n \Rightarrow X$, such that

(i) For every $1 \leq i \leq n$, φ_i expresses a minimally sufficient condition for X; and
(ii) The disjunction $\varphi_1 \vee \ldots \vee \varphi_n$ expresses a minimally necessary condition for X.

Clearly, clause (i) in the MP corresponds to the step of minimizing sufficiency and clause (ii) corresponds to the step of minimizing necessity. Together they entail the soundness of the minimization steps. The question is whether the MP is true. I submit that it is not.

3 Counterexamples to the Minimization Principle

Consider the following example that I will call CHEM. Suppose there are three (and only three) ways to produce chemical E: The combination of chemical A and chemical D will surely result in a reaction to produce E; the combination of chemical B and chemical D will surely result in a reaction to produce E; and the combination of chemical C and chemical X will surely result in a reaction to produce E. Moreover, among several ways to produce chemical D, one way is to have A and Y react that will produce D. As it happens, the reaction between A and Y does not exhaust A so that the reaction between A and the resulting D will in turn take place, which, remember, will surely result in E. Another way is to have B and Z react so that they will produce D. Since this reaction does not exhaust B, the reaction between B and the resulting D will in turn take place, which, remember, will surely result in E.

Let us use each letter to denote the factor of the corresponding chemical being present (of sufficient quantity for the relevant reaction to take place). From the description, I take it uncontroversial that the scenario features the following causal regularities:

$$(A\&Y) \vee (B\&Z) \vee \alpha \Rightarrow D, \text{ and } (A\&D) \vee (B\&D) \vee (C\&X) \Rightarrow E \qquad (2)$$

where α is supposed to catch all the other ways for producing D. As it happens, our factor frame just consists of $\{A, B, C, D, E\}$, and in the background, X, Y, Z happen to be constantly present, and α happens to be constantly unsatisfied. Against this background and relative to the frame, the intuitively correct causal structure is:

$$A \vee B \Rightarrow D, \text{ and } (A\&D) \vee (B\&D) \vee C \Rightarrow E \qquad (3)$$

Moreover, thanks to appropriate variations of some factors in the background that are causally relevant to A, B, and C, we get to observe all logically possible combinations of A, B, and C. So what coincidence list do we get? Clearly, none other than the list in Table 1. As we saw in the previous section, if we feed the list to the Boolean method CNA, its output, given in (1), will not contain (3), the true causal structure, for neither $A\&D$ nor $B\&D$ is minimally sufficient for E. Since the assumptions of determinism, no confounding and empirical exhaustiveness are satisfied, or at least can be made so without contradicting the description of the scenario I have given, this appears to be a counterexample to the soundness of the method. It suggests that the MP is not necessarily true.

Why should we think that in this situation (3) describes the true causal structure (against the said background and relative to the said frame)? Several considerations can reinforce this intuition. First, $A\&D$ and $B\&D$ are specified as two *direct* mechanisms for producing E. Minimizing them to A alone, B alone, or D alone misrepresents the direct mechanisms. Second, the structure as described in (3) correctly answers counterfactual questions about what would happen if the background causal field were altered in a way that constitutes an intervention, in, e.g., James

Woodward's (2003) sense, to realize some configurations that were not actually observed. For example, would there be E in a case where A is present, and B, C are absent if the background were slightly altered so that Y were absent (while α remains unsatisfied), which, according to (2), would lead to the absence of D? Or would there be E in a case where A, B, C are all absent if the background were slightly altered so that (some conjunct in) α were instantiated, which, according to (2), would lead to the presence of D? Under a natural reading of (3), it entails the correct answers to these questions (i.e., "no" to both). By contrast, neither causal structure in (1) gets both questions right. Third, (3) seems to be the most natural restriction of (2) to the said background (where it happens to be the case that X, Y, Z are constantly present, and α is constantly unsatisfied). Since (2) clearly describes the true structure, when not restricted to the given frame and the said background, the structure expressed by (3)— which results from plugging into (2) the supposed values of those background variables—should be regarded as the correct structure relative to the frame and the background.

Despite these considerations, defenders of the MP may bite the bullet and insist that (3) is nonetheless not the correct structure relative to the given frame and the given background, or they may challenge the intelligibility of the scenario, perhaps on the grounds that it is based on stipulations of strange or even impossible chemical reactions. I confess that I do not know how realistic my imagined scenario is, but I would be surprised if it turned out to be chemically impossible. In any case, more mundane examples with the same structure abound. Moreover, to make the same point, we can consider even simpler structures. I now present a simpler example, called PROF, in two versions, one challenging the soundness of minimizing sufficiency, and the other challenging the soundness of minimizing necessity. As I will argue, the two versions together make it even harder to resist the force of the counterexample.

In what I will refer to as the *conjunctive* version of the example, two professors are assigned to evaluate Ph.D. students' qualifying exams, and the rule is clear: a student passes the exam if and only if both professors' verdicts are "pass". Professor 1 grades exams as usual, giving "pass" or "fail" based on her expertly judgement. Professor 2 is unfortunately not confident and is deferential to Professor 1. Before he gives his verdict, he takes a peek at Professor 1's verdict. If he sees Professor 1's verdict, he simply gives the same verdict. Otherwise, he gives a verdict based on his own judgement.

Suppose the given frame is $\mathbf{F} = \{A\} \cup \{B, C\}$, where A denotes the factor of Professor 1's verdict being "pass", B denotes the factor of Professor 2's verdict being "pass", and C denotes the factor of a student passing the exam. According to the description I gave above, I take it to be clear that B and C are regulated by the following simple structure:

$$(A\&X) \vee ({\sim}X\&Y) \Rightarrow B, \text{ and } A\&B \Rightarrow C \tag{4}$$

where X denotes the factor of Professor 2 successfully seeing Professor 1's verdict, and Y denotes the factor of Professor 2's own judgement being "pass".

Table 2 A coincidence list on three factors

	A	*B*	*C*
Case 1	1	1	1
Case 2	0	0	0

Suppose further that the background causal field happens to always instantiate that $X = 1$ (i.e., Professor 2 succeeds at peeking) and $Y = 0$ (i.e., Professor 2's own judgment is harsh). Against this background and relative to the frame, I claim the correct structure is simply

$$A \Rightarrow B, \text{ and } A\&B \Rightarrow C \tag{5}$$

In this case, the coincidence list that satisfies the assumption of empirical exhaustiveness (and can satisfy the assumption of no confounding) is given in Table 2.

It is easy to see from the list that (5) will not be contained in the output of CNA, for $A\&B$ is not minimally sufficient for C. Instead, the CNA algorithm will output two possible regularities for C: either $A \Rightarrow C$, or $B \Rightarrow C$. As a side note, in this case the data alone clearly underdetermine the causal order, for any permutation of the three factors in Table 2 will result in the same list. By contrast, in the previous, more complicated example CHEM, the causal order can be determined from the data, as every endogenous variable has at least two alternative causes in the frame (see Baumgartner 2009b, p. 75 for an elaboration of this point). This otherwise interesting difference does not matter here, for the point is simply that the output of CNA, which may contain many alternative causal structures due to the underdetermination of causal order, will not contain the true causal structure (5).

To simplify the subsequent discussion, let us suppose that the true causal order is given as background knowledge: A causally precedes B and C (which was already indicated when we wrote $\mathbf{F} = \{A\} \cup \{B, C\}$), and B causally precedes C. With this background knowledge, the CNA algorithm outputs two possible causal structures:

$$\begin{array}{lll} & A \Rightarrow B, \text{ and } A \Rightarrow C & \text{(common cause structure)} \\ \text{or} & A \Rightarrow B, \text{ and } B \Rightarrow C & \text{(causal chain structure)} \end{array} \tag{6}$$

As in CHEM, the minimization of sufficient conditions is apparently a mistake in this example.

Before we consider possible responses to this counterexample, let me introduce another version of PROF, which I will refer to as the *disjunctive* version. All is the same as the conjunctive version, except for one change. The rule for passing the qualifying exam is less demanding in this version: a student passes the exam if and only if at least one professor's verdict is "pass". It is trivial to check that, with the same assumptions about the background, the coincidence list remains the same as that in Table 2, and hence the CNA algorithm, with background knowledge of the true causal order, will output (6), whereas the intuitively true causal structure is

$$A \Rightarrow B, \text{and } A \vee B \Rightarrow C \tag{7}$$

Again, CNA apparently misses the target, this time due to the other minimization step, namely the step of minimizing necessity. It misses the structure given in (7) because $A \vee B$ is not minimally necessary for C. The two versions of PROF suggest that both steps of minimization in CNA, or both clauses in the MP, are problematic.

I will consider two possible responses to the apparent counterexample, one in this section, and the other in the next. The response I will discuss presently is to deny that it is a genuine counterexample, by denying that (5) is the true casual structure for the conjunctive version of PROF and denying that (7) is the true causal structure for the disjunctive version. Note that what I said previously in relation to the example CHEM also applies here. The causal regularity for C given in (5) is, I claim, the true one (relative to the frame and against the background) in the conjunctive version, because it corresponds exactly to the specified mechanism (the rule) for determining C, because it answers correctly (as neither structure in the output of CNA does) certain counterfactual questions about consequences of intelligible interventions, and because it is the most natural restriction of (4) to the given frame and the supposed background. And similarly for the causal regularity for Cgiven in (7) in relation to the disjunctive version.

So the response under consideration has the heavy burden of explaining away or debunking these points. It will not do to simply say that the underlying theory of causation is supposed to be a regularity theory, so any appeal to intuitions about mechanisms or about counterfactuals is illegitimate and can be ignored. For any account of causation worth its salt, be it regularity theoretic or not, needs to account for our ordinary locutions and intuitions about mechanisms and interventions, especially if the mechanisms and interventions in question are simple and intelligible. Moreover, even if we agree to suspend the first two reasons I presented above, there is still the third reason that appeals to nothing but a most natural way to restrict the causal structure for a larger frame to a smaller frame against a specified, properly homogeneous background.

How can the response continue? Well, the responder may insist that this natural way of deriving restricted causal structure is simply not right; indeed the responder may point to our example and claim that the example shows that the natural way is not right, for the result violates the MP. In other words, the responder may attempt to resist by stipulation: by definition, the true causal structure satisfies the MP, so in the conjunctive version of the example, (5) cannot be the true causal structure, and in the disjunctive version of the example, (7) cannot be the true causal structure.

It is usually hard to rebut such a question-begging argument, except to say that it, well, begs the question. But there may be a chance to do better here. We can ask: if (5) is not the true causal structure in the conjunctive version of the example, then which structure is? In particular, which of the two structures that satisfy the MP and are hence recommended by CNA is the truth? Note that we are not asking that given only the data (and the assumptions of determinism, no confounding, and empirical exhaustiveness, as well as the background knowledge of causal order), which one we should infer to be the truth. To this question (which we are *not* asking), it is perfectly ok to answer "neither", for it is perfectly normal to have data underdetermine the causal structure. But the question is: given all the description of the conjunctive version of PROF, *as a matter of fact*, which structure in (6) is the true causal structure. To answer "neither" to this question is to say that there is no matter of the fact about causal structure (in a frame that is supposed not to be confounded), which strikes me as far less plausible than the claim that (5) is the true causal structure.

So which one is the truth in the conjunctive version of PROF, the common cause structure or the causal chain structure? I cannot see any good reason to favor one over the other. One may be tempted to choose the common cause structure, on the grounds that from the description of the scenario, A appears to be the ultimate cause or at least more responsible for the result than B is. This temptation results from a conflation of questions. We are not asking about ultimate causes or responsibility, etc., which usually affects (rightly or wrongly) our judgements about singular causation. We are concerned with causal generalizations and are in particular asking about the factors that are *directly* causally relevant to C in the given frame.

And how about the disjunctive version? Presumably any reason in the conjunctive version that favors one minimized structure over the other also applies to the disjunctive version. So the conjunctive version and the disjunctive version would end up featuring the same causal structure for the given frame. I submit that there is a much more plausible alternative here. That is, we should take (5) and (7) to model the conjunctive version and the disjunctive version, respectively, which accurately display the difference between the two versions. I conclude that the counterexample to the MP stands.

4 The Purpose of Minimization

There is another possible response to the counterexample. Instead of resisting it, this response grants the counterexample but claims that the minimization steps in CNA are not meant to be justified by the MP, so the invalidity of the MP does not affect the soundness of the minimization steps. If the CNA algorithm remains sound, despite the fact that (5) and (7) represent respectively the correct causal structures in the conjunctive and disjunctive versions of PROF, then it must be that the output of CNA, though given as (6), does not really eliminate (5) or (7). As far as I can see, this can only be the case if the regularities in (6) are weakly interpreted as follows: $A \Rightarrow C$, for example, says only that A is directly causally relevant to C; it does *not* state that B is not directly causally relevant to C. Similarly for $B \Rightarrow C$. In other words, only causal relevancies are expressed by each "$\Rightarrow$ statement" in the output, but *not* causal irrelevancies. Thus, appearances to the contrary, (6) is actually compatible with more than two causal structures, including (5) and (7).

This weak interpretation of the output of the Boolean method does not seem to fit Mackie's intention with regard to the "N" in "INUS", which corresponds to the minimization of sufficient conditions and which is naturally interpreted as a criterion for distinguishing relevant from irrelevant factors. It is also apparently at odds with some claims in Baumgartner's writings. For example, he wrote (Baumgartner 2009b, pp. 74–75): "A factor A is (positively) directly causally relevant to a factor B iff A is part of a minimal theory of B." The "only if" claim suggests that minimization is intended to filter out causally irrelevant factors. In another interesting paper, Baumgartner (2008) argued that for deterministic systems there is systematic underdetermination of causal chain structures, for observations that are compatible with a causal chain structure are always also compatible with an alternative, common cause structure. One of his main examples features a coincidence list on five factors that is very similar to the list in Table 1, and in (1) we are supposed to witness an exemplary case of underdetermination between a causal chain structure and a common cause structure. This claim of underdetermination seems to presuppose that the two structures given in (1) are rival or incompatible, but if they are interpreted weakly as not claiming causal irrelevancies, then they are compatible and no rival.

No matter. Even if this weak interpretation was not intended before, defenders of the Boolean method are well advised to adopt it now, lest they are forced to take the untenable response discussed in the previous section. There is then the following question: What is the point of minimization if it is not intended to be a way to possibly infer causal irrelevancies? More concretely, if the minimization that results in (6) in our simple example is not supposed to eliminate (5) or (7), what then is the point? Why not explicitly output also (5) and (7) as among the possible causal structures in light of data?

The only answer that occurs to me is that minimization is a way to possibly reveal what is *definitely* causally relevant. To illustrate what I mean, consider a variation of the conjunctive version of PROF, which I shall call PROF-VC (Variation on the Conjunctive version). Suppose Professor 2 is a little less deferential to Professor 1 than in the original story. He is only deferential to Professor 1 when her verdict is "pass" (and he sees it), and will stick to his own judgement when her verdict is "fail". That is, the causal regularity for B is: $(A\&X) \vee Y \Rightarrow B$, where, as before, X represents the factor of Professor 2 successfully seeing Professor 1's verdict, and Y represents the factor of Professor 2's own judgement being "pass". Moreover, suppose this time Y is included in our frame as an exogenous variable: $\mathbf{F} = \{A, Y\} \cup \{B, C\}$, and as in the original story, X happens to be constantly present in the background. Then the causal structure on the given frame relative to the background is:

$$A \vee Y \Rightarrow B, \text{ and } A\&B \Rightarrow C \tag{8}$$

The coincidence list on this frame that is empirically exhaustive is given in Table 3.

As in the original example, assume it is known beforehand that B causally precedes C. The CNA algorithm will then output the following:

$$A \vee Y \Rightarrow B, \text{ and } A \Rightarrow C \tag{9}$$

Table 3 A coincidence list on four factors

	A	*Y*	*B*	*C*
Case 1	1	1	1	1
Case 2	1	0	1	1
Case 3	0	1	1	0
Case 4	0	0	0	0

Again, if (9) is interpreted strongly, as representing a single causal structure that is compatible with the data and the assumptions, the inference is mistaken, for the true structure is given in (8). However, under the weak interpretation, (9), and in particular $A \Rightarrow C$, is supposed to mean only that A is directly causally relevant to C, without indicating that B (or Y) is *not* directly causally relevant to C. In other words, (9) in effect represents multiple causal structures, including the true causal structure that is explicitly given in (8), all of which *agree* that A is directly causally relevant to C. The minimization helps to *reveal* this agreement, which entails that it is definitely the case that A is directly causally relevant to C.

I see nothing wrong with this response to my counterexample to the MP and this modest purpose of minimization, but it is important to stress the implication of this maneuver. Under this weak interpretation of the output of CNA, the role of the minimization steps is insubstantial or inessential, for at least two reasons. First, the minimization steps are not intended to implement substantial or defining constraints on what causal regularities are, such as the MP. Their purpose is not to further narrow down the space of possible causal structures (as a substantial constraint would do at least sometimes), but to reveal common features among (some of) the possible causal structures. For the coincidence data in the example PROF-VC (Table 3), for instance, the minimization reveals that a common feature among all possible causal theories for C is that A is directly causally relevant to C.[4] Second, the minimization is not essential for this modest purpose, for it is just one way to reveal commonalities. It is not the only way. One can, for example, also go through the space of causal structures to see what commonalities there are. In the example PROF-VC (Table 3), for instance, one can infer that A is directly causally relevant to C by observing that every sufficient condition for C contains A as a part.

Let me close this section with an observation. Suppose we still adopt the strong interpretation of the output of CNA; that is, only the "minimal" structures output by the CNA are taken as possible structures. In the case of PROF, for example, the structures in (5) and (7) are taken as eliminated under this interpretation of the output. Consider now using the output to answer questions about interventions (Pearl 2000; Woodward 2003), interventions that could be realized, if needed, by appropriate variations in the background (and so should be acceptable to regularity theorists).

[4]In the original versions of PROF, the minimization amounts to revealing that some (but not all) possible causal structures share the feature that A is directly causally relevant to C and some (but not all) possible causal structures share the feature that B is directly causally relevant to C.

For example, what would be the value of C if it were the case that $A = 1$ and $B = 0$ (because, say, Professor 2 failed to see Professor 1's verdict)? In the standard and natural way of using knowledge about the causal structure to answer questions about interventions (Spirtes et al. 1993; Pearl 2000), the common cause structure in (6) would answer 1, and the causal chain structure in (6) would answer 0. Thus, if the unanimity rule is used, that is, if a definite answer is given if and only if all structures in the output agree on the answer, then the output of the CNA will not give a definite, wrong answer to the question. In this simple case, it is easy to check that whenever the two structures in the output of the CNA agree on the answer to a question about interventions, that answer is true. In other words, given that the two structures are in the output, there is no need to also consider the more complex, "non-minimal" structures described in (5) and (7), for the purpose of answering questions about interventions. This provides a sort of pragmatic justification for employing minimization in this case, and I suspect, more generally when there is such underdetermination, which Baumgartner (2008) showed to be commonplace.

Specifically, Baumgartner (2008) showed that in the Boolean approach to causal discovery, there is systematic underdetermination of causal chain structures, for observations that are compatible with a causal chain structure are always also compatible with a common cause structure. The previous observation suggests that in these otherwise unfortunate cases, minimization is less of a problem even if we adopt a strong interpretation of the output of CNA. By contrast, as Baumgartner pointed out, a common cause structure may sometimes be unambiguously inferred by CNA, as we saw in the case of PROF-VC. In those cases, if we adopt a strong interpretation and use the output of CNA to answer questions about interventions, we are at a higher risk of making mistakes.

5 Conclusion

I have shown that the minimization steps in CNA – a representative Boolean method for inferring deterministic causal structures from coincidence data – are prima facie problematic, for there are situations in which the method's output apparently rules out the true causal structure. I considered two possible responses to my counterexamples. One is to resist the counterexamples by insisting that the structures that I take to be correct in those examples are not true. This response not only faces a number of good reasons to the contrary, but also has a hard time to specify in a principled way which structure is true as a matter of fact.

The other response is one of retreat. It grants the force of the counterexamples against a strong interpretation of the result of the minimization, but defends the minimization steps by adopting a weak interpretation, under which the result is compatible with the truth in those examples. I welcome this retreat, but I stress that under the weak interpretation, the minimization steps play an insubstantial or inessential role.

I hasten to add that I am not suggesting that no minimization principle should play a substantial role in defining what causal regularities are. On the contrary, I believe that causal *irrelevancy* must correspond to some sort of minimality. The problem with the minimization principle criticized in this paper is that it is supposed to apply to any frame and any background that satisfy the assumptions of determinism, no confounding, and empirical exhaustiveness. The trouble arises when some possible configurations of some factors of interest are "locally" uninstantiated, due to certain local, background conditions. A regularity theorist may legitimately reject the idea of globally uninstantiated laws and so ignore considerations of configurations that are globally uninstantiated, but she has no good reason to ignore considerations of configurations that are only locally uninstantiated. My main claim is precisely that it would be a mistake to take the minimization principle as a substantial constraint on the causal structure of a local system when there are locally uninstantiated configurations.

However, when all the logically possible configurations are instantiated,[5] as it is often the case in non-deterministic or probabilistic settings, then some generalized version of the minimization principle will be valid or at least defensible. In fact, the minimization principle in the Boolean framework is analogous to the so-called causal Minimality condition in the framework of (recursive) structural equation model or causal Bayesian network (Spirtes et al. 1993; Pearl 2000; Spohn 2000; Hitchcock 2010; Zhang and Spirtes 2011; Zhang 2013). I will not pursue a formal comparison here, but it is worth noting that the causal Minimality condition has been shown to be extremely well motivated and justified, if the probability distribution over the factors/variables in question is strictly positive – which essentially says that every logically possible configuration has a positive probability of being instantiated – but is problematic otherwise (Zhang and Spirtes 2011). So the standard causal discovery methods in the framework of causal Bayesian network are not straightforwardly applicable to deterministic systems not only because, as Baumgartner (2009b) noted, the so-called causal Faithfulness assumption fails, but also because even the much weaker assumption of causal Minimality condition is called into question. In this connection, it is interesting to note that an adaptation of a standard causal discovery method in that framework to deterministic systems (Glymour 2007) dispenses with the causal Minimality condition; if that method is applied to the example CHEM or the example PROF, it will output the more complex structures as well as the minimal ones. A detailed comparison between the two approaches may yield more insights on this matter, which I will have to leave for another occasion.

Acknowledgements This research was supported in part by the Research Grants Council of Hong Kong under the General Research Fund LU13600715, and by a Faculty Research Grant from Lingnan University.

[5] Or when the uninstantiated configurations are somehow known or assumed to be globally impossible, which is almost never the case.

References

Baumgartner, M. (2008). The causal chain problem. *Erkenntnis, 69*, 201–226.

Baumgartner, M. (2009a). Inferring causal complexity. *Sociological Methods and Research, 38*, 71–101.

Baumgartner, M. (2009b). Uncovering deterministic causal structures: A Boolean approach. *Synthese, 170*, 71–96.

Baumgartner, M., & Epple, R. (2014). A coincidence analysis of a causal chain: The Swiss minaret vote. *Sociological Methods and Research, 43*, 280–312.

Baumgartner, M., & Thiem, A. (2015). Model ambiguities in configurational comparative research. *Sociological Methods and Research*. https://doi.org/10.1177/0049124115610351.

Glymour, C. (2007). Learning the structure of deterministic systems. In A. Gopnik & L. Schulz (Eds.), *Causal Learning. Psychology, Philosophy, and Computation* (pp. 231–240). New York: Oxford University Press.

Hitchcock, C. (2010). Probabilistic causation. In E. Zalta (Ed.), *Stanford Encyclopedia of Philosophy*. https://plato.stanford.edu/entries/causation-probabilistic/.

Mackie, J. L. (1965). Causes and conditions. *American Philosophical Quarterly, 2*(4), 245–264.

Mackie, J. L. (1974). *The Cement of the Universe: A Study of Causation*. Oxford: Clarendon Press.

Mill, J. S. (1843). *Systems of Logic* (Vol. 1). London: John Parker.

Pearl, J. (2000). *Causality: Models, Reasoning, and Inference*. Cambridge, UK: Cambridge University Press.

Ragin, C. C. (1987). *The Comparative Method*. Berkeley: University of California Press.

Ragin, C. C. (2000). *Fuzzy-Set Social Science*. Chicago: University of Chicago Press.

Ragin, C. C., & Alexandrovna Sedziaka, A. (2013). QCA and fuzzy set applications to social movement research. *The Wiley-Blackwell Encyclopedia of Social and Political Movements*. https://doi.org/10.1002/9780470674871.wbespm482.

Spirtes, P., Glymour, C., & Scheines, R. (1993). *Causation, Prediction and Search*. New York: Springer.

Spohn, W. (2000). Bayesian nets are all there is to causal dependence. In M. C. Galavotti, et al. (Eds.), *Stochastic Dependence and Causality* (pp. 157–172). CSLI Publications.

Woodward, J. (2003). *Making Things Happen: A Theory of Causal Explanation*. Oxford and New York: Oxford University Press.

Zhang, J. (2013). A comparison of three Occam's Razors for Markovian causal models. *British Journal for the Philosophy of Science, 64*(2), 423–448.

Zhang, J., & Spirtes, P. (2011). Intervention, determinism, and the causal minimality condition. *Synthese, 182*(3), 335–347.

Contentual and Formal Aspects of Gentzen's Consistency Proofs

Ryota Akiyoshi and Yuta Takahashi

1 Introduction

Gentzen's 1936 consistency proof for first-order arithmetic (Gentzen 1936), located between the first (Gentzen 1974) and third proofs (Gentzen 1938), was a work during his "transition period 1936–1938."[1] (Hereafter, we call the first proof *the 1935 proof*, the second *the 1936 proof* and the third *the 1938 proof*.) On the one hand, the 1936 proof inherited from the 1935 proof the method of "finitist (finit)" interpretation for first-order arithmetical formulas. On the other hand, the 1936 proof was a precursor of the 1938 proof given by the cut-elimination method with an ordinal notation system below ε_0. Because of this intermediate or patchwork feature of the 1936 proof, several researchers have published investigations of its structure. For example, Yasugi (1980) reformulated the 1936 proof within the framework of the 1938 proof and presented some applications of her reformulation. More recently, Buchholz (2015) analyzed the structure of the 1936 proof, using the method of finite notations for infinitary

[1] Arai (2002, p. 438).

We would like to express our deep gratitude to Wilfried Sieg for his valuable comments on our work. In particular, we had fruitful discussions when the first author invited him to Waseda University in 2015. In addition, special thanks is due to the anonymous referees for careful readings, many helpful comments, and suggestions. The first author is partially supported by KAKENHI (Grant-in-Aid for Young Scientists (B)) 16K16690. The second author is supported by KAKENHI (Grant-in-Aid for JSPS Fellows) 16J04925.

R. Akiyoshi (✉)
Waseda Institute for Advanced Study, Tokyo, Japan
e-mail: georg.logic@gmail.com

Y. Takahashi
Research Fellow of Japan Society for the Promotion of Science, Nagoya University, Nagoya, Japan
e-mail: yuuta.taka84@gmail.com

S.C.-M. Yang et al. (eds.), *Philosophical Logic: Current Trends in Asia*,
Logic in Asia: Studia Logica Library, DOI 10.1007/978-981-10-6355-8_6

derivations.[2] While Yasugi and Buchholz have focused on the mathematical side of the 1936 proof, we focus not only on its mathematical side, but also on its conceptual side.[3]

Sieg, in (2012), explained some background of Gentzen's 1936 proof that enables us to understand the intermediate feature of this 1936 proof from another approach. Sieg claimed, "Hilbert's considerations in ["Beweis des tertium non datur" Hilbert (1931)] were a crucial germ for Gentzen's work on consistency."[4] That is to say, Sieg has claimed that Hilbert's 1931 paper "Beweis des Tertium non datur" had great influence on Gentzen's investigation into the consistency of first-order arithmetic, especially on Gentzen's 1936 proof. In the main argument of Hilbert (1931), Hilbert used a concept of *correctness* (Richtigkeit) of formulas to prove the consistency of a first-order arithmetical theory without any induction axiom schema.[5] He formulated a concept of correctness of formulas and then attempted to show, in this sense, that every derivation of the theory has a correct formula as its conclusion.

According to Sieg, Gentzen scrutinized this argument before completing the 1936 proof. By citing some passages from Gentzen's unpublished manuscripts, Sieg explained background of the 1936 proof as follows.[6] Gentzen called a kind of a consistency proof like Hilbert's argument above a *contentual correctness proof* (inhaltlicher Richtigkeitsbeweis) and contrasted this kind with another kind that he called a *formal correctness proof* (formaler Richtigkeitsbeweis). Formal correctness proofs show a theory's consistency by assigning a normal derivation to every derivation of a numeric equation in the theory. A consistency proof by means of Hilbert's epsilon substitution method and Gentzen's 1938 proof are typical examples of such

---

[2]The method of finite notations for infinitary derivations originates from the work by Mints (1978) and has been developed further in subsequent papers by Buchholz (1991, 1997, 2001).

[3]Here, we should explain the literatures in more detail. Historically speaking, the first reconstruction of the 1935 proof was by Bernays (1970) and this proof was further discussed by Kreisel in (1971). More recently, this proof was discussed by von Plato in (2009). As to the connection between the 1935 and 1936 proofs, the paper Negri (1980) by Maurizio Negri should be mentioned here. He reconstructed the 1935 proof using ordinals up to ε_0, which would create a close relationship between the 1935 and 1936 proofs. (A similar analysis was given in Sundholm's Ph.D. thesis Sundholm 1983.) In these papers, authors focused only on mathematical sides of Gentzen's proofs, while we focus also on a conceptual side of the 1936 proof. Moreover, as we will see in Sect. 5.2, we will investigate a uniform idea behind Gentzen's three consistency proofs.

The content of this paper differs from our previous work (Akiyoshi and Takahashi 2013) written in Japanese with respect to the three main points. First, while the aim of Akiyoshi and Takahashi (2013) was to give only a uniform interpretation of Gentzen's proofs, we also considered questions naturally arising from Sieg's paper Sieg (2012). Secondly, we use another version of normalization trees, namely, a version of normalization trees reformulated with (possibly) non-well-founded trees. This notion makes it easier to see a connection between the 1936 proof and some notions of intuitionism like spreads and choice sequences. Finally, Sect. 5.3 of the present paper gives another proof of Kreisel's no-counterexample interpretation, using normalization trees.

[4]Sieg (2012, p. 123).

[5]The theory includes the principle of tertium non datur. Hilbert's aim in this paper was to justify the use of this principle by means of a consistency proof.

[6]In fact, Sieg has explained background of not only the 1936 proof but also the 1935 proof. In the present paper, we concentrate on an analysis of the 1936 proof in the light of Sieg's explanation.

correctness proofs.[7] Then, Gentzen finished the 1936 proof and considered it *intermediate* between these two kinds of consistency proofs.

This explanation by Sieg induces the question whether Gentzen's 1936 proof is both a contentual correctness proof and a formal correctness proof in the sense above, since one might wonder if a consistency proof *in general* can be so. Contentual correctness proofs and formal correctness proofs correspond to semantic consistency proofs and proof-theoretic consistency proofs, respectively, and semantic consistency proofs are usually distinguished from proof-theoretic consistency proofs. (Below we often abbreviate "P is a contentual correctness proof" as "P is contentual." The same abbreviation is used also for formal correctness proofs.) Furthermore, one could ask another question: If the 1936 proof is both contentual and formal, how do its contentual and formal aspects relate to each other?

This paper aims to answer the two questions above. First, we argue that the 1935 proof is a contentual correctness proof and that the 1938 proof is a formal correctness proof. By examining some remarks in Gentzen (1974), we attribute the following interpretation for every sequent of first-order arithmetic: Let Γ be a sequent as a finite set of formulas of first-order arithmetic, then

(GI) Γ is correct if and only if a reduction procedure is statable for Γ.[8]

The main lemma of the 1935 proof shows that every derivation of first-order arithmetic has the correct endsequent in the sense of (GI); thus the 1935 proof is a contentual correctness proof. After arguing so, we briefly see that the 1938 proof is a formal correctness proof because its main lemma assigns a normal derivation to every derivable numeric equation of first-order arithmetic.

Second, we show that the 1936 proof is both contentual and formal because its main lemma implies both the main lemma of the 1935 proof and the main lemma of the 1938 proof. To show this in a uniform way, we formulate the 1936 proof in the framework of *normalization trees*, which were introduced in Akiyoshi (2010), Akiyoshi and Takahashi (2013) through finite notations for infinitary derivations. Here we use a version of normalization trees reformulated in terms of (possibly) non-well-founded trees, to analyze contentual aspects of the 1936 proof more closely. Then, we derive the main lemma of the 1935 proof from the main lemma of the 1936 proof. Next, to analyze formal aspects of the 1936 proof, we utilize not only normalization trees, but also analyses in Buchholz's two papers Buchholz (1997, 2015). We show that the main lemma of the 1938 proof is a special case of the main lemma of the 1936 proof. This means that the main lemma of the 1936 proof assigns a normal derivation to every derivable numeric equation as the main lemma of the 1938 proof does.

Third, we explain how contentual and formal aspects of the 1936 proof relate to each other. Our argument for the claim that the 1936 proof is both contentual and formal enables us to see the following: The correctness of a derivable formula of

[7] As to Hilbert's exposition of the epsilon substitution method, see Hilbert (1928).

[8] Note that Gentzen, in Gentzen (1974), used two-sided sequents. The reason why we formulate (GI) with one-sided sequents is that we use the one-sided sequent calculus in Sects. 3, 4 and 5.

first-order arithmetic in the sense of (GI) is in fact shown by means of syntactic transformation of a given derivation of the formula. This means that contentual aspects of the 1936 proof are formed by its formal aspects.

Finally, we note several consequences of our arguments. One is that we obtain a uniform reading of Gentzen's three consistency proofs: The assignment of a "finitist" sense to each derivable formula of first-order arithmetic is a common method of his three consistency proofs. Furthermore, consequent to our arguments is that Kreisel's no-counterexample interpretation (Kreisel 1951) is obtained from our formulation of the 1936 proof.[9]

This paper is structured as follows. In Sect. 2, we briefly recall Sieg's argument above and then explain that the 1935 proof is a contentual correctness proof and that the 1938 proof is a formal correctness proof. Section 3 gives a formulation of finite notations for infinitary derivations in first-order arithmetic, following Buchholz's paper Buchholz (1997). In Sect. 4, we first define the notion of normalization trees and then argue by means of this notion that the 1936 proof is both contentual and formal. Finally, in Sect. 5, we indicate some consequences of the arguments in the previous sections.

2 Contentual and Formal Correctness Proofs

In this section, we first explain the argument by Sieg for the claim that Hilbert's considerations in "Beweis des Tertium non datur" had great influence on Gentzen's 1936 proof. Sieg's argument provides background for the present paper's contents. Next, preliminarily to our argument after this section, we see that the 1935 proof is a contentual correctness proof and that the 1938 proof is a formal correctness proof.

The following passage from Gentzen's unpublished manuscript *Urdissertation* is key to Sieg's argument.

> The consistency of arithmetic will be proved; in the process, the concept of an infinite sequence of natural numbers will be used, furthermore in one place the principle of the excluded middle. The proof is thus not intuitionist. Perhaps the *tertium non datur* can be eliminated (Sieg 2012, p. 88).

Here, Gentzen made several comments about his tentative proof for the consistency of first-order arithmetic in *Urdissertation*: The concept of an infinite sequence of natural numbers and the principle of tertium non datur were used in that proof. According to Sieg, a connection between Gentzen's investigation into the consistency of first-order arithmetic and Hilbert's considerations in Hilbert (1931) is found here. It was a central feature of the main argument in Hilbert (1931) that both the concept of

[9]Several studies have already pointed out the relationship between Gentzen's 1935 and 1936 proofs and the no-counterexample interpretation. Cf. Kreisel (1971), Sieg and Parsons (1938), Tait (2005). Furthermore, according to Sieg and Parsons (1938), Tait (2001), the idea of the no-counterexample interpretation is found in Gödel's notes for his lecture in 1938.

an infinite number-sequence and the principle of tertium non datur were used in metamathematical investigations.

Hilbert, in (1931), formulated a fragment of arithmetic that lacks an induction axiom schema and the axiom (TND) $\forall x A(x) \vee \exists x \neg A(x)$; then he claimed that the theory can be expanded by means of (TND) without deriving contradiction.[10] An outline of Hilbert's main argument is as follows.[11] First, Hilbert stipulated when a formula of the theory is *correct* (richtig). Next, he tried to prove that every correct formula does not imply contradiction. Finally, he argued that every derivation in the expanded theory by means of (TND) can be transformed into a derivation such that the axioms and inference-rules in it are all correct. As seen from this outline, a notion of correctness is crucial to Hilbert's argument. Later in this section, we will see that a notion of correctness played a crucial role in Gentzen's considerations as well.

Let us see briefly how Sieg explained that the concept of an infinite number-sequence was included in Hilbert's considerations. Hilbert stipulated when a universally quantified formula is correct as follows:

> If the statement $\mathfrak{A}(\mathfrak{z})$ is correct as soon as $\mathfrak{z}$ is a numeral, then the statement $(x)\mathfrak{A}(x)$ holds [$\cdots$] (Hilbert 1931, p.121).[12]

Sieg claimed that this stipulation gave the following rule (HR^*): If $\mathfrak{A}(\mathfrak{z})$ is correct for an arbitrary numeral $\mathfrak{z}$, then the universally quantified formula $(x)\mathfrak{A}(x)$ may be introduced as an *axiom* of theory.[13] Note that the rule (HR^*) differs from standard inference-rules, since standard rules allow us to deduce a conclusion from some premises *within* a formal theory. The rule (HR^*) is for introducing an axiom by means of metatheoretical considerations. According to Sieg, the concept of an infinite sequence of natural numbers is included in this rule. We omit further details of Sieg's explanation.[14]

In addition to this feature of Hilbert's considerations, Sieg argued that Hilbert made metamathematical use of the principle of tertium non datur to show that his notion of correctness actually works:

> This metamathematical statement [any statement either does or does not lead to a contradiction] is an instance of *tnd*, and Hilbert views it as "necessary" for the founding of mathematics. He then uses *tnd* to show that correctness, falsity, and the generalized negation of statements (see Note 29) harmonize in the appropriate way (Sieg 2012, p. 105).

Of course, "tnd" denotes the principle of the excluded middle, so Sieg has claimed here that a metamathematical statement used in Hilbert's argument is an instance

---

[10]In the notation of Hilbert (1931), (TND) is $(x)\mathfrak{A}(x) \vee (Ex)\bar{\mathfrak{A}}(x)$.

[11]We owe this outline to Sieg. Cf. Sieg (2012, pp. 102–105).

[12]This English translation is by Sieg.

[13]Sieg (2012, p. 92, p. 103).

[14]Cf. Sieg (2012, p. 107). Sieg's explanation can be paraphrased as follows: To apply the rule (HR^*), one needs to show that $\mathfrak{A}(\mathfrak{z})$ is correct for an arbitrary numeral $\mathfrak{z}$. Because of Hilbert's finitist attitude, this must be shown by giving an effective method to verify the correctness of $\mathfrak{A}(0)$, $\mathfrak{A}(1)$, $\mathfrak{A}(2)$ and so on. The concept of an infinite sequence of natural numbers is used here.

of the principle of the excluded middle. After quoting the remark from Gentzen's *Urdissertation* (at the beginning of this section), Sieg wrote,

> The connection to Hilbert's considerations in *Beweis des tertium non datur* seems unmistakable, as [Gentzen's] these remarks point exactly to the central features of Hilbert's argument, i.e., the metamathematical use of the rule (HR^*) and *tnd* (Sieg 2012, p. 107).

This connection between Gentzen's *Urdissertation* and Hilbert's argument in Hilbert (1931) is a rationale for Sieg's claim that Hilbert's consideration in Hilbert (1931) had great influence on Gentzen's 1936 proof.

Let us explain the next rationale. After explaining the connection above between Gentzen's *Urdissertation* and Hilbert's argument in Hilbert (1931), Sieg has described other background of Gentzen's 1936 proof as follows. The manuscript INH, written from October 1932 to October 1934 and titled "Die formale Erfassung des Begriffs der inhaltlichen Richtigkeit in der reinen Zahlentheorie, Beziehungen zum Widerspruchsfreiheitsbeweis," starts with a reflection by Gentzen: how a notion of correctness is defined in a given formal theory.[15] Then, Gentzen contrasted a kind of consistency proof he called a *formal correctness proof* with another kind called a *contentual correctness proof*. The following is a quotation by Sieg from INH, p. 2:

> I seek to clarify the questions: what distinguishes a formal correctness or consistency proof from a contentual one, why is the former for certain inferences not even possible by these same inferences (according to Gödel), is a bridge inference involved then, how secure is that [bridge inference, WS], what are the connections with Gödel's proof, what role do the mathematical axioms play? (Sieg 2012, p.114)

In the present paper, what is relevant is the first question, that is, the question of what distinguishes a formal correctness proof from a contentual one.

Citing Gentzen's words from INH, Sieg characterized notions of contentual correctness proofs and formal correctness proofs as follows:

> Calling a proof of a numeric statement a *Normalbeweis* if it contains only numeric statements, Gentzen can now express the difference between (purely) formal and (semi-) contentual correctness proofs by formulating carefully the claim each is to establish. The claim for a *purely formal correctness proof* is, "for every proof of a numeric statement there is a *Normalbeweis* of that statement", and the corresponding claim for the (semi-) contentual correctness proof is, "every proof has a correct result" [where result means end formula, WS] (Sieg 2012, p. 115).

In sum, a formal correctness proof shows the consistency of a theory by assigning a normal derivation (Normalbeweis) to each derivable numeric equation in the theory. Since no normal derivation of $0 = 1$ exists, the consistency follows. The meaning of "normal derivations" and of "numeric equations" may vary according to which theory is considered. For the former notion, it suffices to require that the truth of the conclusion of a normal derivation is verified in a primitive recursive way. On the other hand, a contentual correctness proof shows the consistency of a theory

[15] Sieg (2012, p. 114).

by verifying that every derivation of the theory has a correct conclusion.[16] Then, it follows that no contradiction is derivable in the theory because no contradiction can be correct. Note that Hilbert's argument in Hilbert (1931) is an example of contentual correctness proofs. Gentzen made a further remark about differences between these two kinds of correctness proofs, but we do not enter into its details. The remark was quoted by Sieg from INH, p. 8:

> The semi-contentual proof uses complete induction for a rather complicated statement. This contains Ri erg x [the result of proof x is correct, WS], and this predicate becomes ever more complicated in complicated cases. The formal proof uses complete induction for [the statement] Ey. No y & erg x = erg y [there is a *Normalbeweis* y having the same result as the given proof x, WS]; this is also a statement containing logical signs; it is however of a simple nature, also in more complicated cases (Sieg 2012, p. 115).

As the last two passages from INH show, Gentzen examined the method of Hilbert's argument in Hilbert (1931), namely, the method of contentual correctness proofs before completing the 1936 consistency proof. This is the second rationale for Sieg's claim that Hilbert's consideration in Hilbert (1931) had great influence on Gentzen's 1936 proof. As we said in the introduction, Sieg eventually concluded that Gentzen gave the 1936 proof in an intermediate way between contentual correctness proofs and formal correctness proofs.[17]

In the rest of this section, we argue that Gentzen's 1935 proof is contentual and that his 1938 proof is formal, preliminarily to our argument in Sect. 4.

Gentzen, in the 1935 proof, gave a notion of correctness that was called the *statability of a reduction procedure*, and then remarked that it gives a "finitist" interpretation of arithmetical formulas. He wrote,

> The concept of the 'statability of a *reduction procedure*' (die Angebbarkeit einer Reduziervorschrift) for a sequent, to be defined below, will serve as the formal replacement (formaler Ersatz) of the contentual *concept of correctness* (der inhaltliche Richtigkeitsbegriff); it provides us with a *special finitist interpretation* (finite Deutung) of propositions and takes the place of their *actualist interpretation* [...] (Gentzen 1974, p. 100, Gentzen 1936, p. 536, Gentzen 1969, p. 173, italics original).

From the quotation above, we can extract the following interpretation for every sequent Γ, i.e., every finite set Γ of formulas of first-order arithmetic:

(GI) Γ is correct if and only if a reduction procedure is statable for Γ.

Later, we will define the notion of reduction procedures formally. As Tait pointed out in Tait (2015, Footnote 4), a reduction procedure is statable if and only if both this procedure and a proof for its termination are obtained.[18] But now, it is sufficient

---

[16]The distinction of *purely* formal correctness proofs from formal correctness proofs does not matter to our present concern; neither does the distinction of *semi*-contentual correctness proofs from contentual correctness proofs. Thus we do not consider these distinctions. In the both cases, the former is subsumed into the latter.

[17]Sieg (2012, p. 117).

[18]For Gentzen, the statability of a reduction procedure for Γ gives a meaning to Γ from his "finitist" standpoint. It is not easy to estimate the exact strength of Gentzen's finitist standpoint. As far as we

to recall that the statability of a reduction procedure is a notion of correctness: A sequent Γ is correct when a reduction procedure is statable for Γ. If we stipulate that a formula A is correct if and only if the sequent $\{A\}$ is correct, then we obtain from (GI) the following interpretation for arithmetical formulas:

A is correct if and only if a reduction procedure is statable for $\{A\}$.

The main lemma of the 1935 proof shows that every derivation of a proof-system Z of first-order arithmetic has the correct endsequent in the sense of (GI). The statement of the lemma is as follows.[19]

Main Lemma of the 1935 Proof. For every sequent Γ of Z, if Γ is derivable in Z, then there is a reduction procedure for Γ.

Note that this lemma shows the correctness of each Z-derivable sequent Γ in the sense of (GI) because the proof of the lemma gives not only a reduction procedure for Γ, but also a proof for its termination. The 1935 proof shows the consistency of first-order arithmetic in this way. Thus, according to Sieg's characterization above, the 1935 proof is a contentual correctness proof.

On the other hand, the 1938 proof is a formal correctness proof. As we have said previously, Sieg's characterization of formal correctness proofs is not restricted to a particular theory, so the range of numeric equations and normal derivations could vary according to which theory is considered. To see that Gentzen's 1938 proof is a formal correctness proof, consider the proof-system Z of first-order arithmetic that Gentzen formulated in Gentzen (1938), whose language includes 0 (zero) and S (the successor function) as its all function symbols. Then, the numeric equations we consider are formed by means of these two function symbols and the equality $=$. In addition, we recall Takeuti's formulation of the 1938 proof (Takeuti 1987, Sect. 12). A derivation of Z is called *simple* if and only if it includes no free variable and consists only of mathematical initial sequents and structural inferences except cut-inferences with the non-atomic cut formula.[20] Note that a simple derivation does not include any logical inference and the induction-rule. Below, by "normal derivation" we mean a simple derivation.

Now, the 1938 proof can be outlined as follows. First, it can be verified in a primitive recursive way that there is no normal derivation of a false numeric equation. Next, the following lemma is proved:

Main Lemma of the 1938 Proof. Let h be a derivation of Z for a numeric equation, then there is a reduction sequence of h that terminates with a normal derivation for the equation.

(Footnote 18 continued)
know, his standpoint should be constructive in the following sense. First, all infinite totalities must be generated by some finitary rules (Gentzen 1936, pp. 524–525, 1969, p. 162). For example, the totality of all natural numbers is generated from 0 by the successor rule. Second, one must avoid the use of the principle of the excluded middle for non-decidable predicates (Gentzen 1936, pp. 527–528, 1969, pp. 164–165).

[19] Gentzen (1974, p. 103, 1936, p. 549).

[20] Cf. Takeuti (1987, Definition 12.2.).

There is no derivation of Z for $0 = 1$ by the main lemma and the above fact about normal derivations, so Z is consistent. As this outline shows, the 1938 proof is a formal correctness proof. The main lemma of this proof assigns a normal derivation to every derivable numeric equation.

In Sect. 4, we argue that the 1936 proof is both contentual and formal, by showing that the main lemma of the 1936 proof implies the main lemmas of the 1935 and 1938 proofs. In the next section, we explain the method of *finite notations of infinitary derivations*.

3 Finite Notations for Infinitary Derivations

In this section, we introduce a finitary system Z^* of first-order arithmetic and the corresponding infinitary system Z^∞, then recall some basic properties of finite notations for infinitary derivations. We define these systems in the style of Tait's calculus as explained later, that is, (1) a sequent is a finite set of formulas rather than a finite multi-set of them, and (2) contraction, exchange and weakening rules are implicitly assumed. Although we follow Buchholz' paper Buchholz (1997) with some minor modifications, we repeat and explain important definitions and theorems in it for readers' convenience.

First, we define the basic language L, on which proof systems in Sects. 3 and 4 are defined. The vocabulary consists of the following symbols:

- Predicate Symbols: p for every primitive recursive relation $\mathbf{P}$,
- Function Symbols: 0, S (successor),
- Variables for natural numbers: $x, y, \ldots$,
- Logical connectives: $\forall$, $\exists$, $\wedge$, $\vee$, $\neg$,
- Auxiliary Symbols: (,).

Terms of L are defined in the standard way: x, 0 are terms. if t is a term, then $S(t)$ is also a term, and denoted by t, s. The closed terms are called the *numerals*. *Atomic formulas* are formulas of the form $pt_1 \cdots t_n$ where p is a predicate symbol and $t_1, \ldots, t_n$ are terms. We say an expression is a *literal* if it is an atomic formula or of the form $\neg pt_1 \cdots t_n$. Formulas of L are defined from literals by means of $\forall$, $\exists$, $\wedge$, $\vee$. For every formula A, the *negation* $\neg A$ of A is defined via de Morgan's laws. From this definition it follows that the negation symbol $\neg$ is put only before an atomic formula. For example, $\neg(0 = 0 \wedge 1 = 1)$ denotes the formula $\neg(0 = 0) \vee \neg(1 = 1)$.

The set of all closed true literals is denoted by TRUE_0. For example, $0 = 0$ belongs to this set. We use syntactic variables x, y, z for variables, n, m, k, l for terms, and A, B, C, D for formulas, possibly with suffixes.

Finite sets of formulas are called *sequents* and denoted by Γ, Δ. We often abbreviate $\Gamma \cup \Delta$ (resp. $\Gamma \cup \{A\}$) as Γ, Δ (resp. Γ, A or A, Γ). When θ is a term, a formula or a sequent, we denote the set of the free variables in θ by $FV(\theta)$ and say θ is *closed* if $FV(\theta) = \emptyset$. The expression $\theta(x/t)$ is the result of substituting t for every free occurrence of x in θ after renaming some bound variables in θ if necessary.

A *proof system* $\mathfrak{S}$ is defined by a set of inference symbols. We denote inference symbols by $\mathcal{I}$, the indices of $\mathcal{I}$ by $|\mathcal{I}|$, the principle formulas of $\mathcal{I}$ by $\Delta(\mathcal{I})$ and the premises of $\mathcal{I}$ by $(\Delta_i(\mathcal{I}))_{i\in|\mathcal{I}|}$. By writing

$$(\mathcal{I})\,\frac{\ldots\Delta_i\ldots(i\in I)}{\Delta}$$

we mean that $\mathcal{I}$ is an inference symbol such that $|\mathcal{I}| = I, \Delta(\mathcal{I}) = \Delta$ and $\Delta_i(\mathcal{I}) = \Delta_i$. When $|\mathcal{I}| = \{0, 1, \ldots, n-1\}$, we write simply

$$(\mathcal{I})\,\frac{\Delta_0\ \Delta_1\ \ldots\ \Delta_{n-1}}{\Delta}.$$

For example, in the case of conjunction rule, we write as follows.

$$(\textstyle\bigwedge_{A_0\wedge A_1})\,\frac{A_0\quad A_1}{A_0\wedge A_1}$$

Then $I = \bigwedge_{A_0\wedge A_1}$, $\Delta(I) = \{A_0 \wedge A_1\}$, and $\Delta_i(I) = \{A_i\}$ for $\in \{0, 1\}$. Since the contraction and weakening rules are implicitly assumed, the following example is a correct instance of this rule:

$$\frac{A_0, A_0\wedge A_1\quad A_1}{A_0\wedge A_1, C}$$

Definition 3.1 (*The Inference Symbols of the Finitary Proof System* Z^*) We assume the existence of a primitive recursive set Ax of sequents satisfying the following conditions:

- for all $\Delta \in \mathsf{Ax}$, Δ is a set of literals,
- if there is a substitution instance Δ_0 of Δ, then $\Delta_0 \in \mathsf{Ax}$,
- if $\Delta \in \mathsf{Ax}$ and Δ is closed, then $\Delta \cap \mathsf{TRUE}_0 \neq \emptyset$.

The inference symbols of Z^* are as follows.

$$(\mathsf{Ax}_\Delta)\,\frac{}{\Delta}\quad\text{with }\Delta\in\mathsf{Ax}$$

$$(\textstyle\bigwedge_{A_0\wedge A_1})\,\frac{A_0\quad A_1}{A_0\wedge A_1}\qquad(\bigvee^k_{A_0\vee A_1})\,\frac{A_k}{A_0\vee A_1}\quad\text{with }k\in\{0,1\}$$

$$(\textstyle\bigwedge^y_{\forall xA})\,\frac{A(x/y)}{\forall xA}\qquad(\bigvee^t_{\exists xA})\,\frac{A(x/t)}{\exists xA}$$

$$(\mathsf{Ind}^{y,t}_F)\,\frac{\neg F, F(y/S(y))}{\neg F(y/0), F(y/t)}\qquad(\mathsf{R}_C)\,\frac{C\quad\neg C}{\emptyset}\qquad(\mathsf{E})\,\frac{\emptyset}{\emptyset}$$

Definition 3.2 (Z*-*Derivations*) If $\mathcal{I}$ is an inference symbol of Z^* with $|\mathcal{I}| = \{0, \dots, n-1\}$ and $h_0, \dots, h_{n-1}$ are Z^*-derivations, then $h := \mathcal{I} h_0 \cdots h_{n-1}$ is a Z^*-derivation with

$$\Gamma(h) := \bigcup_{0 \le i \le n-1} (\Gamma(h_i) \setminus \Delta_i(\mathcal{I})) \cup \Delta(\mathcal{I}).$$

A Z^*-derivation h is *closed* if and only if $\Gamma(h)$ is closed. We suppose, if $h := \mathcal{I} h_0 \cdots h_{n-1}$ is a Z^*-derivation, then it satisfies the following Free Variables-Conditions:

- if $\mathcal{I} = \bigwedge^{y}_{\forall x A}$ then $y \notin \mathrm{FV}(\Gamma(h))$,
- if $\mathcal{I} = \mathsf{Ind}^{y,t}_{F}$ then $y \notin \mathrm{FV}(\Gamma(h))$,
- if $\mathcal{I} = \bigvee^{t}_{\exists x A}$ then $\mathrm{FV}(t) \subseteq \mathrm{FV}(\Gamma(h))$,
- if $\mathcal{I} = \mathsf{R}_C$ then $\mathrm{FV}(C) \subseteq \mathrm{FV}(\Gamma(h))$.

The first and second conditions impose the standard proviso of eigenvariables. By the last two conditions, it holds that if h is closed and has the form $\mathcal{I} h_0 \cdots h_{n-1}$ such that $\mathcal{I}$ is neither $\bigwedge^{y}_{\forall x A(x)}$ nor $\mathsf{Ind}^{y,t}_{F}$, then $h_0, \dots, h_{n-1}$ are closed as well. If h is closed and has the form $\bigwedge^{y}_{\forall x A(x)} h_0$ or $\mathsf{Ind}^{y,t}_{F} h_0$, then $FV(\Gamma(h_0)) \subseteq \{y\}$.

Remark 3.1 As to the definition of Ax_Δ, a typical example would be obtained by taking $\Delta = \{0 = 0\}$ or $\{x = x\}$.

Remark 3.2 At a first glance the E-rule is redundant, but this rule is needed for "expressing" the cut-elimination for the corresponding infinitary system inside Z^* (Cf. Theorem 3.2). Similarly, the R-rule corresponds to one-step reduction for it (Cf. Theorem 3.1).

We assign ordinals up to ε_0 to Z^*-derivations. This assignment is motivated by the corresponding ordinal in the infinitary system Z^∞. For a more detailed informal idea behind this, see Remark 3.4. If α, β are ordinals, then $\alpha \sharp \beta$ means the *natural sum* of them.

Definition 3.3 (*Ordinal Assignment of* Z*-*Derivations*)

$$\mathrm{o}(h) := \begin{cases} \mathrm{o}(h_0) \sharp \mathrm{o}(h_1) & \text{if } I = \mathsf{R}_C, \\ \mathrm{o}(h_0) \times \omega & \text{if } I = \mathsf{Ind}^{y,t}_{F}, \\ \omega^{\mathrm{o}(h_0)} & \text{if } I = \mathsf{E}, \\ (\sup_{i \in |I|} \mathrm{o}(h_i)) + 1 & \text{otherwise.} \end{cases}$$

$$\deg(h) := \begin{cases} \max\{\mathrm{rk}(C), \deg(h_0), \deg(h_1)\} & \text{if } I = \mathsf{R}_C, \\ \max\{\mathrm{rk}(F), \deg(h_0)\} & \text{if } I = \mathsf{Ind}^{y,t}_{F}, \\ \deg(h_0) \dot{-} 1 & \text{if } I = \mathsf{E}, \\ \sup_{i \in |I|} \deg(h_i) & \text{otherwise.} \end{cases}$$

We define the system Z as the subsystem of Z^* obtained by omitting the rule E.

Next, we introduce the infinitary proof system Z^∞. Since this system contains Schütte's ω-rule (denoted by $\bigwedge_{\forall x A}$), it suffices to consider only closed terms and formulas. After proving the cut-elimination theorem for it (Theorems 3.1 and 3.2), we recall the canonical embedding from Z^* to Z^∞ by Schütte.

Definition 3.4 (*The Inference Symbols of the Infinitary Proof System* Z^∞) The inference symbols of Z^∞ are as follows:

$$(\mathsf{Ax}_A)\,\frac{}{A}\quad \text{with } A \in \mathsf{TRUE}_0 \qquad (\textstyle\bigwedge_{\forall x A})\,\frac{\ldots A(x/n)\ldots (n \in \mathbb{N})}{\forall x A}$$

$$(\mathsf{Cut}_C)\,\frac{C \quad \neg C}{\emptyset} \qquad (\mathsf{Rep})\,\frac{\emptyset}{\emptyset}$$

and $\bigwedge_{A_0 \wedge A_1}$, $\bigvee^k_{A_0 \vee A_1}$, $\bigvee^t_{\exists x A}$ as in Z^*.

Remark 3.3 At a first glance the Rep-rule is redundant, but this rule is needed for defining finite notations for infinitary derivations in a primitive recursive way (cf. Remark 3.6).

Below, we define the notion of Z^∞-derivations in a precise way. Note that derivations obtained in this way are just ordinary ones using inference rules of Z^∞. As to the ordinal height α of a given Z^∞-derivation, this just means the size of it as a tree. $\deg(\mathtt{d})$ means the cut-rank of it.

The figure

$$\frac{\ldots \Gamma_i \ldots (i \in I)}{\Gamma}\ \mathcal{I}$$

is a *correct* $\mathfrak{S}$-*inference* if and only if $\mathcal{I} \in \mathfrak{S}$, $|\mathcal{I}| = I$, $\Delta(\mathcal{I}) \subseteq \Gamma$ and $\Gamma_i \subseteq \Gamma \cup \Delta_i(\mathcal{I})$ for every $i \in |\mathcal{I}|$.

Definition 3.5 (Z^∞-*Derivations*) If $\mathcal{I}$ is an inference symbol of Z^∞, $(\mathtt{d}_i)_{i\in I}$ is a family of Z^∞-derivations, Γ is a sequent and α is an ordinal such that

$$\frac{\ldots \Gamma(\mathtt{d}_i)\ldots (i \in I)}{\Gamma}\ \mathcal{I}$$

is a correct Z^∞-inference and $\mathrm{o}(\mathtt{d}_i) \prec \alpha$ for every $i \in I$, then the tree $\mathtt{d}$

$$\frac{\ldots \mathtt{d}_i \ldots (i \in I)}{\mathcal{I} : \Gamma : \alpha}$$

is a Z^∞-derivation and

$$\Gamma(\mathtt{d}) := \Gamma,\ \mathrm{last}(\mathtt{d}) := \mathcal{I},\ \mathrm{o}(\mathtt{d}) := \alpha,\ \mathtt{d}(i) := \mathtt{d}_i,$$

$$\deg(\mathrm{d}) := \begin{cases} \max\{\mathrm{rk}(C)+1, \deg(\mathrm{d}_0), \deg(\mathrm{d}_1)\} & \text{if } \mathcal{I} = \mathsf{Cut}_C, \\ \sup_{i \in I} \deg(\mathrm{d}_i) & \text{else.} \end{cases}$$

We write $\mathrm{d} \vdash^{\alpha}_{m} \Gamma$ if $\Gamma(\mathrm{d}) \subseteq \Gamma$, $\mathrm{o}(\mathrm{d}) = \alpha$ and $\deg(\mathrm{d}) \leq m$. In what follows, we may assume $\Gamma(\mathrm{d}) = \Gamma$ unless otherwise stated.

The point of the following theorem is that we can derive Γ from Γ, C and $\Gamma, \neg C$ without increasing the degree of the derivation, under a certain condition.

Theorem 3.1 *For every formula C, there is an operator $\mathcal{R}_C$ such that if $\mathrm{d}_0 \vdash^{\alpha}_{m} \Gamma, C$, $\mathrm{d}_1 \vdash^{\beta}_{m} \Gamma, \neg C$ and $\mathrm{rk}(C) \leq m$, then $\mathcal{R}_C(\mathrm{d}_0, \mathrm{d}_1) \vdash^{\alpha \sharp \beta}_{m} \Gamma$.*

Proof By double induction on d_0 and d_1. Let $\mathcal{I}_0, \mathcal{I}_1$ be the last inference rules of $\mathrm{d}_0, \mathrm{d}_1$ respectively. Then $\mathrm{d}_k = \mathcal{I}_k(\mathrm{d}_{ki})_{i \in |\mathcal{I}_k|}$.

If $C \notin \Delta(\mathcal{I}_0)$, we set

$$\mathcal{R}_C(\mathrm{d}_0, \mathrm{d}_1) := \mathcal{I}_0(\mathcal{R}_C(\mathrm{d}_{0i}, \mathrm{d}_1))_{i \in |\mathcal{I}_0|} \vdash^{\alpha \sharp \beta}_{m} \Gamma.$$

The case $\neg C \notin \Delta(\mathcal{I}_1)$ is treated in the same way.

Hence, we consider the case $C \in \Delta(\mathcal{I}_0)$ and $\neg C \in \Delta(\mathcal{I}_1)$. Although all cases according to the shape of C must be considered, we treat the only case when $C \equiv \forall x A(x)$. In this case, $\mathrm{d}_0 = \bigwedge_{\forall x A(x)}(\mathrm{d}_{0n})_{n \in \omega}$ and $\mathrm{d}_1 = \bigvee^{k}_{\exists x \neg A(x)}(\mathrm{d}_{10})$ for some k. By IH, we have $\mathcal{R}_C(\mathrm{d}_{0k}, \mathrm{d}_1) \vdash^{\alpha_k \sharp \beta}_{m} \Gamma, A(k)$ with $\alpha_k \sharp \beta < \alpha \sharp \beta$. Moreover, again by IH, we have $\mathcal{R}_C(\mathrm{d}_0, \mathrm{d}_{10}) \vdash^{\alpha \sharp \beta_0}_{m} \Gamma, \neg A(k)$ with $\alpha \sharp \beta_0 < \alpha \sharp \beta$. By inserting a cut over $A(k)$, we get the desired derivation:

$$\mathcal{R}_C(\mathrm{d}_0, \mathrm{d}_1) := \mathsf{Cut}_{A(k)}(\mathcal{R}_C(\mathrm{d}_{0k}, \mathrm{d}_1), \mathcal{R}_C(\mathrm{d}_0, \mathrm{d}_{10})) \vdash^{\alpha \sharp \beta}_{m} \Gamma.$$

We remark that $\mathrm{rk}(C) = \mathrm{rk}(\forall x A(x)) > \mathrm{rk}(A(k))$ so that $\deg(\mathcal{R}_C(\mathrm{d}_0, \mathrm{d}_1)) \leq m$ holds. □

Theorem 3.2 *There is an operator $\mathcal{E}$ such that if $\mathrm{d} \vdash^{\alpha}_{m+1} \Gamma$ then $\mathcal{E}(\mathrm{d}) \vdash^{\omega^\alpha}_{m} \Gamma$.*

Proof By induction on d. Let $\mathcal{I}$ be its last inference symbol.

First we consider the crucial case $\mathcal{I} = \mathsf{Cut}_C$ so that $\mathrm{d} = \mathsf{Cut}_C(\mathrm{d}_0, \mathrm{d}_1)$. By IH, we have $\mathcal{E}(\mathrm{d}_0) \vdash^{\omega^{\alpha_0}}_{m} \Gamma, C$ and $\mathcal{E}(\mathrm{d}_1) \vdash^{\omega^{\alpha_1}}_{m} \Gamma, \neg C$ with $\alpha_0, \alpha_1 < \alpha$. By applying Theorem 3.1, we get $\mathcal{R}_C(\mathcal{E}(\mathrm{d}_0), \mathcal{E}(\mathrm{d}_1)) \vdash^{\omega^{\alpha_0} \sharp \omega^{\alpha_1}}_{m} \Gamma$. Since $\omega^{\alpha_0} \sharp \omega^{\alpha_1} < \omega^\alpha$, we may define

$$\mathcal{E}(\mathrm{d}) := \mathsf{Rep}(\mathcal{R}_C(\mathcal{E}(\mathrm{d}_0), \mathcal{E}(\mathrm{d}_1))) \vdash^{\omega^\alpha}_{m} \Gamma.$$

In other cases, we define (using IH)

$$\mathcal{E}(\mathrm{d}) := \mathcal{I}(\mathcal{E}(\mathrm{d}_i))_{i \in |\mathcal{I}|} \vdash^{\omega^\alpha}_{m} \Gamma.$$

For example, if $\mathcal{I} = \bigwedge_{\forall x A(x)}$, then $\mathcal{E}(\mathrm{d}_i) \vdash^{\omega^{\alpha_0}}_{m} \Gamma, A(i)$ with $\alpha_0 < \alpha$ for each $i < \omega$. From this, applying $\bigwedge_{\forall x A(x)}$, we get $\mathcal{I}(\mathcal{E}(\mathrm{d}_i))_{i \in |\mathcal{I}|} \vdash^{\omega^\alpha}_{m} \Gamma$. □

We set $\mathtt{d}_A$ as a fixed cut-free Z^∞-derivation with $\Gamma(\mathtt{d}_A) = \{A, \neg A\}$, $\deg(\mathtt{d}_A) = 0$ and $\mathrm{o}(\mathtt{d}_A) = 2 \cdot \mathrm{rk}(A)$. Such a $\mathtt{d}_A$ may be defined by induction on A.

Next, we define the canonical embedding from Z^*-derivations into Z^∞-derivations.

Definition 3.6 (*Translation of* Z^**-Derivations into* Z^∞*-Derivations*) We define the Z^∞-derivation h^∞ for every closed Z^*-derivation h. Let Γ be $\Gamma(h)$ and α be $\mathrm{o}(h)$.

1. $(\mathsf{Ax}_\Delta)^\infty := \overline{\mathsf{Ax}_A : \Gamma : \alpha}$, where A is the arbitrarily fixed element of $\Delta \cap \mathsf{TRUE}_0$.
2. $(\bigwedge^y_{\forall x A} h)^\infty := \dfrac{\ldots h(y/i)^\infty \ldots (i \in \mathbb{N})}{\bigwedge_{\forall x A} : \Gamma : \alpha}$.
3. $(\mathsf{R}_C h_0 h_1)^\infty := \mathcal{R}_C(h_0^\infty, h_1^\infty)$.
4. $(\mathsf{E}h)^\infty := \mathcal{E}(h^\infty)$.
5. $(\mathsf{Ind}^{y,n}_F h)^\infty := \dfrac{\mathtt{e}_n}{\mathsf{Rep} : \Gamma : \alpha}$ where
$$\mathtt{e}_0 := \mathtt{d}_{F(x/0)},\ \ \mathtt{e}_1 := h(y/0)^\infty,\ \ \mathtt{e}_{i+1} := \mathcal{R}_{F(x/i)}(\mathtt{e}_i, h(y/i)^\infty).$$
6. Otherwise: $\mathcal{I} h_0 \cdots h^\infty_{n-1} := \dfrac{h_0^\infty \ldots h^\infty_{n-1}}{\mathcal{I} : \Gamma : \alpha}$.

Remark 3.4 Now we see that the ordinal assignment for Z^*-derivations (Cf. Definition 3.3) comes from the corresponding theorems for Z^∞-derivations. For example, the clause that $\mathrm{o}(\mathsf{R}_C(h_0, h_1)) = \mathrm{o}(h_0)\sharp\mathrm{o}(h_1)$ in Definition 3.3 is motivated by Theorem 3.1. Following this, we defined $(\mathsf{R}_C h_0 h_1)^\infty := \mathcal{R}_C(h_0^\infty, h_1^\infty)$ as in Clause 3.

Remark 3.5 We insert Rep in Clause 5 for making our definition of finite notations for infinitary derivations primitive recursive. For details, see Remark 3.6.(4) after Definition 3.7.

In what follows, we use the following notation. For every $k \in \mathbb{N}$, we set

$$C[k] := \begin{cases} A(x/k), & \text{if } C \equiv QxA \text{ with } Q \in \{\forall, \exists\}, \\ A_k, & \text{if } C \equiv A_0 \circ A_1 \text{ with } \circ \in \{\wedge, \vee\} \text{ and } k \in \{0, 1\}, \\ \text{undefined}, & \text{if } C = A_0 \circ A_1 \text{ with } \circ \in \{\wedge, \vee\} \text{ and } k \notin \{0, 1\}. \end{cases}$$

We may define h_A as a Z-derivation with $\Gamma(h_A) = \{A, \neg A\}$, $\deg(h_A) = 0$ and $\mathrm{o}(h_A) = 1 + 2 \cdot \mathrm{rk}(A)$ by induction on A. For technical reasons, we treat only Z^*-derivations such that an eigenvariable occurs only at once in them. If a given derivation does not satisfy this condition, then we can obtain another one satisfying it by substituting a fresh variable for such a variable.

Next, we define finite notations for infinitary derivations, that is, the two functions $\mathrm{tp}(h)$ and $h[i]$ with $i \in |\mathrm{tp}(h)|$ for a Z^*-derivation h. To give a simple and short definition, we first define the expression h^+ and then obtain $\mathrm{tp}(h)$ and $h[i]$ from h^+. A definition of $\mathrm{tp}(h)$ and $h[i]$ in this manner can be found in Buchholz (2010).

Definition 3.7 For an arbitrary closed Z^*-derivation h, $\mathrm{tp}(h)$ and $h[i]$ with $i \in |\mathrm{tp}(h)|$ are defined as follows.

1. For every closed Z^*-derivation h of the form Ax_Δ, define

$$h^+ := \mathrm{tp}(h) := \mathsf{Ax}_A,$$

where A is the arbitrarily fixed element of $\Delta \cap \mathsf{TRUE}_0$.

2. Let h be a closed Z^*-derivation with $h \neq \mathsf{Ax}_\Delta$. By induction on the built-up of h, we define h^+ as the expression $\mathcal{I}(h_i)_{i\in|\mathcal{I}|}$ for some inference symbol $\mathcal{I}$ of Z^∞ and some sequence $(h_i)_{i\in|\mathcal{I}|}$ of closed Z^*-derivations. Then, set

$$\mathrm{tp}(h) := \mathcal{I},\ h[i] := h_i \text{ for every } i \in |\mathcal{I}|.$$

   (a) $(\bigwedge_C h_0 h_1)^+ := \bigwedge_C (h_i)_{i\in\{0,1\}}$.
   (b) $(\bigwedge_C^y h_0)^+ := \bigwedge_C (h_0(y/i))_{i\in\mathbb{N}}$.
   (c) $(\bigvee_C^k h_0)^+ := \bigvee_C^k h_0$.
   (d) $(\mathsf{Ind}_F^{y,n} h_0)^+ := \mathsf{Rep}e_n$, where $e_0 := h_{F(x/0)}$, $e_1 := h_0(y/0)$ and $e_{i+1} := \mathsf{R}_{F(x/i)} e_i h_0(y/i)$.
   (e) Let h be $\mathsf{E}h'$ and h'^+ be $\mathcal{I}((h_i)_{i\in|\mathcal{I}|})$, then

$$(\mathsf{E}h')^+ := \begin{cases} \mathsf{Rep}(\mathsf{R}_C \mathsf{E}h_0 \mathsf{E}h_1), & \text{if } \mathcal{I} = \mathsf{Cut}_C, \\ \mathcal{I}((\mathsf{E}h_i)_{i\in|\mathcal{I}|}) & \text{else.} \end{cases}$$

   (f) Let h be $\mathsf{R}_C h_0 h_1$ and h_l^+ be $\mathcal{I}_l((h_{li})_{i\in|\mathcal{I}_l|})$ for $l \in \{0, 1\}$, then

$$(\mathsf{R}_C h_0 h_1)^+ := \begin{cases} \mathcal{I}_0((\mathsf{R}_C h_{0i} h_1)_{i\in|\mathcal{I}_0|}), & \text{if } C \notin \Delta(\mathcal{I}_0), \\ \mathcal{I}_1((\mathsf{R}_C h_0 h_{1i})_{i\in|\mathcal{I}_1|}), & \text{if } \neg C \notin \Delta(\mathcal{I}_1), \\ \mathsf{Cut}_{C[k]}(\mathsf{R}_C h_{0k} h_1 \mathsf{R}_C h_0 h_{10}), & \text{if } C \in \Delta(\mathcal{I}_0), \neg C \in \Delta(\mathcal{I}_1) \text{ and } \mathcal{I}_1 = \bigvee_{\neg C}^k, \\ \mathsf{Cut}_{C[k]}(\mathsf{R}_C h_{00} h_1 \mathsf{R}_C h_0 h_{1k}), & \text{if } C \in \Delta(\mathcal{I}_0), \neg C \in \Delta(\mathcal{I}_1) \text{ and } \mathcal{I}_0 = \bigvee_C^k. \end{cases}$$

Remark 3.6 For readers' convenience, let us make several remarks about Definition 3.7.

1. If $h = \mathsf{Ax}_\Delta$, then $h[i]$ is not defined since $\mathrm{tp}(h) = \mathsf{Ax}_A$ and $|\mathsf{Ax}_A| = \emptyset$ for some A.
2. Clauses (a), (b), (c), (d) are motivated by the definition of $(\ )^\infty$ (Cf. Definition 3.6).
3. In Clause (f), we may suppose that if both $C \notin \Delta(\mathcal{I}_0)$ and $\neg C \notin \Delta(\mathcal{I}_1)$ hold, then $(\mathsf{R}_C h_0 h_1)^+$ is defined according to the first case:

$$(\mathsf{R}_C h_0 h_1)^+ := \mathcal{I}_0((\mathsf{R}_C h_{0i} h_1)_{i\in|\mathcal{I}_0|}).$$

   Moreover, note that if $C \in \Delta(\mathcal{I}_0)$, $\neg C \in \Delta(\mathcal{I}_1)$ and $\mathcal{I}_1 = \bigvee_{\neg C}^k$ hold, then $\mathcal{I}_0$ must be $\bigwedge_C$ because C and $\neg C$ are principal formulas of $\mathcal{I}_0$ and $\mathcal{I}_1$, respectively.

4. Now, we can explain the reason why we inserted Rep in Definition 3.6 when $h = \mathsf{Ind}_F^{y,n} h_0$. If we do not insert Rep, then we might define $\text{tp}(h) := \text{tp}(\mathsf{R}_{F(x/i)}(e_n, h(y/i)))$. However, this is not even recursive because $\text{rk}(\mathsf{R}_{F(x/i)}(e_n, h(y/i)))$ might be bigger than that of h.
5. In Clause (e), we insert Rep for making our definition primitive recursive. The reason is very similar to the last remark.

The next theorem says that the definition of $\text{tp}(h)$ and $h[i]$ satisfies the desired properties; we can recover some useful information of a finitary derivation from the corresponding infinitary one primitive recursively.

Theorem 3.3 (Cf. Buchholz (1997), Theorem 3) *For each closed Z^*-derivation h, the following holds:*

1. $\dfrac{\ldots\Gamma(h[i])\ldots(i \in |\text{tp}(h)|)}{\Gamma(h)}\ \text{tp}(h)$ *is a correct Z^∞-inference.*
2. *If* $\text{tp}(h) = \mathsf{Cut}_C$, *then* $\text{rk}(C) < \deg(h)$.
3. $\deg(h[i]) \le \deg(h)$ *for all* $i \in |\text{tp}(h)|$.
4. $\text{o}(h[i]) \prec \text{o}(h)$ *for all* $i \in |\text{tp}(h)|$.

Proof By induction on h. To explain how the definitions of $\text{tp}(h), h[i]$ work, we focus on the following cases. Other cases are treated in a similar way.

First, let $h = \bigwedge^y_{\forall x A(x)} h_0$. In this case, $\text{tp}(h) = \bigwedge_{\forall x A(x)}$ and $h[i] = h_0(y/i)$. By Definition 3.6, $h^\infty = \bigwedge_{\forall x A(x)}(h_0(y/i)^\infty)$. Let us check the clause (1). Now, $\Gamma(h[i]) = \Gamma, \forall x A(x), A(i)$ for every $i \in \omega$ and $\Gamma(h) = \Gamma, \forall x A(x)$. So, from $\Gamma(h[n]) = \forall x A(x), A(n)$ for all $n \in \omega$, we can infer $\forall x A(x)$ by $\bigwedge_{\forall x A(x)}$. Other clauses are proved easily.

Next, suppose that $h = \mathsf{R}_C h_0 h_1$ with $C \in \Delta(\text{tp}(h_0))$ and $\neg C \in \Delta(\text{tp}(h_1))$ and $C \equiv \forall x A(x)$. From this it follows that $\text{tp}(h_0) = \bigwedge_{\forall x A(x)}$. This is the crucial case of the cut-elimination in Z^∞. By IH, we have $h_i^\infty = \text{tp}(h_i)(h_i[k]^\infty)_{k \in |\text{tp}(h_i)|}$ for $i \in \{0, 1\}$. Therefore,

$$h^\infty = \mathcal{R}_C(h_0^\infty, h_1^\infty) = \mathcal{R}_C\Big(\bigwedge\nolimits_{\forall x A(x)}(h_0[n]^\infty)_{n\in\omega}, \bigvee\nolimits^k_{\exists x \neg A(x)}(h_1[0]^\infty)\Big)$$

Then, according to Theorem 3.1, the last derivation is equivalent to

$$\mathsf{Cut}_{C[k]}\Big(\mathcal{R}_C\big(h_1[0]^\infty, \bigvee\nolimits^k_{\exists x \neg A(x)}(h_1[0]^\infty)\big), \mathcal{R}_C\big(\bigwedge\nolimits_{\forall x A(x)}(h_0[n]^\infty)_{n\in\omega}, h_1[0]^\infty\big)\Big).$$

This is identical with the third case of Definition 3.7.(2f). In particular, $\text{tp}(h) = \mathsf{Cut}_{C[k]}$ is the last inference symbol of h^∞. Now, $h[0]^\infty = (\mathsf{R}_C(h_0[k], h_1))^\infty$ must be the 0th derivation of h^∞, but this is seen from the above formula. In the same way, we see that $h[1]^\infty$ is the first subderivation of h^∞. From these the clause (1) follows.

The clause (2) holds since $\text{rk}(C[k]) < \text{rk}(C) \le \deg(h)$. It is easy to see that the clause (3) holds.

Finally, let us see why the clause (4) holds. By Definition 3.3,

$$\mathrm{o}(h) = \mathrm{o}(h_0)\sharp\mathrm{o}(h_1).$$

On the other hand,

$$\mathrm{o}(h[0]) = \mathrm{o}(\mathsf{R}_C(h_0[k], h_1)) = \mathrm{o}(h_0[k])\sharp\mathrm{o}(h_1).$$

Then, $\mathrm{o}(h[0]) \prec \mathrm{o}(h)$ holds since we have $\mathrm{o}(h_0[k]) \prec \mathrm{o}(h_0)$ by IH. In the same way, we see $\mathrm{o}(h[1]) \prec \mathrm{o}(h)$ holds.

As the last example, let $h = \mathsf{E}h_0$ with $\mathrm{tp}(h_0) = \mathsf{Cut}_C$. In this case, $\mathrm{tp}(h) = \mathsf{Rep}$ and $h[0] = \mathsf{R}_C(\mathsf{E}h_0[0], \mathsf{E}h_0[1])$. Now, $\Gamma(h[0]) = \Gamma(h)$. Note that $\Gamma(h_0[0]) = \Gamma(h), C$ and $\Gamma(h_0[1]) = \Gamma(h), \neg C$. This proves the clause (1). The clause (2) and (3) are trivial. As to the clause (4), we compute

$$\mathrm{o}(h) = \omega^{\mathrm{o}(h_0)},$$

and

$$\mathrm{o}(h[0]) = \omega^{\mathrm{o}(h_0[0])}\sharp\omega^{\mathrm{o}(h_0[1])}.$$

Then, $\mathrm{o}(h[0]) \prec \mathrm{o}(h)$ holds since $\mathrm{o}(h_0) \succ \mathrm{o}(h_0[i])$ holds for $i \in \{0, 1\}$ by IH. □

4 Contentual and Formal Aspects of Gentzen's 1936 Proof

The aim of this section is to show that the 1936 proof is both a contentual correctness proof and a formal correctness proof. First, we formulate and prove the main lemma of the 1936 proof, using *normalization trees*. Next, we show that the main lemma of the 1936 proof implies both the main lemma of the 1935 proof and the main lemma of the 1938 proof. Finally, we argue that the 1936 proof is both contentual and formal because of these consequences.[21]

For every Z^*-derivation h, we use the function $\phi(h)$ in the same sense in Buchholz (1997), i.e., $\phi(h)$ denote the Z-derivation obtained by deleting all E's in h. In addition, define

$$\mathsf{Z}_0^* := \{h \mid h \text{ is a } \mathsf{Z}^*\text{-derivation with } \deg(h) = 0\}.$$

In this section, we often consider only the derivation in Z_0^*, but this is not substantial restriction: We obtain from an arbitrary Z^*-derivation h a derivation $h' \in \mathsf{Z}_0^*$ with the same endsequent as h's by taking h' as

[21] In fact, for the claim that the 1936 proof is both contentual and formal, it is not necessary to show that the main lemma of the 1936 has these consequences. It suffices to show simply that the 1936 proof has both of the features of contentual correctness proofs and formal correctness proofs. Our strategy for the argument in this section comes from the aim of the next section.

$$h' := \mathsf{E}^m h := \underbrace{\mathsf{E} \ldots \mathsf{E}}_{m \text{ times}} h$$

with $\deg(h) = m$.

The following is our formulation of reduction steps of the 1936 proof.

Definition 4.1 (*Reduction Steps of the 1936 Proof*) For every closed Z_0^*-derivation h and every infinite sequence $\langle h_n \rangle_{n \in \mathbb{N}}$ of closed Z_0^*-derivations, the predicate $RED(h, \langle h_n \rangle_{n \in \mathbb{N}})$ holds if and only if either

- $\mathrm{tp}(h) \neq \mathsf{Ax}_A$ for any A, and for all $i \in \mathbb{N}$,
  if $i \in |\mathrm{tp}(h)|$ then $h_i = h[i]$, otherwise $\mathrm{tp}(h_i) = \mathsf{Ax}_{A'}$ for some A',
  or
- $\mathrm{tp}(h) = \mathsf{Ax}_A$ for some A and $h_i = h$ for all i.

As suggested implicitly in Arai (2002) and observed in Buchholz (2015); Akiyoshi (2010), an application of the function $h[i]$ corresponds to an application of a reduction step of the 1936 proof. More precisely, for a closed Z_0^*-derivation h, the step $\phi(h) \mapsto \phi(h[i])$ is a 1936 reduction step to $\phi(h)$.[22] This motivates the above definition: If $RED(h, \langle h_n \rangle_{n \in \mathbb{N}})$ holds, then each of $\phi(h_i)$'s with $i \in |\mathrm{tp}(h)|$ is obtained by an application of a 1936 reduction step to $\phi(h)$. The cases of $i \notin |\mathrm{tp}(h)|$ are inessential.

For example, $RED(h, \langle h_n \rangle_{n \in \mathbb{N}})$ holds if h is a closed Z_0^*-derivation $\bigwedge^y_{\forall x A(x)} h'$ described diagrammatically as

$$\dfrac{\begin{array}{c} \vdots\ h' \\ \Gamma, A(y) \end{array}}{\Gamma, \forall x A(x)} \ \bigwedge^y_{\forall x A(x)}$$

and h_i is $h'(y/i)$ described as

$$\begin{array}{c} \vdots\ h'(y/i) \\ \Gamma, A(i) \end{array}$$

for all $i \in \mathbb{N}$. Here, each $\phi(h_i)$ is obtained from $\phi(h)$ with an application of a reduction step defined in Gentzen (1936, Sect. 14.23). Another example is the following one: $RED(h, \langle h_n \rangle_{n \in \mathbb{N}})$ holds if h is a closed Z_0^*-derivation $\mathsf{E}(\mathsf{R}_{\forall x A(x)}(\bigwedge^y_{\forall x A(x)} h', \bigvee^k_{\exists x \neg A(x)} h''))$ described as

[22] To formulate reduction steps of the 1936 proof by means of finite notations for infinitary derivations, we need to insert sufficiently many E-rules into a given Z-derivation h. Moreover, we also need the function ϕ to delete the inserted E-rules, since, of course, the proof system Gentzen used to provide the 1936 proof does not include the E-rule.

$$\cfrac{\cfrac{\begin{array}{c}\vdots\; h' \\ \Gamma, \forall x A(x), A(y)\end{array}}{\Gamma, \forall x A(x)}\,\bigwedge\nolimits^{y}_{\forall x A(x)} \qquad \cfrac{\begin{array}{c}\vdots\; h'' \\ \Gamma, \exists x \neg A(x), \neg A(k)\end{array}}{\Gamma, \exists x \neg A(x)}\,\bigvee\nolimits^{k}_{\exists x \neg A(x)}}{\cfrac{\Gamma}{\Gamma}\,\mathsf{E}}\,\mathsf{R}_{\forall x A(x)}$$

and h_0 is

$$\mathsf{R}_{A(k)}\{\mathsf{E}(\mathsf{R}_{\forall x A(x)}((\bigwedge\nolimits^{y}_{\forall x A(x)} h')[k], \bigvee\nolimits^{k}_{\exists x \neg A(x)} h'')), \mathsf{E}(\mathsf{R}_{\forall x A(x)}(\bigwedge\nolimits^{y}_{\forall x A(x)} h', (\bigvee\nolimits^{k}_{\exists x \neg A(x)} h'')[0]))\}$$

described as

$$\cfrac{\cfrac{\cfrac{\cfrac{\begin{array}{c}\vdots\; h'(y/k) \\ \Gamma, \forall x A(x), A(k)\end{array} \quad \begin{array}{c}\vdots \\ \Gamma, \exists x \neg A(x)\end{array}}{\Gamma, A(k)}\,\mathsf{R}_{\forall x A(x)}}{\Gamma, A(k)}\,\mathsf{E} \qquad \cfrac{\cfrac{\begin{array}{c}\vdots \\ \Gamma, \forall x A(x)\end{array} \quad \begin{array}{c}\vdots\; h'' \\ \Gamma, \exists x \neg A(x), \neg A(k)\end{array}}{\Gamma, \neg A(k)}\,\mathsf{R}_{\forall x A(x)}}{\Gamma, \neg A(k)}\,\mathsf{E}}{\Gamma}\,\mathsf{R}_{A(k)}$$

and $h_i = \mathsf{Ax}_{\{0=0\}}$ for all $i \in \mathbb{N} \setminus \{0\}$. In this case, $\phi(h_0)$ is obtained from $\phi(h)$ with an application of the reduction step defined in Gentzen (1936, Sect. 14.25).[23]

Let $\mathbb{N}^{<\omega}$ be the set of all finite sequences of natural numbers, and $\vec{u}, \vec{v}, \vec{w}$ be variables for elements of $\mathbb{N}^{<\omega}$. As is known, a primitive recursive coding of elements of $\mathbb{N}^{<\omega}$ as natural numbers can be given, so we treat functions of the domain $(\mathbb{N}^{<\omega})^n$ with $n > 0$ (resp. of the range $(\mathbb{N}^{<\omega})^n$) as functions of the domain $\mathbb{N}^n$ (resp. of the range $\mathbb{N}^n$). Let $\vec{u} * \vec{v}$ be a primitive recursive concatenation function from $(\mathbb{N}^{<\omega})^2$ to $\mathbb{N}^{<\omega}$. Infinite sequences of natural numbers are denoted by α. For an infinite sequence α of natural numbers, the initial segment $\bar{\alpha}(n)$ of length n is defined as

$$\bar{\alpha}(n) := \langle \alpha(0), \ldots, \alpha(n-1) \rangle.$$

We denote by T a function from $\mathbb{N}^{<\omega}$ to the set of all closed Z_0^*-derivations, and abbreviate the value $T(\vec{u})$ of T for $\vec{u}$ as $T_{\vec{u}}$.

Definition 4.2 (*Local Correctness and Well-Foundedness on T*) For every T,

- T is *locally correct* if and only if for every $\vec{u} \in \mathbb{N}^{<\omega}$, $RED(T_{\vec{u}}, \langle T_{\vec{u}*\langle n \rangle} \rangle_{n \in \mathbb{N}})$ holds,
- T is *well-founded* if and only if
  for every infinite sequence α of natural numbers, there is a natural number n such that for some A, $\mathrm{tp}(T_{\bar{\alpha}(n)}) = \mathsf{Ax}_A$ holds.

[23] Strictly speaking, the above step is a special case of the reduction step that Gentzen formulated in Gentzen (1936, Sect. 14.25), since he used not the cut-rule such as R_C but the "chain-rule (Kettenschluss)," one of whose instances is the cut-rule. Recently, Buchholz analyzed the chain-rule by means of finite notations for infinitary derivations (Buchholz 2015).

Roughly speaking, when T is locally correct, T is a tree growing up along reduction steps of the 1936 proof. If T is well-founded, then every branch of T always has a node $T_{\vec{u}}$ such that $\text{tp}(T_{\vec{u}}) = \mathsf{Ax}_A$ holds for some A. Informally, this means that every branch of a well-founded T has a leaf node. Note that if a locally correct T has a node $T_{\vec{u}}$ such that $\text{tp}(T_{\vec{u}}) = \mathsf{Ax}_A$ for some A, then $\text{tp}(T_{\vec{u}*\vec{v}}) = \mathsf{Ax}_A$ for every $\vec{v}$. In other words, if a branch of a locally correct T has a leaf node, then each of the nodes above this leaf node is also a leaf node.

Now we define the main notion of this section.

Definition 4.3 (*Normalization Trees*) For every T and every closed Z_0^*-derivation h, T is a *normalization tree* of h if and only if T is locally correct and $T_{\langle\rangle} = h$.

A typical example T of a normalization tree can be described as follows, if we omit the irrelevant nodes of T, i.e., the nodes $T_{\vec{u}*\langle i\rangle}$ such that $i \notin |\text{tp}(T_{\vec{u}})|$.

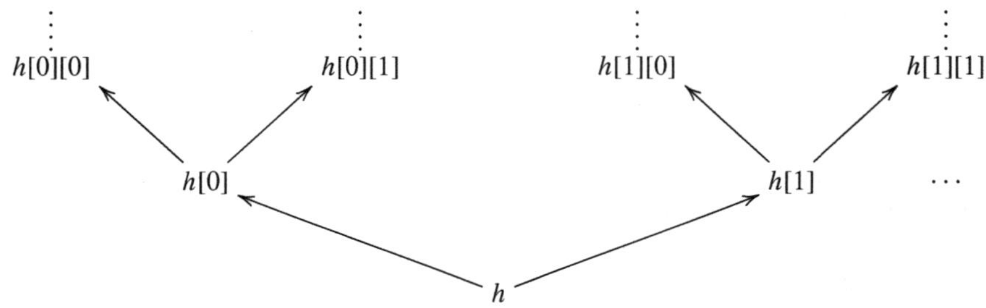

Lemma 4.1 *For every T, every $\vec{u} \in \mathbb{N}^{<\omega}$ and every formula C, $\text{tp}(T_{\vec{u}}) \neq \mathsf{Cut}_C$ holds.*

Proof Suppose that $\text{tp}(T_{\vec{u}}) = \mathsf{Cut}_C$. By Theorem 3.3, $\text{rk}(C) < \deg(T_{\vec{u}})$ holds, but this is impossible because $\deg(T_{\vec{u}}) = 0$ by definition. □

The main lemma of the 1936 proof says that every reduction sequence formed by reduction steps of the 1936 proof terminates with a normal derivation. In our setting, the lemma says that for every closed Z_0^*-derivation h, there is a well-founded normalization tree of h. Note that the truth of the endsequent of a closed Z_0^*-derivation h with $\text{tp}(h) = \mathsf{Ax}_A$ is verified in a primitive recursive way because $\Gamma(h) \cap \mathsf{TRUE}_0 \neq \emptyset$ by Theorem 3.3.(1), which is proved by structural induction on h. For the purpose of Sect. 5.3, we also see that every well-founded normalization tree defined in the proof below is a $< \varepsilon_0$-recursive function. For the definition of $< \varepsilon_0$-recursive functionals of level ≤ 2, see Schwichtenberg (1977, Sect. 4.1).

Proposition 4.1 (Main Lemma of the 1936 Proof) *For every closed Z_0^*-derivation h, there is a well-founded normalization tree T of h such that T is a $< \varepsilon_0$-recursive function.*

Proof If $\text{tp}(h) = \mathsf{Ax}_A$ for some A, then define $T_{\vec{u}} := h$ for every $\vec{u}$. It is obvious that T is both a well-founded normalization tree of h and a primitive recursive function.

Consider the case that $\mathrm{tp}(h) \neq \mathsf{Ax}_A$ for any A. We use transfinite induction up to ε_0. Assume that a well-founded normalization tree of h' is defined for every closed Z_0^*-derivation h' with $\mathrm{o}(h') \prec \mathrm{o}(h) \prec \varepsilon_0$. By Theorem 3.3, there is a well-founded normalization tree $T^{h[i]}$ of $h[i]$ for every $i \in |\mathrm{tp}(h)|$. Set the tree T^{Ax} as

$$T^{\mathsf{Ax}}(\vec{u}) := \mathsf{Ax}_{\{0=0\}} \text{ for every } \vec{u}.$$

Define T as follows:

$$T(\vec{u}) := \begin{cases} h, & \text{if } \vec{u} = \langle\rangle, \\ T^{h[i]}(\vec{v}), & \text{if } \vec{u} = \langle i \rangle * \vec{v} \text{ and } i \in |\mathrm{tp}(h)|, \\ T^{\mathsf{Ax}}(\vec{v}), & \text{if } \vec{u} = \langle i \rangle * \vec{v} \text{ and } i \notin |\mathrm{tp}(h)|. \end{cases}$$

Then, T is a well-founded normalization tree of h. Moreover, it can be shown that T is a $< \varepsilon_0$-recursive function, because $\mathrm{tp}(h)$ and $h[i]$ are primitive recursive. □

Next, we show that Proposition 4.1 implies both the main lemma of the 1935 proof and the main lemma of the 1938 proof. Let us start with our definition for reduction steps of the 1935 proof.

Definition 4.4 (*Reduction Steps of the 1935 Proof*) For every closed sequent Γ and every infinite sequence $\langle \Gamma_n \rangle_{n \in \mathbb{N}}$ of closed sequents, the predicate $RED^{sq}(\Gamma, \langle \Gamma_n \rangle_{n \in \mathbb{N}})$ holds if and only if there is a formula $C \in \Gamma$ such that

1. $C \in \mathsf{TRUE}_0$ and $\Gamma_n = \Gamma$ for every $n \in \mathbb{N}$, or
2. C is of the form $\forall x A(x)$ and $\Gamma_n \subseteq \Gamma \cup \{A(n)\}$ for every $n \in \mathbb{N}$, or
3. C is of the form $A_0 \wedge A_1$ and $\Gamma_i \subseteq \Gamma \cup \{A_i\}$ for every $i \in \{0, 1\}$, or
4. C is of the form $\exists x A(x)$ and $\Gamma_0 \subseteq \Gamma \cup \{A(k)\}$ for some $k \in \mathbb{N}$, or
5. C is of the form $A_0 \vee A_1$ and $\Gamma_0 \subseteq \Gamma \cup \{A_i\}$ for some $i \in \{0, 1\}$.

As Definition 4.1, if $RED^{sq}(\Gamma, \langle \Gamma_n \rangle_{n \in \mathbb{N}})$ holds, then each of Γ_i's is obtained from Γ with an application of a reduction step of the 1935 proof. A typical example is as follows: $RED^{sq}(\Gamma, \langle \Gamma_n \rangle_{n \in \mathbb{N}})$ holds whenever $\Gamma = \{\forall x(x = x)\}$ and $\Gamma_n = \{\forall x(x = x), n = n\}$ hold. Here each of Γ_i's is obtained from Γ with an application of the reduction step being defined in Gentzen (1974, Sect. 13.21). This example can be described as follows.

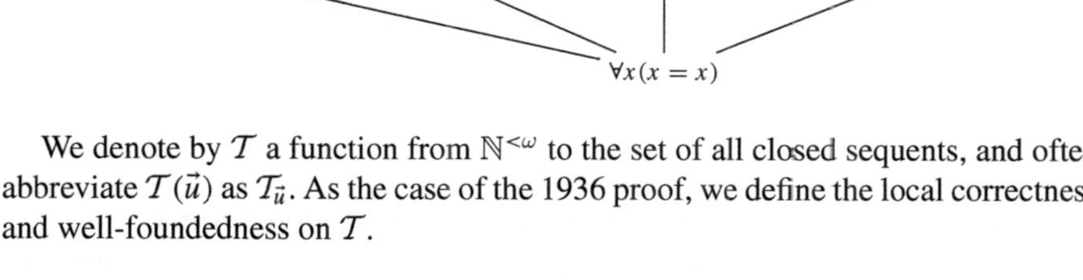

We denote by $\mathcal{T}$ a function from $\mathbb{N}^{<\omega}$ to the set of all closed sequents, and often abbreviate $\mathcal{T}(\vec{u})$ as $\mathcal{T}_{\vec{u}}$. As the case of the 1936 proof, we define the local correctness and well-foundedness on $\mathcal{T}$.

Definition 4.5 (*Local Correctness and Well-Foundedness on $\mathcal{T}$*) For every $\mathcal{T}$,

- $\mathcal{T}$ is *locally correct* if and only if for every $\vec{u} \in \mathbb{N}^{<\omega}$, $RED^{sq}(\mathcal{T}_{\vec{u}}, \langle \mathcal{T}_{\vec{u}*\langle n \rangle} \rangle_{n \in \mathbb{N}})$ holds.
- $\mathcal{T}$ is *well-founded* if and only if for every infinite sequence α of natural numbers, there is a natural number n such that $\mathcal{T}_{\bar{\alpha}(n)} \cap \mathsf{TRUE}_0 \neq \emptyset$.

The main notion of the 1935 proof is the notion of *reduction procedures*, as we have seen in Sect. 2. The following is its definition in our setting.

Definition 4.6 (*Reduction Procedures*) We say $\mathcal{T}$ is a *reduction procedure* for Γ if and only if

- $\mathcal{T}_{\langle\rangle} = \Gamma$,
- $\mathcal{T}$ is locally correct and well-founded.

A reduction procedure is a tree growing up along reduction steps of the 1935 proof, every branch of which always has a node $\mathcal{T}_{\vec{u}}$ that contains at least one element of TRUE_0.

To derive the main lemma of the 1935 proof from Proposition 4.1, we use an induction principle of intuitionistic mathematics called *monotone bar induction*.[24]

Monotone Bar Induction. Let B and P be predicates on $\mathbb{N}^{<\omega}$. If the following four conditions hold:

1. for every infinite sequence α of natural numbers, there is a natual number n such that $B(\bar{\alpha}(n))$ holds,
2. for every $\vec{u}$ and $\vec{v}$, if $B(\vec{u})$ holds then $B(\vec{u} * \vec{v})$ holds,
3. for every $\vec{u}$, if $B(\vec{u})$ holds then $P(\vec{u})$ holds,
4. for every $\vec{u}$, if $P(\vec{u} * \langle n \rangle)$ holds for all n then $P(\vec{u})$ holds,

then $P(\langle\rangle)$ holds.

The idea of our proof for the main lemma of the 1935 proof is as follows.[25] Given a closed Z_0^*-derivation h, a reduction procedure for $\Gamma(h)$ is obtained from a normalization tree T of h by extracting the endsequent of $T_{\vec{u}}$ from $T_{\vec{u}}$ for every $\vec{u}$.

Lemma 4.2 *For every closed Z_0^*-derivation h, if there is a well-founded normalization tree of h, then there is also a reduction procedure for $\Gamma(h)$.*

Proof By monotone bar induction (MBI). Assume that there is a well-founded normalization tree T of h, and define the predicates B and P as follows:

$B(\vec{u})$ if and only if $\mathrm{tp}(T_{\vec{u}}) = \mathsf{Ax}_A$ for some A,
$P(\vec{u})$ if and only if there is a reduction procedure for $\Gamma(T_{\vec{u}})$.

---

[24] As to monotone bar induction, see Troelstra and van Dalen (1988, Chap. 4, Sect. 8).

[25] One can read off this idea in Gentzen's explanation for reduction steps of the 1936 steps. See Gentzen (1936, Sect. 14.2).

Then, Premise 1 of MBI holds by the well-foundedness of T, and Premise 2 of MBI holds by the local correctness of T.

Next, we show that Premise 3 of MBI holds. If $B(\vec{u})$ holds, i.e., $\mathrm{tp}(T_{\vec{u}}) = \mathsf{Ax}_A$ for some A, then

$$\frac{}{\Gamma(T_{\vec{u}})}\ \mathsf{Ax}_A$$

is a correct Z^∞-inference by Theorem 3.3, so $\Gamma(T_{\vec{u}}) \cap \mathsf{TRUE}_0 \neq \emptyset$. Thus, the tree $\mathcal{T}$ being defined as $\mathcal{T}(\vec{v}) := \Gamma(T_{\vec{u}})$ for every $\vec{v}$ is a reduction procedure for $\Gamma(T_{\vec{u}})$.

Finally, we show that Premise 4 of MBI holds. Assume that $P(\vec{u} * \langle n \rangle)$ holds for every n, i.e., there is a reduction procedure $\mathcal{T}^n$ for each $\Gamma(T_{\vec{u}*\langle n \rangle})$.

Consider the case that $\mathrm{tp}(T_{\vec{u}}) = \mathsf{Rep}$. Then, by Theorem 3.3,

$$\frac{\Gamma(T_{\vec{u}*\langle 0 \rangle})}{\Gamma(T_{\vec{u}})}\ \mathsf{Rep}$$

is a correct Z^∞-inference, so $\mathcal{T}^0$ is also a reduction procedure for $\Gamma(T_{\vec{u}})$.

Consider in turn the case that $\mathrm{tp}(T_{\vec{u}}) = \bigwedge_{A_0 \wedge A_1}$. Again, by Theorem 3.3,

$$\frac{\Gamma(T_{\vec{u}*\langle 0 \rangle}) \quad \Gamma(T_{\vec{u}*\langle 1 \rangle})}{\Gamma(T_{\vec{u}})}\ \bigwedge\nolimits_{A_0 \wedge A_1}$$

is a correct Z^∞-inference. Set the tree $\mathcal{T}^{\mathsf{Ax}}$ as

$$\mathcal{T}^{\mathsf{Ax}}(\vec{v}) := \{0 = 0\} \text{ for every } \vec{v}.$$

Then, define $\mathcal{T}$ as follows:

$$\mathcal{T}(\vec{v}) := \begin{cases} \Gamma(T_{\vec{u}}), & \text{if } \vec{v} = \langle\rangle, \\ \mathcal{T}^i(\vec{w}), & \text{if } \vec{v} = \langle i \rangle * \vec{w} \text{ and } i \in \{0, 1\}, \\ \mathcal{T}^{\mathsf{Ax}}(\vec{w}), & \text{if } \vec{v} = \langle i \rangle * \vec{w} \text{ and } i \notin \{0, 1\}. \end{cases}$$

It is obvious that $\mathcal{T}$ is a reduction procedure for $\Gamma(T_{\vec{u}})$. The remaining cases are similar to this case. Note that $\mathrm{tp}(T_{\vec{u}}) \neq \mathsf{Cut}_C$ for any C by Lemma 4.1.

Therefore, $P(\langle\rangle)$ holds by MBI, so there is a reduction procedure for $\Gamma(T_{\langle\rangle}) = \Gamma(h)$. □

Proposition 4.2 (Main Lemma of the 1935 Proof) *For every closed Z_0^*-derivation h, there is a reduction procedure for $\Gamma(h)$.*

Proof By Proposition 4.1 and Lemma 4.2. □

The above proof of Proposition 4.2 shows that the 1936 proof is a contentual correctness proof. Let h be an arbitrary Z-derivation. For the sake of simplicity, we assume that h is closed. A Z_0^*-derivation h' with the same endsequent as h's is obtained from h by inserting sufficiently many E-rules into the bottom of h. By Proposition 4.1, namely, the main lemma of the 1936 proof, there is a well-founded normalization tree T of h'. Then, by Lemma 4.2, we can extract a reduction procedure $\mathcal{T}$ for $\Gamma(h') = \Gamma(h)$ from T with a proof for the termination of $\mathcal{T}$, i.e., a proof for its well-foundedness. Note that we avoided the use of the principle of the excluded middle for non-decidable predicates in our arguments.[26] In Sect. 2, we have seen the following definition of the *statability* of a reduction procedure: For every reduction procedure $\mathcal{T}'$ for Γ, $\mathcal{T}'$ is statable for Γ if and only if both $\mathcal{T}'$ and a proof for its well-foundedness are obtained. Therefore, a reduction procedure is statable for $\Gamma(h)$. In other words, $\Gamma(h)$ is correct in the sense of (GI). In this way, the main lemma of the 1936 proof verifies that every Z-derivation has the correct endsequent, so this proof is contentual.

For our claim that the main lemma of the 1936 proof implies the main lemma of the 1938 proof, the following observation by Buchholz (1997) is crucial: Let h be a closed Z_0^*-derivation for a numeric equation $n = m$, then the step $\phi(h) \mapsto \phi(h[0])$ is a main reduction step of the 1938 proof.[27] For example, the following step, which is called a "operational reduction (Verknüpfungs-Reduktion)," is one of the main reduction steps of the 1938 proof.[28]

$$
\begin{array}{c}
\dfrac{
\begin{array}{c} \vdots\; h' \\ \dfrac{\Delta_0, \forall x A(x), A(y)}{\Delta_0, \forall x A(x)}\bigwedge^{y}_{\forall x A(x)} \\ \vdots \\ \Gamma', \forall x A(x) \end{array}
\qquad
\begin{array}{c} \vdots\; h'' \\ \dfrac{\Delta_1, \exists x \neg A(x), \neg A(k)}{\Delta_1, \exists x \neg A(x)}\bigvee^{k}_{\exists x \neg A(x)} \\ \vdots \\ \Gamma', \exists x \neg A(x) \end{array}
}{\Gamma'}\mathsf{R}_{\forall x A(x)} \\
\vdots \\
\Gamma \\
\vdots \\
n = m
\end{array}
\qquad \triangleright
$$

[26]In the future works, we will investigate the exact strength of Gentzen's finitist standpoint and examine whether our argument can be formalized in this standpoint.

[27]Here, we slightly changed Buchholz's actual observation. The actual observation is that for a supposed Z_0^*-derivation h of the *empty sequent* $\emptyset$, the step $\phi(h) \mapsto \phi(h[0])$ is a main reduction step of the 1938 proof.

[28]For the operational reductions, see Gentzen (1938, Sect. 3.5).

$$
\dfrac{\dfrac{\dfrac{\overset{\vdots\; h'(y/k)}{\Delta_0, A(k), \forall x A(x)} \quad \dfrac{\overset{\vdots\; h''}{\Delta_1, \exists x \neg A(x), \neg A(k)}}{\Delta_1, \exists x \neg A(x)}\, \bigvee^{k}_{\exists x \neg A(x)}}{\overset{\vdots}{\Gamma', A(k), \forall x A(x)} \quad \overset{\vdots}{\Gamma', \exists x \neg A(x)}}}{\overset{}{\Gamma', A(k)}}\, \mathsf{R}_{\forall x A(x)}
$$

$$
\dfrac{\dfrac{\overset{\vdots\; h'}{\Delta_0, \forall x A(x), A(y)}}{\Delta_0, \forall x A(x)}\, \bigwedge^{y}_{\forall x A(x)} \quad \overset{\vdots\; h''}{\Delta_1, \neg A(k), \exists x \neg A(x)}}{\dfrac{\overset{\vdots}{\Gamma', \forall x A(x)} \quad \overset{\vdots}{\Gamma', \neg A(k), \exists x \neg A(x)}}{\Gamma', \neg A(k)}\, \mathsf{R}_{\forall x A(x)}}
$$

$$
\dfrac{\overset{\vdots}{\Gamma, A(k)} \qquad \overset{\vdots}{\Gamma, \neg A(k)}}{\underset{\overset{\vdots}{n = m}}{\Gamma}}\, \mathsf{R}_{A(k)}
$$

Note that all 1938 reduction steps have no branching, in contrast to reduction steps of the 1935 proof and the 1936 proof. In addition, Gentzen introduced the notion of an *end-piece* (Endstück) of a derivation to formulate reduction steps of the 1938 proof.[29] For our purpose, it suffices to explain the notion of an end-piece informally: The end-piece of a Z^*-derivation h consists of the sequents that we encounter when we ascend from the endsequent of h and stop as soon as we arrive at the lower sequent of an inference symbol $\mathcal{I}$ of the form $\bigwedge_A$, $\bigwedge^y_A$ or $\bigvee^k_A$. For example, the end-pieces of the following Z^*-derivations

$$
\dfrac{\dfrac{\dfrac{\overset{\vdots}{\Gamma, A(y), \forall x A(x)}}{\Gamma, \forall x A(x)}\, \bigwedge^{y}_{\forall x A(x)} \quad \dfrac{\overset{\vdots}{\Gamma, \neg A(k), \exists x \neg A(x)}}{\Gamma, \exists x \neg A(x)}\, \bigvee^{k}_{\exists x \neg A(x)}}{\Gamma}\, \mathsf{R}_{\forall x A(x)}}{\Gamma}\, \mathsf{E}
\qquad
\dfrac{\dfrac{}{\Delta, 0 = 0}\, \mathsf{Ax}_{\{\Delta, 0=0\}} \quad \dfrac{}{\Delta, 0 \neq 0}\, \mathsf{Ax}_{\{\Delta, 0\neq 0\}}}{\Delta}\, \mathsf{R}_{0=0}
$$

are

$$
\dfrac{\dfrac{\Gamma, \forall x A(x) \quad \Gamma, \exists x \neg A(x)}{\Gamma}\, \mathsf{R}_{\forall x A(x)}}{\Gamma}\, \mathsf{E}
\qquad
\dfrac{\dfrac{}{\Delta, 0 = 0}\, \mathsf{Ax}_{\{\Delta, 0=0\}} \quad \dfrac{}{\Delta, 0 \neq 0}\, \mathsf{Ax}_{\{\Delta, 0\neq 0\}}}{\Delta}\, \mathsf{R}_{0=0}
$$

respectively.

As seen above, a step $\phi(h) \mapsto \phi(h[i])$ for an *arbitrary* closed Z_0^*-derivation h is a reduction step of the 1936 proof. Thus, one can consider reduction steps of the 1938 proof to be *special cases* of reduction steps of the 1936 proof. That is to say, each of reduction step of the 1938 proof is nothing but a 1936 reduction step to some closed Z-derivation of a *numeric equation.* On the basis of this fact, one can see that the main lemma of the 1938 proof is also a special case of the main lemma of the 1936 proof. In our setting, the former is formulated as the following proposition, which immediately follows from Proposition 4.1, namely, the main lemma of the 1936 proof.

---

[29]Cf. Gentzen (1938, Sect. 3.2). In Gentzen (1969), Szabo translated "Endstück" as "ending".

Proposition 4.3 (Main Lemma of the 1938 Proof) *For every closed* Z_0^**-derivation* h *with the endsequent of the form* $n = m$*, there is a well-founded normalization tree* T *of* h.

In the remaining part of this section, we aim to justify the claim that the main lemma of the 1938 proof can be formulated as above. To achieve this aim, we explain the relationship between normalization trees and reduction steps of the 1938 proof in detail: We explain that each well-founded normalization tree for some closed Z_0^*-derivation with a numeric conclusion corresponds to consecutive applications of 1938 reduction steps.

Let us begin with the definition for the notion of *nominal forms*, which Buchholz introduced to verify his observation we have seen before.[30]

Definition 4.7 (*Nominal Forms*) *Nominal forms* are defined as follows:

1. $*$ is a nominal form and $\mathsf{Cut}(*) := \emptyset$,
2. if $\mathfrak{a}$ is a nominal form and h is a Z^*-derivation, then $\mathsf{E}^m\mathsf{R}_C\mathfrak{a}h$ and $\mathsf{E}^m\mathsf{R}_C h\mathfrak{a}$ are nominal forms, and

$$\mathsf{Cut}(\mathsf{E}^m\mathsf{R}_C\mathfrak{a}h) := \mathsf{Cut}(\mathfrak{a}) \cup \{C\}, \quad \mathsf{Cut}(\mathsf{E}^m\mathsf{R}_C h\mathfrak{a}) := \mathsf{Cut}(\mathfrak{a}) \cup \{\neg C\}.$$

We denote a nominal form by $\mathfrak{a}$, $\mathfrak{b}$, $\mathfrak{c}$. When q is a nominal form or a Z^*-derivation, $\mathfrak{a}\{q\}$ is the result of substituting q for $*$ in $\mathfrak{a}$. For example, the nominal form

$$\mathfrak{a} = \mathsf{E}^m(\mathsf{R}_{C_0}(\mathsf{E}^n(\mathsf{R}_{C_1}(*, h_1)), h_0))$$

is of the following figure:

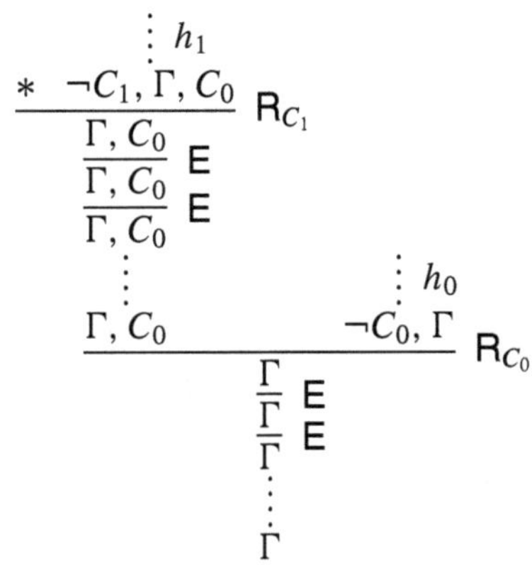

[30] Cf. Buchholz (1997). Buchholz defined also the function $\mathsf{hgt}^*(\mathfrak{a})$ of a nominal form $\mathfrak{a}$, to describe the ordinal assignment in the 1938 proof by means of finite notations for infinitary derivations. Here, we do not need to describe it, so we omit the definition of this function.

It is easily observed that for every nominal form $\mathfrak{a}$ and every Z^*-derivation h, the endsequent of h is included in the end-piece of $\mathfrak{a}\{h\}$. This observation is crucial for our argument below.

For readers' convenience, we carry out the details of the proofs of Lemma 4.3, Propositions 4.4 and 4.5. We suggest a reader being familiar with Buchholz (1997) or not being interested in the details to skip them.

Lemma 4.3 (cf. Buchholz (1997), Lemma 3) *Let h be a closed Z^*-derivation.*

1. *If* $\mathrm{tp}(h) = \mathsf{Rep}$, *then there are* $\mathfrak{a}, h_0, h_1$ *such that* $h = \mathfrak{a}\{h_0\}$, $h[0] = \mathfrak{a}\{h_0[0]\}$ *and either* $h_0 = \mathsf{E}^m \mathsf{Ind}_F^{y,t} h_1$ *or*
   $h_0 = \mathsf{E}^{m+1} h_1$ *and* $\mathrm{tp}(h_1) = \mathsf{Cut}_B$ *for some* B.
2. *If* $\mathrm{tp}(h) = \mathsf{Cut}_B$, *then there are* $\mathfrak{a}, C, h_0, h_1$ *such that* $h = \mathfrak{a}\{\mathsf{R}_C h_0 h_1\}$, $h[i] = \mathfrak{a}\{(\mathsf{R}_C h_0 h_1)[i]\}$ *and either*
   (a) $\mathrm{tp}(h_0) = \bigwedge_C$ & $\mathrm{tp}(h_1) = \bigvee^k_{\neg C}$ & $B = C[k]$ *or*
   (b) $\mathrm{tp}(h_0) = \bigvee^k_C$ & $\mathrm{tp}(h_1) = \bigwedge_{\neg C}$ & $B = C[k]$.
3. *If* $\mathrm{tp}(h) = \bigwedge_{\forall x A(x)}$, *then there are* $\mathfrak{a}, h_0$ *such that* $\forall x A(x) \notin \mathsf{Cut}(\mathfrak{a})$, $h = \mathfrak{a}\{\bigwedge^y_{\forall x A(x)} h_0\}$ *and* $h[i] = \mathfrak{a}\{h_0(y/i)\}$.
4. *If* $\mathrm{tp}(h) = \bigwedge_{A_0 \wedge A_1}$, *then there are* $\mathfrak{a}, h_0, h_1$ *such that* $A_0 \wedge A_1 \notin \mathsf{Cut}(\mathfrak{a})$, $h = \mathfrak{a}\{\bigwedge_C h_0 h_1\}$ *and* $h[i] = \mathfrak{a}\{h_i\}$.
5. *If* $\mathrm{tp}(h) = \bigvee^k_C$, *then there are* $\mathfrak{a}, h_0$ *such that* $C \notin \mathsf{Cut}(\mathfrak{a})$, $h = \mathfrak{a}\{\bigvee^k_C h_0\}$ *and* $h[0] = \mathfrak{a}\{h_0\}$.
6. *If* $\mathrm{tp}(h) = \mathsf{Ax}_A$, *then there is a nominal form* $\mathfrak{a}$ *such that* $A \notin \mathsf{Cut}(\mathfrak{a})$ *and* $h = \mathfrak{a}\{\mathsf{Ax}_\Delta\}$ *with* $A \in \Delta$.

Proof By induction on the length of h.

(1) Since $\mathrm{tp}(h) = \mathsf{Rep}$, h is of either the form $\mathsf{E}^m \mathsf{Ind}_F^{y,t} h'$ or $\mathsf{E}^n h''$ with $\mathrm{last}(h'') \neq \mathsf{E}, \mathsf{Ind}$. If $h = \mathsf{E}^m \mathsf{Ind}_F^{y,t} h'$, then take $\mathfrak{a}$ as $*$, h as h_0 and h' as h_1.

Consider the case that $h = \mathsf{E}^n h''$ with $\mathrm{last}(h'') \neq \mathsf{E}, \mathsf{Ind}$. If $\mathrm{tp}(h'') = \mathsf{Cut}_B$ for some B, then $n > 0$ holds, so take $*$ as $\mathfrak{a}$, h as h_0 and h'' as h_1. If $\mathrm{tp}(h'') \neq \mathsf{Cut}_B$ for any B, then $h'' = \mathsf{R}_C h'_0 h'_1$ and $\mathrm{tp}(h'') = \mathrm{tp}(h'_i) = \mathsf{Rep}$ for some $i \in \{0, 1\}$ by the definition of tp. We consider only the case that $i = 0$, since the other case is similar. By IH, there are $\mathfrak{b}, h''_0, h''_1$ such that $h'_0 = \mathfrak{b}\{h''_0\}$, $h'_0[0] = \mathfrak{b}\{h''_0[0]\}$ and either $h''_0 = \mathsf{E}^m \mathsf{Ind}_F^{y,t} h''_1$ or ($h''_0 = \mathsf{E}^{m+1} h''_1$ & $\mathrm{tp}(h''_1) = \mathsf{Cut}_B$ for some B). Take $\mathsf{E}^n \mathsf{R}_C \mathfrak{b} h'_1$ as $\mathfrak{a}$, h''_0 as h_0 and h''_1 as h_1. Then,

$$h = \mathsf{E}^n \mathsf{R}_C(h'_0, h'_1) = \mathsf{E}^n \mathsf{R}_C(\mathfrak{b}\{h''_0\}, h'_1) = \mathfrak{a}\{h''_0\}$$

and

$$h[0] = (\mathsf{E}^n \mathsf{R}_C(h'_0, h'_1))[0] = \mathsf{E}^n \mathsf{R}_C(h'_0[0], h'_1) = \mathsf{E}^n \mathsf{R}_C(\mathfrak{b}\{h''_0[0]\}, h'_1) = \mathfrak{a}\{h''_0[0]\}.$$

(2) Assume that $\mathrm{tp}(h) = \mathsf{Cut}_B$. Then, $h = \mathsf{R}_C h'_0 h'_1$ and either

$$\begin{array}{l} \mathrm{tp}(h'_0) = \bigwedge_C,\ \mathrm{tp}(h'_1) = \bigvee^k_{\neg C},\ B = C[k], \text{ or} \\ \mathrm{tp}(h'_0) = \bigvee^k_C,\ \mathrm{tp}(h'_1) = \bigwedge_{\neg C},\ B = C[k], \text{ or} \\ \mathrm{tp}(h'_0) = \mathsf{Cut}_B,\ h[i] = \mathsf{R}_C(h'_0[i], h'_1), \text{ or} \\ \mathrm{tp}(h'_1) = \mathsf{Cut}_B,\ h[i] = \mathsf{R}_C(h'_0, h'_1[i]). \end{array}$$

In the first two cases, take $*$ as $\mathfrak{a}$, h'_0 as h_0 and h'_1 as h_1. Below we consider the third case and omit the fourth case, since the latter is similar to the former.

Assume that $\mathrm{tp}(h'_0) = \mathsf{Cut}_B$, $h[i] = \mathsf{R}_C(h'_0[i], h'_1)$. Then, by IH, there are $\mathfrak{b}$, D, h''_0, h''_1 such that $h'_0 = \mathfrak{b}\{\mathsf{R}_D h''_0 h''_1\}$, $h'_0[i] = \mathfrak{b}\{(\mathsf{R}_D h''_0 h''_1)[i]\}$ and either

(a) $\mathrm{tp}(h''_0) = \bigwedge_D$ & $\mathrm{tp}(h''_1) = \bigvee^k_{\neg D}$ & $B = D[k]$ or
(b) $\mathrm{tp}(h''_0) = \bigvee^k_D$ & $\mathrm{tp}(h''_1) = \bigwedge_{\neg D}$ & $B = D[k]$.

Take $\mathsf{R}_C \mathfrak{b} h_1$ as $\mathfrak{a}$, h''_0 as h_0 and h''_1 as h_1. Then,

$$h = \mathsf{R}_C h'_0 h'_1 = \mathsf{R}_C(\mathfrak{b}\{\mathsf{R}_D h''_0 h''_1\}, h'_1) = \mathfrak{a}\{\mathsf{R}_D h''_0 h''_1\}$$

and

$$h[i] = \mathsf{R}_C(h'_0[i], h'_1) = \mathsf{R}_C(\mathfrak{b}\{(\mathsf{R}_D h''_0 h''_1)[i]\}, h'_1) = \mathfrak{a}\{(\mathsf{R}_D h''_0 h''_1)[i]\}.$$

(3) By the definition of $\mathrm{tp}(h)$, either $h = \bigwedge^y_{\forall x A(x)} h'$ or $h = \mathsf{E}^n \mathsf{R}_D h'_0 h'_1$ with $\mathrm{tp}(h'_i) = \bigwedge_{\forall x A(x)}$ for some $i \in \{0, 1\}$. In the former case, take $*$ as $\mathfrak{a}$ and h'_0 as h_0.

In the latter case, we may assume without the loss of generality that $i = 0$, then $D \not\equiv \forall x A(x)$ by the definition of $\mathrm{tp}(h'_i)$. By IH, there are $\mathfrak{b}$, h''_0 such that $\forall x A(x) \notin \mathsf{Cut}(\mathfrak{b})$, $h'_0 = \mathfrak{b}\{\bigwedge^y_{\forall x A(x)} h''_0\}$ and $h'_0[i] = \mathfrak{b}\{h''_0(y/i)\}$. Take $\mathsf{E}^n \mathsf{R}_D \mathfrak{b} h'_1$ as $\mathfrak{a}$ and h''_0 as h_0, then $\forall x A(x) \notin \mathsf{Cut}(\mathfrak{a})$ and $h = \mathsf{E}^n \mathsf{R}_D(h'_0, h'_1) = \mathsf{E}^n \mathsf{R}_D(\mathfrak{b}\{\bigwedge^y_{\forall x A(x)} h''_0\}, h'_1) = \mathfrak{a}\{\bigwedge^y_{\forall x A(x)} h''_0\}$. Moreover,

$$h[i] = \mathsf{E}^n \mathsf{R}_D(h'_0[i], h'_1) = \mathsf{E}^n \mathsf{R}_D(\mathfrak{b}\{h''_0(y/i)\}, h'_1) = \mathfrak{a}\{h''_0(y/i)\}.$$

In the cases of (4), (5) and (6), we can prove the assertion in the similar way to the case of (3). □

We set

$$\mathbf{0}^n := \begin{cases} \underbrace{\langle 0, 0, \ldots, 0 \rangle}_{n\ times}, & \text{if } n > 0, \\ \langle\rangle, & \text{else.} \end{cases}$$

Note that $T_{\mathbf{0}^k}$ denotes the node of T labeled by $\mathbf{0}^k$.

Proposition 4.4 (cf. Buchholz (1997), Theorem 5) *Let T be a normalization tree of some closed $h \in \mathsf{Z}_0^*$ with the endsequent of the form $n = m$. For every $k \in \mathbb{N}$, it holds that* $\mathrm{tp}(T_{\mathbf{0}^k}) = \mathsf{Ax}_{n=m}$ *or* Rep, *and*

1. *if* $\mathrm{tp}(T_{\mathbf{0}^k}) = \mathsf{Ax}_{n=m}$, *then there is a nominal form* $\mathfrak{a}$ *such that* $n = m \notin \mathsf{Cut}(\mathfrak{a})$ *and* $T_{\mathbf{0}^k} = \mathfrak{a}\{\mathsf{Ax}_\Delta\}$ *with* $n = m \in \Delta$, *and*
2. *if* $\mathrm{tp}(T_{\mathbf{0}^k}) = \mathsf{Rep}$, *then one of the following statements holds for some nominal forms* $\mathfrak{a}, \mathfrak{b}, \mathfrak{c}_0, \mathfrak{c}_1$. *(The symmetric cases are suppressed.)*
   - *(a)* $T_{\mathbf{0}^k} = \mathfrak{a}\{\mathsf{E}^m \mathsf{Ind}_F^{y,t} h_0\}$ *and* $T_{\mathbf{0}^{k+1}} = \mathfrak{a}\{\mathsf{E}^m((\mathsf{Ind}_F^{y,t} h_0)[0])\}$.
   - *(b)* $T_{\mathbf{0}^k} = \mathfrak{a}\{\mathsf{E}^{m+1}\mathfrak{b}\{\mathsf{R}_C(\mathfrak{c}_0\{h'_0\}, \mathfrak{c}_1\{h'_1\})\}\}$ *and* $T_{\mathbf{0}^{k+1}} = \mathfrak{a}\{\mathsf{E}^m \mathsf{R}_{C[n]}(\mathsf{E}\mathfrak{b}\{\mathsf{R}_C(\mathfrak{c}_0\{h_0(y/n)\}, \mathfrak{c}_1\{h'_1\})\}, \mathsf{E}\mathfrak{b}\{\mathsf{R}_C(\mathfrak{c}_0\{h'_0\}, \mathfrak{c}_1\{h_1\})\})\}$, *where* $C \equiv \forall x A(x)$, $C \notin \mathsf{Cut}(\mathfrak{c}_0)$, $\neg C \notin \mathsf{Cut}(\mathfrak{c}_1)$, $h'_0 = \bigwedge_C^y h_0$ *and* $h'_1 = \bigvee_{\neg C}^n h_1$.
   - *(c)* $T_{\mathbf{0}^k} = \mathfrak{a}\{\mathsf{E}^{m+1}\mathfrak{b}\{\mathsf{R}_C(\mathfrak{c}_0\{h_0\}, \mathfrak{c}_1\{h_1\})\}\}$ *and* $T_{\mathbf{0}^{k+1}} = \mathfrak{a}\{\mathsf{E}^m \mathsf{R}_{C[i]}(\mathsf{E}\mathfrak{b}\{\mathsf{R}_C(\mathfrak{c}_0\{h_{0i}\}, \mathfrak{c}_1\{h_1\})\}, \mathsf{E}\mathfrak{b}\{\mathsf{R}_C(\mathfrak{c}_C\{h_0\}, \mathfrak{c}_1\{h_{10}\})\})\}$, *where* $C \equiv A_0 \wedge A_1$, $C \notin \mathsf{Cut}(\mathfrak{c}_0)$, $\neg C \notin \mathsf{Cut}(\mathfrak{c}_1)$, $h_0 = \bigwedge_C h_{00}h_{01}$ *and* $h_1 = \bigvee_{\neg C}^i h_{10}$.

Proof By Theorem 3.3.(1) and Lemma 4.1, it follows that for all $\vec{u} \in \mathbb{N}^{<\omega}$, if $\Gamma(T_{\vec{u}}) = \{n = m\}$, then $\mathrm{tp}(T_{\vec{u}}) = \mathsf{Rep}$ or $\mathsf{Ax}_{n=m}$. We show that $\Gamma(T_{\mathbf{0}^k}) = \{n = m\}$ for every $k \in \mathbb{N}$, by induction on k. When $\mathbf{0}^k = \langle\rangle$, we have $\Gamma(T_{\langle\rangle}) = \Gamma(h) = \{n = m\}$. Assume that $\Gamma(T_{\mathbf{0}^k}) = \{n = m\}$. Then, $\mathrm{tp}(T_{\mathbf{0}^k}) = \mathsf{Rep}$ or $\mathsf{Ax}_{n=m}$. In each case, we can show that $\Gamma(T_{\mathbf{0}^{k+1}}) = \{n = m\}$ by the definition of normalization trees and Theorem 3.3.(1). Below we set $\vec{u} := \mathbf{0}^k$ for an arbitrary $k \in \mathbb{N}$.
(1) Assume that $\mathrm{tp}(T_{\vec{u}}) = \mathsf{Ax}_{n=m}$. By Lemma 4.3.(6), there is a nominal form $\mathfrak{a}$ such that $n = m \notin \mathsf{Cut}(\mathfrak{a})$ and $T_{\vec{u}} = \mathfrak{a}\{\mathsf{Ax}_\Delta\}$ with $n = m \in \Delta$.
(2) Assume that $\mathrm{tp}(T_{\vec{u}}) = \mathsf{Rep}$. By Lemma 4.3.(1), there are $\mathfrak{c}$, h''_0, h''_1 such that $T_{\vec{u}} = \mathfrak{c}\{h''_0\}$, $T_{\vec{u}}[0] = T_{\vec{u}*\langle 0\rangle} = \mathfrak{c}\{h''_0[0]\}$ and either $h''_0 = \mathsf{E}^m \mathsf{Ind}_F^{y,t} h''_1$ or ($h''_0 = \mathsf{E}^{m+1} h''_1$ & $\mathrm{tp}(h''_1) = \mathsf{Cut}_B$ for some B). If $h''_0 = \mathsf{E}^m \mathsf{Ind}_F^{y,t} h''_1$, then the statement (a) holds by taking $\mathfrak{c}$ as $\mathfrak{a}$ and h''_0 as h_0. Assume that $h''_0 = \mathsf{E}^{m+1} h''_1$ and $\mathrm{tp}(h''_1) = \mathsf{Cut}_B$ for some B. Then, by Lemma 4.3.(2), there are $\mathfrak{b}_0$, C, h''_{10}, h''_{11} such that $h''_1 = \mathfrak{b}_0\{\mathsf{R}_C h''_{10} h''_{11}\}$, $h''_1[i] = \mathfrak{b}_0\{(\mathsf{R}_C h''_{10} h''_{11})[i]\}$ and either

(†) $\mathrm{tp}(h''_{10}) = \bigwedge_C$ & $\mathrm{tp}(h''_{11}) = \bigvee_{\neg C}^n$ & $B = C[n]$ or
(††) $\mathrm{tp}(h''_{10}) = \bigvee_C^n$ & $\mathrm{tp}(h''_{11}) = \bigwedge_{\neg C}$ & $B = C[n]$.

We consider only the case that (†) holds and $C \equiv \forall x A(x)$. Each of the other cases can be dealt with in a similar way. By Lemma 4.3.(3) and (5), there are $\mathfrak{a}_0$, $\mathfrak{a}_1$, h^*_{10} and h^*_{11} such that

$C \notin \mathsf{Cut}(\mathfrak{a}_0)$, $h''_{10} = \mathfrak{a}_0\{\bigwedge_C^y h^*_{10}\}$, $h''_{10}[n] = \mathfrak{a}_0\{h^*_{10}(y/n)\}$ and
$\neg C \notin \mathsf{Cut}(\mathfrak{a}_1)$, $h''_{11} = \mathfrak{a}_1\{\bigvee_{\neg C}^n h^*_{11}\}$, $h''_{11}[0] = \mathfrak{a}_1\{h^*_{11}\}$.

Thus,

$$T_{\vec{u}} = \mathfrak{c}\{h''_0\} = \mathfrak{c}\{\mathsf{E}^{m+1} h''_1\} = \mathfrak{c}\{\mathsf{E}^{m+1}\mathfrak{b}_0\{\mathsf{R}_C h''_{10} h''_{11}\}\} = \mathfrak{c}\{\mathsf{E}^{m+1}\mathfrak{b}_0\{\mathsf{R}_C(\mathfrak{a}_0\{\bigwedge_C^y h^*_{10}\}, \mathfrak{a}_1\{\bigvee_{\neg C}^n h^*_{11}\})\}\}$$

and

$$T_{\vec{u}*\langle 0\rangle} = \mathfrak{c}\{(\mathsf{E}^{m+1}h_1'')[0]\} = \mathfrak{c}\{\mathsf{E}^m((\mathsf{E}h_1'')[0])\} = \mathfrak{c}\{\mathsf{E}^m(\mathsf{R}_{C[n]}(\mathsf{E}h_1''[0], \mathsf{E}h_1''[1]))\}.$$

Then,

$$\begin{aligned}
&\mathfrak{c}\{\mathsf{E}^m(\mathsf{R}_{C[n]}(\mathsf{E}h_1''[0], \mathsf{E}h_1''[1]))\} = \\
&\mathfrak{c}\{\mathsf{E}^m(\mathsf{R}_{C[n]}(\mathsf{E}\mathfrak{b}_0\{(\mathsf{R}_C h_{10}'' h_{11}'')[0]\}, \mathsf{E}\mathfrak{b}_0\{(\mathsf{R}_C h_{10}'' h_{11}'')[1]\}))\} = \\
&\mathfrak{c}\{\mathsf{E}^m(\mathsf{R}_{C[n]}(\mathsf{E}\mathfrak{b}_0\{\mathsf{R}_C(h_{10}''[n], h_{11}'')\}, \mathsf{E}\mathfrak{b}_0\{\mathsf{R}_C(h_{10}'', h_{11}''[0])\}))\} = \\
&\mathfrak{c}\{\mathsf{E}^m(\mathsf{R}_{C[n]}(\mathsf{E}\mathfrak{b}_0\{\mathsf{R}_C(\mathfrak{a}_0\{h_{10}^*(y/n)\}, \mathfrak{a}_1\{\textstyle\bigvee_{\neg C}^n h_{11}^*\})\}, \mathsf{E}\mathfrak{b}_0\{\mathsf{R}_C(\mathfrak{a}_0\{\textstyle\bigwedge_C^y h_{10}^*\}, \\
&\mathfrak{a}_1\{h_{11}^*\})\}))\}.
\end{aligned}$$

Set

$$\mathfrak{a} := \mathfrak{c},\ \mathfrak{b} := \mathfrak{b}_0,\ \mathfrak{c}_i := \mathfrak{a}_i \text{ for } i \in \{0, 1\},\ h_0 := h_{10}^* \text{ and } h_1 := h_{11}^*.$$

Then, the statement (b) holds. □

Before we proceed, let us make the following remark. As observed in Akiyoshi and Takahashi (2013), reduction steps of the 1936 proof can be reformulated by slightly generalizing Proposition 4.4. Let FO be the set of all Z^∞-inference symbols of the form $\bigwedge_C$ or $\bigvee_C^k$, and FFO be the set of all Z^*-inference symbols of the form $\bigwedge_C^y$, $\bigwedge_C$ or $\bigvee_C^k$.

Proposition 4.5 (cf. Akiyoshi and Takahashi (2013), Theorem 8) *Let T be a normalization tree of some closed $h \in \mathsf{Z}_0^*$. For every $\vec{u}$, one of the following statements holds:*

1. *if $\mathrm{tp}(T_{\vec{u}}) = \mathsf{Ax}_A$ for some A, then there is a nominal form $\mathfrak{a}$ such that $A \notin \mathsf{Cut}(\mathfrak{a})$ and $T_{\vec{u}} = \mathfrak{a}\{\mathsf{Ax}_\Delta\}$ with $A \in \Delta$, and*
2. *if $\mathrm{tp}(T_{\vec{u}}) = \mathsf{Rep}$, then one of the following statements holds for some nominal forms $\mathfrak{a}, \mathfrak{b}, \mathfrak{c}_0, \mathfrak{c}_1$. (The symmetric cases are suppressed.)*
   (a) $T_{\vec{u}} = \mathfrak{a}\{\mathsf{E}^m \mathsf{Ind}_F^{y,t} h_0\}$ *and* $T_{\vec{u}*\langle 0\rangle} = \mathfrak{a}\{\mathsf{E}^m((\mathsf{Ind}_F^{y,t} h_0)[0])\}$.
   (b) $T_{\vec{u}} = \mathfrak{a}\{\mathsf{E}^{m+1}\mathfrak{b}\{\mathsf{R}_C(\mathfrak{c}_0\{h_0'\}, \mathfrak{c}_1\{h_1'\})\}\}$ *and*
   $T_{\vec{u}*\langle 0\rangle} = \mathfrak{a}\{\mathsf{E}^m \mathsf{R}_{C[k]}(\mathsf{E}\mathfrak{b}\{\mathsf{R}_C(\mathfrak{c}_0\{h_0(y/k)\}, \mathfrak{c}_1\{h_1'\})\}, \mathsf{E}\mathfrak{b}\{\mathsf{R}_C(\mathfrak{c}_0\{h_0'\}, \mathfrak{c}_1\{h_1\})\})\}$,
   where $C \equiv \forall x A(x)$, $C \notin \mathsf{Cut}(\mathfrak{c}_0)$, $\neg C \notin \mathsf{Cut}(\mathfrak{c}_1)$, $h_0' = \bigwedge_C^y h_0$ *and* $h_1' = \bigvee_{\neg C}^k h_1$.
   (c) $T_{\vec{u}} = \mathfrak{a}\{\mathsf{E}^{m+1}\mathfrak{b}\{\mathsf{R}_C(\mathfrak{c}_0\{h_0\}, \mathfrak{c}_1\{h_1\})\}\}$ *and*
   $T_{\vec{u}*\langle 0\rangle} = \mathfrak{a}\{\mathsf{E}^m \mathsf{R}_{C[i]}(\mathsf{E}\mathfrak{b}\{\mathsf{R}_C(\mathfrak{c}_0\{h_{0i}\}, \mathfrak{c}_1\{h_1\})\}, \mathsf{E}\mathfrak{b}\{\mathsf{R}_C(\mathfrak{c}_0\{h_0\}, \mathfrak{c}_1\{h_{10}\})\})\}$,
   where $C \equiv A_0 \wedge A_1$, $C \notin \mathsf{Cut}(\mathfrak{c}_0)$, $\neg C \notin \mathsf{Cut}(\mathfrak{c}_1)$, $h_0 = \bigwedge_C h_{00} h_{01}$ *and* $h_1 = \bigvee_{\neg C}^i h_{10}$.
3. *if $\mathrm{tp}(T_{\vec{u}}) \in FO$, then there is a nominal form $\mathfrak{a}$ such that $T_{\vec{u}} = \mathfrak{a}\{\mathcal{I}(h_j)_{j\in|\mathcal{I}|}\}$, $\mathcal{I} \in FFO$, $\Delta(\mathcal{I}) \cap \mathsf{Cut}(\mathfrak{a}) = \emptyset$ and for every $i \in |\mathrm{tp}(T_{\vec{u}})|$,*

$$T_{\vec{u}*\langle i\rangle} = \begin{cases} \mathfrak{a}\{h_0(y/i)\}, & \textit{if } \mathcal{I} = \bigwedge_C^y, \\ \mathfrak{a}\{h_i\}, & \textit{else.} \end{cases}$$

Proof (1) Similar to the proof of Proposition 4.4.(1).
(2) This case is the same as Proposition 4.4.(2).
(3) We consider only the case that $\mathrm{tp}(T_{\vec{u}}) = \bigwedge_{\forall x A(x)}$. By Lemma 4.3.(3), there are $\mathfrak{b}, h^*$ such that $T_{\vec{u}} = \mathfrak{b}\{\bigwedge_{\forall x A(x)}^y h^*\}$, $\Delta(\bigwedge_{\forall x A(x)}^y) \cap \mathsf{Cut}(\mathfrak{b}) = \emptyset$, and $T_{\vec{u}*\langle k\rangle} = (T_{\vec{u}})[k] = \mathfrak{b}\{h^*(y/k)\}$ for all $k \in \mathbb{N}$. We are done. □

Let T be a well-founded normalization tree of some closed $h \in \mathsf{Z}_0^*$ with the endsequent of the form $n = m$. Now we argue that T corresponds to consecutive applications of 1938 reduction steps. First, since $\mathrm{tp}(T_{\mathbf{0}^k}) = \mathsf{Ax}_{n=m}$ or Rep for every $k \in \mathbb{N}$, T essentially has no branching: If we omit its irrelevant nodes, then T is described as follows.

$$T_{\langle\rangle} \longrightarrow T_{\langle 0\rangle} \longrightarrow T_{\langle 0,0\rangle} \longrightarrow \cdots$$

This accords with the fact that reduction steps of the 1938 proof has no branching.

Second, each transition from $\phi(T_{\mathbf{0}^k})$ to $\phi(T_{\mathbf{0}^{k+1}})$ corresponds to a reduction step of the 1938 proof, which is formulated with end-pieces. Consider the case (2a) of Proposition 4.4. There $\phi(T_{\mathbf{0}^k})$ and $\phi(T_{\mathbf{0}^{k+1}})$ are as follows.

$$\dfrac{\begin{matrix}\vdots\, h_0 \\ \neg F(y), F(S(y))\end{matrix}}{\begin{matrix}\neg F(0), F(k) \\ \vdots\, \mathfrak{a} \\ n = m\end{matrix}}\,\mathsf{Ind}_F^{y,k} = \phi(T_{\mathbf{0}^k}) \qquad \rhd$$

$$\dfrac{\dfrac{\begin{matrix}\vdots \\ \Gamma, \neg F(0), F(0)\end{matrix} \quad \begin{matrix}\vdots\, h_0(y/0) \\ \Gamma, \neg F(0), F(1)\end{matrix}}{\Gamma, \neg F(0), F(1)}\,\mathsf{R}_{F(0)} \quad \begin{matrix}\vdots\, h_0(y/1) \\ \Gamma, \neg F(1), F(2)\end{matrix}}{\begin{matrix}\Gamma, \neg F(0), F(2) \\ \vdots \\ \Gamma, \neg F(0), F(k) \\ \vdots\, \mathfrak{a} \\ n = m\end{matrix}}\,\mathsf{R}_{F(1)} = \phi(T_{\mathbf{0}^{k+1}})$$

This transition is done exactly in the end-pieces of $\phi(T_{\mathbf{0}^k})$ and $\phi(T_{\mathbf{0}^{k+1}})$. Thus, it is the same as the "CJ-reduction" of the 1938 proof formulated in Gentzen (1938, Sect. 3.3). Furthermore, in the case (2b) of Proposition 4.4, $\phi(T_{\mathbf{0}^k})$ and $\phi(T_{\mathbf{0}^{k+1}})$ are of the following forms, respectively. Let C be $\forall x A(x)$.

$$
\dfrac{
\begin{array}{c} \vdots\, h_0 \\ \dfrac{\Delta_0, A(y), \forall x A(x)}{\Delta_0, \forall x A(x)}\textstyle\bigwedge^y_C \\ \vdots\, \mathfrak{c}_0 \\ \Gamma', \forall x A(x) \end{array}
\qquad
\begin{array}{c} \vdots\, h_1 \\ \dfrac{\Delta_1, \neg A(k), \exists x \neg A(x)}{\Delta_1, \exists x \neg A(x)}\textstyle\bigvee^k_{\neg C} \\ \vdots\, \mathfrak{c}_1 \\ \Gamma', \exists x \neg A(x) \end{array}
}{
\begin{array}{c} \Gamma' \\ \vdots\, \mathfrak{b} \\ \Gamma \\ \vdots\, \mathfrak{a} \\ n = m \end{array}
}\;\mathsf{R}_C = \phi(T_{0^k}) \qquad \triangleright
$$

$$
\dfrac{
\dfrac{
\begin{array}{c} \vdots\, h_0(y/k) \\ \Delta_0, A(k), \forall x A(x) \\ \vdots\, \mathfrak{c}_0 \\ \Gamma', A(k), \forall x A(x) \end{array}
\quad
\begin{array}{c} \vdots\, h_1 \\ \dfrac{\Delta_1, \neg A(k), \exists x \neg A(x)}{\Delta_1, \exists x \neg A(x)}\textstyle\bigvee^k_{\neg C} \\ \vdots\, \mathfrak{c}_1 \\ \Gamma', \exists x \neg A(x) \end{array}
}{
\begin{array}{c} \Gamma', A(k) \\ \vdots\, \mathfrak{b} \\ \Gamma, A(k) \end{array}
}\mathsf{R}_C
\quad
\dfrac{
\begin{array}{c} \vdots\, h_0 \\ \dfrac{\Delta_0, A(y), \forall x A(x)}{\Delta_0, \forall x A(x)}\textstyle\bigwedge^y_C \\ \vdots\, \mathfrak{c}_0 \\ \Gamma', \forall x A(x) \end{array}
\quad
\begin{array}{c} \vdots\, h_1 \\ \Delta_1, \neg A(k), \exists x \neg A(x) \\ \vdots\, \mathfrak{c}_1 \\ \Gamma', \neg A(k), \exists x \neg A(x) \end{array}
}{
\begin{array}{c} \Gamma', \neg A(k) \\ \vdots\, \mathfrak{b} \\ \Gamma, \neg A(k) \end{array}
}\mathsf{R}_C = \phi(T_{0^k+1})
}{
\begin{array}{c} \Gamma \\ \vdots\, \mathfrak{a} \\ n = m \end{array}
}\;\mathsf{R}_{C[k]}
$$

with $\forall x A(x) \notin \mathsf{Cut}(\mathfrak{c}_0)$, $\exists x \neg A(x) \notin \mathsf{Cut}(\mathfrak{c}_1)$. This is one of the operational reductions formulated in Gentzen (1938, Sect. 3.5). The case (2c) corresponds to another operational reduction.

Third, T gives a normal derivation of $n = m$. By the definition of well-foundedness, T has a node T_{0^l} such that $\mathrm{tp}(T_{0^l}) = \mathsf{Ax}_A$ for some A. By Proposition 4.4.(1), A is of the form $n = m$ and T_{0^l} is of the form $\mathfrak{a}\{\mathsf{Ax}_\Delta\}$ with $n = m \in \Delta$. Since $n = m \notin \mathsf{Cut}(\mathfrak{a})$, $\Gamma(T_{0^l})$ includes $n = m$. In fact, $\Gamma(T_{0^l}) = \{n = m\}$ holds as we have seen in the proof of Proposition 4.4. Thus, T_{0^l} is a Z_0^*-derivation of $n = m$. Moreover, we can consider T_{0^l} a normal derivation, because the endsequent of T_{0^l} is shown to be true in a primitive recursive way.

From these arguments, we conclude that T corresponds to consecutive applications of reduction steps of the 1938 proof and that the formulation of its main lemma as Proposition 4.3 is justified. Proposition 4.3 assigns a normal derivation to every derivable numeric equation of first-order arithmetic, and the main lemma of the 1936 does so as well, because the 1938 proof is a special case of the 1936 proof.

Let us summarize the argument in this section. First, we have argued that the 1936 proof is a contentual correctness proof, by verifying that the main lemma of the 1936 proof (Proposition 4.1) with an additional lemma (Lemma 4.2) implies the main lemma of the 1935 proof (Proposition 4.2). This consequence from the main lemma of the 1936 proof means that the proof shows all axioms and theorems of first-order arithmetic to be correct in the sense of (GI), as the 1935 proof does. Next, we have argued that the 1936 proof is also a formal correctness proof. By utilizing Buchholz's observation in Buchholz (1997), we showed that reduction steps of the 1936 proof for derivations of numeric equations have the same structures as the ones of reductions steps of the 1938 proof. This enabled us to see that the main lemma of the 1938 proof (Proposition 4.3) is a special case of the main lemma of the 1936 proof. Accordingly, the 1936 proof assigns a normal derivation to every derivation of numeric equation in first-order arithmetic as the 1938 proof does.

5 Some Remarks

In this section, we make some remarks about our arguments in previous sections. Section 5.1 explains Gentzen's finitist meaning ascribed to a transfinite proposition in the 1936 proof. In Sect. 5.2, we extract a uniform interpretation of Gentzen's three consistency proofs from our argument in Sect. 4. In particular, we claim that there is a uniform idea behind these proofs. Finally, in Sect. 5.3, we provide another proof of Kreisel's no-counterexample interpretation, using our framework.

5.1 Gentzen's Finitist Interpretation of Transfinite Propositions

According to Gentzen, his reductions steps ascribe a finitist meaning (in his sense) to a transfinite proposition, for example, $\forall x A(x)$ such that x ranges over natural numbers.[31] Gentzen explained the finitist meaning of $\forall x A(x)$ in the following way:

> $\forall \mathfrak{x} \mathfrak{F}(\mathfrak{x})$ may be significantly asserted if $\mathfrak{F}(\mathfrak{x})$ represents a significant (sinnvoll) and true proposition for arbitrary successive replacements of $\mathfrak{x}$ by numbers (Gentzen 1936, p. 526, 1969, p. 163)

Now we see Gentzen's explanation of the finitist meaning of $\forall x A(x)$ using our previous description of Gentzen's reduction steps. Suppose that the following proof-figure $T_{\vec{u}}$ is attached to a node of a normalization tree T for some closed Z_0^*-derivation.

$$\dfrac{\begin{array}{c}\vdots\ h' \\ \Gamma, A(y)\end{array}}{\Gamma, \forall x A(x)} \bigwedge\nolimits^{y}_{\forall x A(x)}$$

To explain the finitist meaning of $\forall x A(x)$, we may suppose that $A(i)$ for each $i \in \mathbb{N}$ is already significant ("sinvoll").

[31] Gentzen (1936, pp. 523–524).

Let us consider the resulting derivations of applying his reduction steps to h. As we have explained in Sect. 4, this node has infinitary many successors $T_{\vec{u}*\langle i\rangle}$ for each $i \in \mathbb{N}$, to which the following derivations are attached:

$$\begin{array}{c} \vdots \; h'(y/i) \\ \Gamma, A(i) \end{array}$$

for all $i \in \mathbb{N}$ with $RED(T_{\vec{u}}, \langle T_{\vec{u}*\langle i\rangle}\rangle_{i\in\mathbb{N}})$. By our assumption stated above, $A(i)$ is already significant for each $i \in \mathbb{N}$. Hence, "If, starting with 1, we substitute for $\mathfrak{x}$ successive natural numbers then, however far we may progress in the formation of numbers, a true proposition results in each case."[32]

Here, let us ask the following natural question: What is the role of the well-foundedness of T in Gentzen's finitist explanation of the meaning of a transfinite proposition? As we have seen above, the finitist explanation of $\forall x A(x)$ is reduced into others of simpler formulas, that is, $A(0)$, $A(1)$, and so on. The situation is the same for $A \wedge B$, $A \vee B$, $\exists x A(x)$. Therefore, in these cases, the finitist meaning of a transfinite proposition A is explained by another of simpler formulas or subformulas of it. Since A is a formula and consists of finitely many symbols, this series of explanations should terminate at a base case.

How about the case that $\mathrm{tp}(T_{\vec{u}}) = \mathsf{Rep}$? If we recall Definition 3.7.(2e), here is the crucial case of cut-elimination. Also, according to Definition 3.7.(2d), this is the case of embedding the mathematical induction axiom. In these cases, the truth of the endsequent of $T_{\vec{u}}$ denoted by $\Gamma(T_{\vec{u}})$ is reduced into the one of $T_{\vec{u}*\langle 0\rangle}$ denoted by $\Gamma(T_{\vec{u}*\langle 0\rangle})$, but we see $\Gamma(T_{\vec{u}}) = \Gamma(T_{\vec{u}*\langle 0\rangle})$. Since there is no subformula property here, then we must guarantee the well-foundedness of this reduction, that is, we need the termination proof of the cut-elimination theorem. Indeed, if T is well-founded, then we will eventually arrive at a "base case," to which true literal like $0 = 0$ is attached. Therefore, if T is well-founded, his finitist explanation of a transfinite proposition is also well-founded, in the sense that it terminates at a base case like $0 = 0$, the truth of which we can see in Gentzen's (or even Hilbert's) finitist standpoint. Otherwise, it could be the case that there is a non-well-founded series of reductions, say $T_{\vec{u}}$, $T_{\vec{u}*\langle 0\rangle}$, $T_{\vec{u}*\langle 0\rangle*\langle 0\rangle}$, and so on. From this argument, the role of the well-foundedness of T in Gentzen's finitist explanation is very clear: The well-foundedness of T guarantees the well-foundedness of his finitist explanation of a transfinite proposition.

Returning to Gentzen's main aim in his papers, that is, the consistency of first-order arithmetic, he proved it by ascribing a finitist meaning to a transfinite proposition. In his lecture "Der Unendlichkeitsbegriff in der Mathematik: ein Vortrag, gehalten in Münster am 27. Juni 1936," this methodology is clearly stated:

[32]Gentzen (1936, p. 526).

> In carrying out a consistency proof for elementary number theory we must consider all conceivable number theoretical proofs and must show that in a certain sense, to be formally defined, each individual proof yields a 'correct' result, in particular, no contradiction. The 'correctness' of a proof depends on the correctness of certain other simpler proofs contained in it as special cases or constituent parts (Gentzen 1969, p. 231).

As Gentzen says, we proved that each individual proof yields a correct result by ascribing a finitist meaning to a transfinite proposition occurring in all conceivable proofs. Additionally, this quotation would be indirect evidence that our strategy to explain the relationship between his consistency proofs is quite natural for him, although he did not explain it in the way we did. The main lemma of the 1938 proof, which belongs to formal correctness proofs, is just a special case of the main lemma of the 1936 proof. Therefore, as we have shown, the 1936 proof is formal as well.

5.2 *Uniform Interpretation of Gentzen's Three Proofs*

Next, we extract a *uniform* interpretation of Gentzen's three consistency proofs from our reconstruction of the 1936 proof in this paper. As noted in the introduction, Gentzen gave three different consistency proofs, that is, the 1935 proof, the 1936 proof, and the 1938 proof. If we read Sect. 17 of the 1936 proof, then it would be more or less obvious that the aim of the 1935 proof is the same as that of the 1936 proof because Gentzen did not change passages to explain their background and motivation. Therefore, it follows that the aim to give a finitist interpretation to a transfinite proposition is uniform behind the 1935 and 1936 proofs. Let us quote his key passage occurring in both:

> The major part of my consistency proof, however, consists precisely in *ascribing a finitist sense* to actualist propositions, viz.: for every arbitrary proposition, as long as it is provable, a *reduction procedure* according to 13.6 *can be stated*, and this fact represents the finitist sense of the proposition concerned and this sense is gained precisely through the consistency proof (Gentzen 1974, p. 118, 1936, p. 564, 1969, p. 163, italics original).

In Sect. 4, we have shown that the 1936 proof is formal, in the sense that reduction steps in the 1938 proof are just special cases of some in the 1936 proof. We remark that this has been shown by the same method, that is, the method of finite notations for infinitary notations. In particular, both reduction steps are recovered from corresponding ones for the infinitary system with the ω-rule. Based on this observation, we claim the aim of the 1936 proof is to be found in the 1938 proof, from which it follows that there is a *uniform idea* behind all three proofs by Gentzen. One might ask why Gentzen could have the same idea in the 1938 proof, since this proof focuses on a derivation of a numeric equation like $n = m$, while he considers arbitrary derivation in the 1936 proof. For this question, we may answer in the following way. In our reconstruction of the 1936 proof, a finitist interpretation in Gentzen's sense is given for a transfinite proposition, which is also given for logical inference rules possibly occurring in the 1938 proof. For example, if we recall our example in Sect. 4, then the interpretation is given for $\bigwedge_{\forall x A(x)}$ occurring in boundary inferences.[33] This interpretation is entirely the same as that in the 1936 proof. In this sense, Gentzen's idea to ascribe a finitist interpretation is *uniform* behind his three proofs.

Here, we might quote a passage from the 1938 paper:

> [O]nly this time the main emphasis will be placed on developing the *fundamental ideas* and on making every single step of the proof as *lucid* as possible (Gentzen 1938, p. 19, 1969, p. 252, italics original).

This quotation would not be direct evidence supporting our reading, but, at least, he seems to suggest that he did not change his fundamental ideas, that is, he just extracted and reformulated some parts of the 1936 proof that were essentially needed for proving the consistency of first-order arithmetic.[34]

5.3 No-Counterexample Interpretation

In this subsection, we prove the theorem called the *no-counterexample interpretation*, following Schwichtenberg's proof Schwichtenberg (1977, Sect. 4) in our setting.[35]

Theorem 5.1 (No-Counterexample Interpretation) *If a closed formula*

$$\exists x_1 \forall y_1 \ldots \exists x_n \forall y_n p(x_1, y_1, \ldots, x_n, y_n)$$

is derivable in Z*, then there are* $< \varepsilon_0$*-recursive functionals* $\mathbf{G}_1, \ldots, \mathbf{G}_n$ *such that for every* k*-ary function* $\mathbf{F}_k$ *on natural numbers* $(1 \leq k \leq n)$*,*

$$\mathbf{P}(\mathbf{G}_1\vec{\mathbf{F}}, \mathbf{F}_1(\mathbf{G}_1\vec{\mathbf{F}}), \mathbf{G}_2\vec{\mathbf{F}}, \mathbf{F}_2(\mathbf{G}_1\vec{\mathbf{F}}, \mathbf{G}_2\vec{\mathbf{F}}), \ldots, \mathbf{G}_n\vec{\mathbf{F}}, \mathbf{F}_n(\mathbf{G}_1\vec{\mathbf{F}}, \ldots, \mathbf{G}_n\vec{\mathbf{F}}))$$

holds, where $\mathbf{G}_k\vec{\mathbf{F}} := \mathbf{G}_k(\mathbf{F}_1, \ldots, \mathbf{F}_n)$ *for all* k $(1 \leq k \leq n)$*.*

For the sake of simplicity, we concentrate on the case that $n = 2$ in the theorem above. Let an arbitrary unary function $\mathbf{F}_1$ and an arbitrary binary function $\mathbf{F}_2$ on natural numbers be given.

First, we extend the language L. Let $L(f_1, f_2)$ be the language obtained by extending L with a unary function-variable f_1 and a binary function-variable f_2. We denote the set of all free number-variables in an expression θ in $L(f_1, f_2)$ by $FV(\theta)$. An expression θ is *closed* if $FV(\theta) = \emptyset$. Let $\mathsf{TRUE}_0^{\mathbf{F}_1,\mathbf{F}_2}$ be the set of all true literals of $L(f_1, f_2)$ under the assignment $f_i \mapsto \mathbf{F}_i$ for $i \in \{1, 2\}$. For every closed term t of $L(f_1, f_2)$, we denote the numeral n such that $t = n \in \mathsf{TRUE}_0^{\mathbf{F}_1,\mathbf{F}_2}$ by $[\![t]\!]$.

[33] As to the definition of a boundary inference, we refer to Takeuti's definition Takeuti (1987, p. 79).

[34] In the future, when we will have more historical resources, we will investigate this point in more detail.

[35] The difference is our use of finite notations for infinitary derivations instead of heavy coding.

On the basis of the language $L(f_1, f_2)$, we define the finitary proof system $\mathsf{Z}^*(\mathbf{F}_1, \mathbf{F}_2)$. We assume the existence of a set $\mathsf{Ax}^{\mathbf{F}_1,\mathbf{F}_2}$ of sequents satisfying the following conditions:

- for all $\Delta \in \mathsf{Ax}^{\mathbf{F}_1,\mathbf{F}_2}$, Δ is a set of literals,
- if there is a substitution instance Δ_0 of Δ, then $\Delta_0 \in \mathsf{Ax}^{\mathbf{F}_1,\mathbf{F}_2}$,
- if $\Delta \in \mathsf{Ax}^{\mathbf{F}_1,\mathbf{F}_2}$ and $FV(\Delta) = \emptyset$, then $\Delta \cap \mathsf{TRUE}_0^{\mathbf{F}_1,\mathbf{F}_2} \neq \emptyset$.

Define the inference symbol Ax_Δ of $\mathsf{Z}^*(\mathbf{F}_1, \mathbf{F}_2)$ as

$$(\mathsf{Ax}_\Delta)\,\overline{\Delta}$$

with $\Delta \in \mathsf{Ax}^{\mathbf{F}_1,\mathbf{F}_2}$. Furthermore, we add the following inference symbol:

$$(\mathsf{S}_s^t)\,\frac{A(x/t)}{A(x/s)}$$

with $[\![t]\!] = [\![s]\!]$. The remaining inference symbols $\bigwedge_{A_0 \wedge A_1}$, $\bigwedge^y_{\forall x A}$, $\bigvee^i_{A_0 \vee A_1}$, $\bigvee^t_{\exists x A}$, $\mathsf{Ind}_F^{y,t}$, R_C and E of $\mathsf{Z}^*(\mathbf{F}_1, \mathbf{F}_2)$ are defined in the same manner as Z^*.

Hereafter, a $\mathsf{Z}^*(\mathbf{F}_1, \mathbf{F}_2)$-derivation is denoted by h possibly with suffixes. Again, we suppose that every $\mathsf{Z}^*(\mathbf{F}_1, \mathbf{F}_2)$-derivation satisfies Free Variables-Conditions. We define the ordinal assignment $\mathrm{o}(h)$ and the degree $\deg(h)$ of a $\mathsf{Z}^*(\mathbf{F}_1, \mathbf{F}_2)$-derivation h with $h = \mathsf{S}_s^t h_0$ as follows: $\mathrm{o}(h) := \mathrm{o}(h_0)$ and $\deg(h) := \deg(h_0)$. The other cases are treated in the same manner as Z^*.

Note that if a closed formula of the form

$$\exists x_1 \forall y_1 \exists x_2 \forall y_2 p(x_1, y_1, x_2, y_2)$$

is derivable in Z, then we can verify only by first-order logic that the formula

$$\exists x_1 \exists x_2 p(x_1, f_1(x_1), x_2, f_2(x_1, x_2))$$

is derivable in $\mathsf{Z}^*(\mathbf{F}_1, \mathbf{F}_2)$. In particular, we have no need of the $\mathsf{Z}^*(\mathbf{F}_1, \mathbf{F}_2)$-axioms about $\mathbf{F}_1$ and $\mathbf{F}_2$ in deriving the formula above.

The infinitary proof system $\mathsf{Z}^\infty(\mathbf{F}_1, \mathbf{F}_2)$ with the language $L(f_1, f_2)$ is defined in the same way as Z^∞ except the inference symbols Ax_A. The inference symbol Ax_A of $\mathsf{Z}^\infty(\mathbf{F}_1, \mathbf{F}_2)$ is defined as

$$(\mathsf{Ax}_A)\,\overline{A}$$

with $A \in \mathsf{TRUE}_0^{\mathbf{F}_1,\mathbf{F}_2}$. We denote a $\mathsf{Z}^\infty(\mathbf{F}_1, \mathbf{F}_2)$-derivation by d possibly with suffixes.

By the following lemma, we define an operator for substitution in $\mathsf{Z}^\infty(\mathbf{F}_1, \mathbf{F}_2)$-derivations that corresponds to the inference symbol S_s^t of $\mathsf{Z}^*(\mathbf{F}_1, \mathbf{F}_2)$.

Lemma 5.1 *For all terms t and s with $[\![t]\!] = [\![s]\!]$, there is an operator $\mathcal{S}^t_s$ such that if $\mathtt{d} \vdash^\alpha_m \Gamma, A(x/t)$, then $\mathcal{S}^t_s(\mathtt{d}) \vdash^\alpha_m \Gamma, A(x/s)$.*

Proof By induction on α. □

The following two theorems can be shown for $\mathsf{Z}^\infty(\mathbf{F}_1, \mathbf{F}_2)$ in a similar way to the case of Z^∞. To prove the first theorem, we use the previous lemma.

Theorem 5.2 *For every formula C, there is an operator $\mathcal{R}_C$ such that if $\mathtt{d}_0 \vdash^\alpha_m \Gamma, C$, $\mathtt{d}_1 \vdash^\beta_m \Gamma, \neg C$ and $\mathrm{rk}(C) \leq m$, then $\mathcal{R}_C(\mathtt{d}_0, \mathtt{d}_1) \vdash^{\alpha\#\beta}_m \Gamma$.*

Theorem 5.3 *There is an operator $\mathcal{E}$ such that if $\mathtt{d} \vdash^\alpha_{m+1} \Gamma$ then $\mathcal{E}(\mathtt{d}) \vdash^{\omega^\alpha}_m \Gamma$.*

We define the translation of $\mathsf{Z}^*(\mathbf{F}_1, \mathbf{F}_2)$ into $\mathsf{Z}^\infty(\mathbf{F}_1, \mathbf{F}_2)$. Let h be a closed $\mathsf{Z}^*(\mathbf{F}_1, \mathbf{F}_2)$-derivation, Γ be $\Gamma(h)$ and α be $\mathrm{o}(h)$.

1. If $h = \mathsf{Ax}_\Delta$, then $(\mathsf{Ax}_\Delta)^\infty := \overline{\mathsf{Ax}_A : \Gamma : \alpha}$, where A is the arbitrarily fixed element of $\Delta \cap \mathsf{TRUE}_0^{\mathbf{F}_1,\mathbf{F}_2}$.
2. If $h = \mathsf{S}^t_s h_0$, then $(\mathsf{S}^t_s h_0)^\infty := \mathcal{S}^t_s(h_0^\infty)$ with $\Gamma(\mathcal{S}^t_s(h_0^\infty)) = \Gamma$.
3. If $h = \mathsf{Ind}^{y,t}_F h_0$ and $[\![t]\!] = n$, then $(\mathsf{Ind}^{y,t}_F h_0)^\infty := \dfrac{\mathcal{S}^n_t(\mathtt{e}_n)}{\mathsf{Rep} : \Gamma : \alpha}$, where
$$\mathtt{e}_0 := \mathtt{d}_{F(x/0)},\ \ \mathtt{e}_1 := h_0(y/0)^\infty,\ \ \mathtt{e}_{i+1} := \mathcal{R}_{F(x/i)}(\mathtt{e}_i, h_0(y/i)^\infty)$$
and $\Gamma(\mathtt{e}_n) = \{\Delta, \neg F(0), F(n)\}$, $\Gamma(\mathcal{S}^n_t(\mathtt{e}_n)) = \{\Delta, \neg F(0), F(t)\} = \Gamma$.
4. Otherwise, define h^∞ in the same manner as the translation of Z^* into Z^∞.

In addition, we define the two functionals $\mathrm{tp}^{\mathbf{F}_1,\mathbf{F}_2}(h)$ and $h(\mathbf{F}_1, \mathbf{F}_2)[i]$ of level ≤ 2 for closed $\mathsf{Z}^*(\mathbf{F}_1, \mathbf{F}_2)$-derivations. Note that these are primitive recursive functional with $\mathbf{F}_1$ and $\mathbf{F}_2$ as function-arguments. For closed terms t and s, the expression $\theta(t \mapsto s)$ is the result of replacing some occurrences of t in θ with s.

Let h be a closed $\mathsf{Z}^*(\mathbf{F}_1, \mathbf{F}_2)$-derivation.

1. If $h = \mathsf{Ax}_\Delta$, then $h^+ := \mathsf{Ax}_A$ and $\mathrm{tp}^{\mathbf{F}_1,\mathbf{F}_2}(h) := \mathsf{Ax}_A$, where A is the arbitrarily fixed element of $\Delta \cap \mathsf{TRUE}_0^{\mathbf{F}_1,\mathbf{F}_2}$.
2. If $h = \mathsf{S}^t_s h'$ and $h'^+ = \mathcal{I}((h_i)_{i\in|\mathcal{I}|})$, then $(\mathsf{S}^t_s h')^+ := \mathcal{I}(t \mapsto s)((\mathsf{S}^t_s h_i)_{i\in|\mathcal{I}(t/s)|})$.
3. If $h = \mathsf{Ind}^{y,t}_F h_0$ and $[\![t]\!] = n$, then $(\mathsf{Ind}^{y,t}_F h_0)^+ := \mathsf{Rep}(\mathsf{S}^n_t e_n)$, where $e_0 := h_{F(x/0)}$, $e_1 := h_0(y/0)$ and $e_{i+1} := \mathsf{R}_{F(x/i)} e_i h_0(y/i)$.
4. If $h = \mathsf{R}_C h_0 h_1$ and $h_l^+ = \mathcal{I}_l((h_{li})_{i\in|\mathcal{I}_l|})$ for $l \in \{0, 1\}$, then

$$(\mathsf{R}_C h_0 h_1)^+ := \begin{cases} \mathcal{I}_0((\mathsf{R}_C h_{0i} h_1)_{i\in|\mathcal{I}_0|}), & \text{if } C \notin \Delta(\mathcal{I}_0), \\ \mathcal{I}_1((\mathsf{R}_C h_0 h_{1i})_{i\in|\mathcal{I}_1|}), & \text{if } \neg C \notin \Delta(\mathcal{I}_1), \\ \mathsf{Cut}_{C[i]}(\mathsf{R}_C h_{0i} h_1 \mathsf{R}_C h_0 h_{10}), & \text{if } C \in \Delta(\mathcal{I}_0), \neg C \in \Delta(\mathcal{I}_1), \mathcal{I}_1 = \bigvee^i_{C_0 \vee C_1}, \\ \mathsf{Cut}_{C[i]}(\mathsf{R}_C h_{00} h_1 \mathsf{R}_C h_0 h_{1i}), & \text{if } C \in \Delta(\mathcal{I}_0), \neg C \in \Delta(\mathcal{I}_1), \mathcal{I}_0 = \bigvee^i_{C_0 \vee C_1}, \\ \mathsf{Cut}_{C[k]}(\mathsf{R}_C h_{0k} h_1 \mathsf{R}_C h_0 \mathsf{S}^t_k h_{10}), & \text{if } C \in \Delta(\mathcal{I}_0), \neg C \in \Delta(\mathcal{I}_1), \mathcal{I}_1 = \bigvee^t_{\exists x C_0(x)} \\ & \text{and } [\![t]\!] = k, \\ \mathsf{Cut}_{C[k]}(\mathsf{R}_C \mathsf{S}^t_k h_{00} h_1 \mathsf{R}_C h_0 h_{1k}), & \text{if } C \in \Delta(\mathcal{I}_0), \neg C \in \Delta(\mathcal{I}_1), \mathcal{I}_0 = \bigvee^t_{\exists x C_0(x)} \\ & \text{and } [\![t]\!] = k. \end{cases}$$

5. Otherwise, define h^+ in the same manner as $\mathrm{tp}(h)$ and $h[i]$.

By induction on the build-up of a given closed $\mathsf{Z}^*(\mathbf{F}_1, \mathbf{F}_2)$-derivation h, we have the following theorem.

Theorem 5.4 (Cf. Theorem 3.3) *Let h be a closed $\mathsf{Z}^*(\mathbf{F}_1, \mathbf{F}_2)$-derivation.*

1. $\dfrac{\ldots \Gamma(h(\mathbf{F}_1, \mathbf{F}_2)[i]) \ldots (i \in |\mathrm{tp}^{\mathbf{F}_1,\mathbf{F}_2}(h)|)}{\Gamma(h)}\ \mathrm{tp}^{\mathbf{F}_1,\mathbf{F}_2}(h)$ *is a correct $\mathsf{Z}^\infty(\mathbf{F}_1, \mathbf{F}_2)$-inference.*
2. *If* $\mathrm{tp}^{\mathbf{F}_1,\mathbf{F}_2}(h) = \mathsf{Cut}_C$, *then* $\mathrm{rk}(C) < \deg(h)$.
3. $\deg(h(\mathbf{F}_1, \mathbf{F}_2)[i]) \leq \deg(h)$ *for all* $i \in |\mathrm{tp}^{\mathbf{F}_1,\mathbf{F}_2}(h)|$.
4. $\mathrm{o}(h(\mathbf{F}_1, \mathbf{F}_2)[i]) \prec \mathrm{o}(h)$ *for all* $i \in |\mathrm{tp}^{\mathbf{F}_1,\mathbf{F}_2}(h)|$.

Define the set $\mathsf{Z}_0^*(\mathbf{F}_1, \mathbf{F}_2)$ as

$$\mathsf{Z}_0^*(\mathbf{F}_1, \mathbf{F}_2) := \{h \mid h \text{ is a } \mathsf{Z}^*(\mathbf{F}_1, \mathbf{F}_2)\text{-derivation with } \deg(h) = 0\}.$$

Hereafter, we denote by T a functional of level ≤ 2 such that it takes $\mathbf{F}_1, \mathbf{F}_2, \vec{u}$ as arguments and outputs a closed $\mathsf{Z}_0^*(\mathbf{F}_1, \mathbf{F}_2)$-derivation. The local correctness, the well-foundedness and normalization trees are defined in a similar manner to Sect. 4. Then, we have the following lemma and proposition.

Lemma 5.2 (Cf. Lemma 4.1) *For every T, every $\vec{u} \in \mathbb{N}^{<\omega}$ and every formula C,* $\mathrm{tp}^{\mathbf{F}_1,\mathbf{F}_2}(T_{\vec{u}}) \neq \mathsf{Cut}_C$ *holds.*

Proposition 5.1 (Cf. Proposition 4.1) *For every closed $\mathsf{Z}_0^*(\mathbf{F}_1, \mathbf{F}_2)$-derivation h, there is a well-founded normalization tree T of h such that T is a $< \varepsilon_0$-recursive functional with $\mathbf{F}_1$ and $\mathbf{F}_2$ as function-arguments.*

In what follows, our proofs for Lemmas 5.3 and 5.4 are inspired by Buchholz' proofs for Buchholz (1991, Lemma 4.4 and Theorem 4.5). We use the following notions:

- $\mathsf{INS}_p := \{\exists x_1 \exists x_2 p(x_1, f_1(x_1), x_2, f_2(x_1, x_2))\} \cup$ $\{\exists x_2 p(t_1, f_1(t_1), x_2, f_2(t_1, x_2)) \mid t_1 \text{ is a closed term}\} \cup$ $\{p(t_1, f_1(t_1), t_2, f_2(t_1, t_2)) \mid t_1, t_2 \text{ are closed terms}\}$.
- For every inference symbol $\mathcal{I}$ of $\mathsf{Z}^\infty(\mathbf{F}_1, \mathbf{F}_2)$,

$$\mathsf{WIT}_p(\mathcal{I}) := \begin{cases} \langle t_1, t_2 \rangle, & \text{if } \mathcal{I} \text{ is of the form } \mathsf{Ax}_{p(t_1, f_1(t_1), t_2, f_2(t_1, t_2))}, \\ \langle \rangle, & \text{else.} \end{cases}$$

- $w_p := \{\langle t_1, t_2 \rangle \mid p(t_1, f_1(t_1), t_2, f_2(t_1, t_2)) \in \mathsf{TRUE}_0^{\mathbf{F}_1,\mathbf{F}_2}\}$.

Lemma 5.3 *Let h be a closed $\mathsf{Z}_0^*(\mathbf{F}_1, \mathbf{F}_2)$-derivation with*

$$\Gamma(h) = \{\exists x_1 \exists x_2 p(x_1, f_1(x_1), x_2, f_2(x_1, x_2))\}.$$

If T is a normalization tree of h and $\mathrm{o}(T_{0^{k+1}}) \not\prec \mathrm{o}(T_{0^k})$, then

$$\mathrm{tp}^{\mathbf{F}_1,\mathbf{F}_2}(T_{0^k}) = \mathsf{Ax}_{p(t_1, f_1(t_1), t_2, f_2(t_1,t_2))}$$

for some closed terms t_1, t_2.

Proof By Lemma 5.2 and Theorem 5.4.(1), it holds that if $\Gamma(T_{0^k}) \subseteq \mathsf{INS}_p$, then

($\dagger$) $\mathrm{tp}^{\mathbf{F}_1,\mathbf{F}_2}(T_{0^k}) \in \{\mathsf{Rep}\} \cup \{\mathsf{Ax}_A \mid A \text{ is of the form } p(t_1, f_1(t_1), t_2, f_2(t_1, t_2))\} \cup$
$\{\bigvee^{t_1}_{\exists x_1 \exists x_2 p(x_1, f_1(x_1), x_2, f_2(x_1,x_2))} \mid t_1 \text{ is a closed term}\} \cup$
$\{\bigvee^{t_2}_{\exists x_2 p(t, f_1(t_1), x_2, f_2(t_1,x_2))} \mid t_1, t_2 \text{ are closed terms}\}$

for all $k \in \mathbb{N}$. We show that $\Gamma(T_{0^k}) \subseteq \mathsf{INS}_p$ by induction on k. The base case is obvious. Assume as IH that $\Gamma(T_{0^k}) \subseteq \mathsf{INS}_p$, then ($\dagger$) holds for T_{0^k}. By Theorem 5.4.(1) again, it holds that $\Gamma(T_{0^{k+1}}) \subseteq \mathsf{INS}_p$.

Assume that $\mathrm{o}(T_{0^{k+1}}) \not\prec \mathrm{o}(T_{0^k})$. Then, $\mathrm{tp}^{\mathbf{F}_1,\mathbf{F}_2}(T_{0^k}) = \mathsf{Ax}_A$ for some $A \in \mathsf{TRUE}_0^{\mathbf{F}_1,\mathbf{F}_2}$ by Theorem 5.4.(4). Since $\Gamma(T_{0^k}) \subseteq \mathsf{INS}_p$, A is of the form $p(t_1, f_1(t_1), t_2, f_2(t_1, t_2))$. □

Lemma 5.4 *Let h be a closed $\mathsf{Z}_0^*(\mathbf{F}_1, \mathbf{F}_2)$-derivation with*

$$\Gamma(h) = \{\exists x_1 \exists x_2 p(x_1, f_1(x_1), x_2, f_2(x_1, x_2))\}.$$

Then, there is a well-founded normalization tree T of h such that the following three statements hold:

1. *For every $k \in \mathbb{N}$, $\mathrm{o}(T_{0^k}) \preceq \mathrm{o}(h)$ holds,*
2. $\mathsf{WIT}_p(\mathrm{tp}^{\mathbf{F}_1,\mathbf{F}_2}(T_{0^m})) \in w_p$, *where $m = \min\{k \mid \mathrm{o}(T_{0^{k+1}}) \not\prec \mathrm{o}(T_{0^k})\}$,*
3. *let $\mathcal{G}_{\mathbf{F}_1,\mathbf{F}_2}$ be the functional of level ≤ 2 with $\mathbf{F}_1$ and $\mathbf{F}_2$ as function-arguments, which is defined as*

$$\mathcal{G}_{\mathbf{F}_1,\mathbf{F}_2}(T_{0^n}) := \begin{cases} \mathcal{G}_{\mathbf{F}_1,\mathbf{F}_2}(T_{0^{n+1}}), & \text{if } \mathrm{o}(T_{0^{n+1}}) \prec \mathrm{o}(T_{0^n}) \preceq \mathrm{o}(h) \prec \varepsilon_0 \\ \mathsf{WIT}_p(\mathrm{tp}^{\mathbf{F}_1,\mathbf{F}_2}(T_{0^n})), & \text{else,} \end{cases}$$

then $\mathcal{G}_{\mathbf{F}_1,\mathbf{F}_2}$ is a $< \varepsilon_0$-recursive functional and $\mathcal{G}_{\mathbf{F}_1,\mathbf{F}_2}(T_{0^0}) \in w_p$ holds.

Proof By Proposition 5.1, there is a normalization tree T of h such that T is a $< \varepsilon_0$-recursive functional with $\mathbf{F}_1$ and $\mathbf{F}_2$ as function-arguments.
(1) By Theorem 5.4.(4), $\mathrm{o}(T_{0^k}) \preceq \mathrm{o}(h) \prec \varepsilon_0$ holds for every $k \in \mathbb{N}$.
(2) By the well-foundedness of T, there is a $k \in \mathbb{N}$ such that $\mathrm{o}(T_{0^{k+1}}) \not\prec \mathrm{o}(T_{0^k})$ holds. The assertion holds by Lemma 5.3.
(3) By (1), (2) and the definition of $\mathcal{G}_{\mathbf{F}_1,\mathbf{F}_2}(T_{0^n})$, we can see that $\mathcal{G}_{\mathbf{F}_1,\mathbf{F}_2}(T_{0^0}) \in w_p$ and that $\mathcal{G}_{\mathbf{F}_1,\mathbf{F}_2}$ is a $< \varepsilon_0$-recursive functional. Note that $\mathrm{tp}^{\mathbf{F}_1,\mathbf{F}_2}(h)$ and $h(\mathbf{F}_1, \mathbf{F}_2)[i]$ are primitive recursive functional and T is a $< \varepsilon_0$-recursive functional. □

Now we prove Theorem 5.1.

Proof Let $\mathcal{G}_{\mathbf{F}_1,\mathbf{F}_2}$ be the functional defined in Lemma 5.4.(3) and $\mathbf{p}_0, \mathbf{p}_1$ be primitive recursive projection functions for pairs. Define

$$\mathbf{G}_0(\mathbf{F}_1, \mathbf{F}_2) := \mathbf{p}_0(\mathcal{G}_{\mathbf{F}_1,\mathbf{F}_2}(T_{\mathbf{0}^0})),\ \mathbf{G}_1(\mathbf{F}_1, \mathbf{F}_2) := \mathbf{p}_1(\mathcal{G}_{\mathbf{F}_1,\mathbf{F}_2}(T_{\mathbf{0}^0})).$$

Then,

$$\mathbf{P}(\mathbf{G}_0(\mathbf{F}_1, \mathbf{F}_2), \mathbf{F}_1(\mathbf{G}_0(\mathbf{F}_1, \mathbf{F}_2)), \mathbf{G}_1(\mathbf{F}_1, \mathbf{F}_2), \mathbf{F}_2(\mathbf{G}_0(\mathbf{F}_1, \mathbf{F}_2), \mathbf{G}_1(\mathbf{F}_1, \mathbf{F}_2)))$$

holds by Lemma 5.4.(3). □

6 Conclusion

First in this paper, we have answered the following question induced by Sieg's paper Sieg (2012): Is Gentzen's 1936 proof both a contentual correctness proof and a formal correctness proof? We have argued that the 1936 proof is both contentual and formal because the main lemma of the 1936 proof implies both the main lemma of the 1935 proof and the main lemma of the 1938 proof. In other words, the main lemma of the 1936 proof not only verifies the correctness of axioms and theorems of first-order arithmetic, but also assigns a normal derivation to each derivable numeric equation of first-order arithmetic. (Here, according to Gentzen, A is correct if and only if a reduction procedure is statable for it.)

Next, we have answered the question of how contentual and formal aspects of the 1936 proof relate to each other: Contentual aspects of the 1936 proof are formed by its formal aspects. In other words, the correctness of a derivable formula A of first-order arithmetic is in fact shown by means of syntactic transformation of a given derivation of the formula.

Finally, we have noted several consequences of our argument for the answers above. Our argument has given a uniform reading of Gentzen's three consistency proofs and another proof of Kreisel's no-counterexample interpretation.

As mentioned in the introduction, we have focused not only on the mathematical side of Gentzen's consistency proofs, but also on their conceptual side. Our whole argument indicates that contentual and formal aspects of Gentzen's consistency proofs are fruitful products of an interaction between philosophical ideas and mathematical proof-theoretical methods.

References

Akiyoshi, R. (2010). Gentzen's first consistency proof revisited. In *CARLS series of advanced study of logic and sensibility* (Vol. 4, pp. 315–324). Tokyo: Keio University.

Akiyoshi, R., & Takahashi, Y. (2013). Reading Gentzen's three consistency proofs uniformly. (in Japanese) *Kagaku kisoron kenkyū (Journal of the Japan Association for Philosophy of Science) 41*, 1–22.

Arai, T. (2002). Review: Three papers on proof theory by W. Buchholz and S. Tupailo. *The Bulletin of Symbolic Logic, 8*, 437–439.

Bernays, P. (1970). On the original Gentzen consistency proof for number theory. In A. Kino, J. Myhill, & R. Vesley (Eds.), *Intuitionism and proof theory*. Amsterdam: North-Holland.

Buchholz, W. (1991). Notation systems for infinitary derivations. *Archive for Mathematical Logic, 30*, 277–96.

Buchholz, W. (1997). Explaining Gentzen's consistency proof within infinitary proof theory. In G. Gottlob, A. Leitsch, & D. Mundici (Eds.), *Computational logic and proof theory: 5th Kurt Gödel colloquium, KGC'97*. Lecture notes in computer science (Vol. 1289, pp. 4–17). Berlin: Springer.

Buchholz, W. (2001). Explaining the Gentzen-Takeuti reduction steps: A second-order system. *Archive for Mathematical Logic, 40*, 255–72.

Buchholz, W. (2010). Another reduction of classical ID_ν to constructive ID^i_ν." In Schindler, R. (Eds.), *Ways of proof theory* (pp. 183–198). Frankfurt: Ontos Verlag.

Buchholz, W. (2015). On Gentzen's first consistency proof for arithmetic. In R. Kahle & M. Rathjen (Eds.), *Gentzen's centenary: The quest for consistency* (pp. 63–87). Switzerland: Springer International Publishing.

Gentzen, G. (1936). Die Widerspruchsfreiheit der reinen Zahlentheorie. *Mathematische Annalen, 112*, 493–565.

Gentzen, G. (1938). Neue Fassung des Widerspruchsfreiheitsbeweises für die reine Zahlentheorie. *Forschungen zur Logik und zur Grundlegung der exakten Wissenschaften, 4*, 19–44.

Gentzen, G. (1969). In M.E. Szabo (Ed.), *The Collected Papers of Gerhard Gentzen*. Amsterdam: North-Holland.

Gentzen, G. (1974). Der erste Widerspruchsfreiheitsbeweis für die klassische Zahlentheorie. *Archiv für mathematische Logik und Grundlagenforschung, 16*, 97–118.

Gödel, K. (1995). In S. Feferman, J. Dawson, S. Kleene, G. Moore, R. Solovay, & J. van Heijenoort (Eds.), *Collected works. III: Unpublished essays and lectures*. Oxford: Oxford University Press

Hilbert, D. (1923). Die logischen Grundlagen der Mathematik. *Mathematische Annalen, 88*, 151–65.

Hilbert, D. (1928). Die Grundlagen der Mathematik. *Abhandlungen aus dem Seminar der Hamburgischen Universität, 6*, 65–85.

Hilbert, D. (1931). Beweis des tertium non datur. *Nachrichten von der Gesellschaft der Wissenschaften zu Göttingen, Mathematische-physikalische Klasse* (pp. 120–125).

Kreisel, G. (1951). On the interpretation of non-finitist proofs - part I. *The Journal of Symbolic Logic, 16*(4), 241–67.

Kreisel, G. (1971). Review: The collected papers of Gerhard Gentzen. by M. E. Szabo. *The Journal of Philosophy, 68*, 238–65.

Mints, G. E. (1978). Finite investigations of transfinite derivations. *Journal of Soviet Mathematics, 10*, 548–96.

Negri, M. (1980). Constructive sequent reduction in Gentzen's first consistency proof for arithmetic. In Maria Luisa Dalla Chiara (Ed.), *Italian studies in the philosophy of science* (pp. 153–68). Dordrecht: D. Reidel Publishing Company.

Schwichtenberg, H. (1977). Proof-theory: Some applications of cut-elimination. In J. Barwise (Ed.), *Handbook of mathematical logic* (pp. 867–95). Amsterdam: North-Holland.

Sieg, W. (2012). In the shadow of incompleteness: Hilbert and Gentzen. In P. Dybjer, S. Lindström, E. Palmgren, & G. Sundholm (Eds.), *Epistemology versus ontology: Essays on the philosophy and foundations of mathematics in honour of Per Martin-Löf* (pp. 87–127). London: Springer.

Sieg, W., & Parsons, C. (1938). Introductory note to 1938*. In [14, pp. 62–85].

Sundholm, G. (1983). *Proof theory: A survey of the omega rule.* Oxford D. Phil. dissertation.
Tait, W. W. (2001). Gödel's unpublished papers on foundations of mathematics. *Philosophia Mathematica 9*, 87–126 (2001).
Tait, W. W. (2005). Gödel's reformulation of Gentzen's first consistency proof for arithmetic: The no-counterexample interpretation. *The Bulletin of Symbolic Logic, 11*, 225–38.
Tait, W. W. (2015). Gentzen's original consistency proof and the Bar Theorem. In R. Kahle & M. Rathjen (Eds.), *Gentzen's centenary: The quest for consistency* (pp. 213–28). Switzerland: Springer International Publishing.
Takeuti, G. (1987). *Proof Theory* (2nd ed.). Amsterdam: North-Holland.
Troelstra, A. S., & van Dalen, D. (1988). *Constructivism in mathematics: An introduction* (Vol. 1). Amsterdam: Elsevier.
von Plato, J. (2009). Gentzen's original proof of the consistency of arithmetic revisited. In G. Primiero & S. Rahman (Eds.), *Acts of knowledge - history, philosophy and logic* (pp. 151–71). London: College Publications.
Yasugi, M. (1980). Gentzen reduction revisited. *Publications of RIMS, 16*, 1–33.

Saving Supervaluationism from the Challenge of Higher-Order Vagueness Argument

Hao-Cheng Fu

Abstract The phenomenon of the higher-order vagueness is one of the most controversial issues of the semantics for vagueness. Timothy Williamson has argued that any theory would inevitably fail if it aims at explicating higher-order vagueness in terms of revising classical logic, for the problem of higher-order vagueness consists not in the inadequacy of classical logic but rather in the limitation of our cognitive ability. Williamson proposes the *epistemic theory* and claims that it is the best option to dissolve the challenge of the phenomenon of higher-order vagueness. Nevertheless, I disagree with Williamson. In this paper, I will argue that the most promising response to the challenge of higher-order vagueness argument is supervaluationism rather than the epistemic theory. However, the orthodox supervaluationism cannot dissolve the phenomenon of higher-order vagueness because the model is static. I suggest that we should turn the static model of supervaluationism into a dynamic one, which, as I will argue, release the pressure of supervaluationism from the challenge of higher-order vagueness argument.

Keywords Vagueness · Higher-order vagueness · Supervaluationism · Timothy Williamson

1 Vagueness and Higher-Order Vagueness

The puzzle of vagueness is initiated by Eubulides of Miletus who was a philosopher of the Megarian school. Eubulides calls the puzzle of vagueness "The heap paradox". Intuitively, all of us would grant that there is a heap if the number of grains of wheat is large enough, for example, if the number of grains of wheat is greater than 10,000. Moreover, we would also grant that the remaining still count as a heap if we just remove one grain of wheat from the given pile of grains of wheat, which we initially judge as a heap. The slight variation would have no effect on our judgments. Conversely, we would also grant that there is a non-heap if the number of grains

H.-C. Fu (✉)
Department of Philosophy, Chinese Culture University, Taipei, Taiwan
e-mail: fhz2@faculty.pccu.edu.tw

S.C.-M. Yang et al. (eds.), *Philosophical Logic: Current Trends in Asia*,
Logic in Asia: Studia Logica Library, DOI 10.1007/978-981-10-6355-8_7

of wheat is too small, for example, if the number of grains of wheat is only 1 or 2. Similarly, the grains of wheat, which we intuitively judge as not a heap, would still be counted as a non-heap when just one more grain of wheat is added. In the same vein, such a slight variation is not sufficient to affect us to turn the thought from a non-heap into a heap. Nevertheless, it follows that we would be puzzled by the concept of "a heap" because the considerations above eventually lead to a clash of intuition.

In general, most philosophers believe that the puzzle of vagueness can be represented adequately by certain forms of sorites paradox. For the sake of simplicity, I divide the argument forms into two directions:

(Upward argument)

1 grain of wheat does not make a heap.
If n grains of wheat do not make a heap then $n + 1$ grains of wheat do not.

So, 10,000 grains of wheat do not make a heap.

(Downward argument)

10,000 grains of wheat make a heap.
If $n + 1$ grains of wheat make a heap then n grains of wheat do.

So, 1 grain of wheat makes a heap.

It is no doubt that neither the conclusion of upward argument nor the conclusion of the downward argument is acceptable. Nonetheless, the premises of each argument seem intuitively reasonable, and both arguments are valid. It is worth mentioning that the second premise of each argument is schematic- they stand for a series of sentences rather than single one. For instance, in the upward argument, we can draw the conclusion "2 grains of wheat do not make a heap" from the first premise "1 grain of wheat does not make a heap" and substitute the number of "n" with 1 to the second premise, i.e. "If 1 grain of wheat does not make a heap then 2 grains of wheat do not". In the next step, we can draw the conclusion that "3 grains of wheat do not make a heap" from the deduced premise "2 grains of wheat do not make a heap" and substitute the number of "n" with 2 to the second premise, i.e. "If 2 grains of wheat do not make a heap then 3 grains of wheat do not". It follows that we would reach the undesirable conclusion that "10,000 grains of wheat do not make a heap" if we carry out the same procedure over and over again. The result violates our intuitive judgments that 10,000 grains of wheat *ipso facto* make a heap. Furthermore, we can also draw the undesirable conclusion "1 grain of wheat makes a heap" by the downward argument in a similar fashion. Again, the result of the downward argument also violates our intuitive judgments that 1 grain of wheat does not make a heap either.

There are three possible strategies to dissolve the puzzle of vagueness. As Sainsbury has suggested (Sainsbury 2009: 47):

(a) Accept the conclusion of the argument.
(b) Reject the reasoning as faulty.

(c) Reject one or more premises.

Among these strategies, (c) is the most popular one. For example, Unger (1979) advocates that we should accept the conclusions of the aforementioned arguments. He argues that the problem with these arguments consists in the vacancy of the concept of heap rather than the faulty of deductive reasoning. Furthermore, he says that there are, strictly, no ordinary things such as tables, stones, etc.:

> The most rational way of responding to the contradictions, I submit, is to deny application for the ordinary concepts: the concepts of stones, tables, feathers and sousaphones. These concepts do not apply. We cannot say, then, 'All feathers are stones', unless we mean it in some peculiar sense of the logicians. But a vivid way of putting our point is to say 'There are no stones' (Unger 1979: 145–6).

Unger's conclusion seems very odd at first glance, for it implies that there are very few (if at all) ordinary things. It seems to me that we should accept Unger's proposal only if all the other options are even less plausible. In other words, we should reject his suggestion if there is a way to explain the phenomenon of vagueness without sacrificing our intuitions about the existence of ordinary things.

Surprisingly, the puzzle of vagueness does not cease to exist even if we provide a tenable solution to first-order vagueness. Why? The difficulty is twofold. Firstly, consider the following case. We all had been a child. But can we point to an exact time when we cease to become a child and turn into an adult? The answer is obviously 'no'. Perhaps someone would argue that the exact time a child turns into an adult is this person's eighteenth birthday. Unfortunately, this answer is unsatisfactory because we can always minimize the limitation of "the exact time" to generate a new puzzle, e.g., is the exact time 6 o'clock a.m. or 6 o'clock p.m. on that day? Even if he insisted that the exact time was a specific moment, e.g. 6 o'clock a.m., we could always continue to inquire. Is it possible that the exact time is one minute more (or less) than the specific moment? How about one second more (or less)? How about one tenth of a second more (or less)? How about one millionth of a second more (or less)? And so on.[1] The sequence of requests shows that the boundary of an exact time is always vague itself.

On the other hand, consider the heap paradox again. Intuitively, we all regard that there are some extreme cases fall into one end of the spectrum, e.g. 10,000 or more grains of wheat indeed make a heap. On the contrary, some extreme cases fall into the opposite side of the spectrum, e.g. 1 or 2 grains of wheat do not make a heap. Between the extreme cases, there are many borderline cases. So far, it seems perfectly fine if we divide all cases into three classes: the *positive extension*, the *negative extension* and the *penumbra*. But a new puzzle appears once we attempt to distribute all cases into these three classes. Can we distribute all cases into one of these classes definitely? Are there sharp boundaries between the positive extension, the penumbra, and the negative extension? Once we start thinking about these issues, it seems obvious that it is hard for us to draw any the sharp boundaries here. Indeed, we are facing the

[1]I have learned the way of this request from the anti-luminosity argument which appears in the chapter four of Timothy Williamson's *Knowledge and Its Limits*.

second-order vagueness in this case. Timothy Williamson characterizes second-order vagueness as follows:

> Second-order vagueness shows itself in an object-language with a 'definitely' operator. If vagueness were only first order, 'Definitely A' would be precise, so 'Either definitely A or definitely not definitely A' would hold. But 'Definitely A' has borderline cases, for otherwise there would be sharp second-order boundaries (Williamson 1994: 157).

D. Hyde proposes another way to represent the phenomenon of second-order vagueness, which appeals to the notion of border cases:

> [T]hat vague predicates not only possess border cases but border cases and border cases of border cases, etc. The worry was that the simple requirement that vague predicates have border cases could not be sufficient for vagueness seemed to have border cases (cases to which the predicate did not appear to determinately apply nor determinately not apply) (Hyde 1994: 39).

If the second-order vagueness is an inevitable consequence resulting from our attempt to comprehend the concept of vagueness, then we are facing the problem of characterizing an infinite series of the hierarchy of different levels of vagueness. This is the so-called *phenomenon of higher-order vagueness*. The phenomenon of higher-order vagueness is puzzling since we can address the same type of question in any different level of vagueness. In my opinion, the phenomenon of higher-order vagueness is indeed a serious challenge, perhaps the most serious challenge to the problem of vagueness. Hence, I take it that there is no way for one to have a proper understanding of vagueness unless the problem of higher-order vagueness is resolved. Thus, it is necessary for any proper theory of vagueness to also provide a proper account to the phenomenon of higher-order vagueness.

2 The Naïve Strategies to Vagueness and Higher-Order Vagueness Challenge

The most intuitive way to solve the puzzle of vagueness is to distribute all cases into three different classes based on the appearance of such cases. From a semantic point of view, the main idea of three-valued logic is that all sentences which include the vague terms map three different values rather than two as adopted by the classical logic. From a three-valued logic point of view, there are some cases which definitely satisfy the predicate, so "a is F" is true if a definitely satisfies the predicate F. On the contrary, "a is F" is false if a definitely does not satisfy the predicate F.Nevertheless, there is an additional value that "a is F" is neither true nor false if the object a indefinitely satisfies or dissatisfies the predicate F.

A crucial question naturally arises. How should we elaborate the idea that an object a indefinitely makes the sentence "a is F" true or false? It seems to me that there are at least two different ways to develop the meaning of the word "indefinitely". On the one hand, one may construe "indefinitely" as saying that the object a definitely belongs to the area that is indefinitely true or false, i.e., the penumbra. In other words, the semantic values of sentential forms such as "*a* is *F*" are definitely indefinite, i.e., such sentential forms are definitely neither true nor false. On the other hand, the meaning of "indefinitely" could be understood as indicating that we are uncertain whether an object *a* belongs to the positive extension, negative extension, or penumbra. In other words, *a* indefinitely makes the sentential form "*a* is *F*" true, neutral or false. Traditionally, the advocators of three-valued logic adopt the first reading of "indefinitely". Unfortunately, this reading has to bite the bullet of the phenomenon of higher-order vagueness. As Williamson aptly points out:

> Second-order vagueness poses a general threat to three-valued logic. If three-valued tables define the standard functors then (a) the three values exhaust all possibilities (otherwise the definition would not cover all cases) and (b) any three-valued truth-table should define some operator (otherwise some constraint on meaning has been left unaccounted for) (Williamson 1994: 112).

The fatal defect of the three-valued logic is the requirement that every sentence has a definite semantic value. It means that there is no second or higher order vagueness since every sentence of the form "*a* is *F*" has a specific semantic value. Put in another way, the existence of the phenomenon of higher-order vagueness threatens the hope of having a *three-valued* logic regarding sentences consisting of vague terms.

The other naïve strategy of solving the puzzle of vagueness appeals to the notion of the degree of truth. The main idea is to assume that the semantic value of every sentence is a positive real number between the closed interval [0, 1]. Let the notation $[p]$ stand for the semantic value of any sentence p. The semantic rules of the standard operators can be shown as the following:

(i) $[p] = d(0 \leq \ d \leq 1)$
(ii) $[\neg p] = 1 - d$
(iii) $[p \wedge q] = \min\{[p], [q]\}$
(iv) $[p \vee q] = \max\{[p], [q]\}$
(v) $[p \supset q] = 1 + \min\{[p], [q]\} - [p]$
(vi) $[p \equiv q] = 1 + \min\{[p], [q]\} - \max\{[p], [q]\}$

Let us put the details aside and concentrate on the problem of the theory of the degree of truth in play. At the very beginning, such a theory suffers from the same difficulty as the three-valued logic. That is, it also presupposes sharp boundaries on the second or higher-order level. Specifically, the theory cannot distinguish the closest cases whose semantic value is less than 1 from the cases whose semantic value is definitely 1. Similarly, the theory is incapable of distinguishing between the closest cases whose semantic value is greater than 0 and the cases whose semantic value is definitely 0. Consider a red-yellow gradient paper. It is hard to draw a sharp boundary between red and yellow, and it is equally hard to draw a sharp boundary between red and non-red. Williamson has put this point nicely:

> Unfortunately, it is not obvious that the notion of degree of truth appropriate for the semantics of comparative and other modifiers is also appropriate for the analysis of vagueness. Of course, that form is relevant only to a very limited range of comparisons. It does not tell us when 'Snow is white' truer than 'Snow is cold', nor does it apply to complex sentences (Williamson 1994: 126).

The naïve strategies, I take it, are unsuccessful. Neither three-valued logic nor the degree of truth is able to explain the phenomenon of second-order vagueness and, *a fortiori*, the phenomenon of higher-order vagueness. This is an undesirable result since the phenomenon of higher-order vagueness is both genuine and legitimate. That is, we can always ask the existence of sharp boundaries at different levels, and, as mentioned above, the $n + 1$ level vagueness will be generated spontaneously whenever we try to account for the n level vagueness. As the result of the failure of the naïve strategies, we should search for an alternative way to solve or dissolve the challenge of higher-order vagueness.

3 Supervaluationism and Williamson's Criticisms

Unlike the naïve strategies, supervaluationism proposes a different solution to the puzzle of vagueness. Why does the phenomenon of vagueness emerge? We may understand the phenomenon of vagueness in terms of the disagreement among people. For example, some people are inclined to claim that Barack Obama is bald, but some are inclined to claim that he is not. How should we react when facing a disagreement? Supervaluationists propose that we should collect the data regarding the disagreement, and deal with them by assigning different levels of values. Such a strategy is different from the naïve strategies, which invite an additional value different from the traditional bivalence at the primary level. Roughly speaking, the advocators of the naïve strategies would say that the controversial cases (e.g., Obama's baldness) fall into a neutral area, which is different from definite truth or definite falsity, or that such cases must be seen as a degree of definiteness between 1 and 0. As noted, the traditional bivalence is unable to account for the phenomenon of vagueness. Nor does the proposal of the degree of truth. Contrary to the naïve strategies, supervaluationism suggests that we should assimilate all disagreements and assign them to certain *admissible interpretations*.[2] Consider the case of Obama's baldness. Some state that he is bald, while others insist that he is not. Whose statement is more reasonable or correct? The answer is neither. The former is no more tenable than the latter, any more than the latter is more tenable than the former. From the supervaluationist point of view, all the interpretations must be treated as equal. At any rate, this is so at the primary level.

The original idea of supervaluation is proposed by Bas van Fraassen (1966, 1969), who exploits the mechanism to deal with non-referential singular terms in free logic. Later, Kit Fine (1975) uses the idea of supervaluation as a solution to the problem of vagueness, which is about the reconciliation between the indefinite truth-values and some other theorems in classical logic such as the law of excluded middle. As Fine puts it:

[2]The name of *admissible interpretations* originated from Kit Fine, and he took the notion of admissibility as primitive. (Fine 1975: 272).

> However, such a model would suffer from familiar difficulties. For example, if the meaning of A is relevant to the meaning of $A \vee -A$, then the vagueness of A should be relevant to the vagueness of $A \vee -A$. Thus $A \vee -A$ should be vague for vague A, though on the super-truth account, say, it is equally and completely precise for all A (Fine 1975:274).

In fact, the struggle of the primary level does not bother supervaluationists because they take it that we have no need to try to dismiss the dispute at the primary level as proposed by the naive strategies. I have already shown in the previous section that one will inevitably suffer from the phenomenon of higher-order vagueness if one tries to draw sharp boundaries between penumbra and positive extension as well as between penumbra and negative extension. Will supervaluationism suffer from the same flaw as the naïve strategies do? To answer, let us start with accounting for the main ideas of supervaluationism. Supervaluationists suggest that we can assimilate all admissible interpretations regarding *supervalues*. This means that whenever we want to decide the semantic value of a statement that includes a certain vague term, we should reflect the admissible interpretations already in hands. The way to assign the semantic value to formulae is as the following:

(a) A statement is *supertrue* if it is true in all admissible interpretations at the primary level.
(b) A statement is *superfalse* if it is false in all admissible interpretations at the primary level.
(c) A statement is neither true nor false otherwise.

At first glance the advantages of supervaluationism are twofold: (1) the theory avoids the primary-level dispute since it takes every admissible interpretation to be equal at this level, and (2) it retains the two-valued semantics at the primary level. So, the phenomenon of vagueness should be regarded as happening at the secondary level rather than the primary level. Unlike the naïve strategies, supervaluationists suggest that we should turn our eyes to the secondary level. Unfortunately, supervaluationism still leaves room for higher-order vagueness since it does not develop a reasonable explanation as to why some interpretations could be counted as admissible but others could not. The framework of supervaluationism should provide a way out to this question. Otherwise, the theory is unable to solve the problem of vagueness.

It is worth mentioning that there are three possible options to meet the higher-order-vagueness challenge. The first is to acknowledge that the phenomenon of higher-order vagueness is fatal to any variant of logical systems, and to admit that we should not seek answer from (classical or non-classical) logic at all. For example, Williamson claims that the phenomenon of vagueness is *ipso facto* caused by our ignorance rather than by the inadequacy of classical logic. The second is to contend that higher-order vagueness is just a pseudo problem. Both D. Hyde and M. Tye adopt this approach. Here is what Hyde says:

> I think that there are higher orders of vagueness, but that this is already entailed by the paradigmatic conception and can be seen to follow when the notion of "border cases" employed therein is properly understood (Hyde 1994: 39).

In addition, Tye has proposed another way to show that higher-order vagueness does not play an important role as what is vaguely vague does.[3] He says:

> In my view, the vaguely vague, unlike the higher-order vague, plays an important role in dissolving the sorites paradoxes, as they apply to common or garden vague predicate (Tye 1994: 45).

Both Hyde and Tye deny that higher-order vagueness has much effect on our understanding of vague predicates. On their view, we have no need to develop any positive account of higher-order vagueness. I agree with the claim that higher-order vagueness does not have much impact on our understanding of vague terms. But I disagree that there is no need to develop a positive account to explain the puzzle of higher-order vagueness. However, it seems to me that the best strategy is to tackle the question of vagueness at the secondary level, i.e. to develop a positive account to eliminate the threat of higher-order vagueness. In the next section, I will amend some stipulations of supervaluationism in order to meet the challenge of higher-order vagueness.

Before proceeding to my positive account, let us briefly review Williamson's criticisms to supervaluationism. Williamson argues that the phenomenon of higher-order vagueness is a serious problem for supervaluationism. Williamson proposes three reasons that show that supervaluationism cannot handle the puzzle of vagueness. More precisely (1) supertruth is elusive; (2) supertruth is disquotational; (3) supervaluationism fails to govern higher-order vagueness, respectively. In my opinion, (3) is the most serious one. Supervaluationism is untenable unless we can provide appropriate semantic rules to govern higher-order vagueness.[4] Williamson's assessment is as follows:

> If a vague term is governed by semantic rules, then they are formulated in equally vague terms. Moreover, it is not plausible that the limits of higher-order vagueness are laid down by a hierarchy of higher-order semantic rules. When admissibility is not pictured as consistency with semantic rules, supervaluationism becomes a less inviting approach (Williamson 1994: 158).

---

[3]The main purpose of Tye's paper is to show that Hyde fails to demonstrate that higher-order vagueness is a real phenomenon but a pseudo one. However, it seems to me that Tye has attempted to dissolve higher-order vagueness and emphasized the importance of vaguely vague, perhaps vaguely vague. Nevertheless, the success of the position depends on the soundness of Sorenson's argument.

[4]Williamson has proposed three difficulties to supervaluationism, the other two are connected to the two-valued logic. As he said, "According to supervaluationism, '*p* or *q*' is sometimes true when no answer to the question 'Which?' is true. For similar reasons, 'Something is *F*' is sometimes true when no answer to the question 'Which thing is *F*?' is true. In this sense supertruth is elusive." (Williamson 1994, p. 153) and "Truth is standardly assumed to have the disquotational property to which Tarski drew attention. ...... Supertruth is not disquotational. If it were, then the supervaluationist would be forced to admit bivalence. Consider any sentence '*A*'. By supervaluationist logic, either *A*or not *A*. Suppose that supertruth is disquotational. Thus '*A*' is supertrue if and only if *A* and 'Not *A*' is supertrue if and only if not *A*. It would then follow, by more supervaluationist logic, that either '*A*' is supertrue or 'Not *A*' is supertrue; in the latter case, '*A*' is superfalse."(Williamson 1994, p. 162) I would ignore these two difficulties here because they are connected the problem of semantic rules for vagueness directly.

I agree with Williamson's assessment of supervaluationism, especially his contention of the inadequacy of admissibility. So, some assumptions of the model of supervaluationism need to be critically reconsidered. There are two options. The first option is to acknowledge that the flaws are fatal and then side with Williamson to abandon this theory. The alternative is to revise supervaluationism to meet Williamson's challenge. In the next section, I would develop a revised version of supervaluationism that dissolves the puzzle of vagueness. I argue that my account offers an adequate account for the phenomenon of higher-order vagueness.

4 Admissibility as Dynamic

We can draw some conclusions based on the previous discussions. First of all, the higher-order vagueness is the core of the problem of vagueness. Moreover, the orthodox supervaluationist solution is unable to resolve the problem of higher-order vagueness. Williamson has argued that many prominent accounts have mistakenly focused on the first-order vagueness, while overlooking the phenomenon of higher-order vagueness, beneath which the real problem lies. The orthodox supervaluationism also fails to take the phenomenon of higher-order vagueness into account. In the previous section, I have mentioned that both Tye and Hyde want to dissolve the problem of the phenomenon of higher-order vagueness by dismissing the effect the phenomenon higher-order vagueness. It seems to me that neither Tye's nor Hyde's solution is appropriate. Tye suggests that "vaguely vague" is the appropriate interpretation of vague terms rather than higher-order vagueness. But it is hard to distinguish "vaguely vague" terms from higher order vagueness. There are two possible ways to specify what Tye means by "vaguely vague". First, we may apply "vaguely vague" only to vague predicates but not to border cases. This application is odd because there is no appropriate way to explain the meaning of "vaguely vague". Second, "vaguely vague" may be applied to border cases. But this means that we need to commit to the existence of border cases. Such a commitment is problematic since we cannot distinguish "vaguely vague" so construed from higher-order vagueness. On the other hand, Hyde argues that the higher-order vagueness is a pseudo question since every vague term must be vague. The supervaluationist R. Keefe has proposed a similar way to defend that supervaluationism is indeed an appropriate way to dissolve the phenomenon of higher-order vagueness. The key point is to reinterpret admissibility in an appropriate way. She proposes three principles for any proper theory of vagueness:

> As a starting point, we can take three things that the earlier chapters suggest we should seek from a theory of vagueness. First, we would like the theory to capture the thought that borderline case predications are neither true nor false, but that they do not each take some specific (numerical) value either. Second, and relatedly, we should avoid the commitment to sharp boundaries with vague predicates: for there is nothing in our language, its use, or the world that determines particular locations for such hidden boundaries. And third, we want a tried and tested logic, avoiding surprising and counter-intuitive consequences concerning principles or inferences that should be unaffected by the acknowledgment of vagueness: classical logic would surely be ideal. (Keefe 2000: 153)

Moreover, Hyde's view of higher-order vagueness is similar with W. C. Salmon's talk about the iterated "why?" question in scientific explanation. They both claim that the infinite regress of the same request is nothing new at the higher level. So the requests to the puzzle of vagueness are meaningful only if they are conducted on the first and the second level:

> Concerning the first order question "Why?". I have raised the second order question "Why ask, 'Why?'?" to which you might naturally respond with the third order question "Why ask, 'Why ask, "Why?"'?" But this way lies madness, to say nothing of an infinite regress. While an infinite sequence of nested intervals may converge upon a point, the series of nested questions just initiated has no point to it, and so we had better cut it off without delay (Salmon 1978:683).

> There are border cases for vague predicates, but this need not be stated as part of the concept of predicate-vagueness any more than, having said that "red" is a predicate, one must then go on state that "'red' is a predicate" is a predicate and that ""red' is a predicate' is a predicate' is a predicate, etc. One is simply repeating oneself and adding nothing new. The vagueness of "vague" can be seen to be built in right form the very start, when we characterised vagueness in terms of border cases, once we have conceded (as we must since "vague" is vague) that the sense of "border case" employed in the characterisation is the vague sense (Hyde 1994: 40).

If the stipulation of supervaluationism is to be restored, the most pressing and critical task is to deliberate the behavior of admissible interpretations. First of all, the admissible interpretations that are collected must be finite based on the *precisification*[5] requirement. Moreover, Williamson complains that supervaluationism is unable to apply to future experiences. Contrary to Williamson, I want to convert the direction into the opposite way to replace the role of future experiences. That is, I propose that future experiences may revise the extent of penumbra rather than behave as admissible interpretations. To be more precise, the future experiences should be counted as the joint interpretation of prior interpretations. In other words, we allow the value of a statement to be changed by future experiences.

To properly characterize the difficulty that supervaluationism faces, we need to examine the model of supervaluation. First of all, an important feature of the model of supervaluationism is that it is a static quadruple model, which can be written as $M =< O, P, f, v >$. On this model, the first factor "O" stands for a set of objects, the second factor "P" for the set of vague predicates, the third factor "f" for a set of precisification functions, and the last factor "v" for the set of semantic values. Every precisification function, say f_i, embraces a sharp boundary between the positive and the negative extensions. So, we could formulate atomic formula with an object in O, say a_i, and any vague predicate, sayP^i into the form of $P^i(a_i)$. And the set of

[5]"Precisification" means that there is a sharp boundary between true and false in every admissible interpretation. So, let us stipulate the *precisification* (or in Sainsbury's word, *sharpening*) $s(w)$ of a vague predicate and the value of a sentence meets the following conditions, (Sainsbury 2009:52)

- If w is definitely true of something, then $s(w)$ is true of it.
- If $w$is definitely false of something, then $s(w)$ is false of it.
- For each object, $s(w)$ is either true of it or false of it.
- $s(w)$ respects the underlying ordering (if there is one). For example, if s("tall") is true of someone $6'$ tall, it is also true of someone $6'1''$.

v consists of three different values, True, False and Indefinite. Any formula $P^i(a_i)$ would be true or false in every precisification function due to the sharp boundary. In brief, any formula of the form $P^i(a_i)$ meets the following conditions,
(1) $P^i(a_i)$ is definitely true if $P^i(a_i)$ is true in every precisification function.
(2) $P^i(a_i)$ is definitely false if $P^i(a_i)$ is false in every precisification function.
(3) $P^i(a_i)$ is neither true nor false if $P^i(a_i)$ is true in some precisification functions and false in others.
The static model can easily provide an account for how we decide the extensions in the spectrum which objects in O falls, i.e. the positive extension, the negative extension, or the penumbra. But how can we justify the given set of precisification functions, or admissibility interpretations? The problem is that supervaluationists claim that their goal is to develop certain 'complete and admissible specifications'[6] that explain the phenomenon of vagueness. There are two scenarios that are related to such 'complete and admissible specifications', and both give rise to some difficulties. The first scene is that it is hard to get 'complete and admissible specifications' if the number of possible admissibilities is infinite. The other is that if the 'complete and admissible specifications' is just meant for one single agent then it is impossible to conciliate the debate about different versions.

Keefe has offered a way out to these problems. On her view, not only is the object language vague, but the metalanguage is vague as well:

> This argument is too quick as it stands, but it does encapsulate a popular line of objection to supervaluationism. We need to examine a key assumption on which it rests, namely that there is a precise and unique set of complete and admissible specifications. If 'complete and admissible specification' is vague, we should reject this assumption. (Keefe 2000: 202)

Let us begin with the first level of vagueness. How does an agent acquire the precisification functions at this level? On an intuitive interpretation, the agent simply collects the judgments from people one by one. Nevertheless, two further questions arise immediately. First, how should we decide whether an interpretation is admissible or not? Secondly, how should we deal with various controversial cases? The static model does not seem to have appropriate answers to these questions. This is the reason why we need to amend the static model into a dynamic one.

What does the term "dynamic" mean? It is based on the eminent idea of dynamic epistemology. P. Gärdenfors is one of the founders who develops a dynamic model that represents the process of the change of epistemic states (or belief sets) (Gärdenfors 2008). The theory is well-known as *AGM* theory (the milestone of this theory is 'On the Logic of Theory Change: Partial Meet Contraction and Revision Functions' published in 1985 and the authors are C. E. Alchourrón, P. Gärdenfors, and D. Makinson). Unlike traditional epistemology, *AGM* theory concentrates on the process of the change of the epistemic states of an agent rather than the sufficient and necessary conditions of knowledge (Alchourrón et al. 1985; Gärdenfors 2008). The essential elements of the dynamic model of belief revision are epistemic states (or belief sets), the notation is K, and the functions operate on K from $\boldsymbol{K} \times \boldsymbol{L}$ to $\boldsymbol{K}$.

---

[6]The phrase 'complete and admissible specifications' is from Keefe (2000): Chap. 8.

There are three ways to change the epistemic states. The first is to add the sentence A in the given epistemic state K directly where A is consistent with K. The belief change should comply with the principle of minimal change. It means that the given set should respond to any disturbance with minimal variations. Thus there is no other procedure needed to proceed except adding the sentence into the given epistemic state when the sentence A is consistent with the given epistemic state K. This procedure is called *expansion*. The second way is to give up some beliefs when we accept the sentence A when the sentence A is inconsistent with the given epistemic state K. The adjustment aims at the consistency of the epistemic state. We call it *revision*. The last way is *contraction*, which occurs when some beliefs in the given epistemic state are retracted but no belief is added.

Let me introduce the dynamic model for supervaluationism briefly. For any agent s, and any vague predicate P, the agent *ipso facto* already has embraced a set of admissible interpretations while she decides to which extension, i.e. positive, negative extension or penumbra, an object belongs. It means that an agent knows a vague predicate if and only if she has a set of precisification functions as the semantic rule when deciding. Then it is easy to see that the agent would feel comfortable to use it if no conflict occurs. In this case, the agent keeps the same set of precisification functions even after the intervention. And we can say that the set of precisification functions expands its interpretative power because it assimilates the new case without revision. However, the agent will be confused if any disagreement occurs. In my opinion, their disagreement is about the set of precisification functions rather than merely about the case. So, what the agent has to do is to revise the set of precisification functions. More precisely, to subtract some items of the set of precisification functions.

The dynamic model adds a new factor to the functions operating on sets of precisification functions. I prefer to use the terms expansion, revision, and contraction functions analogous to Gärdenfors' terminology in the development of dynamic epistemic models. Meanwhile, another factor also needs to be amended. The set of precisification functions should turn into the set of the set of precisification functions. The reason is that every judgment is related to a set of precisification functions rather than depending on a single precisification function only. So the real debate of different judgments is not about whether the case itself is vague or not but rather about the inconsistency of the set of precisification functions.

In general, the dynamic model would be a quintuple model described as $M =< O, P, \boldsymbol{F}, v, R >$ where the elements of $\boldsymbol{F}$ is the set of precisification functions, and R is the set of functions which operate on $\boldsymbol{F}$.

The set of functions R operates on the power set of the set of precisification functions $\boldsymbol{F}$ from $\boldsymbol{F} \times \boldsymbol{F}$ to $\boldsymbol{F}$. This means that a new set of precisification functions will emerge if the former set of precisification functions is disturbed by a new formula $P^i(a_i)$ in $\boldsymbol{L}$. According to the function, $\boldsymbol{F} \times \boldsymbol{F}$ to $\boldsymbol{F}$, the real change of a set of precisification functions is disturbed by another set of precisification functions rather than by a single formula, but the situation that leads us to know about that is the disturbance from a new formula which is incompatible with the original set of precisification functions. For example, suppose Margaret thinks that it is definitely false that Bruce is bald in the given set of precisification functions, i.e., "Bruce is bald" is false in every precisification function. One day when Margaret is chatting with her friends, she finds that some of them do not grant that "Bruce is bald" is false and that they take "Bruce is bald" to be a controversial case, i.e. on their interpretations, "Bruce is bald" is true in some precisification functions but false in others. Assuming Margaret is affected by her friends, it follows that she has to revise her set of precisification functions to assimilate their judgments.

Considering the new information acquired by Margaret, there are three different cases need to be distinguished. The first case is that the new formula $P^i(a_i)$ is compatible with the given set of precisification functions. That is to say, the truth-value of the new formula $P^i(a_i)$ coincides with the formula assessed by the former set of precisification functions. Then the generated set of precisification functions will be the same as former one due to the consistency. In other words, the set of precisification functions directly assimilates the formula without any change of the given set of precisification functions. I call it the expansion function because the formula is compatible with the set of precisification functions or the set of precisification functions is successfully extended to interpret some formulas without contradiction.

Let us turn to the revision function. The revision function is more complicated than expansion. In this case, the revision function will be applied to the precisification functions when the truth-value of a formula $P^i(a_i)$ is incompatible with the assessment from the given set of precisification functions. Consider the above example. Margaret finds out that some of her friends do not agree with her claim that "Bruce is bald" is definitely false, which is based on the set of precisification functions that Margaret holds. Margaret has two options. The first one is to ignore the disagreement from her friends. So the given set of precisification functions has no need to change in this case. The alternative is to revise her set of precisification functions to assimilate the new formula. Nevertheless, it is worth mentioning that what a new formula stands for is a set of precisification functions rather than a judgment. For any set of precisification functions, it should retain the relation of partial order in interpreting the formulas, such as if "6 feet is tall" is definitely true in the interpretation by the set of precisification functions then "6 feet 1 inch is tall" should also be definitely true in the interpretation by the same set of precisification function. At last, the contraction function means that there is no disturbance from outside. It happens when some formulas are released, and the set of precisification functions are affected by the release. Perhaps it narrows the penumbra because some of the precisification functions have been excluded by the procedure.

Here is my preliminary plan to revise the model of supervaluationism, and I believe that it is a defensible response to the challenge of higher-order vagueness. To sum up, the advantages of my proposal are twofold: (1) it is an effective way to realize the admissible interpretations, which is the main difficulty that supervaluationism faces as Williamson points out in his book. Why is it effective? The answer is that the dynamic model emphasizes on the flexible extension of penumbra, which is different from the static model. More precisely, the penumbra is a flexible area which depends on the assimilation process of the new set of precisification functions. (2)

The adjustment also answers the question why the boundary of the penumbra should be counted as vague due to the change of precisification functions. In my opinion, the higher-order vagueness would dismiss if the modification of precisification functions can respond to the disturbance from outside adequately.

References

Alchourrón, C. E., Gärdenfors, P., & Makinson, D. (1985). On the logic of theory change: partial meet contraction and revision functions. *Journal of Symbolic Logic*, *50*, 510–530.

Fine, K. (1975). Vagueness, truth and logic. *Synthese*, *30*, 265–300.

Gärdenfors, P. (2008). *Knowledge in flux*. London: College Publication.

Hyde, D. (1994). Why higher-order is a pseudo-problem. *Mind*, *103*, 35–41.

Keefe, R. (2000). *Theories of vagueness*. Cambridge: Cambridge University Press.

Sainsbury, R. M. (2009). *Paradoxes* (3rd ed.). New York: Cambridge University Press.

Salmon, W. C. (1978). 'Why ask "why?"?'. *Proceedings and Addresses of the American Philosophical Association*, *51*(6), 683–705.

Tye, M. (1994). Why the vague need not be higher-order vague. *Mind*, *103*, 43–45.

Unger, P. (1979). There are no ordinary things. *Synthese*, *41*, 117–54.

van Fraassen, B. (1966). Singular terms, truth-value gap, and free logic. *Journal of Philosophy*, *63*, 481–495.

van Fraassen, B. (1969). Presuppositions, supervaluations and free logic. In Lambert, K. (Ed.), *The logical way of doing things* New Haven: Yale University Press.

Williamson, T. (1994). *Vagueness*. London and New York: Routledge.

Cut Free Labelled Sequent Calculus for Dynamic Logic of Relation Changers

Ryo Hatano, Katsuhiko Sano and Satoshi Tojo

Abstract Dynamic epistemic logic (**DEL**) is known as a large family of logics that extend standard epistemic logic with dynamic operators. Such dynamic operators can be regarded as epistemic actions over Kripke semantics (or its variant). Therefore, **DEL** is often used to model changes of agents' knowledge, belief or preference over Kripke semantics in terms of dynamic operators in many literatures. As a variant of **DEL**, (van Benthem and Liu, J Appl Non-Classical Logics 17(2):157–18 2007; Liu, Reasoning about preference dynamics, Springer Science & Business Media, Berlin 2011) proposed dynamic logic of relation changers (**DLRC**). They provided a general framework to capture many dynamic operators in terms of relation changing operation written in programs of propositional dynamic logic, and they also provided a sound and complete Hilbert-style axiomatization for **DLRC**. While **DLRC** can cover many dynamic operators in a uniform manner, proof theory for **DLRC** is not well-studied except the Hilbert-style axiomatization. Therefore, we propose a cut-free labelled sequent calculus for **DLRC**. We show that our sequent calculus is equipollent with the Hilbert-style axiomatization.

1 Introduction

Dynamic epistemic logic (**DEL**) (van Ditmarsch et al. 2007) is known as a large family of logics that extend standard epistemic logic with dynamic operators. Such operators can be regarded as epistemic actions over Kripke semantics (or its variant).

R. Hatano (✉)
Faculty of Science and Technology, Tokyo University of Science, Noda, Japan
e-mail: r-hatano@rs.tus.ac.jp

K. Sano
Graduate School of Letters, Hokkaido University, Hokkaido, Japan
e-mail: v-sano@let.hokudai.ac.jp

S. Tojo
School of Information Science, Japan Advanced Institute of Science and Technology, Nomi, Japan
e-mail: tojo@jaist.ac.jp

S.C.-M. Yang et al. (eds.), *Philosophical Logic: Current Trends in Asia*,
Logic in Asia: Studia Logica Library, DOI 10.1007/978-981-10-6355-8_8

One of such well-studied logics is public announcement logic (**PAL**) (Plaza 1989) which models changes of agent's knowledge and belief over Kripke semantics. **PAL** has a public announcement operator $[!A]$ whose intuitive meaning is 'after the announcement that A, delete every non-A world from the domain of a Kripke model.' As a result, all links to the non-A world is also deleted from the model.

We may also consider a variant of the public announcement operator, namely, the link-cutting public announcement operator, which keeps the original domain but changes the given accessibility (relation) alone. The link-cutting idea for public announcements was proposed and employed in the context of update semantics by Gerbrandy and Groeneveld (1997), doxastic logic by van Ditmarsch et al. (2007), preference logic and its dynamic extension called dynamic logic of relation changers (**DLRC**) by van Benthem and Liu (2007), Liu (2011). As for the study by van Benthem and Liu, the link-cutting idea is used to upgrade agent's preferences. Their preference logic has the preference modality $[a]$ of agent a, and it is used to describe what is better[1] to the agent. In particular, a formula $[a]A$ reads 'A is true at every world that is the agent a considers as least as good as the current world.' In this logic, dynamic operators can be regarded as preference upgrade operators. For example, the link-cutting operator for public announcements is defined as a preference upgrade operator and called the *link-cutting public update operator*. Intuitively, the update operator means 'after the announcement that A, remove all the links between A-worlds and non-A worlds from a given model.' Moreover, the update operator becomes an instance of *relation changers* in **DLRC**. Relation changers can be specified by programs in propositional dynamic logic (**PDL**) (Harel et al. 2000) without the iteration operator $*$. We may also write many other types of dynamic operators by programs, e.g., the link-adding operator. To acquire such a technically general result, we note that a set of agents in preference logic is regarded as a set of atomic programs.

There are several studies on proof theory of **DEL**s. The first Hilbert-style axiomatization of **PAL** is proposed by Plaza (1989). There is another Hilbert-style axiomatization of **PAL** which is provided by Wang and Cao (2013). Labelled tableau calculus for **PAL** is provided by Balbiani et al. (2010) and its generalized version for non-normal **PAL** is given by Ma et al. (2015). The labelled formalism is also employed in sequent calculi for modal logic such as Negri (2005) and for **PAL** such as Maffezioli and Negri (2010), Balbiani et al. (2014), Nomura et al. (2015). In addition, Frittella et al. (2014) also proposed a sequent calculus for **DEL** based on display calculus which enjoys Belnap-style cut elimination. As for **PDL**, several cut-free sequent calculi are also proposed in Bull (1992), Hill and Poggiolesi (2010), Maffezioli and Naibo (2013). Finally, a Hilbert-style axiomatization for **DLRC** is provided by van Benthem and Liu (2007), Liu (2011). Except for this axiomatization, proof theory for **DLRC** is not well-studied. Therefore, this paper proposes a cut-free labelled sequent

[1] In preference logic by van Benthem and Liu, the notion of *better*-ness for preference is supported by a binary preference relation (i.e., reflexive and transitive relation) over worlds. Common notions of preference play between propositions (or semantically speaking, set of worlds), but their approach emphasizes comparisons of worlds rather than propositions.

calculus to **DLRC**. Similarly to Negri (2005), Balbiani et al. (2014), Nomura et al. (2015), our sequent calculus for **DLRC** also employs the labelled formalism based on Kripke semantics, but it is extended to cover the notion of programs of **PDL** based on Maffezioli and Naibo (2013).

This paper is organized as follows. In Sect. 2, we provide our syntax and semantics for **DLRC** and review the completeness result by van Benthem and Liu (2007), Liu (2011). Then, in Sect. 3, we present our sequent calculus **GDLRC** from a labelled formalism for that. In this section, we establish the following two theorems; all theorems of **HDLRC** are derivable in **GDLRC** (Theorem 1) and our sequent calculus enjoys the cut elimination theorem (Theorem 2). In Sect. 4, we also establish that our sequent is sound for Kripke semantics (Theorem 3). Finally, in Sect. 4, we conclude that our sequent calculus is equipollent with the Hilbert-style axiomatization **HDLRC** (Corollary 1).

2 Dynamic Logic of Relation Changers (DLRC)

2.1 Syntax and Kripke Semantics

Our syntax of dynamic logic of relation changer (**DLRC**) is based on that of propositional dynamic logic without the iteration operator (**PDL**$^-$) (Harel et al. 2000). Given a countably infinite set $\mathsf{Prop} = \{ p, q, \ldots \}$ of *atomic propositional variables* and a finite set $\mathsf{AP} = \{a, b, \ldots\}$ of *atomic programs*, we define the set $\mathsf{Form}_{\mathbf{DLRC}}$ of all formulas, the set PR of all programs, and the set RC of all relation changers by simultaneous induction as follows:

$$\begin{aligned}
&\mathsf{Form}_{\mathbf{DLRC}} \ni A ::= p \mid \neg A \mid A \to A \mid [\alpha]A \mid [\mathsf{r}]A\\
&\mathsf{PR} \ni \alpha ::= a \mid \alpha \cup \alpha \mid \alpha ; \alpha \mid ?A\\
&\mathsf{RC} \ni \mathsf{r} ::= (a := \alpha)_{a \in \mathsf{AP}}
\end{aligned}$$

where $p \in \mathsf{Prop}$ and $a \in \mathsf{AP}$. As usual, we define $\top, \bot, A \wedge B, A \leftrightarrow B$, and $\langle\alpha\rangle A$ as the abbreviations of $A \to A, \neg\top, \neg(A \to \neg B), (A \to B) \wedge (B \to A)$ and $\neg[\alpha]\neg A$, respectively. Then, we also define $\mathsf{Form}_{\mathbf{PDL}^-}$ as the result of dropping the relation changing operator $[\mathsf{r}]$ and the set RC from the above syntax. As we noted in the introduction, we regard each agent's preference as an atomic program. Similarly to propositional dynamic logic (**PDL**), a formula $[\alpha]A$ stands for 'after executing α, it is necessary that A.' A program $\alpha \cup \beta$ is to read 'choose either α or β nondeterministically and execute the chosen one,' a program $\alpha ; \beta$ 'execute α, then execute β' and a program $?A$ 'Test A, proceed if A is true, fail otherwise,' respectively. Based on these programs, we can write the well-known if-then-else program as follows:

$$\text{if } A \text{ then } \alpha \text{ else } \beta \quad \stackrel{\text{def}}{\Leftrightarrow} \quad (?A; \alpha) \cup (?\neg A; \beta).$$

Note that a relation changer r can be regarded as a function from AP to PR. Hence, when $\mathsf{r} = (a := \alpha_a)_{a \in \mathsf{AP}}$, we may use the notation $\mathsf{r}(a)$ to mean α_a. Then a formula $[\mathsf{r}]A$ reads 'After changing an accessibility relation for each programs a by α_a, A holds.' For example, we may consider the following relation changer: given a set $\mathsf{AP} = \{a, b, c\}$ of atomic programs, we set $\mathsf{r} = (a\colon= a, b\colon= c, c\colon= b)$ and so $\mathsf{r}(a) = a$, $\mathsf{r}(b) = c$ and $\mathsf{r}(c) = b$. The intended meaning of this $[\mathsf{r}]$ is that an accessibility relation of the program a is not changed, accessibility relations of b and c is overwritten (or exchanged) by relations of the (original) programs c and b, respectively.

Let us define Kripke semantics with our syntax. A *model* $\mathfrak{M}$ is a tuple $\langle W, (R_a)_{a\in\mathsf{AP}}, V\rangle$ where W is a non-empty set of *possible worlds* (or states), $R_a \subseteq W \times W$ is an *accessibility relation*, and $V\colon\mathsf{Prop} \to \mathscr{P}(W)$ is a *valuation function*. Given any model $\mathfrak{M} = \langle W, (R_a)_{a\in\mathsf{AP}}, V\rangle$ and any possible world $w \in W$, the *satisfaction relation* $\mathfrak{M}, w \models A$ is defined by simultaneous induction as follows:

$$
\begin{array}{lll}
\mathfrak{M}, w \models p & \iff & w \in V(p),\\
\mathfrak{M}, w \models \neg A & \iff & \mathfrak{M}, w \not\models A,\\
\mathfrak{M}, w \models A \to B & \iff & \mathfrak{M}, w \models A \text{ implies } \mathfrak{M}, w \models B,\\
\mathfrak{M}, w \models [\alpha]A & \iff & \mathfrak{M}, v \models A \text{ for all } v \text{ with } wR_\alpha v,\\
wR_{\alpha\cup\beta}v & \iff & wR_\alpha v \text{ or } wR_\beta v,\\
wR_{\alpha;\beta}v & \iff & wR_\alpha u \text{ and } uR_\beta v \text{ for some } u \in W,\\
wR_{?A}v & \iff & w = v \text{ and } \mathfrak{M}, w \models A,\\
\mathfrak{M}, w \models [\mathsf{r}]A & \iff & \mathfrak{M}^{\mathsf{r}}, w \models A,
\end{array}
$$

where $\mathfrak{M}^{\mathsf{r}} = \langle W, (R^{\mathsf{r}}_a)_{a\in\mathsf{AP}}, V^{\mathsf{r}}\rangle$ and R^{r}_a and V^{r} are defined as:

- $R^{\mathsf{r}}_a\colon= R_{\mathsf{r}(a)} = R_{\alpha_a}$ when $\mathsf{r} = (a\colon= \alpha_a)_{a\in\mathsf{AP}}$,
- $w \in V^{\mathsf{r}}(p) \iff w \in V(p)$.

As we described in our syntax, R^{r}_a is changed (or redefined) by that of a (new) program $\mathsf{r}(a)$ $(= \alpha_a)$. On the other hand, $V^{\mathsf{r}}(p)$ is not changed by a relation changer r since $V^{\mathsf{r}}(p)$ is defined by the original $V(p)$. We say that A is *valid on a model* $\mathfrak{M}$ (notation: $\mathfrak{M} \models A$) if $\mathfrak{M}, w \models A$ for all $w \in W$. We also say that A is *valid in a class* $\mathbb{M}$ *of models* (notation: $\mathbb{M} \models A$) if $\mathfrak{M} \models A$ for all $\mathfrak{M} \in \mathbb{M}$.

Example 1 Here we show three examples of relation changers and the first two of them were originally presented in van Benthem and Liu (2007), Liu (2011). The first one is a relation changer for the link-cutting public update operator $[\dagger A]$, as we described in the introduction.[2] Each accessibility relation of the updated model $\mathfrak{M}^{\dagger A}$ is defined as follows: for all $a \in \mathsf{AP}$,

$$R^{\dagger A}_a\colon= \{(u, v) \in R_a \mid \mathfrak{M}, u \models A \iff \mathfrak{M}, v \models A\}$$

The corresponding relation changer can be described as follows:

[2]As we noted in the introduction, the original definitions of atomic programs for these operators are defined as agent's preferences. If we follow the original definition, we have to introduce additional frame properties to our logic. But, in order to make our story simple, we regard atomic programs semantically as ordinary binary relations. Namely, we do not assume any frame property since an accessibility relation might be changed arbitrarily by a dynamic operator.

$$\mathsf{r}_{\dagger A} = (a{:} = (?A; a; ?A) \cup (?\neg A; a; ?\neg A))_{a \in \mathsf{AP}}$$

The second one is a relation changer for a suggestion operator $[\sharp A]$. Intuitively, the suggestion operator removes all links from A-worlds to non-A worlds from the original model $\mathfrak{M}$. Each accessibility relation of the updated model $\mathfrak{M}^{\sharp A}$ is defined as follows: for all $a \in \mathsf{AP}$,

$$R_a^{\sharp A}{:} = R_a \setminus \{ (u, v) \mid \mathfrak{M}, u \models A \text{ and } \mathfrak{M}, v \models \neg A \}.$$

The corresponding relation changer is given as follows:

$$\mathsf{r}_{\sharp A} = (a{:} = (?\neg A; a) \cup (a; ?A))_{a \in \mathsf{AP}}$$

The third one is such a relation changer [if A then b else c] that updates the links of agent a by that of agent b if A holds; otherwise c. Each accessibility relation of the updated model $\mathfrak{M}^{\text{if } A \text{ then } b \text{ else } c}$ is defined as follows: for all $a \in \mathsf{AP}$,

$$R_a^{\text{if } A \text{ then } b \text{ else } c}{:} = (?A; b) \cup (?\neg A; c).$$

The corresponding relation changer is given as follows:

$$\mathsf{r}_{\text{if } A \text{ then } b \text{ else } c} = (a{:} = (?A; b) \cup (?\neg A; c))_{a \in \mathsf{AP}}$$

2.2 *Hilbert-Style Axiomatization* HDLRC

In Table 1, we explicitly provide sound and complete Hilbert-style axiomatization **HDLRC** of dynamic logic of relation changers from van Benthem and Liu (2007), Liu (2011). A *derivation* in **HDLRC** is a sequence of formulas which consist of an instance of an axiom or the result of applying an inference rule to formulas that occur earlier. If a formula A occurs in the last of a derivation, we say A is *derivable* in **HDLRC** (or is a theorem of that) and denote $\vdash_{\mathbf{HDLRC}} A$. The height of a derivation of A is defined as the length of a sequence of the derivation of that.

Lemma 1 *The following three axiom schemes are derivable in* **HDLRC**.

$$\begin{array}{ll} ([\mathsf{r}][\cup]) & [\mathsf{r}][\alpha \cup \beta]A \leftrightarrow [\mathsf{r}][\alpha]A \wedge [\mathsf{r}][\beta]A \\ ([\mathsf{r}][;]) & [\mathsf{r}][\alpha; \beta]A \leftrightarrow [\mathsf{r}][\alpha][\beta]A \\ ([\mathsf{r}][?]) & [\mathsf{r}][?B]A \leftrightarrow [\mathsf{r}](B \to A) \end{array}$$

Proof We only sketch the proof for the case of axiom ($[\mathsf{r}][\cup]$), since the other two cases are shown similarly. First, we can derive the distribution of $[\mathsf{r}]$ over the conjunction $\wedge$, i.e., $[\mathsf{r}](A \wedge B) \leftrightarrow [\mathsf{r}]A \wedge [\mathsf{r}]B$, by axioms ($[\mathsf{r}] \to$) and ($[\mathsf{r}]\neg$). Moreover, we can also derive the distribution of $[\mathsf{r}]$ over the logical equivalence $\leftrightarrow$ by the

Table 1 Hilbert-style axiomatization **HDLRC**

(**Taut**)	All instances of propositional tautologies
($\mathbf{K}_{[\alpha]}$)	$[\alpha](A \to B) \to ([\alpha]A \to [\alpha]B)$
($[\cup]$)	$[\alpha \cup \beta]A \leftrightarrow [\alpha]A \wedge [\beta]A$
($[;]$)	$[\alpha; \beta]A \leftrightarrow [\alpha][\beta]A$
($[?]$)	$[?B]A \leftrightarrow (B \to A)$
($[\mathsf{r}]p$)	$[\mathsf{r}]p \leftrightarrow p$
($[\mathsf{r}]\neg$)	$[\mathsf{r}]\neg A \leftrightarrow \neg[\mathsf{r}]A$
($[\mathsf{r}]\to$)	$[\mathsf{r}](A \to B) \leftrightarrow ([\mathsf{r}]A \to [\mathsf{r}]B)$
($[\mathsf{r}][a]$)	$[\mathsf{r}][a]A \leftrightarrow [\mathsf{r}(a)][\mathsf{r}]A$
(**MP**)	From A and $A \to B$, infer B
($\mathbf{Nec}_{[\alpha]}$)	From A, infer $[\alpha]A$
($\mathbf{Nec}_{[\mathsf{r}]}$)	From A, infer $[\mathsf{r}]A$

distribution of $[\mathsf{r}]$ over $\wedge$ and axiom ($[\mathsf{r}] \to$). Then, by axiom ($[\cup]$) and inference rule ($\mathbf{Nec}_{[\mathsf{r}]}$), we have $[\mathsf{r}]([\alpha \cup \beta] \leftrightarrow [\alpha]A \wedge [\beta]A)$, which implies axiom ($[\mathsf{r}][\cup]$) by the derived distribution of $[\mathsf{r}]$ over $\leftrightarrow$ and $\wedge$. □

Definition 1 The translation $t : \mathsf{Form}_{\mathbf{DLRC}} \to \mathsf{Form}_{\mathbf{PDL}^-}$ is defined by:

$$
\begin{aligned}
t(p) &= p,\\
t(\neg A) &= \neg t(A),\\
t(A \to B) &= t(A) \to t(B),\\
t([a]A) &= [a]t(A),\\
t([\alpha \cup \beta]A) &= t([\alpha]A) \wedge t([\beta]A),\\
t([\alpha; \beta]A) &= t([\alpha][\beta]A),\\
t([?B]A) &= t(B) \to t(A),\\
t([\mathsf{r}]p) &= p,\\
t([\mathsf{r}]\neg A) &= \neg t([\mathsf{r}]A),\\
t([\mathsf{r}](A \to B)) &= t([\mathsf{r}]A) \to t([\mathsf{r}]B),\\
t([\mathsf{r}][a]A) &= [\mathsf{r}(a)]t([\mathsf{r}]A),\\
t([\mathsf{r}][\alpha \cup \beta]A) &= t([\mathsf{r}][\alpha]A) \wedge t([\mathsf{r}][\beta]A),\\
t([\mathsf{r}][\alpha; \beta]A) &= t([\mathsf{r}][\alpha][\beta]A),\\
t([\mathsf{r}][?B]A) &= t([\mathsf{r}]B) \to t([\mathsf{r}]A),\\
t([\mathsf{r}][\mathsf{r}']A) &= t([\mathsf{r}]t([\mathsf{r}']A)).
\end{aligned}
$$

Remark that this translation reflects the idea of 'inside-out' strategy. That is, we start rewriting the innermost occurrences of $[\mathsf{r}]$.

Lemma 2 *Given any formula* $A \in \mathsf{Form}_{\mathbf{DLRC}}$,

$$\vdash_{\mathbf{HDLRC}} A \leftrightarrow t(A).$$

Fact 2 (van Benthem and Liu 2007; Liu 2011) *Let A be a formula in* $\mathsf{Form}_{\mathbf{DLRC}}$ *and $\mathbb{M}$ be the class of all models. The following equivalence holds.*

$$\mathbb{M} \models A \iff \vdash_{\mathbf{HDLRC}} A.$$

3 Labelled Sequent Calculus GDLRC

3.1 Labelled Formalism

Now, we introduce the labelled formalism for our sequent calculus. Let $\mathsf{Var} = \{x, y, z, \dots\}$ be a countably infinite set of *variables* and $L = (\mathsf{r}_1, \mathsf{r}_2, \dots, \mathsf{r}_n)$ be a (possibly empty) finite list of relation changers. If L is empty, we use ε to denote such an empty list. We may regard the list of relation changers as a history or a stack of that. Suppose that $\mathfrak{M}^L, w \models A$ is already defined, then

$$\mathfrak{M}^{\langle L, r \rangle}, w \models A \iff \mathfrak{M}^L, w \models [\mathsf{r}]A.$$

Given variables $x, y \in \mathsf{Var}$, a list L of relation changers, a formula $A \in \mathsf{Form}_{\mathbf{DLRC}}$ and a program $\alpha \in \mathsf{PR}$, we define *labelled expressions* by:

$$\varphi ::= x{:}^L A \mid x\mathsf{R}^L_\alpha y \mid x = y$$

We say that $x{:}^L A$ is a *labelled formula*, $x\mathsf{R}^L_\alpha y$ is a *relational atom* and $x = y$ is an *equality atom*. A labelled formula $x{:}^L A$ means that 'after the successive updates of relation changers in L, A holds at state x,' a relational atom $x\mathsf{R}^L_\alpha y$ means that 'after the successive updates of relation changers in L, there is a link of program α from x to y,' and an equality atom means that 'state x equals to state y.' We note that labelled expressions are counterparts of objects in Kripke semantics. Namely, given a model $\mathfrak{M} = \langle W, (R_a)_{a \in \mathsf{AP}}, V \rangle$ and if we regard $w, v \in W$ as counterparts of $x, y \in \mathsf{Var}$, a labelled formula $x{:}^L A$ corresponds to the satisfaction relation '$\mathfrak{M}^L, w \models A$,' a relational atom $x\mathsf{R}^L_\alpha y$ corresponds to a link '$wR^L_\alpha v$' and an equality atom $x = y$ corresponds to an equality '$w = v$,' respectively. Let Γ and Δ be multisets of labelled expressions. We define a *labelled sequent* $\Gamma \Rightarrow \Delta$ by a pair $\langle \Gamma, \Delta \rangle$ of labelled expressions, whose reading is 'if all of Γ hold, then some of Δ holds.' Based on these definitions, we define our labelled sequent calculus **GDLRC** by all initial sequents and all inference rules in Table 2. We also define the labelled sequent calculus $\mathbf{GDLRC}^-$ by **GDLRC** without the cut rule. In order to refer both left and right rules shortly, we also use simple notation of the rule name without L and R, e.g., (at) means both (Lat) and (Rat) rules. The multisets Γ, Δ etc. in each inference rule in **GDLRC** are called the *context*. In the lower sequent of each inference rule, the labelled expression(s) not in the context is called the *principal labelled expression*(s). A *derivation* $\mathfrak{D}$ in **GDLRC** is inductively defined as a finite tree generated by initial sequents and inference rules in Table 2. A labelled sequent in the root node of $\mathfrak{D}$ is called the *end sequent*. The *height* of a derivation $\mathfrak{D}$ is defined as the maximum length of branches in $\mathfrak{D}$ from the end sequent of that to an initial sequent. We say that $\Gamma \Rightarrow \Delta$ is *derivable* in **GDLRC** (notation: $\vdash_{\mathbf{GDLRC}} \Gamma \Rightarrow \Delta$) if there exists a derivation $\mathfrak{D}$ in **GDLRC** whose end sequent is $\Gamma \Rightarrow \Delta$. A formula A is a theorem of **GDLRC** if the labelled sequent $\Rightarrow x{:}^{\varepsilon} A$ is derivable in **GDLRC** for all $x \in \mathsf{Var}$. These definitions are also used for **GDLRC**$^{-}$.

Remark that all syntactic objects in the labelled sequent calculus **GDLRC** are finite, e.g., labelled expressions, sequents and derivations. We may regard **GDLRC** as a formalized version of Kripke semantics of **DLRC**. The crucial difference between the ordinary Kripke semantics and its formalized version is this finiteness.

Definition 3 The length ℓ of formulas $p \in \mathsf{Prop}$ and $A, B \in \mathsf{Form}_{\mathbf{DLRC}}$, programs $a \in \mathsf{AP}$ and $\alpha, \beta \in \mathsf{PR}$, a relation changer $\mathsf{r} \in \mathsf{RC}$, and labelled expressions are defined as follows:

$$
\begin{array}{ll}
\ell(p) & = 1, \\
\ell(\neg A) & = \ell(A) + 1, \\
\ell(A \to B) & = \ell(A) + \ell(B) + 1, \\
\ell([\alpha]A) & = \ell(\alpha) + \ell(A) + 1, \\
\ell([\mathsf{r}]A) & = \ell(\mathsf{r}) + \ell(A) + 1, \\
\ell(a) & = 1, \\
\ell(\alpha; \beta) & = \ell(\alpha) + \ell(\beta) + 1, \\
\ell(\alpha \cup \beta) & = \ell(\alpha) + \ell(\beta) + 1, \\
\ell(?A) & = \ell(A) + 1, \\
\ell(\varepsilon) & = 0, \\
\ell(\mathsf{r}) & = \Sigma_{a \in \mathsf{AP}} \ell(\alpha_a) \text{ where } \mathsf{r} = (a{:} = \alpha_a)_{a \in \mathsf{AP}}, \\
\ell(L) & = \Sigma_{i=0}^{n} \ell(\mathsf{r}_i) \text{ where } L = (\mathsf{r}_0, \ldots, \mathsf{r}_n), \\
\ell(x{:}^{L} A) & = \ell(L) + \ell(A), \\
\ell(x \mathsf{R}_{\alpha}^{L} y) & = \ell(L) + \ell(\alpha), \\
\ell(x = y) & = 0.
\end{array}
$$

Proposition 4 *Let* $p \in \mathsf{Prop}$, $A \in \mathsf{Form}_{\mathbf{DLRC}}$, $a \in \mathsf{AP}$, $\mathsf{r} \in \mathsf{RC}$ *and* $x, y \in \mathsf{Var}$.

(i) $\ell(x{:}^{\varepsilon}[\mathsf{r}]A) > \ell(x{:}^{\mathsf{r}} A)$.
(ii) $\ell(x{:}^{\mathsf{r}} p) > \ell(x{:}^{\varepsilon} p)$.
(iii) $\ell(x \mathsf{R}_{a}^{\mathsf{r}} y) > \ell(x \mathsf{R}_{\mathsf{r}(a)}^{\varepsilon} y)$.

Proof

(i) The length of the left side is:

$$\ell(x{:}^{\varepsilon}[\mathsf{r}]A) = \ell(\mathsf{r}) + \ell(A) + 1.$$

The right side is:

$$\ell(x{:}^{\mathsf{r}} A) = \ell(\mathsf{r}) + \ell(A).$$

Table 2 Gentzen-style Sequent Calculus **GDLRC**

(Initial sequents)

$$x{:}^L A \Rightarrow x{:}^L A \qquad x\mathsf{R}^L_\alpha y \Rightarrow x\mathsf{R}^L_\alpha y \qquad x = y \Rightarrow x = y$$

(Structural rules)

$$\frac{\Gamma \Rightarrow \Delta}{\Gamma \Rightarrow \Delta, \varphi}\,(Rw) \qquad \frac{\Gamma \Rightarrow \Delta}{\varphi, \Gamma \Rightarrow \Delta}\,(Lw) \qquad \frac{\Gamma \Rightarrow \Delta, \varphi, \varphi}{\Gamma \Rightarrow \Delta, \varphi}\,(Rc) \qquad \frac{\varphi, \varphi, \Gamma \Rightarrow \Delta}{\varphi, \Gamma \Rightarrow \Delta}\,(Lc)$$

(Logical rules)

$$\frac{x{:}^L A, \Gamma \Rightarrow \Delta}{\Gamma \Rightarrow \Delta, x{:}^L \neg A}\,(R\neg) \qquad \frac{\Gamma \Rightarrow \Delta, x{:}^L A}{x{:}^L \neg A, \Gamma \Rightarrow \Delta}\,(L\neg)$$

$$\frac{x{:}^L A_1, \Gamma \Rightarrow \Delta, x{:}^L A_2}{\Gamma \Rightarrow \Delta, x{:}^L A_1 \to A_2}\,(R\to) \qquad \frac{\Gamma \Rightarrow \Delta, x{:}^L A_1 \quad x{:}^L A_2, \Gamma \Rightarrow \Delta}{x{:}^L A_1 \to A_2, \Gamma \Rightarrow \Delta}\,(L\to)$$

$$\frac{x\mathsf{R}^L_\alpha y, \Gamma \Rightarrow \Delta, y{:}^L A}{\Gamma \Rightarrow \Delta, x{:}^L [\alpha] A}\,(R[\alpha])^\dagger \qquad \frac{\Gamma \Rightarrow \Delta, x\mathsf{R}^L_\alpha y \quad y{:}^L A, \Gamma \Rightarrow \Delta}{x{:}^L [\alpha] A, \Gamma \Rightarrow \Delta}\,(L[\alpha])$$

†: y does not appear in the lower sequent.

$$\frac{\Gamma \Rightarrow \Delta, x{:}^{\langle L, \mathsf{r}\rangle} A}{\Gamma \Rightarrow \Delta, x{:}^L [\mathsf{r}] A}\,(R[\mathsf{r}]) \qquad \frac{x{:}^{\langle L, \mathsf{r}\rangle} A, \Gamma \Rightarrow \Delta}{x{:}^L [\mathsf{r}] A, \Gamma \Rightarrow \Delta}\,(L[\mathsf{r}])$$

(Program rules)

$$\frac{\Gamma \Rightarrow \Delta, x\mathsf{R}^L_{\alpha_i} y}{\Gamma \Rightarrow \Delta, x\mathsf{R}^L_{\alpha_1 \cup \alpha_2} y}\,(R\cup)_{i \in \{1,2\}} \qquad \frac{x\mathsf{R}^L_{\alpha_1} y, \Gamma \Rightarrow \Delta \quad x\mathsf{R}^L_{\alpha_2} y, \Gamma \Rightarrow \Delta}{x\mathsf{R}^L_{\alpha_1 \cup \alpha_2} y, \Gamma \Rightarrow \Delta}\,(L\cup)$$

$$\frac{\Gamma \Rightarrow \Delta, x\mathsf{R}^L_{\alpha_1} z \quad \Gamma \Rightarrow \Delta, z\mathsf{R}^L_{\alpha_2} y}{\Gamma \Rightarrow \Delta, x\mathsf{R}^L_{\alpha_1 ; \alpha_2} y}\,(R;) \qquad \frac{x\mathsf{R}^L_{\alpha_1} z, z\mathsf{R}^L_{\alpha_2} y, \Gamma \Rightarrow \Delta}{x\mathsf{R}^L_{\alpha_1 ; \alpha_2} y, \Gamma \Rightarrow \Delta}\,(L;)^\ddagger$$

‡: z does not appear in the lower sequent.

$$\frac{\Gamma \Rightarrow \Delta, x = y \quad \Gamma \Rightarrow \Delta, x{:}^L A}{\Gamma \Rightarrow \Delta, x\mathsf{R}^L_{?A} y}\,(R?) \qquad \frac{x = y, \Gamma \Rightarrow \Delta}{x\mathsf{R}^L_{?A} y, \Gamma \Rightarrow \Delta}\,(L?_1) \qquad \frac{x{:}^L A, \Gamma \Rightarrow \Delta}{x\mathsf{R}^L_{?A} y, \Gamma \Rightarrow \Delta}\,(L?_2)$$

(Equality rules)

$$\frac{\Gamma \Rightarrow \Delta, x = y \quad \Gamma \Rightarrow \Delta, x{:}^\varepsilon p}{\Gamma \Rightarrow \Delta, y{:}^\varepsilon p}\,(= at) \qquad \frac{x = x, \Gamma \Rightarrow \Delta}{\Gamma \Rightarrow \Delta}\,(= ref)$$

$$\frac{\Gamma \Rightarrow \Delta, x = y \quad y = x, \Gamma \Rightarrow \Delta}{\Gamma \Rightarrow \Delta}\,(= sym) \qquad \frac{\Gamma \Rightarrow \Delta, x = y \quad \Gamma \Rightarrow \Delta, y = z \quad x = z, \Gamma \Rightarrow \Delta}{\Gamma \Rightarrow \Delta}\,(= tra)$$

$$\frac{\Gamma \Rightarrow \Delta, x = y \quad \Gamma \Rightarrow \Delta, x\mathsf{R}^L_a z \quad y\mathsf{R}^L_a z, \Gamma \Rightarrow \Delta}{\Gamma \Rightarrow \Delta}\,(= rel_1)$$

$$\frac{\Gamma \Rightarrow \Delta, x = y \quad \Gamma \Rightarrow \Delta, z\mathsf{R}^L_a x \quad z\mathsf{R}^L_a y, \Gamma \Rightarrow \Delta}{\Gamma \Rightarrow \Delta}\,(= rel_2)$$

(Relation changer rules)

$$\frac{\Gamma \Rightarrow \Delta, x{:}^L p}{\Gamma \Rightarrow \Delta, x{:}^{\langle L, \mathsf{r}\rangle} p}\,(Rat) \qquad \frac{x{:}^L p, \Gamma \Rightarrow \Delta}{x{:}^{\langle L, \mathsf{r}\rangle} p, \Gamma \Rightarrow \Delta}\,(Lat)$$

$$\frac{\Gamma \Rightarrow \Delta, x\mathsf{R}^L_{\mathsf{r}(a)} y}{\Gamma \Rightarrow \Delta, x\mathsf{R}^{(L, \mathsf{r})}_a y}\,(Rrel) \qquad \frac{x\mathsf{R}^L_{\mathsf{r}(a)} y, \Gamma \Rightarrow \Delta}{x\mathsf{R}^{(L, \mathsf{r})}_a y, \Gamma \Rightarrow \Delta}\,(Lrel)$$

(Cut rule)

$$\frac{\Gamma \Rightarrow \Delta, \varphi \quad \varphi, \Sigma \Rightarrow \Pi}{\Gamma, \Sigma \Rightarrow \Delta, \Pi}\,(Cut)$$

Therefore, $\ell(\mathsf{r}) + \ell(A) + 1 > \ell(\mathsf{r}) + \ell(A)$.

(ii) The length of the left side is:

$$\begin{aligned}\ell(x{:}^{\mathsf{r}}p) &= \ell(\mathsf{r}) + \ell(p),\\ &= \ell(\mathsf{r}) + 1.\end{aligned}$$

The right side is:

$$\begin{aligned}\ell(x{:}^{\varepsilon}p) &= \ell(p),\\ &= 1.\end{aligned}$$

Since $\ell(\mathsf{r}) > 0$, we obtain $\ell(\mathsf{r}) + 1 > 1$.

(iii) The length of the left side is:

$$\begin{aligned}\ell(x\mathsf{R}^{\mathsf{r}}_{a}y) &= \ell(\mathsf{r}) + \ell(a),\\ &= \ell(\mathsf{r}) + 1.\end{aligned}$$

The right side is:

$$\ell(x\mathsf{R}^{\varepsilon}_{\mathsf{r}(a)}y) = \ell(\alpha_a).$$

Since $\ell(\mathsf{r}) \geq \ell(\alpha_a)$, we obtain $\ell(\mathsf{r}) + 1 > \ell(\alpha_a)$.

□

3.2 *All Theorems of* HDLRC *are Derivable in* GDLRC

Lemma 3 *For any $n \in \mathbb{N}$ and any list $L \in (\mathsf{RC})^*$ of relation changers,*

(i) *if $\ell(x{:}^{L}A) \leq n$, then $\vdash_{\mathbf{GDLRC}} x = y,\ x{:}^{L}A \Rightarrow y{:}^{L}A$ for any $x, y \in \mathsf{Var}$ and any $A \in \mathsf{Form}_{\mathbf{DLRC}}$.*

(ii) *if $\ell(x\mathsf{R}^{L}_{\alpha}y) \leq n$, then $\vdash_{\mathbf{GDLRC}} x = y,\ y\mathsf{R}^{L}_{\alpha}z \Rightarrow x\mathsf{R}^{L}_{\alpha}z$ for any $x, y, z \in \mathsf{Var}$ and any $\alpha \in \mathsf{PR}$.*

Proof The proofs of (i) and (ii) are done simultaneously by induction on $n \in \mathbb{N}$. First, we show base cases for both (i) and (ii).

Basis for (i) where $x :^{L} A$ is of the form $x :^{\varepsilon} p$ ($p \in \mathsf{Prop}$):

$$\dfrac{\dfrac{x = y \Rightarrow x = y}{x = y, x{:}^{\varepsilon}p \Rightarrow x = y}\,(Lw) \qquad \dfrac{x{:}^{\varepsilon}p \Rightarrow x{:}^{\varepsilon}p}{x = y, x{:}^{\varepsilon}p \Rightarrow x{:}^{\varepsilon}p}\,(Lw)}{x = y, x{:}^{\varepsilon}p \Rightarrow y{:}^{\varepsilon}p}\,(= at)$$

Basis for (ii) where $x\mathsf{R}^{L}_{\alpha}y$ is of the form $x\mathsf{R}^{\varepsilon}_{a}y (a \in \mathsf{AP})$:

First, we construct the following derivation $\mathfrak{D}$ where double line in $\mathfrak{D}$ means finite applications of (w) rules.

$$\dfrac{\dfrac{y = x \Rightarrow y = x}{y = x, y\mathsf{R}_a^{\varepsilon}z \Rightarrow x\mathsf{R}_a^{\varepsilon}z, y = x}\,(w) \quad \dfrac{y\mathsf{R}_a^{\varepsilon}z \Rightarrow y\mathsf{R}_a^{\varepsilon}z}{y = x, y\mathsf{R}_a^{\varepsilon}z \Rightarrow x\mathsf{R}_a^{\varepsilon}z, y\mathsf{R}_a^{\varepsilon}z}\,(w) \quad \dfrac{x\mathsf{R}_a^{\varepsilon}z \Rightarrow x\mathsf{R}_a^{\varepsilon}z}{x\mathsf{R}_a^{\varepsilon}z, y = x, y\mathsf{R}_a^{\varepsilon}z \Rightarrow x\mathsf{R}_a^{\varepsilon}z}\,(w)}{y = x, y\mathsf{R}_a^{\varepsilon}z \Rightarrow x\mathsf{R}_a^{\varepsilon}z}\,(= rel_1)$$

Next, we obtain the following derivation by $\mathfrak{D}$.

$$\dfrac{\dfrac{x = y \Rightarrow x = y}{x = y, y\mathsf{R}_a^{\varepsilon}z \Rightarrow x\mathsf{R}_a^{\varepsilon}z, x = y}\,(w) \quad \dfrac{\mathfrak{D}}{y = x, x = y, y\mathsf{R}_a^{\varepsilon}z \Rightarrow x\mathsf{R}_a^{\varepsilon}z}\,(Lw)}{x = y, y\mathsf{R}_a^{\varepsilon}z \Rightarrow x\mathsf{R}_a^{\varepsilon}z}\,(= sym)$$

Next, we show inductive steps for both (i) and (ii).

Case (i-a) where $x :^{L} A$ is of the form $x :^{\langle L, r\rangle} p$:

$$\dfrac{\begin{matrix}\vdots\ \mathfrak{D}\\ x = y, x{:}^{L}p \Rightarrow y{:}^{L}p\end{matrix}}{x = y, x{:}^{\langle L, \mathsf{r}\rangle}p \Rightarrow y{:}^{\langle L, \mathsf{r}\rangle}p}\,(at)$$

where $\mathfrak{D}$ is obtained by induction hypothesis.

Case (i-b) where $x :^{L} A$ is of the form $x :^{L} \neg B$:

$$\dfrac{\dfrac{\dfrac{x = y \Rightarrow x = y}{x = y, y{:}^{L}B \Rightarrow x{:}^{L}B, x = y}\,(w) \quad \dfrac{\begin{matrix}\vdots\ \mathfrak{D}\\ y = x, y{:}^{L}B \Rightarrow x{:}^{L}B\end{matrix}}{y = x, x = y, y{:}^{L}B \Rightarrow x{:}^{L}B}\,(Lw)}{x = y, y{:}^{L}B \Rightarrow x{:}^{L}B}\,(= sym)}{x = y, x{:}^{L}\neg B \Rightarrow y{:}^{L}\neg B}\,(\neg)$$

where $\mathfrak{D}$ is obtained by induction hypothesis.

Case (i-c) where $x :^{L} A$ is of the form $\mathrm{x} :^{\mathrm{L}} \mathrm{A}_1 \rightarrow \mathrm{A}_2$:

$$\dfrac{\dfrac{\dfrac{\dfrac{\dfrac{x = y \Rightarrow x = y}{y{:}^{L}A_1, x = y \Rightarrow x{:}^{L}A_1, x = y}\,(w) \quad \dfrac{\begin{matrix}\vdots\ \mathfrak{D}_1\\ y = x, y{:}^{L}A_1, \Rightarrow x{:}^{L}A_1\end{matrix}}{y = x, y{:}^{L}A_1, x = y \Rightarrow x{:}^{L}A_1}\,(Lw)}{y{:}^{L}A_1, x = y \Rightarrow x{:}^{L}A_1}\,(= sym)}{y{:}^{L}A_1, x = y \Rightarrow y{:}^{L}A_2, x{:}^{L}A_1}\,(Rw) \quad \dfrac{\begin{matrix}\vdots\ \mathfrak{D}_2\\ x{:}^{L}A_2, x = y \Rightarrow y{:}^{L}A_2\end{matrix}}{x{:}^{L}A_2, y{:}^{L}A_1, x = y \Rightarrow y{:}^{L}A_2}\,(Lw)}{y{:}^{L}A_1, x = y, x{:}^{L}A_1 \rightarrow A_2 \Rightarrow y{:}^{L}A_2}\,(L \rightarrow)}{x = y, x{:}^{L}A_1 \rightarrow A_2 \Rightarrow y{:}^{L}A_1 \rightarrow A_2}\,(R \rightarrow)$$

where $\mathfrak{D}_1$ and $\mathfrak{D}_2$ are obtained by induction hypothesis.

Case (i-d) where $x :^{L} A$ is of the form $x :^{L} [\alpha]B$:

$$\dfrac{\dfrac{\begin{array}{c}\vdots\ \mathfrak{D}\\ y\mathsf{R}^L_\alpha z, x = y \Rightarrow x\mathsf{R}^L_\alpha z\end{array}}{y\mathsf{R}^L_\alpha z, x = y \Rightarrow z{:}^L B, x\mathsf{R}^L_\alpha z}\,(Rw) \quad \dfrac{z{:}^L B \Rightarrow z{:}^L B}{z{:}^L B, y\mathsf{R}^L_\alpha z, x = y \Rightarrow z{:}^L B}\,(w)}{\dfrac{y\mathsf{R}^L_\alpha z, x = y, x{:}^L[\alpha]B \Rightarrow z{:}^L B}{x = y, x{:}^L[\alpha]B \Rightarrow y{:}^L[\alpha]B}\,(R[\alpha])}\,(L[\alpha])$$

where $\mathfrak{D}$ is obtained by induction hypothesis.

Case (i-e) where $x :^L A$ is of the form $x :^L [\mathsf{r}]B$:

$$\dfrac{\begin{array}{c}\vdots\ \mathfrak{D}\\ x = y, x{:}^{\langle L,r\rangle}A \Rightarrow y{:}^{\langle L,r\rangle}A\end{array}}{x = y, x{:}^L[\mathsf{r}]B \Rightarrow y{:}^L[\mathsf{r}]B}\,([\mathsf{r}])$$

where $\mathfrak{D}$ is obtained by induction hypothesis. Note that $\ell(x{:}^{\langle L,r\rangle}A) < \ell(x{:}^L[\mathsf{r}]A)$.

Case (ii-a) where $y\mathsf{R}^L_\alpha z$ is of the form $y\mathsf{R}^L_{\alpha_1\cup\alpha_2}z$:

$$\dfrac{\dfrac{\begin{array}{c}\vdots\ \mathfrak{D}_1\\ x = y, y\mathsf{R}^L_{\alpha_1}z \Rightarrow x\mathsf{R}^L_{\alpha_1}z\end{array}}{x = y, y\mathsf{R}^L_{\alpha_1}z \Rightarrow x\mathsf{R}^L_{\alpha_1\cup\alpha_2}z}\,(R\cup) \quad \dfrac{\begin{array}{c}\vdots\ \mathfrak{D}_2\\ x = y, y\mathsf{R}^L_{\alpha_2}z \Rightarrow x\mathsf{R}^L_{\alpha_2}z\end{array}}{x = y, y\mathsf{R}^L_{\alpha_2}z \Rightarrow x\mathsf{R}^L_{\alpha_1\cup\alpha_2}z}\,(R\cup)}{x = y, y\mathsf{R}^L_{\alpha_1\cup\alpha_2}z \Rightarrow x\mathsf{R}^L_{\alpha_1\cup\alpha_2}z}\,(L\cup)$$

where $\mathfrak{D}_1$ and $\mathfrak{D}_2$ are obtained by induction hypothesis.

Case (ii-b) where $y\mathsf{R}^L_\alpha z$ is of the form $y\mathsf{R}^L_{\alpha_1;\alpha_2}z$:

$$\dfrac{\dfrac{\dfrac{\begin{array}{c}\vdots\ \mathfrak{D}\\ x = y, y\mathsf{R}^L_{\alpha_1}w \Rightarrow x\mathsf{R}^L_{\alpha_1}w\end{array}}{x = y, y\mathsf{R}^L_{\alpha_1}w, w\mathsf{R}^L_{\alpha_2}z \Rightarrow x\mathsf{R}^L_{\alpha_1}w}\,(Lw) \quad \dfrac{w\mathsf{R}^L_{\alpha_2}z \Rightarrow w\mathsf{R}^L_{\alpha_2}z}{x = y, y\mathsf{R}^L_{\alpha_1}w, w\mathsf{R}^L_{\alpha_2}z \Rightarrow w\mathsf{R}^L_{\alpha_2}z}\,(Lw)}{x = y, y\mathsf{R}^L_{\alpha_1}w, w\mathsf{R}^L_{\alpha_2}z \Rightarrow x\mathsf{R}^L_{\alpha_1;\alpha_2}z}\,(R;)}{x = y, y\mathsf{R}^L_{\alpha_1;\alpha_2}z \Rightarrow x\mathsf{R}^L_{\alpha_1;\alpha_2}z}\,(L;)$$

where $\mathfrak{D}$ is obtained by induction hypothesis.

Case (ii-c) where $y\mathsf{R}^L_\alpha z$ is of the form $y\mathsf{R}^L_{?A}z$: First, we construct the following derivation $\mathfrak{D}_1$.

$$\dfrac{\dfrac{x = y \Rightarrow x = y}{x = y, y = z \Rightarrow x = z, x = y}\,(w) \quad \dfrac{y = z \Rightarrow y = z}{x = y, y = z \Rightarrow x = z, y = z}\,(w) \quad \dfrac{x = z \Rightarrow x = z}{x = z, x = y, y = z \Rightarrow x = z}\,(Lw)}{x = y, y = z \Rightarrow x = z}\,(= tra)$$

Second, we also construct the following derivation $\mathfrak{D}_2$.

$$\dfrac{\dfrac{x = y \Rightarrow x = y}{x = y, y{:}^L A \Rightarrow x{:}^L A, x = y}(w) \quad \dfrac{\dfrac{\vdots\ \mathfrak{D}_3}{y = x, y{:}^L A \Rightarrow x{:}^L A}}{y = x, x = y, y{:}^L A \Rightarrow x{:}^L A}(Lw)}{x = y, y{:}^L A \Rightarrow x{:}^L A}(= sym)$$

where $\mathfrak{D}_3$ is obtained by induction hypothesis. Finally, we obtain the following derivation by $\mathfrak{D}_1$ and $\mathfrak{D}_2$.

$$\dfrac{\dfrac{\mathfrak{D}_1}{x = y, y\mathsf{R}^L_{?A}z \Rightarrow x = z}(L?_1) \quad \dfrac{\mathfrak{D}_2}{x = y, y\mathsf{R}^L_{?A}z \Rightarrow x{:}^L A}(L?_2)}{x = y, y\mathsf{R}^L_{?A}z \Rightarrow x\mathsf{R}^L_{?A}z}(R?)$$

Case (ii-d) where $y\mathsf{R}^L_\alpha z$ is of the form $y\mathsf{R}^L_{\alpha_1 \cup \alpha_2} z$:

$$\dfrac{\dfrac{\vdots\ \mathfrak{D}}{x = y, y\mathsf{R}^L_{\mathsf{r}(a)}z \Rightarrow x\mathsf{R}^L_{\mathsf{r}(a)}z}}{x = y, y\mathsf{R}^{\langle L,r\rangle}_a z \Rightarrow x\mathsf{R}^{\langle L,r\rangle}_a z}(rel)$$

where $\mathfrak{D}$ is obtained by induction hypothesis. □

Theorem 1 *For any $A \in \mathsf{Form}_{\mathbf{DLRC}}$ and any $x \in \mathsf{Var}$,*

$$\text{if } \vdash_{\mathbf{HDLRC}} A, \text{ then } \vdash_{\mathbf{GDLRC}} \Rightarrow x{:}^\varepsilon A.$$

Proof We show our goal by induction of the height of the derivation of A in **HDLRC**. Here, we show only the cases of axiom ([?]), inference rule (**MP**), and axioms and rules for relation changers of **HDLRC** listed in Table 1, i.e., the axioms from ($[\mathsf{r}]p$) to ([r]?) and the rule ($\mathbf{Nec}_{[\mathsf{r}]}$) in the table. In the following, we show the base cases.

Case of axiom ([?])

The direction from left to right is shown as follows.

$$\dfrac{\dfrac{\dfrac{\dfrac{\dfrac{x = x \Rightarrow x = x}{\Rightarrow x = x}(= Ref)}{x{:}^\varepsilon B \Rightarrow x{:}^\varepsilon A, x = x}(Lw) \quad \dfrac{x{:}^\varepsilon B \Rightarrow x{:}^\varepsilon B}{x{:}^\varepsilon B \Rightarrow x{:}^\varepsilon A, x{:}^\varepsilon B}(Rw)}{x{:}^\varepsilon B \Rightarrow x{:}^\varepsilon A, x\mathsf{R}^\varepsilon_{?B}x}(R?) \quad \dfrac{x{:}^\varepsilon A \Rightarrow x{:}^\varepsilon A}{x{:}^\varepsilon A, x{:}^\varepsilon B \Rightarrow x{:}^\varepsilon A}(Lw)}{x{:}^\varepsilon B, x{:}^\varepsilon [?B]A \Rightarrow x{:}^\varepsilon A}(L[?B])}{\dfrac{x{:}^\varepsilon [?B]A \Rightarrow x{:}^\varepsilon B \rightarrow A}{\Rightarrow x{:}^\varepsilon [?B]A \rightarrow (B \rightarrow A)}(R\rightarrow)}(R\rightarrow)$$

The direction from right to left is shown as follows.

$$\dfrac{\dfrac{\dfrac{x{:}^{\varepsilon} B \Rightarrow x{:}^{\varepsilon} B}{x{:}^{\varepsilon} B \Rightarrow y{:}^{\varepsilon} A, x{:}^{\varepsilon} B}\,(Rw)}{x\mathsf{R}^{\varepsilon}_{?B}y \Rightarrow y{:}^{\varepsilon} A, x{:}^{\varepsilon} B}\,(L?_2) \qquad \dfrac{\begin{array}{c}\text{Lemma 3}\\ \vdots \\ x{:}^{\varepsilon} A, x = y \Rightarrow y{:}^{\varepsilon} A\end{array}}{x{:}^{\varepsilon} A, x\mathsf{R}^{\varepsilon}_{?B}y \Rightarrow y{:}^{\varepsilon} A}\,(L?_1)}{\dfrac{\dfrac{x\mathsf{R}^{\varepsilon}_{?B}y, x{:}^{\varepsilon} B \to A \Rightarrow y{:}^{\varepsilon} A}{x{:}^{\varepsilon} B \to A \Rightarrow x{:}^{\varepsilon}[?B]A}\,(R[?B])}{\Rightarrow x{:}^{\varepsilon}(B \to A) \to [?B]A}\,(R\to)}\,(L\to)$$

Case of axiom ([r]p)]

$$\dfrac{\dfrac{\dfrac{\dfrac{x{:}^{\varepsilon} p \Rightarrow x{:}^{\varepsilon} p}{x{:}^{\mathsf{r}} p \Rightarrow x{:}^{\varepsilon} p}\,(Lat)}{x{:}^{\varepsilon}[\mathsf{r}]p \Rightarrow x{:}^{\varepsilon} p}\,(L[\mathsf{r}])}{\Rightarrow x{:}^{\varepsilon}[\mathsf{r}]p \to p}\,(R\to) \qquad \dfrac{\dfrac{\dfrac{x{:}^{\varepsilon} p \Rightarrow x{:}^{\varepsilon} p}{x{:}^{\varepsilon} p \Rightarrow x{:}^{\mathsf{r}} p}\,(Rat)}{x{:}^{\varepsilon} p \Rightarrow x{:}^{\varepsilon}[\mathsf{r}]p}\,(R[\mathsf{r}])}{\Rightarrow x{:}^{\varepsilon} p \to [\mathsf{r}]p}\,(R\to)}{\Rightarrow x{:}^{\varepsilon}[\mathsf{r}]p \leftrightarrow p}\,(R\wedge)$$

Case of axiom ([r]¬)

$$\dfrac{\dfrac{\dfrac{\dfrac{\dfrac{x{:}^{\mathsf{r}} A \Rightarrow x{:}^{\mathsf{r}} A}{x{:}^{\varepsilon}[\mathsf{r}]A \Rightarrow x{:}^{\mathsf{r}} A}\,(L[\mathsf{r}])}{x{:}^{\mathsf{r}}\neg A \Rightarrow x{:}^{\varepsilon}\neg[\mathsf{r}]A}\,(\neg)}{x{:}^{\varepsilon}[\mathsf{r}]\neg A \Rightarrow x{:}^{\mathsf{r}}\neg[\mathsf{r}]A}\,(L[\mathsf{r}])}{\Rightarrow x{:}^{\varepsilon}[\mathsf{r}]\neg A \to \neg[\mathsf{r}]A}\,(R\to) \qquad \dfrac{\dfrac{\dfrac{\dfrac{x{:}^{\mathsf{r}} A \Rightarrow x{:}^{\mathsf{r}} A}{x{:}^{\mathsf{r}} A \to x{:}^{\varepsilon}[\mathsf{r}]A}\,(R[\mathsf{r}])}{x{:}^{\varepsilon}\neg[\mathsf{r}]A \Rightarrow x{:}^{\mathsf{r}}\neg A}\,(\neg)}{x{:}^{\varepsilon}\neg[\mathsf{r}]A \Rightarrow x{:}^{\varepsilon}[\mathsf{r}]\neg A}\,(R[\mathsf{r}])}{\Rightarrow x{:}^{\varepsilon}\neg[\mathsf{r}]A \to [\mathsf{r}]\neg A}\,(R\to)}{\Rightarrow x{:}^{\varepsilon}[\mathsf{r}]\neg A \leftrightarrow \neg[\mathsf{r}]A}\,(R\wedge)$$

Case of axiom ([r] →)

$$\dfrac{\dfrac{\dfrac{\dfrac{\dfrac{\dfrac{x{:}^{\mathsf{r}} A \Rightarrow x{:}^{\mathsf{r}} A}{x{:}^{\mathsf{r}} A \Rightarrow x{:}^{\mathsf{r}} B, x{:}^{\mathsf{r}} A}\,(Rw) \quad \dfrac{x{:}^{\mathsf{r}} B \Rightarrow x{:}^{\mathsf{r}} B}{x{:}^{\mathsf{r}} B, x{:}^{\mathsf{r}} A \Rightarrow x{:}^{\mathsf{r}} B}\,(Lw)}{x{:}^{\mathsf{r}} A \to B, x{:}^{\mathsf{r}} A \Rightarrow x{:}^{\mathsf{r}} B}\,(L\to)}{x{:}^{\varepsilon}[\mathsf{r}](A \to B), x{:}^{\varepsilon}[\mathsf{r}]A \Rightarrow x{:}^{\varepsilon}[\mathsf{r}]B}\,([\mathsf{r}])}{x{:}^{\varepsilon}[\mathsf{r}](A \to B) \Rightarrow x{:}^{\varepsilon}[\mathsf{r}]A \to [\mathsf{r}]B}\,(R\to)}{\Rightarrow x{:}^{\varepsilon}[\mathsf{r}](A \to B) \to ([\mathsf{r}]A \to [\mathsf{r}]B)}\,(R\to) \quad \dfrac{\dfrac{\dfrac{\dfrac{\dfrac{x{:}^{\mathsf{r}} A \Rightarrow x{:}^{\mathsf{r}} A}{x{:}^{\mathsf{r}} A \Rightarrow x{:}^{\mathsf{r}} B, x{:}^{\mathsf{r}} A}\,(Rw) \quad \dfrac{x{:}^{\mathsf{r}} B \Rightarrow x{:}^{\mathsf{r}} B}{x{:}^{\mathsf{r}} B, x{:}^{\mathsf{r}} A \Rightarrow x{:}^{\mathsf{r}} B}\,(Lw)}{x{:}^{\mathsf{r}} A \to B, x{:}^{\mathsf{r}} A \Rightarrow x{:}^{\mathsf{r}} B}\,(L\to)}{x{:}^{\varepsilon}[\mathsf{r}](A \to B), x{:}^{\varepsilon}[\mathsf{r}]A \Rightarrow x{:}^{\varepsilon}[\mathsf{r}]B}\,([\mathsf{r}])}{x{:}^{\varepsilon}[\mathsf{r}](A \to B) \Rightarrow x{:}^{\varepsilon}[\mathsf{r}]A \to [\mathsf{r}]B}\,(R\to)}{\Rightarrow x{:}^{\varepsilon}([\mathsf{r}]A \to [\mathsf{r}]B) \to [\mathsf{r}](A \to B)}\,(R\to)}{\Rightarrow x{:}^{\varepsilon}[\mathsf{r}](A \to B) \leftrightarrow ([\mathsf{r}]A \to [\mathsf{r}]B)}\,(R\wedge)$$

Case of axiom ([r][a])
The direction from left to right is shown as follows.

$$
\dfrac{
  \dfrac{
    \dfrac{
      \dfrac{
        \dfrac{x\mathsf{R}^{\varepsilon}_{\mathsf{r}(a)}y \Rightarrow x\mathsf{R}^{\varepsilon}_{\mathsf{r}(a)}y}{x\mathsf{R}^{\varepsilon}_{\mathsf{r}(a)}y \Rightarrow x\mathsf{R}^{\mathsf{r}}_{a}y}\,(Rrel)
      }{x\mathsf{R}^{\varepsilon}_{\mathsf{r}(a)}y \Rightarrow y{:}^{\mathsf{r}}A, x\mathsf{R}^{\mathsf{r}}_{a}y}\,(Rw)
      \quad
      \dfrac{y{:}^{\mathsf{r}}A \Rightarrow y{:}^{\mathsf{r}}A}{y{:}^{\mathsf{r}}A, x\mathsf{R}^{\varepsilon}_{\mathsf{r}(a)}y \Rightarrow y{:}^{\mathsf{r}}A}\,(Lw)
    }{x\mathsf{R}^{\varepsilon}_{\mathsf{r}(a)}y, x{:}^{\mathsf{r}}[a]A \Rightarrow y{:}^{\mathsf{r}}A}\,(L[a])
  }{
    \dfrac{x\mathsf{R}^{\varepsilon}_{\mathsf{r}(a)}y, x{:}^{\mathsf{r}}[a]A \Rightarrow y{:}^{\varepsilon}[\mathsf{r}]A}{
      \dfrac{x{:}^{\mathsf{r}}[a]A \Rightarrow x{:}^{\varepsilon}[\mathsf{r}(a)][\mathsf{r}]A}{
        x{:}^{\varepsilon}[\mathsf{r}][a]A \Rightarrow x{:}^{\varepsilon}[\mathsf{r}(a)][\mathsf{r}]A
      }\,(L[\mathsf{r}])
    }\,(R[\mathsf{r}(a)])
  }\,(R[\mathsf{r}])
}{\Rightarrow x{:}^{\varepsilon}[\mathsf{r}][a]A \to [\mathsf{r}(a)][\mathsf{r}]A}\,(R\to)
$$

The direction from right to left is shown as follows.

$$
\dfrac{
  \dfrac{
    \dfrac{
      \dfrac{
        \dfrac{x\mathsf{R}^{\varepsilon}_{\mathsf{r}(a)}y \Rightarrow x\mathsf{R}^{\varepsilon}_{\mathsf{r}(a)}y}{x\mathsf{R}^{\mathsf{r}}_{a}y \Rightarrow x\mathsf{R}^{\varepsilon}_{\mathsf{r}(a)}y}\,(Lrel)
      }{x\mathsf{R}^{\mathsf{r}}_{a}y \Rightarrow y{:}^{\mathsf{r}}A, x\mathsf{R}^{\varepsilon}_{\mathsf{r}(a)}y}\,(Rw)
      \quad
      \dfrac{\dfrac{y{:}^{\mathsf{r}}A \Rightarrow y{:}^{\mathsf{r}}A}{y{:}^{\varepsilon}[\mathsf{r}]A \Rightarrow y{:}^{\mathsf{r}}A}\,(L[\mathsf{r}])}{y{:}^{\varepsilon}[\mathsf{r}]A, x\mathsf{R}^{\mathsf{r}}_{a}y \Rightarrow y{:}^{\mathsf{r}}A}\,(Lw)
    }{x\mathsf{R}^{\mathsf{r}}_{a}y, x{:}^{\varepsilon}[\mathsf{r}(a)][\mathsf{r}]A \Rightarrow y{:}^{\mathsf{r}}A}\,(L[\mathsf{r}(a)])
  }{
    \dfrac{x{:}^{\varepsilon}[\mathsf{r}(a)][\mathsf{r}]A \Rightarrow x{:}^{\mathsf{r}}[a]A}{
      x{:}^{\varepsilon}[\mathsf{r}(a)][\mathsf{r}]A \Rightarrow x{:}^{\varepsilon}[\mathsf{r}][a]A
    }\,(R[\mathsf{r}])
  }\,(R[a])
}{\Rightarrow x{:}^{\varepsilon}[\mathsf{r}(a)][\mathsf{r}]A \to [\mathsf{r}][a]A}\,(R\to)
$$

In the inductive step, we show the cases of inference rules in **HDLRC**. Here, we show only the case of inference rules (**MP**) and ($\mathbf{Nec}_{[\mathsf{r}]}$).

Case of inference rule (MP):

Fix any $x \in \mathsf{Var}$. Suppose derivations $\mathfrak{D}_1$ for $\Rightarrow x{:}^{\varepsilon}A \to B$ and $\mathfrak{D}_2$ for $\Rightarrow x{:}^{\varepsilon}A$. We show that there is a derivation of $\Rightarrow x{:}^{\varepsilon}B$. This is shown by:

$$
\dfrac{
  \begin{matrix}\vdots\ \mathfrak{D}_2\\ \Rightarrow x{:}^{\varepsilon}A\end{matrix}
  \quad
  \dfrac{
    \begin{matrix}\vdots\ \mathfrak{D}_1\\ \Rightarrow x{:}^{\varepsilon}A \to B\end{matrix}
    \quad
    \dfrac{
      \dfrac{x{:}^{\varepsilon}A \Rightarrow x{:}^{\varepsilon}A}{x{:}^{\varepsilon}A \Rightarrow x{:}^{\varepsilon}B, x{:}^{\varepsilon}A}\,(Rw)
      \quad
      \dfrac{x{:}^{\varepsilon}B \Rightarrow x{:}^{\varepsilon}B}{x{:}^{\varepsilon}B, x{:}^{\varepsilon}A \Rightarrow x{:}^{\varepsilon}B}\,(Lw)
    }{x{:}^{\varepsilon}A \to B, x{:}^{\varepsilon}A \Rightarrow x{:}^{\varepsilon}B}\,(L\to)
  }{x{:}^{\varepsilon}A \Rightarrow x{:}^{\varepsilon}B}\,(Cut)
}{\Rightarrow x{:}^{\varepsilon}B}\,(Cut)
$$

Case of inference rule ($Nec_{[\mathsf{r}]}$):

Fix any $x \in \mathsf{Var}$ and any $\mathsf{r} \in \mathsf{RC}$. Suppose a derivation $\mathfrak{D}$ of $\Rightarrow x{:}^{\varepsilon}A$. Our goal is to show the derivation $\mathfrak{D}'$ of $\Rightarrow x{:}^{\varepsilon}[\mathsf{r}]A$. Using rule ($R[\mathsf{r}]$) in **GDLRC** we can obtain the derivation $\mathfrak{D}'$ by $\mathfrak{D}$ where r is inserted into the head of the list of relation changers of each node in $\mathfrak{D}$.

□

3.3 *Cut Elimination of* GDLRC

Definition 5 Let $w, x, y, z \in \mathsf{Var}$, $A \in \mathsf{Form}_{\mathbf{DLRC}}$, $\alpha \in \mathsf{PR}$, $L \in (\mathsf{RC})^*$ and φ be a sequent. A substitution $\varphi[y/x]$ (the result of substituting x in φ uniformly with y) is defined by:

$$z[y/x] \equiv \begin{cases} y & \text{if } z = x \\ z & \text{if } z \neq x \end{cases}$$
$$(z{:}^L A)[y/x] \equiv z[y/x]{:}^L A$$
$$(w\mathsf{R}^L_\alpha z)[y/x] \equiv w[y/x]\mathsf{R}^L_\alpha z[y/x]$$
$$(w = z)[y/x] \equiv w[y/x](= z[y/x])$$

A substitution $\Gamma[y/x]$ is also defined by:

$$\Gamma[y/x] = \{\varphi[y/x] \mid \varphi \in \Gamma\}.$$

Lemma 4 *If* $\vdash_{\mathbf{GDLRC}} \Gamma \Rightarrow \Delta$ *by a derivation* $\mathfrak{D}$*, then* $\vdash_{\mathbf{GDLRC}} \Gamma[y/x] \Rightarrow \Delta[y/x]$ *by a derivation* $\mathfrak{D}'$ *where the height of* $\mathfrak{D}$ *is equal to the height of* $\mathfrak{D}'$*.*

Definition 6 (*Extended cut rule*) Let $m, n \geq 0$ and φ be a labelled expression.

$$\frac{\Gamma \Rightarrow \Delta, \varphi^m \quad \varphi^n, \Sigma \Rightarrow \Pi}{\Gamma, \Sigma \Rightarrow \Delta, \Pi}\ (Ecut)$$

We say that φ is an *Ecut labelled expression.*

We define $\mathbf{GDLRC}^*$ as $\mathbf{GDLRC}^-$ with the rule of $(Ecut)$. Note that this form of extended cut rule is a variant of Gentzen's *mix rule* (Gentzen 1964). For $(Ecut)$, it is allowed to leave some occurrences of cut labelled expressions φ in either Δ or Σ of upper sequents, while it is not for the mix, i.e., we have to eliminate all occurrences of φ from either Δ or Σ in a mix. It is also allowed the case where $m = 0$ or $n = 0$ for $(Ecut)$, while it is not for the mix. Kashima introduced this form of extended cut rule in his text book to show the cut elimination of a sequent calculus for first-order logic (Kashima 2009).

Definition 7 (*Ecut-bottom form for* $\mathbf{GDLRC}^*$) A derivation $\mathfrak{D}$ in $\mathbf{GDLRC}^*$ is *Ecut-bottom form* if it has the following form:

$$\dfrac{\dfrac{\vdots\, \mathfrak{D}_L}{\Gamma \Rightarrow \Delta, \varphi^m}\ \mathtt{rule}(\mathfrak{D}_L) \quad \dfrac{\vdots\, \mathfrak{D}_R}{\varphi^n, \Sigma \Rightarrow \Pi}\ \mathtt{rule}(\mathfrak{D}_R)}{\Gamma, \Sigma \Rightarrow \Delta, \Pi}\ (Ecut)$$

where $\mathtt{rule}(\mathfrak{D})$ is the last applied rule of a given derivation $\mathfrak{D}$ and there are no-application of $(Ecut)$ in $\mathfrak{D}_L$ nor $\mathfrak{D}_R$.

Definition 8 Let $\mathfrak{D}$ be a derivation in **GDLRC*** which has the Ecut-bottom form. The *weight* and the *complexity* of $\mathfrak{D}$ are defined by:

$$w(\mathfrak{D}) = |\mathfrak{D}_L| + |\mathfrak{D}_R| \ (\geq 2) \text{ where } |\cdot| \text{ is the number of nodes as a tree,}$$
$$c(\mathfrak{D}) = \ell(\varphi) \text{ where } \varphi \text{ is the labelled expressions in } \mathfrak{D}.$$

Definition 9 Let $c(\mathfrak{D}) = i$ and $w(\mathfrak{D}) = j$.

$$(i, j) \geq (i', j') \iff i = i' \text{ and } j \geq j'.$$

Lemma 5 *For any Ecut-bottom form $\mathfrak{D}$ in* **GDLRC***, *there is a derivation $\mathfrak{D}'$ in* **GDLRC**$^-$ *such that the end sequent of $\mathfrak{D}$ is the same as that of $\mathfrak{D}'$.*

Proof Since $\mathfrak{D}$ is the Ecut-bottom form in **GDLRC***, it has the following form:

$$\dfrac{\dfrac{\vdots\, \mathfrak{D}_L}{\Gamma \Rightarrow \Delta, \varphi^m}\ \texttt{rule}(\mathfrak{D}_L) \qquad \dfrac{\vdots\, \mathfrak{D}_R}{\varphi^n, \Sigma \Rightarrow \Pi}\ \texttt{rule}(\mathfrak{D}_R)}{\Gamma, \Sigma \Rightarrow \Delta, \Pi}\ (Ecut)$$

Let $c(\mathfrak{D}) = i \ (\geq 0)$ and $w(\mathfrak{D}) = j \ (\geq 2)$. We show the proof by double induction on $(c(\mathfrak{D}), w(\mathfrak{D}))$.[3] Our goal is to obtain a derivation $\mathfrak{D}'$ in **GDLRC**$^-$ such that the end sequent of $\mathfrak{D}$ is equal to that of $\mathfrak{D}'$. Our proof is organized as follows: first, we consider the cases where $m = 0$ (for φ^m) or $n = 0$ (for φ^n) in $\mathfrak{D}$. Since we can easily show such cases by (Lw) and (Rw), we skip them. Second, we consider the cases where $m > 0$ and $n > 0$ in $\mathfrak{D}$. These cases can be divided into the following five sub cases.

(1) At least one of $\texttt{rule}(\mathfrak{D}_L)$ and $\texttt{rule}(\mathfrak{D}_R)$ is an initial sequent.
(2) At least one of $\texttt{rule}(\mathfrak{D}_L)$ and $\texttt{rule}(\mathfrak{D}_R)$ is a structural rule.
(3) At least one of $\texttt{rule}(\mathfrak{D}_L)$ and $\texttt{rule}(\mathfrak{D}_R)$ is a logical rule, a program rule, or a relation changer rule where the cut labelled expression is not principal.
(4) Both $\texttt{rule}(\mathfrak{D}_L)$ and $\texttt{rule}(\mathfrak{D}_R)$ are logical rules, program rules, or relation changer rules for the same connectives where the cut labelled expression is principal for both rules.
(5) At least one of $\texttt{rule}(\mathfrak{D}_L)$ and $\texttt{rule}(\mathfrak{D}_R)$ is an equality rule where the cut labelled expression is not principal.

For case (5), the equality rule $(= at)$ can be regarded as a right rule, but we do not have any corresponding left rule for that. In addition, there is no principal labelled expression in the other equality rules. Therefore we do not need to consider the case of 'both $\texttt{rule}(\mathfrak{D}_L)$ and $\texttt{rule}(\mathfrak{D}_R)$ are equality rules where the cut labelled expression is principal for both rules.'

[3]Ono and Komori (1985) showed the cut elimination theorem of sequent calculi without contraction rules by double induction on complexity and weight, while Kashima (2009) also employed the same measures to show the cut elimination for first-order logic with contraction rules. We note that Gentzen (1964) used measures of complexity and 'grade.'

Now, let us see the proof for each cases. We skip cases (1) and (2) since they are rather easy.

Case (3):

Suppose that at least one of $\texttt{rule}(\mathfrak{D}_L)$ and $\texttt{rule}(\mathfrak{D}_R)$ is a logical rule, a program rule, or a relation changer rule where the cut labelled expression is not principal. Here, we only show the sub cases for the logical rule $(R[\mathsf{r}])$ and the relation changer rules (Rat) and $(Rrel)$.

Case (3-1) where $\texttt{rule}\ (\mathfrak{D}_L)$ **is** $(R[\mathsf{r}])$**:**

The Ecut-bottom form of this case can be written as follows:

$$\dfrac{\dfrac{\begin{array}{c}\vdots\ \mathfrak{D}_{L'}\\ \Gamma \Rightarrow \Delta, x{:}^{\langle L,r\rangle}A, \varphi^m\end{array}}{\Gamma \Rightarrow \Delta, x{:}^{L}[\mathsf{r}]A, \varphi^m}(R[\mathsf{r}]) \qquad \begin{array}{c}\vdots\ \mathfrak{D}_R\\ \varphi^n, \Sigma \Rightarrow \Pi\end{array}}{\Gamma, \Sigma \Rightarrow \Delta, \Pi, x{:}^{L}[\mathsf{r}]A}(Ecut)$$

We can transform the above derivation into the following one:

$$\dfrac{\dfrac{\begin{array}{c}\vdots\ \mathfrak{D}_{L'}\\ \Gamma \Rightarrow \Delta, x{:}^{\langle L,r\rangle}A, \varphi^m\end{array} \qquad \begin{array}{c}\vdots\ \mathfrak{D}_R\\ \varphi^n, \Sigma \Rightarrow \Pi\end{array}}{\Gamma, \Sigma \Rightarrow \Delta, \Pi, x{:}^{\langle L,r\rangle}A}(Ecut)}{\Gamma, \Sigma \Rightarrow \Delta, \Pi, x{:}^{L}[\mathsf{r}]A}(R[\mathsf{r}])$$

Since $\ell(x{:}^{L}[\mathsf{r}]A) > \ell(x{:}^{\langle L,r\rangle}A)$ (cf. Proposition 4 (i)), the complexity of the above transformed derivation is less than that of the original one. Therefore, we can obtain a derivation for $\Gamma, \Sigma \Rightarrow \Delta, \Pi, x{:}^{L}[\mathsf{r}]A$ in $\mathbf{GDLRC}^{-}$ by induction hypothesis.

Case (3-2) where $\texttt{rule}\ (\mathfrak{D}_L)$ **is** (Rat)**:**

The Ecut-bottom form of this case can be written as follows:

$$\dfrac{\dfrac{\begin{array}{c}\vdots\ \mathfrak{D}_{L'}\\ \Gamma \Rightarrow \Delta, x{:}^{L}p, \varphi^m\end{array}}{\Gamma \Rightarrow \Delta, x{:}^{\langle L,r\rangle}p, \varphi^m}(Rat) \qquad \begin{array}{c}\vdots\ \mathfrak{D}_R\\ \varphi^n, \Sigma \Rightarrow \Pi\end{array}}{\Gamma, \Sigma \Rightarrow \Delta, \Pi, x{:}^{\langle L,r\rangle}p}(Ecut)$$

We can transform the above derivation into the following one:

$$\dfrac{\dfrac{\begin{array}{c}\vdots\ \mathfrak{D}_{L'}\\ \Gamma \Rightarrow \Delta, x{:}^{L}p, \varphi^m\end{array} \qquad \begin{array}{c}\vdots\ \mathfrak{D}_R\\ \varphi^n, \Sigma \Rightarrow \Pi\end{array}}{\Gamma, \Sigma \Rightarrow \Delta, \Pi, x{:}^{L}p}(Ecut)}{\Gamma, \Sigma \Rightarrow \Delta, \Pi, x{:}^{\langle L,r\rangle}p}(Rat)$$

Since $\ell(x{:}^{\langle L,r\rangle}p) > \ell(x{:}^{L}p)$ (cf. Proposition 4 (ii)), the complexity of the above transformed derivation is less than that of the original one. Therefore, we can obtain a derivation for $\Gamma, \Sigma \Rightarrow \Delta, \Pi, x{:}^{\langle L,r\rangle}p$ in **GDLRC**$^{-}$ by induction hypothesis.

Case (3-3) where `rule` **($\mathfrak{D}_L$) is (*Rrel*):**

The Ecut-bottom form of this case can be written as follows:

$$\dfrac{\dfrac{\begin{array}{c}\vdots\ \mathfrak{D}_{L'}\\ \Gamma \Rightarrow \Delta, x\mathsf{R}^{L}_{\mathsf{r}(a)}y, \varphi^m\end{array}}{\Gamma \Rightarrow \Delta, x\mathsf{R}^{\langle L,r\rangle}_{a}y, \varphi^m}(Rrel) \quad \begin{array}{c}\vdots\ \mathfrak{D}_R\\ \varphi^n, \Sigma \Rightarrow \Pi\end{array}}{\Gamma, \Sigma \Rightarrow \Delta, \Pi, x\mathsf{R}^{\langle L,r\rangle a}_{y}}(Ecut)$$

We can transform the above derivation into the following one:

$$\dfrac{\dfrac{\begin{array}{c}\vdots\ \mathfrak{D}_{L'}\\ \Gamma \Rightarrow \Delta, x\mathsf{R}^{L}_{\mathsf{r}(a)}y, \varphi^m\end{array} \quad \begin{array}{c}\vdots\ \mathfrak{D}_R\\ \varphi^n, \Sigma \Rightarrow \Pi\end{array}}{\Gamma, \Sigma \Rightarrow \Delta, \Pi, x\mathsf{R}^{L}_{\mathsf{r}(a)}y}(Ecut)}{\Gamma, \Sigma \Rightarrow \Delta, \Pi, x\mathsf{R}^{\langle L,r\rangle}_{a}y}(Rrel)$$

Since $\ell(x\mathsf{R}^{\langle L,r\rangle}_{a}y) > \ell(x\mathsf{R}^{\mathsf{r}}_{\mathsf{r}(a)}y)$ (cf. Proposition 4 (iii)), the complexity of the above transformed derivation is less than that of the original one. Therefore, we can obtain a derivation for $\Gamma, \Sigma \Rightarrow \Delta, \Pi, x\mathsf{R}^{\langle L,r\rangle a}_{y}$ in **GDLRC**$^{-}$ by induction hypothesis.

Case (4):

Suppose that both `rule`($\mathfrak{D}_L$) and `rule`($\mathfrak{D}_R$) are logical rules, program rules, or relation changer rules for the same connectives where the cut labelled expression is principal for both rules. Here, we only show the sub cases for the logical rules of ($[\mathsf{r}]$) and relation changer rules of (rel) and (at).

Case (4-1) where `rule` **($\mathfrak{D}_L$) is (*R*[r])and** `rule` **($\mathfrak{D}_R$) is (*L*[r]):**

The Ecut-bottom form of this case can be written as follows:

$$\dfrac{\dfrac{\begin{array}{c}\vdots\ \mathfrak{D}_{L'}\\ \Gamma \Rightarrow \Delta, x{:}^{\langle L,r\rangle}A, (x{:}^{L}[\mathsf{r}]A)^{m-1}\end{array}}{\Gamma \Rightarrow \Delta, (x{:}^{L}[\mathsf{r}]A)^m}(R[\mathsf{r}]) \quad \dfrac{\begin{array}{c}\vdots\ \mathfrak{D}_{R'}\\ (x{:}^{L}[\mathsf{r}]A)^{n-1}, x{:}^{\langle L,r\rangle}A, \Sigma \Rightarrow \Pi\end{array}}{(x{:}^{L}[\mathsf{r}]A)^n, \Sigma \Rightarrow \Pi}(L[\mathsf{r}])}{\Gamma, \Sigma \Rightarrow \Delta, \Pi}(Ecut)$$

From this derivation, we can construct the following derivation $\mathfrak{D}'_1$:

$$\dfrac{\begin{array}{c}\vdots\ \mathfrak{D}_{L'}\\ \Gamma \Rightarrow \Delta, x{:}^{\langle L,r\rangle}A, (x{:}^{L}[\mathsf{r}]A)^{m-1}\end{array} \quad \dfrac{\begin{array}{c}\vdots\ \mathfrak{D}_{R'}\\ (x{:}^{L}[\mathsf{r}]A)^{n-1}, x{:}^{\langle L,r\rangle}A, \Sigma \Rightarrow \Pi\end{array}}{(x{:}^{L}[\mathsf{r}]A)^n, \Sigma \Rightarrow \Pi}(L[\mathsf{r}])}{\Gamma, \Sigma \Rightarrow \Delta, \Pi, x{:}^{\langle L,r\rangle}A}(Ecut)$$

Similarly, we can also construct the following derivation $\mathfrak{D}_2'$:

$$\dfrac{\dfrac{\begin{matrix}\vdots\ \mathfrak{D}_{L'}\\ \Gamma \Rightarrow \Delta, x{:}^{\langle L,r\rangle}A, (x{:}^{L}[\mathsf{r}]A)^{m-1}\end{matrix}}{\Gamma \Rightarrow \Delta, (x{:}^{L}[\mathsf{r}]A)^{m}}\,(R[\mathsf{r}]) \qquad \begin{matrix}\vdots\ \mathfrak{D}_{R'}\\ (x{:}^{L}[\mathsf{r}]A)^{n-1}, x{:}^{\langle L,r\rangle}A, \Sigma \Rightarrow \Pi\end{matrix}}{x{:}^{\langle L,r\rangle}A, \Gamma, \Sigma \Rightarrow \Delta, \Pi}\,(Ecut)$$

Since the weight of the Ecut-bottom form in $\mathfrak{D}_1'$ and $\mathfrak{D}_2'$ are smaller than the original one, we can obtain derivations for $\Gamma, \Sigma \Rightarrow \Delta, \Pi, x{:}^{\langle L,r\rangle}A$ and $x{:}^{\langle L,r\rangle}A, \Gamma, \Sigma \Rightarrow \Delta, \Pi$ in **GDLRC**$^-$ by induction hypothesis. Then, by $\mathfrak{D}_1'$ and $\mathfrak{D}_2'$, we can transform the original derivation into the following one:

$$\dfrac{\dfrac{\mathfrak{D}_1' \quad \mathfrak{D}_2'}{\Gamma, \Gamma, \Sigma, \Sigma \Rightarrow \Delta, \Delta, \Pi, \Pi}\,(Ecut)}{\Gamma, \Sigma \Rightarrow \Delta, \Pi}\,(c)$$

Since $\ell(x{:}^{L}[\mathsf{r}]A) > \ell(x{:}^{\langle L,r\rangle}A)$ (cf. Proposition 4 (i)), the complexity of the above transformed derivation is less than that of the original one. Therefore, we can obtain a derivation for $\Gamma, \Sigma \Rightarrow \Delta, \Pi$ in **GDLRC**$^-$ by induction hypothesis.

Case (4-2) where `rule` ($\mathfrak{D}_L$) is (*Rat*)and `rule` ($\mathfrak{D}_R$) is (*Lat*):

The Ecut-bottom form of this case can be written as follows:

$$\dfrac{\dfrac{\begin{matrix}\vdots\ \mathfrak{D}_{L'}\\ \Gamma \Rightarrow \Delta, x{:}^{L}p, (x{:}^{\langle L,r\rangle}p)^{m-1}\end{matrix}}{\Gamma \Rightarrow \Delta, (x{:}^{\langle L,r\rangle}p)^{m}}\,(Rat) \qquad \dfrac{\begin{matrix}\vdots\ \mathfrak{D}_{R'}\\ (x{:}^{\langle L,r\rangle}p)^{n-1}, x{:}^{L}p, \Sigma \Rightarrow \Pi\end{matrix}}{(x{:}^{\langle L,r\rangle}p)^{n}, \Sigma \Rightarrow \Pi}\,(Lat)}{\Gamma, \Sigma \Rightarrow \Delta, \Pi}\,(Ecut)$$

From this derivation, we can construct the following derivation $\mathfrak{D}_1'$:

$$\dfrac{\begin{matrix}\vdots\ \mathfrak{D}_{L'}\\ \Gamma \Rightarrow \Delta, x{:}^{L}p, (x{:}^{\langle L,r\rangle}p)^{m-1}\end{matrix} \qquad \dfrac{\begin{matrix}\vdots\ \mathfrak{D}_{R'}\\ (x{:}^{\langle L,r\rangle}p)^{n-1}, x{:}^{L}p, \Sigma \Rightarrow \Pi\end{matrix}}{(x{:}^{\langle L,r\rangle}p)^{n}, \Sigma \Rightarrow \Pi}\,(Lat)}{\Gamma, \Sigma \Rightarrow \Delta, \Pi, x{:}^{L}p}\,(Ecut)$$

Similarly, we can also construct the following derivation $\mathfrak{D}_2'$:

$$\dfrac{\dfrac{\begin{matrix}\vdots\ \mathfrak{D}_{L'}\\ \Gamma \Rightarrow \Delta, x{:}^{L}p, (x{:}^{\langle L,r\rangle}p)^{m-1}\end{matrix}}{\Gamma \Rightarrow \Delta, (x{:}^{\langle L,r\rangle}p)^{m}}\,(Rat) \qquad \begin{matrix}\vdots\ \mathfrak{D}_{R'}\\ (x{:}^{\langle L,r\rangle}p)^{n-1}, x{:}^{L}p, \Sigma \Rightarrow \Pi\end{matrix}}{x{:}^{L}p, \Gamma, \Sigma \Rightarrow \Delta, \Pi}\,(Ecut)$$

Since the weight of the Ecut-bottom form in $\mathfrak{D}'_1$ and $\mathfrak{D}'_2$ are smaller than the original one, we can obtain derivations for $\Gamma, \Sigma \Rightarrow \Delta, \Pi, x{:}^{L} p$ and $x{:}^{L} p, \Gamma, \Sigma \Rightarrow \Delta, \Pi$ in **GDLRC**$^-$ by induction hypothesis. Then, by $\mathfrak{D}'_1$ and $\mathfrak{D}'_2$, we can transform the original derivation into the following one:

$$\dfrac{\dfrac{\mathfrak{D}'_1 \quad \mathfrak{D}'_2}{\Gamma, \Gamma, \Sigma, \Sigma \Rightarrow \Delta, \Delta, \Pi, \Pi}\,(Ecut)}{\Gamma, \Sigma \Rightarrow \Delta, \Pi}\,(c)$$

Since $\ell(x{:}^{\langle L, r\rangle} p) > \ell(x{:}^{L} p)$ (cf. Proposition 4 (ii)), the complexity of the above transformed derivation is less than that of the original one. Therefore, we can obtain a derivation for $\Gamma, \Sigma \Rightarrow \Delta, \Pi$ in **GDLRC**$^-$ by induction hypothesis.

Case (4-3) where `rule` **($\mathfrak{D}_L$) is ($Rrel$)and** `rule` **($\mathfrak{D}_R$) is ($Lrel$):**

The Ecut-bottom form of this case can be written as follows:

$$\dfrac{\dfrac{\begin{matrix}\vdots\, \mathfrak{D}_{L'} \\ \Gamma \Rightarrow \Delta, x\mathsf{R}^{L}_{\mathsf{r}(a)}y, (x\mathsf{R}^{\langle L,r\rangle}_{a}y)^{m-1}\end{matrix}}{\Gamma \Rightarrow \Delta, (x\mathsf{R}^{\langle L,r\rangle}_{a}y)^{m}}\,(Rrel) \quad \dfrac{\begin{matrix}\vdots\, \mathfrak{D}_{R'} \\ (x\mathsf{R}^{\langle L,r\rangle}_{a}y)^{n-1}, x\mathsf{R}^{L}_{\mathsf{r}(a)}y, \Sigma \Rightarrow \Pi\end{matrix}}{(x\mathsf{R}^{\langle L,r\rangle}_{a}y)^{n}, \Sigma \Rightarrow \Pi}\,(Lrel)}{\Gamma, \Sigma \Rightarrow \Delta, \Pi}\,(Ecut)$$

From this derivation, we can construct the following derivation $\mathfrak{D}'_1$:

$$\dfrac{\begin{matrix}\vdots\, \mathfrak{D}_{L'} \\ \Gamma \Rightarrow \Delta, x\mathsf{R}^{L}_{\mathsf{r}(a)}y, (x\mathsf{R}^{\langle L,r\rangle}_{a}y)^{m-1}\end{matrix} \quad \dfrac{\begin{matrix}\vdots\, \mathfrak{D}_{R'} \\ (x\mathsf{R}^{\langle L,r\rangle}_{a}y)^{n-1}, x\mathsf{R}^{L}_{\mathsf{r}(a)}y, \Sigma \Rightarrow \Pi\end{matrix}}{(x\mathsf{R}^{\langle L,r\rangle}_{a}y)^{n}, \Sigma \Rightarrow \Pi}\,(Lrel)}{\Gamma, \Sigma \Rightarrow \Delta, \Pi, x\mathsf{R}^{L}_{\mathsf{r}(a)}y}\,(Ecut)$$

Similarly, we can also construct the following derivation $\mathfrak{D}'_2$:

$$\dfrac{\dfrac{\begin{matrix}\vdots\, \mathfrak{D}_{L'} \\ \Gamma \Rightarrow \Delta, x\mathsf{R}^{L}_{\mathsf{r}(a)}y, (x\mathsf{R}^{\langle L,r\rangle}_{a}y)^{m-1}\end{matrix}}{\Gamma \Rightarrow \Delta, (x\mathsf{R}^{\langle L,r\rangle}_{a}y)^{m}}\,(Rrel) \quad \begin{matrix}\vdots\, \mathfrak{D}_{R'} \\ (x\mathsf{R}^{\langle L,r\rangle}_{a}y)^{n-1}, x\mathsf{R}^{L}_{\mathsf{r}(a)}y, \Sigma \Rightarrow \Pi\end{matrix}}{x\mathsf{R}^{L}_{\mathsf{r}(a)}y, \Gamma, \Sigma \Rightarrow \Delta, \Pi}\,(Ecut)$$

Since the weight of the Ecut-bottom form in $\mathfrak{D}'_1$ and $\mathfrak{D}'_2$ are smaller than the original one, we can obtain derivations for $\Gamma, \Sigma \Rightarrow \Delta, \Pi, x\mathsf{R}^{L}_{\mathsf{r}(a)}y$ and $x\mathsf{R}^{L}_{\mathsf{r}(a)}y, \Gamma, \Sigma \Rightarrow \Delta, \Pi$ in **GDLRC**$^-$ by induction hypothesis. Then, by $\mathfrak{D}'_1$ and $\mathfrak{D}'_2$, we can transform the original derivation into the following one:

$$\dfrac{\dfrac{\mathfrak{D}_1' \quad \mathfrak{D}_2'}{\Gamma, \Gamma, \Sigma, \Sigma \Rightarrow \Delta, \Delta, \Pi, \Pi}\ (Ecut)}{\Gamma, \Sigma \Rightarrow \Delta, \Pi}\ (c)$$

Since $\ell(x\mathsf{R}_a^{\langle L,r\rangle}y) > \ell(x\mathsf{R}_{\mathsf{r}(a)}^{\mathsf{r}}y)$ (cf. Proposition 4 (iii)), the complexity of the above transformed derivation is less than that of the original one. Therefore, we can obtain a derivation for $\Gamma, \Sigma \Rightarrow \Delta, \Pi$ in **GDLRC**$^-$ by induction hypothesis.

Case (5):

Suppose that at least one of `rule`$(\mathfrak{D}_L)$ and `rule`$(\mathfrak{D}_R)$ is an equality rule where the cut labelled expression is not principal. Here, we only show the sub cases for equality rules $(= at)$ and $(= ref)$.

Case (5-1) where `rule` $(\mathfrak{D}_L)$ is $(= at)$:

The Ecut-bottom form can be written as follows:

$$\dfrac{\dfrac{\overset{\vdots\ \mathfrak{D}_{L_1'}}{\Gamma \Rightarrow \Delta, x = y, \varphi^m} \quad \overset{\vdots\ \mathfrak{D}_{L_2'}}{\Gamma \Rightarrow \Delta, x{:}^{\varepsilon} p, \varphi^m}}{\Gamma \Rightarrow \Delta, y{:}^{\varepsilon} p, \varphi^m}\ (= at) \quad \overset{\vdots\ \mathfrak{D}_R}{\varphi^n, \Sigma \Rightarrow \Pi}}{\Gamma, \Sigma \Rightarrow \Delta, \Pi, y{:}^{\varepsilon} p}\ (Ecut)$$

From this derivation, we can construct the following derivation $\mathfrak{D}_1'$:

$$\dfrac{\overset{\vdots\ \mathfrak{D}_{L_1'}}{\Gamma \Rightarrow \Delta, x = y, \varphi^m} \quad \overset{\vdots\ \mathfrak{D}_R}{\varphi^n, \Sigma \Rightarrow \Pi}}{\Gamma, \Sigma \Rightarrow \Delta, \Pi, x = y}\ (Ecut)$$

Similarly, we can also construct the following derivation $\mathfrak{D}_2'$:

$$\dfrac{\overset{\vdots\ \mathfrak{D}_{L_2'}}{\Gamma \Rightarrow \Delta, x{:}^{\varepsilon} p, \varphi^m} \quad \overset{\vdots\ \mathfrak{D}_R}{\varphi^n, \Sigma \Rightarrow \Pi}}{\Gamma, \Sigma \Rightarrow \Delta, \Pi, x{:}^{\varepsilon} p}\ (Ecut)$$

Since the weight of the Ecut-bottom form in $\mathfrak{D}_1'$ and $\mathfrak{D}_2'$ are smaller than the original one, we can obtain derivations for $\Gamma, \Sigma \Rightarrow \Delta, \Pi, x = y$ and $\Gamma, \Sigma \Rightarrow \Delta, \Pi, x{:}^{\varepsilon} p$ in **GDLRC**$^-$ by induction hypothesis. By $\mathfrak{D}_1'$ and $\mathfrak{D}_2'$, we can transform the original derivation into the following one:

$$\dfrac{\mathfrak{D}_1' \quad \mathfrak{D}_2'}{\Gamma, \Sigma \Rightarrow \Delta, \Pi, y{:}^{\varepsilon} p}\ (= at)$$

Since the transformed derivation does not contain $(Ecut)$, we can obtain a derivation for $\Gamma, \Sigma \Rightarrow \Delta, \Pi, y{:}^{\varepsilon} p$ in **GDLRC**$^-$.

Case (5-2) where `rule` $(\mathfrak{D}_L)$ is $(= ref)$:

The Ecut-bottom form can be written as follows:

$$\dfrac{\dfrac{\begin{array}{c}\vdots\ \mathfrak{D}_{L'}\\ x = x, \Gamma \Rightarrow \Delta, \varphi^m\end{array}}{\Gamma \Rightarrow \Delta, \varphi^m}\,(= ref) \qquad \begin{array}{c}\vdots\ \mathfrak{D}_R\\ \varphi^n, \Sigma \Rightarrow \Pi\end{array}}{\Gamma, \Sigma \Rightarrow \Delta, \Pi}\,(Ecut)$$

We can transform the above derivation into the following one:

$$\dfrac{\dfrac{\begin{array}{c}\vdots\ \mathfrak{D}_{L'}\\ x = x, \Gamma \Rightarrow \Delta, \varphi^m\end{array} \quad \begin{array}{c}\vdots\ \mathfrak{D}_R\\ \varphi^n, \Sigma \Rightarrow \Pi\end{array}}{x = x, \Gamma, \Sigma \Rightarrow \Delta, \Pi}\,(Ecut)}{\Gamma, \Sigma \Rightarrow \Delta, \Pi}\,(= ref)$$

Since the weight of the Ecut-bottom form in the above transformed derivation is smaller than the original one, we can obtain a derivation for $\Gamma, \Sigma \Rightarrow \Delta, \Pi$ in **GDLRC**$^-$ by induction hypothesis. □

Theorem 2 (Cut Elimination) *For any multisets Γ and Δ of labelled expressions, if* $\vdash_{\mathbf{GDLRC}} \Gamma \Rightarrow \Delta$, *then* $\vdash_{\mathbf{GDLRC}^-} \Gamma \Rightarrow \Delta$.

Proof Since (Cut) is a special form of $(Ecut)$, we have that if $\vdash_{\mathbf{GDLRC}} \Gamma \Rightarrow \Delta$, then $\vdash_{\mathbf{GDLRC}^*} \Gamma \Rightarrow \Delta$. Thus, it suffices to show that if $\vdash_{\mathbf{GDLRC}^*} \Gamma \Rightarrow \Delta$, then $\vdash_{\mathbf{GDLRC}^-} \Gamma \Rightarrow \Delta$. Let $\mathfrak{D}$ be a derivation for $\Gamma \Rightarrow \Delta$ in **GDLRC***. Since **GDLRC*** contains $(Ecut)$, $\mathfrak{D}$ might contain some Ecut-bottom forms. If $\mathfrak{D}$ does not contain $(Ecut)$, it is trivial. Therefore, we suppose that $\mathfrak{D}$ contains some Ecut-bottom forms. Now, let us focus on the nearest uppermost occurrence of Ecut-bottom form $\mathfrak{D}'$ from an initial sequent in $\mathfrak{D}$. If we apply Lemma 5 to this $\mathfrak{D}'$, we can get a derivation $\mathfrak{D}''$ in **GDLRC**$^-$ whose initial sequent is the same as that of $\mathfrak{D}'$ in **GDLRC***. If we replace such $\mathfrak{D}'$ in $\mathfrak{D}$ by $\mathfrak{D}''$, we can eliminate an occurrence of $(Ecut)$ from $\mathfrak{D}$. Similarly, if we repeat the above procedure to each uppermost occurrence of Ecut-bottom form in $\mathfrak{D}$, we can obtain a derivation for $\Gamma \Rightarrow \Delta$ that does not contain $(Ecut)$, therefore $\vdash_{\mathbf{GDLRC}^-} \Gamma \Rightarrow \Delta$. □

4 Soundness of GDLRC for Kripke Semantics

To show the soundness of labelled sequent calculus **GDLRC** for Kripke semantics, we need to give a semantic interpretation of labelled expressions in terms of Kripke semantics. Thus, we lift our Kripke semantics to labelled expressions by an assignment function for variables.

Definition 10 Let $\mathfrak{M} = \langle W, (R_a)_{a \in \mathsf{AP}}, V \rangle$ be a model. We say $f{:}\mathsf{Var} \to W$ is an *assignment* on $\mathfrak{M}$. Given any model $\mathfrak{M}$ and any assignment f on $\mathfrak{M}$, we define $\mathfrak{M}, f \models \varphi$ by:

$$\mathfrak{M}, f \models x{:}^L A \iff \mathfrak{M}^L, f(x) \models A,$$
$$\mathfrak{M}, f \models x \mathsf{R}^L_\alpha y \iff f(x) R^L_\alpha f(y),$$
$$\mathfrak{M}, f \models x = y \iff f(x) = f(y).$$

Definition 11 A labelled sequent $\Gamma \Rightarrow \Delta$*holds in* $\mathfrak{M}$ *under* f (notation: $\mathfrak{M}, f \models \Gamma \Rightarrow \Delta$) if, whenever $\mathfrak{M}, f \models \gamma$ for all $\gamma \in \Gamma$, $\mathfrak{M}, f \models \delta$ for some $\delta \in \Delta$. We say that $\Gamma \Rightarrow \Delta$ is *valid in a model* $\mathfrak{M}$ (notation: $\mathfrak{M} \models \Gamma \Rightarrow \Delta$) if $\mathfrak{M}, f \models \Gamma \Rightarrow \Delta$ for any assignment f on $\mathfrak{M}$. We also say that $\Gamma \Rightarrow \Delta$ is *valid in a class* $\mathbb{M}$ *of models* (notation: $\mathbb{M} \models \Gamma \Rightarrow \Delta$) if $\mathfrak{M} \models \Gamma \Rightarrow \Delta$ for all models $\mathfrak{M} \in \mathbb{M}$.

Lemma 6 *Let* $\mathbb{M}_{\text{all}}$ *be the class of all models. For any multisets* Γ *and* Δ *of labelled expressions,*

$$\textit{if } \vdash_{\mathbf{GDLRC}} \Gamma \Rightarrow \Delta, \textit{ then } \mathbb{M}_{\text{all}} \models \Gamma \Rightarrow \Delta.$$

Proof Let φ be a labelled expression. We proceed the proof by induction on the height n of a derivation of $\Gamma \Rightarrow \Delta$ in **GDLRC**. Since the base case for initial sequents is trivial, we skip it. Then, for inductive steps, we show only the cases for $(L[\alpha])$, $(R[\alpha])$, $(L[\mathsf{r}])$, (Lat) and $(Lrel)$.

Case where the last applied rule of our derivation is $(L[\alpha])$:

Suppose that $\mathbb{M}_{\text{all}} \models \Gamma \Rightarrow \Delta, x\mathsf{R}^L_\alpha y$ and $\mathbb{M}_{\text{all}} \models y{:}^L A, \Gamma \Rightarrow \Delta$. We show that $\mathbb{M}_{\text{all}} \models x{:}^L[\alpha]A, \Gamma \Rightarrow \Delta$, i.e., $\mathfrak{M}, f \models x{:}^L[\alpha]A, \Gamma \Rightarrow \Delta$ for all $\mathfrak{M}$ and all f on $\mathfrak{M}$. Fix any $\mathfrak{M} = \langle W, (R_a)_{a\in\mathsf{AP}}, V \rangle$ and f such that $\mathfrak{M}, f \models x{:}^L[\alpha]A$ and $\mathfrak{M}, f \models \gamma$ for all $\gamma \in \Gamma$. Our goal is to show that

$$\mathfrak{M}, f \models \delta \text{ for some } \delta \in \Delta.$$

By induction hypothesis, we have $\mathfrak{M}, f \models \Gamma \Rightarrow \Delta, x\mathsf{R}^L_\alpha y$. Since $\mathfrak{M}, f \models \gamma$ for all $\gamma \in \Gamma$, it follows that $\mathfrak{M}, f \models \delta$ for some $\delta \in \Delta$ or $\mathfrak{M}, f \models x\mathsf{R}^L_\alpha y$. If $\mathfrak{M}, f \not\models x\mathsf{R}^L_\alpha y$, we immediately obtain our goal. Hence, we consider the case where $\mathfrak{M}, f \models x\mathsf{R}^L_\alpha y$, i.e., $f(x) R^L_\alpha f(y)$. By definition of $\mathfrak{M}, f \models x{:}^L[\alpha]A$ and $f(x) R^L_\alpha f(y)$, we get $\mathfrak{M}^L, f(y) \models A$, i.e., $\mathfrak{M}, f \models y{:}^L A$. By induction hypothesis, we have $\mathfrak{M}, f \models y{:}^L A, \Gamma \Rightarrow \Delta$, which implies our goal.

Case where the last applied rule of our derivation is $(R[\alpha])$:

Suppose that $\mathbb{M}_{\text{all}} \models x\mathsf{R}^L_\alpha y, \Gamma \Rightarrow \Delta, y{:}^L A$ where y does not appear in Γ, Δ nor $x{:}^L[\alpha]A$. We show that $\mathbb{M}_{\text{all}} \models \Gamma \Rightarrow \Delta, x{:}^L[\alpha]A$, i.e., $\mathfrak{M}, f \models \Gamma \Rightarrow \Delta, x{:}^L[\alpha]A$ for all $\mathfrak{M}$ and all f on $\mathfrak{M}$. Fix any $\mathfrak{M} = \langle W, (R_a)_{a\in\mathsf{AP}}, V \rangle$ and f such that $\mathfrak{M}, f \models \gamma$ for all $\gamma \in \Gamma$. Our goal is to show that $\mathfrak{M}, f \models \delta$ for some $\delta \in \Delta$ or $\mathfrak{M}, f \models x{:}^L[\alpha]A$. Suppose that $\mathfrak{M}, f \not\models \delta$ for all $\delta \in \Delta$. We show that $\mathfrak{M}, f \models x{:}^L[\alpha]A$, i.e., if $f(x) R^L_\alpha v$, then $\mathfrak{M}^L, v \models A$ for all $v \in W$. Fix any $v \in W$ such that $f(x) R^L_\alpha v$. Our goal is to show that $\mathfrak{M}^L, v \models A$. Define a new assignment function $f' : \mathsf{Var} \to W$ by

$$f'(z) = \begin{cases} v & \text{if } z = y, \\ f(z) & \text{otherwise.} \end{cases}$$

Since $f'(y)$ is equal to v, it suffices to show that $\mathfrak{M}^L, f'(y) \models A$. Since y does not appear in Γ nor Δ, our suppositions implies $\mathfrak{M}, f' \models \gamma$ for all $\gamma \in \Gamma$ and $\mathfrak{M}, f' \not\models \delta$ for all $\delta \in \Delta$. Since $x \neq y$ and $f(x)R^L_\alpha v$, we have $f'(x)R^L_\alpha f'(y)$, i.e., $\mathfrak{M}, f' \models x\mathsf{R}^L_\alpha y$. By induction hypothesis, we have $\mathfrak{M}, f' \models x\mathsf{R}^L_\alpha y, \Gamma \Rightarrow \Delta, y{:}^L A$, which implies $\mathfrak{M}, f' \models y{:}^L A$.

Case where the last applied rule of our derivation is $(L[\mathsf{r}])$:

Suppose that $\mathbb{M}_{\text{all}} \models x{:}^{\langle L,\mathsf{r}\rangle} A, \Gamma \Rightarrow \Delta$. We show that $\mathbb{M}_{\text{all}} \models x{:}^L[\mathsf{r}]A, \Gamma \Rightarrow \Delta$, i.e., $\mathfrak{M}, f \models x{:}^L[\mathsf{r}]A, \Gamma \Rightarrow \Delta$ for all $\mathfrak{M}$ and all f on $\mathfrak{M}$. Fix any $\mathfrak{M}$ and f such that $\mathfrak{M}, f \models x{:}^L[\mathsf{r}]A$ and $\mathfrak{M}, f \models \gamma$ for all $\gamma \in \Gamma$. Our goal is to show that $\mathfrak{M}, f \models \delta$ for some $\delta \in \Delta$. Since $\mathfrak{M}, f \models x{:}^L[\mathsf{r}]A$, we have $\mathfrak{M}, f \models x{:}^{\langle L,\mathsf{r}\rangle} A$. By induction hypothesis, we have $\mathfrak{M}, f \models x{:}^{\langle L,\mathsf{r}\rangle} A, \Gamma \Rightarrow \Delta$, which implies our goal.

Case where the last applied rule of our derivation is (Lat):

Suppose that $\mathbb{M}_{\text{all}} \models x{:}^L p, \Gamma \Rightarrow \Delta$. We show that $\mathbb{M}_{\text{all}} \models x{:}^{\langle L,\mathsf{r}\rangle} p, \Gamma \Rightarrow \Delta$, i.e., $\mathfrak{M}, f \models x{:}^{\langle L,\mathsf{r}\rangle} p, \Gamma \Rightarrow \Delta$ for all $\mathfrak{M}$ and all f on $\mathfrak{M}$. Fix any $\mathfrak{M} = \langle W, (R_a)_{a\in\mathsf{AP}}, V \rangle$ and f such that $\mathfrak{M}, f \models x{:}^{\langle L,\mathsf{r}\rangle} p$ and $\mathfrak{M}, f \models \gamma$ for all $\gamma \in \Gamma$. Our goal is to show that $\mathfrak{M}, f \models \delta$ for some $\delta \in \Delta$. Then,

$$\begin{aligned}
\mathfrak{M}, f \models x{:}^{\langle L,\mathsf{r}\rangle} p &\iff \mathfrak{M}^{\langle L,\mathsf{r}\rangle}, f(x) \models p,\\
&\iff f(x) \in V^{\langle L,\mathsf{r}\rangle}(p),\\
&\iff f(x) \in V^L(p),\\
&\iff \mathfrak{M}^L, f(x) \models p,\\
&\iff \mathfrak{M}, f \models x{:}^L p.
\end{aligned}$$

So, $\mathfrak{M}, f \models x{:}^L p$. By induction hypothesis, we have $\mathfrak{M}, f \models x{:}^L p, \Gamma \Rightarrow \Delta$, which implies our goal.

Case where the last applied rule of our derivation is $(Lrel)$:

Suppose that $\mathbb{M}_{\text{all}} \models x\mathsf{R}^L_{\mathsf{r}(a)} y, \Gamma \Rightarrow \Delta$. We show that $\mathbb{M}_{\text{all}} \models x\mathsf{R}^{\langle L,\mathsf{r}\rangle}_a y, \Gamma \Rightarrow \Delta$, i.e., $\mathfrak{M}, f \models x\mathsf{R}^{\langle L,\mathsf{r}\rangle}_a y, \Gamma \Rightarrow \Delta$ for all $\mathfrak{M}$ and all f on $\mathfrak{M}$. Fix any $\mathfrak{M} = \langle W, (R_a)_{a\in\mathsf{AP}}, V \rangle$ and f such that $\mathfrak{M}, f \models x\mathsf{R}^{\langle L,\mathsf{r}\rangle}_a y$ and $\mathfrak{M}, f \models \gamma$ for all $\gamma \in \Gamma$. Our goal is to show that $\mathfrak{M}, f \models \delta$ for some $\delta \in \Delta$. Then,

$$\begin{aligned}
\mathfrak{M}, f \models x\mathsf{R}^{\langle L,\mathsf{r}\rangle}_a y &\iff f(x)R^{\langle L,\mathsf{r}\rangle}_a f(y),\\
&\iff f(x)R^L_{\mathsf{r}(a)} f(y),\\
&\iff \mathfrak{M}, f \models x\mathsf{R}^L_{\mathsf{r}(a)} y.
\end{aligned}$$

So, $\mathfrak{M}, f \models x\mathsf{R}^L_{\mathsf{r}(a)} y$. By induction hypothesis, we have $\mathfrak{M}, f \models x\mathsf{R}^L_{\mathsf{r}(a)} y, \Gamma \Rightarrow \Delta$, which implies our goal. □

Theorem 3 (Soundness) *If* $\vdash_{\mathbf{GDLRC}^-} \Rightarrow x{:}^\varepsilon A$ *for all* $x \in \mathsf{Var}$, *then* A *is valid on all models.*

Proof Suppose $\vdash_{\mathbf{GDLRC}^-} \Rightarrow x{:}^\varepsilon A$ for all $x \in \mathsf{Var}$. We show that A is valid on all models, i.e., $\mathfrak{M}, w \models A$ for all $\mathfrak{M} = \langle W, (R_a)_{a\in\mathsf{AP}}, V \rangle$ and $w \in W$. Fix any $\mathfrak{M}$ and

$w \in W$. Define a new assignment function f':$\mathsf{Var} \to W$ by $f'(y) = w$ if $y = x$; $f'(y) = v$ otherwise. By Lemma 6, we get that if $\vdash_{\mathbf{GDLRC}} \Gamma \Rightarrow \Delta$, then $\mathfrak{M}, f' \models \Gamma \Rightarrow \Delta$ for any multisets of Γ and Δ of labelled expressions. Since $\vdash_{\mathbf{GDLRC}^-} \Rightarrow x{:}^{\varepsilon} A$ for all $x \in \mathsf{Var}$, we get $\mathfrak{M}, f' \models\Rightarrow x{:}^{L} A$, which implies $\mathfrak{M}, w \models A$. □

Finally, by Fact 2 and our Theorems 1–3, we obtain the following corollary.

Corollary 1 *Given any formula A, the following are equivalent:*

(i) *A is valid on all Kripke models,*
(ii) $\vdash_{\mathbf{HDLRC}} A$,
(iii) $\vdash_{\mathbf{GDLRC}} \Rightarrow x{:}^{\varepsilon} A$ *for all* $x \in \mathsf{Var}$,
(iv) $\vdash_{\mathbf{GDLRC}^-} \Rightarrow x{:}^{\varepsilon} A$ *for all* $x \in \mathsf{Var}$.

Proof Firstly, the direction from (i) to (ii) is established by Fact 2 (**HDLRC** is sound and complete). Secondly, the direction from (ii) to (iii) is shown by Theorem 1 (all theorems of **HDLRC** are derivable in **GDLRC**). Thirdly, the direction from (iii) to (iv) is shown by Theorem 2 (cut elimination holds for **GDLRC**). Finally, the direction from (iv) to (i) is shown by Theorem 3 (**GDLRC** is sound for Kripke semantics). □

5 Conclusion

We have studied a cut-free labelled sequent calculus for dynamic logic of relation changers (**DLRC**). Based on a labelled formalism for Kripke semantics of **DLRC**, we have defined our sequent calculus (with the cut rule, Table 2). We have shown that all theorems of the known Hilbert-style axiomatization of **DLRC** by van Benthem and Liu (2007), Liu (2011) are derivable in our sequent calculus (Theorem 1) and our sequent calculus enjoys the cut elimination theorem (Theorem 2). In addition, we have also shown that our sequent calculus holds soundness for Kripke semantics (Theorem 3) since we have defined equivalence between the ordinary Kripke semantics and labelled expressions of our calculus by the notion of an assignment function. Finally, we have obtained that our sequent calculus is equipollent with the above sound and complete Hilbert-style axiomatization of this logic (Corollary 1).

As for further directions, we are planning to provide a direct proof of completeness of our sequent calculus without the cut rule. In particular, there must be a semantic proof of cut elimination. In addition, we can also consider another complete Hilbert-style axiomatization for **DLRC** by the method of Wang and Cao (2013). Their method was used to obtain the complete axiomatization for public announcement logic without reduction axioms. Is it possible to apply their method to obtain the complete axiomatization for **DLRC** without reduction axioms?

As for the variant of our sequent calculus, we may consider restricted version of ours, e.g., the intuitionistic labelled system, or extended version of that, e.g., the labelled system with the iteration operator $*$. In order to investigate such extensions or restrictions effectively, implementation of theorem prover might also be helpful to obtain rules or conditions for them.

Acknowledgements We would like to thank an anonymous reviewer for his/her helpful comments to revise our manuscript. We would be grateful to the audiences and organizing committee of the joint conference of the 3rd Asian workshop on philosophical logic and the 3rd Taiwan philosophical logic colloquium (AWPL-TPLC 2016). During the conference, the first author was supported by travel grants from the organizing committee of AWPL-TPLC 2016. The work of the second author was partially supported by the Japan Society for the Promotion of Science (JSPS) KAKENHI Grant-in-Aid for Young Scientists (B) Grant Number 15K21025. The work of the second and the third authors were also supported by JSPS Core-to-Core Program (A. Advanced Research Networks).

References

Balbiani, P., Van Ditmarsch, H., Herzig, A., & De Lima, T. (2010). Tableaux for public announcement logic. *Journal of Logic and Computation*, *20*(1), 55–76.

Balbiani, P., Demange, V., & Galmiche, D. (2014). A sequent calculus with labels for Public Announcement Logic. *Advances in Modal Logic (AiML 2014), 6*.

Bull, R. (1992). Cut elimination for propositional dynamic logic without *. *Zeitschrift fuer Mathematische Logik und Grundlagen der Mathematik*, *38*, 85–100.

Frittella, S., Greco, G., Kurz, A., Palmigiano, A., & Sikimíc, V. (2014). Multi-type display calculus for dynamic epistemic logic. *Journal of Logic and Computation*, *26*(6), 2017–2065.

Gentzen, G. (1964). Investigations into logical deduction. *American Philosophical Quarterly*, *1*(4), 288–306.

Gerbrandy, J., & Groeneveld, W. (1997). Reasoning about information change. *Journal of Logic, Language and Information*, *6*(2), 147–169.

Harel, D., Kozen, D., & Tiuryn, J. (2000). *Dynamic logic*. Cambridge: MIT press.

Hill, B., & Poggiolesi, F. (2010). A contraction-free and cut-free sequent calculus for propositional dynamic logic. *Studia Logica*, *94*(1), 47–72.

Kashima, R. (2009). *Mathematical logic*. Asakura Publishing Co. Ltd. (in Japanese).

Liu, F. (2011). *Reasoning about preference dynamics* (Vol. 354). Synthese Library, Berlin: Springer Science & Business Media.

Ma, M., Sano, K., Schwarzentruber, F., & Velázquez-Quesada, F. R. (2015). Tableaux for non normal public announcement logic. *Logic and Its Applications (ICLA2015), Lecture Notes in Computer Science*, *8923*, 132–145.

Maffezioli, P., & Naibo, A. (2013). Proof theory of epistemic logic of programs. *Logic and Logical Philosophy*, *23*(3), 301–328.

Maffezioli, P., & Negri, S. (2010). A Gentzen-style analysis of public announcement logic. *Proceedings of the International Workshop on Logic and Philosophy of Knowledge, Communication and Action*, 293–313.

Negri, S. (2005). Proof analysis in modal logic. *Journal of Philosophical Logic*, *34*(5), 507–544.

Nomura, S., Sano, K., & Tojo, S. (2015). Revising a sequent calculus for public announcement logic. In *Structural Analysis of Non-classical Logics: Proceedings of the Second Taiwan Philosophical Logic Colloquium (TPLC-2014)*, 131–157.

Ono, H., & Komori, Y. (1985). Logics without the contraction rule. *The Journal of Symbolic Logic*, *50*(01), 169–201.

Plaza, J. (1989). Logics of public communications. In M. L. Emrich, M. S. Pfeifer, M. Hadzikadic & Z. W. Ras (Eds.), *Proceedings of the 4th International Symposium on Methodologies for Intelligent Systems*, 201–216.

van Benthem, J., & Liu, F. (2007). Dynamic logic of preference upgrade. *Journal of Applied Non-Classical Logics*, *17*(2), 157–182.
van Ditmarsch, H., van der Hoek, W., & Kooi, B. P. (2007). *Dynamic epistemic logic* (Vol. 337). Synthese Library, Berlin: Springer.
Wang, Y., & Cao, Q. (2013). On axiomatizations of public announcement logic. *Synthese*, *190*(1), 103–134.

On Second Order Propositional Intuitionistic Logics

Ryo Kashima

Abstract We give an alternative proof of (a slightly strong form of) the completeness of two second order propositional intuitionistic logics with respect to Kripke models. One is the logic having the full comprehension axiom, and the other has the constant domain axiom in addition. We also show that, if the language does not contain disjunction as a primitive symbols, then the constant domain axiom is not needed for the completeness with respect to constant domain models. To show the completeness, we use the technique of nested sequent calculi.

Keywords Second order propositional intuitionistic logic · Kripke model
Completeness · Cut-elimination

1 Introduction

Second order propositional intuitionistic logic (SOPIL) is obtained from the propositional intuitionistic logic by adding quantifiers which bind propositional variables. There are some variants of SOPIL. Among them, this paper mainly treats the logic which is axiomatized by the following inference rules in the sequent calculus:

$$\frac{A(B), \Gamma \Rightarrow C}{\forall x A(x), \Gamma \Rightarrow C}\ (\forall\ \text{left}) \qquad \frac{\Gamma \Rightarrow A(p)}{\Gamma \Rightarrow \forall x A(x)}\ (\forall\ \text{right})$$

$$\frac{A(p), \Gamma \Rightarrow C}{\exists x A(x), \Gamma \Rightarrow C}\ (\exists\ \text{left}) \qquad \frac{\Gamma \Rightarrow A(B)}{\Gamma \Rightarrow \exists x A(x)}\ (\exists\ \text{right})$$

where p does not occur in the lower sequent. We call this logic SOPIL_0. (SOPIL_0 is also characterized by the *full comprehension axiom*: $\exists x (x \leftrightarrow A)$.)

R. Kashima (✉)
Department of Mathematical and Computing Science, Tokyo Institute of Technology,
Ookayama, Meguro, Tokyo 152-8552, Japan
e-mail: kashima@is.titech.ac.jp

S.C.-M. Yang et al. (eds.), *Philosophical Logic: Current Trends in Asia*,
Logic in Asia: Studia Logica Library, DOI 10.1007/978-981-10-6355-8_9

Recall that a Kripke model for propositional intuitionistic logic is of the form $\mathcal{M} = \langle \mathcal{W}, \mathcal{R}, \mathcal{I} \rangle$ where $\langle \mathcal{W}, \mathcal{R} \rangle$ is a partially ordered set of "worlds" and $\mathcal{I}$ is an interpretation which assigns an $\mathcal{R}$-upper closed subset to each propositional variable; that is, $\mathcal{I} = \{\mathcal{I}_p \in \mathrm{UP}(\mathcal{W}, R) \mid p \text{ is a propositional variable}\}$ where $\mathrm{UP}(\mathcal{W}, \mathcal{R}) = \{\alpha \subseteq \mathcal{W} \mid \forall a, b \in \mathcal{W}[\text{ if } a \in \alpha \text{ and } a\mathcal{R}b, \text{ then } b \in \alpha]\}$. Then, for a formula A and a world $a \in \mathcal{W}$, the relation $a \models A$ ("A holds at a") is inductively defined as follows.

$$a \models p \Leftrightarrow a \in \mathcal{I}_p. \quad a \not\models \bot.$$
$$a \models A \to B \Leftrightarrow (b \not\models A) \text{ or } (b \models B) \text{ for all } b \text{ such that } a\mathcal{R}b.$$
$$a \models A \wedge B \Leftrightarrow (a \models A) \text{ and } (a \models B). \quad a \models A \vee B \Leftrightarrow (a \models A) \text{ or } (a \models B).$$

A formula A is said to be *valid* in the model if $a \models A$ for all $a \in \mathcal{W}$.

A Kripke model for SOPIL_0 is obtained from the above by adding the "domain" $\mathcal{D} = \{\mathcal{D}_a \subseteq \mathrm{UP}(\mathcal{W}, \mathcal{R}) \mid a \in \mathcal{W}\}$ that satisfies the *hereditary condition*: $a\mathcal{R}b \Rightarrow \mathcal{D}_a \subseteq \mathcal{D}_b$. The quantifiers are interpreted as follows.

$$a \models \forall x A(x) \Leftrightarrow b \models A(\alpha) \text{ for all } b \text{ and all } \alpha \text{ such that } a\mathcal{R}b \text{ and } \alpha \in \mathcal{D}_b.$$
$$a \models \exists x A(x) \Leftrightarrow a \models A(\alpha) \text{ for some } \alpha \in \mathcal{D}_a.$$

In this paper, we prove the completeness theorem: *A formula is provable in* SOPIL_0 *if and only if it is valid in any full model*, where the condition "full" is defined as follows. Given a formula $A(p_1, p_2, \ldots, p_n)$, there is an n-array function f on $\mathrm{UP}(\mathcal{W}, \mathcal{R})$ such that $f(\alpha_1, \alpha_2, \ldots, \alpha_n) = \{a \in \mathcal{W} \mid a \models A(\alpha_1, \alpha_2, \ldots, \alpha_n)\}$. We say that this function f is *induced by* the formula A. For example, the formula $p \vee q$ induces the function $f(\alpha, \beta) = \alpha \cup \beta$. Then the domain $\mathcal{D}_a$ is said to be *full* if the set $\mathcal{D}_a$ is closed under any function that is induced by a formula, and the Kripke model is said to be *full* if each $\mathcal{D}_a$ is full. Note that these definitions are informal; precise definitions are given in Sect. 4. Note also that the condition full is stronger than the condition "full" by Sørensen and Urzyczyn (2006). Therefore our completeness theorem is slightly stronger than the theorem by Sobolev (1977) and Sørensen and Urzyczyn.

We also prove the completeness theorem for the constant domain case. Like the first order intuitionistic logic, the axiom schema $\forall x(A \vee B(x)) \to A \vee \forall x B(x)$ is called **CD**. Then we prove the completeness theorem: *A formula is provable in* $\mathrm{SOPIL}_0 +$ **CD** *if and only if it is valid in any constant domain full model*, where the model is said to be *constant domain* if $\mathcal{D}_a = \mathcal{D}_b$ for any $a, b \in \mathcal{W}$. This completeness theorem is already proved by Gabbay (1974). In this paper, we give an alternative proof using "*sequent tree calculi*" (or, often called "*nested sequent*"; see e.g. Fitting (2014) for the history of nested sequent calculi). Note that the axiom **CD** is not provable in SOPIL_0. Indeed we show that an instance

$$\forall x\big(p \vee ((p \to x) \vee \neg x)\big) \to \big(p \vee \forall x((p \to x) \vee \neg x)\big) \tag{1}$$

of the axiom **CD** is not valid in some full model.

Incidentally it is well known that the disjunction can be defined by

$$A \vee B \;=\; \forall x((A \to x) \to (B \to x) \to x) \tag{2}$$

in SOPIL_0. Then one may think that we do not have to make a distinction between the two situations: "$\vee$ is a primitive symbols" and "$\vee$ is an abbreviation as above." But this is wrong. In this paper we show that an apparently strong claim

> *a formula A is provable in* SOPIL_0 *(*without *the axiom* **CD***) if and only if A is valid in any* constant domain *full model*

holds *if the language does not contain* $\vee$ *as a primitive symbol.* In the case of the language with $\vee$, the above claim does not hold even if A is $\vee$-free — a counter example is obtained from the formula (1) by erasing $\vee$ by the conversion (2). A point is that the condition "$\mathscr{D}_a$ is closed under any function that is induced by a formula", which is the definition of "full", is weakened by the lack of $\vee$ (since $\vee$-free formulas do not require that $\mathscr{D}_a$ is closed under $\cup$).

The structure of this paper is as follows. In Sect. 2 we introduce the sequent calculi. In Sect. 3 we study the the conversion (2) of $\vee$. In Sect. 4 we give the definition of Kripke models; then the soundness of sequent calculi are shown in Sect. 5. In Sect. 6 we introduce sequent tree calculi and we prove their completeness. In Sect. 7 we show translations from sequent tree calculi to sequent calculi. Finally in Sect. 8 we give precise statements of the main theorems of this paper; moreover we make additional remarks and give problems for future study.

Detailed proofs of all the results of this paper are given in the author's research report (Kashima 2016).

2 Sequent Calculi

We fix two countably infinite sets **PropPara** (of *propositional parameters*) and **PropVar** (of *propositional variables*) such that **PropPara** $\cap$ **PropVar** $= \emptyset$. Then *formulas* are inductively defined as follows.

> Propositional parameters and variables, and $\bot$ are formulas. If A and B are formulas and x is a propositional variable, then $(A \to B)$, $(A \wedge B)$, $(A \vee B)$, $(\forall x A)$, and $(\exists x A)$ are formulas.

Note that propositional parameters cannot be bound by quantifiers. Propositional parameters are denoted by $p, q, \ldots$, propositional variables are denoted by $x, y, \ldots$, and formulas are denoted by $A, B, \ldots$. We consider the logical operator $\neg$ to be the abbreviations as $\neg A = A \to \bot$. Parentheses are omitted by the convention that $\forall x$, $\exists x$ and $\neg$ bind more strongly than binary operators; $\wedge$ and $\vee$ bind more strongly than $\to$; and that $A_1 \to A_2 \to \cdots A_n$ is $(A_1 \to (A_2 \to \cdots (A_{n-1} \to A_n) \cdots))$.

As usual, FreeVar(A) denotes the set of propositional variables that occur free in A. Moreover PropPara(A) denotes the set of propositional parameters that occur in A. We call a formula A is *closed* if FreeVar$(A) = \emptyset$. If A and B are formulas and x is a propositional variable, then $A[x := B]$ denotes the formula that is obtained from A by replacing all the free occurrences of x by B. Note that if B is a closed formula, then this substitution does not cause undesired binding of variables in B; hence we do not need the renaming of bound variables. This is the reason why we introduce propositional parameters into the language.

If Γ is a finite set of *closed* formulas and A is a *closed* formula, then the expression $\Gamma \Rightarrow A$ is called a *sequent*. As usual, if $\Gamma = \{B_1, B_2, \dots, B_n\}$, then $\Gamma \Rightarrow A$ is written as $B_1, B_2, \dots, B_n \Rightarrow A$ instead of $\{B_1, B_2, \dots, B_n\} \Rightarrow A$.

LJ2 is a system which deduces sequents. Note that all the formulas occurring below must be closed since the definition of sequents. Axioms (initial sequents) of **LJ2** are of the forms

$$A \Rightarrow A \quad \text{and} \quad \bot \Rightarrow A.$$

Inference rules of **LJ2** are (cut), (weakening), ($\to$ left), ..., ($\exists$ right) as follows.

$$\frac{\Gamma \Rightarrow A \quad A, \Delta \Rightarrow B}{\Gamma, \Delta \Rightarrow B}\ (\text{cut}) \qquad \frac{\Gamma \Rightarrow A}{B, \Gamma \Rightarrow A}\ (\text{weakening})$$

$$\frac{\Gamma \Rightarrow A \quad B, \Delta \Rightarrow C}{A \to B, \Gamma, \Delta \Rightarrow C}\ (\to \text{left}) \qquad \frac{A, \Gamma \Rightarrow B}{\Gamma \Rightarrow A \to B}\ (\to \text{right})$$

$$\frac{A, \Gamma \Rightarrow C}{A \wedge B, \Gamma \Rightarrow C}\ (\wedge \text{left}) \quad \frac{B, \Gamma \Rightarrow C}{A \wedge B, \Gamma \Rightarrow C}\ (\wedge \text{left}) \quad \frac{\Gamma \Rightarrow A \quad \Gamma \Rightarrow B}{\Gamma \Rightarrow A \wedge B}\ (\wedge \text{right})$$

$$\frac{A, \Gamma \Rightarrow C \quad B, \Gamma \Rightarrow C}{A \vee B, \Gamma \Rightarrow C}\ (\vee \text{left}) \quad \frac{\Gamma \Rightarrow A}{\Gamma \Rightarrow A \vee B}\ (\vee \text{right}) \quad \frac{\Gamma \Rightarrow B}{\Gamma \Rightarrow A \vee B}\ (\vee \text{right})$$

$$\frac{A[x := B], \Gamma \Rightarrow C}{\forall x A, \Gamma \Rightarrow C}\ (\forall \text{left})(\dagger 1) \qquad \frac{\Gamma \Rightarrow A[x := p]}{\Gamma \Rightarrow \forall x A}\ (\forall \text{right})(\dagger 2)$$

$$\frac{A[x := p], \Gamma \Rightarrow C}{\exists x A, \Gamma \Rightarrow C}\ (\exists \text{left})(\dagger 2) \qquad \frac{\Gamma \Rightarrow A[x := B]}{\Gamma \Rightarrow \exists x A}\ (\exists \text{right})(\dagger 1)$$

(†1) Proviso: B is a closed formula.

(†2) Proviso: p is a propositional parameter which does not occur in the lower sequent.

Note that the inference rules

$$\frac{A, A, \Gamma \Rightarrow B}{A, \Gamma \Rightarrow B}\ (\text{contraction}) \qquad \frac{\Gamma, A, B, \Delta \Rightarrow C}{\Gamma, B, A, \Delta \Rightarrow C}\ (\text{exchange})$$

are available because $\{A, A, \Gamma\} = \{A, \Gamma\}$ and $\{\Gamma, A, B, \Delta\} = \{\Gamma, B, A, \Delta\}$.

We will use the symbol "$\vdash$" to mean the provability; that is, "$L \vdash \Gamma \Rightarrow A$" means the fact that the sequent $\Gamma \Rightarrow A$ is provable in the sequent calculus L.

Sequent calculus **LJ2CD** is obtained from **LJ2** by adding the "*constant domain axiom*" which is any sequent of the form

$$\forall x(A \vee B) \Rightarrow A \vee \forall x B.$$

Note that $x \notin \text{FreeVar}(A)$ since $A \vee \forall x B$ is a closed formula by the definition of sequents. The constant domain axiom is not provable in **LJ2**; this will be shown in Sect. 5.

3 Deleting Disjunction

We say that a formula (or a sequent) is *disjunction free* if it contains no occurrence of $\vee$. Sequent calculus $\mathbf{LJ2}^{\text{df}}$ is obtained from **LJ2** by imposing the restriction that all the sequents occurring in the proofs are disjunction free. Of course neither ($\vee$ left) nor ($\vee$ right) is available in $\mathbf{LJ2}^{\text{df}}$.

Now $A + B$ denotes the formula $\forall x\big((A \to x) \to (B \to x) \to x\big)$ provided that $x \notin \text{FreeVar}(A, B)$, and A^{df} denotes the formula that is obtained from A by replacing each subformula $P \vee Q$ in A by $P + Q$. For example, both

$$\forall x\big((\forall y((p \to y) \to (q \to y) \to y) \to x\big) \to (r \to x\big) \to x\big)$$

and

$$\forall z\big((\forall z((p \to z) \to (q \to z) \to z) \to z\big) \to (r \to z\big) \to z\big)$$

are $((p \vee q) \vee r)^{\text{df}}$. As this example shows, all the formulas $((p \vee q) \vee r)^{\text{df}}$ are obtained from each other by renaming bound variables. Hence, given A, we will consider A^{df} to be a unique formula.

The following lemmas are easily proved.

Lemma 1 $\mathbf{LJ2} \vdash A \Rightarrow A^{\text{df}}$ *and* $\mathbf{LJ2} \vdash A^{\text{df}} \Rightarrow A$ *for any closed formula* A.

Lemma 2 *If* $\mathbf{LJ2} \vdash \Gamma \Rightarrow A$, *then* $\mathbf{LJ2}^{\text{df}} \vdash \Gamma^{\text{df}} \Rightarrow A^{\text{df}}$, *where* $\Gamma^{\text{df}} = \{C^{\text{df}} \mid C \in \Gamma\}$.

Thus we have:

Theorem 1 *For any sequent* $\Gamma \Rightarrow A$, *the following three conditions are equivalent. (1)* $\mathbf{LJ2} \vdash \Gamma \Rightarrow A$. *(2)* $\mathbf{LJ2} \vdash \Gamma^{\text{df}} \Rightarrow A^{\text{df}}$. *(3)* $\mathbf{LJ2}^{\text{df}} \vdash \Gamma^{\text{df}} \Rightarrow A^{\text{df}}$.

Proof ($1 \Rightarrow 3$) is implied by Lemma 2. ($3 \Rightarrow 2$) is trivial. ($2 \Rightarrow 1$) is implied by Lemma 1 using the cut rule. □

Corollary 1 *For any disjunction free sequent* $\Gamma \Rightarrow A$, *the following two conditions are equivalent. (1)* $\mathbf{LJ2} \vdash \Gamma \Rightarrow A$. *(2)* $\mathbf{LJ2}^{\text{df}} \vdash \Gamma \Rightarrow A$.

4 Kripke Model

If $\langle \mathscr{W}, \mathscr{R} \rangle$ is a partially ordered set, then UP($\mathscr{W}, \mathscr{R}$) denotes the set of $\mathscr{R}$-upward closed subsets of $\mathscr{W}$; that is

$$\mathrm{UP}(\mathscr{W}, \mathscr{R}) = \{\alpha \subseteq \mathscr{W} \mid \forall a, b \in \mathscr{W} [\text{ if } a \in \alpha \text{ and } a\mathscr{R}b, \text{ then } b \in \alpha.]\}$$

We define a *Kripke model* (for second order propositional intuitionistic logic) to be a tuple $\mathscr{M} = \langle \mathscr{W}, \mathscr{R}, \mathscr{D}, \mathscr{I} \rangle$ as follows.

- $\mathscr{W}$ is a non-empty set (of "*worlds*").
- $\mathscr{R}$ is a partial order ("*accessibility relation*") on $\mathscr{W}$.
- $\mathscr{D} = \{\mathscr{D}_a \mid a \in \mathscr{W}\}$ where each $\mathscr{D}_a$ is a non-empty subset of UP($\mathscr{W}, \mathscr{R}$) satisfying the condition: $a\mathscr{R}b \Rightarrow \mathscr{D}_a \subseteq \mathscr{D}_b$. ($\mathscr{D}_a$ is the *domain* of quantification at the world a).
- $\mathscr{I} = \{\mathscr{I}_p \mid p \in \mathbf{PropPara}\}$ where each $\mathscr{I}_p$ is an element of UP($\mathscr{W}, \mathscr{R}$). ($\mathscr{I}_p$ is the "interpretation" of the propositional parameter p.)

$\mathscr{M}$ is said to be *constant domain* if $\mathscr{D}_a = \mathscr{D}_b$ for all $a, b \in \mathscr{W}$. From now on, $\mathscr{M}$ always denotes a Kripke model $\langle \mathscr{W}, \mathscr{R}, \mathscr{D}, \mathscr{I} \rangle$.

Given a Kripke model $\mathscr{M}$, we extend the language by adding "propositional constant" $\overline{\alpha}$ for each $\alpha \in \mathscr{D}_a$. We call $\overline{\alpha}$ the *name* of α, and define $\mathbf{Name}(\mathscr{M}) = \{\overline{\alpha} \mid \alpha \in \mathscr{D}_a \text{ for some } a \in \mathscr{W}\}$. The definition of formulas are extended by the clause: "Each name $\overline{\alpha}$ is a formula." We call a formula of this extended language an "$\mathscr{M}$-*formula*"; in other words, an $\mathscr{M}$-formula is a formula in which the names of elements of the domains can be used like propositional parameters.

For a Kripke model $\mathscr{M}$, $a \in \mathscr{W}$, and a closed $\mathscr{M}$-formula A, the relation $\mathscr{M}, a \models A$ ("A holds at a world a in $\mathscr{M}$") is defined inductively as follows ("$\mathscr{M}$" is often omitted).

$$\begin{array}{l}
a \models p \iff a \in \mathscr{I}_p, \ \text{for } p \in \mathbf{PropPara}. \\
a \models \overline{\alpha} \iff a \in \alpha, \ \text{for } \overline{\alpha} \in \mathbf{Name}(\mathscr{M}). \\
a \not\models \bot. \\
a \models A \to B \iff [(b \not\models A) \text{ or } (b \models B)] \text{ for all } b \text{ such that } a\mathscr{R}b. \\
a \models A \wedge B \iff (a \models A) \text{ and } (a \models B). \\
a \models A \vee B \iff (a \models A) \text{ or } (a \models B). \\
a \models \forall x A \iff b \models A[x := \overline{\alpha}] \text{ for all } b \text{ such that } a\mathscr{R}b \text{ and all } \alpha \in \mathscr{D}_b. \\
a \models \exists x A \iff a \models A[x := \overline{\alpha}] \text{ for some } \alpha \in \mathscr{D}_a.
\end{array}$$

We define $[\![A]\!]_{\mathscr{M}}$, which is a subset of $\mathscr{W}$, by

$$[\![A]\!]_{\mathscr{M}} = \{a \in \mathscr{W} \mid \mathscr{M}, a \models A\}.$$

We can easily show that $[\![A]\!]_{\mathscr{M}} \in \mathrm{UP}(\mathscr{M}, \mathscr{R})$ by induction on A. Moreover the following equations are easily verified.

$$[\![p]\!]_{\mathscr{M}} = \mathscr{I}_p. \quad [\![\overline{\alpha}]\!]_{\mathscr{M}} = \alpha. \quad [\![\bot]\!]_{\mathscr{M}} = \emptyset.$$
$$[\![A \wedge B]\!]_{\mathscr{M}} = [\![A]\!]_{\mathscr{M}} \cap [\![B]\!]_{\mathscr{M}}. \quad [\![A \vee B]\!]_{\mathscr{M}} = [\![A]\!]_{\mathscr{M}} \cup [\![B]\!]_{\mathscr{M}}.$$

We say that a closed $\mathscr{M}$-formula A is *interpretable at* a world a if the following two conditions hold for all $p \in$ **PropPara** and all $\overline{\alpha} \in$ **Name**$(\mathscr{M})$:

If p occurs in A, then $\mathscr{I}_p \in \mathscr{D}_a$. If $\overline{\alpha}$ occurs in A, then $\alpha \in \mathscr{D}_a$.

We say that A is *valid* in $\mathscr{M}$ if $\mathscr{M}, a \models A$ *for any world a at which A is interpretable.*

In the following, we will define two conditions "full" and "df-full" on Kripke models. $\mathscr{M}$ is said to be *full* if the following condition holds for any closed $\mathscr{M}$-formula A and any world $a \in \mathscr{W}$.

($\dagger$) *If A is interpretable at a, then $[\![A]\!]_{\mathscr{M}} \in \mathscr{D}_a$.*

If $\mathscr{M}$ is full, then each domain $\mathscr{D}_a$ contains $\emptyset$ ($\because$ take $\bot$ as A in ($\dagger$)), and $\mathscr{D}_a$ is closed under $\cap$ and $\cup$ ($\because$ take $\overline{\alpha} \wedge \overline{\beta}$ and $\overline{\alpha} \vee \overline{\beta}$ as A in ($\dagger$)). On the other hand, $\mathscr{M}$ is said to be *df-full* if the above condition ($\dagger$) holds for any *disjunction free* closed $\mathscr{M}$-formula A and any world a. Any full model is df-full, but the converse is not true in general because a domain $\mathscr{D}_a$ might not be closed under $\cup$ in a df-full model.

Note that the term "full" is due to Sørensen and Urzyczyn (2006). However the definition is slightly different. $\mathscr{M}$ is said to be *full in the sense of Sørensen and Urzyczyn* if the following condition holds for any closed $\mathscr{M}$-formula A and any world $a \in \mathscr{W}$.

If A is interpretable at a, then there is an element $\alpha \in \mathscr{D}_a$ such that $[\![A]\!]_{\mathscr{M}}$ and α coincide above a; that is, $\{a' \mid a' \in [\![A]\!]_{\mathscr{M}} \text{ and } a\mathscr{R}a'\} = \{a' \quad a' \in \alpha \text{ and } a\mathscr{R}a'\}$.

Our definition of "full" is stronger and simpler than this. Therefore our statement of the completeness (i.e., if a closed formula A is valid in any full Kripke model, then ($\Rightarrow A$) is provable in **LJ2**) is stronger than the completeness in Sørensen and Urzyczyn (2006).

5 Soundness of Sequent Calculi

We say that a sequent $B_1, B_2, \ldots, B_n \Rightarrow A$ is *valid* in a Kripke model $\mathscr{M} = \langle \mathscr{W}, \mathscr{R}, \mathscr{D}, \mathscr{I} \rangle$ if the following condition holds for any world $a \in \mathscr{W}$:

If $B_1, B_2, \ldots, B_n$ and A are interpretable at a and if $\mathscr{M}, a \models B_1$, $\mathscr{M}, a \models B_2$, $\ldots$, and $\mathscr{M}, a \models B_n$, then $\mathscr{M}, a \models A$.

By the standard method of induction, we can show the following results:

Theorem 2 (Soundness of **LJ2**) *If* **LJ2** $\vdash \Gamma \Rightarrow A$, *then this sequent is valid in any full Kripke model.*

Theorem 3 (Soundness of $\mathbf{LJ2}^{\mathrm{df}}$) *If* $\mathbf{LJ2}^{\mathrm{df}} \vdash \Gamma \Rightarrow A$, *then this sequent is valid in any df-full Kripke model.*

Theorem 4 (Soundness of **LJ2CD**) *If* **LJ2CD** $\vdash \Gamma \Rightarrow A$, *then this sequent is valid in any constant domain full Kripke model.*

Theorem 2 implies the fact that **LJ2CD** is strictly stronger than **LJ2**:

Theorem 5 **LJ2** *does not prove the constant domain axiom* $\forall x(A \vee B) \Rightarrow A \vee \forall x B$ *in general.*

Proof We show

$$\mathbf{LJ2} \nvdash \forall x(p \vee ((p \to x) \vee \neg x)) \Rightarrow p \vee \forall x((p \to x) \vee \neg x)$$

by constructing a counter-model $\mathscr{M} = \langle \mathscr{W}, \mathscr{R}, \mathscr{D}, \mathscr{I} \rangle$ as follows. $\mathscr{W} = \{a, b, c\}$. $a\mathscr{R}b\mathscr{R}c$. $\mathscr{D}_a = \{\alpha, \beta, \emptyset\}$ and $\mathscr{D}_b = \mathscr{D}_c = \{\alpha, \beta, \gamma, \emptyset\}$ where $\alpha = \{a, b, c\}$, $\beta = \{b, c\}$, and $\gamma = \{c\}$. We can easily verify that this is a full Kripke model and that $a \models \forall x(\overline{\beta} \vee ((\overline{\beta} \to x) \vee \neg x))$ ($\because$ $a \models \overline{\beta} \to \overline{\alpha}$, $a \models \overline{\beta} \to \overline{\beta}$, $a \models \neg\overline{\emptyset}$, $b \models \overline{\beta}$, $c \models \overline{\beta}$), and $a \not\models \overline{\beta} \vee \forall x((\overline{\beta} \to x) \vee \neg x)$ ($\because$ $a \not\models \overline{\beta}$, $b \not\models \overline{\beta} \to \overline{\gamma}$, $b \not\models \neg\overline{\gamma}$). □

6 Sequent Tree Calculi and Their Completeness

If $n_1, n_2, \ldots, n_k$ are natural numbers, then the finite sequence $\langle n_1, n_2, \ldots, n_k \rangle$ is called a *label*. In the following, $\langle \mathbf{n} \rangle$ denotes a label. We define a partial order $\preceq$ between labels by $\langle \mathbf{m} \rangle \preceq \langle \mathbf{n} \rangle$ if and only if $\langle \mathbf{m} \rangle$ is a prefix of $\langle \mathbf{n} \rangle$, that is, $\langle \mathbf{n} \rangle = \langle \mathbf{m}, a_1, a_2, \ldots, a_k \rangle$ for some $k \geq 0$ and some a_1, a_2,..., a_k. We say that a (finite or infinite) set T of labels is a *label tree* if T contains the empty label $\langle\rangle$ and if T is closed under prefix, that is, if $\langle \mathbf{m} \rangle \preceq \langle \mathbf{n} \rangle \in T$, then $\langle \mathbf{m} \rangle \in T$. For example, $\{\langle\rangle, \langle 1 \rangle, \langle 2 \rangle, \langle 2, 1 \rangle, \langle 2, 2 \rangle\}$ is a label tree, where $\langle\rangle$ is the root node, nodes $\langle 1 \rangle$ and $\langle 2 \rangle$ are the children of the root node, and nodes $\langle 2, 1 \rangle$ and $\langle 2, 2 \rangle$ are the children of $\langle 2 \rangle$, as illustrated in Fig. 1.

If Γ is a set of closed formulas and Δ is a non-empty set of closed formulas, then the expression $\Gamma \Rightarrow \Delta$ is called a *c-sequent* (Γ and Δ may be infinite sets). A *sequent tree* is a label tree each node of which is associated with a pair of a c-sequent and a set of propositional parameters. Figure 2 is an example of a sequent tree, in which the root is associated with the c-sequent $(p, q \Rightarrow \neg p, \neg q)$ and the set $\{p, q\}$, and the inner node $\langle 2 \rangle$ is associated with the c-sequent $(r \to p \Rightarrow q)$ and the set $\{p, q, r\}$. By a *"finite sequent tree"*, we mean a sequent tree whose label tree is a finite set and all the c-sequents and the sets of propositional parameters are finite.

Fig. 1 Label tree

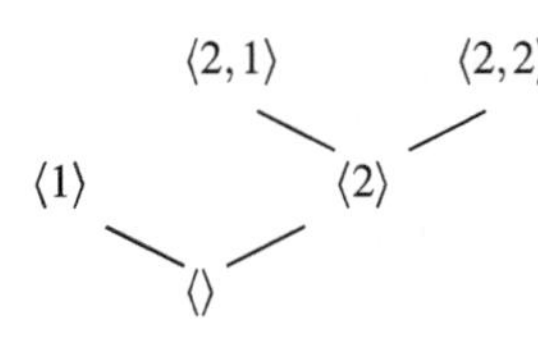

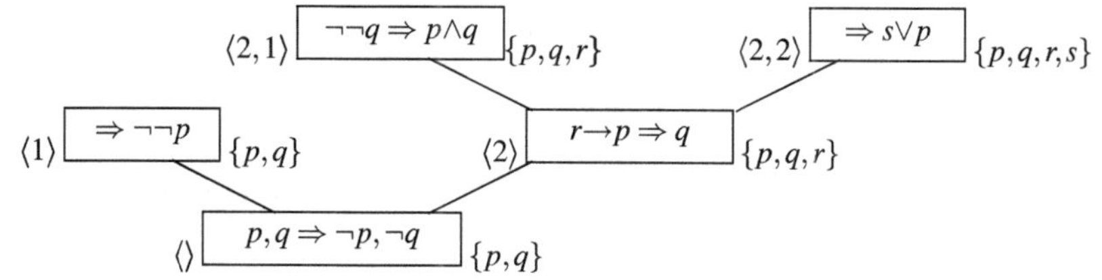

Fig. 2 Sequent tree

A *labelled formula* is an expression

$$\langle \mathbf{n} \rangle : A$$

where $\langle \mathbf{n} \rangle$ is a label and A is a formula. In the following, Γ, Δ, .. will denote (finite or infinite) sets of labelled formulas. Using labelled formulas, we will write any sequent tree in compact form as

$$\Gamma \overset{T,f}{\Rightarrow} \Delta$$

where T is its label tree, f is a function from T to $\mathfrak{P}(\mathbf{PropPara})$ such that $f(\langle \mathbf{n} \rangle)$ is the set of propositional parameters that is associated with the node $\langle \mathbf{n} \rangle$, and Γ and Δ are sets of labelled formulas such that $(\langle \mathbf{n} \rangle : A) \in \Gamma$ (or Δ) if and only if the formula A is in the left (or right, respectively) part of the c-sequent that is associated with the node $\langle \mathbf{n} \rangle$. For example, the sequent tree of Fig. 2 is written as

$$\langle\rangle : p,\ \langle\rangle : q,\ \langle 2 \rangle : r \rightarrow p,\ \langle 2,1 \rangle : \neg\neg q \overset{T,f}{\Rightarrow} \langle\rangle : \neg p,\ \langle\rangle : \neg q,\ \langle 1 \rangle : \neg\neg p,\ \langle 2 \rangle : q,\ \langle 2,1 \rangle : p \wedge q,\ \langle 2,2 \rangle : s \vee p$$

where $T = \{\langle\rangle, \langle 1 \rangle, \langle 2 \rangle, \langle 2,1 \rangle, \langle 2,2 \rangle\}$, $f(\langle\rangle) = f(\langle 1 \rangle) = \{p, q\}$, $f(\langle 2 \rangle) = f(\langle 2,1 \rangle) = \{p, q, r\}$, and $f(\langle 2,2 \rangle) = \{p, q, r, s\}$.

We say that a sequent tree $\Gamma \overset{T,f}{\Rightarrow} \Delta$ is *legal* if it satisfies the following two conditions. (1) If $(\langle \mathbf{n} \rangle : A) \in \Gamma \cup \Delta$, then $\mathrm{PropPara}(A) \subseteq f(\langle \mathbf{n} \rangle)$; in other words, $f(\langle \mathbf{n} \rangle)$ denotes the set of propositional parameters that can be used in the formulas in the node $\langle \mathbf{n} \rangle$. (2) If $\langle \mathbf{m} \rangle \preceq \langle \mathbf{n} \rangle$, then $f(\langle \mathbf{m} \rangle) \subseteq f(\langle \mathbf{n} \rangle)$; therefore, available propositional parameters are inherited (sometimes new elements are added) by the children of the node. All the sequent tree that we treat are legal; so we will use the term "sequent tree" to mean "legal sequent tree" from now on.

We introduce a system **TLJ2**, which deduces finite sequent trees. Axioms of **TLJ2** are finite sequent trees of the forms

$$\langle \mathbf{n} \rangle : A,\ \Gamma \overset{T,f}{\Rightarrow} \Delta,\ \langle \mathbf{n} \rangle : A \qquad \text{and} \qquad \langle \mathbf{n} \rangle : \bot,\ \Gamma \overset{T,f}{\Rightarrow} \Delta,\ \langle \mathbf{n} \rangle : A.$$

Inference rules of **TLJ2** are the following eleven ((heredity), ($\wedge$ left), ..., ($\exists$ right)). Note that the cut rule is not there.

$$\frac{\langle \mathbf{n}, a\rangle : A,\ \Gamma \overset{T,f}{\Rightarrow} \Delta}{\langle \mathbf{n}\rangle : A,\ \Gamma \overset{T,f}{\Rightarrow} \Delta}\ (\text{heredity})(\dagger 1)$$

($\dagger$1) Proviso: $\text{PropPara}(A) \subseteq f(\langle \mathbf{n}\rangle)$ (hence the lower sequent tree is legal).

$$\frac{\Gamma \overset{T,f}{\Rightarrow} \Delta, \langle \mathbf{n}\rangle : A \quad \langle \mathbf{n}\rangle : B, \Gamma \overset{T,f}{\Rightarrow} \Delta}{\langle \mathbf{n}\rangle : A \to B,\ \Gamma \overset{T,f}{\Rightarrow} \Delta}\ (\to\text{left}) \qquad \frac{\langle \mathbf{n}, a\rangle : A, \Gamma \overset{T^+,f^+}{\Rightarrow} \Delta, \langle \mathbf{n}, a\rangle : B}{\Gamma \overset{T,f}{\Rightarrow} \Delta,\ \langle \mathbf{n}\rangle : A \to B}\ (\to\text{right})(\dagger 2)$$

($\dagger$2) Proviso: $\langle \mathbf{n}, a\rangle$ is a leaf of T^+, the c-sequent associated with $\langle \mathbf{n}, a\rangle$ is just $(A \Rightarrow B)$ (this node has no other formula than A, B), $T = T^+ \setminus \{\langle \mathbf{n}, a\rangle\}$, $f^+(\langle \mathbf{n}, a\rangle) = f^+(\langle \mathbf{n}\rangle)$, and $f = f^+ \restriction_T$.

$$\frac{\langle \mathbf{n}\rangle : A,\ \langle \mathbf{n}\rangle : B,\ \Gamma \overset{T,f}{\Rightarrow} \Delta}{\langle \mathbf{n}\rangle : A \wedge B,\ \Gamma \overset{T,f}{\Rightarrow} \Delta}\ (\wedge\ \text{left}) \qquad \frac{\Gamma \overset{T,f}{\Rightarrow} \Delta,\ \langle \mathbf{n}\rangle : A \quad \Gamma \overset{T,f}{\Rightarrow} \Delta,\ \langle \mathbf{n}\rangle : B}{\Gamma \overset{T,f}{\Rightarrow} \Delta,\ \langle \mathbf{n}\rangle : A \wedge B}\ (\wedge\ \text{right})$$

$$\frac{\langle \mathbf{n}\rangle : A,\ \Gamma \overset{T,f}{\Rightarrow} \Delta \quad \langle \mathbf{n}\rangle : B,\ \Gamma \overset{T,f}{\Rightarrow} \Delta}{\langle \mathbf{n}\rangle : A \vee B,\ \Gamma \overset{T,f}{\Rightarrow} \Delta}\ (\vee\ \text{left}) \qquad \frac{\Gamma \overset{T,f}{\Rightarrow} \Delta,\ \langle \mathbf{n}\rangle : A,\ \langle \mathbf{n}\rangle : B}{\Gamma \overset{T,f}{\Rightarrow} \Delta,\ \langle \mathbf{n}\rangle : A \vee B}\ (\vee\ \text{right})$$

$$\frac{\langle \mathbf{n}\rangle : A[x := B],\ \Gamma \overset{T,f}{\Rightarrow} \Delta}{\langle \mathbf{n}\rangle : \forall x A,\ \Gamma \overset{T,f}{\Rightarrow} \Delta}\ (\forall\ \text{left})(\dagger 3) \qquad \frac{\Gamma \overset{T^+,f^+}{\Rightarrow} \Delta,\ \langle \mathbf{n}, a\rangle : A[x := p]}{\Gamma \overset{T,f}{\Rightarrow} \Delta,\ \langle \mathbf{n}\rangle : \forall x A}\ (\forall\ \text{right})(\dagger 4)$$

($\dagger$3) Proviso: B is a closed formula.

($\dagger$4) Proviso: p is a propositional parameter which does not occur in the lower sequent tree, $\langle \mathbf{n}, a\rangle$ is a leaf of T^+, the c-sequent associated with $\langle \mathbf{n}, a\rangle$ is just $(\Rightarrow A[x := p])$, $T = T^+ \setminus \{\langle \mathbf{n}, a\rangle\}$, $f^+(\langle \mathbf{n}, a\rangle) = f^+(\langle \mathbf{n}\rangle) \cup \{p\}$, and $f = f^+ \restriction_T$.

$$\frac{\langle \mathbf{n}\rangle : A[x := p],\ \Gamma \overset{T,f^+}{\Rightarrow} \Delta}{\langle \mathbf{n}\rangle : \exists x A,\ \Gamma \overset{T,f}{\Rightarrow} \Delta}\ (\exists\ \text{left})(\dagger 5) \qquad \frac{\Gamma \overset{T,f}{\Rightarrow} \Delta,\ \langle \mathbf{n}\rangle : A[x := B]}{\Gamma \overset{T,f}{\Rightarrow} \Delta,\ \langle \mathbf{n}\rangle : \exists x A}\ (\exists\ \text{right})(\dagger 6)$$

($\dagger$5) Proviso: p is a propositional parameter which does not occur in the lower sequent tree, and f^+ and f satisfy the following equation.

$$f^+(\langle \mathbf{m}\rangle) = \begin{cases} f(\langle \mathbf{m}\rangle) \cup \{p\} & \text{if } \langle \mathbf{m}\rangle \text{ is equal to or a descendant of } \langle \mathbf{n}\rangle, \\ f(\langle \mathbf{m}\rangle) & \text{otherwise.} \end{cases}$$

($\dagger$6) Proviso: B is a closed formula.

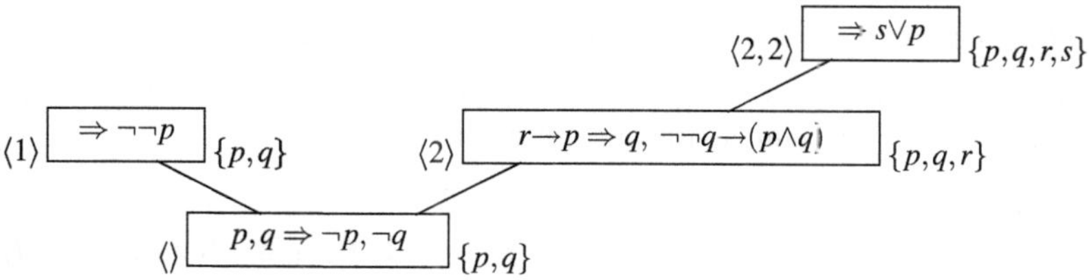

Fig. 3 A conclusion of (→ right) for Fig. 2

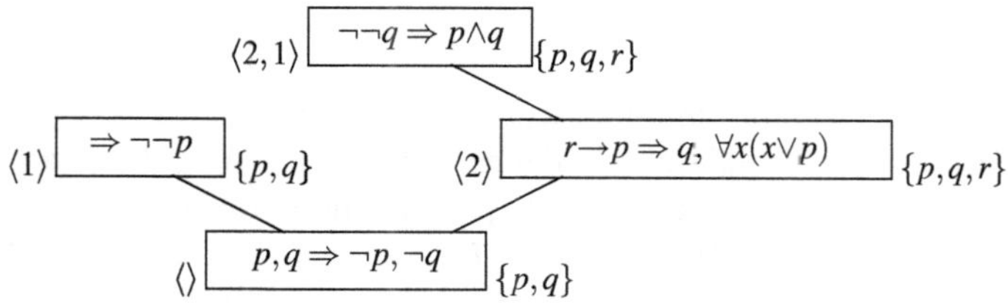

Fig. 4 A conclusion of (∀ right) for Fig. 2

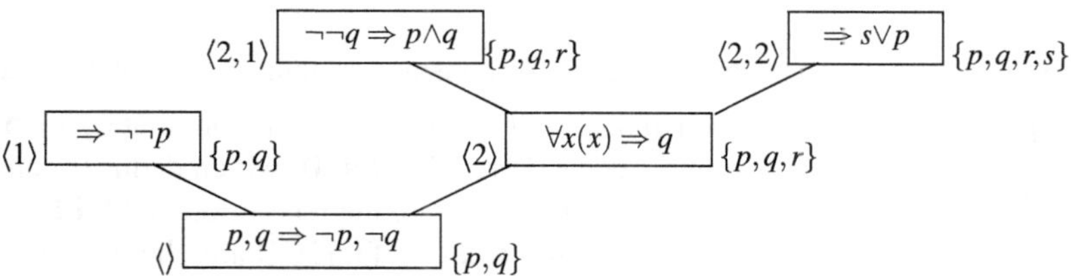

Fig. 5 A conclusion of (∀ left) for Fig. 2

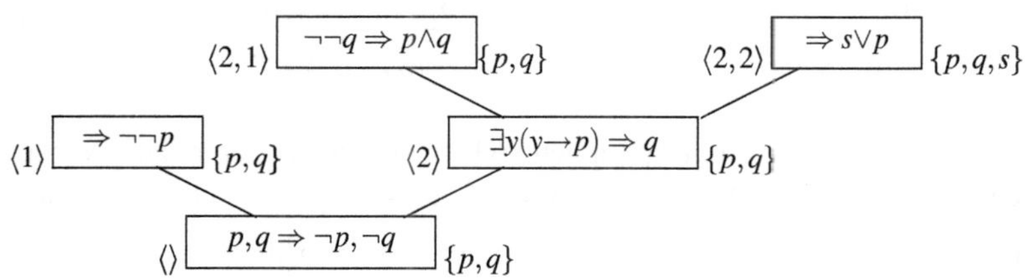

Fig. 6 A consequence of (∃ left) for Fig. 2

For example, if we apply the rules (→ right), (∀ right), (∀ left), and (∃ left) to the sequent tree of Fig. 2, the resulting sequent trees are Figs. 3, 4, 5, and 6, respectively.

Theorem 6 (Completeness Theorem of **TLJ2**) *If a closed formula A is valid in any full Kripke model, then the sequent tree* ($\overset{T,f}{\Rightarrow}$ $\langle\rangle : A$) *is provable in* **TLJ2** *where* $T = \{\langle\rangle\}$ *and* $f(\langle\rangle) = \mathrm{PropPara}(A)$.

Proof (*Sketch*) We show the contraposition. Suppose $\mathbf{TLJ2} \nvdash (\overset{T,f}{\Rightarrow} \langle\rangle : A)$. We define an infinite sequence $(\Gamma_0 \overset{T_0,f_0}{\Rightarrow} \Delta_0)$, $(\Gamma_1 \overset{T_1,f_1}{\Rightarrow} \Delta_1)$, $(\Gamma_2 \overset{T_2,f_2}{\Rightarrow} \Delta_2), \ldots$ of finite sequent trees so as to satisfy certain conditions, for example:

- $(\Gamma_0 \overset{T_0,f_0}{\Rightarrow} \Delta_0) = (\overset{T,f}{\Rightarrow} \langle\rangle : A)$.
- $T_i \subseteq T_{i+1}$, $\Gamma_i \subseteq \Gamma_{i+1}$, $\Delta_i \subseteq \Delta_{i+1}$, and $f_i(\langle\mathbf{n}\rangle) \subseteq f_{i+1}(\langle\mathbf{n}\rangle)$ for all $\langle\mathbf{n}\rangle \in T_i$.
- If $\langle\mathbf{n}\rangle : B \wedge C \in \Gamma_i$, then $\langle\mathbf{n}\rangle : B \in \Gamma_{i+1}$ and $\langle\mathbf{n}\rangle : C \in \Gamma_{i+1}$.
- If $\langle\mathbf{n}\rangle : \forall x B \in \Delta_i$, then there are a child $\langle\mathbf{n}, a\rangle$ of $\langle\mathbf{n}\rangle$ and a closed formula C such that $\langle\mathbf{n}, a\rangle : B[x := C] \in \Delta_{i+1}$.
- $\mathbf{TLJ2} \nvdash (\Gamma_i \overset{T_i,f_i}{\Rightarrow} \Delta_i)$.

Then we define a (possibly infinite) sequent tree $(\Gamma^+ \overset{T^+,f^+}{\Rightarrow} \Delta^+) = \bigcup_{i\in\mathbb{N}}(\Gamma_i \overset{T_i,f_i}{\Rightarrow} \Delta_i)$. For each closed formula X, we define $\text{Left}(X) = \{\langle\mathbf{n}\rangle \in T^+ \mid \langle\mathbf{n}\rangle : X \in \Gamma^+\}$ and $\overline{\text{Right}}(X) = \{\langle\mathbf{n}\rangle \in T^+ \mid \langle\mathbf{n}\rangle : X \notin \Delta^+\}$. Using these, we define a Kripke-model $\mathcal{M}^+ = \langle \mathcal{W}, \mathcal{R}, \mathcal{D}, \mathcal{I} \rangle$ as follows. $\mathcal{W} = T^+$. $\mathcal{R} = \preceq$. $\mathcal{D}_{\langle\mathbf{n}\rangle} = \{\alpha \in \text{UP}(\mathcal{W}, \mathcal{R}) \mid$ there is a closed formula X such that $\text{PropPara}(X) \subseteq f^+(\langle\mathbf{n}\rangle)$ and $\text{Left}(X) \subseteq \alpha \subseteq \overline{\text{Right}}(X)\}$. $\mathcal{I}_p = \text{Left}(p)$. Then we have $(\langle\mathbf{n}\rangle : X \in \Gamma^+ \implies \mathcal{M}^+, \langle\mathbf{n}\rangle \models X)$ and $(\langle\mathbf{n}\rangle : X \in \Delta^+ \implies \mathcal{M}^+, \langle\mathbf{n}\rangle \not\models X)$ for any closed formula X. This implies $\mathcal{M}^+$, $\langle\rangle \not\models A$. □

Next we define sequent tree calculus for constant domain model. We say that a sequent tree $\Gamma \overset{T,f}{\Rightarrow} \Delta$ is *constant domain* if f is a constant function (that is, $f(\langle\mathbf{n}\rangle) = f(\langle\mathbf{m}\rangle)$ for all $\langle\mathbf{n}\rangle$, $\langle\mathbf{m}\rangle$). We introduce a system **TLJ2CD**, which deduces constant domain finite sequent trees. Axioms of **TLJ2CD** are the restriction of **TLJ2**-axioms to constant domain sequent trees. Inference rules of **TLJ2CD** are obtained from the rules of **TLJ2** by replacing ($\forall$ right) and ($\exists$ left) as follows.

$$\frac{\Gamma \overset{T,f^+}{\Rightarrow} \Delta,\ \langle\mathbf{n}\rangle : A[x := p]}{\Gamma \overset{T,f}{\Rightarrow} \Delta,\ \langle\mathbf{n}\rangle : \forall x A}\ (\forall\ \text{right}_{\text{cd}})(\dagger) \qquad \frac{\langle\mathbf{n}\rangle : A[x := p],\ \Gamma \overset{T,f^+}{\Rightarrow} \Delta}{\langle\mathbf{n}\rangle : \exists x A,\ \Gamma \overset{T,f}{\Rightarrow} \Delta}\ (\exists\ \text{left})(\dagger)$$

(†) Proviso: p is a propositional parameter which does not occur in the lower sequent tree, and f^+ and f satisfy the equation $f^+(\langle\mathbf{m}\rangle) = f(\langle\mathbf{m}\rangle) \cup \{p\}$ for all $\mathbf{m}$.

Similar to the above Completeness Theorem 6, we can show:

Theorem 7 (Completeness Theorem of **TLJ2CD**) *If a closed formula A is valid in any constant domain full Kripke model, then the sequent tree* $(\overset{T,f}{\Rightarrow} \langle\rangle : A)$ *is provable in* **TLJ2CD** *where* $T = \{\langle\rangle\}$ *and* $f(\langle\rangle) = \text{PropPara}(A)$.

Moreover this result can be restricted to disjunction free formulas. $\mathbf{TLJ2CD}^{\text{df}}$ is obtained from **TLJ2CD** by imposing restriction that all the sequent tree are disjunction free.

Theorem 8 (Completeness Theorem of $\mathbf{TLJ2CD}^{\text{df}}$) *If a disjunction free closed formula A is valid in any constant domain df-full Kripke model, then the sequent tree* $(\overset{T,f}{\Rightarrow} \langle\rangle : A)$ *is provable in* $\mathbf{TLJ2CD}^{\text{df}}$ *where* $T = \{\langle\rangle\}$ *and* $f(\langle\rangle) = \text{PropPara}(A)$.

Fig. 7 Sequent tree $\mathscr{S}$

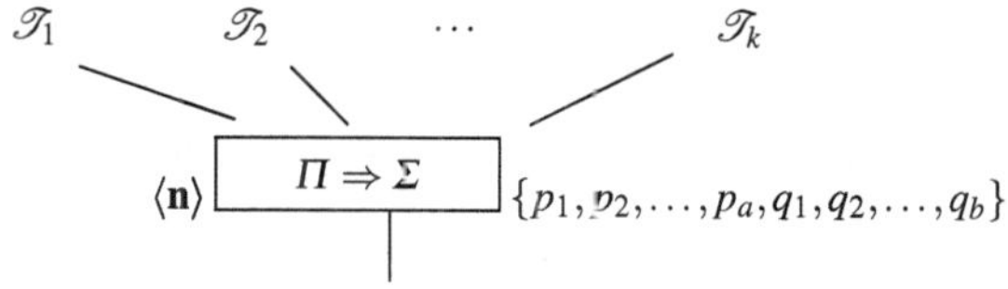

7 From Sequent Trees to Formulas

For any sequent tree $\Gamma \overset{T,f}{\Rightarrow} \Delta$, we will define a formula $\#(\Gamma \overset{T,f}{\Rightarrow} \Delta)$, which is considered to be an interpretation of $\Gamma \overset{T,f}{\Rightarrow} \Delta$, by induction on the height of the tree T. In general, a subtree $\mathscr{S}$ of a finite sequent tree $\Gamma \overset{T,f}{\Rightarrow} \Delta$ is illustrated as Fig. 7. Suppose that $q_1, q_2, \ldots, q_b$ are the propositional parameters that are newly available in this node; that is, if $\langle \mathbf{m} \rangle$ is the parent of $\langle \mathbf{n} \rangle$, then $p_i \in f(\langle \mathbf{m} \rangle)$ and $q_j \notin f(\langle \mathbf{m} \rangle)$ for all i, j. Then the formula $\#(\mathscr{S})$ is defined as

$$\#(\mathscr{S}) = \forall q_1 \forall q_2 \cdots \forall q_b \Big((\bigwedge \Pi) \to \big((\bigvee \Sigma) \vee \#(\mathscr{T}_1) \vee \#(\mathscr{T}_2) \vee \cdots \vee \#(\mathscr{T}_k) \big) \Big).$$

Note that $q_1, \ldots, q_b$ are not propositional variables but propositional parameters; therefore, precisely speaking, the above $\#(\mathscr{S})$ is not a formula (because only propositional variables can be bound by quantifiers in our definition). For the correct definition, we have to replace $q_1, \ldots, q_b$ by appropriate variables. But here we do not use such a correct definition for the sake of simplicity. An example is as follows. If $\mathscr{S}$ is the sequent tree in Fig. 2, then $\#(\mathscr{S})$ is the formula

$$\forall p \forall q \Big[(p \wedge q) \to \Big[\neg p \vee \neg q \vee \neg\neg p \vee \forall r \Big((r \to p) \to \big(q \vee (\neg\neg q \to p \wedge q) \vee \forall s (s \vee p) \big) \Big) \Big] \Big].$$

Theorem 9 *(1) If* **TLJ2** $\vdash \mathscr{S}$*, then* **LJ2** $\vdash\Rightarrow \#(\mathscr{S})$.
(2) If **TLJ2CD** $\vdash \mathscr{S}$*, then* **LJ2CD** $\vdash\Rightarrow \#(\mathscr{S})$.

Proof By induction on the proof of $\mathscr{S}$ in **TLJ2** and **TLJ2CD**. □

Corollary 2 *Let A be a closed formula.*

(1) If **TLJ2** $\vdash \overset{\{\langle\rangle\},f}{\Rightarrow} \langle\rangle : A$*, then* **LJ2** $\vdash\Rightarrow A$.
(2) If **TLJ2CD** $\vdash \overset{\{\langle\rangle\},f}{\Rightarrow} \langle\rangle : A$*, then* **LJ2CD** $\vdash\Rightarrow A$.

Proof The sequent $(\#(\overset{\{\langle\rangle\},f}{\Rightarrow} \langle\rangle : A)) \Rightarrow A$ is provable in **LJ2**. Thus Corollary 2 is obtained from Theorem 9 using the cut rule. □

If the language does not contain disjunction, the above Corollary 2 (2) can be remarkably strengthened as follows: If $\mathbf{TLJ2CD}^{\mathrm{df}} \vdash \overset{\{\langle\rangle\},f}{\Rightarrow} \langle\rangle : A$, then $\mathbf{LJ2}^{\mathrm{df}}_{\mathrm{cutfree}} \vdash\Rightarrow A$, where $\mathbf{LJ2}^{\mathrm{df}}_{\mathrm{cutfree}}$ is the sequent calculus obtained from $\mathbf{LJ2}^{\mathrm{cf}}$ by deleting the cut rule. Note that $\mathbf{LJ2}^{\mathrm{df}}_{\mathrm{cutfree}}$ has neither the cut rule nor the constant domain axiom although $\mathbf{TLJ2CD}^{\mathrm{df}}$ is the system corresponding to constant domain models. To show this fact, we define a set $\Gamma_{\langle \mathbf{n} \rangle \downarrow}$ (of formulas), for given sequent tree $\Gamma \overset{T,f}{\Rightarrow} \Delta$ and label $\langle \mathbf{n} \rangle$, as follows.

$$\Gamma_{\langle \mathbf{n} \rangle \downarrow} = \{A \mid \langle \mathbf{m} \rangle : A \in \Gamma \text{ for some label } \langle \mathbf{m} \rangle \in T \text{ such that } \langle \mathbf{m} \rangle \preceq \langle \mathbf{n} \rangle\}.$$

For example, if $\Gamma \overset{T,f}{\Rightarrow} \Delta$ is the sequent tree of Fig. 2, then $\Gamma_{\langle 2,1 \rangle \downarrow} = \{\neg\neg q, r \to p, p', q\}$ and $\Gamma_{\langle 2 \rangle \downarrow} = \{r \to p, p, q\}$.

Lemma 3 *If a sequent tree* $\Gamma \overset{T,f}{\Rightarrow} \Delta$ *is provable in* $\mathbf{TLJ2CD}^{\mathrm{df}}$, *then there exists a labelled formula* $\langle \mathbf{n} \rangle : A \in \Delta$ *such that the sequent* $\Gamma_{\langle \mathbf{n} \rangle \downarrow} \Rightarrow A$ *is provable in* $\mathbf{LJ2}^{\mathrm{df}}_{\mathrm{cutfree}}$.

Proof By induction on the proof of $\Gamma \overset{T,f}{\Rightarrow} \Delta$ in $\mathbf{TLJ2CD}^{\mathrm{df}}$. □

Corollary 3 *Let A be a disjunction free closed formula. If* $\mathbf{TLJ2CD}^{\mathrm{df}} \vdash \overset{\{\langle\rangle\},f}{\Rightarrow} \langle\rangle : A$, *then* $\mathbf{LJ2}^{\mathrm{df}}_{\mathrm{cutfree}} \vdash \Rightarrow A$.

8 Conclusions and Future Study

Combining all the results of Sects. 2–7, we obtain main theorems of this paper.

Theorem 10 (Completeness of **LJ2**) *For any closed formula A, the following four conditions are equivalent.*

(1) $\mathbf{LJ2} \vdash \Rightarrow A$.
(2) $\mathbf{LJ2}^{\mathrm{df}} \vdash \Rightarrow A^{\mathrm{df}}$.
(3) $\mathbf{TLJ2} \vdash \overset{\{\langle\rangle\},f}{\Rightarrow} \langle\rangle : A$ *where* $f(\langle\rangle) = \mathrm{PropPara}(A)$.
(4) A is valid in any full Kripke model.

Proof By Theorems 1, 2 and 6, and Corollary 2(1). □

Theorem 11 (Completeness of **LJ2CD**) *For any closed formula A, the following three conditions are equivalent.*

(1) $\mathbf{LJ2CD} \vdash \Rightarrow A$.
(2) $\mathbf{TLJ2CD} \vdash \overset{\{\langle\rangle\},f}{\Rightarrow} \langle\rangle : A$ *where* $f(\langle\rangle) = \mathrm{PropPara}(A)$.
(3) A is valid in any constant domain full Kripke model.

Proof By Theorems 4 and 7, and Corollary 2(2). □

Theorem 12 (Completeness and cut-elimination for disjunction free formulas) *For any disjunction free closed formula A, the following seven conditions are equivalent.*

(1) $\mathbf{LJ2} \vdash \Rightarrow A$.
(2) $\mathbf{LJ2}^{\mathrm{df}} \vdash \Rightarrow A$.
(3) $\mathbf{LJ2}^{\mathrm{df}}_{\mathrm{cutfree}} \vdash \Rightarrow A$.
(4) $\mathbf{TLJ2CD}^{\mathrm{df}} \vdash \overset{\{\langle\rangle\},f}{\Rightarrow} \langle\rangle : A$ *where* $f(\langle\rangle) = \mathrm{PropPara}(A)$.
(5) A is valid in any full Kripke model.
(6) A is valid in any df-full Kripke model.
(7) A is valid in any constant domain df-full Kripke model.

Proof (1 $\Leftrightarrow$ 2) is shown by Corollary 1. (1 $\Leftrightarrow$ 5) is shown by Theorem 10. (2 $\Rightarrow$ 6) is shown by Theorem 3. (6 $\Rightarrow$ 5), (6 $\Rightarrow$ 7), and (3 $\Rightarrow$ 2) are trivial. (7 $\Rightarrow$ 4) is shown by Theorem 8. (4 $\Rightarrow$ 3) is shown by Corollary 3. □

Finally we give some remarks and problems for future study.

Theorem 12 implies the semantical cut-elimination for $\mathbf{LJ2}^{\mathrm{df}}$: If $\mathbf{LJ2}^{\mathrm{df}} \vdash \Rightarrow A$, then $\mathbf{LJ2}^{\mathrm{df}}_{\mathrm{cutfree}} \vdash \Rightarrow A$, which is already proved (for the $\rightarrow$-$\forall$-fragment) by Prawitz (1970) using Beth semantics instead of Kripke semantics. For the full language with disjunction, the semantical cut-elimination for **LJ2** is already proved by Okada (2002) using phase semantics. However I do not know how to prove it by using Kripke semantics. Thus we present:

Problem 1 *Prove one of the following by not using the cut-elimination theorem for* **LJ2**.

(a) If $\mathbf{LJ2}^{\mathrm{df}}_{\mathrm{cutfree}} \vdash \Rightarrow A^{\mathrm{df}}$, *then* $\mathbf{LJ2}_{\mathrm{cutfree}} \vdash \Rightarrow A$.
(b) If $\mathbf{TLJ2} \vdash \overset{\{\langle\rangle\},f}{\Rightarrow} \langle\rangle : A$ *where* $f(\langle\rangle) = \mathrm{PropPara}(A)$, *then* $\mathbf{LJ2}_{\mathrm{cutfree}} \vdash \Rightarrow A$.

The cut-elimination for **LJ2** is obviously implied by Theorems 10, 12 and Problem 1(a), as well as by Theorem 10 and Problem 1(b).

Let $C = \forall x\big(p \vee ((p \to x) \vee \neg x)\big) \to \big(p \vee \forall x((p \to x) \vee \neg x)\big)$. In Sect. 5, we show an example of full Kripke model in which the formula C is not valid. Then, by Theorems 10 and 12, there is a constant domain df-full Kripke model in which C^{df} is not valid.

Problem 2 *Find a simple constant domain df-full Kripke model in which the formula* C^{df} *is not valid.*

Note that C^{df} is valid in any constant domain full Kripke model. A solution to this problem should be a good example to show the distinction between $A \vee B$ and $\forall x((A \to x) \to (B \to x) \to x)$.

In the study of nonclassical logics, there have been a lot of results on propositional and predicate *intermediate* (between classical and intuitionistic) logics. However, I do not know substantial studies on second order propositional intermediate logics. (Theorem 11 — completeness of the constant domain intermediate logic — is the only result on second order propositional intermediate logics I know.) Thus we give:

Problem 3 *Develop studies on second order propositional intermediate logics.*

References

Fitting, M. (2014). Nested sequents for intuitionistic logics. *Notre Dame Journal of Formal Logic*, *55*(1), 41–61.
Gabbay, D. M. (1974). On 2nd order intuitionistic propositional calculus with full comprehension. *Archive for Mathematical Logic*, *16*, 177–186.
Kashima, R. (2016). Completeness of second order propositional intuitionistic logics. Technical report, Tokyo Institute of Technology, from http://www.is.titech.ac.jp/~kashima/pub/C284.pdf.
Okada, M. (2002). A uniform semantic proof for cut-elimination and completeness of various first and higher order logics. *Theoretical Computer Science*, *281*, 471–498.
Prawitz, D. (1970). Some results for intuitionistic logic with second order quantification rules. In A. Kino, J. Myhill, & R. E. Vesley (Eds.), *Intuitionism and Proof Theory, Proceedings of the Summer Conference at Buffalo N.Y. 1968* (pp. 259–269): North-Holland. chapter 17.
Sobolev, S. K. (1977). The intuitionistic propositional calculus with quantifiers. *Mathematical Notes*, *22*, 528–532. (English version of Mat. Zametki 22:69–76 (1977)).
Sørensen, M. H., & Urzyczyn, P. (2006). *Lectures on the curry-howard isomorphism*. 0: Elsevier.

Classical Model Existence Theorem in Subclassical Predicate Logics. II

Jui-Lin Lee

Abstract This paper is a sequel to Lee (2007) and Lee (2009). In Lee (2009) it is proved that there are weak subclassical predicate logics (which are classically sound but weaker than the first-order logic) which also satisfy the Classical Model Existence property (*CME* for short): Every consistent set has a classical model. In this paper we improve the result in Lee (2007) by showing that some weak propositional logics (extensions of the implicational linear logic *BCI*) also satisfy *CME* in propositional case. For subclassical predicate logics two types of approaches (by prenex normal form construction or by Hintikka style construction) will be considered. By this we prove that *CME* for predicate logics also holds for some subclassical logics with weak propositional part.

1 Introduction

Classical model existence theorem, usually called the strong completeness theorem, states that every consistent set has a classical model. In Lee (2007) it is proved that in propositional logics there are a lot of logics which satisfy *CME* (as a property) (for example, the intuitionistic propositional logic, or even some paraconsistent logic). In Lee (2009) it is proved that there are some subclassical predicate logics which also satisfy *CME*.

This paper is a sequel to Lee (2007, 2009). In Sect. 2 we improve the result in Lee (2007) by showing that some weak propositional logics ($\mathscr{H}_3$ and $\mathscr{H}_4$, which are extensions of the implicational linear logic *BCI*) also satisfy *CME*. Glivenko's theorem is proved for some weak propositional logics by using *CME* and taking axioms carefully.

J.-L. Lee (✉)
Center for General Education and Department of CSIE, National Formosa University, No.64, Wunhua Rd., Yunlin 63201, Taiwan
e-mail: jlleelogician@gmail.com

S.C.-M. Yang et al. (eds.), *Philosophical Logic: Current Trends in Asia*, Logic in Asia: Studia Logica Library, DOI 10.1007/978-981-10-6355-8_10

Then we study the predicate logic case. In Sect. 3 we take Herbrand-Henkin style approach (i.e., prenex normal form style approach) to construct proof systems which satisfy prenex normal form theorem and *CME*. The construction is based on $\mathscr{H}_3$ and $\mathscr{H}_4$, which allow only weak deduction theorem. In Sect. 4 we take Hintikka style approach to construct proof systems which satisfy *CME*, without converting sentences into prenex normal form first.

At the end we propose some research problems for further study.

2 Classical Model Existence Theorem in Extensions of *BCI*

In this section we modify the proof in Lee (2007, 2009) to show that, in propositional logics, some weak axiomatic extensions of *BCI* logic also satisfy *CME* w.r.t. $\bot$-consistency. First we will sketch the proof idea in Lee (2007) and then give the modified axioms which suffice to prove *CME* in propositional cases. Note that a weaker version of deduction theorem has been used. And in the extensions of *BCI* the notion of negation completeness needs to be modified, too.

Here we rename two axioms[1] in Lee (2009) (the axioms (K) and (S) are previously called $(DT1)$ and $(DT2)$, respectively). We denote the Hilbert proof system $\mathscr{H}_1 = \mathscr{H}(K, S, ECQ_\bot; MP)$, which is the $\{\to, \bot\}$-fragment of *IPL*, and its subsystem $\mathscr{H}_2 = \mathscr{H}(K, S, DA^*_\bot; MP)$, which is a paraconsistent logic. These two systems have only one inference rule *MP*: $\alpha, \alpha \to \beta$ infer β. Their axioms are

$$(K)\quad \alpha \to (\beta \to \alpha)$$
$$(S)\quad [\alpha \to (\beta \to \gamma)] \to [(\alpha \to \beta) \to (\alpha \to \gamma)]$$
$$(ECQ_\bot)\quad \bot \to \alpha$$
$$(DA^*_\bot)\quad (\alpha \to \bot) \to \{[(\alpha \to \beta) \to \bot] \to \bot\}$$

Now we sketch the proof idea of that $\mathscr{H}_1$ and $\mathscr{H}_2$ both satisfy *CME* (i.e., every $\bot$-consistent set Σ has a classical model). Assume that Σ is *$\bot$-consistent*, i.e., $\Sigma \nvdash \bot$ (w.r.t. $\mathscr{H}_1$ or $\mathscr{H}_2$). To show that Σ has a classical model, we first enumerate all sentences $\psi_0, \ldots, \psi_n, \ldots$ and define $\Delta_0 = \Sigma$, $\Delta_{n+1} = \Delta_n \cup \{\psi_n\}$ if $\bot$-consistent; otherwise $\Delta_{n+1} = \Delta_n$. By Lindenbaum's theorem (every $\bot$-consistent set has a maximal $\bot$-consistent extension, see Lee (2007)), $\Delta = \bigcup_{n\in\omega} \Delta_n$ is a maximal $\bot$-consistent extension of Σ. Then Δ is proved to be *deductively closed* (i.e., $\Delta \vdash \varphi$ implies $\varphi \in \Delta$ for any φ, see Lee (2007)) and *negation complete* (i.e., for any sentence φ exactly one of φ and $\neg\varphi (= \varphi \to \bot)$ is in Δ, see Lee (2007)). Note that the negation completeness of Δ is proved by *MP* (to show that not-both $\varphi, \neg\varphi$ in Δ) and the deduction theorem (to show that at-least-one φ, $\neg\varphi$ in Δ). Here we recall the proof that if $\varphi \notin \Delta$, then $\neg\varphi \in \Delta$: If $\varphi = \psi_n \notin \Delta$, then $\Delta \cup \{\varphi\} \vdash \bot$. By deduction theorem $\Delta \vdash \varphi \to \bot$. Since Δ is deductively closed, $\neg\varphi \in \Delta$.

To show that the truth-functionality of $\to$ (i.e., $(\varphi \to \psi) \in \Delta$ iff $\varphi \notin \Delta$ or $\psi \in \Delta$, for any φ, ψ) holds in Δ, it is done by *MP* (from left to right), by (K) (from $\psi \in \Delta$ to

[1] Actually here we use axiom schemes, but call them axioms for simplicity.

$\varphi \to \psi \in \Delta$), and by ($ECQ_\perp$) or ($DA^*_\perp$) (from $\varphi \notin \Delta$ to $\varphi \to \psi \in \Delta$). Note that in $\mathscr{H}_1$ the axiom ($DA_\perp$): $(\alpha \to \perp) \to (\alpha \to \beta)$ is derivable and it can be used to show $\varphi \to \psi \in \Delta$ from $\varphi \notin \Delta$. In $\mathscr{H}_2$ the weaker axiom ($DA^*_\perp$) will suffice to show the truth-functionality of $\to$ on Δ by the following argument: "if $\varphi \notin \Delta$ and $\varphi \to \psi \notin \Delta$, then $\Delta \vdash \perp$." This is by an indirect proof. However, it shows that the weaker axiom ($DA^*_\perp$) is a sufficient condition for the truth-functionality of $\to$ on Δ. Note that by taking ($DA^*_\perp$) instead of ($ECQ_\perp$) the system $\mathscr{H}_2$ is paraconsistent.

With the axioms we added (so that $\perp \notin \Delta$ and Δ satisfies truth functionality of $\to$ on Δ), the maximal $\perp$-consistent Δ has a classical model: Let μ be the truth assignment which sends atomic sentence p_i to be true iff $p_i \in \Delta$. One can easily prove by induction that, $\varphi \in \Delta$ implies $\mu \models \varphi$ for any sentence φ. (Later we will deal with first order sentences, for which the corresponding models will be constructed by truth assignment of atomic sentences over closed terms.)

Note that we use only general proof properties of Hilbert systems to prove Lindenbaum's theorem with respect to $\perp$-consistency and Δ being deductively closed, i.e., these results hold for any Hilbert system (see Lee (2007)). For proving negation completeness of Δ (i.e., two-valued) and truth functionality on Δ (i.e., truth functional), we employ deduction theorem and add axioms which are sufficient conditions for the proof. However, the deduction theorem in classical logic is not a necessary condition for proving *CME*. A weaker version of deduction theorem will suffice.

To see this, we need a more careful treatment on derivation. We modify Troelstra's S-derivation (see Paoli (2002)), p. 76) as follows[2]: Consider any Hilbert system $\mathscr{H}$ with axioms and only one inference rule *MP*. A (finite) *multiset* Λ is a (possibly empty) list of formulas $\varphi_1, \ldots, \varphi_n$. Two multisets Λ_1, Λ_2 are the same iff there is a rearrangement (of formulas) which permutes Λ_1 into Λ_2. An $\mathscr{H}$-proof tree $\mathscr{D}$ is a labelled finite binary tree whose leaves are labelled by expressions either of the form $\alpha \Rightarrow \alpha$ (where α is a formula, and it means "from assumption α infer α"), or of the form $\Rightarrow \beta$ (where β is an axiom of $\mathscr{H}$), and all the other nodes of $\mathscr{D}$ (from leaves down to the root) are labelled according to *MP* (where Λ_1, Λ_2 are multisets of formulas):

$$\frac{\Lambda_1 \Rightarrow \alpha \qquad \Lambda_2 \Rightarrow \alpha \to \beta}{\Lambda_1, \Lambda_2 \Rightarrow \beta}$$

Note that when $\Lambda \Rightarrow \alpha$ has an $\mathscr{H}$-proof tree, it means that every $\varphi \in \Lambda$ are actually a *used* assumption in that proof tree.

Let Σ be a (possibly infinite) set of formulas and Λ be a finite multiset of formulas. We define that $\Lambda \Rightarrow \varphi$ *is provable in* $\mathscr{H}$ iff there is an $\mathscr{H}$-proof tree $\mathscr{D}$ with root labelled by $\Lambda \Rightarrow \varphi$. And define that $\Sigma \vdash \varphi$ *in* $\mathscr{H}$ iff there is a finite multiset Λ which, as a set, is a subset of Σ and also that $\Lambda \Rightarrow \varphi$ has an $\mathscr{H}$-proof tree.

We then take the following definition: Let Σ be a set and φ, ψ be formulas. We say that *in* $\mathscr{H}$ *the formula* ψ *is provable from* $\Sigma \cup \{\varphi\}$ *with* φ *used* (in symbols, $\Sigma \cup \{\varphi\}^+ \vdash \psi$ *in* $\mathscr{H}$) iff there is an $\mathscr{H}$-proof tree $\mathscr{D}$ of $\Lambda \Rightarrow \psi$ for some finite

[2]Here we omit, in Paoli (2002), the adjunction rule "from α and β infer $\alpha \wedge \beta$" and replace $\vdash$ by $\Rightarrow$. We preserve the symbol $\vdash$ for the standard provability in Hilbert systems.

multiset Λ with $\varphi \in \Lambda \subseteq \Sigma \cup \{\varphi\}$. We also define the nested implication as follows: let $\varphi \to^1 \psi$ be $\varphi \to \psi$ and $\varphi \to^{n+1} \psi$ be $\varphi \to (\varphi \to^n \psi)$ for any positive integer n.

We recall the following axioms:

(*B*) $(\alpha \to \beta) \to [(\gamma \to \alpha) \to (\gamma \to \beta)]$
(*C*) $[\alpha \to (\beta \to \gamma)] \to [\beta \to (\alpha \to \gamma)]$
(*I*) $\alpha \to \alpha$
(*W*) $[\alpha \to (\alpha \to \beta)] \to (\alpha \to \beta)$

The Hilbert system *BCI* is $\mathscr{H}(B, C, I; MP)$. A Hilbert system $\mathscr{H}$ is an *axiomatic extension* of *BCI* iff $\mathscr{H}$ is obtained by adding axioms to *BCI*. For example, *BCIW* is *BCI* + (*W*).

Then we have the following weak deduction theorem:

Theorem 2.1 (Weak Deduction Theorem)

(1) Let $\mathscr{H}$ be an axiomatic extension of BCI and Λ be a finite multiset of formulas. If $\Lambda, \varphi \Rightarrow \psi$ is provable in $\mathscr{H}$, then $\Lambda \Rightarrow \varphi \to \psi$ is provable in $\mathscr{H}$.
(2) Let $\mathscr{H}$ be an axiomatic extension of BCI and Σ be a set of formulas. If $\Sigma \cup \{\varphi\}^+ \vdash \psi$ in $\mathscr{H}$, then $\Sigma \vdash \varphi \to^n \psi$ in $\mathscr{H}$ for some integer $n > 0$.
(3) Let $\mathscr{H}$ be an axiomatic extension of BCIW and Σ be a set of formulas. If $\Sigma \cup \{\varphi\}^+ \vdash \psi$ in $\mathscr{H}$, then $\Sigma \vdash \varphi \to \psi$ in $\mathscr{H}$.

Note that in (2) the n of $\Sigma \vdash \varphi \to^n \psi$ is actually the multiplicity number of φ which is used in a proof P of $\Sigma \cup \{\varphi\}^+ \vdash \psi$ in $\mathscr{H}$, and we actually construct the proof of $\Sigma \vdash \varphi \to^n \psi$ inductively based on this proof P.

The proof of Theorem 2.1(1) is straightforward, by induction on the depth of $\mathscr{H}$-proof tree of $\Lambda, \varphi \Rightarrow \psi$ (also see Paoli (2002)): For $\psi \Rightarrow \psi$ we use axiom (*I*). If the last step of the $\mathscr{H}$-proof tree for $\Lambda, \varphi \Rightarrow \psi$ is by *MP*, we use induction hypothesis and axioms (*B*) or (*C*) to construct the required $\mathscr{H}$-proof tree. (2) is proved by (1), and (3) is proved by (2) to obtain $\Sigma \vdash \varphi \to^n \psi$, and one can use axiom (*W*) to do the contraction: $\{\varphi \to^n \psi\} \vdash \varphi \to^{n-1} \psi$. Then repeat this to achieve $\varphi \to \psi$.

Now we modify the proof in Lee (2007) to some axiomatic extension of *BCIW*, call it $\mathscr{H}_3$. We will show what kind of axioms will suffice for $\mathscr{H}_3$ to satisfy *CME*. Assume that $\Sigma \nvdash \bot$ in $\mathscr{H}_3$. Enumerate all sentences $\psi_0, \ldots, \psi_n, \ldots$ and do Lindenbaum extension according to $\bot$-consistency. Now we show that Δ is negation complete. Not-both is by *MP*: Recall that $\neg\alpha$ is defined by $\alpha \to \bot$. At-least-one is by Theorem 2.1(3): If $\alpha \notin \Delta$, then $\Delta \cup \{\alpha\}^+ \vdash \bot$. By Theorem 2.1(3) and Δ being deductively closed, $\alpha \to \bot \in \Delta$. For the truth functionality of $\to$ on Δ, we take the following axioms:

($K^*_\bot$) $\beta \to \{[(\alpha \to \beta) \to \bot] \to \bot\}$
($DA^*_\bot$) $(\alpha \to \bot) \to \{[(\alpha \to \beta) \to \bot] \to \bot\}$

Here by an indirect proof we can use the axiom ($K^*_\bot$), instead of (*K*), to show that $\psi \in \Delta$ implies $\varphi \to \psi \in \Delta$. ($DA^*_\bot$) is then used to prove that $\alpha \notin \Delta$ implies $(\alpha \to \beta) \in \Delta$. (If the reader feels puzzling, she/he can take all the axioms requested above, and restart to check above proof, so that starting from a $\bot$-consistent set Σ one can get a maximal $\bot$-consistent extension Δ from Σ, with $\bot \notin \Delta$ and truth functionality of $\to$ on Δ, and this Δ will induce a classical model of Σ.)

We then conclude that

Theorem 2.2 (Classical Model Existence Theorem for $\mathscr{H}_3$) *In $\mathscr{H}_3$ every $\bot$-consistent set has a classical model. (Here $\mathscr{H}_3 = \mathscr{H}(B, C, I, W, K^*_\bot, DA^*_\bot; MP)$.)*

Note that $\mathscr{H}_3$ is weaker than $\mathscr{H}_2$. We use the following 4-valued semantics to prove this. Let $V = \{0, 1, 2, 3\}$ and the set of designated values $D = \{2, 3\}$. We interpret $\bot$ to be 0 and take the truth function f for $\to$ as follows:

f	0	1	2	3
0	2	1	2	1
1	0	2	2	2
2	0	1	2	1
3	0	1	2	3

Then MP is valid, the axioms (B), (C), (I), (W), $(K^*_\bot)$, $(DA^*_\bot)$ are tautologies, but (K) is not a tautology.

One can easily extend $\mathscr{H}_3$ by adding for $\wedge, \vee$ the following six weak axioms (which are $(Ax1')$-$(Ax6')$ in Lee (2007) with $\neg$ interpreted by $\to \bot$):

$(\wedge I^*_\bot)$ $\quad \alpha \to (\beta \to \{[(\alpha \wedge \beta) \to \bot] \to \bot\})$
$(\wedge E1^*_\bot)$ $\quad (\alpha \wedge \beta) \to [(\alpha \to \bot) \to \bot]$
$(\wedge E2^*_\bot)$ $\quad (\alpha \wedge \beta) \to [(\beta \to \bot) \to \bot]$
$(\vee I1^*_\bot)$ $\quad \alpha \to \{[(\alpha \vee \beta) \to \bot] \to \bot\}$
$(\vee I2^*_\bot)$ $\quad \beta \to \{[(\alpha \vee \beta) \to \bot] \to \bot\}$
$(\vee E^*_\bot)$ $\quad (\alpha \vee \beta) \to \{(\alpha \to \bot) \to [(\beta \to \bot) \to \bot]\}$

Let $\mathscr{H}_3'$ be the system $\mathscr{H}_3$ plus above six axioms. By checking truth functionality of $\wedge, \vee$, it is clear that $\mathscr{H}_3'$ satisfies CME (w.r.t. connectives $\to, \_, \wedge, \vee$).

For Glivenko's theorem we add to $\mathscr{H}_3$ or $\mathscr{H}_3'$ the following axiom

$(ECQ^*_\bot)$ $\quad \bot \to [(\alpha \to \bot) \to \bot]$

Theorem 2.3 (Glivenko's Theorem for $\mathscr{H}_3 + (ECQ^*_\bot), \mathscr{H}_3' + (ECQ^*_\bot)$)

(1) If $\Sigma \vdash \varphi$ in the $\{\to, \bot\}$-fragment of classical propositional logic, then $\Sigma \vdash \neg\neg\varphi$ in $\mathscr{H}_3 + (ECQ^*_\bot)$.
(2) If $\Sigma \vdash \varphi$ in the $\{\to, \bot, \wedge, \vee\}$-fragment of classical propositional logic, then $\Sigma \vdash \neg\neg\varphi$ in $\mathscr{H}_3' + (ECQ^*_\bot)$.

Proof (1): If $\Sigma \vdash \bot$ in $\mathscr{H}_3 + (ECQ^*_\bot)$, by $(ECQ^*_\bot)$ and MP we have $\Sigma \vdash \neg\neg\varphi$ in $\mathscr{H}_3$. Now assume that Σ is $\bot$-consistent, i.e., $\Sigma \nvdash \bot$ in $\mathscr{H}_3 + (ECQ^*_\bot)$. If $\Sigma \cup \{\neg\varphi\} \vdash \bot$ in $\mathscr{H}_3 + (ECQ^*_\bot)$, then it must be the case that $\Sigma \cup \{\neg\varphi\}^+ \vdash \bot$ in $\mathscr{H}_3 + (ECQ^*_\bot)$. By Theorem 2.1(3), $\Sigma \vdash \neg\neg\varphi$ in $\mathscr{H}_3 + (ECQ^*_\bot)$. Else if $\Sigma \cup \{\neg\varphi\} \nvdash \bot$ in $\mathscr{H}_3 + (ECQ^*_\bot)$, by Theorem 2.2 there is a classical model $\mathscr{M} \models \Sigma \cup \{\neg\varphi\}$, i.e., $\Sigma \not\models \varphi$, a contradiction. (2) is proved in a similar way. □

Note that (K) is not provable in $\mathscr{H}_3' + (ECQ^*_\perp)$. This is done by above 4-valued semantics together with interpreting $\wedge$ by *min* and $\vee$ by *max*.

Now we consider *BCI* extension. Instead of negation completeness,[3] we take the following

Definition 2.4 A set Δ is $\rightarrow^+ \perp$ *complete* iff for any α, either ($\alpha \in \Delta$ and $(\alpha \rightarrow^n \perp) \notin \Delta$ for every $n > 0$), or ($\alpha \notin \Delta$ and $(\alpha \rightarrow^n \perp) \in \Delta$ for some $n > 0$).

Because of the nature of weak deduction theorem for *BCI*, we have to modify the negation completeness for *BCI* extension. We may think $\rightarrow^+ \perp$ as a kind of generalized negation. It is clear that in (any axiomatic extension of) *BCI* if $\Sigma \nvdash \perp$ then the Δ obtained by Lindenbaum extension from Σ is $\rightarrow^+ \perp$ complete. Now we construct a Hilbert system $\mathscr{H}_4$ which, as an axiomatic extension of *BCI*, satisfies *CME*. The axioms we are going to add for truth functionality of $\rightarrow$ on Δ are

$(K^+_\perp)$ $\beta \rightarrow \{[(\alpha \rightarrow \beta) \rightarrow^+ \perp] \rightarrow \perp\}$
$(DA^+_\perp)$ $(\alpha \rightarrow^+ \perp) \rightarrow \{[(\alpha \rightarrow \beta) \rightarrow^+ \perp] \rightarrow \perp\}$

Here the axiom $(DA^+_\perp)$: $(\alpha \rightarrow^+ \perp) \rightarrow \{[(\alpha \rightarrow \beta) \rightarrow^+ \perp] \rightarrow \perp\}$ means $(\alpha \rightarrow^n \perp) \rightarrow \{[(\alpha \rightarrow \beta) \rightarrow^m \perp] \rightarrow \perp\}$ for all $n, m > 0$ (and $(K^+_\perp)$ is interpreted in a similar way). We should explain why in these axioms some $\rightarrow$ are with $+$ and some are not. The reason is that those $\rightarrow^+$ are actually obtained by using weak deduction theorem. For example, in $(K^+_\perp)$, since we may not be able to know which n it is on the obtainable $(\alpha \rightarrow \beta) \rightarrow^n \perp$ (by weak deduction theorem), we add more axioms (by using $\rightarrow^+$) so that (part of) the truth functionality of $\rightarrow$ does hold.

One may think that modifying $(K^+_\perp)$ by replacing also the first $\rightarrow$ and the last $\rightarrow$ by $\rightarrow^+$. Taking this more general axiom (which includes $(K^+_\perp)$) is not really necessary, because taking $(K^+_\perp)$ will suffice to prove the truth functionality part we need.

By axiom $(K^+_\perp)$ and Theorem 2.1(2) one can *indirectly* prove that $\beta \in \Delta$ implies $\alpha \rightarrow \beta \in \Delta$. Similarly one can prove that $\alpha \notin \Delta$ implies $\alpha \rightarrow \beta \in \Delta$ by axiom $(DA^+_\perp)$ and Theorem 2.1(2). We then conclude that

Theorem 2.5 (Classical Model Existence Theorem for $\mathscr{H}_4$) *In $\mathscr{H}_4$ every $\perp$-consistent set has a classical model. (Here $\mathscr{H}_4 = \mathscr{H}(B, C, I, K^+_\perp, DA^+_\perp; MP)$.)*

Let Δ be the maximal consistent Lindenbaum extension. Though $\rightarrow^+ \perp$ completeness can only show that $\alpha \notin \Delta$ implies $\alpha \rightarrow^n \perp \in \Delta$ for some $n > 0$ (by Theorem 2.1(2)), actually $\alpha \rightarrow^n \perp \in \Delta$ for all $n > 0$: By the truth functionality of $\rightarrow$ on Δ, all $\alpha \rightarrow^n \perp$ are true by the truth assignment μ with $\mu(p_i) = T$ iff $p_i \in \Delta$. Since Δ is maximal and the fact that the proof system is classically sound, $\alpha \rightarrow^n \perp \in \Delta$ for all $n > 0$.

$\mathscr{H}_4$ is a sublogic of $\mathscr{H}_3$: For any $m > 0$, $\varphi \rightarrow^m \perp \Rightarrow \varphi \rightarrow \perp$ is provable in $\mathscr{H}_3$. With this, axioms $(K^*_\perp)$, $(DA^*_\perp)$, and Theorem 2.1(1), one can easily derive $(K^+_\perp)$, $(DA^+_\perp)$ in $\mathscr{H}_3$.

[3] Precisely speaking, the negation completeness here is $\rightarrow \perp$ completeness, which may not be equivalent to $\neg$ completeness while $\neg$ is a primitive logic connective.

Similarly we define $\mathscr{H}_4'$ to be $\mathscr{H}_4$ plus the following six axioms:

$(\wedge I_\bot^+)$ $\alpha \to (\beta \to \{[(\alpha \wedge \beta) \to^+ \bot] \to \bot\})$
$(\wedge E1_\bot^+)$ $(\alpha \wedge \beta) \to [(\alpha \to^+ \bot) \to \bot]$
$(\wedge E2_\bot^+)$ $(\alpha \wedge \beta) \to [(\beta \to^+ \bot) \to \bot]$
$(\vee I1_\bot^+)$ $\alpha \to \{[(\alpha \vee \beta) \to^+ \bot] \to \bot\}$
$(\vee I2_\bot^+)$ $\beta \to \{[(\alpha \vee \beta) \to^+ \bot] \to \bot\}$
$(\vee E_\bot^+)$ $(\alpha \vee \beta) \to \{(\alpha \to^+ \bot) \to [(\beta \to^+ \bot) \to \bot]\}$

It is clear that $\mathscr{H}_4'$ satisfies *CME* and $\mathscr{H}_4'$ is a sublogic of $\mathscr{H}_3'$. Similarly, for Glivenko's theorem we add to $\mathscr{H}_4$ or $\mathscr{H}_4'$ two axioms: $(ECQ_\bot^*)$ and

$(Ax19)$ $[(\alpha \to \bot) \to^+ \bot] \to [(\alpha \to \bot) \to \bot]$

Theorem 2.6 (Glivenko's theorem for $\mathscr{H}_4 + (ECQ_\bot^*), \mathscr{H}_4' + (ECQ_\bot^*)$)

(1) If $\Sigma \vdash \varphi$ in the $\{\to, \bot\}$-fragment of classical propositional logic, then $\Sigma \vdash \neg\neg\varphi$ in $\mathscr{H}_4 + (ECQ_\bot^*) + (Ax19)$.
(2) If $\Sigma \vdash \varphi$ in the $\{\to, \bot, \wedge, \vee\}$-fragment of classical propositional logic, then $\Sigma \vdash \neg\neg\varphi$ in $\mathscr{H}_4' + (ECQ_\bot^*) + (Ax19)$.

The proof of Theorem 2.6 is similar to that of Theorem 2.3. Note that $(Ax19)$ serves for contraction of negated sentences.

Finally we show that axiom (W) is not provable in $\mathscr{H}_4'$. With this, $\mathscr{H}_4$ (or $\mathscr{H}_4'$) is weaker than $\mathscr{H}_3$ (or $\mathscr{H}_3'$). We use the following 4-valued semantics to prove this.

Let $V = \{0, 1, 2, 3\}$ and the set of designated values $D = \{2, 3\}$. We interpret $\bot$ as 2, $\wedge$ by *min*, $\vee$ by *max*, and take the truth function g for $\to$ as follows:

g	0	1	2	3
0	3	0	3	2
1	2	3	2	3
2	0	1	3	2
3	0	1	2	3

Then *MP* is valid, and the axioms (B), (C), (I), $(K_\bot^+)$, (DA^-), $(\wedge I_\bot^+)$, $(\wedge E1_\bot^+)$, $(\wedge E2_\bot^+)$, $(\vee I1_\bot^+)$, $(\vee I2_\bot^+)$, $(\vee E_\bot^+)$, $(ECQ_\bot^*)$, $(Ax19)$ are tautologies, but axiom (W) is not a tautology. Note that although we assign to $\bot$ the truth value 2 (which also means true in this 4-valued semantics), and those axioms with $\bot$ (in $\mathscr{H}_4$ or $\mathscr{H}_4'$) are not interpreted as expected (i.e., $\bot$ is expected to be false), this 4-valued semantics suffices to show the unprovability of axiom (W) in $\mathscr{H}_4'$.

Also note that in this semantics $\varphi \to^n \bot$ and $\varphi \to^{n+2} \bot$ have the same truth function. Then to compute the truth values for axiom $(K_\bot^+)$ (which means the truth values for $\beta \to \{[(\alpha \to \beta) \to^n \bot] \to \bot\}$ with $n \geq 1$), it suffices to just compute the truth values for $\beta \to \{[(\alpha \to \beta) \to \bot] \to \bot\}$ and $\beta \to \{[(\alpha \to \beta) \to^2 \bot] \to \bot\}$ (which cover all cases).

3 Herbrand-Henkin Style Proof Systems for *CME* in *BCI*-Based Predicate Logics

For simplicity we will focus only on the $\{\rightarrow, \bot\}$-fragment. And $\forall$ and $\exists$ are both primitive symbols, that is, $\exists x$ is not defined by $\neg\forall x\neg$.

In this section first we recall predicate proof systems $\mathcal{Q}_1\mathcal{H}_1$, $\mathcal{Q}_1\mathcal{H}_2$, $\mathcal{Q}_1{}^{=}\mathcal{H}_1$, $\mathcal{Q}_1{}^{=}\mathcal{H}_2$ and some setting from Lee (2009). Then we explain how *CME* is proved for these subclassical predicate logics. Finally we modify these systems into first order quantificational extensions of $\mathcal{H}_3$, $\mathcal{H}_4$ which satisfy *CME* in the predicate sense.

Definition 3.1 (*from* Lee (2009)) Let $i \in \{1, 2\}$. The equality-free proof system $\mathcal{Q}_1\mathcal{H}_i$ has only one inference rule *MP*, and the axioms of $\mathcal{Q}_1\mathcal{H}_i$ are axioms from $\mathcal{H}_i$ (in the predicate sense), $(Ax1)$–$(Ax15)$, and all their universal generalizations.[4] Here all formulas used in $\mathcal{Q}_1\mathcal{H}_i$ are equality-free.

The proof system $\mathcal{Q}_1{}^{=}\mathcal{H}_i$ are defined in a similar way (in this case the equality symbol can be used in formula construction). Axioms of $\mathcal{Q}_1{}^{=}\mathcal{H}_i$ are: the axioms of $\mathcal{H}_i$, the axioms $(Ax1)$–$(Ax18)$, and all their universal generalizations. Below we list the axioms $(Ax1)$–$(Ax18)$:

$(Ax1)$ $\forall x\varphi(x) \rightarrow \varphi(t)$, where the term t is free for x in $\varphi(x)$.
$(Ax2)$ $\varphi(t) \rightarrow \exists x\varphi(x)$, where the term t is free for x in $\varphi(x)$.
$(Ax3)$ $\varphi \rightarrow \forall x\varphi$, where $x \notin FV(\varphi)$ (x is not a free variable of φ).
$(Ax4)$ $\forall x(\varphi \rightarrow \psi) \rightarrow [\forall x\varphi \rightarrow \forall x\psi]$
$(Ax5)$ $\forall x(\varphi \rightarrow \psi) \rightarrow [\exists x\varphi \rightarrow \exists x\psi]$
$(Ax6)$ $\forall x(\varphi \rightarrow \psi) \rightarrow [\varphi \rightarrow \forall x\psi]$, where $x \notin FV(\varphi)$.
$(Ax7)$ $[\varphi \rightarrow \forall x\psi] \rightarrow \forall x(\varphi \rightarrow \psi)$, where $x \notin FV(\varphi)$.
$(Ax8)$ $\exists x(\varphi \rightarrow \psi) \rightarrow [\varphi \rightarrow \exists x\psi]$, where $x \notin FV(\varphi)$.
$(Ax9)$ $[\varphi \rightarrow \exists x\psi] \rightarrow \exists x(\varphi \rightarrow \psi)$, where $x \notin FV(\varphi)$.
$(Ax10)$ $\forall x(\varphi \rightarrow \psi) \rightarrow [\exists x\varphi \rightarrow \psi]$, where $x \notin FV(\psi)$.
$(Ax11)$ $[\exists x\varphi \rightarrow \psi] \rightarrow \forall x(\varphi \rightarrow \psi)$, where $x \notin FV(\psi)$.
$(Ax12)$ $\exists x(\varphi \rightarrow \psi) \rightarrow [\forall x\varphi \rightarrow \psi]$, where $x \notin FV(\psi)$.
$(Ax13)$ $[\forall x\varphi \rightarrow \psi] \rightarrow \exists x(\varphi \rightarrow \psi)$, where $x \notin FV(\psi)$.
$(Ax14)$ $\forall x\varphi(x) \rightarrow \forall y\varphi(y)$, where y does not occur in $\varphi(x)$.
$(Ax15)$ $\exists x\varphi(x) \rightarrow \exists y\varphi(y)$, where y does not occur in $\varphi(x)$.
$(Ax16)$ $x = x$
$(Ax17)$ $x = y \rightarrow t(v_0 \dots v_{i-1}xv_{i+1} \dots v_n) = t(v_0 \dots v_{i-1}yv_{i+1} \dots v_n)$
$(Ax18)$ $x = y \rightarrow [R(v_0 \dots v_{i-1}xv_{i+1} \dots v_n) \rightarrow R(v_0 \dots v_{i-1}yv_{i+1} \dots v_n)]$

In $(Ax17)$,$(Ax18)$ the term t and the relation symbol R are from the predicate language in concern. Note that all of these axioms are classically valid.

[4]That is, if φ is an axiom of $\mathcal{Q}_1\mathcal{H}_i$, then so is $\forall\vec{x}\,\varphi$.

Definition 3.2 (*from* Lee (2009)) $\mathscr{Q}_1$ is the proof system with inference rule *MP* and axioms (*K*),(*S*),(*Ax*1)–(*Ax*15), and all their universal generalizations. A proof system $\mathscr{Q}$ is called an *axiomatic extension* of $\mathscr{Q}_1$ iff $\mathscr{Q}$ is obtained by adding some axioms and all their universal generalizations to $\mathscr{Q}_1$. Let $\mathscr{Q}_1^=$ is $\mathscr{Q}_1$+(*Ax*16)–(*Ax*18).

We recall the following setting from Lee (2009). We assume that any first order language in concern contains at least one constant symbol. $\mathscr{L}(\Gamma)$ is the set of language symbols in Γ (including predicate symbols, function symbols, and constant symbols[5]). "φ is a Π sentence" means that φ is of prenex normal form and it is a Π_m sentence for some $m \geq 0$. $\forall\overrightarrow{x}$ means $\forall x_1 \dots \forall x_k$ for some k. $CT(\Gamma)$ is the set of all closed terms of $\mathscr{L}(\Gamma)$. Σ sentences, $\mathscr{L}(\varphi)$, $\exists\overrightarrow{x}$, $CT(\varphi(\overrightarrow{x}))$ are defined in a similar way. The word "countable" means finite or infinitely countable.

Now we want to modify the proof in Lee (2009) to construct first order quantificational extensions of $\mathscr{H}_3$ and $\mathscr{H}_4$ so that *CME* holds for these extensions. Before doing so, it is better to explain how $\mathscr{Q}_1$ and $\mathscr{Q}_1^=$ are designed. In Lee (2009), $\mathscr{Q}_1$ and $\mathscr{Q}_1^=$ are used to convert, provably equivalently, any first order sentence into a prenex normal form sentence. To do so, some basic properties like Theorem 4.1 in Lee (2009), schema of alphabetic change of a bound variable, and substitutivity of equivalence (Theorem 4.2 in Lee (2009)) are proved in $\mathscr{Q}_1$ and its axiomatic extensions.

After converting all sentences into prenex normal form, the construction of classical model for a $\perp$-consistent set in Lee (2009) is done by witnessing $\forall$ with all closed terms "by now" (i.e., adding to the consistent set with all sentences of $\varphi(t)$ with all closed term t while dealing with $\forall x\varphi(x)$), and then witnessing $\exists$ with new constant symbol "by now" (i.e., adding to the $\perp$-consistent set a sentence $\psi(c)$ with new constant symbol c while dealing with $\exists x\psi(x)$).

Note that in these two types of extensions the $\perp$-consistency is preserved. In Lee (2009), this is proved in Theorem 3.1, *CME* for Π_1 and Σ_1 cases. However, in extensions of $\mathscr{H}_4$, it requires some different axioms. Therefore we will rephrase the consistency preservation theorem for extensions of $\mathscr{H}_3$ and $\mathscr{H}_4$ and give a proof in this section (see Theorem 3.7).

Also note that while extending consistent sets by witnessing $\forall$ and $\exists$ alternatively, after introducing new constant symbols it will damage the previous effort of adding $\varphi(t)$ with "all" closed terms to witness $\forall x\varphi(x)$ (because now not all closed terms are treated), we can redo these two types of extensions in the next level, and then do upto ω many levels. In this way eventually we will have a consistent set which witnesses $\forall$ by all closed terms and $\exists$ by the corresponding constant symbol. Finally the classical model is constructed with domain of all closed terms (or modulo equivalence relation generated by equality if it contains the equality symbol) at the quantifier-free level (i.e., assigning truth values to atomic sentence like $R(t_1, \dots, t_n)$).

However, $\mathscr{Q}_1$ and $\mathscr{Q}_1^=$ are based on (*K*), (*S*) and deduction theorem. But in $\mathscr{H}_3$ and $\mathscr{H}_4$ we can only use weak deduction theorem. Hence we need the following modification.

[5] Add one if there is none.

Definition 3.3 Let $\mathcal{Q}_4$ be the proof system with inference rule *MP* and axioms (*B*),(*C*),(*I*),($Ax1$)–($Ax15$),($Ax23^+$) and all their universal generalizations. $\mathcal{Q}_3$ is the proof system with inference rule *MP* and axioms (*B*),(*C*),(*I*),(*W*),($Ax1$)–($Ax15$) and all their universal generalizations. $\mathcal{Q}_4{}^=$ is the proof system with inference rule *MP* and axioms (*B*),(*C*),(*I*),($Ax1$)–($Ax18$),($Ax23^+$) and all their universal generalizations. $\mathcal{Q}_3{}^=$ is the proof system with inference rule *MP* and axioms (*B*), (*C*), (*I*), (*W*), ($Ax1$)–($Ax18$) and all their universal generalizations. $\mathcal{Q}_3\mathcal{H}_3$ is the proof system with rule *MP* and axioms from $\mathcal{H}_3$ and $\mathcal{Q}_3$ and all their universal generalizations. $\mathcal{Q}_3{}^=\mathcal{H}_3$, $\mathcal{Q}_4\mathcal{H}_4$, $\mathcal{Q}_4{}^=\mathcal{H}_4$ are defined in a similar way.

Here we introduce axiom ($Ax23^+$), which is used for consistency preservation in *BCI* case:

($Ax23^+$) $\quad \exists x\varphi(x) \to [\forall x(\varphi(x) \to^+ \bot) \to \bot]$

Here $\mathcal{Q}_i$ deals with the quantifier part. Its axioms are chosen so that weak deduction theorem and prenex normal form theorem are applicable and also that axioms for witnessing quantifiers are included. However, when dealing with the quantifier-free part, which is essentially the propositional part, we need axioms from $\mathcal{H}_i$ for the construction of classical models.

Recall from Lee (2009) that φ and ψ are *provably equivalent* in $\mathcal{Q}$ iff $\vdash_{\mathcal{Q}} \varphi \to \psi$ and $\vdash_{\mathcal{Q}} \psi \to \varphi$.

Let $FV(\Sigma)$ be the collection of all free variables in φ for some $\varphi \in \Sigma$. We also recall from Lee (2009) the following convention: In any derivation of $\Sigma \vdash_{\mathcal{Q}} \varphi$, the universal generalization on variable x must satisfy that $x \notin FV(\Sigma)$. And to avoid the trouble that Σ uses all free variables, we take the view that "There are always countably many un-used free variables (to Σ)."

Now we check carefully to show that Theorem 4.1 in Lee (2009) also holds in $\mathcal{Q}_3$ and its axiomatic extensions. We rephrase the theorem as follows:

Theorem 3.4 *Let $\mathcal{Q}$ be a proof system with only one rule MP and at least axioms (Ax3),(Ax4), and it is closed under all universal generalizations of axioms. If $\Sigma \vdash_{\mathcal{Q}} \varphi$ and $x \notin FV(\Sigma)$, then $\Sigma \vdash_{\mathcal{Q}} \forall x\varphi$.*

Proof This is proved by induction on the length of proof. Let $\psi_1, \ldots, \psi_n$ be the proof of φ ($= \psi_n$) from Σ and $x \notin FV(\Sigma)$. If ψ_i is from Σ, then $x \notin FV(\psi_i)$ and we use ($Ax3$). If ψ_i is an axiom, then $\forall x\psi_i$ is also an axiom (since $\mathcal{Q}$ is closed under universal generalization of axioms). If ψ_i is obtained by *MP*, we use induction hypotheses, ($Ax4$) and *MP*. □

The *schema of alphabet change of a bound variable* for $\mathcal{Q}$ is that in $\mathcal{Q}$ the following sentences are provable: $\forall x\varphi(x) \to \forall y\varphi(y)$, $\forall y\varphi(y) \to \forall x\varphi(x)$, $\exists x\varphi(x) \to \exists y\varphi(y)$, $\exists y\varphi(y) \to \exists x\varphi(x)$ (where y does not occur in $\varphi(x)$). This does hold in $\mathcal{Q}$ if we simply take axioms ($Ax14$), ($Ax15$) into it.

Now we need to modify Theorem 4.2 in Lee (2009) because the statement there does not work for logics with weak deduction theorem.

Theorem 3.5 (Substitutivity of Equivalence) *Let $\mathscr{Q}$ be a proof system with only MP and axioms at least (Ax1),(Ax3),(Ax4),(Ax5),(B),(C),(I), and it is closed under universal generalization of axioms. Let $\varphi(\overrightarrow{x})$, $\psi(\overrightarrow{x})$ be formulas with free variables among $\overrightarrow{x}$. Suppose that B results from A by replacing zero or more occurrences of $\varphi(\overrightarrow{x})$ in A by $\psi(\overrightarrow{x})$. Then*

$$\{\forall\overrightarrow{x}[\varphi(\overrightarrow{x}) \to \psi(\overrightarrow{x})], \forall\overrightarrow{x}[\psi(\overrightarrow{x}) \to \varphi(\overrightarrow{x})]\} \vdash_{\mathscr{Q}} [A \to B]$$

and

$$\{\forall\overrightarrow{x}[\varphi(\overrightarrow{x}) \to \psi(\overrightarrow{x})], \forall\overrightarrow{x}[\psi(\overrightarrow{x}) \to \varphi(\overrightarrow{x})]\} \vdash_{\mathscr{Q}} [B \to A]$$

Proof (In this proof we call $\{\forall\overrightarrow{x}[\varphi(\overrightarrow{x}) \to \psi(\overrightarrow{x})], \forall\overrightarrow{x}[\psi(\overrightarrow{x}) \to \varphi(\overrightarrow{x})]\}$ the assumption.) If A,B are identical, use axiom (I). If A is $\varphi(\overrightarrow{x})$ and B is $\psi(\overrightarrow{x})$, use assumption, ($Ax1$), MP.

Now we prove this by induction on the complexity of A. We may assume that B results from A by replacing only one occurrence $\varphi(\overrightarrow{x})$ to $\psi(\overrightarrow{x})$. The general case can be done by a sequence of single occurrence replacement: $A = A_1, A_2, \ldots, A_n = B$. Since $A_i \to A_{i+1}, A_{i+1} \to A_i$ for $1 \le i \le n-1$ are all provable from assumption after the equivalence of single occurrence replacement is established, we can then use axiom (B) to prove the equivalence of A and B.

If A is $\alpha_1 \to \beta$ and B is $\alpha_2 \to \beta$, where α_2 results from α_1 by single occurrence replacement. By induction hypothesis, $\alpha_1 \to \alpha_2$ and $\alpha_2 \to \alpha_1$ are provable from assumption. Note that $\alpha_1 \to \beta, \alpha_2 \to \alpha_1 \Rightarrow \alpha_2 \to \beta$ is provable in BCI using $\alpha_1 \to \beta$ exactly one time. By Theorem 2.1(3), $\alpha_2 \to \alpha_1 \Rightarrow (\alpha_1 \to \beta) \to (\alpha_2 \to \beta)$ is provable in BCI. From this we have that $(\alpha_1 \to \beta) \to (\alpha_2 \to \beta)$ is provable in $\mathscr{Q}$ from assumption. Hence $(\alpha_1 \to \beta)$ and $(\alpha_2 \to \beta)$ are provably equivalent in $\mathscr{Q}$ under assumption. The case that $A = \alpha \to \beta_1$ and $B = \alpha \to \beta_2$ is similar.

If A is $\forall xA'$ and B is $\forall xB'$, this is done by Theorem 3.4 and ($Ax4$). If A is $\exists xA'$ and B is $\exists xB'$, this is done by Theorem 3.4 and ($Ax5$). □

Note that the crucial part of above proof is the BCI provability of $(\alpha_1 \to \beta) \to (\alpha_2 \to \beta)$ from $\alpha_2 \to \alpha_1$.

Finally we have the prenex normal form theorem for $\mathscr{Q}_4$ and its axiomatic extensions.

Theorem 3.6 *Let $\mathscr{Q}$ be a proof system with only MP and axioms at least (Ax1), (Ax3)-(Ax15),(B),(C),(I), and it is closed under universal generalization of axioms. In $\mathscr{Q}$ every formula is provably equivalent to a formula in prenex normal form.*

Similar to prenex normal form theorem in Lee (2009), this is proved by

(1) Change all bound variables so that they are all distinct (using Theorems 3.4 and 3.5 and the schema of alphabetic change of a bound variable).
(2) Use ($Ax6$),($Ax7$),($Ax8$),($Ax9$),($Ax10$),($Ax11$),($Ax12$),($Ax13$) to proof-theoretically convert the formula into a provably equivalent prenex normal form formula.

Now it suffices to prove *CME* for prenex normal form sentences. To do so, first we prove that consistency is preserved by adding instantiation sentences (where $\forall$ is by closed terms and $\exists$ is by new constant symbols).

Theorem 3.7 *(1) Let the proof system $\mathcal{Q}$ be a proof system with only MP and axioms at least (Ax1), and it is closed under universal generalization of axioms. If $\Sigma \cup \{\forall x\varphi(x)\}$ is $\bot$-consistent, then so is $\Sigma \cup \{\forall x\varphi(x), \varphi(t)\}$, where t is a closed term.*

(2) Let the proof system $\mathcal{Q}$ be a proof system with only MP and axioms at least (Ax1), (Ax3), (Ax4), (Ax10), (Ax14), (Ax15), (B), (C), (I), (W), and it is closed under universal generalization of axioms. If $\Sigma \cup \{\exists x\varphi(x)\}$ is $\bot$-consistent, then so is $\Sigma \cup \{\exists x\varphi(x), \varphi(c)\}$, where c is a new constant symbol (w.r.t. $\Sigma \cup \{\exists x\varphi(x)\}$).

(3) Let the proof system $\mathcal{Q}$ be a proof system with only MP and axioms at least (Ax1),(Ax3),(Ax4),(Ax23$^+$)(Ax14),(Ax15),(B),(C),(I), and it is closed under universal generalization of axioms. If $\Sigma \cup \{\exists x\varphi(x)\}$ is $\bot$-consistent, then so is $\Sigma \cup \{\exists x\varphi(x), \varphi(c)\}$, where c is a new constant symbol (w.r.t. $\Sigma \cup \{\exists x\varphi(x)\}$). Recall that (Ax23$^+$) is $\exists x\varphi(x) \to [\forall x(\varphi(x) \to^+ \bot) \to \bot]$.

Proof (1) If $\Sigma \cup \{\forall x\varphi(x)\}$ is $\bot$-consistent but $\Sigma \cup \{\forall x\varphi(x), \varphi(t)\} \vdash_{\mathcal{Q}} \bot$, then $\varphi(t)$ must be used in proving $\bot$. By (*Ax*1), $\forall x\varphi(x) \vdash_{\mathcal{Q}} \varphi(t)$. By this, $\Sigma \cup \{\forall x\varphi(x)\}$ is $\bot$-inconsistent, a contradiction.

(2) If $\Sigma \cup \{\exists x\varphi(x)\}$ is $\bot$-consistent but $\Sigma \cup \{\exists x\varphi(x), \varphi(c)\} \vdash_{\mathcal{Q}} \bot$, where c is a new constant, then $\varphi(c)$ must be used in proving $\bot$. By Theorem 2.1(3), $\Sigma \cup \{\exists x\varphi(x)\} \vdash_{\mathcal{Q}} \varphi(c) \to \bot$. Since c is a new constant, we may replace c in the whole proof by a new variable, say, v. Hence $\Sigma \cup \{\exists x\varphi(x)\} \vdash_{\mathcal{Q}} \varphi(v) \to \bot$. By Theorem 3.4, $\Sigma \cup \{\exists x\varphi(x)\} \vdash_{\mathcal{Q}} \forall v(\varphi(v) \to \bot)$. By alphabetic change of a bound variable, (*Ax*10), we have $\Sigma \cup \{\exists x\varphi(x)\} \vdash_{\mathcal{Q}} \exists x\varphi(x) \to \bot$. This leads to the inconsistency of $\Sigma \cup \{\exists x\varphi(x)\}$, a contradiction.

(3) Similar to (2). However, since the proof system is just an extension of *BCI*, we can use only the weak deduction for *BCI*, i.e., $\Sigma \cup \{\exists x\varphi(x)\} \vdash_{\mathcal{Q}} \varphi(c) \to^k \bot$ for some k. Replace c by v, use Theorem 3.4, use alphabetic change of a bound variable, then use (*Ax*23$^+$), we have $\Sigma \cup \{\exists x\varphi(x)\} \vdash_{\mathcal{Q}} \bot$, which leads to a contradiction. □

Now we explain the witnessing process from a consistent set Γ which contains only sentences in prenex norm form. The witness process will enlarge Γ by adding sentences which witness $\forall x\varphi(x)$ or $\exists x\varphi(x)$, and the enlarged set still preserves the consistency. There are ω many levels. Let $\Gamma_{0,0} = \Gamma$. $n \geq 0$. At level L_n we have at most countably many stages $\Gamma_{n,0}, \Gamma_{n,1}, \ldots$ (where $\Gamma_{n,i}$ is the consistent extension upto this stage). At stage $\Gamma_{n,2k}$ we are going to add all $\varphi(t)$ with any closed term t if $\forall x\varphi(x)$ is in $\Gamma_{n,2k-1}$. At stage $\Gamma_{n,2k+1}$ we are going to add $\varphi(c)$ with new constant symbol c if $\exists x\varphi(x)$ is in $\Gamma_{n,2k}$. Since every sentence has only finitely many quantifiers, at this level n every sentence will be witnessed by quantifier-free sentences at stage $\Gamma_{n,k}$ for some k.

However, the problem is that some new constant symbols are introduced so that the $\forall$ is not fully witnessed by all closed terms. To solve this problem, we go to the next level L_{n+1} (let $\Gamma_{n+1,0} = \bigcup_{i\in\omega} \Gamma_{n,i}$) and repeat the whole process. In doing so, upto countably many levels, we will be able to fully witness $\forall$ by all closed terms, and witness $\exists$ by some constant symbols.

Finally we set $\Gamma' = \bigcup_{n=0}^{\infty} \Gamma_{n,0}$ (which is the union of the whole extension), and let Γ'' be the quantifier-free part of Γ'. As a subset, Γ'' is $\perp$-consistent.

Now we have the canonical model existence theorem for Γ''.

Theorem 3.8 (Canonical model existence theorem for quantifier-free case) *Consider proof systems* $\mathcal{Q}_3\mathcal{H}_3$, $\mathcal{Q}_3{}^{=}\mathcal{H}_3$, $\mathcal{Q}_4\mathcal{H}_4$, $\mathcal{Q}_4{}^{=}\mathcal{H}_4$.

(1) *(Without equality, for* $\mathcal{Q}_3\mathcal{H}_3$, $\mathcal{Q}_4\mathcal{H}_4$*) If* Γ'' *is a* $\perp$*-consistent set of quantifier-free sentences, then* Γ'' *has a classical model with domain* $CT(\Gamma'')$.

(2) *(With equality, for* $\mathcal{Q}_3{}^{=}\mathcal{H}_3$, $\mathcal{Q}_4{}^{=}\mathcal{H}_4$*) If* Γ'' *is a* $\perp$*-consistent set of quantifier-free sentences, then* Γ'' *has a classical model with domain* $CT(\Gamma'')/\sim$, *where* $\sim$ *is an equivalence relation (on* $CT(\Gamma'')$*) obtained by the Lindenbaum maximal extension construction.*

Since the quantifier-free model construction is similar to *CME* in $\mathcal{H}_3$ and $\mathcal{H}_4$, we skip the proof here. Note that while without equality, we take the domain $CT(\Gamma'')$ and construct the propositional model over atomic sentences like $R(t_1, \ldots, t_n)$ with $t_1, \ldots, t_n \in CT(\Gamma'')$. And while with equality, we take the domain $CT(\Gamma'')/\sim$, where the equivalence relation $\sim$ is defined by: $t_1 \sim t_2$ if and only if $t_1 = t_2 \in \Delta$ (here Δ is the Lindenbaum extension from Γ'').

Now let Γ be a consistent set of prenex normal form sentences and we construct Γ' by witnessing process and let Γ'' be its quantifier-free part. By Theorem 3.8, Γ'' has a classical model which uses $CT(\Gamma'')$ (without equality) or $CT(\Gamma'')/\sim$ (with equality). Since $\forall$ is for all closed terms and $\exists$ is for some constant symbol, this model is also for Γ' and Γ. Now we have prenex normal form *CME* for $\mathcal{Q}_3\mathcal{H}_3$, $\mathcal{Q}_3{}^{=}\mathcal{H}_3$, $\mathcal{Q}_4\mathcal{H}_4$, $\mathcal{Q}_4{}^{=}\mathcal{H}_4$.

Theorem 3.9 (*CME* for prenex normal form sentences) *Let the proof system* $\mathcal{Q}$ *be one of* $\mathcal{Q}_3\mathcal{H}_3$, $\mathcal{Q}_3{}^{=}\mathcal{H}_3$, $\mathcal{Q}_4\mathcal{H}_4$, $\mathcal{Q}_4{}^{=}\mathcal{H}_4$. *If* Γ *is a* $\perp$*-consistent set of prenex normal form sentences, then* Γ *has a classical model.*

Finally we have the general *CME* for $\mathcal{Q}_3\mathcal{H}_3$, $\mathcal{Q}_3{}^{=}\mathcal{H}_3$, $\mathcal{Q}_4\mathcal{H}_4$, $\mathcal{Q}_4{}^{=}\mathcal{H}_4$.

Theorem 3.10 (Classical model existence theorem for $\mathcal{Q}_3\mathcal{H}_3$, $\mathcal{Q}_3{}^{=}\mathcal{H}_3$, $\mathcal{Q}_4\mathcal{H}_4$, $\mathcal{Q}_4{}^{=}\mathcal{H}_4$) *Let the proof system* $\mathcal{Q}$ *be one of* $\mathcal{Q}_3\mathcal{H}_3$, $\mathcal{Q}_3{}^{=}\mathcal{H}_3$, $\mathcal{Q}_4\mathcal{H}_4$, $\mathcal{Q}_4{}^{=}\mathcal{H}_4$. *If* Σ *is a* $\perp$*-consistent set, then* Σ *has a classical model.*

This is proved as follows: (1) Use prenex normal form theorem to convert each sentence in Σ into a provably equivalent prenex normal sentence. Let the collection of prenex normal form sentences be Γ. (2) Use *CME* for prenex normal form theorem and we have a classical model for Γ. Since Γ is provably equivalent to Σ, it is also a model for Σ.

4 Hintikka Style Proof Systems for *CME* in *BCI*-Based Predicate Logics

In this section we also construct predicate logics which satisfy *CME*, without the aid of prenex normal form theorem. Here the primitive logic symbols are $\forall, \exists, \rightarrow, \bot$. Suppose that we have a $\bot$-consistent set Σ. What we will do is to enlarge this set so that $\forall$ means "for (all) closed terms" and $\exists$ means "there is a (new) constant symbol." However, what is different in Hintikka style construction is that sentences are not converted into prenex normal form sentences in advance, and the proof system may not have sufficient axioms to convert sentences into prenex normal forms. Therefore we extend Σ in a mixed way, i.e., we need to simultaneously treat $\forall x$, $\exists x$, $\alpha \rightarrow \beta$ (including $(\forall x\varphi(x)) \rightarrow \bot$, $(\exists x\varphi(x)) \rightarrow \bot$, $(\gamma \rightarrow \delta) \rightarrow \bot$).

First we list the axioms needed for preserving consistency.

($Ax20^+$) $[(\bot \rightarrow \delta) \rightarrow \bot)] \rightarrow^+ \bot$
($Ax21^+$) $\forall x\varphi(x) \rightarrow [(\varphi(t) \rightarrow^+ \bot) \rightarrow \bot]$, where t is a closed term.
($Ax22^+$) $(\forall x\varphi(x) \rightarrow \bot) \rightarrow \{[\exists x(\varphi(x) \rightarrow \bot) \rightarrow^+ \bot] \rightarrow \bot\}$
($Ax23^+$) $\exists x\varphi(x) \rightarrow [\forall y(\varphi(y) \rightarrow^+ \bot) \rightarrow \bot]$, where x is free for y in $\varphi(y)$ and y is free for x in $\varphi(x)$.
($Ax24^+$) $(\exists x\varphi(x) \rightarrow \bot) \rightarrow \{[\forall x(\varphi(x) \rightarrow \bot) \rightarrow^+ \bot] \rightarrow \bot\}$
($Ax25^+$) $(\beta \rightarrow \gamma) \rightarrow \{(\gamma \rightarrow^+ \bot) \rightarrow [(\beta \rightarrow \bot) \rightarrow^+ \bot) \rightarrow \bot]\}$
($Ax26^+$) $[(\gamma \rightarrow \delta) \rightarrow \bot] \rightarrow [(\gamma \rightarrow^+ \bot) \rightarrow \bot]$
($Ax27^+$) $[(\gamma \rightarrow \delta) \rightarrow \bot] \rightarrow \{[(\delta \rightarrow \bot) \rightarrow^+ \bot] \rightarrow \bot\}$
($Ax28^+$) $[(\gamma \rightarrow \delta) \rightarrow \bot] \rightarrow [\{[\gamma \rightarrow^+ [(\delta \rightarrow \bot) \rightarrow^+ \bot]\} \rightarrow \bot]$

Now we enlarge the $\bot$-consistent set Σ. Let $\Gamma_{0,0} = \Sigma$. We pick up sentence α from $\Gamma_{0,0}$ and add sentences according to the following way:

(1) If α is $\forall x\varphi(x)$: add $\varphi(t)$ for all closed term t. (Axiom ($Ax21^+$) preserves the consistency of this extension.)
(2) If α is $\exists x\varphi(x)$: add $\varphi(c)$ for new constant symbol c. (We need Theorem 3.7(3) and Axiom ($Ax23^+$) to preserve the consistency of this extension. Therefore we also need to take axioms from Sect. 3: ($Ax1$), ($Ax3$), ($Ax4$), ($Ax13$), ($Ax14$), ($Ax15$), and also the setting that the proof system is closed under universal generalization of axioms.)
(3) If α is $\bot \rightarrow \gamma$: add nothing. ($\bot \rightarrow \gamma$ is always true.)
(4) If α is $\beta \rightarrow \gamma$ and β, γ are not $\bot$: add γ if this preserves consistency, else add $\beta \rightarrow \bot$. (Axiom ($Ax25^+$) preserves the consistency of extension, even when the second case happens.)
(5) If α is $\forall x\varphi(x) \rightarrow \bot$: add $\exists x(\varphi(x) \rightarrow \bot)$. (Axiom ($Ax22^+$) preserves the consistency of this extension.)
(6) If α is $\exists x\varphi(x) \rightarrow \bot$: add $\forall x(\varphi(x) \rightarrow \bot)$. (Axiom ($Ax24^+$) preserves the consistency of this extension.)
(7) If α is $(\gamma \rightarrow \delta) \rightarrow \bot$ and γ is not $\bot$: add γ and $\delta \rightarrow \bot$. (Axioms ($Ax26^+$), ($Ax27^+$), ($Ax28^+$) preserve the consistency of adding these two sentences.)

(8) If α is $(\bot \to \delta) \to \bot$: by Axiom $(Ax20^+)$, this will make the current set inconsistent. (Note that $(\bot \to \delta) \to \bot$ is always false. So we add Axiom $(Ax20^+)$ to avoid this situation.)

Note that Axiom $(Ax20^+)$ can be used to trace back inconsistency. For example, if we start with $\beta \to \gamma = (\bot \to \delta) \to [(\bot \to \delta) \to \bot]$, we cannot add γ because of Axiom $(Ax20^+)$, and then we also cannot add $\beta \to \bot$, which is $(\bot \to \delta) \to \bot$. Now use Axiom $(Ax25^+)$ from $\beta \to \gamma$ one can derive $\bot$, i.e., $\beta \to \gamma$ is inconsistent.

Also note that in (7) we need 3 axioms to show consistency preservation because we are dealing with *BCI* extension, which is sensitive on what assumptions with multiplicity are used in that proof.

Now from $\Gamma_{0,0}$ to $\Gamma_{0,7}$, we do extensions of types from (1) to (7): from $\Gamma_{0,i-1}$ to $\Gamma_{0,i}$, we do extension of type (i) for all sentences in $\Gamma_{0,i-1}$ which satisfy the condition of (i), $1 \le i \le 7$. We repeat these seven types of extensions upto countably many stages at level L_0, so that all sentences in $\Gamma_{0,0}$ will be decomposed into atomic sentences.

Let $\Gamma_{1,0} = \bigcup_{i=0}^{\infty} \Gamma_{0,i}$, and we repeat the same process at level L_1, and keep doing this on countably many levels. Then let $\Gamma' = \bigcup_{n=0}^{\infty} \Gamma_{n,0}$. We then take the quantifier-free part of Γ', say, Γ''. We then construct canonical model for quantifier-free Γ'' with $\mathscr{H}_4$ over domain $CT(\Gamma')$ if without equality, and with $\mathscr{H}_4 + (Ax16) - (Ax18)$ over domain $CT(\Gamma')/\sim$ if with equality. Since $\forall x\varphi(x)$ is true when $\varphi(t)$ is true for all closed term t (which ranges over all domain), and $\exists x\varphi(x)$ is true when $\varphi(c)$ is true, and in above extension that α being true is granted by the added sentences being true, we can conclude the following classical model existence theorem:

Theorem 4.1 *Let Hin_4 be the proof system with MP and axioms from $\mathscr{H}_4$ and $(Ax20^+) - (Ax28^+)$, $(Ax1)$, $(Ax3)$, $(Ax4)$, $(Ax13) - (Ax15)$, and it is closed under universal generalization of axioms. Similarly $Hin_4^=$ is $Hin_4 + (Ax16) - (Ax18)$. In Hin_4 or $Hin_4^=$ every $\bot$-consistent set has a classical model.*

We list the axioms for Hin_3 by removing $+$.

$(Ax20)$	$[(\bot \to \delta) \to \bot)] \to \bot$
$(Ax21)$	$\forall x\varphi(x) \to [(\varphi(t) \to \bot) \to \bot]$, where t is a closed term.
$(Ax22)$	$(\forall x\varphi(x) \to \bot) \to \{[\exists x(\varphi(x) \to \bot) \to \bot] \to \bot\}$
$(Ax23)$	$\exists x\varphi(x) \to [\forall y(\varphi(y) \to \bot) \to \bot]$, where x is free for y in $\varphi(y)$, y is free for x in $\varphi(x)$, and y does not occur in $\varphi(x)$. (This is actually a special case of variation of $(Ax10)$.)
$(Ax24)$	$(\exists x\varphi(x) \to \bot) \to \{[\forall x(\varphi(x) \to \bot) \to \bot] \to \bot\}$
$(Ax25)$	$(\beta \to \gamma) \to \{(\gamma \to \bot) \to [(\beta \to \bot) \to \bot) \to \bot]\}$
$(Ax26)$	$[(\gamma \to \delta) \to \bot] \to [(\gamma \to \bot) \to \bot]$
$(Ax27)$	$[(\gamma \to \delta) \to \bot] \to \{[(\delta \to \bot) \to \bot] \to \bot\}$
$(Ax28)$	$[(\gamma \to \delta) \to \bot] \to [\{[\gamma \to [(\delta \to \bot) \to \bot]\} \to \bot]$

By a similar argument we have:

Theorem 4.2 *Let Hin_3 be the proof system with MP and axioms from $\mathscr{H}_3$ and* ($Ax20$) – ($Ax28$), ($Ax1$), ($Ax3$), ($Ax4$), ($Ax13$) – ($Ax15$), *and it is closed under universal generalization of axioms. $Hin_3^=$ is Hin_3 +* ($Ax16$) – ($Ax18$). *In Hin_3 or $Hin_3^=$ every $\perp$-consistent set has a classical model.*

5 Concluding Remarks

Here we propose some questions for further research along this direction.

Firstly, the models/semantics for these logics $\mathscr{H}_3$, $\mathscr{H}_4$, $\mathscr{H}_3'$, $\mathscr{H}_4'$, $\mathscr{Q}_3\mathscr{H}_3$, $\mathscr{Q}_4\mathscr{H}_4$, $\mathscr{Q}_3^=\mathscr{H}_3$, $\mathscr{Q}_4^=\mathscr{H}_4$ need further investigation.

Secondly, Hin_3 and Hin_4 (introduced in Sect. 4) are constructed from different ideas, if compared with $\mathscr{Q}_3\mathscr{H}_3$ and $\mathscr{Q}_4\mathscr{H}_4$ (introduced in Sect. 3). A natural question one should ask is: In Hin_3 or Hin_4, is prenex normal form theorem provable? If not, the differences between $\mathscr{Q}_3\mathscr{H}_3$ and Hin_3 deserve further study.

Finally, converting Hilbert proof systems like Hin_3 and $\mathscr{Q}_3\mathscr{H}_3$ into sequent calculus style proof systems (see Lee (2007a), where we present weakest system for proving *CME* in propositional logics with primitive symbol $\neg$) may be helpful in study of weak predicate proof systems satisfying *CME*.

References

Lee, J. -L. (2007). Classical model existence theorem in propositional logics. In J. -Y. Béziau & A. Costa-Leite (Eds.), *Perspectives on Universal Logic* (pp. 179–97). Italy: Polimetrica, Monza.

Lee, J. -L. (2007a). Classical model existence and left resolution. *Logic and Logical Philosophy*, *16*(4), 333–52.

Lee, J. -L. (2009) Classical model existence theorem in subclassical predicate logics. I. In D. Makinson, J. Malinowski, H. Wansing (eds.), *Trends in Logic: Towards mathematical philosophy* Trends in logic—Studia Logica Library, vol. 28, (pp. 187-99). Dordrecht: Springer.

Paoli, F. (2002). *Substructural Logics: A Primer*, Trends in logic—Studia Logica Library: Kluwer Academic Publishers.

On Incorporating Reasoning Time into Epistemic Logic

Ren-June Wang

Abstract In this paper I argue for the naturality and importance of the concept of *reasoning-based knowledge*, in contrast with the more familiar *information-based knowledge*, which is interpreted plainly by the possible world semantic setting. In search of a formal system for the concept, I propose two extensions of the newly introduced epistemic logical framework, tMEL, which has formulas of form $K^i\phi$ to mean that ϕ is known by the agent at the time i. The two introduced extensions are $\mathsf{tMEL}^{\mathsf{K}}$ and tMEL^{∞}, with each of them having a device representing the information-based knowledge and reasoning-based knowledge respectively. These extensions will be investigated and compared, and then I will demonstrate an application of tMEL^{∞}, with a detour to discuss Moore's paradox from the reasoning-based knowledge point of view.

1 Introduction

Normally we trace the development of modern epistemic logic back to von Wright's writing in which we find the first axiomatization of epistemic logic. But what is rarely discussed about the work is that the epistemic concept formulated therein is actually "being verified." Or to put it in another way, von Wright had a definite epistemic interpretation when the system was laid down. "$\mathbf{V}\phi$," the primitive form of the proposed epistemic formulas, is suggested to read as ϕ *is known* or directly ϕ *is verified*.[1] Since von Wright didn't say much about the meaning of "being verified," we might understand it in its ordinary sense, i.e., to say a sentence *is known* is to say that there is a procedure for verifying the truth of the sentence. From the proposed axioms in the system, it is fair to say that the systematic natures that was observed by von Wright of the usage of the epistemic term and then formulated in the form of

[1] See Wright (1951).

R.-J. Wang (✉)
Department of Philosophy, National Chung-Cheng University, Chiayi County, Taiwan
e-mail: wrenjune@gmail.com

S.C.-M. Yang et al. (eds.), *Philosophical Logic: Current Trends in Asia*,
Logic in Asia: Studia Logica Library, DOI 10.1007/978-981-10-6355-8_11

modal logic are that verification leads to the truth of the verified sentence, and that the body of human knowledge is closed under verification.[2]

The semantic of epistemic logic, whose story is relatively well-known, appeared late in Hintikka's work, and, in its modern form, is still a standard formal mechanism in present days. The idea behind the semantics is the old adage that "*knowledge is the elimination of uncertainty*" (Hintikka, 1986), and thus, according to the analysis, "a knows that p" means p is true in all the epistemic alternatives with respect to the knower or agent a's current situation; or in other words, a is in a position to be certain that p is the case. The original intent for the semantic study is, as Hintikka himself describes, to supplement what has been the endless example-counterexample type of debates on the adequacy of proposed syntactical epistemic principles at the time, with a concrete and even pictorial device to work with Hintikka (1968). From today's vantage point of view, Hintikka's project is of great success, especially given its enormous applications in the science of computing and artificial intelligence Fagin et al. (1995), Meyer and Van Der Hoek (2004). But I still find the analysis unsatisfactory. Something is just missing in the picture. Basically, an epistemic event such as someone or some agent holding some knowledge only can be a *temporal event*. A known sentence can become unknown to an agent because the truth value of the sentence itself, like the state of weather, changes along with the change of the time, e.g. that a knows it is raining can't longer be true if it stops raining; or because the agent under discussion loses memory of what is originally known: I no longer know where my key is if I forgot it.

From a theoretical point of view, we have good reasons to focus our epistemic reasoning on scenarios in which the basic events to know about are general events, like that *the earth is round*, whose truth-value won't change at times, and the agents in discussion won't forget what have been known, as we might try to understand the behavior of an inanimate machine or concern with a committing agent who won't lose any relevant information when devoted to completing a task. But even so, we are beings able to enrich our life by expanding our knowledge, that means turning something unknown into known. We learn things from our contact with the external world, or getting information from others. But finally, even if we limit our attention to *close systems* where the agents are not supposed to communicate with the outside world, the agents can still increase their knowledge just by activating their inner reasoning faculty, to figure out something unknown previously.

Thus my complaint about Hintikka's semantic analysis is that it doesn't take *reasoning time* into account. The concept of knowing itself should not be understood only as the static condition of the truth-values of the known formula in the epistemic alternatives. It is also related to how the agent under discussion figuring out the truth of the known formula. Sentences such as "a knows that p," according to the analysis, involves some temporal measurement which tells us *when* the sentence p is known by the agent, and the magnitude of the measurement is reflecting the reasoning ability of the agent a and the difficulty of the sentence p to be reasoned out. This interpretation

[2]Ibid. The axiomatization is called M in von Wright's essay, which in today's nomenclature is the modal logical system T.

also accords with von Wright's original investigation; anyway, no matter concerning with the individual knowledge or the body of human being's knowledge, verification is a time-consuming procedure, which can't be found in Hintikka's analysis.

I call the concept of knowledge that is defined plainly by the possible-world semantics the *information-based knowledge*, and the concept I emphasize here the *reasoning-based knowledge.* In Wang (2011), I proposes an epistemic logical framework tMEL, *timed Modal Epistemic Logic*, in which the basic epistemic formulas have the form of $K^i\phi$ with the intended meaning that ϕ is known by the agent under discussion at the time i.[3] It is then a natural step to extend the systems in the framework with an epistemic connective to reflect the reasoning-based knowledge; and this is the main goal of this paper.

To achieve the goal, we will do several things in this paper. First in the next section, we will introduce the needed preliminary machineries, including the systems and semantics of MEL, Modal Epistemic Logic, and tMEL, which, for the expository purpose, will be presented in details. tMEL, just like MEL, is an umbrella name including many logics. To simplify the discussion, we will focus on S4, the epistemic logic discussed in Hintikka (1962), and its corresponding counterpart tS4 in tMEL. Then we introduce two logical frameworks extended from tMEL, $\mathsf{tMEL}^{\mathsf{K}}$ and tMEL^{∞}, with each of them including a device representing information-based knowledge and reasoning-based knowledge respectively. These frameworks will be introduced in Sects. 3 and 4, respectively, and, following that, a section is devoted to their comparison. Finally, in Sect. 6, we will utilize tMEL^{∞} to take some issue in epistemic logic which couldn't be handled by the plain possible world semantic setting; but before that a detour is made to discuss the Moore's paradox from the perspective of reasoning-based knowledge.

2 From MEL to tMEL

The formal languages we are going to see in this paper will all be extended from the language of propositional logic, built up from a set of propositional letters and a set of truth-functional connectives, which, for convenience, will only comprise connectives $\sim$ (*negation*) and $\rightarrow$ (*implication*). One single non-truth-functional modal connective will be added to these languages, which is denoted as K. Nonetheless, concerning with the modal connective, two types of modal formulas will be constructed. One is the *simple modal formulas*, constructed based on the rule that if ϕ is a formula, so is $K\phi$; and the other is the *labelled modal formulas*, based on the rule that if ϕ is a formula, so is $K^i\phi$ with i in a set of labels. Taken as examples, the *language of* MEL is extended from the language of propositional logic by the simple modal

[3]In Wang (2011), I have written the formulas as $K\phi^i$, instead of $K^i\phi$ to prevent the misinterpretation that treats the system as a multi-modal system with K^i as modalities for every i. But it has been suggested that the formulas of the original form are difficult to read, and hence now I adopt the form of $K^i\phi$.

formulas and the *language of* tMEL is extended by labelled modal formulas with natural numbers as labels.

In the standard *possible world semantics*, a *structure* $M = \langle W, R, \mathcal{V}\rangle$ is composed of three components: a set of *possible worlds* W, a binary *accessibility relation*, $R \subseteq W \times W$, and a function $\mathcal{V}$ assigning to each propositional letter a set of possible worlds. In this paper we will only consider the modal logic S4 and, in a way, its various variants, so the accessibility relation is taken to be *reflexive* and *transitive*, if not stated otherwise. But as you can see the method practiced here can be generalized to be concerned with other modal epistemic logics. The truth-values of MEL formulas is recursively given as follows:

$(M, w) \vDash p$ for a propositional letter p iff $w \in \mathcal{V}(p)$,,
$(M, w) \vDash \sim\phi$ iff $(M, w) \nvDash \phi$,
$(M, w) \vDash \phi \to \psi$ iff $(M, w) \nvDash \phi$ or $(M, w) \vDash \psi$,
$(M, w) \Vdash K\phi$ iff $(M, w') \Vdash \phi$ for all $w' \in W$ with wRw'.

So $K\phi$ is suggested to read as "*it is known that* ϕ", or "*a knows that* ϕ" for the agent a. In this paper we will only discuss the case of a single agent. The corresponding complete axiom system of S4 is then the following:

Axioms:

A0 classical propositional axiom schemes,
A1 $K(\phi \to \psi) \to (K\phi \to K\psi)$,
A2 $K\phi \to K(K\phi)$,
A3 $K\phi \to \phi$,

Inference Rules

R1 if $\vdash \phi \to \psi$ and $\vdash \phi$, then $\vdash \psi$,
R2 if $\vdash \phi$, then $\vdash K\phi$.

The provability and validity for any logics discussed in this paper are treated in the usual way.

Next we consider the semantics of tMEL logics, which is augmented from the standard possible world semantics, and we will focus on the semantics of tS4.[4] One thing distinctive of tMEL logics is that they are *base-parameterized*. By a base, formally we mean a tuple $\mathcal{A} = \langle \mathbf{A}, \mathbf{f}\rangle$, with $\mathbf{A}$ a set of tMEL formulas, and $\mathbf{f}\colon \mathbf{A} \mapsto \mathbb{N}$ (natural numbers). We will explain the significance of bases later. Now given a base $\mathcal{A}$, an *$\mathcal{A}$-awareness function* α is defined as a partial function mapping tMEL formulas to natural numbers and satisfying the following conditions ($\alpha(\phi)\downarrow$ denotes $\alpha(\phi)$ is defined):

[4]tMEL is derived from the study of Justification Logic Artemov (2001); Artemov and Fitting (2015). For details, see Wang (2011).

0 If $A \in \mathbf{A}$, then $\alpha(A) \leq \mathbf{f}(A)$

(Initial Condition)

1 If $\alpha(\phi \to \psi)\downarrow$ and $\alpha(\phi)\downarrow$, then
$\alpha(\psi) \leq \max(\alpha(\phi \to \psi), \alpha(\phi)) + 1$

(Deduction by Modus Ponens)

2 If $A \in \mathbf{A}$ and $f(A) \leq i$, then
$\alpha(K^i A) \leq i + 1$

(Deduction by $\mathcal{A}$-Epistemization)

3 If $\alpha(\phi)\downarrow$ and $\alpha(\phi) \leq i$, then
$\alpha(K^i \phi) \leq i + 1$

(Inner Positive Introspection)

The goal of these awareness functions is nothing but chronologically recording the deductive reasoning that the modeled agent practices, and these conditions reflect the logical rules that the depicted agent is assumed to perform, with each application of the rules taking one unit of time.[5] In order to document the reasoning process of the agent, we need to assume the reasoning system employed by the agent. The simplest way to do that is to assume the agent is reasoning axiomatically. This assumption, which is not that realistic, might be improved by later research, but at this moment we rely on it. Then a *logical* base, whose formal definition will be given in the following, plays the role to give us the information of what axioms are used by the agent, and when at the latest s/he became aware of the truths of the axioms, since at the time the agent definitely will interact with a source of the information. Now corresponding to each condition of an $\mathcal{A}$-awareness function with $\mathcal{A} = \langle \mathbf{A}, \mathbf{f} \rangle$, the rule that is supposed to be able to be performed by the agent is the following:

0. *Initial Condition:* This condition says nothing about any special kind of deduction rule to be performed. It indicates that at the very latest the agent will be aware of an axiom A employed by the agent at that time $\mathbf{f}(A)$.

1. *Deduction by Modus Ponens*: in one unit of time, the agent is able to perform *modus ponens*.

2. *Deduction by $\mathcal{A}$-Epistemization*: in one unit of time, the agent is able to perform $\mathcal{A}$-Epistemization, which means to deduce the agent's knowledge of an axiom that is employed by the agent.

3. *Inner Positive Introspection*: in one unit of time, the agent is able to perform some kind of introspection in order to be in a state of awareness of that, right before the time of the performance, the agent knows the formula which s/he has been aware of.

[5] Fagin and Halpern (1987) might be the earliest work in which an awareness function is introduced. To call the functions defined here also *awareness function* is indeed inspired by their work, and from purely technical point view, the setting given here could be regarded as an extension of their work. But there is a significant ideological difference between our work and theirs. An awareness function as we deal with it here does not mean to tell us when the agent starts to recognize a formula, as if the agent wasn't aware of the formula before the time indicated by the awareness function; rather, the function employed here is telling us *at what time the agent first time notices the truth-value of the formula is possible to be true in relation to the formulas that s/he has deduced beforehand.*.

Notice that these conditions are for a tS4 awareness function. Obviously, the condition **2** is subsumed under the condition **3**. The reason for this separation is for the consideration of generality; the condition **3** is typical of a tS4 agent, not of an agent with weaker reasoning ability, and the condition **2** is something we want to reserve for the weaker agents.[6]

Fix a base $\mathcal{A}$. By a tS4($\mathcal{A}$)-*structure* we mean a tuple $M = \langle W, R, \mathfrak{A}, \mathcal{V}\rangle$, where $\langle W, R, \mathcal{V}\rangle$ is an S4 structure and $\mathfrak{A}$ is a collection of $\mathcal{A}$-awareness functions α_w, one for each world $w \in W$, satisfying the *monotonicity condition*[7]: given a tMEL formula ϕ, if $\alpha_w(\phi)\downarrow$, then $\alpha_{w'}(\phi) \leq \alpha_w(\phi)$. Informally, the condition, which is not needed for weaker tMEL logics, is stating that a tS4 agent can only imagine a formula known equally or earlier in time at the accessible world, which means tS4 agents are so firm about their knowledge that in the possibilities, only new sources of information can make them have the knowledge early. So the basic idea behind this semantics is that at each world, there is a static world view, like before, which includes the truth-values of those basic events, represented by propositional letters, and the reasoning process taken by the agent at the world.

Now we can define the truth-value for tMEL formulas. For the propositional cases, they are treated in the same way as those of MEL formulas, and for a labelled modal formulas, its truth-value is given as follows:

$$(M, w) \models K^i\phi \quad \text{iff} \quad (M, w') \models \phi \text{ for all } w' \in W \text{ with } wRw', \text{ and } \alpha_w(\phi) \leq i.$$

According to this analysis, an agent knows at the time i that ϕ is the case if and only if it is true at all accessible possible worlds and the agent is able to, by activating her/his deductive reasoning faculty on the information the agent has at the world, be aware of the truth of the formula before that time.

To define a tS4 *logical base*, we need some efforts, since the definition of the logical truths, i.e. the valid formulas, will depend on the base to be worked with; however, we also need every formula in the logical base to be logical truths. So we start from the empty base, which means the agent at the beginning is not assumed as being aware of any logical truth. But still with this base we can deduce some interesting logical truths, which include the classical tautologies and the ones reflect the conditions in the definition of awareness functions, including axiom instances of the following systems. But no formulas of the form $K^i\phi$ are logical truths; that is, an agent has no logical knowledge with empty base.

Some terminology is in order. Given bases $\mathcal{A} = \langle \mathbf{A}, f\rangle$ and $\mathcal{B} = \langle \mathbf{B}, g\rangle$, $\mathcal{B} \subseteq \mathcal{A}$ means $\mathbf{B} \subseteq \mathbf{A}$ and $f(B) \leq g(B)$ for all $B \in \mathbf{B}$. A collection of bases $\{\mathcal{A}_i (= \langle \mathbf{A}_i, f_i\rangle)\}_{i\in\mathbb{N}}$ is an *ascending chain* if $\mathcal{A}_1 \subseteq \mathcal{A}_2 \subseteq \ldots$, and a base $\mathcal{A}$ is the *limit* of the chain if $\mathcal{A} = \bigcup \mathcal{A}_i$, i.e., $\mathbf{A} = \bigcup \mathbf{A}_i$ and $f(A) = \min\{f_i(A) : f_i(A)\downarrow\}$.

Definition 2.1 A base $\mathcal{A}$ is tS4 *logical* if one of the following is true:
(1) $\mathcal{A}$ is empty,
(2) in $\mathcal{A} = \langle \mathbf{A}, f\rangle$, $\mathbf{A}$ consists of tS4($\mathcal{B}$) valid formulas with $\mathcal{B}$ a tS4 logical base,

[6] See Wang (2011) for these weaker logics.

[7] Don't confuse this condition with the *monotonicity axiom* that we will see later.

(3) $\mathcal{A}$ is the limit of an ascending chain of tS4 logical bases $\{\mathcal{A}_i\}_{i\in\mathbb{N}}$, with $\mathcal{A}_{i+1}$ consisting of tS4($\mathcal{A}_i$) valid formulas.

Lemma 2.2 *If $\mathcal{A} = \langle \mathbf{A}, \mathbf{f}\rangle$ is a tS4 logical base, every formula in* $\mathbf{A}$ *is tS4($\mathcal{A}$) valid.*

Let $\mathcal{A}$ be a tS4 logical base. The axiom system of tS4($\mathcal{A}$) is as follows:

Axioms:

A0 Classical propositional axiom schemes,
A1 $K^i(\phi \to \psi) \to (K^j\phi \to K^k\psi)$ $i, j < k$,
A1′ $K^iA \to K^j(K^iA)$ $i < j$ if $A \in \mathbf{A}$ and $f(A) \leq i$,
A2 $K^i\phi{\to}K^j(K^i\phi)$ $i < j$,
A3 $K^i\phi \to \phi$,
A4 $K^i\phi \to K^j\phi$ $i < j$, (*Monotonicity Axiom*)

Inference Rules

R1 if $\vdash \phi \to \psi$ and $\vdash \phi$, then $\vdash \psi$,
R2 if $A \in \mathbf{A}$ and $f(A) \leq i$, then $\vdash K^iA$, (*$\mathcal{A}$-Epistemization*)
for all $i, j, k \in \mathbb{N}$.

Theorem 2.3 (Theorem 4.5 in Wang (2011)) *Given a* tS4 *logical base $\mathcal{A}$, a* tMEL *formula is* tS4*($\mathcal{A}$) valid if and only if it is provable in the axiom system of* tS4*($\mathcal{A}$).*

So corresponding to an MEL logic, such as S4, there is in fact a family of tMEL logics, tS4($\mathcal{A}$), introduced, with each tS4 logical base $\mathcal{A}$ to indicate the basic logical truths to be used by the agent. We have seen the empty base. We can also have a *comprehensive logical base* which include all valid formulas.[8] One logical base especially interests us. We call a tMEL logic, also the logics introduced later on, has the *internalization property* if for any valid formula ϕ, there is an $i \in \mathbb{N}$ such that $K^i\phi$ is valid. Then we call a logical base $\mathcal{A}$ *full*, if, using the base, the tMEL logic, as well as the logics defined later, have the internalization property. Certainly a comprehensive base is full, but so is a base containing all the axiom instances of the schemes listed in the above system. Given the completeness result, this can be proved by induction on a proof of the valid formula, with only axioms A1 and A1′ applied.[9] We can understand agents with full or stronger logical bases as having enough basic logical truths from which they can have *complete* epistemic logical knowledge about themselves. Logics with full or stronger logical bases normally have nice properties, which include the following one concerning the formal relation between MEL logics and their tMEL counterparts. Here's the version for S4 and tS4:

Theorem 2.4 (Temporalization Theorem, Corollary 4.10 of Wang (2013)) *Given a full* tS4 *logical base $\mathcal{A}$, an* MEL *formula ϕ is a* S4 *valid formula if and only if there*

[8]If $\mathcal{A}$ is the limit of an ascending chain of tS4 logical bases $\{\mathcal{A}_i\}_{i\in\mathbb{N}}$, with $\mathcal{A}_{i+1}$ consisting of all tS4($\mathcal{A}_i$) valid formulas, then $\mathcal{A}$ is comprehensive. For the details, see Wang (2011).

[9]This is the reason we keep the condition **2** of awareness function for weaker tMEL logics.

is a suitable label for each of the epistemic modalities K in ϕ such that turning all the epistemic modalities in ϕ into modalities with their suitable labels K^i, we have a tS4($\mathcal{A}$) *valid formula.*[10]

3 Logic of tS4^{K}

The first logical framework we are going to introduce is rather a direct combination of MEL and tMEL. *The language for* tMEL^{K}, and hence *for* tS4^{K}, is extended from the language of tMEL by having the simple modal formulas of the form $K\phi$.

A tMEL^{K} base is a tuple $\mathcal{A} = \langle \mathbf{A}, \mathbf{f} \rangle$, with $\mathbf{A}$ a set of tMEL^{K} formulas, and $\mathbf{f}\colon \mathbf{A} \mapsto \mathbb{N}$. Given a tMEL^{K} base $\mathcal{A}$, $M = \langle W, R, R_t, \mathfrak{A}, \mathcal{V} \rangle$ is a tS4^{K}($\mathcal{A}$)*-structure*, if $R \subseteq R_t$, $\langle W, R, \mathcal{V} \rangle$ is an S4-structure, and $\langle W, R_t, \mathfrak{A}, \mathcal{V} \rangle$ is a tS4($\mathcal{A}$)-structure with the domains of the awareness functions α_w in $\mathfrak{A}$ consisting of tMEL^{K} formulas, instead of just tMEL formulas. The truth-values of the tMEL^{K} formulas for the modal cases are defined as follows:

$$(M, w) \models K\phi \quad \text{iff} \quad (M, w') \models \phi \text{ for all } w' \in W \text{ with } wRw',$$

and for a labelled modal formula:

$$(M, w) \models K^i\phi \quad \text{iff} \quad (M, w') \models \phi \text{ for all } w' \in W \text{ with } wR_tw', \text{ and } \alpha_w(\phi) \leq i.$$

Then a tS4^{K} *logical base* is recursively defined in the same way as the one for tS4, except that it is defined on the language of tMEL^{K}. Given a tS4^{K} logical base $\mathcal{A}$, the axiom system tS4^{K}($\mathcal{A}$) is the following:

I.

Axioms:

A0 Classical propositional axiom schemes,
A1 $K^i(\phi \to \psi) \to (K^j\phi \to K^k\psi) \quad i, j < k$,
A1′ $K^iA \to K^j(K^iA) \quad i < j$ if $A \in \mathbf{A}$ and $f(A) \leq i$,
A2 $K^i\phi \to K^j(K^i\phi) \quad i < j$,
A3 $K^i\phi \to \phi$,
A4 $K^i\phi \to K^j\phi \quad i < j \quad K^i\phi \to K^j\phi \quad i < j$,

Inference Rules

R1 if $\vdash \phi \to \psi$ and $\vdash \phi$, then $\vdash \psi$,
R2 if $A \in \mathbf{A}$ and $f(A) \leq i$, then $\vdash K^iA$, ($\mathcal{A}$*-Epistemization*)

[10] A constructive proof of this theorem can be found in Wang (2014).

II.

Axioms:

A0 classical propositional axiom schemes
A1 $K(\phi \to \psi) \to (K\phi \to K\psi)$,
A2 $K\phi \to K(K\phi)$,
A3 $K\phi \to \phi$,

Inference Rules

R1 if $\vdash \phi \to \psi$ and $\vdash \phi$, then $\vdash \psi$,
R2 if $\vdash \phi$, then $\vdash K\phi$, (*Necessitation Rule*)

III. (Connection Axiom)

A1 $K^i\phi \to K\phi$, for any i.

The system, as it is presented, is composed of three parts, which explicitly shows that it is a combination of the systems S4 and tS4, though some of the axioms and rules are listed repeatedly. The only exception is *Connection Axiom*, which reflects the idea that the agent can only reason to know what is determined by the static possible world setting. The semantic condition which corresponds to this axiom is that $R \subseteq R_t$. On the contrary, reversing the relation, we can't obtain any connection theorem. Observe that even though we can infer from $K\phi$ is true in a world that ϕ is true in all R_t-related worlds, given that R_t included in R, it is not guaranteed that the agent is aware of the truth of ϕ in the world.

The discussed condition can also be set to $R = R_t$, with the resulting semantics still validating the same set of formulas. But to keep them separated, we can introduce more of this type of tMEL-MEL combination logics, e.g. the tS4+S5 logic, whose semantics is the same as what is introduced here, except that R needs to satisfy the equivalence relation, and whose axiom system is the same as the one listed here but with an additional 5 axiom for the modality K. Although this type of logics are not the focus of this paper, but they have their own theoretical interest worth further examination.

Theorem 3.1 *Given a* $\mathsf{tS4}^{\mathsf{K}}$ *logical base* $\mathcal{A}$, *a* $\mathsf{tMEL}^{\mathsf{K}}$ *formula is* $\mathsf{tS4}^{\mathsf{K}}(\mathcal{A})$ *valid if and only if it is* $\mathsf{tS4}^{\mathsf{K}}(\mathcal{A})$ *provable.*

Proof The completeness results for these logics can be proved based on the standard canonical model method. We sketch the proof. For any $\mathsf{tS4}^{\mathsf{K}}(\mathcal{A})$ maximal consistent set Γ, let α_Γ be the partial function such that for any $\mathsf{tMEL}^{\mathsf{K}}$ formulas ϕ, $\alpha_\Gamma(\phi) = i$ if and only if $K^i\phi \in \Gamma$, and for $j<i$, $K^j\phi \notin \Gamma$, and $\Gamma^\# = \{\phi\ \Box\phi \in \Gamma\}$. Let $M = \langle W, R, R_t, \mathfrak{A}, \mathcal{V}\rangle$ be the canonical model, in which W is the set of all $\mathsf{tS4}^{\mathsf{K}}(\mathcal{A})$ maximal consistent sets, $R \subseteq W \times W$ such that $\Gamma R \Gamma'$ if and only if $\Gamma^\# \subseteq \Gamma'$, $R_t = R$, $\mathfrak{A} = \{\alpha_\Gamma\}$ for $\Gamma \in W$, and $\Gamma \in \mathcal{V}(p)$ if and only if $p \in \Gamma$, for a propositional letter p. Then it can be proved that this is a $\mathsf{tS4}^{\mathsf{K}}(\mathcal{A})$-structure and Truth Lemma holds for M. $\Box$

4 Logic of tS4$^\infty$

In this section we introduce a system extended from tS4 by having a device representing the reasoning-based concept of knowledge. This can be done in a similar way as we did above by introducing an epistemic modality K into tS4 and endowing it with the reason-based knowledge interpretation. But since in tS4 we have the *Monotonicity Axiom*: $K^i\phi \to K^j\phi$ for $i < j$, we will take advantage of it. If an agent won't forget what s/he knows, then to say the agent knows something *at some time* is the same as describing the agent as knowing that *eventually*. Hence, instead of introducing a new modality, we will extend the label set from natural numbers $\mathbb{N}$ to extended natural numbers $\mathbb{N}^\infty$ and regard formulas of the form $K^\infty\phi$ as the ones we expect. By doing so, as we will see, the axiom systems can be simplified.

Now, formally the language of tMEL$^\infty$ is extended from the language of tMEL by adopting the extended natural numbers $\mathbb{N}^\infty$ as the labels for the labelled modal formulas, where the order of ∞ is that $x<\infty$ for any $x \in \mathbb{N}$ and $\infty \leq \infty$. A tMEL$^\infty$ base is a tuple $\mathcal{A} = \langle \mathbf{A}, \mathbf{f} \rangle$, with $\mathbf{A}$ a set of tMEL$^\infty$ formulas, and $\mathbf{f} : \mathbf{A} \mapsto \mathbb{N}$ (noticing that it is $\mathbb{N}$, not $\mathbb{N}^\infty$). Given a tMEL$^\infty$ base $\mathcal{A}$, a tS4$^\infty$($\mathcal{A}$)-*structure* is a tS4 *structure* $M = \langle W, R, \mathfrak{A}, \mathcal{V} \rangle$, with the language of tMEL$^\infty$ as the domain of awareness functions $\alpha_w \in \mathfrak{A}$.

The conditions for the truth-values of tS4$^\infty$ formulas are exactly the same as the conditions for those of tS4 formulas, including the following:

$$(M, w) \models K^i\phi \text{ iff } (M, w') \models \phi \text{ for all } w' \in W \text{ with } wRw', \text{ and } \alpha_w(\phi) \leq i.$$

The labelled modal formula $K^\infty\phi$ gets its truth-value through this condition, and notice that the range of the awareness functions for the tMEL$^\infty$ semantics is the same as the range of the functions for the tMEL semantics, which is $\mathbb{N}$, and hence whenever $K^\infty\phi$ is true at some world then there is a $K^i\phi$ true at the world with $i \in \mathbb{N}$.

A tS4$^\infty$ *logical base* is recursively defined in the same way as the one for tS4, except that it is defined on the language of tS4$^\infty$. Given a tS4$^\infty$ logical base $\mathcal{A}$, the axiom system of tS4$^\infty$($\mathcal{A}$) is the following:

Axioms:

A0 Classical propositional axiom schemes,

A1 $K^i(\phi \to \psi) \to (K^j\phi \to K^k\psi) \quad i, j < k$,
A1′ $K^iA \to K^j(K^iA) \quad i < j$ if $A \in \mathbf{A}$ and $f(A) \leq i$,
A2 $K^i\phi \to K^j(K^i\phi) \quad i < j$,
A3 $K^i\phi \to \phi$,
A4 $K^i\phi \to K^j\phi \quad i < j$, (*Monotonicity Axiom*)

Inference Rules

R1 if $\vdash \phi \to \psi$ and $\vdash \phi$, then $\vdash \psi$,
R2 if $A \in \mathbf{A}$ and $f(A) \leq i$, then $\vdash K^iA$,

for all $i, j, k \in \mathbb{N}^\infty$.

These systems are almost the same as the axiom systems of tS4, except that the label variables i, j, k range over $\mathbb{N}^\infty$ instead of $\mathbb{N}$. But exactly because of this exception, more axioms are permitted by this system. For example, $K^5(\phi \to \psi) \to (K^7\phi \to K^\infty\psi)$ and $K^\infty(\phi \to \psi) \to (K^\infty\phi \to K^\infty\psi)$ are both instances of the axiom A1, and, corresponding to Connection Axiom in the $\mathsf{tS4}^K$ system, $K^i\phi \to K^\infty\phi$ is an instance of A4 axiom.

To proceed to prove the completeness theorem for $\mathsf{tS4}^\infty$ logics, one should notice that the compactness theorem does not hold for these systems, and hence the canonical model method can't be used. Consider the following sequence of formulas $\sim K^0\phi, \sim K^1\phi, \ldots, K^\infty\phi$. This is a consistent sequence and finitely satisfiable but not satisfiable, for if $K^\infty\phi$ is true at a world of a structure, then there must be some formulas $K^i\phi$ with finite i also true at the world. So alternatively we will prove this theorem by constructing a structure based on the set of relative maximal consistent sets. And it turns out, as you can see, the method is general enough to be adapted to give alternative proofs for the completeness theorem of tS4 logics.

Theorem 4.1 *Given a* $\mathsf{tS4}^\infty$ *logical base* $\mathcal{A}$, *a* $\mathsf{tS4}^\infty$ *formula is* $\mathsf{tS4}^\infty(\mathcal{A})$ *valid if and only if it is* $\mathsf{tS4}^\infty$ *provable.*

Proof For a tMEL^∞ formula ϕ, let $\mathrm{Sub}^+(\phi)$ be the set of all subformulas of ϕ and their negations. Given a $\mathsf{tS4}^\infty$ logical base $\mathcal{A} = \langle \mathbf{A}, \mathbf{f}\rangle$ and a tMEL formula ϕ, we say a set $\Gamma \subseteq \mathrm{Sub}^+(\phi)$ is a $\mathsf{tS4}^\infty(\mathcal{A})$ maximal consistent set relative to ϕ if 1) it is consistent, that is, from Γ we won't prove contradiction, and 2) for any formula ψ in $\mathrm{Sub}^+(\phi)$ but not in Γ, $\Gamma \cup \{\psi\}$ is inconsistent. For a set of tMEL^∞ formulas Γ, we define $\Gamma^\sharp = \{\phi \mid K^i\phi \in \Gamma, \text{and } i \in \mathbb{N}^\infty\}$, $\Gamma^\sharp_{\mathrm{Fin}} = \{\phi \mid K^i\phi \in \Gamma, \text{and } i \in \mathbb{N}\}$, and $\kappa_\Gamma = \max\{i \in \mathbb{N} \mid \sim K^i\phi \in \Gamma\}$. Now we inductively create a set of tMEL^∞ formulas U relative to Γ, stage by stage. At stage 0, $\psi \in U$ if $\psi \in \mathbf{A}$ and $\mathbf{f}(A) = 0$ or $K^0\psi \in \Gamma$. At the stage $n \neq 0$, we put formula ψ into U, if ψ is not in U and satisfies one of the following conditions:

1. if $\psi = A \in \mathbf{A}, \mathbf{f}(A) = n$,
2. if there are formulas $\theta \to \psi, \theta$ in U,
3. if $\psi = K^{n-1}A$, where $A \in \mathbf{A}, \mathbf{f}(A) \leq n - 1$,
4. if $\psi = K^{n-1}\phi$, where $\phi \in U$,
5. if ψ is in $\Gamma^\sharp_{\mathrm{Fin}}$, and $n = \min\{i \mid K^i\psi \in \Gamma\}$,
6. if ψ is in $\Gamma^\sharp$ but not in $\Gamma^\sharp_{\mathrm{Fin}}$, and $n = \kappa_\Gamma + 1$.

Let W be the collection of relative maximal consistent sets, R be the binary relation such that for any Γ, Γ' in W, $\Gamma R \Gamma'$ if and only if $\Gamma^\sharp \subseteq \Gamma'$, $\mathcal{V}$ maps propositional letters to subsets of W such that $\Gamma \in \mathcal{V}(p)$ if and only if $p \in \Gamma$ for a propositional letter p, and $\alpha_\Gamma(\phi) = i$ if and only if ϕ is added to the set U_Γ at the ith stage, and let $M = \langle W, R, \mathfrak{A}, \mathcal{V}\rangle$. Then it can be checked that this is a $\mathsf{tS4}^\infty(\mathcal{A})$ -structure and the Truth Lemma: for every formula $\psi \in \mathrm{Sub}^+(\phi)$, $M, \Gamma \models \psi$ if and only if $\psi \in \Gamma$, holds. Now if ϕ is consistent then ϕ will belong to some relative maximal consistent set and hence it is satisfiable. We leave the reader to check that M is a $\mathsf{tS4}^\infty(\mathcal{A})$-structure. Here we prove the Truth Lemma for the case of labelled modal formulas.

Suppose $K^i\psi \in \text{Sub}^+(\phi)$ and $M, \Gamma \models K^i\psi$ for $i \in \mathbb{N}$, then $\psi \in U_\Gamma$ at a stage earlier than or equal to i. But if $\sim K^i\psi \in \Gamma$, then ψ can't be added into U_Γ before or equal to i, a contradiction, so $K^i\psi \in \Gamma$. The case for $K^\infty\psi$ is similar, since $M, \Gamma \models K^\infty\psi$, then $M, \Gamma \models K^i\psi$ for some i. For the other direction, suppose $K^i\psi$ is in Γ for $i \in \mathbb{N}^\infty$, then ψ is in Γ' for all $\Gamma R \Gamma'$, and $\alpha_\Gamma(\psi) \leq i$, for $i \in \mathbb{N}$, and $\alpha_\Gamma(\psi) \leq j$ for some $j \in \mathbb{N}$ when $i = \infty$, so $M, w \models K^i\psi$. □

5 Comparisons

Now we undertake to compare the logics $\mathsf{tS4}^K$ and $\mathsf{tS4}^\infty$. We will compare them in the way of putting formulas of the form $K\phi$ in $\mathsf{tS4}^K$ and $K^\infty\phi$ in $\mathsf{tS4}^\infty$ on a par,[11] since they are the formulas introduced to represent information-based knowledge and reasoning-based knowledge respectively. The main difference we try to capture is that for an $\mathsf{tS4}^\infty$ agent, whether $K^\infty\phi$ is true to her/him or not completely depend on the reasoning ability of the agent, i.e. in a formal term, the logical base of the system describing the epistemic strength of the agent. But this is not always the case for $K\phi$ in $\mathsf{tS4}^K$.

Lemma 5.1 *For any* $\mathsf{tS4}^\infty$ *logical base* $\mathcal{A}$, *if formula* $K^\infty\phi$ *is* $\mathsf{tS4}^\infty(\mathcal{A})$ *valid, then there is an* $i \in \mathbb{N}$ *such that* $K^i\phi$ *is* $\mathsf{tS4}^\infty(\mathcal{A})$ *valid.*

Proof Given a $\mathsf{tS4}^\infty$ logical base $\mathcal{A}$, we can create a $\mathsf{tS4}^\infty(\mathcal{A})$ awareness function $\alpha^*_\mathcal{A}$ such that for any $\mathsf{tS4}^\infty(\mathcal{A})$ awareness function α, if $\alpha^*_\mathcal{A}(\phi)\downarrow$ then $\alpha(\phi) \leq \alpha^*_\mathcal{A}(\phi)$. The construction of $\alpha^*_\mathcal{A}$ is similar to the construction of α_Γ in the above proof. We will inductively create a set of tMEL^∞ formulas U stage by stage. At stage 0, $\psi \in U$ if $\psi \in \mathbf{A}$ and $\mathbf{f}(A) = 0$. At the stage $n \neq 0$, we put formula ψ into U, if ψ has not been in U yet and satisfies one of the following conditions:

1. if $\psi = A \in \mathbf{A}$, $\mathbf{f}(A) = n$.
2. if there are formulas $\theta \to \psi, \theta$ in U,
3. if $\psi = K^{n-1}A$, where $A \in \mathbf{A}$, $\mathbf{f}(A) \leq n-1$.
4. if $\psi = K^{n-1}\phi$, where $\phi \in U$.

Let $\alpha^*_\mathcal{A}$ to be the partial function such that for any tMEL^∞ formula ϕ, $\alpha^*_\mathcal{A}(\phi)\downarrow$ if $\phi \in U$, and $\alpha^*_\mathcal{A}(\phi) = i$ if ϕ is added into U at the stage i. Then it is not difficult to show that this awareness function satisfies our requirement, including that if $\alpha^*_\mathcal{A}(\phi)\downarrow$ then $\alpha(\phi) \leq \alpha^*_\mathcal{A}(\phi)$, for any $\mathsf{tS4}^\infty(\mathcal{A})$ awareness function α. Now suppose $K^\infty\phi$ is $\mathsf{tS4}^\infty(\mathcal{A})$ valid, then $\alpha^*_\mathcal{A}(\phi)$ must be defined. So $K^i\phi$ is $\mathsf{tS4}^\infty(\mathcal{A})$ valid, if $\alpha^*_\mathcal{A}(\phi) \leq i$. □

But this is not generally true for $\mathsf{tS4}^K$ logics.

[11] This is kind of a misleading way to put it, since ϕ in $K\phi$ as a formula in $\mathsf{tS4}^K$, and ϕ in $K^\infty\phi$ as a formula in $\mathsf{tS4}^\infty$ cannot always mean the same formula. But for convenience, and for comparison, we will still write them as such, and readers should remember they do not have to mean the same formula.

Lemma 5.2 *There is a* $\mathsf{tS4}^K$ *logical base* $\mathcal{A}$*, such that* $K\phi$ *is* $\mathsf{tS4}^K(\mathcal{A})$ *valid, but no* $i \in \mathbb{N}$ *such that* $K^i\phi$ *is* $\mathsf{tS4}^K(\mathcal{A})$ *valid.*

Proof Consider the extreme case to take $\mathcal{A}$ as the empty set. Then for a classical tautology ϕ, $K\phi$ is $\mathsf{tS4}^K(\mathcal{A})$ valid, but no formula of the form $K^i\phi$ is logical valid. □

So in $\mathsf{tS4}^K$, the formulas $K\phi$ like the final goals, waiting for the agent who has enough reasoning ability to reach. In the literature, it is sometime called the *implicit knowledge* of the agent Levesque (1984). But on the other hand, the concept of knowledge captured by $K^\infty\phi$ is, with any assumption of the reasoning strength of the agent, what is actually known by the agent, but *with the omission of the exact time of knowing*. Of course if we assume that the agent has enough reasoning ability, then things will become different. Consider the situation within $\mathsf{tS4}^K$ logics:

Lemma 5.3 *Given a full* $\mathsf{tS4}^K$ *logical base* $\mathcal{A}$*, such that* $K\varphi$ *is* $\mathsf{tS4}^K(\mathcal{A})$ *valid, then there is an* $i \in \mathbb{N}$ *such that* $K^i\phi$ *is* $\mathsf{tS4}^K(\mathcal{A})$ *valid.*

Proof Since $K\phi$ is $\mathsf{tS4}^K(\mathcal{A})$ valid, so is ϕ. Then since $\mathcal{A}$ is full, there is an $i \in \mathbb{N}$ such that $K^i\phi$ is $\mathsf{tS4}^K(\mathcal{A})$ valid. □

Now consider the situation between logics. If in $\mathsf{tS4}^\infty(\mathcal{A})$, $\mathcal{A}$ is supposed to be full, then in a sense $K^\infty\phi$ is no difference with the $K\phi$ in $\mathsf{tS4}^K(\mathcal{A})$, and also the $K\phi$ in $\mathsf{S4}$. Let $(\,)^\infty$ be the function mapping tMEL^K formulas to tMEL^∞ formulas by the following recursive conditions:

$(p)^\infty = p$ if p is a propositional variable,
$(\sim\phi)^\infty = \sim(\phi)^\infty$,
$(\phi \to \psi)^\infty = (\phi)^\circ \to (\psi)^\infty$,
$(K^i\phi)^\infty = K^\infty(\phi)^\infty$ for any an $i \in \mathbb{N}$,
$(K\phi)^\infty = K^\infty(\phi)^\infty$.

Theorem 5.4 *Let* ϕ *be an* MEL *formula,* $\mathcal{A}$ *a full* $\mathsf{tS4}^\infty$ *logical base, and* $\mathcal{B}$ *any* $\mathsf{tS4}^K$ *logical base.*
(1) ϕ *is* $\mathsf{S4}$ *valid if and only if* ϕ *is* $\mathsf{tS4}^K(\mathcal{B})$ *valid;*
(2) ϕ *is* $\mathsf{tS4}^K(\mathcal{B})$ *valid if and only if* $(\phi)^\infty$ *is* $\mathsf{tS4}^\infty(\mathcal{A})$ *valid.*

Proof The easiest way to prove it is by comparing the proofs in the different proof systems. Here we demonstrate the direction from left to right of (2). Suppose ϕ has a $\mathsf{tS4}^K(\mathcal{B})$ proof, then it is easy to check by induction that for every ψ in the proof, $(\psi)^\infty$ is either a $\mathsf{tS4}^\infty(\mathcal{A})$ axiom or $\mathsf{tS4}^\infty(\mathcal{A})$ theorem. For the case that either $K\psi$ is derived by necessitation, or $K^j\psi$ is derived by $\mathcal{B}$-Epistemization, from ψ in $\mathsf{tS4}^K(\mathcal{B})$, since $(\psi)^\infty$ is provable in $\mathsf{tS4}^\infty(\mathcal{A})$, and $\mathcal{A}$ a full logical base, so there is an $i \in \mathbb{N}$ such that $K^i(\psi)^\infty$ is provable, and hence $(K\psi)^\infty$, or $(K^j\psi)^\infty$, is $\mathsf{tS4}^\infty(\mathcal{A})$ provable. □

6 Some Discussions

In this final section, I will use an example to demonstrate that tMEL^{∞} could give us insights into problems that could not be brought about by epistemic logics plainly equipped with the possible world semantics. But before that, we will deviate from the formal spirit of this paper to discuss a more pragmatic issue with regard to the approach of our analysis of knowledge. As what has been said in the beginning of the paper, a knowledge event is temporal, hence moving out of the logical territory to decide the truth-value of an utterance of a knowledge claim, we need to determine what's the time the event denoted by the knowledge claim is supposed to take place. That is, when people say "*a knows that* ...," or "*a does not know that* ...," to determine the truth-values of these utterances, we need to, first, from the context, determine when is the time of the event that the utterer refers to. Normally it is the time of the utterance, but that's not mandatory; and for the purpose of discussion, we don't take it for granted.[12] The pragmatic issue we are going to discuss is Moore's paradox.

6.1 *Moore's Paradox*

Moore's paradox has attracting many discussions. An anthology has been published on the issue Green and Williams (2007). Moore has an opinion on it, so is Wittgenstein, and Wittgenstein even claims that the discovery of the paradox is the most important philosophical contribution of Moore Malcolm (1984). So if someone says to us that "*it's raining, and I don't know it's raining*," we sense some kind of absurdity.[13] But what is said is not against any semantic rules, and it is quite possible to be true. Then what's the source of the sense of absurdity? Roughly speaking, Moore thought it's our asserting implying our knowledge about what is asserted, and the one who utters the Moore's sentence makes what is implied by the assertion, knowing that it's raining, go against what is asserted, not knowing that it's raining Moore (1942); Baldwin (1993). On the other hand, Wittgenstein thought the sense of absurdity indicates that in everyday practice, our first personal claim about our own inner state of mind is a way of speaking about something outside us. Thus the second part of an utterance of a Moorean sentence is saying something of the world, *it's not raining*, which is against what is said in the first part of the sentence, *it's raining* Wittgenstein (1953, 1980).

[12]Of course, in English, the verb form of a sentence indicates its tense. So no problem to judge if the utterer is saying something in English about, e.g., the past. But for the purpose of argument, one could suppose the utterer speaks in some language in which the tense is not built into the grammar of sentences, such as Chinese. Our conclusion does not rely on this supposition.

[13]The most discussed Moore's paradox is the belief version that "*it's raining, and I don't believe it's raining*," but Moore himself did discuss the knowledge version, which is in Moore (1962), published after his death.

I only very briefly demonstrate their ideas, which might miss some refined points, and it is properly better to understand these analyses as being about belief, rather than knowledge.[14] But my object here is to give the idea of what a possible answer to the paradox could be. There is no judgement about whose analysis is better, and my personally attitude to a paradox as interesting as this, is that we can get different insights from our different analyses about it; and in this paper I try to offer another one.

So to determine the truth-value of an utterance such as that *"p and I don't know that p,"* we have to determine what's the time of the non-knowing event that the utterer refers to. It has long been known, if what the utterer intends to say is *"p and I didn't know that p* in the past," i.e., the intended time of the part of the no-knowledge claim of the utterance is earlier than the time of the utterance, then there is no absurdity in the case.[15] It's a normal utterance, with the content that could be true and could be false. Now what if what the utterer tries to say is about the future, such as *"p and I will not know that p* forever"? In literature, as far as I know, there is no discussion about this case yet, but I think there is no suspense in this one – it is simply false. There is no sense of absurdity arising either, if this is what the utterer means. The difference is that this time the utterer just says something false. But we can continue to ponder what if what the utter tries to say is, within the Moorean utterance, that *"I won't know p* tomorrow"? "six hours later"? or "10 min later"? Does the utterer say something absurd? I would say "no" to these cases. Then there must be a case in which the content of the utterance is false, and the intended time of the non-knowing claim is the earliest among such cases. Let's call that time as the *t-time, the turning time*, of the Moorean utterance. We would say the *t*-time in our raining example is, of course, not in the past, but very close to the time of the utterance. My claim is that it is this *closeness* that is the source of absurdity. Back to the normal circumstance. If no further information is provided or needed to be concerned, a Moorean utterance is normally interpreted as meaning that *"it's raining, and I don't know it's raining* now"; but the closeness makes the truth-value of the content of the utterance in a shaky place.

To make my point clearer, consider the following example. If what the utterer says is that *"p and I don't know q,"* with q a logical consequence of p but very far away in the sense that it takes a long derivation to get q from p. Then do we still sense any absurdity? To judge that the utterer says something definitely false, we probably understand what the utterer tries to mean is that *"p and I don't know q forever."* The t-time for this utterance is far in the future. Hence for most cases we will just understand the utterance as a normal utterance. On the other hand, consider this: *"I don't know what the weather is, but as a matter of fact it is raining."* Absurd! So if my analysis is correct, then the source of the sense of the absurdity of the utterance

[14] Since our main results here focus on knowledge, I try to keep all the discussions uniformly about knowledge. But it needs not to be so. Hence if you like, you could change the discussions in this part to be about belief. Of course knowledge and belief have some essential differences which could make the discussions of knowledge version and belief version of Moore's paradox go in different directions. However, as you can see, our analysis here is not dependent on these differences.

[15] See Baldwin (1993), Hintikka (1962).

of Moorean sentence is in a way just like the absurdity we sense when we sincerely judge someone says "now *the time is ...,*" or simply *"the time is ...,"* to be very specific to the point of second. Whenever it is said, the truth-value of it is hard to tell. Whenever the utterance is completed, it is almost a false utterance.

6.2 Moorean Sentences

A Moorean sentence not only causes problems when it is uttered; it also generates puzzlement when put into an epistemic logic system that concerns knowledge of the sentence. It turns out that a Moorean sentence, which is not a contradiction, can't be consistently known. That is, $\sim K(p \,\&\!\sim\! Kp)$ is logically valid. Why after all we can't know a non-contradictory sentence? What prevents us from knowing it? Researchers have not yet tried to figure out why this is the case, but instead have directly utilized it in other philosophical programs. Hintikka takes it as the source of the absurdity of Moore's paradox Hintikka (1962), contending that it's a pragmatic norm not to utter something that couldn't be consistently known, and Fitch uses it to derive the knowability paradox, arguing that innocent verification principle leads to that all truth are known Fitch (1963), Salerno (2009). I think the reason why the Moorean sentences can't be consistently known haven't been serious discussed is due to lack of machinery. Here I will use tMEL^{∞}, trying to figure out what's going on out there.

First, it is possible to know Moorean types of sentences (in the following i, j are natural numbers):

Lemma 6.1 *There is a* $\mathsf{tS4}^{\infty}$ *logical base* $\mathcal{A}$ *weak enough such that* $K^i(p \,\&\!\sim\! K^j p)$ *is* $\mathsf{tS4}^{\infty}(\mathcal{A})$ *satisfiable.*[16]

Proof Consider the empty base $\mathcal{A}$. But actually any logical base with which the agent could not simplify a conjunction can do. Then we can create a set of U just like the one in the proof of Lemma 5.1, but adding $p \,\&\!\sim\! K^j p$ for some j to the set at some stage. Notice that since $\mathcal{A}$ is so weak that p won't be in U. Let M be a single world $\mathsf{tS4}^{\infty}(\mathcal{A})$-structure where the $\mathcal{A}$ awareness function at the world is the one created from U, and p is true at the world. $(M, w) \vDash p \,\&\!\sim\! K^j p$, and hence $(M, w) \vDash K^i(p \,\&\!\sim\! K^j p)$ for some i. □

But suppose the agent is smart enough to be able to simplify an conjunction. Then still the agent can know some Moorean sentences. Let's call a $\mathsf{tS4}^{\infty}$ logical base $\mathcal{A}$ such that for every $\mathcal{A}$ awareness function α, if $\alpha(\phi \,\&\, \psi) = n$, then $\alpha(\phi) = m$ for some m with $n < m$ a *smart enough* logical base.

Lemma 6.2 *For any smart enough* $\mathsf{tS4}^{\infty}$ *logical base* $\mathcal{A}$*, there are* i, j *such that* $K^i(p \,\&\!\sim\! K^j p)$ *is* $\mathsf{tS4}^{\infty}(\mathcal{A})$ *satisfiable.*

[16]Just consider & as a derived connective in our system.

Proof Again define U and the structure M just like the way they are defined in the above lemma. But now p will be in U at some stage, say m later than the stage, say n, to which the formula $p\,\&{\sim}K^j p$ is added. Now even so, $(M, w) \models p\,\&{\sim}K^j p$ for $j \le n$, hence $(M, w) \models K^i(p\,\&{\sim}K^j p)$, for $i \ge m$ be larger or equal to the stage number in which $p\,\&{\sim}K^j p$ is added to U. □

Now is a result of inconsistent knowledge.

Lemma 6.3 *For any smart enough* tS4$^\infty$ *logical base $\mathcal{A}$, $K^i(p\,\&{\sim}K^\infty p)$ is not* tS4$^\infty(\mathcal{A})$ *satisfiable.*

Proof Let $M = \langle W, R, \mathfrak{A}, \mathcal{V}\rangle$ be the tS4$^\infty(\mathcal{A})$-structure such that for some $w \in W$ $(M, w) \models K^i(p\,\&{\sim}K^\infty p)$. Then $\alpha_w(p\,\&{\sim}K^\infty p) \le i$ for the $\mathcal{A}$ awareness function α_w. But then $\alpha_w(p){\downarrow}$, and $p\,\&{\sim}K^\infty p$ is not satisfiable at w. Neither is $K^i(p\,\&{\sim}K^\infty p)$. □

Here is a general result concerning tS4$^\infty$.

Lemma 6.4 *For any* tS4$^\infty$ *logical base $\mathcal{A}$, if for any i, ${\sim}K^i\phi$ is* tS4$^\infty(\mathcal{A})$ *valid, then ${\sim}K^\infty\phi$ is* tS4$^\infty(\mathcal{A})$ *valid.*

From these lemmas, we can derive that ${\sim}K^\infty(p\,\&{\sim}K^\infty p)$ is tS4$^\infty(\mathcal{A})$ valid for smart enough logical bases $\mathcal{A}$.

In all the arguments given here, we don't really need the possible world machinery but mostly play with the awareness functions. So in a way, the statement is valid because of the relation of the labels. It is because the label of the first epistemic modality (∞ on the left) of the sentence is smaller than the label of the second modality (∞ on the right); or we can consider ${\sim}K^\infty(p\,\&{\sim}K^\infty p)$ is the limit of the sequences of formulas ${\sim}K^i(p\,\&{\sim}K^\infty p)$. In either way, it means at any time we could not correctly predict that something is true but we could not know it in the distant future. A meaningful statement.

Some more words on the issue. When we set a limitation on the domain of epistemic logic such that the basic events, events represented by propositional letters, are general events, whose truth-values would not change over time, it simplifies things a lot. But even so, when we take seriously that knowledge will evolve, the truth-values of knowledge sentences vary with time; so things couldn't be too simplified. Consider a higher-order knowledge sentence $K\phi$ in which ϕ contains other knowledge sentences. Then the variation of $K\phi$ in the truth-value as the time goes by depends not only on the agent's change of her/his state of mind, but also on the variations of the truth-values of ϕ. Now regarding the knowledge of Moorean sentences, which is certainly higher-order knowledge, in the course of the agent's reasoning, we have different cases to examine. Suppose p is false, $p\,\&{\sim}Kp$ can't be true and hence ${\sim}K(p\,\&{\sim}Kp)$ is true of the agent under discussion. Then suppose p is true, and in the beginning of the reasoning the agent doesn't know p and doesn't know the Moorean sentence $p\,\&{\sim}Kp$, we have the following three cases of what's happening in the course of the agent's reasoning:

Case 1. The agent doesn't learn the truth of p, and doesn't learn that $p \& {\sim} Kp$, which will be true in the whole course of reasoning.

So in the whole course of reasoning the agent doesn't know that $p \& {\sim} Kp$.

Case 2. Suppose the agent at some point learns that p is true, but not that $p \& {\sim} Kp$, even if it is true.

Then before the point, $p \& {\sim} Kp$ is true, but since the agent doesn't learn this, so ${\sim} K(p \& {\sim} Kp)$. After that point, since $p \& {\sim} Kp$ is false, so again ${\sim} K(p \& {\sim} Kp)$. Thus at the end, the agent doesn't know that $p \& {\sim} Kp$.

Case 3. Suppose the agent at some point learns that $p \& {\sim} Kp$ but doesn't learn that p at the point yet.

Before this point, ${\sim} K(p \& {\sim} Kp)$, and just right after this point the agent knows that $p \& {\sim} Kp$. However, assume the agent is smart enough. Then p will be known at some other point later. Then before this latter point, still $K(p \& {\sim} Kp)$, but after that, since ${\sim} Kp$ is false, so at the end, the agent doesn't know that $p \& {\sim} Kp$.

Remember that in the argument of the invalidity of knowing a Moorean sentence with the possible world semantics, we argue that since if $(M, w) \models K(p \& {\sim} Kp)$ for some structure and some world, then $(M, w') \models p \& {\sim} Kp$ for all accessible worlds w' from w, so it is impossible $(M, w') \models {\sim} Kp$, given that p is true in all the accessible worlds. However, even given that p is true, there are still cases to consider about the conditions of the agent's reasoning.

In all these cases we discussed above, agents don't know that $p \& {\sim} Kp$ at the end, but they don't know it for different reasons. Furthermore, in Case 3, the agent indeed knows the Moorean sentence for some time. Also notice that a sentence of the form $p \& {\sim} Kq$, where q is a logical consequence of p, can't be consistently known either. The above analysis can be applied to this sentences as well. Consider Case 3. Then even though eventually the agent doesn't know that $p \& {\sim} Kq$, the agent can know this for quite some time in the course of the reasoning; the magnitude of the period of time depends on how long for the agent to reason in order to know q after the agent learns that p is the case. This is where we know Moorean sentences, which are not self-falsified.

The technical details in this paper is not that dramatic. But the goal of this paper is to bring out the idea that a logical analysis of knowledge should include the temporal aspects that reflects the amount of time that an agent needs in order to deduce to have new knowledge. For this idea, we introduce systems of epistemic logic of this kind, and show you with examples that this analysis can shed some light on some old riddles and puzzles in epistemic reasoning.

References

Artemov, S., & Fitting, M. (2015). Justification logic. In E. N. Zalta (Ed.), *The Stanford encyclopedia of philosophy*. Fall 2015. http://plato.stanford.edu/entries/behaviorism/.

Artemov, S. N. (2001). Explicit provability and constructive semantics. *Bulletin of Symbolic logic*, 1–36.

Baldwin, T. (1993). *G.E. Moore: selected writings*. Routledge.

Fagin, R., & Halpern, J. Y. (1987). Belief, awareness, and limited reasoning. *Artificial Intelligence*, *34*(1), 39–76.

Fagin, R., Halpern, J. Y., Moses, Y., & Vardi, M. Y. (1995). *Reasoning about knowledge*. Cambridge: MIT Press.

Fitch, F. (1963). A logical analysis of some value concepts. *Journal of Symbolic Logic*, *28*(2), 135–142.

Green, M. S., & Williams, J. (2007). *Moore's paradox: new essays on belief, rationality, and the first person*. Oxford: Oxford University Press.

Hintikka, J. (1962). *Knowledge and belief: An introduction to the logic of the two notions*. New York: Cornell University Press.

Hintikka, J. (1968). Epistemic logic and the methods of philosophical analysis. *Australasian Journal of Philosophy*, *46*(1), 37–51.

Hintikka, J. (1986). Reasoning about knowledge in philosophy: The paradigm of epistemic logic. In *Proceedings of the 1986 Conference on Theoretical aspects of reasoning about knowledge* (pp. 63–80). Monterey: Morgan Kaufmann Publishers Inc.

Levesque, H. J. (1984). A logic of implicit and explicit belief. In *Proceedings of the National Conference on Artificial Intelligence* (pp. 198–202). Austin.

Malcolm, N. (1984). *Ludwig wittgenstein: A memoir*. New York: Oxford University Press.

Meyer, J.-J. C., Van Der Hoek, W. (2004). *Epistemic logic for AI and computer science* (Vol. 41). Cambridge: Cambridge University Press.

Moore, G. E. (1942). *A reply to my critics* (pp. 535–677). Evanston: Northwestern University Press.

Moore, G. E. (1962). *The commonplace book: 1919–1953*. London: Allen & Unwin.

Salerno, J. (2009). *New essays on the knowability paradox*. USA: Oxford University Press.

Wang, R.-J. (2011). Knowledge, time, and the problem of logical omniscience. *Fundamenta Informaticae*, *106*(2–4), 321–338.

Wang, R.-J. (2013). Temporalizing modal epistemic logic. In *International symposium on logical foundations of computer science* (pp. 359–371). Berlin: Springer.

Wang, R.-J. (2014). Non-circular proofs and proof realization in modal logic. *Annals of Pure and Applied Logic*, *165*(7), 1318–1338.

Wittgenstein, L. (1953). *Philosophical investigations*. Oxford: B. Blackwell.

Wittgenstein, L. (1980). *Remarks on the philosophy of psychology* (Vols. 1 and 2).

Wright, G. H. (1951). *An essay in modal logic* (Vol. 5). North-Holland Publishing Company.

Proof-Theoretic Embedding from Visser's Basic Propositional Logic to Modal Logic K4 via Non-labelled Sequent Calculi

Sakiko Yamasaki and Katsuhiko Sano

Abstract Albert Visser introduced a subintuitionistic logic called Basic Propositional Logic (**BPL**) by dropping the requirement of reflexivity from Kripke semantics for intuitionistic logic, and he showed that **BPL** can be embedded into modal logic **K4** by a semantic method. This paper provides a contraction-free non-labelled sequent calculus for **BPL** and shows that the calculus enjoys the admissibility of cut. Moreover, we establish a proof-theoretic embedding from **BPL** into **K4** via a Gödel–McKinsey–Tarski translation.

1 Introduction

Intuitionistic logic (**Int**) is semantically complete for Kripke models satisfying the persistency condition of a valuation and the frame properties of reflexivity and transitivity. The persistency condition means that a formula remains true once it has been established to be true. By dropping the property of reflexivity from Kripke models for **Int**, we can obtain a logic called *basic propositional logic* (**BPL**), which is introduced by Visser (1981). The syntax of **BPL** is the same as intuitionistic logic (we use P, Q, etc. for atomic variables). Neither $P\&(P \supset Q) \supset Q$ nor $(P \supset (P \supset Q)) \supset (P \supset Q)$ are theorems of **BPL**, but these formulas are theorems of intuitionistic logic. Hence the set of theorems of **BPL** is a proper subset of that of **Int**. In this sense, **BPL** can be regarded as a *subintuitionistic logic* (see Restall 1994 for more details). Since the implication of **BPL** is weaker than that of **Int**, **BPL** enables us to study the more basic property of implication than intuitionistic logic. In Ruitenburg (1991, 1993), **BPL** is employed as a basis for naive set theory, because it is considered as a system with a non-circular interpretation of connectives more than classical logic and **Int**.

S. Yamasaki (✉)
Graduate School of Humanities, Tokyo Metropolitan University, Tokyo, Japan
e-mail: megumegu.world8008@gmail.com

K. Sano
Graduate School of Letters, Hokkaido University, Hokkaido, Japan
e-mail: v-sano@let.hokudai.ac.jp

S.C.-M. Yang et al. (eds.), *Philosophical Logic: Current Trends in Asia*,
Logic in Asia: Studia Logica Library, DOI 10.1007/978-981-10-6355-8_12

There are many varieties of proof system for **BPL**. Visser introduced a natural deduction system for **BPL**. Hilbert-style axiomatizations were proposed in Suzuki and Ono (1997); Suzuki et al. (1998); Kikuchi (2001). Gentzen-style sequent calculi were also given in Ardeshir (1995); Ishii et al. (2001); Yamasaki and Sano (2016). Ardeshir (1995) provided a cut-free sequent calculus for **BPL**, though the calculus does not enjoy the subformula property. In Ishii et al. (2001), a cut-free sequent calculus for **BPL** was presented, and their system enjoys the subformula property. Yamasaki and Sano (2016) introduced a cut-free **G3**-style labelled sequent calculus for **BPL** by generalizing the labelled sequent system in Dyckhoff and Negri (2012). At the same time, they also presented labelled sequent systems for logics of strict implication which cover weak logics with strict implication (Corsi 1987) and intermediate logics (Dyckhoff and Negri 2012). As far as the authors know, there seems no contraction-free and non-labelled sequent calculus for **BPL** such that it enjoys the subformula property.

It is well-known that **Int** can be embedded into modal logic **S4** by the following Gödel-McKinsey-Tarski translation sending a formula of **Int** to a formula of modal logic **S4**:

$$\begin{aligned} P^{\Box} &:= \Box P \\ \bot^{\Box} &:= \bot \\ (A \& B)^{\Box} &:= A^{\Box} \& B^{\Box} \\ (A \vee B)^{\Box} &:= A^{\Box} \vee B^{\Box} \\ (A \supset B)^{\Box} &:= \Box(A^{\Box} \supset B^{\Box}). \end{aligned}$$

By this translation, we can establish the following:

A is a theorem of **Int** if, and only if, $A^{\Box}$ is a theorem of **S4**.

The left-to-right direction is shown by Gödel (1933) with a proof-theoretic method. Here, a proof-theoretic method means that we can effectively construct a derivation of $A^{\Box}$ in **S4** from a derivation of A in **Int**. In Gödel (1933), Gödel conjectured that the right-to-left direction (*faithfulness*) also holds. In 1948 McKinsey and Tarski (1948) completed the direction of faithfulness with a semantic method. That is, their proof does not provide an effective procedure for rewriting a derivation of $A^{\Box}$ in **S4** into a corresponding derivation of A in **Int**. Although McKinsey and Tarski gave the proof of faithfulness for embedding of **Int** with a semantic method, now many proof theoretic methods have been proposed for the proof of faithfulness. For there are also merits for a proof-theoretic method. For example, once we obtain a proof-theoretic embedding from **Int** to **S4**, we can reduce the cut admissibility, the decidability, etc. of **Int** into those of **S4**. Thus, many authors gave a proof-theoretic argument for faithfulness (Maehara 1954; Prawitz and Malmnäs 1968; Troelstra and Schwichtenberg 2000; Dyckhoff and Negri 2012; Mints 2012; Yamasaki and Sano 2016). For example, Maehara (1954) employed a non-labelled sequent calculus to establish faithfulness by cut-elimination, but he used another translation which prefixes $\Box$ to all subformulas of A. For the same purpose, Troelstra and Schwichtenberg (2000) employed the non-labelled **G3**-style sequent calculi and the above translation $A^{\Box}$. Dyckhoff and Negri (2012) employed the **G3**-style labelled systems for a proof-theoretic embedding from intermediate logics to modal logics extending **S4** with the translation $A^{\Box}$. Yamasaki and Sano (2016) also employed the labelled system, but extended Dyckhoff and Negri (2012)'s result to obtain a proof-theoretic embedding from logics of strict implication into modal logics in a uniform manner. They changed the above translation $A^{\Box}$ into the one sending P to $P\&\Box P$. This translation was already introduced by Visser (1981).

Visser (1981) showed that **BPL** is embedded into **K4** with a semantic method via a modified Gödel–McKinsey–Tarski translation such that it sends atomic variable P to $P\&\Box P$. Although Yamasaki and Sano (2016)'s result included the proof-theoretic embedding from Visser's basic propositional logic into modal logic **K4**, as far as the authors know, there are no studies such that a *non*-labelled sequent system is employed to demonstrate a proof-theoretic embedding from **BPL** into **K4**.

In this paper, we provide a cut-free **G3**-style (i.e., contraction-free) non-labelled sequent calculus for **BPL** with the help of the system in Ishii et al. (2001). Moreover, we follow basic ideas of Troelstra and Schwichtenberg (2000); Mints (2012) to establish a proof-theoretic embedding from **BPL** into **K4** via the above translation sends atomic variable P to $P\&\Box P$. There are at least two reasons why we would like to employ the non-labelled system instead of the labelled system based on Kripke semantics. First, our non-labelled sequent system enjoys the subformula property, while a labelled system (say, a system in Yamasaki and Sano 2016) does not. For in the labelled system the expressions xRy (corresponding to an accessibility relation) explicitly appear in a derivation of a provable sequent. Second, we would like to claim that non-labelled sequent systems enable us to capture the 'behavior' of the implication more directly than labelled systems. Via the proof-theoretic embedding, we can see how the behavior of the implication in **BPL** is simulated in the corresponding modal logic **K4**. However, if we employ a labelled system based on Kripke semantics, we may fail to capture such aspect. This is because the inference rules of the implication, say in the labelled system in Yamasaki and Sano (2016), are common to all weak logics of strict implications, and so the differences of the implication in several logics consist only in additional frame properties of Kripke models.

The following is the outline of this paper. In Sect. 2, we first present the syntax for **BPL** and its Kripke semantics. Section 3 introduces a **G3**-style non-labelled sequent calculus for **BPL**, and explains Ishii et al. (2001)'s **LBP** which is the non-**G3**-style non-labelled sequent calculus for **BPL**. Section 4 establishes the admissibility of the cut rule for **G3BPL**, and shows that **G3BPL** is sound and complete for Kripke semantics by referring to the soundness and completeness of **LBP** for Kripke semantics. In Sect. 5, we introduce a **G3**-style non-labelled sequent calculus for **K4** (**G3K4**), and then establish our proof-theoretic (constructive) embedding from **G3BPL** into **G3K4**.

2 Visser's Basic Propositional Logic

The syntax $\mathscr{L}$ of **BPL** is the same as the syntax of intuitionistic logic. That is, $\mathscr{L}$ consists of a countably infinite set Atom of atomic variables (P, Q, etc.) as well as the logical connectives &, $\vee$, $\supset$, and $\bot$. The set $\mathsf{Form}_{\mathscr{L}}$ of all *$\mathscr{L}$-formulas* is inductively defined as follows:

$$\mathsf{Form}_{\mathscr{L}} \ni A ::= P \mid \bot \mid A\&A \mid A \vee A \mid A \supset A,$$

where $P \in \mathsf{Atom}$. We denote $\mathscr{L}$-formulas by A, B, C, etc. Given finite multisets Γ and Δ of formulas, we say that an expression $\Gamma \Rightarrow \Delta$ is a *sequent*, where the Γ and Δ are called the *antecedent* and the *succedent* of a sequent $\Gamma \Rightarrow \Delta$ respectively. Before introducing a **G3**-style non-labelled sequent calculus for **BPL** we first explain Kripke semantics for **BPL**.

We say that $\mathfrak{F} = (S, R)$ is a *frame* if S is a non-empty set and R ($\subseteq S \times S$) is a transitive binary relation on S. A tuple $\mathfrak{M} = (S, R, V)$ is a *model* if (S, R) is a frame and V is a mapping $\mathsf{Atom} \to \mathscr{P}(S)$, called a *valuation*. We say that a valuation V is *monotone* if sRs' and $s \in V(P)$ jointly imply $s' \in V(P)$ for all $s, s' \in S$ and $P \in \mathsf{Atom}$. A model $\mathfrak{M} = (S, R, V)$ is said to be *monotone* if a valuation V is monotone. Given a model $\mathfrak{M} = (S, R, V)$, a state $s \in S$ and a formula A, the *satisfaction relation* $\mathfrak{M}, s \models A$ is defined by:

$$\begin{array}{ll}
\mathfrak{M}, s \models P & \text{iff } s \in V(P), \\
\mathfrak{M}, s \models \bot & \quad\text{Never}, \\
\mathfrak{M}, s \models A\&B & \text{iff } \mathfrak{M}, s \models A \text{ and } \mathfrak{M}, s \models B, \\
\mathfrak{M}, s \models A \vee B & \text{iff } \mathfrak{M}, s \models A \text{ or } \mathfrak{M}, s \models B, \\
\mathfrak{M}, s \models A \supset B & \text{iff for all } s' \in S \text{ with } sRs' : \mathfrak{M}, s' \models A \text{ implies } \mathfrak{M}, s' \models B.
\end{array}$$

Given a model $\mathfrak{M} = (S, R, V)$, we say that A is *valid in* $\mathfrak{M}$ if $\mathfrak{M}, s \models A$ for all states $s \in S$ and, that $\Gamma \Rightarrow \Delta$ is *valid in* $\mathfrak{M}$ if, whenever $\mathfrak{M}, s \models A$ for any $A \in \Gamma$, $\mathfrak{M}, s \models C$ for some $C \in \Delta$, for all states $s \in S$.

3 G3-Style Non-labelled Sequent Calculus for BPL

We define that a *rule* is of the following form:

$$\frac{\Gamma_1 \Rightarrow \Delta_1 \quad \dots \quad \Gamma_n \Rightarrow \Delta_n}{\Gamma \Rightarrow \Delta}\, r,$$

where $\Gamma_i \Rightarrow \Delta_i$ and $\Gamma \Rightarrow \Delta$ are called the *premise(s)* and the *conclusion* of the rule r, respectively. In what follows, we may also consider a rule possibly not in **G3BPL**.

Table 1 G3-style sequent calculus: **G3BPL**

(Axioms)

$$\frac{}{P, \Gamma \Rightarrow \Delta, P}\ (Id) \qquad \frac{}{\bot, \Gamma \Rightarrow \Delta}\ (L\bot)$$

(Logical Rules)

$$\frac{A, B, \Gamma \Rightarrow \Delta}{A\&B, \Gamma \Rightarrow \Delta}\ (L\&) \qquad \frac{\Gamma \Rightarrow \Delta, A \quad \Gamma \Rightarrow \Delta, B}{\Gamma \Rightarrow \Delta, A\&B}\ (R\&)$$

$$\frac{A, \Gamma \Rightarrow \Delta \quad B, \Gamma \Rightarrow \Delta}{A \vee B, \Gamma \Rightarrow \Delta}\ (L\vee) \qquad \frac{\Gamma \Rightarrow \Delta, A, B}{\Gamma \Rightarrow \Delta, A \vee B}\ (R\vee)$$

$$\frac{C_1 \supset D_1, \ldots, C_n \supset D_n, \Delta_1, \Sigma, A \Rightarrow B, \Gamma_1 \quad \ldots \quad C_1 \supset D_1, \ldots, C_n \supset D_n, \Delta_{2^n}, \Sigma, A \Rightarrow B, \Gamma_{2^n}}{\Sigma, C_1 \supset D_1, \ldots, C_n \supset D_n \Rightarrow A \supset B, \Theta}\ (\supset)\dagger$$

†: $n \geq 0$, $\Gamma_i = \{C_j \mid j \in \gamma(i)\}$, $\Delta_i = \{D_j \mid j \in \delta(i)\}$, $1 \leqslant i \leqslant 2^n$, and $\gamma(i)$ runs through the subsets of $\{1, \ldots, n\}$ and $\delta(i) = \{1, \ldots, n\} \setminus \gamma(i)$.

In Table 1, we provide a **G3**-style non-labelled sequent calculus for **BPL**. Let us give some examples of the rule $(\supset)$ for the implication. When $n = 0, 1, 2$, the forms of the rule are as follows:

$$\frac{\Sigma, A \Rightarrow B}{\Sigma \Rightarrow A \supset B, \Theta}\ (\supset), \qquad \frac{\Sigma, C_1 \supset D_1, A \Rightarrow B, C_1 \quad D_1, \Sigma, C_1 \supset D_1, A \Rightarrow B}{\Sigma, C_1 \supset D_1 \Rightarrow A \supset B, \Theta}\ (\supset),$$

$$\frac{\Sigma, \Phi, A \Rightarrow B, C_1, C_2 \quad D_1, \Sigma, \Phi, A \Rightarrow B, C_2 \quad D_2, \Sigma, \Phi, A \Rightarrow B, C_1 \quad D_1, D_2, \Sigma, \Phi, A \Rightarrow B}{\Sigma, C_1 \supset D_1, C_2 \supset D_2 \Rightarrow A \supset B, \Theta}\ (\supset),$$

where $\Phi = C_1 \supset D_1, C_2 \supset D_2$.

We note that no structural rules of contraction and weakening are found in **G3BPL** but these rules are shown to be admissible in the next section, i.e., the addition of these rules does not change all the derivable sequents.

It is immediate to obtain the following subformula property of **G3BPL**.

Proposition 1 **G3BPL** *enjoys the subformula property, i.e., if a sequent* $\Gamma \Rightarrow \Delta$ *is derivable in* **G3BPL** *then there is a derivation all of whose sequents are constructed by subformulas of the sequent* $\Gamma \Rightarrow \Delta$.

Definition 1 (*Context and Principal Formula*) The Γ and Δ in an inference rule of **G3BPL** except for $(\supset)$ are called the *context*. In the rule $(\supset)$, Σ and Θ are the *context*. In the conclusion of each rule of **G3BPL**, the formula(s) not in the context is called the *principal formula*(*s*).

Definition 2 (*Derivation*) A *derivation* $\mathscr{D}$ in **G3BPL** is a finite tree generated from the axioms by the rules of **G3BPL**. We say that the *end sequent* of $\mathscr{D}$ is the sequent in the root node of $\mathscr{D}$. The *height* of a derivation is the maximum length of branches in the derivation from the end sequent to an axiom. A sequent $\Gamma \Rightarrow \Delta$ is *derivable* in **G3BPL** (notation: **G3BPL** $\vdash \Gamma \Rightarrow \Delta$) if it has a derivation $\mathscr{D}$ in **G3BPL** whose end sequent is $\Gamma \Rightarrow \Delta$. We write **G3BPL** $\vdash_n \Gamma \Rightarrow \Delta$ to mean that $\Gamma \Rightarrow \Delta$ has a derivation whose height is at most n.

Proposition 2 *Any sequent of the form $A, \Gamma \Rightarrow \Delta, A$ is derivable in* **G3BPL**.

Proof By induction on the complexity of A. If A is an atomic variable or $\bot$, it is obvious. If $n > 0$, then we can prove simply all cases by applying induction hypothesis. Here we present the case of implication. If $A \equiv B \supset C$, then the last step of the derivation is

$$\dfrac{\begin{array}{cc}\vdots & \vdots \\ B\supset C, \Gamma, B \Rightarrow C, \Delta, B & C, B\supset C, \Gamma, B \Rightarrow C, \Delta\end{array}}{B\supset C, \Gamma \Rightarrow \Delta, B\supset C}\ (\supset),$$

where $B\supset C, \Gamma, B \Rightarrow C, \Delta, B$ and $C, B\supset C, \Gamma, B \Rightarrow C, \Delta$ are clearly derivable by induction hypothesis. □

As we have already mentioned in the introduction, Ishii et al. (2001) also provided a cut-free sequent calculus for **BPL** which is called "**LBP**" (see Table 2).

Note that Ishii et al. (2001) defined a sequent as a pair of finite {¥em sets} of formulas. Therefore, their system has no rules of contraction (or exchange). The form of axioms of **LBP** are different from that of **G3BPL**. The rules of & and $\vee$ are the same as these of **G3BPL**. But the rule of $\supset$ is not the same as that of **G3BPL**. When we compare the rule of $\supset$ of **LBP** with that of **G3BPL**, the rule of $\supset$ of **LBP** is simpler than that of **G3BPL**. The rule of $\supset$ of **G3BPL** has the complex form, because the rules of weakening and contraction are admissible in **G3BPL**.

Table 2 Sequent calculus for BPL: **LBP** (Ishii et al. 2001)

(Axioms)

$$\overline{A \Rightarrow A} \quad \overline{\bot \Rightarrow}$$

(Inference Rules)

$$\dfrac{\Gamma \Rightarrow \Delta}{A, \Gamma \Rightarrow \Delta}\ (LW) \qquad \dfrac{\Gamma \Rightarrow \Delta}{\Gamma \Rightarrow \Delta, A}\ (RW)$$

$$\dfrac{A, B, \Gamma \Rightarrow \Delta}{A\&B, \Gamma \Rightarrow \Delta}\ (L\&) \qquad \dfrac{\Gamma \Rightarrow \Delta, A \quad \Gamma \Rightarrow \Delta, B}{\Gamma \Rightarrow \Delta, A\&B}\ (R\&)$$

$$\dfrac{A, \Gamma \Rightarrow \Delta \quad B, \Gamma \Rightarrow \Delta}{A \vee B, \Gamma \Rightarrow \Delta}\ (L\vee) \qquad \dfrac{\Gamma \Rightarrow \Delta, A, B}{\Gamma \Rightarrow \Delta, A \vee B}\ (R\vee)$$

$$\dfrac{\Delta_1, \Sigma, A \Rightarrow B, \Gamma_1 \quad \dots \quad \Delta_{2^n}, \Sigma, A \Rightarrow B, \Gamma_{2^n}}{\Sigma, C_1\supset D_1, \dots, C_n\supset D_n \Rightarrow A\supset B}\ (\supset)\dagger$$

$\dagger$: $n \geq 0$, $\Gamma_i = \{C_j \mid j \in \gamma(i)\}$, $\Delta_i = \{D_j \mid j \in \delta(i)\}$, $1 \leqslant i \leqslant 2^n$, and $\gamma(i)$ runs through the subsets of $\{1, \dots, n\}$ and $\delta(i) = \{1, \dots, n\} \setminus \gamma(i)$.

The notions of context, principal formula and derivation for **LBP** are the same as **G3BPL**. We will show that **G3BPL** is equivalent to **LBP** later.

4 Admissibility of Cut

In this section, we establish the admissibility of the cut rule in **G3BPL**. As for the proof, we follow the standard syntactic argument for cut-elimination in contraction-free sequent system (see e.g. Negri and von Plato 2001; Troelstra and Schwichtenberg 2000), though the 'combinatorial' nature of the rule ($\supset$) makes our argument more involved.

Definition 3 The rule of cut is defined as follows:

$$\frac{\Gamma \Rightarrow \Delta, A \quad A, \Pi \Rightarrow \Sigma}{\Gamma, \Pi \Rightarrow \Delta, \Sigma}\ (Cut),$$

where we say that A is the *cut formula*.

Definition 4 (*Admissibility*) A rule r is said to be *admissible* in **G3BPL** if, whenever the premise(s) of the rule is derivable in **G3BPL**, the conclusion of the rule is also derivable in **G3BPL**. A rule r is said to be *height-preserving admissible* (*hp-admissible*) in **G3BPL** if, whenever the premise(s) of the rule is derivable in **G3BPL** with height at most n, the conclusion of the rule is also derivable in **G3BPL** with height at most n.

Lemma 1 (Weakening) *The rules of weakening are hp-admissible in* **G3BPL**, *i.e.,*

(i) *If* $\vdash_n \Gamma \Rightarrow \Delta$, *then* $\vdash_n A, \Gamma \Rightarrow \Delta$.
(ii) *If* $\vdash_n \Gamma \Rightarrow \Delta$, *then* $\vdash_n \Gamma \Rightarrow \Delta, A$.

Proof We show each item by induction on height n of derivation. Here we focus on the second item (the first item is obvious by induction hypothesis). Suppose that there is a derivation $\mathcal{D}$ of $\Gamma \Rightarrow \Delta$. When $\mathcal{D}$ is a derivation of height zero, i.e., an axiom, then so is $A, \Gamma \Rightarrow \Delta$.

If the last applied rule of $\mathcal{D}$ is one of ($L\&$), ($R\&$), ($L\vee$), and ($R\vee$), then we apply induction hypothesis to the premise(s), and then apply the rule to the result of induction hypothesis. If the last applied rule is ($\supset$), then the original derivation is

$$\frac{\begin{array}{c}\vdots\\ C_1{\supset}D_1,\ldots,C_n{\supset}D_n,\Delta_1,\Sigma,E \Rightarrow F,\Gamma_1\end{array} \quad \ldots \quad \begin{array}{c}\vdots\\ C_1{\supset}D_1,\ldots,C_n{\supset}D_n,\Delta_{2^n},\Sigma,E \Rightarrow F,\Gamma_{2^n}\end{array}}{\Sigma, C_1{\supset}D_1,\ldots,C_n{\supset}D_n \Rightarrow E{\supset}F,\Theta}\ (\supset),$$

where $n \geq 0$, Γ_i and Δ_i are defined as in Table 1. In order to obtain our goal, we apply ($\supset$) to all premises of the original derivation, where we regard the context as Θ, A. □

Definition 5 (*Invertibility*) A rule r is said to be *height-preserving invertible* (*hp-invertible*) in **G3BPL** if, whenever the conclusion of the rule is derivable in **G3BPL** with height at most n, the premise(s) of the rule is also derivable in **G3BPL** with height at most n.

Lemma 2 (Inversion) *All rules of* **G3BPL** *except for the rule* ($\supset$) *are hp-invertible.*

Proof We have four cases: the left and right rules of & and $\vee$. For example, in the case of $(L\&)$, it is enough to show that $\vdash_n A\&B, \Gamma \Rightarrow \Delta$ implies $\vdash_n A, B, \Gamma \Rightarrow \Delta$. If $A\&B, \Gamma \Rightarrow \Delta$ is an axiom, then $A, B, \Gamma \Rightarrow \Delta$ is also an axiom. If $n > 0$, 1) if $A\&B$ is the principal formula, then it is obvious. 2) Otherwise, apply induction hypothesis to the premise(s) of the original derivation, and then apply the last applied rule in the original derivation. □

Lemma 3 (Contraction) *The rules of contraction are hp-admissible in* **G3BPL**, *i.e.*,

(i) *If* $\vdash_n A, A, \Gamma \Rightarrow \Delta$, *then* $\vdash_n A, \Gamma \Rightarrow \Delta$.
(ii) *If* $\vdash_n \Gamma \Rightarrow \Delta, A, A$, *then* $\vdash_n \Gamma \Rightarrow \Delta, A$.

Proof We show both items by simultaneous induction on height n of derivations. If $n = 0$, then each sequent assumed is an axiom. It is clear that the desired sequents are also axioms. Let $n > 0$. We consider each item in turn.

(i) If $n > 0$, then we have two cases, depending on the form of A. First, if A is not a principal formula, then we consider which rules are applied to the derivation $A, A, \Gamma \Rightarrow \Delta$. In this case, we can straightforwardly use induction hypothesis to obtain $A, \Gamma \Rightarrow \Delta$. Otherwise, we distinguish further cases, depending on the last applied rule, which can introduce a principal formula in the antecedent of the conclusion of the rule. The candidates are these three rules: $(L\&)$, $(L\vee)$ and $(\supset)$. In the case of $(L\&)$ or $(L\vee)$, our argument proceeds easily by using hp-invertibility for $(L\&)$ or $(L\vee)$, and induction hypothesis. In the case of $(\supset)$, the same reasoning cannot be applied, because there are possibly many principal formulas of $(\supset)$ in the antecedent. Therefore, there are two possibilities: (1) one occurrences of A is principal and the other is the context, and (2) both occurrences of A are principal. For the case of (1), consider that one of occurrences of A (where $A \equiv C_1 \supset D_1$) is in the context. The original derivation is as follows:

$$\dfrac{\begin{array}{c}\vdots\\ C_1\supset D_1,\ldots,C_n\supset D_n,\Delta_1,\Sigma,C_1\supset D_1,E\Rightarrow F,\Gamma_1\end{array}\quad\ldots\quad\begin{array}{c}\vdots\\ C_1\supset D_1,\ldots,C_n\supset D_n,\Delta_{2^n},\Sigma,C_1\supset D_1,E\Rightarrow F,\Gamma_{2^n}\end{array}}{\Sigma,C_1\supset D_1,C_1\supset D_1,\ldots,C_n\supset D_n\Rightarrow E\supset F,\Theta}\,(\supset),$$

where $n \geq 1$, Γ_i and Δ_i are defined as in Table 1. Then we apply induction hypothesis to all premises of the original derivation, and then apply $(\supset)$. Thus, we can obtain the following derivation:

$$\dfrac{\begin{array}{c}\vdots\\ C_1\supset D_1,\ldots,C_n\supset D_n,\Delta_1,\Sigma,E\Rightarrow F,\Gamma_1\end{array}\quad\ldots\quad\begin{array}{c}\vdots\\ C_1\supset D_1,\ldots,C_n\supset D_n,\Delta_{2n},\Sigma,E\Rightarrow F,\Gamma_{2n}\end{array}}{\Sigma,C_1\supset D_1,\ldots,C_n\supset D_n\Rightarrow E\supset F,\Theta}\,(\supset).$$

For the case of (2), consider that both occurrences of A are principal formulas of ($\supset$). Suppose, for example, that the original derivation is of the following form

$$\dfrac{\begin{matrix}\vdots & & \vdots\\ C_1\supset D_1,\ldots,C_n\supset D_n,C_n\supset D_n,\Delta_1,\Sigma,E\Rightarrow F,\Gamma_1 & \ldots & C_1\supset D_1,\ldots,C_n\supset D_n,C_n\supset D_n,\Delta_{2^{n+1}},\Sigma,E\Rightarrow F,\Gamma_{2^{n+1}}\end{matrix}}{\Sigma,C_1\supset D_1,\ldots,C_n\supset D_n,C_n\supset D_n\Rightarrow E\supset F,\Theta}\,(\supset),$$

where $n \geq 1$, $\Gamma_i = \{C_j \mid j \in \gamma(i)\}$, $\Delta_i = \{D_j \mid j \in \delta(i)\}$, $1 \leqslant i \leqslant 2^{n+1}$, $\gamma(i)$ runs through the subsets of $\{1,\ldots,n+1\}$ and $\delta(i) = \{1,\ldots,n+1\} \setminus \gamma(i)$. Note that $C_{n+1} = C_n$ and $D_{n+1} = D_n$.
In the premises of ($\supset$) of the original derivation there are 2^{n-1} sequents such that Γ_i contains two occurrences of C_n, and there are also 2^{n-1} sequents such that Δ_i contains two occurrences of D_n. Now, we contract two occurrences of C_n or that of D_n from these sequents by applying induction hypothesis, and contract also two occurrences of $C_n\supset D_n$. Next, we apply ($\supset$) to all these resulting sequents. Hence, we can obtain the following desired derivation:

$$\dfrac{\begin{matrix}\vdots & & \vdots\\ C_1\supset D_1,\ldots,C_n\supset D_n,\Delta'_1,\Sigma,E\Rightarrow F,\Gamma'_1 & \ldots & C_1\supset D_1,\ldots,C_n\supset D_n,\Delta'_{2^n},\Sigma,E\Rightarrow F,\Gamma'_{2^n}\end{matrix}}{\Sigma,C_1\supset D_1,\ldots,C_n\supset D_n\Rightarrow E\supset F,\Theta}\,(\supset),$$

where note that $n \geq 0$, $\Gamma'_i = \{C_j \mid j \in \gamma(i)\}$, $\Delta'_i = \{D_j \mid j \in \delta(i)\}$, $1 \leqslant i \leqslant 2^n$, and $\gamma(i)$ runs through the subsets of $\{1,\ldots,n\}$ and $\delta(i) = \{1,\ldots,n\} \setminus \gamma(i)$.

(ii) Similarly to (i), we only consider the case where A is principal. If the last applied rule to obtain $\Gamma \Rightarrow \Delta, A, A$ is $(R\&)$ or $(R\vee)$, it is easy. If the outermost connective of A is $\supset$ (so $A \equiv E\supset F$), then one of occurrences of A is always in the context, because the number of occurrences of a principal formula is only one in the succedent of the conclusion of ($\supset$). The original derivation ends with

$$\dfrac{\begin{matrix}\vdots & & \vdots\\ C_1\supset D_1,\ldots,C_n\supset D_n,\Delta_1,\Sigma,E\Rightarrow F,\Gamma_1 & \ldots & C_1\supset D_1,\ldots,C_n\supset D_n,\Delta_{2^n},\Sigma,E\Rightarrow F,\Gamma_{2^n}\end{matrix}}{\Sigma,C_1\supset D_1,\ldots,C_n\supset D_n\Rightarrow E\supset F,E\supset F,\Theta}\,(\supset),$$

where $n \geq 0$, Γ_i and Δ_i are defined as in Table 1.
We apply ($\supset$) to all premises of ($\supset$) of the original derivation, and can obtain the following derivation:

$$\dfrac{\begin{matrix}\vdots & & \vdots\\ C_1\supset D_1,\ldots,C_n\supset D_n,\Delta_1,\Sigma,E\Rightarrow F,\Gamma_1 & \ldots & C_1\supset D_1,\ldots,C_n\supset D_n,\Delta_{2^n},\Sigma,E\Rightarrow F,\Gamma_{2^n}\end{matrix}}{\Sigma,C_1\supset D_1,\ldots,C_n\supset D_n\Rightarrow E\supset F,\Theta}\,(\supset),$$

where note that we do not introduce $E\supset F$ in the context. □

Definition 6 The *weight* of formula A is the number of logical connectives in A, and the *cut-height* of (Cut) is the sum of heights of derivations of the two premises of (Cut).

Theorem 1 *The rule of* (Cut) *is admissible in* **G3BPL**.

Proof Recall that the rule of (Cut) is defined as follows:

$$\frac{\Gamma \Rightarrow \Delta, A \quad A, \Pi \Rightarrow \Sigma}{\Gamma, \Pi \Rightarrow \Delta, \Sigma}\,(Cut).$$

We prove the admissibility of (Cut) by induction on the weight of the cut formula A, with subinduction on the cut-height of (Cut). Our proof is organized as follows. First, we consider the cases ((i) and (ii)) where at least one of the premises of (Cut) is an axiom. For the rest, there are three cases: (iii) the cut formula is not principal in the left premise; (iv) the cut formula is principal in the left premise only; (v) the cut formula is principal in both premises of (Cut).

(i) The left premise of (Cut) is an axiom: We omit the proof of this case.

(ii) The right premise of (Cut) is an axiom: First, suppose that the right premise $A, \Pi \Rightarrow \Sigma$ is the axiom (Id). That is, we have one of the following cases: the right premise is of the form $A, P, \Pi' \Rightarrow \Sigma', P$ or of the form $P, \Pi \Rightarrow \Sigma', P$, where $A \equiv P$. For the former case, we note that $\Gamma, \Pi \Rightarrow \Delta, \Sigma$ is also an axiom (Id). For the latter case, we need to obtain $\Gamma, \Pi \Rightarrow \Delta, \Sigma', P$, which is derivable from the left premise $\Gamma \Rightarrow \Delta, P$ by hp-weakening. Second, suppose that the right premise is the axiom $(L\bot)$. If $A \not\equiv \bot$ in the cut formula A, we can find a $\bot$ in Π and so $\Gamma, \Pi \Rightarrow \Delta, \Sigma$ is also an axiom $(L\bot)$. Otherwise, i.e., if $A \equiv \bot$, we need to check the last rule of the left premise $\Gamma \Rightarrow \Delta, \bot$. If the last rule is an axiom, this case is reduced to the case (i). Otherwise, this case becomes a special case of (iii).

(iii) The cut formula is not principal in the left premise: We divide our argument into cases, depending on the last applied rule of the left premise of (Cut). That is, there are five cases including all logical rules. First, we demonstrate the case of &. We have the following derivation:

$$\frac{\dfrac{\vdots \atop B, C, \Gamma \Rightarrow \Delta, A}{B\&C, \Gamma \Rightarrow \Delta, A}\,(L\&) \quad \vdots \atop A, \Pi \Rightarrow \Sigma}{B\&C, \Gamma, \Pi \Rightarrow \Delta, \Sigma}\,(Cut).$$

This derivation is transformed into the following one:

$$\dfrac{\dfrac{\vdots \atop B, C, \Gamma \Rightarrow \Delta, A \quad \vdots \atop A, \Pi \Rightarrow \Sigma}{B, C, \Gamma, \Pi \Rightarrow \Delta, \Sigma}\,(Cut)}{B\&C, \Gamma, \Pi \Rightarrow \Delta, \Sigma}\,(L\&).$$

In this case, the cut-height of (Cut) in the transformed derivation becomes smaller than that of (Cut) in the original derivation. The other cases except for $(\supset)$ are similar to this case. For the case of $(\supset)$, our argument proceeds as follows. We have the original derivation:

$$\dfrac{\dfrac{\begin{matrix}\vdots\\ C_1\supset D_1,\ldots,C_n\supset D_n,\Delta_1,\Gamma,E\Rightarrow F,\Gamma_1\end{matrix}\quad\ldots\quad\begin{matrix}\vdots\\ C_1\supset D_1,\ldots,C_n\supset D_n,\Delta_{2^n},\Gamma,E\Rightarrow F,\Gamma_{2^n}\end{matrix}}{C_1\supset D_1,\ldots,C_n\supset D_n,\Gamma\Rightarrow E\supset F,\Delta,A}(\supset)\quad\begin{matrix}\vdots\\ A,\Pi\Rightarrow\Sigma\end{matrix}}{C_1\supset D_1,\ldots,C_n\supset D_n,\Gamma,\Pi\Rightarrow E\supset F,\Delta,\Sigma}(Cut),$$

where $n\geq 0$, Γ_i and Δ_i are defined as in Table 1.

Since the cut formula A is in the context of the conclusion in the rule $(\supset)$, we can obtain the following derivation:

$$\dfrac{\dfrac{\begin{matrix}\vdots\\ C_1\supset D_1,\ldots,C_n\supset D_n,\Delta_1,\Gamma,E\Rightarrow F,\Gamma_1\end{matrix}\quad\ldots\quad\begin{matrix}\vdots\\ C_1\supset D_1,\ldots,C_n\supset D_n,\Delta_{2^n},\Gamma,E\Rightarrow F,\Gamma_{2^n}\end{matrix}}{C_1\supset D_1,\ldots,C_n\supset D_n,\Gamma\Rightarrow E\supset F,\Delta,\Sigma}(\supset)}{\overline{C_1\supset D_1,\ldots,C_n\supset D_n,\Gamma,\Pi\Rightarrow E\supset F,\Delta,\Sigma}},$$

where the double line means finitely many applications of weakening.

(iv) The cut formula is principal in the left premise only: We divide our argument into cases, depending on the last applied rule of the left premise of (Cut), where the cut formula A is *principal*. We have three cases: $(R\&)$, $(R\vee)$ and $(\supset)$. If the last applied rule is $(R\&)$ or $(R\vee)$, these cases are similar to item (iii). If the last applied rule is $(\supset)$, we need to look at the last applied rule of the right premise of (Cut), where the cut formula A is *not* principal. The only non-trivial case is a case where the last applied rule of the *right* premise of (Cut) is $(\supset)$, because the other cases are shown similar to item (iii). If the last applied rule of the right premise of (Cut) is $(\supset)$, the cut formula A is of the form $C\supset D$, then we have the following original derivation:

$$\dfrac{\dfrac{\begin{matrix}\vdots\\ \Psi,\Gamma,\Theta_1,C\Rightarrow D,\Xi_1\end{matrix}\ \ldots\ \begin{matrix}\vdots\\ \Psi,\Gamma,\Theta_{2^n},C\Rightarrow D,\Xi_{2^n}\end{matrix}}{\Psi,\Gamma\Rightarrow\Delta,C\supset D}(\supset)\quad\dfrac{\begin{matrix}\vdots\\ C\supset D,\Phi,\Pi,\Delta_1,E\Rightarrow F,\Gamma_1\end{matrix}\ \ldots\ \begin{matrix}\vdots\\ C\supset D,\Phi,\Pi,\Delta_{2^m},E\Rightarrow F,\Gamma_{2^m}\end{matrix}}{C\supset D,\Phi,\Pi\Rightarrow E\supset F,\Sigma}(\supset)}{G_1\supset H_1,\ldots,G_n\supset H_n,\Gamma,C_1\supset D_1,\ldots,C_m\supset D_m,\Pi\Rightarrow\Delta,E\supset F,\Sigma}(Cut),$$

where $\Psi=G_1\supset H_1,\ldots,G_n\supset H_n, n\geq 0, \Xi_i=\{G_j\mid j\in\gamma(i)\}, \Theta_i=\{H_j\mid j\in\delta(i)\}$, $1\leqslant i\leqslant 2^n$, and $\gamma(i)$ runs through the subsets of $\{1,\ldots,n\}$ and $\delta(i)=\{1,\ldots,n\}\setminus\gamma(i)$, and $\Phi=C_1\supset D_1,\ldots,C_m\supset D_m$, $m\geq 0$, $\Gamma_k=\{C_j\mid j\in\gamma'(k)\}$, $\Delta_k=\{D_j\mid j\in\delta'(k)\}$, $1\leqslant k\leqslant 2^m$, and $\gamma'(k)$ runs through the subsets of $\{1,\ldots,m\}$ and $\delta'(k)=\{1,\ldots,m\}\setminus\gamma'(k)$.

Our process of the transformation proceeds as follows. First, we apply $(\supset)$ to all premises of the applied rule $(\supset)$ of the left premise of (Cut):

$$\frac{\begin{array}{ccc}\vdots & & \vdots \\ \Psi, \Gamma, \Theta_1, C \Rightarrow D, \Xi_1 & \dots & \Psi, \Gamma, \Theta_{2^n}, C \Rightarrow D, \Xi_{2^n}\end{array}}{\Psi, \Gamma \Rightarrow C \supset D} (\supset),$$

where note that the context in the succedent of the conclusion of $(\supset)$ is now empty. Next, based on this result, we eliminate the occurrences of $C \supset D$ in all premises of the applied rule $(\supset)$ of the right premise of (Cut):

$$\mathscr{D}_k = \frac{\dfrac{\begin{array}{ccc}\vdots & & \vdots \\ \Psi, \Gamma, \Theta_1, C \Rightarrow D, \Xi_1 & \dots & \Psi, \Gamma, \Theta_{2n}, C \Rightarrow D, \Xi_{2n}\end{array}}{\Psi, \Gamma \Rightarrow C \supset D} (\supset) \quad \begin{array}{c}\vdots \\ C \supset D, \Phi, \Pi, \Delta_k, E \Rightarrow F, \Gamma_k\end{array}}{\Psi, \Gamma, \Phi, \Pi, \Delta_k, E \Rightarrow F, \Gamma_k} (Cut),$$

where $1 \leqslant k \leqslant 2^m$ and this subscript k is the number of the premises of $(\supset)$ of the right premise of (Cut) in the original derivation.

Finally, from all derivations $\mathscr{D}_k$ $(1 \leqslant k \leqslant 2^m)$, we can obtain the following derivation:

$$\frac{\begin{array}{ccc}\mathscr{D}_1 & & \mathscr{D}_k \\ \Psi, \Gamma, \Phi, \Pi, \Delta_1, E \Rightarrow F, \Gamma_1 & \dots & \Psi, \Gamma, \Phi, \Pi, \Delta_k, E \Rightarrow F, \Gamma_k\end{array}}{\Psi, \Gamma, \Phi, \Pi \Rightarrow \Delta, E \supset F, \Sigma} (\supset).$$

(v) The cut formula is principal in both premises of (Cut). We concentrate on the case where $A \equiv C_1 \supset D_1$. Before presenting a general argument, let us see one example. Suppose that the original derivation is as follows:

$$\frac{\dfrac{\begin{array}{c}\mathscr{D}_{l1} \\ \Gamma, C_1 \Rightarrow D_1\end{array}}{\Gamma \Rightarrow \Delta, C_1 \supset D_1} (\supset) \quad \dfrac{\begin{array}{cc}\mathscr{D}_{r1} & \mathscr{D}_{r2} \\ C_1 \supset D_1, \Pi, E \Rightarrow F, C_1 & D_1, C_1 \supset D_1, \Pi, E \Rightarrow F\end{array}}{C_1 \supset D_1, \Pi \Rightarrow E \supset F, \Sigma} (\supset)}{\Gamma, \Pi \Rightarrow \Delta, E \supset F, \Sigma} (Cut).$$

We divide our argument into three steps. In the first step, we eliminate the occurrences of $C_1 \supset D_1$ in all premises of $(\supset)$ of the right premise of (Cut). From $\mathscr{D}_{l1}$ and $\mathscr{D}_{r1}$, we construct the following derivation $\mathscr{D}_1$:

$$\mathscr{D}_1 = \frac{\dfrac{\begin{array}{c}\mathscr{D}_{l1} \\ \Gamma, C_1 \Rightarrow D_1\end{array}}{\Gamma \Rightarrow C_1 \supset D_1} (\supset) \quad \begin{array}{c}\mathscr{D}_{r1} \\ C_1 \supset D_1, \Pi, E \Rightarrow F, C_1\end{array}}{\Gamma, \Pi, E \Rightarrow F, C_1} (Cut).$$

From $\mathscr{D}_{l1}$ and $\mathscr{D}_{r2}$, we construct the following derivation $\mathscr{D}_2$:

$$\mathscr{D}_2 = \frac{\dfrac{\begin{array}{c}\mathscr{D}_{l1} \\ \Gamma, C_1 \Rightarrow D_1\end{array}}{\Gamma \Rightarrow C_1 \supset D_1} (\supset) \quad \begin{array}{c}\mathscr{D}_{r2} \\ C_1 \supset D_1, D_1, \Pi, E \Rightarrow F\end{array}}{D_1, \Gamma, \Pi, E \Rightarrow F} (Cut).$$

Note that in each $\mathscr{D}_1$ and $\mathscr{D}_2$ we can eliminate (Cut) since the cut-height becomes smaller.

In the second step, we eliminate both occurrences of subformulas (C_1 or D_1) of the cut formula $C_1 \supset D_1$ from the premise of $(\supset)$ of the left premise of (Cut). From $\mathscr{D}_1$, $\mathscr{D}_2$ and $\mathscr{D}_{l1}$, we construct the following derivation:

$$\dfrac{\dfrac{\dfrac{\dfrac{\overset{\mathscr{D}_1}{\Gamma, \Pi, E \Rightarrow F, C_1} \quad \overset{\mathscr{D}_{l1}}{C_1, \Gamma \Rightarrow D_1}}{\Gamma, \Pi, E, \Gamma \Rightarrow F, D_1}\,(Cut)}{\Gamma, \Pi, E \Rightarrow F, D_1} \qquad \overset{\mathscr{D}_2}{D_1, \Gamma, \Pi, E \Rightarrow F}}{\Gamma, \Pi, E, \Gamma, \Pi, E \Rightarrow F, F}\,(Cut)}{\Gamma, \Pi, E \Rightarrow F},$$

where the double line means finitely many applications of contraction and note that the application of (Cut) is possible because the weight of the cut formulas C_1 or D_1 is smaller than that of the original cut formula $C_1 \supset D_1$.

In the final step, we apply $(\supset)$ to the conclusion of the second to obtain the following derivation:

$$\dfrac{\begin{array}{c}\vdots \\ \Gamma, \Pi, E \Rightarrow F\end{array}}{\Gamma, \Pi \Rightarrow \Delta, E \supset F, \Sigma}\,(\supset).$$

Now we have checked how we can eliminate (Cut) in the simple example. In what follows, we present a general argument. Let the original derivation be the following:

$$\dfrac{\dfrac{\overset{\mathscr{D}_{l1}}{\Psi, \Gamma, \Theta_1, C_1 \Rightarrow D_1, \Xi_1} \quad \dots \quad \overset{\mathscr{D}_{l2^n}}{\Psi, \Gamma, \Theta_{2^n}, C_1 \Rightarrow D_1, \Xi_{2^n}}}{\Psi, \Gamma \Rightarrow \Delta, C_1 \supset D_1}\,(\supset) \quad \dfrac{\overset{\mathscr{D}_{r1}}{C_1 \supset D_1, \Phi, \Pi, \Delta_1, E \Rightarrow F, \Gamma_1} \quad \dots \quad \overset{\mathscr{D}_{r2^m}}{C_1 \supset D_1, \Phi, \Pi, \Delta_{2^m}, E \Rightarrow F, \Gamma_{2^m}}}{C_1 \supset D_1, \Phi, \Pi \Rightarrow E \supset F, \Sigma}\,(\supset)}{G_1 \supset H_1, \dots, G_n \supset H_n, \Gamma, C_2 \supset D_2, \dots, C_m \supset D_m, \Pi \Rightarrow \Delta, \Sigma, E \supset F}\,(Cut),$$

where $\Psi = G_1 \supset H_1, \dots, G_n \supset H_n, n \geq 0, \Xi_i = \{G_j \mid j \in \gamma(i)\}, \Theta_i = \{H_j \mid j \in \delta(i)\}$, $1 \leqslant i \leqslant 2^n$, and $\gamma(i)$ runs through the subsets of $\{1, \dots, n\}$ and $\delta(i) = \{1, \dots, n\} \setminus \gamma(i)$, and $\Phi = C_2 \supset D_2, \dots, C_m \supset D_m$, $m \geq 1$, $\Gamma_k = \{C_j \mid j \in \gamma'(k)\}$, $\Delta_k = \{D_j \mid j \in \delta'(k)\}$, $1 \leqslant k \leqslant 2^m$, and $\gamma'(k)$ runs through the subsets of $\{1, \dots, m\}$ and $\delta'(k) = \{1, \dots, m\} \setminus \gamma'(k)$. Recall that the cut formula is principal in both premises of (Cut). As we have seen in the previous example, this argument is divided into three steps.

(Step 1) We eliminate the occurrences of $C_1 \supset D_1$ in each premise of $(\supset)$ of the right premise of (Cut) as follows:

$$\dfrac{\dfrac{\begin{matrix}\mathscr{D}_{l1} & & \mathscr{D}_{l2^n}\\ \Psi,\Gamma,\Theta_1,C_1\Rightarrow D_1,\Xi_1 & \ldots & \Psi,\Gamma,\Theta_{2^n},C_1\Rightarrow D_1,\Xi_{2^n}\end{matrix}}{\Psi,\Gamma\Rightarrow C_1\supset D_1}\,(\supset) \qquad \begin{matrix}\mathscr{D}_{rk}\\ C_1\supset D_1,\Phi,\Pi,\Delta_k,E\Rightarrow F,\Gamma_k\end{matrix}}{\Psi,\Gamma,\Phi,\Pi,\Delta_k,E\Rightarrow F,\Gamma_k}\,(Cut).$$

(Step 2) Before going to the detail of our argument, we introduce a useful notion in this step: *a nice pair of sequents*, to take care of the occurrences of C_1 and D_1 in premises of $(\supset)$ of the right premises of (Cut). We define $\Gamma'_k = \{C_j \mid j \in \gamma''(k)\}$, $\Delta'_k = \{D_j \mid j \in \delta''(k)\}$, $1 \leqslant k \leqslant 2^{m-1}$ where $\gamma''(k)$ runs through the subsets of $\{2, \ldots, m\}$ and $\delta''(k) = \{2, \ldots, m\} \setminus \gamma''(k)$. By (Step 1), the number of derivations whose root is a sequent of the following form is 2^{m-1}:

$$\Psi, \Gamma, \Phi, \Pi, \Delta'_k, E \Rightarrow F, \Gamma'_k, C_1$$

and the number of derivations whose root is a sequent of the following form is also 2^{m-1}:

$$D_1, \Psi, \Gamma, \Phi, \Pi, \Delta'_k, E \Rightarrow F, \Gamma'_k.$$

A nice pair of sequents is defined as a pair of sequents of the form specified just above. So the sequents in the nice pair differ only in the occurrence of C_1 in the succedent and the occurrence of D_1 in the antecedent.

Let us fix the subscript i ($1 \leqslant i \leqslant 2^n$) in the left premise of (Cut) of the original derivation such that $\Psi, \Gamma, \Theta_i, C_1 \Rightarrow D_1, \Xi_i$ is derivable. First, we eliminate, by (Cut), the occurrence of C_1 in the antecedent of $\Psi, \Gamma, \Theta_i, C_1 \Rightarrow D_1, \Xi_i$ by using all the resulting sequents $\Psi, \Gamma, \Phi, \Pi, \Delta'_k, E \Rightarrow F, \Gamma'_k, C_1$. Furthermore, from each of all these results obtained by eliminating the occurrences of C_1, we eliminate also the occurrences of D_1 by (Cut) by using the other sequent $D_1, \Psi, \Gamma, \Phi, \Pi, \Delta'_k, E \Rightarrow F, \Gamma'_k$ in the nice pair of sequents than the sequent used to eliminate C_1. To sum up, for each k ($1 \leqslant k \leqslant 2^{m-1}$) and i ($1 \leqslant i \leqslant 2^n$), we define a derivation:

$$\mathscr{D}_{(k,i)} = \begin{array}{c}
\begin{array}{cc}\vdots & \mathscr{D}_{li}\\ \Psi,\Gamma,\Phi,\Pi,\Delta'_k,E\Rightarrow F,\Gamma'_k,C_1 & C_1,\Phi,\Gamma,\Theta_i\Rightarrow D_1,\Xi_i\end{array}\\ \hline
\Psi,\Gamma,\Phi,\Pi,\Delta'_k,E,\Phi,\Gamma,\Theta_i\Rightarrow F,\Gamma'_k,D_1,\Xi_i \quad (Cut)\\ \hline\hline
\Psi,\Gamma,\Phi,\Pi,\Delta'_k,E,\Theta_i\Rightarrow F,\Gamma'_k,\Xi_i,D_1 \qquad \begin{array}{c}\vdots\\ D_1,\Psi,\Gamma,\Phi,\Pi,\Delta'_k,E\Rightarrow F,\Gamma'_k\end{array}\\ \hline
\Psi,\Gamma,\Phi,\Pi,\Delta'_k,E,\Theta_i,\Psi,\Gamma,\Phi,\Pi,\Delta'_k,E\Rightarrow F,\Gamma'_k,\Xi_i,F,\Gamma'_k \quad (Cut),\\ \hline\hline
\Psi,\Gamma,\Phi,\Pi,\Delta'_k,\Theta_i,E\Rightarrow F,\Gamma'_k,\Xi_i
\end{array}$$

where the double line means finitely many applications of contraction.

(Step 3) In this step, we want to apply $(\supset)$ to all the resulting sequent

$$\Psi, \Gamma, \Phi, \Pi, \Delta'_k, \Theta_i, E \Rightarrow F, \Gamma'_k, \Xi_i$$

of (Step 2). The number of such sequents is $2^n \cdot 2^{m-1} = 2^{n+m-1}$. Then we obtain

$$\frac{\mathscr{D}_{(1,1)} \quad \mathscr{D}_{(1,2)} \quad \cdots \quad \mathscr{D}_{(1,2^n)} \quad \mathscr{D}_{(2,1)} \quad \cdots \quad \mathscr{D}_{(2,2^n)} \quad \cdots \quad \mathscr{D}_{(2^{m-1},2^n)}}{G_1 \supset H_1, \ldots, G_n \supset H_n, \Gamma, C_2 \supset D_2, \ldots, C_m \supset D_m, \Pi \Rightarrow E \supset F, \Delta, \Sigma} (\supset)$$

Why can we obtain the desired conclusion in the above application of $(\supset)$? Let us define $G_{n+1} \supset H_{n+1}, \ldots, G_{n+m-1} \supset H_{n+m-1}$ as $C_2 \supset D_2, \ldots, C_m \supset D_m$. Then, it is easy from this identification to see that the principal formulas of the antecedent of the conclusion of $(\supset)$ are appropriately introduced, where recall that $\Psi = G_1 \supset H_1, \ldots, G_n \supset H_n$ and $\Phi = C_2 \supset D_2, \ldots, C_m \supset D_m$. □

In the next section, we also provide another syntactic proof of this theorem by reducing the admissibility of cut of **G3BPL** to that of a sequent calculus for modal logic **K4** with the help of our proof-theoretic (or constructive) embedding.

4.1 Soundness and Completeness

We can show the equivalence of **G3BPL** and **LBP**. Note that there is a difference between **G3BPL** and **LBP** of concepts of a sequent. Recall in **G3BPL** a sequents is a pair of multisets, but in **LBP** a sequent is a pair of sets. To establish the equivalence of **G3BPL** and **LBP**, we need to define the function $(-)^*$ sending a multiset to the corresponding set. For the direction from **LBP** to **G3BPL**, we use the admissibility of weakening of **G3BPL**.

Proposition 3 *Let Γ and Δ be finite multisets of formulas. Then* **LBP** $\vdash \Gamma^* \Rightarrow \Delta^*$ *iff* **G3BPL** $\vdash \Gamma \Rightarrow \Delta$.

From the soundness and completeness of **LBP** for Kripke semantics and the equivalence of **G3BPL** and **LBP**, it follows that **G3BPL** is sound and complete for the class of all transitive Kripke frames. First of all, we state that the soundness and completeness for **LBP** in Ishii et al. (2001). Note that the concept of validity of a sequent in **LBP** is the same as that of **G3BPL**.

Fact 1 (Ishii et al. 2001) *Let Γ and Δ be finite sets of formulas. Then* **LBP** $\vdash \Gamma \Rightarrow \Delta$ *iff $\Gamma \Rightarrow \Delta$ is valid in any model.*

Theorem 2 (Soundness and Completeness) *Let Γ and Δ be finite multisets of formulas. Then* **G3BPL** $\vdash \Gamma \Rightarrow \Delta$ *iff $\Gamma \Rightarrow \Delta$ is valid in any model.*

Proof For the left-to-right direction, we first assume that **G3BPL** $\vdash \Gamma \Rightarrow \Delta$. From Proposition 3 we obtain **LBP** $\vdash \Gamma^* \Rightarrow \Delta^*$. Then, from Fact 1 it follows that $\Gamma^* \Rightarrow \Delta^*$ is valid in any model. Hence, it is easily seen that $\Gamma \Rightarrow \Delta$ is valid in any model. For the opposite direction, we assume that $\Gamma \Rightarrow \Delta$ is valid in any model. It is obvious that $\Gamma^* \Rightarrow \Delta^*$ is valid in any model. From Fact 1 it follows that **LBP** $\vdash \Gamma^* \Rightarrow \Delta^*$. Then, from Proposition 3 we obtain **G3BPL** $\vdash \Gamma \Rightarrow \Delta$. □

5 Proof-Theoretic Embedding from BPL to Modal Logic K4

In this section, we establish that **G3BPL** can be embedded into **G3K4**. We first present a **G3**-style non-labelled sequent calculus for **K4**.

5.1 G3-Style Non-labelled Sequent Calculus for K4

The modal syntax $\mathscr{ML}$ is obtained by adding to $\mathscr{L}$ the modal operator $\Box$, where we keep the same set Atom of atomic variables as $\mathscr{L}$. Given finite multisets Γ and Δ of modal formulas, we say that the expression of the form $\Gamma \Rightarrow \Delta$ is a *sequent*. The notions of derivability, admissibility, etc. in **G3K4** are defined similarly to **G3BPL**. The sequent calculus for modal logic **K4** is obtained by adding the following rule to **LK**:

$$\frac{\Gamma, \Box\Gamma \Rightarrow A}{\Box\Gamma \Rightarrow \Box A}\ .$$

This sequent calculus for **K4** is first presented by Sambin and Valentini (1982). We note that Sambin and Valentini's system enjoys the cut-elimination theorem (see Sambin and Valentini 1982, Theorem 2.4).

It is easy to define a **G3**-style non-labelled sequent calculus for **K4**, denoted by **G3K4** as in Table 3. That is, we add the following rule:

$$\frac{\Box\Gamma, \Gamma \Rightarrow A}{\Pi, \Box\Gamma \Rightarrow \Box A, \Sigma}\ (\Box_4)$$

to a **G3**-style non-labelled sequent calculus for classical logic in Troelstra and Schwichtenberg (2000). This rule for $\Box$ is easily seen to be obtained by revising the above rule by Sambin and Valentini in Sambin and Valentini (1982).

The notion of context in all rules of **G3K4** except for $(\Box_4)$ is similarly defined to that of **G3BPL**. In $(\Box_4)$, Π and Σ are the *context*. In the conclusion of each rule of **G3K4** except for $(\Box_4)$, the formula(s) not in the context is called the *principal formula(s)*. For $(\Box_4)$, $\Box\Gamma$ and $\Box A$ in the conclusion are the *principal formulas*. In **G3K4** the admissibility of hp-weakening, hp-invertibility for all rules except for $(\Box_4)$, and the admissibility of hp-contraction hold. The system **G3K4** also enjoys the cut-elimination theorem as follows.

Fact 2 *The rule of cut is admissible in* **G3K4**.

$$\frac{\Gamma \Rightarrow \Delta, A \quad A, \Pi \Rightarrow \Sigma}{\Gamma, \Pi \Rightarrow \Delta, \Sigma}\ (Cut).$$

Table 3 G3-style sequent calculus: **G3K4**

(Axioms)

$$\frac{}{P, \Gamma \Rightarrow \Delta, P}\ (Id) \qquad \frac{}{\bot, \Gamma \Rightarrow \Delta}\ (L\bot)$$

(Logical Rules)

$$\frac{A, B, \Gamma \Rightarrow \Delta}{A \& B, \Gamma \Rightarrow \Delta}\ (L\&) \qquad \frac{\Gamma \Rightarrow \Delta, A \quad \Gamma \Rightarrow \Delta, B}{\Gamma \Rightarrow \Delta, A \& B}\ (R\&)$$

$$\frac{A, \Gamma \Rightarrow \Delta \quad B, \Gamma \Rightarrow \Delta}{A \vee B, \Gamma \Rightarrow \Delta}\ (L\vee) \qquad \frac{\Gamma \Rightarrow \Delta, A, B}{\Gamma \Rightarrow \Delta, A \vee B}\ (R\vee)$$

$$\frac{\Gamma \Rightarrow \Delta, A \quad B, \Gamma \Rightarrow \Delta}{A \supset B, \Gamma \Rightarrow \Delta}\ (L\supset) \qquad \frac{A, \Gamma \Rightarrow \Delta, B}{\Gamma \Rightarrow \Delta, A \supset B}\ (R\supset)$$

$$\frac{\Box\Gamma, \Gamma \Rightarrow A}{\Pi, \Box\Gamma \Rightarrow \Box A, \Sigma}\ (\Box_4)$$

5.2 *Embedding Theorem*

Now let us define our version of Gödel–McKinsey–Tarski translation as follows (as noted in the introduction, this version of translation already appeared in Visser 1981).

Definition 7 (*Translation* $\boxplus$)

$$\begin{array}{ll} P^{\boxplus} & := P \& \Box P, \\ \bot^{\boxplus} & := \bot, \\ (A \supset B)^{\boxplus} & := \Box(A^{\boxplus} \supset B^{\boxplus}), \\ (A * B)^{\boxplus} & := A^{\boxplus} * B^{\boxplus} \ (* \in \{\&, \vee\}). \end{array}$$

For a finite multiset $\Gamma \equiv A_1, ..., A_n$, we define $\Gamma^{\boxplus} := A_1^{\boxplus}, ..., A_n^{\boxplus}$.

We note that this translation was also employed to establish the proof-theoretic embedding from **BPL** into **K4** within the labelled system (see Yamasaki and Sano 2016).

In what follows, we are going to establish the following equivalence with our *non*-labelled sequent calculus:

$$\mathbf{G3BPL} \vdash \Gamma \Rightarrow \Delta \text{ iff } \mathbf{G3K4} \vdash \Gamma^{\boxplus} \Rightarrow \Delta^{\boxminus}.$$

First, we establish the left-to-right direction (Lemma 6). To establish this, we need to prepare two lemmas.

Lemma 4 **G3K4** $\vdash A^{\boxplus} \Rightarrow \Box A^{\boxplus}$.

Proof By induction on the complexity of A. If $A \equiv P$, then we show that $P\&\Box P \Rightarrow \Box(P\&\Box P)$.

$$\dfrac{\dfrac{\dfrac{P, \Box P \Rightarrow P \quad \dfrac{P, \Box P \Rightarrow P}{P, \Box P \Rightarrow \Box P}(\Box_4)}{P, \Box P \Rightarrow P\&\Box P}(R\&)}{P, \Box P \Rightarrow \Box(P\&\Box P)}(\Box_4)}{P\&\Box P \Rightarrow \Box(P\&\Box P)}(L\&) \quad ,$$

where note that each occurrence of Ps in the antecedent of the conclusion of the applications of $(\Box_4)$ is in the context.

If $A \equiv B\&C$, then we show that $B^{\boxplus}\&C^{\boxplus} \Rightarrow \Box(B^{\boxplus}\&C^{\boxplus})$. Assume that we have the derivations of $B^{\boxplus} \Rightarrow \Box B^{\boxplus}$ and $C^{\boxplus} \Rightarrow \Box C^{\boxplus}$ by induction hypothesis. Then, we obtain the following derivation:

$$\dfrac{\dfrac{\dfrac{\dfrac{\begin{matrix}\vdots \\ B^{\boxplus} \Rightarrow \Box B^{\boxplus}\end{matrix}}{B^{\boxplus}, C^{\boxplus} \Rightarrow \Box B^{\boxplus}}(hp-weak) \quad \dfrac{\begin{matrix}\vdots \\ C^{\boxplus} \Rightarrow \Box C^{\boxplus}\end{matrix}}{B^{\boxplus}, C^{\boxplus} \Rightarrow \Box C^{\boxplus}}(hp-weak)}{B^{\boxplus}, C^{\boxplus} \Rightarrow \Box B^{\boxplus}\&\Box C^{\boxplus}}(R\&) \quad \dfrac{\dfrac{\dfrac{\dfrac{\begin{matrix}\vdots \\ \Box B^{\boxplus}, \Box C^{\boxplus}, B^{\boxplus}, C^{\boxplus} \Rightarrow B^{\boxplus}\end{matrix} \quad \begin{matrix}\vdots \\ \Box B^{\boxplus}, \Box C^{\boxplus}, B^{\boxplus}, C^{\boxplus} \Rightarrow C^{\boxplus}\end{matrix}}{\Box B^{\boxplus}, \Box C^{\boxplus}, B^{\boxplus}, C^{\boxplus} \Rightarrow B^{\boxplus}\&C^{\boxplus}}(R\&)}{\Box B^{\boxplus}, \Box C^{\boxplus} \Rightarrow \Box(B^{\boxplus}\&C^{\boxplus})}(\Box_4)}{\Box B^{\boxplus}\&\Box C^{\boxplus} \Rightarrow \Box(B^{\boxplus}\&C^{\boxplus})}(L\&)}{B^{\boxplus}, C^{\boxplus} \Rightarrow \Box(B^{\boxplus}\&C^{\boxplus})}(Cut)}{B^{\boxplus}\&C^{\boxplus} \Rightarrow \Box(B^{\boxplus}\&C^{\boxplus})}(L\&) \quad .$$

where $(hp-weak)$ means the applications of hp-weakening.

If $A \equiv B \vee C$, then the proof is similar to that of the case &. If $A \equiv B \supset C$, then it is obvious, because $\Box(B^{\boxplus} \supset C^{\boxplus}) \Rightarrow \Box\Box(B^{\boxplus} \supset C^{\boxplus})$ is an instance of axiom schema: $\Box A \Rightarrow \Box\Box A$. This axiom can be derivable in **G3K4**. □

Lemma 5 *Let* $\Theta^{\boxplus}$ *be the multiset of the translated formulas. The following rule is admissible in* **G3K4**.

$$\dfrac{\Theta^{\boxplus}, \Box\Gamma, \Gamma \Rightarrow A}{\Theta^{\boxplus}, \Box\Gamma, \Pi \Rightarrow \Box A, \Sigma}(\Box') \quad .$$

Proof It is sufficient to show that **G3K4** $\vdash \Theta^{\boxplus}, \Box\Gamma, \Gamma \Rightarrow A$ implies **G3K4** $\vdash \Theta^{\boxplus}, \Box\Gamma, \Pi \Rightarrow \Box A, \Sigma$. Assume that $\vdash \Theta^{\boxplus}, \Box\Gamma, \Gamma \Rightarrow A$ in **G3K4**. From $\vdash \Theta^{\boxplus}, \Box\Gamma, \Gamma \Rightarrow A$ it follows that $\vdash \Box\Theta^{\boxplus}, \Theta^{\boxplus}, \Box\Gamma, \Gamma \Rightarrow A$ by applying hp-weakening. Let $\Box\Theta^{\boxplus} = \{\Box B^{\boxplus} \mid B \in \Theta\}$. Then we apply $(\Box_4)$ to $\vdash \Box\Theta^{\boxplus}, \Theta^{\boxplus}, \Box\Gamma, \Gamma \Rightarrow A$ to obtain $\vdash \Box\Theta^{\boxplus}, \Box\Gamma,$ ¥$Pi \Rightarrow \Box A, \Sigma$. From Lemma 4 we have $B^{\boxplus} \Rightarrow \Box B^{\boxplus}$ for any B. Now it suffices to apply (Cut) to each member $\Box B^{\boxplus}$ of $\Box\Theta^{\boxplus}$ with $B^{\boxplus} \Rightarrow \Box B^{\boxplus}$. Consequently, we obtain $\vdash \Theta^{\boxplus}, \Box\Gamma, \Pi \Rightarrow \Box A, \Sigma$. □

Lemma 5 means that we may put the context $\Theta^{\boxplus}$ in the application of $(\Box_4)$.

Lemma 6 **G3BPL** $\vdash \Gamma \Rightarrow \Delta$ *implies* **G3K4** $\vdash \Gamma^{\boxplus} \Rightarrow \Delta^{\boxplus}$.

Proof By induction on height n of a derivation of $\Gamma \Rightarrow \Delta$ in **G3BPL**. Assume that there is a derivation of $\Gamma \Rightarrow \Delta$ in **G3BPL**. If the height of this derivation is 0, then $\Gamma \Rightarrow \Delta$ is an axiom. If $\Gamma \Rightarrow \Delta$ is an axiom (that is, (Id) or $(L\bot)$), then the translation $\Gamma^{\boxplus} \Rightarrow \Delta^{\boxplus}$ is clearly derivable.

Let us consider the case where the height of the derivation is more than 0. Suppose that the last applied rule is $(\supset)$, i.e., we have the following derivation:

$$\dfrac{\begin{matrix}\vdots & & \vdots\\ C_1\supset D_1, \dots, C_m\supset D_m, \Delta_1, \Gamma', A \Rightarrow B, \Gamma_1 & \dots & C_1\supset D_1, \dots, C_m\supset D_m, \Delta_{2^m}, \Gamma', A \Rightarrow B, \Gamma_{2^m}\end{matrix}}{\Gamma', C_1\supset D_1, \dots, C_m\supset D_m \Rightarrow A\supset B, \Delta'}\,(\supset),$$

where $m \geq 0$, $\Gamma_i = \{C_j \mid j \in \gamma(i)\}$, $\Delta_i = \{D_j \mid j \in \delta(i)\}$, $1 \leqslant i \leqslant 2^m$, and $\gamma(i)$ runs through the subsets of $\{1, \dots, m\}$ and $\delta(i) = \{1, \dots, m\} \setminus \gamma(i)$.

By induction hypothesis, we obtain the sequents $\{\Box(C_i^{\boxplus}\supset D_i^{\boxplus})\}_{1\leqslant i\leqslant m}, \Delta_i^{\boxplus}, \Gamma'^{\boxplus}, A^{\boxplus} \Rightarrow B^{\boxplus}, \Gamma_i^{\boxplus}$, where $1 \leqslant i \leqslant 2^m$. For these sequents we apply finite many applications of $(L\supset)$ and construct. Furthermore, we apply $(R\supset)$ to the result of applications of $(L\supset)$ and then apply $(\Box_4')$. Thus, we obtain the following derivation in **G3K4**:

$$\dfrac{\dfrac{\dfrac{\begin{matrix}\vdots & \vdots\\ \Gamma'^{\boxplus}, \{\Box(C_i^{\boxplus}\supset D_i^{\boxplus})\}_{1\leqslant i\leqslant m}, \{C_i^{\boxplus}\supset D_i^{\boxplus}\}_{1\leqslant i\leqslant m-1}, A^{\boxplus} \Rightarrow B^{\boxplus}, C_m^{\boxplus} & D_m^{\boxplus}, \Gamma'^{\boxplus}, \{\Box(C_i^{\boxplus}\supset D_i^{\boxplus})\}_{1\leqslant i\leqslant m}\ \{C_i^{\boxplus}\supset D_i^{\boxplus}\}_{1\leqslant i\leqslant m-1}, A^{\boxplus} \Rightarrow B^{\boxplus}\end{matrix}}{\Gamma'^{\boxplus}, \{\Box(C_i^{\boxplus}\supset D_i^{\boxplus})\}_{1\leqslant i\leqslant m}, \{C_i^{\boxplus}\supset D_i^{\boxplus}\}_{1\leqslant i\leqslant m}, A^{\boxplus} \Rightarrow B^{\boxplus}}\,(L\supset)}{\Gamma'^{\boxplus}, \{\Box(C_i^{\boxplus}\supset D_i^{\boxplus})\}_{1\leqslant i\leqslant m}, \{C_i^{\boxplus}\supset D_i^{\boxplus}\}_{1\leqslant i\leqslant m} \Rightarrow A^{\boxplus}\supset B^{\boxplus}}\,(R\supset)}{\Gamma'^{\boxplus}, \{\Box(C_i^{\boxplus}\supset D_i^{\boxplus})\}_{1\leqslant i\leqslant m} \Rightarrow \Box(A^{\boxplus}\supset B^{\boxplus}), \Delta'^{\boxplus}}\,(\Box_4'),$$

whose end sequent is the result of the translation $\Gamma', C_1\supset D_1, \dots, C_m\supset D_m \Rightarrow A\supset B, \Delta'$. The argument for the remaining other cases is similar to that for the case of $(\supset)$. $\Box$

Second, we establish the right-to-left direction by the following Lemma 7. This is because we can show the right-to-left direction of Theorem 3 as a special case of Lemma 7 by putting $\Pi = \Sigma = \emptyset$.

Lemma 7 *Let Γ and Δ be finite multisets of formulas and let Π and Σ be finite multisets of atomic formulas. Then,*

$$\mathbf{G3K4} \vdash \Gamma^{\boxplus}, \Pi, \Box\Pi \Rightarrow \Sigma, \Delta^{\boxplus} \ \textit{implies}\ \mathbf{G3BPL} \vdash \Gamma, \Pi \Rightarrow \Sigma, \Delta.$$

Proof By induction on height n of a derivation of $\Gamma^{\boxplus}, \Pi, \Box\Pi \Rightarrow \Sigma, \Delta^{\boxplus}$ in **G3K4**. If $n = 0$, $\Gamma^{\boxplus}, \Pi, \Box\Pi \Rightarrow \Sigma, \Delta^{\boxplus}$ is an axiom (there are just two cases: (Id) and $(L\bot)$) in **G3K4**, so $\Gamma, \Pi \Rightarrow \Sigma, \Delta$ is also an axiom in **G3BPL**.

If $n > 0$, we divide our argument into cases depending on the last rule of the derivation. Since Π and Σ are finite multisets of atomic formulas for the syntax $\mathscr{L}$, the outermost logical connective of a formula in the translations $\Gamma^{\boxplus}$ and $\Delta^{\boxplus}$ never be the implication symbol $\supset$. So, the last applied rule must be other than the rules for $\supset$. In what follows, we consider the following cases: (i) the last applied rule is $(L\vee)$ or $(R\vee)$; (ii) the last applied rule is $(L\&)$ or $(R\&)$; (iii) the last applied rule is $(\Box_4)$.

(i) The last applied rule is $(L\vee)$ or $(R\vee)$: the straightforward application of induction hypothesis gives us the required derivation in **G3BPL**.

(ii) The last applied rule is $(L\&)$ or $(R\&)$: we distinguish two further cases: 1) $P^{\boxplus} \equiv P\&\Box P$ is the principal formula of the rule, and 2) $(A\&B)^{\boxplus} \equiv A^{\boxplus}\&B^{\boxplus}$ is the principal formula of the rule. The latter case 2) is similarly proved to the case (i). Here we concentrate on the case of 1). We first suppose that the last applied rule is $(L\&)$, i.e., the derivation in **G3K4** is of the following form:

$$\dfrac{\vdots \atop P, \Box P, \Gamma'^{\boxplus}, \Pi, \Box\Pi \Rightarrow \Sigma, \Delta^{\boxplus}}{P\&\Box P, \Gamma'^{\boxplus}, \Pi, \Box\Pi \Rightarrow \Sigma, \Delta^{\boxplus}}\ (L\&).$$

By induction hypothesis, we obtain, from the premise of $(L\&)$, the following derivation in **G3BPL** as required:

$$\vdots \atop P, \Gamma', \Pi \Rightarrow \Sigma, \Delta.$$

Second, we suppose that the last applied rule is $(R\&)$. Then, the last step of this derivation looks like:

$$\dfrac{\vdots \atop \Gamma^{\boxplus}, \Pi, \Box\Pi \Rightarrow \Sigma, \Delta'^{\boxplus}, P \quad \vdots \atop \Gamma^{\boxplus}, \Pi, \Box\Pi \Rightarrow \Sigma, \Delta'^{\boxplus}, \Box P}{\Gamma^{\boxplus}, \Pi, \Box\Pi \Rightarrow \Sigma, \Delta'^{\boxplus}, P\&\Box P}\ (R\&).$$

Then, we apply induction hypothesis to the *left* premise of $(R\&)$ to obtain the desired derivation:

$$\vdots \atop \Gamma, \Pi \Rightarrow \Sigma, \Delta', P.$$

(iii) Let the last applied rule be $(\Box_4)$ in **G3K4**. Then the original derivation is of the following form $(m \geq 0)$:

$$\dfrac{\vdots \atop \Pi'', \Box\Pi'', C_1^{\boxplus}\supset D_1^{\boxplus}, \ldots, C_m^{\boxplus}\supset D_m^{\boxplus}, \Box(C_1^{\boxplus}\supset D_1^{\boxplus}), \ldots, \Box(C_m^{\boxplus}\supset D_m^{\boxplus}) \Rightarrow A^{\boxplus}\supset B^{\boxplus}}{\Gamma'^{\boxplus}, \Pi', \Box\Pi', \Pi'', \Box\Pi'', \Box(C_1^{\boxplus}\supset D_1^{\boxplus}), \ldots, \Box(C_m^{\boxplus}\supset D_m^{\boxplus}) \Rightarrow \Box(A^{\boxplus}\supset B^{\boxplus}), \Delta'^{\boxplus}, \Sigma}\ (\Box_4),$$

where we note that $\Gamma'^{\boxplus}$, Π', $\Box\Pi'$, Π'' and $\Delta'^{\boxplus}$, Σ are the context of the conclusion of the rule ($\Box_4$), Γ' and Δ' are multisets of $\mathscr{L}$-formulas of **BPL**, and all of Π', Π'' and Σ are multisets of atomic formulas. What we want to show is that the following sequent is derivable in **G3BPL**:

$$\Gamma', \Pi', \Pi'', C_1 \supset D_1, \ldots, C_m \supset D_m \Rightarrow A \supset B, \Delta', \Sigma.$$

Then it suffices for us to derive all the premises of the following application to the rule ($\supset$):

$$\dfrac{\begin{array}{c}\vdots \\ \Delta_1, \Gamma', \Pi', \Pi'', C_1{\supset}D_1, \ldots, C_m{\supset}D_m, A \Rightarrow B, \Gamma_1\end{array} \quad \ldots \quad \begin{array}{c}\vdots \\ \Delta_{2^m}, \Gamma', \Pi', \Pi'', C_1{\supset}D_1, \ldots, C_m{\supset}D_m, A \Rightarrow B, \Gamma_{2^m}\end{array}}{\Gamma', \Pi', \Pi'', C_1{\supset}D_1, \ldots, C_m{\supset}D_m \Rightarrow A{\supset}B, \Sigma, \Delta'}\ (\supset),$$

where $m \geq 0$, $\Gamma_i = \{C_j \mid j \in \gamma(i)\}$, $\Delta_i = \{D_j \mid j \in \delta(i)\}$, $1 \leqslant i \leqslant 2^m$, and $\gamma(i)$ runs through the subsets of $\{1, \ldots, m\}$ and $\delta(i) = \{1, \ldots, m\} \setminus \gamma(i)$. We note that Γ', Π', Π'', Σ and Δ' are the context in the conclusion of ($\supset$). Let $1 \leqslant i \leqslant 2^m$. So it suffices to show that the following sequent is derivable in **G3BPL**:

$$\Delta_i, \Gamma', \Pi', \Pi'', C_1{\supset}D_1, \ldots, C_m{\supset}D_m, A \Rightarrow B, \Gamma_i.$$

Now let us recall the original derivation above. We apply hp-invertibility for ($R{\supset}$) to the premise of ($\Box_4$) in the original derivation to obtain that

$$\Pi'', \Box\Pi'', C_1^{\boxplus}{\supset}D_1^{\boxplus}, \ldots, C_m^{\boxplus}{\supset}D_m^{\boxplus}, \Box(C_1^{\boxplus}{\supset}D_1^{\boxplus}), \ldots, \Box(C_m^{\boxplus}{\supset}D_m^{\boxplus}), A^{\boxplus} \Rightarrow B^{\boxplus}$$

is also derivable in **G3K4**. Furthermore, we apply hp-invertibility for ($L{\supset}$) to this sequent to have that:

$$\Delta_j^{\boxplus}, \Pi'', \Box\Pi'', \Box(C_1^{\boxplus}{\supset}D_1^{\boxplus}), \ldots, \Box(C_m^{\boxplus}{\supset}D_m^{\boxplus}), A^{\boxplus} \Rightarrow B^{\boxplus}, \Gamma_j^{\boxplus}$$

is derivable in **G3K4** for all $1 \leqslant j \leqslant 2^m$. Recall that we have fixed i such that $1 \leqslant i \leqslant 2^m$. Since we can derive

$$\Delta_i^{\boxplus}, \Pi'', \Box\Pi'', \Box(C_1^{\boxplus}{\supset}D_1^{\boxplus}), \ldots, \Box(C_m^{\boxplus}{\supset}D_m^{\boxplus}), A^{\boxplus} \Rightarrow B^{\boxplus}, \Gamma_i^{\boxplus}$$

in **G3K4** with at most the same height as the premise of ($\Box_4$) in the original derivation, we can still apply induction hypothesis to this sequent to obtain that:

$$\Delta_i, \Pi'', C_1{\supset}D_1, \ldots, C_m{\supset}D_m, A \Rightarrow B, \Gamma_i$$

is derivable in **G3BPL**. Then hp-weakening to the antecedent of this sequent (we add Γ' and Π') enables us to conclude that our desired sequent is derivable in **G3BPL**. □

From Lemmas 6 and 7, we can establish the proof-theoretic embedding from **G3BPL** into **G3K4**.

Theorem 3 **G3BPL** $\vdash \Gamma \Rightarrow \Delta$ iff **G3K4** $\vdash \Gamma^{\boxplus} \Rightarrow \Delta^{\boxplus}$.

Let us mention two important consequences of Theorem 3. As a consequence of the proof-theoretic embedding, we can reduce the cut admissibility and the decidability of **BPL** to those of **K4**. Although we used the admissibility of (Cut) for **G3K4** in Lemma 6, we can prove Lemma 7 without (Cut). Hence, we can reduce the cut admissibility of **G3BPL** to that of **G3K4**. Let us say that a logic **L** is *decidable* if there exists an effective algorithm such that, for any input formula A, the algorithm can decide whether A is provable in **L** or not. Here we focus on the reduction of the admissibility of cut, since the reduction of the decidability is shown similarly.

Theorem 4 *The following rule of cut is admissible in* **G3BPL**.

$$\frac{\Gamma \Rightarrow \Delta, A \quad A, \Pi \Rightarrow \Sigma}{\Gamma, \Pi \Rightarrow \Delta, \Sigma}\ (Cut).$$

Proof For the reduction of the admissibility of cut, it suffices to show that $\vdash \Gamma \Rightarrow \Delta, A$ and $\vdash A, \Pi \Rightarrow \Sigma$ jointly imply $\vdash \Gamma, \Pi \Rightarrow \Delta, \Sigma$ in **G3BPL**. We proceed as follows: assume $\Gamma \Rightarrow \Delta, A$ and $A, \Pi \Rightarrow \Sigma$ are derivable in **G3BPL**. Then by these assumptions and Theorem 3, $\Gamma^{\boxplus} \Rightarrow \Delta^{\boxplus}, A^{\boxplus}$ and $A^{\boxplus}, \Pi^{\boxplus} \Rightarrow \Sigma^{\boxplus}$ are also derivable in **G3K4**. By the admissibility of cut in **G3K4**, $\Gamma^{\boxplus}, \Pi^{\boxplus} \Rightarrow \Delta^{\boxplus}, \Sigma^{\boxplus}$ is derivable in **G3K4**. Finally, we apply Lemma 7 to obtain that $\Gamma, \Pi \Rightarrow \Delta, \Sigma$ is derivable in **G3BPL**. □

6 Extending Our Proof-Theoretic Embedding

Our argument for proof-theoretic embedding can be extended to a pair of **DNT**(= **BPL** $\oplus \neg\neg\top$) and **KD4**(= **K4** $\oplus \Diamond\top$),[1] if the **G3**-style sequent calculus for these logics are provided. Both **DNT** and **KD4** are semantically characterized by the class of all serial (for any $s \in S$, for some $s' \in S$, sRs') and transitive Kripke frames. Let us consider how we can obtain each of **G3**-style sequent calculi for **DNT** and **KD4**. The cut-free sequent calculus for **DNT** was presented in Ishigaki and Kashima (2008). We can obtain the **G3**-style sequent calculus for **DNT** by referring to Ishigaki and Kashima (2008). If we want to have a **G3**-style sequent calculus for **DNT** (**G3DNT**), then it suffices to add to **G3BPL** the following rule:

[1]The embedding from **DNT** into **KD4** can be found in Ma and Sano (2015). Note that they use the semantic method from Chagrov and Zakharyaschev (1997) and the translation sends P into $\Box P$.

$$\frac{C_1 \supset D_1, \ldots, C_n \supset D_n, \Delta_1, \Sigma \Rightarrow \Gamma_1 \quad \ldots \quad C_1 \supset D_1, \ldots, C_n \supset D_n, \Delta_{2^n}, \Sigma \Rightarrow \Gamma_{2^n}}{\Sigma, C_1 \supset D_1, \ldots, C_n \supset D_n \Rightarrow \Theta} (\supset_{\mathrm{D}}),$$

where $n \geq 0$, $\Gamma_i = \{C_j \mid j \in \gamma(i)\}$, $\Delta_i = \{D_j \mid j \in \delta(i)\}$, $1 \leqslant i \leqslant 2^n$, and $\gamma(i)$ runs through the subsets of $\{1, \ldots, n\}$ and $\delta(i) = \{1, \ldots, n\} \setminus \gamma(i)$. Since the sequent calculus for **KD4** is already presented in Goble (1974), we can easily obtain the **G3**-style sequent calculus **G3KD4** for **KD4** as well as the case of **K4**. For **G3KD4**, we add to **G3K4** the following rule:

$$\frac{\Box\Gamma, \Gamma \Rightarrow}{\Sigma, \Box\Gamma \Rightarrow \Theta} \Box_{\mathrm{D}}.$$

We can establish the admissibility of the cut rule for both **G3DNT** and **G3KD4** by the standard argument for a **G3**-style sequent calculus. For **G3DNT**, the similar argument for **G3BPL** works. We can establish the following extension of our embedding argument to **G3DNT** and **G3KD4**:

Theorem 5 **G3DNT** $\vdash \Gamma \Rightarrow \Delta$ iff **G3KD4** $\vdash \Gamma^{\boxplus} \Rightarrow \Delta^{\boxplus}$.

7 Further Direction

There are several further directions of this work. The first direction is to extend our embedding argument to other pairs of logics. In Sect. 6, we presented the case of **DNT** and **KD4**, but there are more embedding relations, e.g. logics of strict implication and the corresponding modal logics (Yamasaki and Sano 2016). In Yamasaki and Sano (2016), they established the proof-theoretic embedding from logics of strict implication into the corresponding modal logics in a uniform manner via a labelled system, because as we have mentioned in the introduction, each labelled system can be obtained easily. For a non-labelled **G3**-style sequent calculus, according to the manner of our proof-theoretic embedding argument, we may establish also the proof-theoretic embedding for other pairs of logics. Since it seems that we need to provide a non-labelled **G3**-style sequent calculi for each logic individually, it would be a nice result to establish the proof-theoretic embedding for many logical systems in a uniform manner at the same time.

The second direction is to establish Craig interpolation property (Craig 1957) for **G3BPL**. In general, when we established cut-elimination for some logical systems, the subformula property of this system is one of the important corollaries of cut-elimination. One of the other possible applications of cut-elimination is Craig interpolation property. It seems difficult to establish Craig interpolation property for **BPL**, because the implication rule for **BPL** is more complex than that of **Int**. But the interpolation property for **BPL** has already been shown in Ardeshir (1995); Ardeshir and Ruitenburg (2001); Aghaei and Ardeshir (2001). In Aghaei and Ardeshir (2006), the algebraic counterpart for interpolation property for **BPL** is presented. Is it possible to establish Craig interpolation property for **G3BPL** in terms of proof-theoretic methods?

Acknowledgements We would like to thank an anonymous reviewer for his/her invaluable comments and suggestions. In particular we owe the reviewer for the simple notation used in (Step 2) and (Step 3) of the case (v) in the proof of Theorem 1. We are grateful to Ryo Kashima for setting opportunities for the first author to give the presentation on this topic at Tokyo Institute of Technology. We would like to thank the audiences of the events where the content of this paper is presented, including the 50th MLG meeting at Kyoto, Japan, The Joint Conference of The Third Asian Workshop on Philosophical Logic (AWPL 2016) and The Third Taiwan Philosophical Logic Colloquium (TPLC 2016) in Taiwan. The first author's visit to Taiwan was supported by the grant from travel awards for students and young researchers of AWPL-TPLC 2016. The work of the first author was supported by JSPS KAKENHI Grant Number JP 15J07255. The work of the second author was partially supported by JSPS KAKENHI Grant-in-Aid for Young Scientists (B) Grant Number 15K21025 and JSPS Core-to-Core Program (A. Advanced Research Networks).

References

Aghaei, M., & Ardeshir, M. (2001). Gentzen-style axiomatizations for some conservative extentions of basic propositional logic. *Studia Logica*, *68*(2), 263–285.

Alizadeh, M., & Ardeshir, M. (2006). Amalgamation property for the class of basic algebras and some of its natural subclasses. *Archive for Mathematical Logic*, *45*, 913–930.

Ardeshir, M. (1995). *Aspects of Basic Logic*. Ph.D thesis, Marquette University.

Ardeshir, M., & Ruitenburg, W. (2001). Basic propositional calculus II, interpolation. *Archive for Mathematical Logic*, *40*, 349–364.

Chagrov, A., & Zakharyaschev, M. (1997). *Modal logic*. Oxford: Oxford University Press.

Corsi, G. (1987). Weak logics with strict implication. *Mathematical Logic Quarterly*, *33*, 389–406.

Craig, W. (1957). Linear reasoning. A new form of the Herbrand-Gentzen theorem. *The Journal of Symbolic Logic*, *22*(3), 250–268.

Dyckhoff, R., & Negri, S. (2012). Proof analysis in intermediate logics. *Archive for Mathematical Logic*, *51*, 71–92.

Goble, L. (1974). Gentzen systems for modal logics. *Notre Dame Journal of Formal Logic*, *15*, 455–461.

Gödel, K. (1933). Eine interpretation des intuitionistischen Aussagenkalküls. *Ergebnisse Eines Mathematischen Kolloquiums*, *4*, 39–40.

Ishigaki, R., & Kashima, R. (2008). Sequent calculi for some strict implication logics. *Logic Journal of the IGPL*, *16*(2), 155–174.

Ishii, K., Kashima, R., & Kikuchi, K. (2001). Sequent calculi for Visser's propositional logics. *Notre Dame Journal of Formal Logic*, *42*(1), 1–22.

Kikuchi, K. (2001). Relationships between basic propositional calculus and substructural logics. *Bulletin of the Section of Logic*, *30*(1), 15–20.

Ma, M., & Sano, K. (2015). Extensions of Basic Propositional Logic. In X. Zhao, X. Feng, B. Kim, & L. Yu (Eds.), *Proceedings of the 13th Asian Logic Conference*, 13th Asian Logic Conference, Guangzhou, China, 16–20 September 2013, World Scientific (pp. 170–200).

Maehara, S. (1954). Eine Darstellung der intuitionistischen Logik in der klassischen. *Nagoya Mathematical Journal*, *7*, 45–64.

McKinsey, J. C. C., & Tarski, A. (1948). Some theorems about the sentential calculi of Lewis and Heyting. *Journal of Symbolic Logic*, *13*, 1–15.

Mints, G. (2012). The Gödel-Tarski Translations of Intuitionistic Propositional Formulas. In E. Erdem, J. Lee, Y. Lierler, & D. Pearce (Eds.), *Correct Reasoning*. Lecture Notes in Computer Science (Vol. 7265, pp. 487–491). Berlin, Heidelberg: Springer.

Negri, S., & von Plato, J. (2001). *Structural proof theory*. Cambridge: Cambridge University Press.

Prawitz, D., & Malmnäs, P.-E. (1968). A survey of some connections between classical, intuitionistic and minimal logic. *Studies in Logic and the Foundations of Mathematics*, *50*, 215–229.

Restall, G. (1994). Subintuitionistic logics. *Notre Dame Journal of Formal Logic*, *35*, 116–129.

Ruitenburg, W. (1991). Constructive logic and the paradoxes. *Modern Logic*, *1*(4), 271–301.

Ruitenburg, W. (1993). Basic logic and fregean set theory. *Dirk van Dalen Festschrift, Quaestiones Infinitae*, *5*, 121–142.

Sambin, G., & Valentini, S. (1982). The modal logic of provability. The sequential approach. *Journal of Philosophical Logic*, *11*(3), 311–342.

Suzuki, Y., & Ono, H. (1997). Hilbert-style proof system for BPL. Technical Report IS-RR-97-0040F, Japan Advanced Institute of Science and Technology.

Suzuki, Y., Wolter, F., & Zakharyaschev, M. (1998). Speaking about transitive frames in propositional languages. *Journal of Logic, Language and Information*, *7*(3), 317–339.

Troelstra, A.S., & Schwichtenberg, H. (2000). *Basic proof theory*, 2nd edn. Cambridge: Cambridge University Press.

Visser, A. (1981). A propositional logic with explicit fixed points. *Studia Logica*, *40*(2), 155–175.

Yamasaki, S., & Sano, K. (2016). Constructive embedding from extensions of logics of strict implication into modal logics. In S. M. Yang, D. M. Deng, & H. Lin (Eds.), *Structural analysis of non-classical logics*, Logic in Asia: Studia Logica Library (pp. 223–251). Berlin, Heidelberg: Springer.

Varieties of Parthood

Paul Hovda

Abstract We assume that there are multiple varieties of parthood, or multiple relations that are importantly similar to parthood, and that are related to each other by specific principles we consider. We are motivated by the idea that the class of these relations is closed under certain operations, so that, for example, the transitive closure of one variety of parthood is itself a variety of parthood. Another type of principle we embrace is that parthood relations can be merged to form new ones. We show that our principles about this class yield well-defined formal patterns that connect in clear ways to familiar forms of mereology, including Classical Extensional Mereology. We introduce a novel closure condition on the class, involving what we will call the overlap-inclusion of a given relation. We connect this notion to the classical mereological axiom of Strong Supplementation. We introduce a notion of sameness of structure that is slightly weaker than the standard notion of isomorphism, that we call quasi-isomorphism. We then study some formal patterns that govern the merging of a new domain and part-like relation with an old domain and relation, together with our closure conditions. Our main result is that, under a natural general assumption about how new things relate to old ones, we can show that, when we ignore the differences between quasi-isomorphic structures, there is basically only one way to extend the old structure: all additions are quasi-isomorphic to a substructure of a certain natural extension of the original, which is itself quasi-isomorphic to a Classical Mereology.

Keywords Mereology · Classical mereology · Mereological pluralism · Compositional pluralism · Parthood

1 Introduction

The basic intuitive idea driving the work in this paper is that there are multiple relations that are *like* the relation of part-to-whole, and various general principles

P. Hovda (✉)
Department of Philosophy, Reed College, 3203 Southeast Woodstock Boulevard, Portland, OR 97202, USA
e-mail: hovdap@reed.edu

S.C.-M. Yang et al. (eds.), *Philosophical Logic: Current Trends in Asia*,
Logic in Asia: Studia Logica Library, DOI 10.1007/978-981-10-6355-8_13

that govern and organize these relations. A variation of this idea is that there are multiple ways of being a part; another is that there are multiple relations that are equally good candidates for being called "the" part-whole relation.

To illustrate one version of this idea, we might think that there are distinct primitive manners of being a part: for example, Socrates' nose is a part of Socrates in one manner (call it R), and Socrates is a part of the set {Socrates} in a different manner (call it S). And it seems plausible that Socrates' nose is not a part of the set in any primitive manner. But it is a part in some derivative manner. We might say that there is a derivative form of parthood that is the "union" of R and S, call it $R \cup S$. Now the nose does not bear even this relation to the set. But it does bear the ancestral of this relation to the set. So we have a fourth, more inclusive or "tolerant" form of being a part.

There is a much more detailed discussion of informal versions of the ideas pursued here in Hovda (2016). The basic outlook is deeply indebted to Kit Fine's (Fine 2010), though there are some differences of approach. A key difference throughout is that we do not assume that all forms of parthood are anti-symmetric. As a brief motivation, we suggest that it is a reasonable view that a statue and a lump of clay might be parts of one another, in a suitably tolerant sense, without being identical. Another difference from Fine is that we do not here take mereological summation or fusion as a primitive "building operation."

Some of the main informal ideas that motivate our work are these general closure principles: if P is a part-like relation, so is its ancestral, and so is the relation (defined below) we call the *overlap-inclusion* of its ancestral; if P and Q are part-like relations, so is there union. Another general principle that motivates us is that given objects typically bear primitive parthood relations to abstract objects (e.g., sets) "constructed out of" them.

Here is an over-view of what is to come: we introduce the notion of the overlap-inclusion of a given relation, and show that it is intimately connected with the traditional mereological axiom or principle of *strong supplementation*. We then discuss a notion of "sameness of structure" that is weaker than the standard notion of isomorphism, that we call *quasi-isomorphism*. The basic idea is quite simple: structures are quasi-isomorphic if they would be isomorphic when failures of anti-symmetry are "factored out."

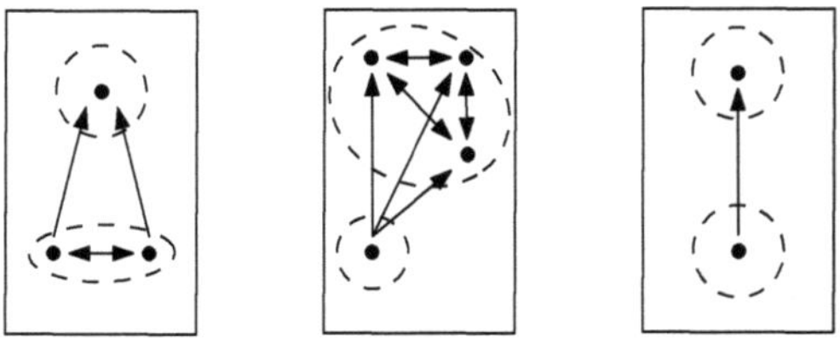

three non-isomorphic, quasi-isomorphic reflexive relations

We connect overlap-inclusion to the notions of (mereological) fusion and the axioms of Classical Mereology. Next, we look at some formal structure that concerns the "merging" of relations, which corresponds to the step in the informal presentation above where we went from the separate primitive ways of being a part (R and S) to their union, and then to the ancestral of that union. We show that under a natural general assumption about how "new" things relate to "old" ones, we can show that, when we ignore the differences between quasi-isomorphic structures, there is really only one way to extend the "old" structure: all additions are quasi-isomorphic to a substructure of a certain natural extension of the original, which is itself quasi-isomorphic to a Classical Mereology. The following figure gives a quick rough impression of the result, applied to a simple illustration:

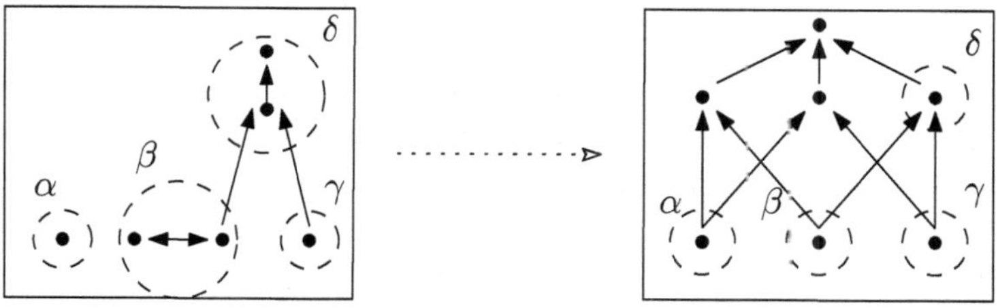

Roughly put, if the structure on the left shows an original domain and some primitive parthood relations on it, then, under a certain main variety of parthood, every "completion" of it is quasi-isomorphic to a substructure of the structure on the right (where the relational links on the right include the transitive closure of those shown).

Most of the arguments below are easy in the sense that they should be accessible to anyone with a basic undergraduate-level background in logic and set theory. The reader should be able to skip any given proof and pick up later, although some forms of argument that are spelled out in detail in one proof will be compressed or left implicit in later proofs.

One reason for a bit of the complexity of the details here is that we attempt to remain fairly neutral about reflexivity. In Hovda (2016), the reflexive closure of a part-like relation is taken to be itself part-like, and we do not disagree with that here, but we do not stress it, and our definitions are slightly different, in order to accommodate, as much as possible, any part-like relations there might be that are neither reflexive nor irreflexive.

2 Transitivity

As an introduction to our notation and style, let us consider the general idea of the *ancestral of* or *transitive closure of* a relation. Let R be any relation. Officially, we will work in set theory, and we will think of relations as being sets of ordered pairs. Almost everything we say would be straightforward to adapt to a different background setting, and we may unofficially talk a little differently about relations at times. (In particular, much that we say could be adapted to a setting in which a relation is "too big" to be represented as a *set* of ordered pairs.) As is standard, we will write $x\ R\ y$ to indicate that $\langle x, y\rangle \in R$. And we will sometimes speak of x "bearing R to" y.

Now the ancestral of R, call it $\tau(R)$ or ${}^{\tau}R$ for short, is that relation such that for arbitrary x and y

$x\ {}^{\tau}R\ y$ iff there is a sequence of things $(z_1, \ldots, z_n)$, such that for each $i < n$, $z_i\ R\ z_{i+1}$, and $x = z_1$ and $y = z_n$.

We may indicate such a sequence as a "chain"

$x\ R\ z_2\ R\ z_3\ R\ \ldots z_{n-1}\ R\ z_n.$

Now it is not hard to see that ${}^{\tau}T$ is a transitive relation, and that it includes R in the sense that if $x\ R\ y$ then $x\ {}^{\tau}R\ y$. It can be shown that ${}^{\tau}T$ is the "smallest" such relation, in the sense that any transitive relation that includes R in this sense also includes ${}^{\tau}T$. Accordingly, we know that if R is itself transitive, then ${}^{\tau}T$ is simply R itself (given, that is, that a relation is just a set of ordered pairs).

As a result, instead of saying that R is transitive, we could simply say $R = {}^{\tau}T$, and we will sometimes do this in the following.

To remind ourselves of our motivation, we conjecture that for every part-like relation R, ${}^{\tau}R$ is also a part-like relation.

3 Overlap-Inclusion and Strong Supplementation

3.1 Overlap

For any relation R, let the "field" of R, or $fld(R)$ for short, be the set of all things x such that there is a y with either $x\ R\ y$ or $y\ R\ x$.

Notational convention: for any relation S, let us write

$x\ S^{=}\ y$

for $x\ S\ y \vee x = y$.

Definition 1 Given any R, define the relation of (left-) R-overlap as follows:

$$\forall x \forall y (x \circ_R y \leftrightarrow (x \in fld(R) \wedge y \in fld(R) \wedge \exists z (z\ R^{=}\ x \wedge z\ R^{=}\ y)))$$

We use the word "overlap," and use this notation because of the connection with parthood and standard formal treatments of parthood.

The following four lemmas hold no matter what R, x, and y are:

Lemma 1 *if $x\ R\ y$, then $x \circ_R y$.*

Lemma 2 *$x \in fld(R)$ iff $x \circ_R x$ iff $x \in fld(\circ_R)$.*

Lemma 3 *$x \circ_R y$ iff $y \circ_R x$.*

Lemma 4 *If R is transitive, then: if* $x\ R\ y$ *and* $w \circ_R x$*, then* $w \circ_R y$.

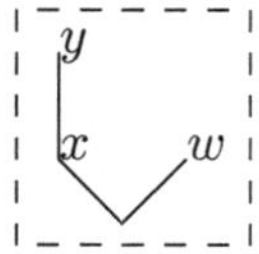

These are all easy to see; above is a visual "proof" for the last. ⊣

3.2 Overlap-Inclusion: General

Definition 2 Let R be any relation. We define the "(left-)overlap-inclusion" of R, which we will here denote as $\sigma(R)$, or ${}^\sigma R$ for short, as follows:

$$\forall x \forall y(x\ {}^\sigma R\ y \leftrightarrow (x \in fld(R) \wedge y \in fld(R) \wedge \forall z(z \circ_R x \rightarrow z \circ_R y))).$$

We will write "$x\ {}^\sigma R^=\ y$" for "$x\ {}^\sigma R\ y \vee x = y$." Note these three facts about overlap-inclusions in general:

Lemma 5 $x \in fld(R)$ *iff* $x\ {}^\sigma R\ x$ *iff* $x \in fld({}^\sigma R)$.

Lemma 6 ${}^\sigma R$ *is transitive.*

Lemma 7 *if* $x \in fld(R)$*, then:* $x\ {}^\sigma R^=\ y \rightarrow x \circ_R y$.

Perhaps the third of these is not immediately obvious, but it follows easily once we note that $x \circ_R x$. ⊣

3.3 Overlap-Inclusion of a Transitive Relation

Let T be any relation that is transitive. Then we get a number of nice results about ${}^\sigma T$.

Lemma 8 *For* $x, y \in fld(T)$*: if* $x\ T^=\ y$ *then* $x\ {}^\sigma T\ y$.

This is easy to prove by appeal to Lemmas 4 and 5.

Lemma 9 $x \circ_T y$ *iff* $x \circ_{{}^\sigma T} y$.

Proof Left-to-right: get z witnessing $x \circ_T y$; using Lemma 8 it is easy to get that z witnesses $x \circ_{{}^\sigma T} y$. Right-to-left: get z witnessing $x \circ_{{}^\sigma T} y$. Since $z\ {}^\sigma T^=\ x$ we get $z \circ_T x$ (recall Lemma 7); let w witness this, so we have $w\ T^=\ z$ and $w\ T^=\ x$. Now note that since $w \circ_T z$ and $z\ {}^\sigma T^=\ y$, we may get that $w \circ_T y$; let v witness this.

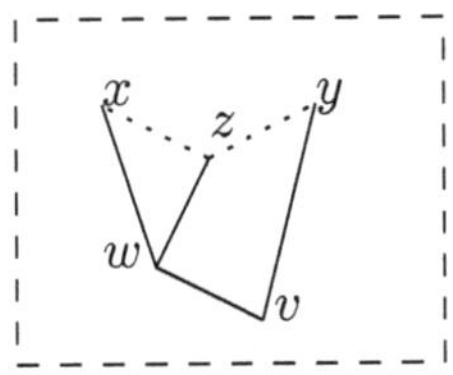

Now we have $v\ T^{=}\ w$ and $w\ T^{=}\ x$, so, by transitivity of T we have $v\ T^{=}\ x$; but also $v\ T^{=}\ y$, so v witnesses $x \circ_T y$. ⊣

3.4 Strong Supplementation

Definition 3 We will say that a relation Q "obeys strong supplementation" when

$$(\forall a, b \in fld(Q))\ (\forall x(x \circ_Q a \rightarrow x \circ_Q b) \rightarrow a\ Q\ b)$$

The name "strong supplementation" is standard in the literature; see Simons (1987) as well as Varzi (2016) and the references therein. (There is also a principle standardly called "weak supplementation," but, despite the names, the two principles are, strictly speaking, independent. The one called "strong" does not imply the "weak" one without further assumptions, not all of which we are making here.)

Theorem 1 (where T is transitive) *${}^{\sigma}T$ obeys strong supplementation; i.e., we have that*

$$(\forall a, b \in fld({}^{\sigma}T))\ (\forall x(x \circ_{{}^{\sigma}T} a \rightarrow x \circ_{{}^{\sigma}T} b) \rightarrow a\ {}^{\sigma}T\ b).$$

Proof Suppose the antecedent. Now, to show that $a\ {}^{\sigma}T\ b$, we suppose that $x \circ_T a$ and need to show that $x \circ_T b$. Apply Lemma 9 to get $x \circ_{{}^{\sigma}T} a$, and our hypothesis to get $x \circ_{{}^{\sigma}T} b$; applying Lemma 9 again gets us $x \circ_T b$. ⊣

There is an alternate form of strong supplementation:

Lemma 10 $\forall x(x\ {}^{\sigma}T\ a \rightarrow x \circ_{{}^{\sigma}T} b) \rightarrow a\ {}^{\sigma}T\ b$.

Proof Left to the reader; do not forget that T is assumed to be transitive. ⊣

The ancestral of a transitive relation R is (extensionally) identical with R; we show that a similar result obtains for overlap-inclusion.

Lemma 11 *If T is transitive and obeys strong supplementation, then $a\ T\ b \leftrightarrow a\ {}^{\sigma}T\ b$.*

Proof We may assume $a, b \in fld(T)$ (since otherwise we are done by Lemma 5). Assume the hypothesis, so $\forall x(x \circ_T a \rightarrow x \circ_T b) \rightarrow a\ T\ b$. Now $a\ T\ b \rightarrow a\ {}^{\sigma}T\ b$ is immediate from Lemma 8. And for $a\ {}^{\sigma}T\ b$ to hold is for $\forall x(x \circ_T a \rightarrow x \circ_T b)$, hence we get $a\ T\ b$ by hypothesis. ⊣

Lemma 11 with Theorem 1 obviously yield

Lemma 12 (for any transitive relation T)

(A) ${}^{\sigma}T = \sigma(\sigma(T))$, *i.e.,* $x\ \sigma(T)\ y \leftrightarrow x\ \sigma(\sigma(T))\ y$
(B) T *obeys strong supplementation iff* $T = {}^{\sigma}T$.

Hence, in the following, instead of saying "R obeys strong supplementation," we may, when it is known that R is transitive, say simply $R = {}^{\sigma}R$. And note that we may now conclude that for any relation R whatsoever, ${}^{\sigma}({}^{\tau}R)$ obeys strong supplementation, etc. Accordingly, for arbitrary relations Q, let us write ${}^{\Sigma}Q$ for ${}^{\sigma}({}^{\tau}Q)$. Note that we have

$$ {}^{\Sigma}R = R \longleftrightarrow (R \text{ is transitive and } R \text{ obeys strong supplementation}). $$

In Hovda (2016), an argument is given for the conclusion that (using our notation), if R is a (transitive) part-like relation, then so is ${}^{\sigma}R$. Given the principle that the transitive closure of a part-like relation is itself part-like, this is equivalent to saying that for any part-like R, ${}^{\Sigma}R$ is part-like.

3.5 Reflexivity

While it is common in formal mereology to work with a relation that is taken to be reflexive or irreflexive, we are allowing that a given parthood relation could be neither. (For a brief example motivating this idea, imagine that we take set-theoretic membership to be a variety of parthood, and we allow for there to be a set that is a member of itself, as well as one that is not.) This causes some slight deviations in some of our formulations, such as in our definition of overlap above.

Let us look at the situation in a little more detail. Supposing that R is a variety of parthood, one natural but quite strict conception of overlap would say that x overlaps y iff there is a z that bears R to both x and y. But it is also very natural to consider a notion of overlap that ensures that if $x\ R\ y$, then x overlaps y (relative to R), even if x fails to bear R to itself. The notion we have focussed on ensures this, but it is in fact even more permissive, since we have that $x \circ_R x$ iff $x \in fld(R)$. So our notion automatically includes, as cases of x "overlapping" y, two types of case that are not necessarily included on the strictest notion of overlap: the case in which $x\ R\ y$ and the case in which $x = y$. This might seem somewhat artificial, but it is necessary for the way our notion yields nice definitions, of ${}^{\sigma}R$, and of the principle of strong supplementation.

However we define our terms, we are motivated by the thought that if R is a variety of parthood, then so is (the relation we have called) ${}^{\Sigma}R$, and, remarkably, this relation is guaranteed to satisfy a principle very much like that of the classical strong supplementation axiom; our definitions have the virtue of letting us state all this fairly smoothly. Whatever you call it, the relation ${}^{\Sigma}R$ is, we want to say, is a parthood-like relation, though a tolerant (rather than strict), and perhaps derivative (rather than primitive) one.

In fact, however, there may be reason to consider a broader conception of a tolerant, derivative form of parthood, based on a given form of parthood R. For it may be that R, roughly speaking, naturally "goes with" a certain domain D that is in fact a proper superset of $fld(R)$. For one example, on some views, (primitive) parthood among concrete objects is a relation that runs from part x to concrete whole y only when x is among some things that collectively behave in a unified manner (and these will be the parts of y). On this sort of view, if b is a concrete thing that has no parts, and does not join with any other things in unified behavior, then b is not part of anything. Yet, suppose that partless z is primitively a part of something; in this case, given our definitions, and denoting primitive concrete parthood with R, z will bear ${}^{\Sigma}R$ to itself, but b will not bear ${}^{\Sigma}R$ to itself. We should consider the relation Q that x bears to y just when either $x\ {}^{\Sigma}R\ y$ or ($x = y$ and x is a concrete thing). There is a sense in which primitive concrete parthood is "appropriate" for object b even though b is not in its field, while it is not "appropriate" for an abstract thing like the number 3, which is also not in its field. One way to reflect this is to regard the relation Q as a variety of parthood, while denying that the relation $Q \cup \langle 3, 3\rangle$ is a variety of parthood, or at least holding that the former is more natural than the latter. In fact, Q may be regarded as more natural than ${}^{\Sigma}R$.

In the following, we will continue to make a certain amount of room for relations that are neither reflexive nor irreflexive, but we will also bear in mind situations like the example of the R and Q just discussed. In general form, we have a domain D, a relation R whose field is a subset of D and we wish to speak of the relation that results from "expanding" ${}^{\Sigma}R$ to include all the $\langle x, x\rangle$ pairs where $x \in D$. We could achieve this effect in a number of equivalent ways. To illustrate, let $=_D$ be the relation of identity restricted to D, so that our target is $({}^{\Sigma}R) \cup =_D$. There are many ways we could reach this target. It can be shown that

$$^{\tau}(R) \cup =_D \quad = \quad ^{\tau}(R \cup =_D)$$

and in fact

$$({}^{\Sigma}R) \cup =_D \quad = \quad ^{\sigma}(^{\tau}(R)) \cup =_D \quad = \quad ^{\sigma}(^{\tau}(R \cup =_D)) \quad = \quad ^{\sigma}(^{\tau}(R) \cup =_D).$$

Thus $=_D$ can be "added," so to speak, at any of a few stages along the way from our given parthood relation R, to our eventual target. In the example recently discussed it might well be thought that (the correlate of) $R \cup =_D$ is itself a natural part-like relation, since we might say that each concrete thing itself behaves in a unified manner, hence, in a limit-case way, is automatically among some "things" (namely it and itself) that together behave in a unified manner. Hence there is some justification for looking at situations where we have a relation R that is reflexive on a given associated domain D (i.e., $x\ R\ x$ for all $x \in D$). This is effectively what we will do in the final section of the paper.

3.6 Fusion

We will define a notion of fusion that is basically standard, but (as is less familiar) explicitly relativized to a relation, and which is slightly more tolerant of irreflexive or non-reflexive relations.

Definition 4 Define, for any relation Q

$$Fu_Q(b, [x|\phi_x])$$

as

$$\forall x\,(\phi_x \to x\ Q^= b) \wedge \forall w\,(w\ Q^= b \to \exists x\,(\phi_x \wedge x \circ_Q w)).$$

We formulate our definition schematically, with the idea that ϕ_z could be any formula of any language whatsoever (provided that language includes the logical apparatus used here, and a relation-symbol Q, or suitable surrogates). If we wished to be more rigorous, of course, we could specify some formal language instead. As an abbreviation device, when it is clear that X is a set, we will write $Fu_Q(b, X)$ to mean $Fu_Q(b, [x|x \in X])$.

Lemma 13 *If* $Fu_Q(b, [x|\phi_x])$, *then:*

(A) $\exists x\,\phi_x$
(B) $b \in fld(Q)$
(C) $\forall x\,(\phi_x \to x \in fld(Q))$.

Proof Straightforward; note that $b\ Q^= b$ by logic alone, and $fld(Q) = fld(\circ_Q)$. ⊣

Lemma 14 *If* $T =^{\tau} T$ *then:*

(A) $Fu_T(b, [x|\phi_x]) \;\to\; Fu_{\circ^\sigma T}(b, [x|\phi_x])$
(B) $Fu_T(b, [x|\phi_x]) \leftrightarrow$
$[b \in fld(T) \wedge \forall x\,(\phi_x \to x\ T^= b) \wedge \forall w\,(w \circ_T b \to \exists x\,(\phi_x \wedge x \circ_T w))]$
(C) $Fu_{\circ^\sigma T}(b, [x|\phi_x]) \leftrightarrow$
$[b \in fld(T) \wedge \forall w\,(w \circ_{\circ^\sigma T} b \leftrightarrow \exists x\,(\phi_x \wedge x \circ_{\circ^\sigma T} w))]$
(D) *if both* $Fu_{\circ^\sigma T}(b, [x|\phi_x])$ *and* $Fu_{\circ^\sigma T}(b', [x|\phi_x])$, *then both* $b\ {}^\sigma T\ b'$ *and* $b'\ {}^\sigma T\ b$
(E) *if both* $Fu_T(b, [x|\phi_x])$ *and* $Fu_T(b', [x|\phi_x])$, *then both* $b\ {}^\sigma T\ b'$ *and* $b'\ {}^\sigma T\ b$
(F) *if* $b \in fld(T)$, *then* $Fu_T(b, [x|x\ T^= b])$
(G) *if both* $b\ {}^\sigma T\ b'$ *and* $b'\ {}^\sigma T\ b$, *and also* $Fu_{\circ^\sigma T}(b, [x|\phi_x])$, *then* $Fu_{\circ^\sigma T}(b', [x|\phi_x])$.

Proof Now for (A), suppose $\forall x\,(\phi_x \to x\ T^= b) \wedge \forall w\,(w\ T^= b \to \exists x\,(\phi_x \wedge x \circ_T w))$. Then $\forall x\,(\phi_x \to x\ {}^\sigma T^= b)$ is immediate from Lemma 8 (and the fact that $fld(T) = fld({}^\sigma T)$). Now suppose that $w\ {}^\sigma T^= b$; we need to show that $\exists x\,(\phi_x \wedge x \circ_{\circ^\sigma T} w))$. By Lemma 7, $w \circ_T b$; let z witness this, so $z\ T^= w$ and $z\ T^= b$. The latter and our hypothesis get us an $x \in G$ with $(\phi_x \wedge x \circ_T z))$. Apply Lemma 4 and conclude $x \circ_T w$; then Lemma 9 yields $x \circ_{\circ^\sigma T} w$, and we have (A). (B) is clear from applications of Lemmas 1, 2, and 4.

For (C), the left-to-right direction is similar to (B). For the right-to-left: supposing the right side, we first show that $\forall x\,(\phi_x \rightarrow x\ {}^{\sigma}T^{=}\ b)$. Now if we have an x with ϕ_x, we get that for any z, if $z \circ_{\sigma T} x$ then, by supposition $z \circ_{\sigma T} b$. But Theorem 1 then yields $x\ {}^{\sigma}T\ b$. And it is easy to confirm the other fusion condition, so we have $Fu_{\sigma T}(b, [x|\phi_x])$.

To get (D): supposing the hypothesis and that $z \circ_T b$, we want to show that $z \circ_T b'$. We get $z \circ_{\sigma T} b$, and we may then activate the fusion condition on b (and, in view of (B)) get an x such that ϕ_x and $z \circ_{\sigma T} x$ Given the fusion condition on b', and (C), we have $z \circ_{\sigma T} b'$, so $z \circ_T b'$, as desired. Hence $b\ {}^{\sigma}T\ b'$ and the same reasoning yields $b'\ {}^{\sigma}T\ b$. (E) is immediate from (A) and (D). (F) and (G) are easy to confirm. ⊣

Note that parts (B) and (C) show us, in effect, alternative ways to define fusion (given background assumptions of transitivity or obedience of strong supplementation). See Hovda (2009) and Varzi (2016) for discussion of various definitions of fusion.

Definition 5 For any relation R and sets A and B, write

$$A \cong_{\circ_R} B$$

for $A, B \subseteq fld(R)$, and $\forall x((\exists y \in A)\, x \circ_R y \leftrightarrow (\exists y \in B)\, x \circ_R y)$.

Lemma 15 *Suppose $S = {}^{\Sigma}S$ and that A and B are sets. Then*

$$Fu_S(b, A) \rightarrow (Fu_S(b, B) \leftrightarrow A \cong_{\circ_S} B).$$

Proof Supposing $Fu_S(b, A)$, the left-to-right direction is easy (cf. Lemma 14(C)). For the right-to-left, assume $A \cong_{\circ_S} B$ and let $x \in B$. We will argue that $x\ S\ b$ by way of strong supplementation (noting that $x, b \in fld(S)$). Suppose $z \circ_S x$; by hypothesis, get $y \in A$ with $z \circ_S y$, hence $z \circ_S b$, and we thus deduce $x\ S\ b$. Next, we need to show that if $z\ S^{=}\ b$, then $(\exists y \in B)\, z \circ_S b$. But this is straightforward from our hypotheses. ⊣

4 Quasi-isomorphism

Later, we will show that under certain general and natural assumptions, there is a sense in which our principles for finding new part-like relations, given old ones, are guaranteed to stop generating any novel structure after only a few applications. Roughly, a certain structure arises quickly, as a "fixed point." To make this more precise, we will use a notion of "quasi-isomorphism," which we will now define and investigate several formal features of, especially in connection to part-like relations and the axioms of Classical Mereology.

First, let us articulate the standard idea of an isomorphism between two relational structures:

Let R and S be relations whose fields are subsets of D, and E, respectively. Then f is an *isomorphism* from $\langle D, R\rangle$ to $\langle E, S\rangle$ iff

$f : D \to E$ (i.e., f is a function with domain D and whose range is a subset of E); and
f is one-to-one (i.e., $f(x) = f(y) \;\to\; x = y$); and
f is onto E (i.e., $(\forall y \in E)(\exists x \in D)\, y = f(x)$); and
$(\forall x, y \in D)\ (x\ R\ y \leftrightarrow f(x)\ S\ f(y))$.

We say that two structures are isomorphic when there exists an isomorphism from one to the other. Such pairs of structures are "structurally identical;" R is transitive iff S is transitive, etc.

Now, the intuitive version of our notion of quasi-isomorphism is this: suppose we are considering relations R and S that are transitive. Now if, for some x, y with $x \neq y$, $x\ R\ y$ and $y\ R\ x$, then x and y are "equivalent" relative to R: in particular, $x\ R\ a \leftrightarrow y\ R\ a$ and $a\ R\ x \leftrightarrow a\ R\ y$, no matter what a is. Now it could be that $\langle D, R\rangle$ and $\langle E, S\rangle$ are not isomorphic, but "would be if we identify equivalent things." This is roughly what we mean when we say that they are "quasi-isomorphic."

To make this more precise, we will first define the idea of a "reduction" of a given structure, which will be a substructure of it, that it is quasi-isomorphic to, and which is quasi-isomorphic to no "smaller" substructure. The reduction will basically just choose one out of every collection of equivalent things, and throw the rest of the things out, and leave the remaining relational links intact. (At times it would have been easier to work with new relations relating equivalence-classes of our old things, rather than working with restrictions of the old relations, but, for our eventual application, we wish our approach to be more directly adaptable to situations in which the equivalence-classes in question would be "too big to be a set.")

First, a couple abbreviations:

we will write $x \equiv_R y$ to mean that $(x\ R\ y \wedge y\ R\ x)$;
we will write $x \equiv_R^= y$ to mean that $(x \equiv_R y \vee x = y)$.

Definition 6 $\langle D^-, R^-\rangle$ is a substructure of $\langle D, R\rangle =_{def} D^- \subseteq D$ and $R^- = R \restriction D^-$ (i.e., $(\forall x, y)(x\ R^-\ y \leftrightarrow (x \in D^- \wedge y \in D^- \wedge x\ R\ y)))$.

Now for our next notions:

Definition 7 Say that $c : D \to D$ "reduces" $\langle D, R\rangle$ (to $\langle D^-, R^-\rangle$) iff R is a relation with $fld(R) \subseteq D$, meeting these conditions:

(i) $(\forall x \in D)\ x \equiv_R^= c(x)$ and
(ii) $(\forall x, y \in D)\ (x \equiv_R^= y \;\leftrightarrow\; c(x) = c(y))$ and
(iii) $D^- = ran(c)$ and $R^- = R \restriction D^-$.

Say that $\langle D^-, R^-\rangle$ is a "reduction" of $\langle D, R\rangle$ if there is a c that reduces $\langle D, R\rangle$ to $\langle D^-, R^-\rangle$.

Say that $\langle D, R\rangle$ is "quasi-isomorphic" to $\langle E, S\rangle$ if there is a reduction of the one that is isomorphic to a reduction of the other.

Now, this definition makes quasi-isomorphism non-transitive. For example, consider the structures pictured below, where a solid arrow indicates a holding of the relation, and a circle around a point indicates the holding of the relation from that point to itself.

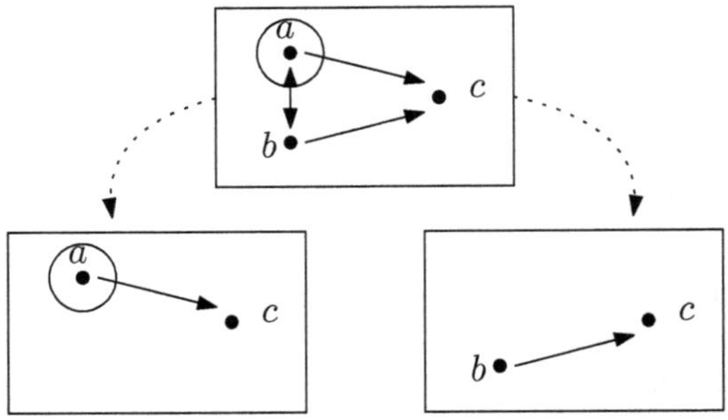

The structure at the top may be reduced, by our definition, to either of the ones below. But those two are not isomorphic, yet each reduces to itself, and hence would be quasi-isomorphic to itself, so we would have a failure of transitivity of "quasi-isomorphism".

Yet this kind of situation will only arise with relations that are not transitive, as we will show with Theorem 2.

Lemma 16 *Let c reduce $\langle D, R\rangle$ to $\langle D^-, R^-\rangle$, and let $a, b \in D^-$. Then:*

(A) $a\ R\ b \leftrightarrow a\ R^-\ b$
(B) $a = c(a)$
(C) R^- *is anti-symmetric (i.e.,* $(a\ R^-\ b \wedge b\ R^-\ a) \rightarrow a = b)$
(D) if R is anti-symmetric, then $(\forall x \in D)\, c(x) = x$.

Proof (A) is obvious. For (B), if $a = c(x)$, then, since $x \equiv^{=}_{R} a$, $c(x) = c(a)$. For (C), if $(a\ R^-\ b \wedge b\ R^-\ a)$, then $a \equiv^{=}_{R} b$ so $c(a) = c(b)$, but then, applying (B), we have $a = c(a) = c(b) = b$. For (D), we are supposing that $\forall x, y\ (x \equiv_R y \rightarrow x = y)$. So $(x \equiv^{=}_{R} y \rightarrow x = y)$. Now for $x \in D$, $x \equiv^{=}_{R} c(x)$, so $x = c(x)$. ⊣

From (D) we see that if R is anti-symmetric, then $\langle D, R\rangle$ is "irreducible" in the sense that the one and only reduction of it is the identity map. Then with (C) we see that for any $\langle D, R\rangle$ whatsoever, every reduction of it is irreducible.

4.1 Reduction of a Transitive Relation

We have seen some properties of reductions that do not depend on transitivity, but our main use of the notion is for transitive relations. We first note that for these, we may state the condition of something's being a reduction in somewhat simpler terms:

Lemma 17 *Suppose $R =^{T} R$ (i.e., R is transitive), with $fld(R) \subseteq D$. Then $c : D \rightarrow D$ reduces $\langle D, R\rangle$ iff*

(i) $(\forall x \in D)\ x \equiv_R^= c(x)$ and
(ii) $(\forall x, y \in D)\ (x \equiv_R^= y \ \rightarrow\ c(x) = c(y))$.

Proof Given transitivity of R and that condition (i) is met, we may easily verify that $c(x) = c(y) \ \rightarrow\ x \equiv_R^= y$, since then we have $x \equiv_R^= c(x) \equiv_R^= c(y) \equiv_R^= y$. ⊣

Lemma 18 *Suppose $R =^T R$, and let c reduce $\langle D, R\rangle$, to $\langle D^-, R^-\rangle$. Then:*

(A) R^- *is transitive*
(B) $(\forall x, y \in D):\ x \circ_R y \ \leftrightarrow\ c(x) \circ_R y$
(C) $(\forall x, y \in D):\ x\ R\ y \ \leftrightarrow\ c(x)\ R\ c(y) \ \leftrightarrow\ c(x)\ R^-\ c(y)$
(D) $(\forall a, b \in D^-):\ a \circ_R b \ \leftrightarrow\ a \circ_{R^-} b$
(E) $(\forall x \in D)(\forall a \in D^-):\ x \circ_R a \ \leftrightarrow\ c(x) \circ_{R^-} a$
(F.1) $c(x)\ R^-\ c(x) \ \leftrightarrow\ x\ R\ x \ \leftrightarrow\ (\exists y \in D)\ x \equiv_R y$
(F.2) *if either* $\neg x\ R\ x$ *or* $\neg c(x)\ R^-\ c(x)$, *then* $c(x) = x$
(G) $fld(R^-) \subseteq fld(R)$, *and* $(\forall x \in D)\ (x \in fld(R) \ \leftrightarrow\ c(x) \in fld(R^-))$.

Proof Note that if $\alpha\ R^=\ \beta$ and $\beta\ R^=\ \gamma$ then $\alpha\ R^=\ \gamma$. Now (A) is obvious, given what R^- is. (B) is easy (recalling Lemma 4) since $x\ R^=\ c(x)$ and vice-versa. For the first biconditional of (C), if $x\ R\ y$, we have $c(x)\ R^=\ x\ R\ y\ R^=\ c(y)$; and if $c(x)\ R\ c(y)$, we have $x\ R^=\ c(x)\ R\ c(y)\ R^=\ y$. The second biconditional of (C) is trivial. For (D), left-to-right: suppose $z\ R^=\ a$ and $z\ R^=\ b$ and $a, b \in fld(R)$; then $a, b \in fld(R^-)$ (cf. part (G)), and we have $c(z)\ R^=\ z\ R^=\ a$ and $c(z)\ R^=\ z\ R^=\ b$. So $c(z)\ R^{-=}\ a$ and $c(z)\ R^{-=}\ b$, as desired. The right-to-left part of (D) is easy. (E) can be obtained easily from (B) and (D). The first biconditional in (F.1) is easy from (C), and the second biconditional is elementary. (F.2) is easy and (G) is straightforward, recalling (C). ⊣

(C) makes explicit the way the reduction of $\langle D, R\rangle$ "represents" the structure of R, and helps to justify our term "quasi-isomorphic," as does

Theorem 2 *Suppose R is transitive, and let c reduce $\langle D, R\rangle$, to $\langle D^-, R^-\rangle$. Then every reduction of $\langle D, R\rangle$ is isomorphic to $\langle D^-, R^-\rangle$.*

Proof Suppose that d reduces $\langle D, R\rangle$ to $\langle D_-, R_-\rangle$; we show that $g = d \restriction D^-$ is an isomorphism from $\langle D^-, R^-\rangle$ to $\langle D_-, R_-\rangle$. (g is the function with domain D^- such that for all $x \in D^-$, $g(x) = d(x)$.) Now $(\forall x, y \in D^-)\ (x \neq y \rightarrow \neg x \equiv_R^= y)$, so g is one-to-one. And if $y \in D_-$, $c(y) \in D^-$, and $g(c(y)) \in D_-$. But we also have $y \equiv_R^= c(y) \equiv_R^= g(c(y))$, so $y = g(c(y))$; hence g is onto D_-. And, appealing to Lemma 18(C) (applied to d), we see that for any $x, y \in D^-$, $x\ R^-\ y \ \leftrightarrow\ x\ R\ y \ \leftrightarrow\ g(x)\ R_-\ g(y)$. ⊣

An obvious corollary is

Lemma 19 *If $\langle D, R\rangle$ is quasi-isomorphic to $\langle E, S\rangle$, and both R and S are transitive, then every reduction of the one is isomorphic to every reduction of the other.*

4.2 Features Preserved Across Reductions

Lemma 20 *Suppose* $R =^{\tau} R$ *and that c reduces* $\langle D, R\rangle$, *to* $\langle D^{-}, R^{-}\rangle$. *Then:*

$$R = ^{\sigma}R \leftrightarrow R^{-} = ^{\sigma}(R^{-})$$

Proof We will make repeated appeal to parts of Lemma 18, which we will refer to by their letters. For left-to-right: we are supposing that $(\forall x, y \in fld(R))(\forall z(z \circ_R x \to z \circ_R y) \to x\ R\ y)$. Now suppose that for some $a, b \in fld(R^{-})$, we have that $\forall z(z \circ_{R^-} a \to z \circ_{R^-} b)$; we must show that $a\ R^{-}\ b$. By (G), $a, b \in fld(R)$, and so, by our main supposition, we will be done if we can show that $\forall z(z \circ_R a \to z \circ_R b)$. Now if $z \circ_R a$, then $z \in fld(R)$, and so by (E), $c(z) \circ_{R^-} a$. So, by supposition, $c(z) \circ_{R^-} b$. But now by (D), $c(z) \circ_R b$, so by (B), $z \circ_R b$ and we are done.

For right-to-left: we are supposing that $(\forall a, b \in fld(R^{-}))(\forall z(z \circ_{R^-} a \to z \circ_{R^-} b) \to a\ R^{-}\ b)$. Now suppose that for some $x, y \in fld(R)$, we have that $\forall z(z \circ_R x \to z \circ_R y)$; we must show that $x\ R\ y$. Now we know, by (G), that $c(x), c(y) \in fld(R^{-})$. Suppose now that $z \circ_{R^-} c(x)$. Then $z \in fld(R^{-}) \subseteq D^{-}$ and so $z \circ_R c(x)$, and so $z \circ_R x$. Now, by supposition, $z \circ_R y$, so $z \circ_R c(y)$, and so $z \circ_{R^-} c(y)$. z was arbitrary, so, by our main supposition, $c(x)\ R^{-}\ c(y)$, and hence $x\ R\ y$ and we are done. ⊣

4.2.1 Fusions and Reduction

Recall our abbreviation device that for a set X $Fu_Q(b, X)$ means $Fu_Q(b, [x|x \in X])$.

Lemma 21 *Suppose* $R = {}^{\tau}R$, *and let c reduce* $\langle D, R\rangle$, *to* $\langle D^{-}, R^{-}\rangle$. *Suppose* $X \subseteq D$, $Y \subseteq D^{-}$, $(\forall x \in X)\ c(x) \in Y$ *and* $(\forall y \in Y)(\exists x \in X)\ c(x) = y$. *Then:*

$$Fu_R(b, X) \leftrightarrow Fu_{R^-}(c(b), Y)$$

Proof Left-to-right: if $y \in Y$, get $x \in X$ with $c(x) = y$, so $x\ R^{=}\ b$ by hypothesis, so $c(x)\ R^{-=}\ c(b)$, recalling Lemma 18(C). For the second fusion condition, note $b \in fld(R)$, and confirm $c(b) \in fld(R^{-})$, recalling Lemma 18(G). Now, bearing in mind Lemma 14(B), we argue: if $z \circ_{R^-} c(b)$, then note that $z \in D^{-}$, and we have $z \circ_R c(b)$, so $z \circ_R b$, hence there is an $x \in X$ with $z \circ_R x$; but then $z \circ_{R^-} c(x)$. So we have that if $z \circ_{R^-} c(b)$, then $(\exists y \in Y)\ z \circ_{R^-} y$, as desired.

For the other direction, similarly: if $x \in X$, $c(x) \in Y$, so $c(x)\ R^{-=}\ c(b)$, so $x\ R^{=}\ b$. Note $c(b) \in fld(R^{-})$ so $b \in fld(R)$. And, supposing $z \circ_R b$, we deduce $c(z) \circ_{R^-} c(b)$, hence there is a $y \in Y$ with $c(z) \circ_{R^-} y$. But then there is an $x \in X$ with $c(x) = y$, and we get, appealing to Lemma 18(E), that $c(z) \circ_R x$ and hence $z \circ_R x$ with $x \in X$, as desired. ⊣

4.2.2 Fusion-Completeness

Definition 8 D is R-fusion complete $=_{def}$

D is non-empty and, for every non-empty $X \subseteq D$, $(\exists b \in D)\ Fu_R(b, X)$.

Lemma 22 *If D is R-fusion complete, then* $D \subseteq fld(R)$.

Proof Given $z \in D$, get $b \in D$ with $Fu_R(b, [x|x = z])$, deduce $\exists x(x = z \wedge x \circ_R z)$, hence $z \circ_R z$, so $z \in fld(R)$. ⊣

Lemma 23 *Suppose* $R = {}^{\tau}R$, *and let c reduce* $\langle D, R\rangle$, *to* $\langle D^-, R^-\rangle$. *Then:*

D is R-fusion complete iff D^- *is* R^-*-fusion complete.*

Proof Left-to-right: if Y is a non-empty subset of D^-, then, by hypothesis, there is some $b \in D$ such that $Fu_R(b, Y)$. But Lemma 21 yields $Fu_{R^-}(c(b), Y)$. For the other direction: for any non-empty $X \subseteq D$, let Y be $\{y \in D^- : y = c(x)$ for some $x \in X\}$. By hypothesis, there is a $b \in D^-$ with $Fu_{R^-}(b, Y)$. Since $b = c(b)$, we may deduce, using Lemma 21 again, $Fu^D_R(b, X)$. ⊣

Lemma 24 *Suppose that R and S are transitive and that* $\langle D, R\rangle$ *is quasi-isomorphic to* $\langle E, S\rangle$. *Then:*

(A) ${}^{\sigma}R = R$ *iff* ${}^{\sigma}S = S$
(B) D is R-fusion complete iff E is S-fusion complete.

Proof Get reductions $\langle D^-, R^-\rangle$ and $\langle E^-, S^-\rangle$ of $\langle D, R\rangle$ and $\langle E, S\rangle$. By Lemma 19, $\langle D^-, R^-\rangle$ and $\langle E^-, S^-\rangle$ are isomorphic. Hence, if one of these obeys strong supplementation, so does the other; similarly for fusion completeness. (We leave it to the reader to confirm this elementary result.) Now apply Lemmas 20 and 23. ⊣

4.3 The Existence of Reductions and Quasi-isomorphisms

Next we note that reductions exist, in general, for our desired application:

Lemma 25 *Let R be a transitive relation with* $fld(R) \subseteq D$. *Then there exists a function* $c : D \to D$ *that reduces* $\langle D, R\rangle$.

Proof For each $x \in D$, let $[x]_R = \{y \in D : y \equiv^{=}_R x\}$. $X \subseteq D$ is an equivalence-class under $\equiv_R$ iff there is $x \in D$ such that $X = [x]_R$. By the axiom of choice, there is a function e whose domain is the set of these equivalence-classes and such that for every such class X, $e(X) \in X$. Now let $c : D \to D$ be the function $c(x) = e([x]_R)$ and it is easy to verify that c reduces $\langle D, R\rangle$. (We need transitivity of R to ensure that if $x \equiv^{=}_R y$ then $[x]_R = [y]_R$.) ⊣

Theorem 2 and Lemma 25 tell us that for each transitive relational structure, there are reductions of it, but they are unique "up to isomorphism." In the language of isomorphism types, each transitive relational structure has a unique reduction type. The reduction type is (of) a transitive, anti-symmetric relation.

Lemma 26 *Suppose $\langle C, Q\rangle$ is a substructure of $\langle D, R\rangle$ and $R =^{\tau} R$, and $fld(R) \subseteq D$. Suppose $(\forall x \in D)(\exists y \in C)\ x \equiv_R^{=} y$. Then $\langle D, R\rangle$ is quasi-isomorphic to $\langle C, Q\rangle$, and in particular, there is a $\langle C^-, Q^-\rangle$ that is a reduction of both.*

Proof By Lemma 25, get a $c : C \to C$ that reduces $\langle C, Q\rangle$ to $\langle C^-, Q^-\rangle$. Now, since for each $x \in D$ there is at least one $y \in C$ with $x \equiv_R^{=} y$, we may use the axiom of choice to obtain a function $d : D \to C$ that chooses one of these; i.e., we have $(\forall x \in D)\ d(x) \equiv_R^{=} x$. Then set $b : D \to C$ as $b(x) = c(d(x))$ and it is easy to show that b reduces $\langle D, R\rangle$ to $\langle C^-, Q^-\rangle$. ⊣

4.4 Unique Fusions and Classical Mereology

Definition 9 Say that D is R-fusion !-complete iff:

D is non-empty and for every non-empty $X \subseteq D$, $(\exists! b \in D)\ Fu_R(b, X)$.

Lemma 27 *Suppose $R = {}^{\Sigma}R$, $fld(R) \subseteq D$, and that D is R-fusion complete. Then (by Lemma 22) we have $fld(R) = D$, and:*

(A) R is anti-symmetric iff D is R-fusion !-complete
(B) if c reduces $\langle D, R\rangle$, to $\langle D^-, R^-\rangle$, then D^- is R^--fusion !-complete.

Proof For (A), the left-to-right direction is straightforward from Lemma 14(D) and the anti-symmetry of ${}^{\Sigma}R$ $(=^{\sigma} R)$. The right-to-left direction may be obtained when we observe (Lemma 14(F)) that under the supposed conditions every $y \in D$ is an R-fusion of the things $R^{=} y$, i.e., $Fu_R(y, [x|xR^{=}y])$. For it is easy to see that if $y \equiv^R z$, then also $Fu_R(z, [x|xR^{=}y])$. For (B), combine Lemma 23 with Lemma 16(C).⊣

Structures with a transitive relation R that obeys strong supplementation and where $fld(R)$ is R-fusion !-complete are Classical Mereologies in the sense that the relation R on the domain $fld(R)$ obeys all the axioms of Classical Mereology, interpreting "parthood" or "$\leq$" as R, and where we assume that the fusion-existence axiom (or axiom scheme) is formulated in such a way as to guarantee the actual "completeness" of the structure: basically, for *every* non-empty subset of the domain, there is a fusion for that subset (not just for the subsets that are expressed by an open formula of some language). (Hovda 2009 has an extended discussion of the axioms of Classical Mereology and a brief discussion of the issue of how to ensure the "completeness" in question.) The characteristic features are that parthood is transitive and every non-empty condition has a unique fusion. And so we have that if R is transitive, obeys strong supplementation, and is fusion complete on $D = fld(R)$, then $\langle D, R\rangle$ reduces to (and is thus quasi-isomorphic with) a Classical Mereology.

Let us call a structure $\langle D, R\rangle$ a *sub-Classical Mereology* if $D = fld(R)$, R obeys the axioms of transitivity, strong supplementation, and D is R-fusion-complete. (We ignore here the issue of how fusion-completeness would be expressed axiomatically.) Then we may restate some of the content of Lemma 27 as

Theorem 3 *Every sub-Classical Mereology reduces to (and hence is quasi-isomorphic with) a Classical Mereology.*

This is basically to say that if we add anti-symmetry to the axioms for sub-Classical Mereology, we get Classical Mereology.

4.5 Groundedness

We will now show a number of results that involve situations in which we have a set B that is "grounded" in a set A in the sense that every $y \in B$ has something in A that bears some relevant relation to y.

Definition 10 For any set C and relation R, for any y:

$[R]\,y$ (y's left R-extension) $=_{def} \{x : x\ R\ y\}$ (assuming this set exists)
$[R^{=}]\,y =_{def} \{x : x\ R^{=}\ y\}$ (again, assuming it exists)
$[R_C]\,y$ (y's R-extension in C) $=_{def} \{x \in C : x\ R\ y\}$
$[R^{=}_{C}]\,y =_{def} \{x \in C : x\ R^{=}\ y\}$.

(The qualifications about existence are only relevant if we are working with a relation R that is itself "too big" to be a set. In most of our development, we are officially ignoring such relations, but we do wish many of our results to extend to such relations.)

Definition 11 For any sets C, D and relation R, say that D is R-grounded in C when:

$(\forall y \in D)\ [R_C]\,y \neq \emptyset$, i.e., $(\forall y \in D)(\exists x \in C)\ x\ R\ y$.

Say that D is $R^{=}$-grounded in C when

$(\forall y \in D)\ [R^{=}_{C}]\,y \neq \emptyset$, i.e., $(\forall y \in D)(\exists x \in C)\ x\ R^{=}\ y$.

Lemma 28 *If* $R = {}^{\tau}R$ *and* $x\ R\ y$*, then:*

(A) $[R]\,x \subseteq [R]\,y$, $[R^{=}]\,x \subseteq [R^{=}]\,y$, $[R_C]\,x \subseteq [R_C]\,y$, *and* $[R^{=}_{C}]\,x \subseteq [R^{=}_{C}]\,y$*;*
(B) if, in addition, either $[R^{=}]\,x$ *or* $[R^{=}]\,y$ *(exists and) is* $R^{=}$*-grounded in* C*, and* $a\ R^{=}\ x$ *and* $b\ R^{=}\ y$*, then, if* $a \circ_R b$*, there is a* $w \in C$ *that witnesses the overlap, i.e.,* $w\ R^{=}\ a$ *and* $w\ R^{=}\ b$.

Proof Part (A) is obvious; for part (B), observe that we have a z such that $z\ R^{=}\ a$ and $z\ R^{=}\ b$; but then $z\ R^{=}\ x$ and $z\ R^{=}\ y$, hence, one way or the other, there is a $w \in C$ with $w\ R^{=}\ z$. But then w is the desired witness. ⊣

Lemma 29 *If* $R = {}^{\Sigma}R$ *then:*

(A) $x\ R\ y \longleftrightarrow [R]\,x \subseteq [R]\,y$
(B) if, in addition, $[R]\,x$ *is* $R^{=}$*-grounded in* C*, then*

(B.1) $x\ R\ y \longleftrightarrow [R_C]\,x \subseteq [R_C]\,y$
(B.2) $x\ R\ y \longleftrightarrow (\forall z \in [R_C]\,x)(\exists w \in [R^=]y)\, z \circ_R w$
(B.3) $x\ R\ y \longleftrightarrow (\forall z \in [R_C]\,x)(\exists w \in [R_C{}^=]y)\, z \circ_R w$
(B.4) $x\ R\ y \longleftrightarrow (\forall z \in [R_C]\,x)\, z \circ_R y.$

Proof The left-to-right directions are covered by Lemma 28(A) or are easy. For the right-to-left direction of (A), check that if $a\ R\ x$, then $a\ R\ y$, hence $a \circ_R y$; hence (cf. Lemma 10) $x\ R\ y$. For the right-to-left direction of (B.2): if $a\ R\ x$, then, by hypothesis, there is a $b \in C$ with $b\ R\ a$. Then $b \in [R_C]x$, so, by our other hypothesis, we get $w \in [R^=]\,y$ with $b \circ_R w$. We also have $w\ R^=\ y$, $b\ R\ a$, hence $a \circ_R y$. Apply Lemma 10. The right-to-left directions of all the others follow easily from (B.2). ⊣

Theorem 4 *Suppose $R = {}^{\Sigma}R$, $fld(R) \subseteq D$, $C \subseteq D$ and let $Q = R \restriction C$ and suppose that C is Q-fusion !-complete and that D is $R^=$-grounded in C. Then:*

(A) $\langle C, Q\rangle$ *is a reduction of* $\langle D, R\rangle$
(B) $\langle C, Q\rangle$ *is a Classical Mereology.*

Proof First, note that for any $y \in D$, by groundedness, there is a $z \in C$ such that $z\ R^=\ y$; then, recalling Lemma 22, we have $z \in fld(Q) \subseteq fld(R)$, from which we may deduce that $y \in fld(R)$. So $D = fld(R)$.

Now, for a notion we will use repeatedly below, for $y \in D$, let $y^* = [R_C^=]\,y$, i.e., $\{a \in C : a\ R^=\ y\}$. Now let $f(y)$ be the unique $b \in C$ such that $fu_Q(b, [x | x \in y^*])$. By Lemma 28, if $z\ R^=\ y$ then $z^* \subseteq y^*$.

Now we claim that (i) for all $y \in D$, $y \equiv_R f(y)$. First, suppose that $z \circ_R f(y)$; we will argue that $z \circ_R y$ and use strong supplementation (on R, whose field includes D) to conclude that $f(y)\ R\ y$.

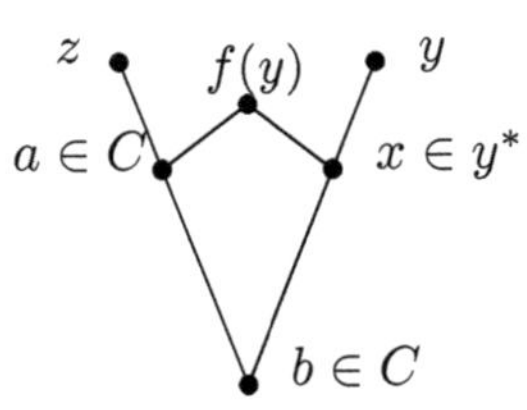

By Lemma 28(B), get $a \in C$ with $a\ R^=\ z$ and $a\ R^=\ f(y)$. Given the fusion condition on $f(y)$, get $x \in y^*$ with $a \circ_Q x$; let $b \in C$ witness this. It should be clear from the figure now that $z \circ_R y$. For the second half of (i), we argue that if $z \circ_R y$, then $z \circ_R f(y)$ and use strong supplementation to conclude that $y\ R\ f(y)$. Let w witness $z \circ_R y$ and by Lemma 28(A) $w^* \subseteq y^*$. By groundedness, get $x \in w^*$, hence $x \in y^*$. So $x\ R^=\ f(y)$, while $x\ R^=\ z$, so $z \circ_R f(y)$ and we are done.

Next we claim (ii) $(\forall y, z \in D)\ (y \equiv_R^= z \rightarrow f(y) = f(z))$. This is clear from Lemma 28(A), since if $y \equiv_R^= z$ then $y^* \subseteq z^*$ and vice-versa, so $y^* = z^*$. Putting together (i) and (ii) with Lemma 17, we have that f reduces $\langle D, R\rangle$ to $\langle C, Q\rangle$, hence we have part (A).

To get (B), given our hypotheses, we need only confirm that $Q =^{\tau} Q$ and $Q =^{\sigma} Q$. The former (which we have implicitly used) is trivial (cf. Lemma 18(A)), and we may get the latter by our proof of part (A), together with Lemma 20. (The fact that we did not need to assume strong supplementation for Q in our hypotheses may be seen as an aspect of Tarski's showing that Classical Mereology can be axiomatized as transitivity plus unique fusion existence; see Hovda (2009) for discussion.) ⊣

We elevate this fact to the status of Theorem because the reasoning in the proof we give contains the basic germ or pattern for characterizing a number of aspects of part-like relations under certain assumptions. Very roughly, if we imagine that there is some sub-domain of the universe that "looks like" a Classical Mereology under some (suitably tolerant) part-like relation, then, if everything in the rest of the universe can be traced back to this sub-domain along that part-like relation, then the universe is quasi-isomorphic with the sub-domain under that relation. This is born out in the following three Lemmas, the later Lemmas 36 and 39–42, and, especially, Theorem 6.

Lemma 30 *Suppose $\langle C, Q\rangle$ is a substructure of $\langle D, R\rangle$, $R =^{-} R$, $fld(R) \subseteq D$, and that D is $R^{=}$-grounded in C. Then $(\forall b \in fld(R))\ Fu_R(b, [R^{=}_{C}]\, b)$.*

Proof It is trivial that $(\forall x \in [R^{=}_{C}]\, b)\ x\ R^{=}\ b$. Now suppose $z\ R^{=}\ b$; since $b \in fld(R) \subseteq D$, $z \in D$, and, by groundedness, there is a $w \in [R^{=}_{C}]\, z$; but then $w\ R^{=}\ b$, so we have $z \circ_R w$, and also $w \in [R^{=}_{C}]\, b$. ⊣

Lemma 31 *Suppose $R =^{\Sigma} R$, $fld(R) = D$, and that D is $R^{=}$-grounded in $C \subseteq D$. In general, let $Y^* = \bigcup\{[R_C]\, y : y \in Y\}$. Then: $(\forall b \in D)$ and $Y \subseteq D$:*

$$Fu_R(b, Y^*) \longleftrightarrow Fu_R(b, Y).$$

Proof Deduce (recalling Lemma 28(B) and noting that $Y, Y^* \subseteq fld(R)$) that $Y \cong_{\circ_R} Y^*$ and apply Lemma 15. ⊣

Lemma 32 *Suppose $\langle C, Q\rangle$ is a substructure of $\langle D, R\rangle$, $fld(R) \subseteq D$, and $R =^{\Sigma} R$. Suppose also that $Q =^{\sigma} Q$, C is Q-fusion complete, and that D is $R^{=}$-grounded in C. Then $fld(R) = D$, $\langle D, R\rangle$ is quasi-isomorphic to $\langle C, Q\rangle$, and they reduce to a Classical Mereology.*

Proof Q is transitive and so by Lemmas 25 and 27 we get that there is a c that reduces $\langle C, Q\rangle$ to a Classical Mereology $\langle C^-, Q^-\rangle$. Now, for any $x \in D$ there is a $y \in C$ with $y\ R^{=}\ x$. (To see that $fld(R) = D$, recall Lemma 22 and note that if $x = y$, then $x \in fld(Q) \subseteq fld(R)$, and if $x \neq y$, $x \in fld(R)$.) But then $c(y)\ R^{=}\ y$ while $c(y) \in C^-$; hence D is $R^{=}$-grounded in C^- and we may apply Theorem 4 to conclude that $\langle D, R\rangle$ reduces to $\langle C^-, Q^-\rangle$. ⊣

5 Merging Domains

In this section we will be considering situations in which we have a domain of objects with a relation whose field is a subset of that domain, and a second, disjoint domain, with a relation whose field is a subset of the union of the two domains.

Notational convention for Sect. 5: let D and E be disjoint non-empty sets, and let $F = D \cup E$; and let R be a relation on D (i.e., $fld(R) \subseteq D$), and S be a relation on F. We let P be $R \cup S$ (i.e., $\forall x \forall y(x\ P\ y \leftrightarrow (x\ R\ y \vee x\ S\ y))$).

One application to bear in mind is this: think of D as the set of concrete (physical) objects or similar, and R as being the union of the primitive ways such things can be parts of one another. Then think of E as containing all the non-empty subsets of D, and take membership to be the primal parthood relation for things in E, so that S would be the membership relation, (restricted to $D \cup E$). This application will nicely fit many of the general formal results we now work through. Alternatively, try thinking of E as the class of all (non-empty) sets that can be iteratively "built-up from" members of D (not including the empty set). Of course E is then not a set, but there are ways to adapt our results nonetheless – one simple way would be to stop the iteration at some point and let E be the set of sets generated up to that point.

5.1 Non-interference

On the condition that S never relates things within or into D, we get some nice results about induced part-like relations.

Lemma 33 *If* $\neg(\exists x \in F)(\exists y \in D)\, x\ S\ y$ *then:*

(0) $P \restriction D = R$
(1) $\neg(\exists x \in E)(\exists y \in D)\, x\ {}^{\tau}P\ y$
(2) $({}^{\tau}P) \restriction D = {}^{\tau}R$
(3) $(\circ_{{}^{\tau}P}) \restriction D = \circ_{{}^{\tau}R}$
(4) $(\forall x \in E)(\forall y \in D)\ (x \circ_{{}^{\tau}P} y \rightarrow (\exists z \in D)(z\ {}^{\tau}P^{=}\ y \wedge z\ {}^{\tau}P\ x))$
(5) $({}^{\Sigma}P) \restriction fld(R) = {}^{\Sigma}R$

Proof Clause (0) is easy. (1) is straightforward: any chain of P links that starts with a thing in E must stay in E, since each P link is either a R link or an S link, and R is confined to D, so to speak, while S never takes us into D, by hypothesis. (2) is clear for similar reasons: any chain of P links from $x \in D$ to $y \in D$ must stay within D and consist of R links. Hence $({}^{\tau}P) \restriction D \subseteq {}^{\tau}R$; the other direction is easy, since a chain of R links is a chain of P links, all within D. (3) is then easy. To get (4), if w witnesses $x \circ_{{}^{\tau}P} y$, then, with $y \in D$, by (1), $w \in D$ as well, and witnesses the desired result. For (5): supposing $x \in fld(R)(= fld({}^{\tau}R))$ and $y \in fld(R)$ and $x\ {}^{\Sigma}P\ y$, we want to show that $\forall z(z \circ_{{}^{\tau}R} x \rightarrow z \circ_{{}^{\tau}R} y)$, so suppose $z \circ_{{}^{\tau}R} x$. By (3), $z \circ_{{}^{\tau}P} x$, and so by hypothesis $z \circ_{{}^{\tau}P} y$, and another application of (3) gets us what we need. The other half of (5) is similar. $\dashv$

Note that with our hypotheses, even though we get $P \upharpoonright D = R$, and clearly $fld(R) \subseteq fld(P)$, we are not guaranteed that $fld(P) \cap D \subseteq fld(R)$, since it could be that $x\ S\ y$ for $x \in D$ and $y \in E$ with $x \notin fld(R)$.

Some limited results obtain with a weaker stricture.

Lemma 34 *Suppose that* $E = E^+ \cup E^-$ *and that* $(\forall x, y \in F)(x\ S\ y \to \neg[(x \in E^+ \wedge y \in D) \vee (x \in E^+ \wedge y \in E^-) \vee (x \in D \wedge y \in E^-)]$. *Then:*

(0) $P \upharpoonright D = R$
(1) $\neg(\exists x \in E^+)(\exists y \in (D \cup E^-))\, x\ {}^{\tau}P\ y$
(2) $\neg(\exists x \in D)(\exists y \in E^-)\, x\ {}^{\tau}P\ y$
(3) $({}^{\tau}P) \upharpoonright D = {}^{\tau}R$

These are all straightforward to obtain. ⊣

On the condition that S never relates things within or out of D, we need a more stringent requirement to get similar results to Lemma 33, and even then there are limitations.

Lemma 35 *If* $\neg(\exists x \in D)(\exists y \in F)\, x\ S\ y$ *and* $(\forall x, y \in D)\,(x \circ_{\tau P} y \to x \circ_{\tau R} y)$, *then:*

(0) $P \upharpoonright D = R$
(1) $\neg(\exists x \in D)(\exists y \in E)\, x\ {}^{\tau}P\ y$
(2) $({}^{\tau}P) \upharpoonright D = {}^{\tau}R$
(3) $(\circ_{\tau P}) \upharpoonright D = \circ_{\tau R}$
(4) $fld(P) \cap D = fld(R)$
(5) $({}^{\Sigma}P) \upharpoonright D \subseteq {}^{\Sigma}R$

(0)–(2) are already covered by Lemma 34, (3) is immediate from our hypothesis. (4) also follows from our hypothesis, since, if $x \in D$ and $x \in fld(P)$, then $x \circ_{\tau P} x$, hence $x \circ_{\tau R} x$. (5) is then straightforward, but we cannot conclude that ${}^{\Sigma}R \subseteq ({}^{\Sigma}P) \upharpoonright D$. ⊣

5.2 *Articulation and Fusion*

Now we may show that if the domain E articulates or "comprehends," the domain D, in a certain sense, then there are fusions for arbitrary non-empty subsets of D, under both the ${}^{\tau}P$ relation and the ${}^{\Sigma}P$ relation, with respect to domain F.

There are various conditions on E that give fusion results. Here is a simple one, that fits the idea of letting E be the set of non-empty subsets of D and S be membership (restricted to F) (and assuming that no subset of D is itself a member of D).

Lemma 36 *Suppose that for every non-empty* $X \subseteq D$, *there is at least one* $y \in E$ *such that (1)* $(\forall x \in D)\,(x\ S\ y \leftrightarrow x \in X)$ *and (2)* $(\forall z \in E)\,(z\ S\ y \to z = y)$. *Then:*

(A) for $X \subseteq D$ *with* $X \neq \emptyset$*:*

(A.1) $(\exists y \in E)\, Fu_{{}^{\tau}P}(y, X)$; *and*
(A.2) $(\exists y \in E)\, Fu_{{}^{\Sigma}P}(y, X)$.

(B) If E is ${}^{\tau}S^{=}$*-grounded in D, then F is* ${}^{\Sigma}P$*-fusion complete.*

Proof For (A), fix non-empty $X \subseteq D$. Given y that witnesses the supposition, we may show that it also serves as both the desired fusions. For (A.1), the only non-obvious step is to show that, given $w \in F$ with $w\ {}^{\tau}P^{=}\ y$, $\exists x \in X\ x \circ_{{}^{\tau}P} w$. We may assume $w \neq y$ (since if $w = y$ then we are easily done). For any chain of P links that runs from w to y, we may consider the first link that runs to y; this link has to be either an R link or an S link; but by hypothesis it must be an S link running from some $x \in X$ to y; but then $w\ {}^{\tau}P^{=}\ x$ and hence $w \circ_{{}^{\tau}P} x$. For (A.2), invoke Lemma 14(A).

For (B), we first argue that $fld({}^{\Sigma}P) = F$: given $z \in F$, if $z \in E$, then, since E is ${}^{\tau}S^{=}$-grounded in D, get $x \in D$ with $x\ {}^{\tau}S\ z$, hence $z \in fld(S) \subseteq fld(P) = fld({}^{\Sigma}P)$; if $z \in D$, then by hypothesis get $y \in E$ with $x\ S\ y$ (consider $\{z\} \subseteq D$) hence again $z \in fld(S)$. We also may deduce that F is ${}^{\Sigma}P^{=}$-grounded in D. Now, given non-empty $Y \subseteq F$, letting $Y^{*} = \bigcup\{[{}^{\Sigma}P_{D}]y : y \in Y\}$, note that Y^{*} is non-empty, and use (A.2) to get a $b \in F$ such that $Fu_{{}^{\Sigma}P}(b, Y^{*})$. We may then invoke Lemma 31 to get that $Fu_{{}^{\Sigma}P}(b, Y)$. ⊣

Definition 12 *(For any sets* A, B, *and relation* Q*)* B Q-flows from $A =_{def}$

B is ${}^{\tau}Q$-grounded in A and $\neg(\exists x \in A)(\exists y)\, y\ Q\ x$.

Note that, in general, if B Q-flows from A, then $B \cap A = \emptyset$ and $B \subseteq fld(Q)$.

Lemma 37 *If E S-flows from D, then* $\langle fld(R), {}^{\Sigma}R\rangle$ *is a substructure of* $\langle F, {}^{\Sigma}P\rangle$.

Proof Check that the conditions of Lemma 33 part (5) apply, so $({}^{\Sigma}P) \restriction fld(R) = {}^{\Sigma}R$. ⊣

Lemma 38 *Suppose that E S-flows from D and* $y \in E$. *Then* $[{}^{\tau}S_{D}]\,y$ *is non-empty, and:*

(A) $(\forall z \in D)(z\ {}^{\tau}P\ y \leftrightarrow (\exists x \in [{}^{\tau}S_{D}]\,y)\, z\ {}^{\tau}P^{=}\ x)$
(B) $(\forall z \in E)(z\ {}^{\tau}P^{=}\ y \rightarrow [{}^{\tau}S_{D}]\,z \subseteq [{}^{\tau}S_{D}]\,y)$
(C) $Fu_{{}^{\tau}P}(y, [{}^{\tau}S_{D}]\,y)$
(D) $Fu_{{}^{\Sigma}P}(y, [{}^{\tau}S_{D}]\,y)$

Proof Note that for any $z_0 \in D$ such that $z_0\ {}^{\tau}P\ y$, there must be a chain $z_0\ P\ z_1$, $z_1\ P\ z_2$, $\ldots z_{n-1}\ P\ z_n$, where $z_n = y$ (and with the possibility that $n = 1$). For any such chain, there must be an $i \leq n$ such that for $1 \leq j \leq i$, $z_j \in D$, and $z_{i+1} \in E$. Since R is confined to D and no member of E bears S to a member of D, we have that for $i + 1 \leq j \leq n$, $z_j \in E$, and, moreover $z_i\ S\ z_{i+1}$, $z_{i+1}\ S\ z_{i+2}$, and so forth; so $z_i\ {}^{\tau}S\ y$; hence $z_i \in [{}^{\tau}S_{D}]\,y$. This gives us (A). Lemma 28 gives us (B). Lemma 30 gives us (C), and (D) is then immediate from Lemma 14(A).

Lemma 39 *Suppose that E S-flows from D and that for every non-empty $X \subseteq D$ there is a $y \in E$ such that $X = [{}^{\tau}S_D]\,y$. Then F is ${}^{\Sigma}P$-fusion complete.*

Proof We may deduce that $fld({}^{\Sigma}P) = F$ by an argument like that used for Lemma 36(B). We may deduce that F is ${}^{\Sigma}P$-grounded in D. And given a non-empty $Y \subseteq F$, letting $Y^* = \bigcup\{[{}^{\Sigma}P_D]y : y \in Y\}$, check that Y^* is non-empty; then our hypothesis gets us a $y \in E$ with $Y^* = [{}^{\tau}S_D]\,y$; may then invoke Lemmas 38(D) and 31 to get that $Fu_{\Sigma P}(y, Y)$. ⊣

Lemma 40 *Suppose that E S-flows from D and that for every non-empty $X \subseteq D$ there is an $X' \subseteq D$ with $X \cong_{\circ\Sigma_P} X'$ and $y \in E$ such that $X' = [{}^{\tau}S_D]\,y$. Then F is ${}^{\Sigma}P$-fusion complete.*

Proof This follows easily from the previous Lemma together with Lemma 15. ⊣

This gives us a very general condition for the ${}^{\Sigma}P$-fusion completeness of F.

Lemma 41 *If D is ${}^{\Sigma}R$-fusion complete and E S-flows from D, then $\langle F, {}^{\Sigma}P\rangle$ is quasi-isomorphic to $\langle D, {}^{\Sigma}R\rangle$.*

Proof Deduce that $fld({}^{\Sigma}R) = D$, recalling Lemma 22. Get that ${}^{\Sigma}P \restriction D = {}^{\Sigma}R$ from Lemma 33 part (5). Then invoke Lemmas 32 and 37. (For the applicability of Lemma 32, note that of course ${}^{\sigma}({}^{\Sigma}R) = {}^{\Sigma}R$.) ⊣

Lemma 42 *Suppose D is ${}^{\Sigma}R$-fusion complete, E S-flows from D, and for every non-empty $X \subseteq D$ there is a $y \in E$ such that $X = [{}^{\tau}S_D]\,y$. Then $\langle F, {}^{\Sigma}P\rangle$, $\langle E, ({}^{\Sigma}P) \restriction E\rangle$, and $\langle D, {}^{\Sigma}R\rangle$ are all quasi-isomorphic.*

Proof Given the previous two Lemmas, it will suffice to show that $\langle F, {}^{\Sigma}R\rangle$ is quasi-isomorphic to $\langle E, ({}^{\Sigma}P) \restriction E\rangle$, and then appeal to the transitivity of quasi-isomorphism (for transitive relational structures), which is straightforward to deduce from Theorem 2. To show this, it is enough to note that $(\forall z \in F)(\exists y \in E)\ z \equiv^{=}_{\Sigma_P} y$, and then apply Lemma 26. Given $z \in D$, then we note that by hypothesis there is a $y \in E$ with $[{}^{\tau}S_D]\,y = \{z\}$ and note that $Fu_{\Sigma P}(z, [x|x = z])$, and $Fu_{\Sigma P}(y, [x|x = z])$ (by Lemma 38(D)), and hence $y \equiv_{\Sigma P} z$ (recalling Lemma 14(E))⊣.

5.3 Discussion

In one intuitive application of the above results, we let D be the set of concrete objects and R be the union of the primitive ways such things can be parts of one another. (Officially, we place no constraints on R, though there are some reasonable ones we could try). And we then let E be the set of all the non-empty subsets of D, and take membership to be the primal parthood relation for things in E, so that S would be the membership relation, (restricted to $D \cup E$). Then we have that E S-flows from D, and we can apply our recent Lemmas to deduce that $\langle F, {}^{\Sigma}P\rangle$ is quasi-isomorphic to a Classical Mereology, among other things.

Moreover, suppose we iterate the basic process that took us from $\langle D, R\rangle$ to $\langle F, {}^{\Sigma}P\rangle$: we take E^2 to be the set of non-empty subsets of F, and let S^2 be the membership relation restricted to $F^2 = F \cup E^2$. And we set $P^2 = {}^{\Sigma}P \cup S^2$. Then E^2 S^2-flows from F, and we may apply our Lemmas to get that $\langle F^2, {}^{\Sigma}P^2\rangle$ is quasi-isomorphic to $\langle F, {}^{\Sigma}P\rangle$. Hence, if we differentiate structures only "up to quasi-isomorphism," he find that iterating this process does not alter structure. In effect, iterating just places more nodes within clusters of "equivalent" items: e.g., if $z \in E^2$, then there is an $x \in F$ such that $x \equiv_{{}^{\Sigma}P^2} z$ and for all $y \in F$ $y \equiv_{{}^{\Sigma}P^2} z \leftrightarrow y \equiv_{{}^{\Sigma}P} x$.

The basic phenomena generalize. To illustrate, let $E^1 = E = \{w : w \neq \emptyset \wedge w \in \mathcal{P}(D)\}$, $E^2 = \{w : w \neq \emptyset \wedge w \in \mathcal{P}(D \cup E^1)\}$ and so forth, and let E^{ω} be $\{x : x \in E^n$ for some finite $n\}$. Let G be any subset of E^{ω}, let S^G be the membership relation restricted to $H = D \cup G$, and let P^G be $R \cup S^G$. Then H is ${}^{\tau}P^G$-grounded in D, and G ${}^{\Sigma}P^G$-flows from D. We will show in a moment (Theorem 6) that $\langle G, {}^{\Sigma}({}^{\Sigma}P \cup S^G)\rangle$ is quasi-isomorphic to a substructure of $\langle F, {}^{\Sigma}P\rangle$. We note that if G includes "enough" sets from E^{ω}, then the conditions of Lemma 39 apply, and (as we will show), $\langle G, {}^{\Sigma}({}^{\Sigma}P \cup S^G)\rangle$ is quasi-isomorphic to $\langle F, {}^{\Sigma}P\rangle$.

And there is nothing peculiar about *sets*, as opposed to other objects that "flow from" our base. Roughly: any additional objects under a relation on which they flow from our base D are structurally bound either to "join in" to existing equivalence clusters, or to fill in "gaps" in places where there is a cluster in the structure that arises when enough objects are added to yield fusion-completeness.

6 Completions and General Structure

We now turn to making this last remark more precise.

Definition 13 Say that $\langle F, P\rangle$ is a completion of $\langle D, R\rangle$ when

(i) there are relations R° and P°, with $D = fld(R^{\circ})$, $F = fld(P^{\circ})$, $R =^{\Sigma} R^{\circ}$ and $P =^{\Sigma} P^{\circ}$; and
(ii) $\langle D, R\rangle$ is a substructure of $\langle F, P\rangle$; and
(iii) F is P-fusion complete; and
(iv) F is P-grounded in D.

Theorem 5 *For any relation R°, there is a completion of $\langle fld(R^{\circ}), {}^{\Sigma}R^{\circ}\rangle$.*

Proof Our result is basically evident from our discussion thus far. Officially, we are assuming here that R° is a set of ordered pairs. Given that $D = fld(R^{\circ})$ is a set, we can consider letting $E = \{X : X \neq \emptyset \wedge X \subseteq D\}$, $F = D \cup E$, $P^{\circ} =$ the union of R° with membership restricted to F, and $P =^{\Sigma} P^{\circ}$. Assuming that $E \cap D = \emptyset$, we have seen (cf. Lemmas 36 and 37) that then $\langle F, P\rangle$ is a completion of $\langle D, {}^{\Sigma}R^{\circ}\rangle$. If some subsets of D are members of D, then we will need some other construction. But we should still be able to find some set of objects E' and relation S' to play the roles played by E and S in Lemma 36 or 39 or 42. If we are working with ZFC, then we can use sets whose rank is higher than that of D. In other set theories, there should be similar resources available, unless R° is "too big" to be a set anyway. ⊣

Theorem 6 *For any relation* R°, *with* $D = fld(R^\circ)$ *and* $R =^\Sigma R^\circ$, *if* $\langle F, P\rangle$ *is a completion of* $\langle D, R\rangle$ *then:*

> *for any* $\langle G, Q\rangle$ *of which* $\langle D, R\rangle$ *is a substructure, where* $Q =^\Sigma Q^\circ$ *(with* $G = fld(Q^\circ)$*) and where* G *is* ${}^{\mathcal{T}}Q^{\circ=}$*-grounded in* D*:*

(A) $\langle G, Q\rangle$ *is quasi-isomorphic to a substructure of* $\langle F, P\rangle$*; and*
(B) if G *is* Q*-fusion complete, then* $\langle G, Q\rangle$ *is quasi-isomorphic to* $\langle F, P\rangle$*.*

Proof By Theorem 3, we can get a c that reduces $\langle F, P\rangle$ to a Classical Mereology $\langle F^-, P^-\rangle$. Let d reduce $\langle G, Q\rangle$ to $\langle G^-, Q^-\rangle$. Our main goal is to find an isomorphism h from $\langle G^-, Q^-\rangle$ to some substructure of $\langle F^-, P^-\rangle$. Given $y \in G^-$, since G is ${}^{\mathcal{T}}Q^{\circ=}$-grounded in D, G is $Q^=$-grounded in D, (and $y \in fld(Q)$), and we may apply Lemma 30 to get that $Fu_Q(y, [Q_D]\, y)$. (Note that for any $y \in G$, $[Q_D]y = [Q^={}_D]y$, since Q is reflexive on $G = fld(Q)$.) Now, since $\langle F, P\rangle$ is P-fusion complete, we get a $y' \in F$ with $Fu_P(y', [Q_D]\, y)$, and we may set $h(y) = c(y')$. This is well-defined, since for any $y'' \in F$ with $Fu_P(y'', [Q_D]\, y)$, $y' \equiv_P y''$, hence $c(y') = c(y'')$.

Now for any $y, z \in G^-$, we get $y', z' \in F$ with $Fu_P(y', [Q_D]\, y)$ and $Fu_P(z', [Q_D]\, z)$ and $h(y) = c(y')$ and $h(z) = c(z')$. We claim that:

$$y\ Q^-\ z \longleftrightarrow h(y)\ P^-\ h(z).$$

This follows from the correctness of these four biconditionals:

$$\begin{aligned} y\ Q^-\ z &\longleftrightarrow_{(1)} y\ Q\ z \\ &\longleftrightarrow_{(2)} (\forall x \in [Q_D]y)(\exists w \in [Q_D]z)\ x \circ_Q w \\ &\longleftrightarrow_{(3)} y'\ P\ z' \\ &\longleftrightarrow_{(4)} c(y')\ P^-\ c(z'). \end{aligned}$$

Biconditionals (1) and (4) are covered by Lemma 18. (2) is covered by Lemma 29. For the left-to-right direction of (3), bearing in mind both (B.1) and (B.3) of Lemma 29, we have that $[Q_D]\, y \subseteq [Q_D]\, z$, and so if $a\ P\ y'$, then there is $x \in [Q_D]z$ with $a \circ_P x$ and hence $a \circ_P z'$; now appeal to Lemma 10. For the right-to-left direction: given $x \in [Q_D]\, y$, we have $x\ P\ y'\ P\ z'$. Thus there is a $w \in [Q_D]\, z$ such that $x \circ_P w$. Bearing in mind that F is P-grounded in D, by Lemma 28(B), we get a $b \in D$ that witnesses this overlap: $b\ P\ x$ and $b\ P\ w$. But then $b\ R\ x$ and $b\ R\ w$, and hence $b\ Q\ x$ and $b\ Q\ w$. Thus $x \circ_Q w$, and we are done.

Now, since P^- is reflexive, if $h(y) = h(z)$, $h(y) \equiv_{P^-} h(z)$, so $y \equiv_{Q^-} z$. Since Q^- is anti-symmetric, we then have $y = z$. Hence h is one-to-one.

Letting E be the range of h, we may conclude that $h : G^- \to E$ is an isomorphism from $\langle G^-, Q^-\rangle$ to $\langle E, P^- \restriction E\rangle$, which is a substructure of $\langle F^-, P^-\rangle$, hence of $\langle F, P\rangle$. This completes the argument for part (A).

For part (B), let $x \in F^-$. $[P_D]\, x$ is non-empty; by our additional hypothesis, there is a $y \in G$ with $Fu_Q(y, [P_D]x)$. As we noted above, we know $Fu_Q(y, [Q_D]y)$, so, recalling Lemma 15, we have $[P_D]x \cong_{\circ_Q} [Q_D]y$. Now we argue that $[P_D]x \cong_{\circ_P}$

$[Q_D]y$: if $b \in [P_D]x$ and $a \circ_P b$, then (Lemma 28) get $w \in D$ with $w\ P^= a$ and $w\ P^= b$. But since b also $\in D$, we have $w\ R^= b$ and $w\ Q^= b$. So $w \circ_Q b$, so get $e \in [Q_D]y$ with $w \circ_Q e$. Deduce $w \circ_P e$, hence $a \circ_P e$. Parallel reasoning runs the other way to yield $[P_D]x \cong_{\circ_P} [Q_D]y$. But we also have $Fu_P(h(d(y)), [Q_D]\,y)$; appealing again to Lemma 15, we get $Fu_P(h(d(y)), [P_D]x)$. But also $Fu_P(x, [P_D]\,x)$, hence $x \equiv_P h(d(y))$. But x and $h(d(y))$ are both in the range of c, so $x = c(x) = c(h(d(y))) = h(d(y))$. So x is in the range of h, and thus $E = F^-$. ⊣

6.1 Discussion

Imagine again that we start with some domain D and some relation R° on D (with $fld(R^\circ) = D$ that we take to capture some or all primitive parthood connections among things in D (including identity on D; recall our discussion of reflexivity above). We consider ${}^{\mathcal{T}}R^\circ$ and ${}^{\Sigma}R^\circ$ to be derivative parthood relations on D. Now we note that there may be "new" (abstract, perhaps) objects E that could be "built out of" the things in D, and we imagine that the primitive parthood relation S° that connects these new things to D, and might also connect them to one another, "flows from" D in basically the sense made precise above. We might think of S° as marking exactly the genesis of a new thing: $x\ S^\circ\ y$ just when y was immediately generated by, or built from, (possibly among other things) x. Then ${}^{\mathcal{T}}S^\circ$ traces the ancestry of a new thing and is itself a parthood relation. When we take view of the domain $F = E \cup D$, we see that $P^\circ = R^\circ \cup S^\circ$ represents the sum of the primitive parthood relations on this domain, and, accordingly ${}^{\mathcal{T}}P^\circ$ and ${}^{\Sigma}P^\circ$ represent derivative parthood relations. Within the domain D, these relations are no different from ${}^{\mathcal{T}}R^\circ$ and ${}^{\Sigma}R^\circ$ (cf. Lemma 29).

Let us say that two quasi-isomorphic structures have the same q-type, and that one q-type is a sub-type of another if any structure with the first q-type is quasi-isomorphic to a sub-structure of a structure with the second q-type. Then Theorem 6 tells us that there is a unique q-type (fixed by the q-type of the initial $\langle D, R\rangle$) that the q-type of any such $\langle F, {}^{\Sigma}P^\circ\rangle$ must be a sub-type of: this is the q-type of a completion of $\langle D, R\rangle$.

For another perspective, consider the addition of a single new object y to a given $\langle D, R^\circ\rangle$, letting $R =^{\Sigma} R^\circ$. Think of y as being generated by some thing or things in D, and let S° mark this generation, so that $[S^\circ_D]y$ is precisely the set of things that generated y. Now letting $F = D \cup \{y\}$ and $P^\circ = R^\circ \cup S^\circ$ and $P =^{\Sigma} P$, there are only two possibilities for the q-type of $\langle F, P\rangle$, *vis a vis* $\langle D, R\rangle$. We know (Lemmas 37 and 30) that $Fu_P(y, [P_D]\,y)$. The two possibilities are whether or not there is already $z \in D$ such that $Fu_R(z, [P_D]\,y)$. If there is such a z, then (Lemma 14(D)) $z \equiv_P y$, and it is easy to see that $\langle F, P\rangle$ is quasi-isomorphic to $\langle D, R\rangle$, so that the new structure has exactly the same q-type as the old one. If not, then there is no $z \in D$ such that $z \equiv_P y$. This is because if there were such a z, we would have $Fu_P(z, [P_D]\,y)$, hence (cf. Lemma 28(B)) $Fu_R(z, [P_D]\,y)$. In this case, $\langle F, P\rangle$ is not quasi-isomorphic to its substructure $\langle D, R\rangle$, and y occupies a new position, so to speak. The q-type of $\langle D, R\rangle$ is then a proper sub-type of the q-type of $\langle F, P\rangle$. Thus we can think of $\langle D, R\rangle$ as having "gaps" precisely where it lacks R-fusions for various subsets of D.

We might imagine now adding a new thing y' to F in just the manner in which y was added to D, and so forth. (One can check that each of the resulting structures flows from $\langle D, R\rangle$, and it does not really matter whether we add objects and generate structures cumulatively, in this manner, or all at once as in our previous discussion.) At each step in the cumulative process, when a new object is added, it may fill in a "gap" in the pattern of the previous structure. If we reach a stage without gaps, then any further additions will generate nothing new *modulo* quasi-isomorphism. We will have reached a fixed point: all further additions will have the same q-type.

Again, we were thinking of the S° relations as being the primitive ways in which the new things have parts. It is the less discerning derivative relations, especially the ones in the role of P, for which we were able to give a strong structural characterization.

6.2 *Expansion Without Groundedness*

We have been focusing on what happens when we "extend" a given domain (with part-like relations on it) to a larger domain, where the new objects are born new primitive part-like relations *from* the old ones. Our main result was roughly that, relative to one particular sort of tolerant parthood relation, there is an aspect in which there is a unique pattern that can result, fixed by the pattern in the given seed. But what if some new objects are not connected to the old ones in such a manner? We made some very brief points about such situations in Lemmas 34 and 35, but did not explore very far.

There are two very different cases: (1) a new object might simply be disconnected from the old domain, with no parthood relations running from it to the old domain, nor to it, from the old domain; or (2) a new object might be a primitive part of an old object. Roughly speaking, in case (1) the new object is guaranteed not to "disturb" the old domain and its parthood relations, in a fairly strong sense: we can merge the new object and its relation into the old domain and relation, take the transitive closure and overlap-inclusion of the merged relations, and the restrictions of these relations will be the corresponding old ones. But in case (2), there is no such guarantee: for example, when we merge, old objects that did not overlap under the old relation might overlap under the new one, and hence the restriction, to the old domain, of the overlap-inclusion of the transitive closure of the merged relation might not be identical with its old counterpart. Now there is some reason to insist on metaphysical grounds that non-concrete objects cannot bear genuine parthood relations to concrete ones; this is natural when we think of the non-concrete ones as either independent of (think "Platonic") or generated by (think "Aristotelean") concrete ones. But there may be some reason to think otherwise; (see section X of Fine 2010 for some motivations). There is not room for further discussion here, but we may hope that some of the analytical tools developed above could be of use in exploring the formal patterns that can emerge, given some reasonable constraints.

References

Fine, K. (2010). Towards a theory of part. *Journal of Philosophy*, *107*(11), 559–589.

Hovda, P. (2009). What is classical mereology? *Journal of Philosophical Logic*, *38*(1), 55–82.

Hovda, P. (2016). Parthood-like relations: Closure principles and connections to some axioms of classical mereology. *Philosophical Perspectives*, *30*(1), 183–197.

Simons, P. (1987). *Parts: A study in ontology*. Oxford: Oxford University Press.

Varzi, A. (2016). Mereology. In E. N. Zalta (Ed.), *The Stanford Encyclopedia of Philosophy*. Metaphysics Research Lab, Stanford University, winter 2016 ed.

Infinite "Atomic" Mereological Structures

Hsing-Chien Tsai

Abstract Mereology is the theory based on a binary relation "being a part of". Intuitively, using such a relation, notions such as "being an atom" and "being composed of" can be defined, and indeed, there are relevant axioms which can be found in the literature, such as the atomicity axiom and the fusion axiom schema. In this light, it might be hoped that some axiomatized mereological theory can secure an atomic domain, that is, can guarantee that everything in the domain is composed of atoms only. However, several negative results have been shown previously. This paper will give more negative results by showing that the strongest first-order atomic mereological theory which can be generated by axioms found in the literature, that is, General Extensional Mereology with the atomicity axiom, as well as a natural second-order extension of such a theory, that is, Classical Mereology with the atomicity axiom, cannot secure an atomic domain either.

1 Introduction

It is known that mereology is the theory (or rather a class of theories) based on a binary relation "being a part of". Quite a few notions can be defined using such a relation. In particular, "being an atom" as well as "being composed of" can be defined as follows: something is an atom if and only if it contains no part other than itself; something is composed of a collection of objects if and only if each object in that collection is a part of that thing and anything which overlaps that thing also overlaps some object in that collection, where "overlapping" means "sharing a part".[1]

[1]The definition of "being composed of" here, which is in effect also the definition of "being a fusion of", is slightly different from the definition of "being a sum of" mentioned in (Tsai and Varzi 2016) in that the former does not require objects in the collection to be pairwise disjoint whereas the latter does. But this does not matter for my purpose, for atoms are of course pairwise disjoint.

H.-C. Tsai (✉)
Department of Philosophy, National Chung-Cheng University,
168 University Road, Min-Hsiung, Chia-Yi 62102, Taiwan
e-mail: pythc@ccu.edu.tw

S.C.-M. Yang et al. (eds.), *Philosophical Logic: Current Trends in Asia*,
Logic in Asia: Studia Logica Library, DOI 10.1007/978-981-10-6355-8_14

Indeed, relevant mereological axioms, such as the atomicity axiom and the fusion axiom schema, can be found in the literature. In this light, one might hope to find an axiomatized mereological theory which is sufficient enough to secure an atomic domain or in other words, to guarantee that everything in the domain is composed of atoms only.[2] However, some negative results have been shown previously (Tsai and Varzi 2016; Tsai 2015). In this paper, more negative results will be given. More precisely, we are going to show that the strongest first-order atomic mereological theory that can be generated by axioms found in the literature as well as a natural second-order extension of such a theory still cannot guarantee that everything in the domain is composed of atoms only.

To make this paper more or less self-contained, we will as usual give a brief review of the formal setting.

The signature of the first-order mereological language contains only one binary predicate P, where the intended meaning of Pxy will be "x is a part of y". In order to simplify complicated expressions, five additional predicates are defined as follows:

(Proper Part) $PPxy = Pxy \wedge \neg Pyx$
(Overlap) $Oxy = \exists z(Pzx \wedge Pzy)$
(Underlap) $Uxy = \exists z(Pxz \wedge Pyz)$
(Atom) $Atom(x) = \forall y \neg PPyx$
(Gunk) $Gunk(x) = \forall y(Pyx \rightarrow \exists z PPzy)$

The following are some mereological axioms which can be found in the literature (Simons 1987; Casati and Varzi 1999).

Reflexivity (P1): $\forall x Pxx$
Anti-symmetry (P2): $\forall x \forall y((Pxy \wedge Pyx) \rightarrow x = y)$
Transitivity (P3): $\forall x \forall y \forall z((Pxy \wedge Pyz) \rightarrow Pxz)$
Strong Supplementation Principle (SSP): $\forall x \forall y(\neg Pyx \rightarrow \exists z(Pzy \wedge \neg Ozx))$
Finite Sum (FS): $\forall x \forall y(Uxy \rightarrow \exists z \forall w(Owz \leftrightarrow (Owx \vee Owy)))$
Finite Product (FP): $\forall x \forall y(Oxy \rightarrow \exists z \forall w(Pwz \leftrightarrow (Pwx \wedge Pwy)))$
Complement (C): $\forall x(\neg \forall y Pyx \rightarrow \exists z \forall w(Pwz \leftrightarrow \neg Owx))$
Atomicity (A): $\forall x \exists y(Pyx \wedge Atom(y))$
Greatest Member (G): $\exists x \forall y Pyx$
Fusion (F): For any formula α where z and y do not occur free and where variables other than x might occur free, $\exists x\alpha \rightarrow \exists z \forall y(Oyz \leftrightarrow \exists x(\alpha \wedge Oyx))$. Note that (F) is an axiom schema which defines infinitely many axioms.

Also let (Infinity) stand for the list of infinitely many axioms each of which says that there are at least n members, for $n \geq 2$. It should be clear that each axiom on this list is first-order expressible.

Mereological theories can be formed by using axioms listed above. The following are some theories named in the literature (Casati and Varzi 1999).

[2]An atomic domain contains a collection of "atoms" (an atom is something which has no proper parts) and each member there is composed of some atoms and nothing else. If any model of a mereological theory has an atomic domain, we say that such a theory can "secure an atomic domain" or can "guarantee that everything in the domain is composed of atoms only".

Ground Mereology **GM** = (P1) + (P2) + (P3)
Extensional Mereology **EM** = **GM** + (SSP)
Extensional Closure Mereology **CEM** = **EM** + (FS) + (FP)
General Extensional Mereology **GEM** = **EM** + (F)

It is not difficult to see that the ordering of the foregoing theories in terms of proving strength is as follows.

GM < **EM** < **CEM** < **GEM**

The atomicity axiom (A) says that everything has a part which is an atom. Of course, if everything is composed of atoms only, (A) should be the case. Conversely, it is hoped that working with some mereological theory, (A) can really guarantee that everything is composed of atoms only. However, it has been shown that **EM** + (A) cannot secure an atomic domain, for that theory has a model whose domain contains some members which are mixtures of atoms and gunks, or in other words, which have "gunky parts" (Tsai and Varzi 2016). Moreover, **CEM** + (A) cannot secure an atomic domain either, for we can still find that kind of model for such a theory (Tsai 2015).

This paper will give two more negative results.

(1) **GEM** + (A), which is the strongest first-order atomic mereological theory that can be generated by axioms found in the literature, cannot secure an atomic domain either.[3]
(2) A natural second-order extension of **GEM** + (A), that is, classical mereology with the atomicity axiom, still cannot secure an atomic domain.

2 The Case of GEM + (A)

In order to show that **GEM** + (A) cannot secure an atomic domain, we have to come up with a model of **GEM** + (A) whose domain contains members with gunky parts. Such a model must be infinite since any gunky part can be divided indefinitely. Hence we actually have to find such a model of **GEM** + (A) + (Infinity), which is indeed a consistent theory.[4] It has been shown previously that **CEM** + (G) + (C) + (A) + (Infinity), a subtheory of **GEM** + (A) + (Infinity), is a complete theory (Tsai 2013), and hence the two theories are equivalent. Therefore, it suffices to show that

[3]An atomic mereological theory is a theory whose language has only one binary predicate P and whose axioms contain the atomicity axiom. **GEM** + (A) is the strongest first-order atomic theory that can be axiomatized by axioms found in the literature, for **GEM** can prove all other traditional axioms which are compatible with (A). For this issue, please refer to the ordering of theories mentioned above and to the axioms listed in (Simons 1987) and in (Casati and Varzi 1999). By the way, after showing this result, I realized that I had made a careless mistake in (Tsai 2015) by saying that **GEM** + (A) can secure an atomic domain.

[4]The consistency of **GEM** + (A) + (Infinity) can be easily checked. Also see M_4 below.

CEM + (G) + (C) + (A) + (Infinity) has a model whose domain contains members with gunky parts.

To see what an "intended model" of **CEM** + (G) + (C) + (A) + (Infinity) looks like, let M_1 be the mereological structure whose domain is $\{S \subseteq \omega$: S is nonempty and S is either finite or cofinite$\}$ and whose interpretation of P is the set inclusion, that is, $M_1 \vDash Pxy[a, b]$ if and only if $a \subseteq b$. It is easy to check that M_1 satisfies **CEM** + (G) + (C) + (A) + (Infinity). Apparently, atoms in M_1 are those singletons $\{n\}$, for $n \in \omega$, and each member there is composed of exactly the singletons contained in that member. Furthermore, it is easy to see that M_1 is the smallest infinite model which satisfies **CEM** + (G) + (C) + (A) + (Infinity).

Now we will define a model M_2 of **CEM** + (G) + (C) + (A) + (Infinity) such that there are members with gunky parts in $Dom(M_2)$ (henceforth we will use $Dom(M)$ to stand for the domain of M). We will first define a set D of mixtures of atoms and gunks by induction on ω and then we will use D with a set of atoms as the starting point to define $Dom(M_2)$ by induction on ω again.[5]

Let $D_0 = \{\omega \cup (0, 1)\}$.
$D_{n+1} = \{A \cup G$: for some $A' \subseteq \omega$ and some $G' \subseteq (0, 1)$, such that $A' \cup G' \in D_n$, either $A = \{n \in A'$: n appears in an odd position of the usual linear ordering of $A'\}$ and G is the left half subinterval generated by cutting through the middle point p of G' such that $p \in G$, or $A = \{n \in A'$: n appears in an even position of the usual linear ordering of $A'\}$ and G is the right half subinterval generated by cutting through the middle point p of G' such that $p \notin G\}$.

For example, $D_1 = \{\{0, 2, 4, \dots\} \cup (0, 1/2], \{1, 3, 5, \dots\} \cup (1/2, 1)\}$; $D_2 = \{\{0, 4, 8, \dots\} \cup (0, 1/4], \{2, 6, 10, \dots\} \cup (1/4, 1/2], \{1, 5, 9, \dots\} \cup (1/2, 3/4], \{3, 7, 11, \dots\} \cup (3/4, 1)\}$.

Let $D = \bigcup_{n\in\omega} D_n$. Intuitively, each member in D is composed of a piece of gunk and infinitely many atoms and here each atom is a singleton $\{n\}$, for some $n \in \omega$. Now let $D' = D \cup \{\{n\} : n \in \omega\}$. Also let $\sim X$ stand for $(\omega \cup (0, 1)) \setminus X$, that is, $\{x \in \omega \cup (0, 1) : x \notin X\}$. Define each D'_n by induction on ω as follows.

$D'_0 = D'$
$D'_{n+1} = D'_n \cup \{X \cup Y : X \in D'_n \text{ and } Y \in D'_n\} \cup \{\sim X : X \in D'_n \text{ and } X \neq \omega \cup (0, 1)\}$

Finally let $Dom(M_2) = \bigcup_{n\in\omega} D'_n$. The interpretation of P in M_2 is as usual the set inclusion. The following are some useful facts about M_2.

(Fact 1) For any $X \in Dom(M_2)$, X is of the form $A \cup G$, where $A \subseteq \omega$ and $G \subseteq (0, 1)$.
(Fact 2) $\varnothing \notin Dom(M_2)$.
(Fact 3) For any $X \in Dom(M_2)$, if $\omega \subseteq X$, then $X = \omega \cup (0, 1)$, that is, X is the greatest member in $Dom(M_2)$.

[5]Here the construction of M_2 is based on an idea similar to that in (Tsai and Varzi 2016).

(Fact 4) The $A \subseteq \omega$ mentioned in (Fact 1) cannot be empty.
(Fact 5) For any two members $X = A \cup G$, $Y = A' \cup G'$ in $Dom(M_2)$, if A is a subset of A', then G is a subset of G'.

(Fact 1) and (Fact 2) are trivial.

For (Fact 3), we will first show that each $x \in Dom(M_2)$ is of the form $(Y_{11} \cup Y_{12} \cup \cdots \cup Y_{1k_1}) \cap (Y_{21} \cup Y_{22} \cup \cdots \cup Y_{2k_2}) \cap \cdots \cap (Y_{m1} \cup Y_{m2} \cup \cdots \cup Y_{mk_m})$, where each $Y_{ij} \in Dom(M_2)$ belongs to one of the following three kinds: (a) a finite subset of ω; (b) a member of $D \setminus \{\omega \cup (0, 1)\}$; (c) a member of the form $A \cup G$, where A is a cofinite subset of ω and $G = (0, 1)$. The proof will be an induction on ω using the distribution laws and De Morgan laws, that is, $(A \cup B) \cap C = (A \cap C) \cup (B \cap C)$, $(A \cap B) \cup C = (A \cup C) \cap (B \cup C)$, $\sim(A \cup B) = \sim A \cap \sim B$ and $\sim(A \cap B) = \sim A \cup \sim B$. If $x \in D'_0$, x is obviously of the form specified. Suppose that each $x \in D'_n$ is of the form specified. Consider any $x \in D'_{n+1}$. There are two possibilities. (1) For some $y, z \in D'_n$, $x = y \cup z$. By the inductive hypothesis, y and z are of the form specified, but then by the distribution laws mentioned above, it is easy to see that x is also of the form specified. For example, suppose $y = (Y_1 \cup Y_2) \cap (Y_3 \cup Y_4)$ and $z = (Z_1 \cup Z_2) \cap (Z_3 \cup Z_4)$. Then $x = ((Y_1 \cup Y_2) \cap (Y_3 \cup Y_4)) \cup ((Z_1 \cup Z_2) \cap (Z_3 \cup Z_4)) = (Y_1 \cup Y_2 \cup Z_1 \cup Z_2) \cap (Y_3 \cup Y_4 \cup Z_1 \cup Z_2) \cap (Y_1 \cup Y_2 \cup Z_3 \cup Z_4) \cap (Y_3 \cup Y_4 \cup Z_3 \cup Z_4)$. This simple example can be easily generalized. (2) For some $y \in D'_n$, $x = \sim y$. By the inductive hypothesis, y is of the form specified. For example, suppose $y = (Y_1 \cup Y_2) \cap (Y_3 \cup Y_4)$. Then by De Morgan laws and the distribution laws, $\sim y = (\sim Y_1 \cap \sim Y_2) \cup (\sim Y_3 \cap \sim Y_4) = (\sim Y_1 \cup \sim Y_3) \cap (\sim Y_1 \cup \sim Y_4) \cap (\sim Y_2 \cup \sim Y_3) \cap (\sim Y_2 \cup \sim Y_4)$. If Y_i, for $i = 1$ to 4, belongs to (a)/(c), then $\sim Y_i$ belongs to (c)/(a). Now suppose that Y_i belongs to (b). Then for some $n \in \omega$ and $n \neq 0$, $Y_i \in D_n$. By definition, $\bigcup D_n = \omega \cup (0, 1)$ and D_n contains only finitely many, in fact, 2^n, pairwise disjoint subsets of $\omega \cup (0, 1)$ each of which is of the form $A \cup G$, where A is an infinite but non-cofinite subset of ω and G is a nonempty proper subinterval of $(0, 1)$. Hence $\sim Y_i$ is the union of $2^n - 1$ pairwise disjoint members each of which belongs to (b). Again this simple example can be easily generalized.

Then in order to show (Fact 3), it suffices to show that for any $X = Y_1 \cup Y_2 \cup \cdots \cup Y_k$, for some $k \geq 1$, where each Y_i belongs to one of the said three kinds, ω cannot be a subset of X if $X \neq \omega \cup (0, 1)$. Now if some Y_i belongs to (c), then $(0, 1) \subseteq X$ and hence $X = \omega \cup (0, 1)$ if $\omega \subseteq X$. So we can assume without loss of generality that no Y_i belongs to (c). Furthermore, we can also assume that no Y_i belongs to (a) because of the following two facts. First, the union of a finite subset of ω and an infinite non-cofinite subset of ω will still be an infinite non-cofinite subset, which must be a proper subset, of ω. Second, for any nonempty finite subset S of D, if $\bigcup S$ does not include ω, $\omega \setminus \bigcup S$ must be an infinite non-cofinite subset of ω (see below). Therefore, assume $Y_1, Y_2, \ldots$ and Y_k belong to (b). By the definition of D, we can further assume that each $Y_i \in D_n$, for some $n \in \omega \setminus \{0\}$. Let $m = Max\{n$: for some $i \leq k$, $Y_i \in D_n\}$. Then it is easy to see that each Y_i is the union of some members from D_m (it can be shown by an easy induction that, if $m \geq n$, then each member from D_n is the union of some members from D_m). Hence $X = Y_1 \cup Y_2 \cup \cdots \cup Y_k$ will also be the union of some members from D_m. If $X \neq \omega \cup (0, 1)$, then for some

$Y \in D_m$, $Y \cap X = \varnothing$. This is because $\bigcup D_m = \omega \cup (0, 1)$ and members in D_m are pairwise disjoint. But Y contains a nonempty subset of ω and hence ω is not included in X (actually, by the definition of D, Y must contain an infinite non-cofinite subset of ω and we can see from this the aforementioned second fact).

For (Fact 4), it suffices to show that every nonempty subset of (0, 1) is not in $Dom(M_2)$. If some nonempty $X \subseteq (0, 1)$ is in $Dom(M_2)$, then its complement $Y = (\omega \cup (0, 1)) \setminus X$ will be such that $Y \neq \omega \cup (0, 1)$ and $\omega \subseteq Y$, which cannot be the case by (Fact 3).

For (Fact 5), suppose that there are $X = A \cup G$ and $Y = A' \cup G'$ in $Dom(M_2)$ such that A is a subset of A' but G is not a subset of G'. If $A = A'$, by (Fact 3), A cannot be ω (otherwise, X will be equal to Y since both will be $\omega \cup (0, 1)$) and hence by (Fact 4), A must be a nonempty proper subset of ω. So $\sim X$ exists in $Dom(M_2)$. However, since G is not a subset of G', $G \setminus G' \neq \varnothing$, and therefore $\omega \subseteq \sim X \cup Y \subset \omega \cup (0, 1)$, which contradicts (Fact 3). Therefore, $A \subset A'$. But then X cannot be $\omega \cup (0, 1)$ and hence $\sim X$ exists in $Dom(M_2)$. Moreover, since G is not a subset of G', $\sim X \cup Y$ cannot be $\omega \cup (0, 1)$, and hence $\sim(\sim X \cup Y)$ exists in $Dom(M_2)$. But $\sim(\sim X \cup Y) = \sim(\sim X) \cap \sim Y = X \cap \sim Y = G \setminus G'$, which is a nonempty subset of (0, 1), and this contradicts (Fact 4).

Now we will check that M_2 satisfies **CEM** + (G) + (C) + (A) + (Infinity). It is obvious that M_2 satisfies (P1), (P2), (P3), (G) and (Infinity). By (Fact 4), M_2 satisfies (A). By (Fact 5), M_2 satisfies (SSP), for if $X = A \cup G$ is not a subset of $Y = A' \cup G'$, then A cannot be a subset of A', which implies that there is some $n \in \omega$ such that $n \in X \setminus Y$.

To show that M_2 satisfies (C), consider any $X \in Dom(M_2)$ which is not the greatest member. By definition, $\sim X$ is in $Dom(M_2)$ and obviously, every part of $\sim X$ does not overlap X. Assume that Y does not overlap X. We shall show that Y is a part of $\sim X$. Suppose $Y = A \cup G$ and $X = A' \cup G'$. Then $A \cap A' = \varnothing$. For otherwise, since neither X nor Y is the greatest member, $\sim X$ as well as $\sim Y$ exists in $Dom(M_2)$, and furthermore, since $A \cap A' \neq \varnothing$, $\sim X \cup \sim Y$ cannot be the greatest member either and therefore $\sim(\sim X \cup \sim Y) = X \cap Y$ exists in $Dom(M_2)$, which contradicts the assumption that Y does not overlap X. So $A \cap A' = \varnothing$ and therefore, $A \subseteq \omega \setminus A'$. That is, any atom of Y will be in $\sim X$. Hence, by (Fact 5), Y is a part of $\sim X$.

To show that M_2 satisfies (FP), consider, for any $X, Y \in Dom(M_2)$, the following two cases. (1) $X \subseteq Y$ or $Y \subseteq X$. Then obviously X and Y will have a greatest common part, that is, X if $X \subseteq Y$ or Y if $Y \subseteq X$. (2) $X \cap Y \neq \varnothing$ but neither is included in the other. Then both $\sim X$ and $\sim Y$ exist in $Dom(M_2)$. Furthermore, $\sim X \cup \sim Y$ cannot be the greatest member since $X \cap Y \neq \varnothing$. So $\sim(\sim X \cup \sim Y) = X \cap Y$ exists in $Dom(M_2)$, and it is obvious that this is the greatest common part of X and Y.

To show that M_2 satisfies (FS), it suffices to show that for any $X, Y, Z \in Dom(M_2)$, if Z overlaps $X \cup Y$, Z will overlap X or Y. Suppose that Z overlaps $X \cup Y$. So $Z \cap (X \cup Y) \neq \varnothing$ and hence $Z \cap X$ and $Z \cap Y$ cannot both be empty. Assume without loss of generality that $Z \cap X \neq \varnothing$. We can furthermore assume that neither Z nor X is the greatest member, for otherwise it must be the case that Z overlaps X. Then both $\sim Z$ and $\sim X$ exist in $Dom(M_2)$. Moreover, $\sim Z \cup \sim X$

cannot be the greatest member since $Z \cap X \neq \varnothing$. So $\sim(\sim Z \cup \sim X) = Z \cap X$ exists in $Dom(M_2)$, which means that Z overlaps X.

Hence M_2 is a model of **CEM** + (G) + (C) + (A) + (Infinity). But as mentioned earlier, **CEM** + (G) + (C) + (A) + (Infinity) is a complete theory, and therefore is equivalent to **GEM** + (A) + (Infinity). Hence M_2 also satisfies **GEM** + (A) + (Infinity). However, M_2 has members with gunky parts.

3 The Case of Classical Mereology with the Atomicity Axiom

So first-order theories axiomatized by traditional mereological axioms cannot work. But it might still be hoped that some second-order theories can rule out members with gunky parts from the domain. For example, classical mereology is a second-order theory which guarantees that every nonempty subset of the domain has a fusion. Formally, let capital S stand for a set variable and lower-case x, y and z for individual variables. Then on top of **EM**, classical mereology has the following axiom.[6]

(F2) $\forall S(\exists x Sx \rightarrow \exists z Sum(z, S))$, where Sum is a binary predicate whose definition is as follows.

$Sum(z, S) =_{df} \forall x(Sx \rightarrow Pxz) \wedge \forall x(Pxz \rightarrow \exists y(Sy \wedge Oxy))$

It is fair to say that classical mereology is a natural second-order extension of **GEM**. For the latter guarantees that every first-order definable nonempty subset of the domain has a fusion, which is the best a first-order axiomatizable mereological theory can do, while the former extends the guarantee to all nonempty subsets of the domain.

Intuitively, the model M_2 defined in the previous section is not a model of classical mereology since there are some nonempty subsets of ω whose corresponding fusions are not in M_2. But we can modify M_2 a little bit so as to make it satisfy classical mereology. Define M_3 as follows.

$Dom(M_3) = Dom(M_2) \cup \{A \cup G \subseteq \omega \cup (0, 1) : A \subseteq \omega,\ A \neq \varnothing$, for any $A' \cup G' \in Dom(M_2)$, where $A' \subseteq \omega$ and $G' \subseteq (0, 1)$, $A \neq A'$ and $G = \bigcup\{G' \subseteq (0, 1) :$ for some $A' \subseteq \omega$, $A' \cup G' \in Dom(M_2)$ and $A' \subseteq A\}\}$. The interpretation of P is still the set inclusion.

It goes without saying that M_3 satisfies (P1), (P2) and (P3) and it is also trivial that M_3 satisfies (A). Before checking the satisfaction of other axioms, observe that any nonempty finite subset of ω is in $Dom(M_2)$ and that for any cofinite subset A of ω, $A \cup (0, 1) \in Dom(M_2)$. Therefore, by the definition of $Dom(M_3)$, every member in $Dom(M_3) \setminus Dom(M_2)$ must be of the form $A \cup G$, where A is an infinite non-cofinite subset of ω and $G \subseteq (0, 1)$.

Now we will show that M_3 satisfies (SSP). Suppose $x, y \in Dom(M_3)$ is such that x is not a part of y, that is, x is not included in y. There are four possibilities.

---

[6]The name "classical mereology" follows the nomenclature in Pietruszczak (2015) and the formulation here is a rephrasing of the definition of classical mereology given there.

(1) Both $x, y \in Dom(M_2)$. Then x must have a part z which does not overlap y, for M_2 satisfies (SSP).
(2) Both x, y are not from $Dom(M_2)$. Suppose $x = A \cup G$ and $y = A' \cup G'$, where $A \subseteq \omega$, $G \subseteq (0, 1)$, $A' \subseteq \omega$, and $G' \subseteq (0, 1)$. By the definition of $Dom(M_3)$, if $A \subseteq A'$, it must be the case that $G \subseteq G'$. So A cannot be a subset of A', and hence there must be some $n \in \omega$ such that $n \in x \setminus y$.
(3) x is from $Dom(M_2)$ but y is not. Let $x = A \cup G$, where $A \subseteq \omega$ and $G \subseteq (0, 1)$. There are three possibilities. (3.1) A is a nonempty finite subset of ω. Then by (Fact 5) and the fact that every nonempty finite subset of ω is in $Dom(M_2)$, it must be the case that $x = A$ and therefore there must be some $n \in \omega$ such that $n \in x \setminus y$ since x is not included in y. (3.2) A is a cofinite subset of ω. But then there must be some $n \in \omega$ such that $n \in x \setminus y$, for as mentioned earlier, $\{n \in \omega : n \in y\}$ must be an infinite non-cofinite subset of ω. (3.3) A is an infinite non-cofinite subset of ω. Let $y = A' \cup G'$, for some $A' \subseteq \omega$ and some $G' \subseteq (0, 1)$. If $A \subseteq A'$, then by the definition of $Dom(M_3)$, $G \subseteq G'$ and hence x will be included in y, which contradicts the assumption. Therefore, A is not a subset of A' and hence there must be some $n \in \omega$ such that $n \in x \setminus y$.
(4) y is from $Dom(M_2)$ but x is not. Suppose $x = A \cup G$ and $y = A' \cup G'$, where $A \subseteq \omega$, $G \subseteq (0, 1)$, $A' \subseteq \omega$ and $G' \subseteq (0, 1)$. If $A \subseteq A'$, it must be the case that $G \subseteq G'$, for by definition, $G = \bigcup\{Y \subseteq (0, 1)$: for some $X \subseteq \omega$, $X \cup Y \in Dom(M_2)$ and $X \subseteq A\}$, but by (Fact 5), each $Y \subseteq (0, 1)$ in the said set will be included in G'. But then $x \subseteq y$, which contradicts the assumption. So A is not included in A', and therefore there must be some $n \in \omega$ such that $n \in x \setminus y$.

Finally, let's check that M_3 satisfies (F2). Let S be any nonempty subset of $Dom(M_3)$ and let $A = \bigcup\{A' \subseteq \omega$: for some $G' \subseteq (0, 1)$, $A' \cup G' \in S\}$. By the definition of $Dom(M_3)$, there must be some $G \subseteq (0, 1)$ such that $A \cup G \in Dom(M_3)$ and we will show that $A \cup G$ is the fusion of S, or formally, $M_3 \vDash Sum(x, Y)[A \cup G, S]$.

First, we will show that each member in S is included in $A \cup G$. There are two possibilities.

(1) $A \cup G \in Dom(M_2)$. Consider any $y = A' \cup G' \in S$. If $y \in Dom(M_2)$, by (Fact 5), $y \subseteq A \cup G$ since $A' \subseteq A$. Otherwise, $y \in Dom(M_3) \setminus Dom(M_2)$. By definition, $G' = \bigcup\{Y \subseteq (0, 1)$: for some $X \subseteq \omega$, $X \cup Y \in Dom(M_2)$ and $X \subseteq A'\}$. Then since $A' \subseteq A$, by (Fact 5), each $Y \subseteq (0, 1)$ in the said set must be included in G and hence $G' \subseteq G$.
(2) $A \cup G \in Dom(M_3) \setminus Dom(M_2)$. Consider any $y = A' \cup G' \in S$. By the definition of $Dom(M_3)$, it is trivial that $G' \subseteq G$.

Second, we will show that each part z of $A \cup G$ overlaps some $y \in S$. Observe that no nonempty subset of $(0, 1)$ is in $Dom(M_3)$. This can be easily seen by (Fact 4) and the fact that any member in $Dom(M_3) \setminus Dom(M_2)$ must contain a nonempty subset of ω. Consider any $z = A' \cup G' \in Dom(M_3)$ such that $z \subseteq A \cup G$. As just mentioned, A' cannot be empty. But $A' \subseteq A$ and $A = \bigcup\{X \subseteq \omega$: for some $Y \subseteq (0, 1)$, $X \cup Y \in S\}$. Hence for some X in the said set, $A' \cap X \neq \varnothing$ and therefore z must overlap some $y \in S$.

The foregoing shows that M_3 satisfies classical mereology with the atomicity axiom, that is, **EM** + (F2) + (A), but M_3 still has members with gunky parts.

4 Concluding Remarks

Can there be any second-order theory stronger than **EM** + (F2) + (A) which can secure an atomic domain? This is unlikely. Consider the following model M_4. $Dom(M_4) = \{A \subseteq \omega : A \neq \varnothing\}$ and P is interpreted in M_4 as usual as the set inclusion. It is trivial that M_4 satisfies **EM** + (F2) + (A) and moreover, M_4 can be thought of as an "intended model" which satisfies **EM** + (F2) + (A). If a second-order mereological theory T which is stronger than **EM** + (F2) + (A) can secure an atomic domain, it should be the case that M_4 satisfies T, for we can hardly imagine what else a model could be. However, it is not difficult to see that M_3 and M_4 are isomorphic. From the fact that M_3 satisfies **EM** and (A), it is easy to see that for each nonempty subset A of ω, $Dom(M_3)$ contains exactly one member which is $A \cup G$, for some $G \subseteq (0, 1)$. Therefore, an isomorphism f from M_3 to M_4 can be defined as follows. For any $x = A \cup G \in Dom(M_3)$, $f(x) = A$. It is easy to check that f is indeed an isomorphism. So any second-order mereological theory T satisfied by M_4 will also be satisfied by M_3. That is, T still cannot rule out members with gunky parts from the domain.

One more thing is that the theories considered in this paper are based on traditional mereological axioms. But can we find a theory with some new mereological axioms which help to secure an atomic domain? Let's look into such a possibility a little bit now. Assume that there is some theory T which can secure an atomic domain. We shall also assume that T has infinite models, for otherwise T will be uninteresting (since any finite model must be atomic).[7] We may further assume that T is stronger than **EM** + (A), for we have to put some basic requirements on T anyway and **EM** + (A)

[7] An anonymous referee seems to suggest that a finite model can also have members with gunky parts. But this can hardly make sense. Consider the following finite model whose domain contains three members $\{2, 3\} \cup (0, 1)$, $\{2\}$ and $\{3\}$ and whose interpretation of P is the set inclusion. In this model, $\{2, 3\} \cup (0, 1)$ is composed of $\{2\}$ and $\{3\}$, or in other words, $\{2, 3\} \cup (0, 1)$ is the "fusion" of $\{2\}$ and $\{3\}$. One might think that the fusion of $\{2\}$ and $\{3\}$ should be $\{2, 3\}$ instead of $\{2, 3\} \cup (0, 1)$, but $\{2, 3\}$ is not in the domain (or to put it more drastically, it does not "exist") and hence in that model the fusion will be $\{2, 3\} \cup (0, 1)$. Furthermore, one cannot sensibly say that $\{2, 3\} \cup (0, 1)$ is a member with a gunky part, for $(0, 1)$ has not been divided indefinitely in that model. It is not tenable either to say that $(0, 1)$ is "necessarily" a piece of gunk, for it won't be a piece of gunk in a model whose domain contains all nonempty subsets of the set of real numbers. In short, whether something is a piece of gunk depends on the model where it is situated. In my opinion, here an important issue is: what kind of models can be accepted by metaphysicians as counterexamples to the traditional definition of composition. A model as simple as the one mentioned above can hardly be acceptable. Some philosophical concerns must be considered. Even though we have made a point previously that the traditional definition of composition seems to be defective (Tsai and Varzi 2016), for me it is still unclear how far a mathematical model can go when it comes to a metaphysical issue. This paper is aimed at giving two further technical results and I do not intend to address the aforementioned philosophical issue, but I will probably deal with it somewhere else.

seems to be a reasonable starting point. Based on the result given in the second section, we know that any infinite model of T cannot satisfy all of (FS), (FP), (G) and (C), for otherwise such a model will be elementarily equivalent to M_2 (recall that **CEM** + (G) + (C) + (A) + (Infinity) is complete) and hence M_2 will also satisfy T. But it is not difficult to see that **EM** + (FS) + (G) + (C) implies (FP) and therefore T + (Infinity) must imply $\neg$(FS) $\vee$ $\neg$(G) $\vee$ $\neg$(C). By compactness, this implies that some finite model of T satisfies $\neg$(FS) $\vee$ $\neg$(G) $\vee$ $\neg$(C), if T has finite models. But it is also not difficult to see that any finite model of **EM** + (FS) + (G) satisfies (C), and therefore, some finite model of T must satisfy $\neg$(FS) $\vee$ $\neg$(G), which implies that T cannot have both (FS) and (G). The foregoing preliminary observation gives the following moral: if a first-order axiomatizable theory can secure an atomic domain, it must put some restrictions on (FS), (G) and (C).

Acknowledgements This paper is a product of a research project (104-2410-H-194 -098 -MY3) funded by Ministry of Science and Technology, Taiwan.

References

Casati, R., & Varzi, A. C. (1999). *Parts and Places*. Cambridge: MIT Press.

Pietruszczak, A. (2015). Classical mereology is not elementarily axiomatizable. *Logic and Logical Philosophy*, *24*, 485–98.

Simons, P. (1987). *Parts: A Study in Ontology*. Oxford: Clarendon Press.

Tsai, H.-C. (2013). Decidability of general extensional mereology. *Studia Logica*, *101*(3), 619–36.

Tsai, H.-C. (2015). Notes on models of first-order mereological theories. *Logic and Logical Philosophy*, *24*, 469–82.

Tsai, H.-C., & Varzi, A. C. (2016). The Limits of 'Composition'. *Erkenntnis*, *81*, 231–35.

MIX
Papier aus verantwortungsvollen Quellen
Paper from responsible sources
FSC® C105338

If you have any concerns about our products, you can contact us on
ProductSafety@springernature.com

In case Publisher is established outside the EU, the EU authorized representative is:
Springer Nature Customer Service Center GmbH
Europaplatz 3, 69115 Heidelberg, Germany

Printed by Libri Plureos GmbH
in Hamburg, Germany

The Romance of Science: Essays in Honour of Trevor H. Levere

Archimedes

NEW STUDIES IN THE HISTORY AND PHILOSOPHY OF SCIENCE AND TECHNOLOGY

VOLUME 52

Archimedes has three fundamental goals; to further the integration of the histories of science and technology with one another: to investigate the technical, social and practical histories of specific developments in science and technology; and finally, where possible and desirable, to bring the histories of science and technology into closer contact with the philosophy of science. To these ends, each volume will have its own theme and title and will be planned by one or more members of the Advisory Board in consultation with the editor. Although the volumes have specific themes, the series itself will not be limited to one or even to a few particular areas. Its subjects include any of the sciences, ranging from biology through physics, all aspects of technology, broadly construed, as well as historically-engaged philosophy of science or technology. Taken as a whole, *Archimedes* will be of interest to historians, philosophers, and scientists, as well as to those in business and industry who seek to understand how science and industry have come to be so strongly linked.

More information about this series at http://www.springer.com/series/5644

Jed Buchwald • Larry Stewart
Editors

The Romance of Science: Essays in Honour of Trevor H. Levere

Springer

Editors
Jed Buchwald
Caltech
Pasadena, CA, USA

Larry Stewart
College of Arts & Science
University of Saskatchewan
Saskatoon, Saskatchewan, Canada

ISSN 1385-0180 ISSN 2215-0064 (electronic)
Archimedes
ISBN 978-3-319-58435-5 ISBN 978-3-319-58436-2 (eBook)
DOI 10.1007/978-3-319-58436-2

Library of Congress Control Number: 2017944446

Printed on acid-free paper

This Springer imprint is published by Springer Nature
The registered company is Springer International Publishing AG
The registered company address is: Gewerbestrasse 11, 6330 Cham, Switzerland

Contents

List of Contributors

Robert G.W. Anderson spent most of his career working in UK national museums, first as a history of science curator and then latterly as director: the Royal Scottish Museum (Edinburgh), the Science Museum (London), the National Museum of Scotland, and the British Museum. He has acted as president of the Chemical Heritage Foundation, Philadelphia, and as chairman of the Society for the History of Alchemy and Chemistry. His main interest is chemistry in the Scottish Enlightenment.

Victor D. Boantza is assistant professor in the Program for History of Science, Technology, and Medicine at the University of Minnesota. He works on the history of the early modern physical sciences, with interests in the relations between theory and practice in physics and chemistry in the seventeenth and eighteenth centuries. *Matter and Method in the Long Chemical Revolution: Laws of Another Order* (Ashgate, 2013) is his first book

Jed Buchwald is Doris and Henry Dreyfuss Professor of history at Caltech and is finishing a book on the decipherment of Egyptian hieroglyphs. His most recent publications are *The Zodiac of Paris* (Princeton, 2010, coauthor Diane Greco Josefowicz), *Newton and the Origin of Civilization* (Princeton, 2012, coauthor Mordechai Feingold), and "Kirchhoff's theory for optical diffraction" (*Archive for History of Exact Sciences*, 70 (2016):464–511, coauthor Chen-Pang Yeang).

Andrew Ede is a historian of science specializing in twentieth-century chemistry and physical sciences, teaching in the Department of History and Classics at the University of Alberta. He is also the director of the Science, Technology and Society Program. Along with Lesley B. Cormack, he is the coauthor of *A History of Science in Society: From Philosophy to Utility*.

Gregory A. Good publishes mostly on the history of geophysics and other Earth sciences. He has been a fellow of the Smithsonian Institution and the Alexander von Humboldt-Stiftung and has received several grants from the National Science

Foundation. He taught for 25 years at West Virginia University and has been director of the Center for History of Physics at the American Institute of Physics since 2009. He is a fellow of the Geological Society of America and a life member of Clare Hall, Cambridge, UK. He is currently writing a book titled *John Herschel's Planet: Earth in the Cosmos*.

Ernie Hamm is a historian of science and associate professor in the Department of Science and Technology Studies, York University, Toronto, Canada. He works on the history of geology and on aspects of Enlightenment and Romantic science.

Jennifer Hubbard is associate professor of the history of science and technology at Ryerson University and specializes in the political, environmental, institutional, and economic history of fisheries and ocean science. Through the University of Toronto Press, she published in 2006 *A Science on the Scales: The Rise of Canadian Atlantic Fisheries Biology 1898–1939* and, in 2016, *A Century of Maritime Science: The St. Andrews Biological Station 1908–2008*, which includes chapters by scientists, technologists, and historians of science, which she coedited with Robert Stephenson and David Wildish. She has also published articles in *ISIS*, the *ICES Journal of Marine Science*, and *Environmental History*. She continues to find inspiration from the archives and influence of Archibald Gowanlock Huntsman, first director of the then-named Atlantic Biological Station.

David Knight Reading chemistry at Oxford, **David Knight** was tempted into history by Alistair Crombie and put to work on Humphry Davy with Sir Harold Hartley as my guide. Ignorantly supposing that the Romantics were hostile to science, he was astonished to find that Davy was a close friend of S. T. Coleridge and through him of William Wordsworth and Walter Scott: they all hoped for a dynamic science, especially chemistry. Coleridge and Davy had been brought together by the enthusiast for chemical medicine, Thomas Beddoes, and Trevor's researches and David's were complementary; they met and made friends. In 1964, he was appointed to start history of science in the Philosophy Department at Durham University, while Trevor went to Canada; both settled happily and have kept in close contact ever since. Both have focused especially on the Revolutionary and Romantic periods, but ranged more widely as one thing led to another. David remarks, "I'm very grateful to have had so good a friend and helpful critic as Trevor right through my career."

Janis Langins is an emeritus professor of history of technology at the Institute for the History and Philosophy of Science and Technology at the University of Toronto, where he also received his doctorate in the history of science and technology under Trevor Levere with a thesis on the early history of the Ecole Polytechnique during the French Revolution. Previously he graduated from McGill University with a master's degree in chemical engineering. His research interests are in the social history of engineering, in the history of the military engineering corps of Old Regime France and engineers in the late nineteenth and early twentieth centuries.

Gordon McOuat is Professor in the History of Science and Technology and the Contemporary Studies Programmes at the University of King's College/Dalhousie University, the Director of the international partnership project, "Cosmopolitanism and the Local in Science and Nature: East and West", and the past Director of the Canadian national knowledge cluster, "Situating Science". His research centres on the history and philosophy of the life sciences, classification and logic, including the formation of language rules and the origins of modern quantified logic. His recent publications include the *Origins of Natural Kinds, Cataloguing Power, and Bentham's Logic*, along with the co-edited volumes, *Circulation of Knowledge between England, India and China* (Brill) and *Narratives of Nature and Science: East and West* (Routledge).

David Philip Miller is an emeritus professor of history of science at the University of New South Wales and a fellow of the Australian Academy of the Humanities. He continues to work and publish on the life of James Watt. He also investigates the nature of discovery and invention, concerning which a recent publication is "Of Patents, Principles and the Construction of Heroic Invention: The Case of Neilson's Hot Blast in Iron Production," *Proceedings of the American Philosophical Society* **160** (December 2016), 1–62. As coeditor of *Annals of Science*, he was proud to succeed Trevor Levere.

David Pantalony is curator of physical sciences and medicine at the Canada Science and Technology Museum, Ottawa, Canada. His main research interest is in the history of scientific instruments. As adjunct faculty at the University of Ottawa, he also teaches a collection-based seminar for history students.

Margaret Schabas is a professor of philosophy at the University of British Columbia. She works primarily in the history and philosophy of economics and has published over 40 articles or book chapters. She is the author of two monographs. *A World Ruled by Number* (Princeton, 1990) and *The Natural Origins of Economics* (Chicago, 2005), both of which are in paperback. She is the recipient of a UBC Killam Research Prize (2015) and has given many keynote addresses, to the Hume Society, HOPOS, CPA, and as president of the HES (2013–2014).

Larry Stewart is professor of history at the University of Saskatchewan and is currently writing a study of experiment during the late Enlightenment and Industrial Revolution. His most recent work, edited with Erika Dyck, is *The Uses of Humans in Experiment. Perspectives from the 17th to the 20th Century* (Brill/Rodopi, 2016).

Suzanne Zeller is professor of history at Wilfrid Laurier University. Her research focuses on the history of science and environmental history in Canada.

Chapter 1
Trevor Levere, Affinities That Matter

Ernie Hamm

Abstract Trevor Levere's manifold contributions to the history of science are rooted in the history of chemistry and find important early expression in his foundational study of nineteenth-century ideas of affinity. Theories of matter took Levere to the all-encompassing romanticism and science of Coleridge and *Naturphilosophie*. From there he has proceeded to leave an important mark on many others interconnected areas of the history of science, including the study of scientific instruments, science and the Arctic, the history of science in Canada, and the utilitarian and political interactions of Enlightenment chemistry.

From the perspective of the early twenty-first century, it is all too easy to gaze wistfully upon the 1960s as a most propitious time to embark upon a life as an historian of science. The founding of departments, institutes and research centres for the history of science, the flourishing of conferences, societies and other scholarly fora, the dramatic expansion of universities over the past half century can take on a sense of inevitability when looking backward. Things must have looked rather different to a young English academic in 1968 moving from Oxford to the University of Toronto for a 1-year appointment as Lecturer in a field that had relatively few practitioners and limited institutional support. Fortunately, his new colleagues took the long view and recognized Trevor's outstanding preparation and promise. He was appointed Assistant Professor in 1969 and soon thereafter rose through the ranks, accumulating national and international honours along the way, and playing a crucial role in building an institutional base for the history of science at the University of Toronto. While his self-effacing style can belie such accomplishments, those who know him well—among them the colleagues, students and friends who have contributed to this collection—recognize a most unusual and endearing affinity between Trevor's unassuming manner and the insight, imagination and determination that characterize his intellectual life.

E. Hamm (✉)
Department of Science and Technology Studies, York University, Toronto, Canada
e-mail: ehamm@yorku.ca

J. Buchwald, L. Stewart (eds.), *The Romance of Science: Essays in Honour of Trevor H. Levere*, Archimedes 52, DOI 10.1007/978-3-319-58436-2_1

It is hard to resist the temptation to see in Trevor's early work the seed that would grow into the work of the mature scholar. In his first monograph, *Martinus van Marum and the Introduction of Lavoisier's Chemistry into the Netherlands*, one can now see a number of fundamental themes and concerns that run through much of Trevor's work, including a focus on chemistry, especially pneumatic chemistry, and the elaborate scientific instruments that were developed alongside it and experimental philosophy more generally during the in so many ways revolutionary decades of the late eighteenth and early nineteenth centuries.[1] Van Marum and his location at Teyler's Museum in Haarlem, a linguistic and cultural crossroads of Europe, was ideally suited for someone with facility in languages and who eschewed the notion that the history of science could be fundamentally delineated by linguistic or national borders. However, this early work should also be looked at from the other direction, as the product of a young man who as a boy had the advantage of a good chemistry set and an excellent chemistry teacher at St. Paul's School in London before going up to Oxford to pursue his historical, philosophical and scientific interests. Remarkably, his work on van Marum was first presented as an *undergraduate* thesis—clearly this was a student who was going places.

Trevor went on to do a D.Phil. (1969) in Oxford's Department of Modern History under the supervision of A. C. Crombie. His thesis, published with substantial modifications as *Affinity and Matter: Elements of Chemical Philosophy 1800–1865* (Oxford: Clarendon, 1971), effectively reoriented our picture of much of nineteenth-century science, showing that affinity was the unifying concept of chemistry and was the basis for uniting "chemistry as a whole with physics." (*Affinity and Matter*, p. vi). Affinity, a protean idea with roots reaching well back into Renaissance hermeticism, seemed an unlikely candidate for such a prominent place in nineteenth-century science. However, through a study of the key figures such as Davy, Faraday, Berthollet and Berzelius, and other more unexpected ones such as Coleridge and Oersted, Trevor showed that affinity appealed to those who sought models for explaining chemical composition and decomposition, and that it was deeply intertwined with theories of matter. While *Affinity and Matter* is very much a contribution to the history of chemistry, it also challenges fundamentally the notion of a disciplinary history, for its balanced treatment of affinity reaches well beyond any particular discipline and involves natural philosophy and theology, physics and philosophy.

Trevor's eye for the interconnections of seemingly disparate branches of knowledge was ideally suited for his next major project, *Poetry Realized in Nature: Samuel Taylor Coleridge and Early Nineteenth-Century Science* (Cambridge: Cambridge University Press, 1981). This book removed any remaining doubts about the seriousness of Coleridge's engagement with science and made a compelling case that romanticism could no longer be dismissed as an anti-scientific movement, as a gigantic loss of intellectual nerve by people unable or unwilling to accept mech-

[1] "Martinus van Marum and the Introduction of Lavoisier's Chemistry into the Netherlands," in *Martinus van Marum: Life and Work*, Volume I, ed. R. J. Forbes (Haarlem: H. D. Tjeenk Willink & Zoon, 1969), 158–286.

anistic explanations of the Enlightenment. At the time, this was something of a revelation to literary historians; for historians of science it was an exciting development, an important demonstration that presentist concerns were an unreliable guide when grappling with the central concerns of different times and places.

Anyone who first encountered his work through *Poetry Realized in Nature*, as did I, might think of Trevor primarily as a historian of science with a focus on romanticism, rather than a historian of chemistry. Yet chemistry is a science with all sorts of unexpected connections. It turns out Trevor was first drawn to Coleridge through his connection with Davy, which then led him to seek the insight of his Toronto colleague Kathleen Coburn, one of the outstanding Coleridge scholars of the twentieth century. Coburn recognized that she had found just the person to make sense of Coleridge's unpublished notebooks, which are rife with chemical and natural philosophical entries. Trevor seized this opportunity and followed the notebooks where they led, including to libraries all over Britain and North America tracking down Coleridge's marginalia, at that point still unpublished. This was the work of a scholar's scholar and was recognized as such by his contribution to *Coleridge's Notebooks* and *The Collected Works of Samuel Taylor Coleridge*, definitive editions published in Princeton's Bollingen Series. Immersing himself in Coleridge's sources took him elsewhere as well, into the at turns marvelous, arcane and remarkable world of *Naturphilosophie*. This was heady stuff and Trevor's exemplary treatment of this material is always insightful and even-handed. *Poetry Realized in Nature* gives us a Coleridge with a dynamic and active view of nature and who saw unity in the various parts of science, a deep connection between mind and nature, and rejected simplistic distinctions between empirical science and speculative philosophy. This was inspiring work for those who would later contribute to the now large body of work on science and romanticism.

By the early 1980s Trevor was a full Professor, Fellow of the Royal Society of Canada and a much admired teacher and colleague in the University of Toronto's Institute for the History and Philosophy of Science. His intellectual breadth was much sought after in the classroom, particularly in seminars. His graduate course on the intellectual context of nineteenth century science was a favourite of many students, but the scope of his interests and knowledge already then reached well beyond Victorian science, from early modern to well into the twentieth century—as students in his seminars on the history of chemistry or on the history of science in Canada can attest.

Not many years after arriving in Toronto, Trevor saw and met a need to foster the study of science in Canadian history. Perhaps this was to be expected from an avid birder who loves spending time outdoors, including camping trips (summer and winter) with graduate students during his early years in Toronto. While doing his work on Coleridge he and a former graduate student, Richard Jarrell, co-edited *A Curious Field-book: Science & Society in Canadian History* (Toronto: Oxford University Press, 1974), a source book for the social history of Canadian science. Jarrell went on to take a leading role developing the history of Canadian science as a field of study, a field in which Trevor would also make a foundational contribution.

Somehow, in between writing his book on Coleridge, teaching, and exploring the history of science in Canada, Trevor developed his administrative skills and became an obvious choice for the directorship of the Institute for the History and Philosophy of Science, a role in which he served 2 five-year terms. The director of a small and young academic unit faces many challenges, not least securing resources in an environment where larger well-established departments have more institutional clout. One of the chief tasks facing Trevor was finding a new home for the Institute, and here he succeeded in securing an elegant suite of offices, including space for graduate students (on which he insisted) in Victoria College, a home that has lasted to this day. The only thing missing was air-conditioning, a problem the director broached with great success at the end of a lengthy meeting, carefully scheduled on a midsummer afternoon, with the more senior administration. Trevor maintained a most amiable, collegial tone and the festive end of term potluck dinners were a highlight for grad students, faculty, their partners and children, and did much more to make the Institute an amiable intellectual environment than any "self-study", external review, or any of the other inventions of high-level university bureaucrats. Trevor deserves nothing but admiration for walking the fine line between the demands of a large university and the needs of a small academic unit, all the while fostering a spirit of collegiality.

Trevor supervised at least 17 PhD students and many MA students, and as one of the former I can say that he is a most congenial *Doktorvater*: always prepared to offer guidance when needed, encouragement when appropriate, a nudge when required and, when the time is right, to let a student focus on research and writing. His aim is always to encourage his students to develop their interests and do good work, not create disciples or epigones, and in my experience he never sees supervision as an opportunity to promote a particular view or approach to the history of science.

One of the delights of my graduate education was having Trevor guide me through a number of the texts foundational for an understanding of *Naturphilosophie*. After working through those texts it became clear I needed to know more about the Enlightenment thinkers that seemed to be the foil for a generation of romantics and proposed as much to Trevor, who encouraged me to give him a reading list. By the time I did, I was pretty pleased with my list (which was probably not much more than the usual suspects). He looked it over, nodded, paused and told me since I was interested in the Enlightenment I needed to pay attention to technology. This not being the comment I was hoping for, I proceeded with what I imagined to be a vigorous defense of the Enlightenment as fundamentally about ideas, but what more likely sounded like a long-winded recapitulation of the hoary trope that history of science is about thinkers not tinkers. Trevor nodded, this time with his characteristic gentle smile, and said "Take a look at the *Encyclopédie*. The plates."

I did as I was told and eventually the lesson of the rich technological tableaux of the *Encyclopédie*, a work emblematic of the Enlightenment, took hold. Long before it had become a commonplace to say that the history of science was profoundly connected with technological change, Trevor knew that technology mattered. The practices and ideas of natural philosophers, chemists and natural historians, among others, were deeply intertwined with technologies of measurement, observation, classification and experimentation, to name but a few.[2] This was true, I would learn, even for many things that seemed purely intellectual, or were presented as such. Arguments for the utility, including technological utility, of science needed to be taken seriously, though not uncritically. Paying attention to technology, I learned, is good for thinking about the history of science.

After completing his work on Coleridge, Trevor emerged from the world of *Naturphilosophie* and romanticism prepared for something more bracing. *Science and the Canadian Arctic: A Century of Exploration, 1818–1918* (Cambridge: Cambridge University Press, 1993) once again showed its author was well ahead of the field, this time giving exploration a central place in the history of science some years before that became a trend. Trevor looked at the Canadian arctic as a field for collaboration and competition, usually the latter, by British, Scandinavian, American, German, Austro-Hungarian and, in the latter stages of the story he tells, Canadian (i.e. those who came from the south) expeditions. Scientific expeditions throughout the century he covers were bound up with imperial ambitions, national pride, international competition, commercial activity and sovereignty. Trevor is not the sort of historian who puts his craft to the service of activism, but in his understated way he made the case for the continuing relevance of these themes. The events of the intervening years have shown Trevor to be more prescient than ever. In his Afterword he noted there is "a real sense in which northern regions around the pole have colonial economies in relation to southern governments and industries." (427) Yet the Inuit, he suggested, could provide important evidence of Canadian sovereignty. *Science and the Canadian Arctic* is not a study of what is now known as indigenous knowledge, but Trevor noted that throughout the period he covers and beyond, those expeditions with the best chance for survival (a basic measure of success in the Arctic) were those that fostered contact with and made a point of trying to learn from the ways of the Inuit.

By the 1990s Trevor had mastered at least three distinct fields in the history of science and was on a trajectory to the highest echelons of the profession. For many professors this is the stage where they can with some justification focus on research and leave matters of professional and editorial service to others. This was never a possibility for Trevor, whose quiet sense of obligation to colleagues, students and the profession seems almost quaintly out of place in an era where advancement is increasingly measured by increments not by the quality of a contribution. During these years he served his second term as director and, among other things, devoted his editorial acumen to a number of project, including a book co-edited with William Shea, *Nature, Experiment and the Sciences: Essays on Galileo and the History of Science in Honour of Stillman Drake* (Dordrecht: Kluwer, 1990) and a three volume collection of Stillman Drake's essays, co-edited with Noel Swerdlow, *Essays on*

[2] For an early example of this see Gerard L'E. Turner and Trevor H. Levere, *Van Marum's Scientific Instruments in Teyler's Museum*, vol. 4 of *Martinus van Marum: Life and Work* (Leyden: Noordhoff, 1973), more recently in Frederic L. Holmes and Trevor H. Levere, eds, *Instruments and Experimentation in the History of Chemistry* (Cambridge, MA: MIT Press, 2000).

Galileo and the History and Philosophy of Science (Toronto: University of Toronto Press, 1999). In 1999 he also took over as editor of *Annals of Science*, the first and oldest history of science journal published in England, a position he held for a full 14 years.

Annals of Science gave Trevor the opportunity to encourage the work of young scholars, something in which he takes great pleasure. The same held true for the Ashgate book series "Science, Technology and Culture, 1700–1945," which he co-edited with David Knight. Over the years a number of colleagues and friends from Europe (including Britain) and North America have told me how grateful they are for his quiet and steadfast support.

Lest there be any doubt, editing did not keep Trevor from other writing projects. *Transforming Matter: A History of Chemistry from Alchemy to the Buckyball* (Baltimore: Johns Hopkins University Press, 2001) is a shrewd introductory study aimed at a non-specialist audience that begins with the premise that chemistry was and remains a practical science. Trevor's long experience teaching the history of chemistry allows him to cover the ground from a discussion of alchemy that will surely surprise the intended readership, through to environmental chemistry at the close of the twentieth century.

The practical character of chemistry, its concern with developing tools and instruments for manipulating matter, makes it a subject of considerable social and political import. This importance is amplified during times of revolutionary social change. *Discussing Chemistry and Steam: The Minutes of a Coffee House Philosophical Society, 1780–1787*, co-authored and edited by Trevor and Gerard L'E. Turner with contributions by Jan Golinski and Larry Stewart (Oxford: Oxford University Press, 2002) gives us a place at the table, as it were, of an eighteenth-century society devoted to promoting natural knowledge. In Trevor's pithy assessment, "The appetite of members for profitable science was intense." (7).

Of the chemists who were drawn to more than industrial revolution, one of the foremost was Thomas Beddoes, a crucial figure in the later stages of the English Enlightenment who, thanks to his connections with Davy and Coleridge, has long attracted Trevor's attention. Beddoes's interest in medicine, chemistry, geology and education, and his reforming tendencies and democratic politics made him a controversial figure at a time when, as Trevor has pointed out, the British authorities treated democracy with about as much enthusiasm as McCarthy treated communism. *The Enlightenment of Thomas Beddoes: Science, Medicine and Reform*, co-authored by Trevor, Larry Stewart and Hugh Torrens with Joseph Wachelder (London: Routledge, 2016), presents us with a Beddoes who saw medicine and chemistry as tools for pulling the poor out of poverty and who stayed true to his radical tendencies in an often hostile environment.

This brief tour of Trevor's life in the history of science is very far from complete. Space does not permit a discussion of all of his edited volumes, much less his scores of articles, or of the many encounters related by the contributors (students, colleagues and friends) to this volume. As one might expect, he has a long list of honours. Besides those already mentioned, he has held numerous visiting fellowships in North America, Europe and Japan, has been elected to the International Academy of the History of Science in Paris and the Royal Holland Society of Sciences, and is a Fellow of the Royal Geographical Society in London. He has held Killam and Guggenheim Fellowships and was awarded a D.Litt. by Oxford in 1999. In 2006 the University of Toronto appointed him University Professor and in 2009 the American Chemical Society recognized him with the Edelstein Award "for the breadth and depth of his historical interests, his research productivity, his understanding of the intellectual and contextual aspects of the history of chemistry, and his promotion of the history of science in Canada."

These stellar achievements reflect Trevor's place among the world's leading historians of science, but it is more appropriate to conclude with the man rather than his honours. Those of us fortunate enough to socialize with him admire him for his affinity for good conversation on any number of topics—the arts, politics, literature, science, family, nature—always enlived with his subtle and sly wit (akin to that of W.C. Fields, it has been said). Such qualities, together with his keen awareness that intellectual life is by no means limited to the academy, make him a most amiable dinner companion. No doubt his intellectual breadth serves him well, as does the way he has balanced his outstanding career with his devotion to his wife, children and grandchildren. Trevor knows what matters, and it is a genuine pleasure to present him this volume, a token of friendship and collegiality!

Chapter 2
Elements, Instruments, and Menstruums: Boerhaave's Imponderable Fire Between Chemical Masterpiece and Physical Axiom

Victor D. Boantza

> *Fire is a Body, and a Body in motion. Its Motion is argued from its expanding the Air; which is not effected without communicating Motion thereto: And its Corporeity is proven by Experiment. For pure Mercury being enclosed in a Phial with a long Neck, and kept in a gentle Heat for the space of a Year, is reduced into a Solid; and its Weight considerably increased; which Increase can arise from nothing but the accession of Fire ... The Nature of Fire is so wonderful, and abstruse, that the Ancients generally revered it as a God: Among the moderns, we can scarce name one Point in all Philosophy of more importance, or less understood. ("Fire," Ephraim Chambers, Cyclopædia, 1728)*

> *Hence, in all physics, it is found exceedingly difficult perfectly to distinguish the very action of fire from that of other concurring causes; while yet the nature of fire is so very different from theirs, that to confound them together will introduce the utmost uncertainty and disorder. (Herman Boerhaave, A New Method of Chemistry, 1741)*

2.1 Introduction

In 1796, in the aftermath of the Chemical Revolution, 2 years after Lavoisier's execution and Priestley's flight to America, Lavoisier's close collaborator Antoine-François de Fourcroy (1755–1809) wrote:

An earlier version of this essay was presented at a workshop held at McGill University in 2010: "The Maker's Universe: Science, Art, and Instruments in Early Modern Europe." Trevor Levere acted as commentator and I thank him for his insightful questions and remarks.

V.D. Boantza (✉)
History of Science Technology and Medicine, Tate Lab of Physics,
University of Minnesota, 116 Church Street SE, Minneapolis, MN 55455, USA
e-mail: vboantza@umn.edu

J. Buchwald, L. Stewart (eds.), *The Romance of Science: Essays in Honour of Trevor H. Levere*, Archimedes 52, DOI 10.1007/978-3-319-58436-2_2

> In his noble work, entitled *Physica subterranea*, [Johann Joachim Becher] had collected all the chemical phenomena at that time known, and has described them with amazing accuracy … Stahl, who was born with a natural fondness for chemistry, undertook to illustrate the doctrine of Becher by commentary. His favourite subject, to which he more particularly directed his attention, was to demonstrate the existence of that inflammable earth which he denominated Phlogiston … His treatises on sulphur and salts, together with that which is entitled *Trecenta Experimenta*, have acquired him immortal honour … Boerhaave, amid his various engagements, found leisure to cultivate chemistry, and composed a very profound work on this science, which has long been highly celebrated. The treatises on the four elements, particularly that on fire, which form a part of this work, are each of them masterpieces, containing all that was at that time known on their particular subjects.[1]

The "masterpieces" on the "four elements, particularly that on fire" were part of Herman Boerhaave's *Elementa Chemiae* of 1732 (Fig. 2.1).[2] These were the developments, authors, and works Fourcroy regarded as most central to what he framed as the "Rise and Progress of Philosophical Chemistry, from the middle of the seventeenth to the middle of the eighteenth Century."[3] A few decades earlier, in the early 1750s, the Montpellier chemist Gabriel François Venel (1723–1775) mentioned the very same works and names, but in a tellingly different way and context. Fourcroy's words were meant as a historical reflection; Venel's allusions were part of an argument he advanced in his article "chimie" that appeared in the third volume (1753) of Diderot and d'Alembert's *Encyclopédie*.

Both Fourcroy and Venel, separated by a generation, recognized the importance of chemical research on fire. Where the two chemists differed, importantly, was in their understanding of what constituted chemical—versus physical—research. Drawing on fire's double role as a physical agent of repulsion and as a chemical substance whose definition was similar to traditional phlogiston, Venel remarked:

> As for fire, and its essential quality of rarifying all bodies by which Boerhaave, who has best treated it physically, characterizes this fluid, this property is clearly related to fire as

---

[1] Antoine-François Fourcroy, *Elements of Chemistry, and Natural History. To which is Prefixed the Philosophy of Chemistry*, 4th ed. trans. R. Heron (London: J. Murray et al., 1796), vol. 1, 31–2. This is the fourth edition. The first three editions appeared, with slightly different titles in 1782, 1788, and 1790.

[2] All the references and quotations in this essay are from the 1741 edition: *A New Method of Chemistry, Including the History, Theory, and Practice of the Art*. For details on Boerhaave's work and life see Arie G. Lindeboom, *Herman Boerhaave: The Man and His Work* (London: Methuen, 1968); Arie G. Lindeboom, ed., *Boerhaave and his Time: Papers Read at the International Symposium in Commemoration of the Tercentenary of Boerhaave's Birth* (Leiden: Brill, 1970). An insightful study of Boerhaave's chemistry in the context of his pedagogy and practice is John C. Powers, *Inventing Chemistry: Herman Boerhaave and the Reform of the Chemical Arts* (Chicago, 2012). See also Rina Knoeff, *Herman Boerhaave (1668–1738): Calvinist Chemist and Physician* (Amsterdam: Koninklijke Nederlandse Akademie van Wetenschappen, 2002); Mi Gyung Kim, *Affinity, That Elusive Dream: A Genealogy of the Chemical Revolution* (Cambridge, MA, 2003), 177–88. For a more narrowly focused study see Ursula Klein, "Experimental History and Herman Boerhaave's Chemistry of Plants," *Studies in History and Philosophy of Science*, **34** (2003), 533–67.

[3] Fourcroy, *Elements*, 30.

a

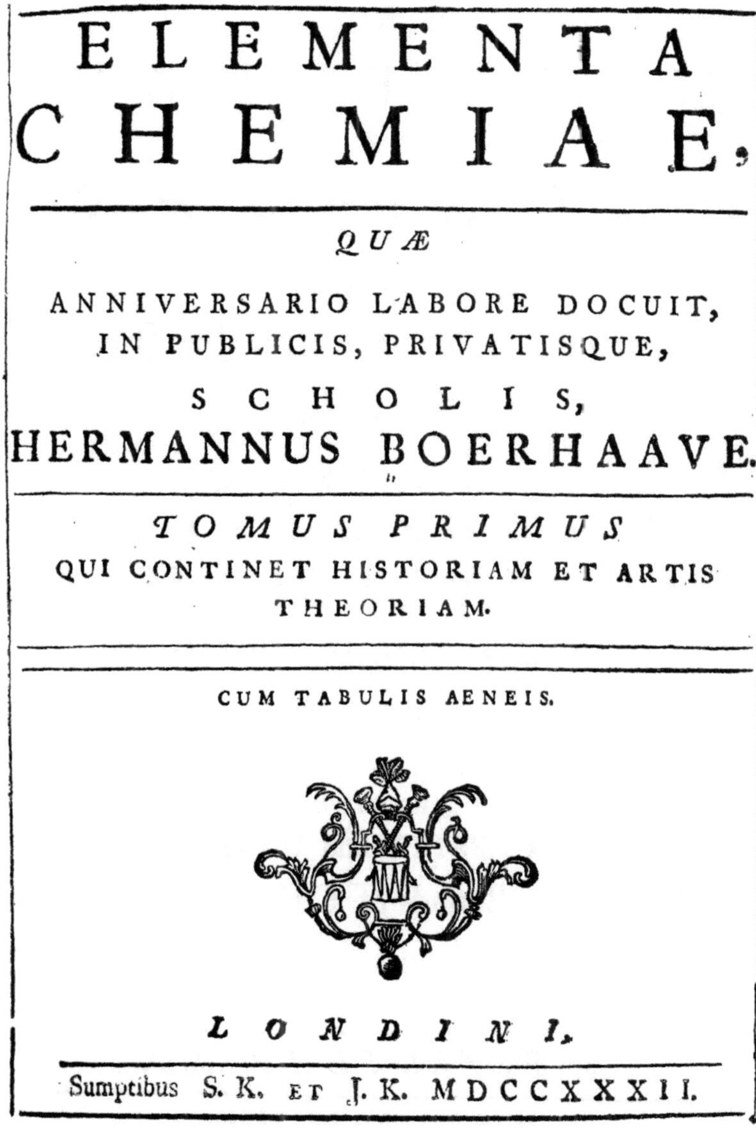

ELEMENTA CHEMIAE,

QUÆ

ANNIVERSARIO LABORE DOCUIT, IN PUBLICIS, PRIVATISQUE, SCHOLIS, HERMANNUS BOERHAAVE.

TOMUS PRIMUS QUI CONTINET HISTORIAM ET ARTIS THEORIAM.

CUM TABULIS AENEIS.

LONDINI,

Sumptibus S. K. ET J. K. MDCCXXXII.

Fig. 2.1 Boerhaave's 1732 *Elementa chemiae above* and the second English edition *below*, translated by Peter Shaw and published in 1741 with his commentaries. The first English edition appeared in 1735, 3 years before Boerhaave's death; it was translated by Timothy Dallowe and published as *Elements of Chemistry, being the lectures of Herman Boerhaave, M.D. Formerly Professor of Botany, And at present, Professor of Physick in the University of Leyden*. Both editions are in two volumes

b

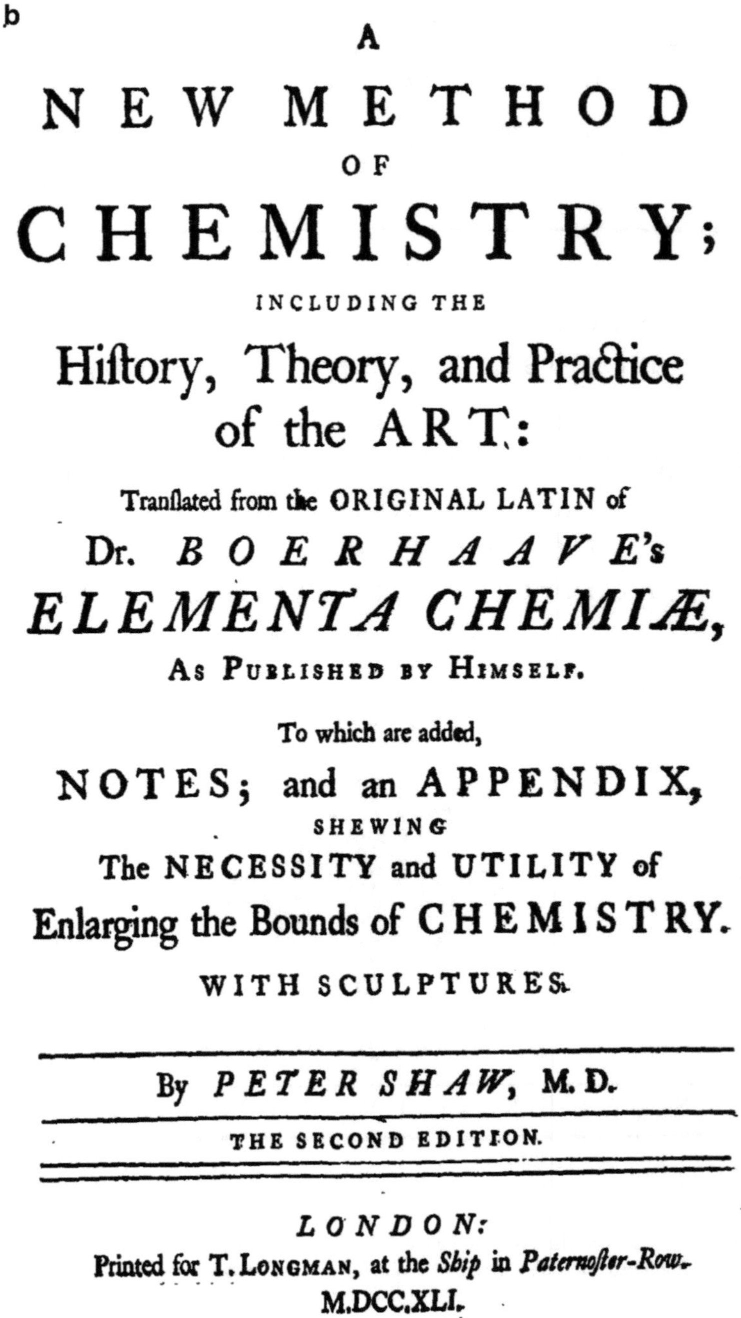
A

NEW METHOD

OF

CHEMISTRY;

INCLUDING THE

Hiſtory, Theory, and Practice of the ART:

Tranſlated from the ORIGINAL LATIN of

Dr. *BOERHAAVE*'s

ELEMENTA CHEMIÆ,

As Published by Himself.

To which are added,

NOTES; and an APPENDIX,

SHEWING

The NECESSITY and UTILITY of Enlarging the Bounds of CHEMISTRY.

WITH SCULPTURES.

By *PETER SHAW*, M.D.

THE SECOND EDITION.

LONDON:

Printed for T. Longman, at the *Ship* in *Paternoſter-Row*.

M.DCC.XLI.

Fig. 2.1 (continued)

> mass or aggregate; we can thus confirm that all but five or six lines of Boerhaave's treatise on fire deal with physics ... [yet] fire, considered as a particular chemical object, is a principle capable of combination and precipitation, conferring to those mixts of which it is a principle color, inflammability, metallicity, etc., hence the treatise on fire, known as Stahl's *Trecenta* [*experimenta*], is fundamentally chemical.[4]

According to Venel, Georg Ernst Stahl (1659–1734) perceived fire "as a particular chemical object" whereas Boerhaave (1668–1738), the famous Leiden chemist and scientific educator, "treated it physically." We should of course guard against reading too much into Fourcroy's and Venel's generalized statements. But as the following analysis will demonstrate, their words embodied key features of the dialectics between scientific disciplines and styles of research in the physical sciences of the early eighteenth century. A careful examination of Boerhaave's influential account of fire in theoretical, experimental, and institutional contexts is at the center of the ensuing discussion.

Lavoisier began his *Elements of Chemistry* (1789) with a nod to Boerhaave that confirmed Venel's view. "That every body," the book's opening line read, "whether solid or fluid, is augmented in all its dimensions by any increase in its sensible heat, was long ago established as a physical axiom, or universal proposition, by the celebrated Boerhaave."[5] In the early 1750s Venel expressed considerable concern over the fact that the single most influential part of Boerhaave's celebrated chemical treatise—the chapter on fire—dealt with physics more than it did with chemistry. By the late 1780s, Lavoisier touted Boerhaave's study of heat as a staple of chemical theory. Despite momentous shifts in disciplinary boundaries, however, ontological challenges persisted. The significance of Boerhaave's "physical axiom" notwithstanding, Lavoisier still included in his new system a chemical and substantial form of heat—the vaguely defined element he named caloric. Both Lavoisier's chemical caloric and Boerhaave's fire were material and imponderable. They differed in that Lavoisier considered heat as an accumulation of caloric whereas Boerhaave believed that heat resulted from the motion of the particles of both ponderable matter and imponderable fire.[6]

Fourcroy, Lavoisier's ally in matters of experimental physics, grouped Stahl (alongside Becher) and Boerhaave together as the masterminds of early eighteenth-century chemistry. Venel, by contrast, situated them on the opposing sides of the physical–chemical divide. Although not rigid or impermeable, this divide was characterized by a stark imbalance, as Venel disconcertedly recognized. "Some chemistry can be found in the works of true physicists, for indeed many of them

---

[4] Gabriel François Venel, "Chymie ou Chimie," in Denis Diderot and Jean le Rond D'Alembert, eds., *Encyclopédie, ou dictionnaire raisonné des sciences, des arts et des métiers, etc.*, vol. 3, 414. Retrieved from <University of Chicago: ARTFL Encyclopédie Project (Spring 2011 Edition), Robert Morrissey,ed., http://encyclopedie.uchicago.edu/>.

[5] Antoine Laurent Lavoisier, *Elements of Chemistry, In a New Systematic Order; Containing all the Modern Discoveries*, trans. J. Kerr (Edinburgh, 1790 [1789]), 1.

[6] Robert Fox, *The Caloric Theory of Gases from Lavoisier to Regnault* (Oxford, 1971), 12.

deal with chemistry, including views on general systems and fundamental doctrinal principles … [but] what Boyle, Newton, Keill, Freind, Boerhaave, etc. have written is marked by lack of experimentation. Thus we cannot expect to get a true picture of chemistry by studying such people." The works of these "physicists," Venel added, were widely known, "quoted and praised, whereas the greater ideas and unique notions of Stahl, in both areas, are recognized by only a few chemists."[7] Physical theories and methods dominated contemporary natural philosophy; so much so, that even chemists, ignorant of their own scientific heritage, had been led astray.

An important disseminator of Stahlian ideas in France, Venel was also the most prominent writer on chemical topics for the *Encyclopédie*.[8] In his landmark article "chimie" he attempted to carve out an autonomous theoretical and practical space for a midcentury chemistry he sensed was misunderstood, misrepresented, and threatened by the rising influence of Newtonian physics. Venel's "chimie" became an important manifesto throughout the second half of the century; in 1796, the same year Fourcroy wrote the passage with which we have begun, he reprinted Venel's article in the *Encyclopédie Méthodique*, a revised and expanded version of the *Encyclopédie*.[9] Fourcroy's close acquaintance with Venel's midcentury assessment rendered his late eighteenth-century delineation of the disciplinary boundaries all the more instructive.

Chemistry's disciplinary struggle was not new. Venel traced its origins to Boyle's late seventeenth-century attempts to interpret chemical phenomena along mechanistic principles. In order to "assume the form of true science," Venel lamented, chemistry "accepted the prevalent physical systems, as it has become successively Cartesian, corpuscular, Newtonian, academic or experimental."[10] The latter denoted the growing authority of experimental physics in the early eighteenth century, a trend best represented by the first generations of Anglo-Dutch Newtonians. Singling out scientific demonstrators and popularizers like John Freind, John Keill, John Theophilus Desaguliers, Pieter van Musschenbroek, as well as Boerhaave, Venel remarked: "we cannot raise chemistry to the level it deserves by demonstrating its

[7] Venel, "Chymie ou Chimie," 408.

[8] Venel contributed more than 200 entries on various topics in chemistry (out of more than 800 dedicated to the subject), including entries on calcination, precipitation, mixts, elements, combustion, decomposition, fermentation, solvents, distillation, and niter. For the best study on chemistry in the *Encyclopédie*, including a detailed analysis of the "chemical corpus" see Jean-Claude Guedon, "The Still Life of a Transition: Chemistry in the Encyclopédie" (Ph.D. Thesis, University of Wisconsin, 1974); see also Rémi Franckowiak, "La chimie dans l'Encyclopédie: une branche tour à tour dépréciée, réévaluée et autonome," *Recherches sur Diderot et sur l'Encyclopédie* **40/41** (2006), 59–70.

[9] See Antoine-François Fourcroy, *Encyclopédie méthodique*, "chimie" (Paris, 1796), vol. 3, 262–303. Venel's association with the *Encyclopédie* is also discussed in Jacques Proust, *L'Encyclopédisme dans le Bas-Languedoc au XVIIIe siècle* (Montpellier: Faculté des Lettres et Sciences Humaines de Montpellier, 1968), 23–7 and 33–5.

[10] Venel "Chymie ou Chimie," 408–9.

philosophical side … we cannot do for chemistry what elegant machines, optics, and electricity have done for physics.".[11]

For Venel, experimental physics and chemistry offered causal explanations based on experimental inquiries. But he also drew a firm distinction between the two that rested on a sophisticated fusion of matter theoretical considerations and an operative epistemology. Tracing out the contours of physics and chemistry, Venel argued: "two sciences that consider objects from such different points of view will inevitably produce particular and distinct knowledge … and each will produce a different language, method, and operating procedure."[12] At its core, however, the distinction did not turn on linguistic or methodological divergences but on ontological concerns, as signified by Venel's reference to "objects." Important as methods and operating procedures might be, they were fundamentally dependent on a matter theoretical distinction—between the chemical *mixt* and the physical *aggregate* modes of matter, a distinction found in Stahl. Physics and chemistry conceived of and therefore manipulated matter in profoundly different ways. Physicists studied aggregates, or materially homogenous masses governed by universal forces of attraction and repulsion; chemists dealt with mixts, or materially heterogeneous compounds brought together by the force of affinity. The attempts to unite these two approaches, Venel stressed, have been equally detrimental to both sciences. Like Fourcroy, Venel considered Stahl as one of the pillars of eighteenth-century chemistry, even though Fourcroy's—post Chemical Revolution—association of Stahl with "that inflammable earth which he denominated Phlogiston" already foreshadowed the biases of a much later, enduring historiography.[13] Boerhaave, Venel emphasized, viewed "fire as mass or aggregate."

Boerhaave can be seen as a link between Boyle and Venel: chronologically, but also from a thematic and disciplinary standpoint. An examination of Boerhaave's views on elements, focusing on his treatise on fire, which became the standard account on the subject during much of the eighteenth century, provides valuable insights into the processes of disciplinary negotiation between chemistry and physics from Boyle to the midcentury crisis depicted by Venel. In his 1771 *Dictionary of Chemistry*, Venel's contemporary, the French chemist Pierre-Joseph Macquer portrayed Boerhaave's treatise on fire as "an astonishing master-piece, so complete, that the human understanding can scarcely make any addition to it." Although his language had likely inspired Fourcroy's depiction, Macquer also added an evocative juxtaposition: "Next to *Stahl* we place the immortal *Boerhaave*; though he excelled in a different way."[14] By highlighting their contributions to "Philosophical Chemistry," Fourcroy would later obscure this very contrast.

[11] Ibid., 410.

[12] Ibid., 416.

[13] For a comprehensive survey of this see John G. McEvoy, *The Historiography of the Chemical Revolution: Patterns of Interpretation in the History of Science* (London: Pickering & Chatto, 2010).

[14] Pierre Joseph Macquer, *A Dictionary of Chemistry, Containing the Theory and Practice of that Science*, trans. J. Keir (London: T. Cadell and P. Elmsly, 1771), vol. 1, xi.

Stahl's phlogiston, and its various theoretical ramifications, has received ample scholarly attention, particularly in the context of the Chemical Revolution. Given how influential Boerhaave's account of fire was throughout the eighteenth century, it has been the subject of several studies, most of which center on the role of and effect on Newtonianism.[15] In what follows I am not interested in what earned Boerhaave the title of "immortal"—be it chemist or physicist. His clear aim, the culmination of which was the *Elementa Chemiae*, was to reform chemistry. And yet, contemporary chemists questioned his contributions to their science, as they perceived it. The view, which goes back to the eighteenth century, that Boerhaave introduced physicalist considerations into chemistry, is basically correct. My attempt here is to understand *why*, and in particular *how*, it is so. Following Venel's judgment, I explore the conflicts, challenges, and inconsistencies inherent in that endeavor. Tracing the origins of these ontological, epistemological, and institutional tensions back to the 1660s, I examine influences on Boerhaave's account of elements and instruments—fire was both—as well as his ideas about menstruums and solution chemistry.[16] The contributions of Samuel Cottereau Duclos and Wilhelm Homberg, two chemists and founding members of the French Royal Academy of Sciences, are considered as part of the story. I pay particular attention to Boerhaave's view of fire as an imponderable agent of material change in relation to various mechanistic and materialistic trends in matter theory. Reconstructing these contexts reveals diverging conceptions of the nature of elements, instruments, analysis, composition, and material change more generally, as these entities and categories straddled the shifting perimeters of the physical–chemical divide around the turn of the eighteenth century.

---

[15] An exception is Rosaleen Love, "Some Sources of Herman Boerhaave's Concept of Fire," *Ambix* **19** (1972), 157–74. Otherwise see Héléne Metzger, "La théorie du feu d'après Boerhaave," *Revue Philosophique de la France Et de L'Étranger* **109** (1930), 253–85; I. Bernard Cohen, *Franklin and Newton: An Inquiry Into Speculative Newtonian Experimental Science and Franklin's Work in Electricity as an Example Thereof* (Cambridge, MA: Harvard University Press, 1966), 214–34; David R. Dyck, "The Nature of Heat and Its Relationship to Chemistry in the Eighteenth Century" (Ph.D. Thesis, University of Wisconsin, 1967), 101–17; Deirdre M. La Porte, "Theories of Fire and Heat in the First Half of the Eighteenth Century" (Ph.D. Thesis, Harvard University, 1970), 155–78; Bernard Joly, "Voltaire chimiste: l'influence des théories de Boerhaave sur sa doctrine du feu," *Revue de nord* **77** (1995), 817–43; Arnold Thackray, *Atoms and Powers: An Essay on Newtonian Matter-Theory and the Development of Chemistry* (Cambridge, MA: Harvard University Press, 1970), 111–113.

[16] For an excellent recent study of Boerhaave's 'skeptical' views of chemical elements and analysis, relating them to Boyle's early critiques as well as to chemistry in the second half of the century, see John C. Powers, "Fire Analysis in the Eighteenth Century: Herman Boerhaave and Scepticism about the Elements," *Ambix* **61** (2014), 385–406.

2.2 The Elements and Instruments of Chemistry

The relationship between chemical analysis, composition, and the nature of elements was a controversial issue in early modern chemistry. Practicing chemists faced a variety of questions in this context. What were the primary constituents of bodies? Could substances be broken down into a fixed and limited number of elements? Could such elements be obtained through chemical analysis? Could one be certain that the end products of chemical analysis—whether using fire (distillation) or solvents—did preexist in any given substance and were not created anew during the process? Could one know whether the products of analysis were elementary substances, the genuine building blocks of matter, and not just limits of a particular analytical setup?

Such questions were at the center of Boyle's critical manifesto, *The Sceptical Chymist* (1661). With respect to chemical composition, elements, and matter theory Boyle denounced the traditionally accepted Aristotelian Four Elements, the Paracelsian *tria prima* (salt, sulfur, mercury) as well as practically informed combinations thereof, most commonly represented by the *tria prima*, to which water and earth would be added occasionally as passive elements. Concerning chemical analysis, Boyle charged the traditional distillation practices, or fire analysis, that grew out of the iatrochemical and pharmaceutical traditions. The two sets of critiques, on elements and analysis, were clearly interlinked, as Boyle cautioned against "relying too much upon the Fire alone, for the resolving of Bodies, so without fruitlessly contending to force them into more Elements than Nature made Them up of, or strip the sever'd Principles so naked, as by making Them Exquisitely Elementary to make them almost useless."[17] In order to elevate chemistry to a more 'philosophical' status, Boyle looked to dissociate it from its technical and artisanal roots while introducing corpuscular and mechanical explanations.

The tenor of Boyle's criticism can be gleaned from the phrases he chose to convey his reservations. Chemical discourse was "obscure, Ambiguous and Aenigmatical"; chemists failed to write "intelligibly enough" and exhibited an "over great-reservedness"; "without proving" their arguments, the "validity" of their interpretations was questionable; given the "unreasonable liberty they give themselves of playing with Names at pleasure," chemical entities lacked fixed referents; finally, the "Un-Philosophical" chemists lacked "Clear and Distinct Notions" of the elements.[18]

Nearly a century later, Venel complained about the disciplinary state of chemistry. "Chemistry is scarcely cultivated among us," he lamented, "not even among scientists, despite the claim for universal knowledge, which is presently so dominant. Chemists made up a distinct group of people, not very numerous, having their own language, laws, mysteries, and living rather isolated within a large community

[17] Robert Boyle, *The Sceptical Chymist*, in Michael Hunter and Edward B. Davis, eds., *The Works of Robert Boyle* (London: Pickering and Chatto, 1999–2000), vol. 2, 375.

[18] Ibid., 209, 211, 213, 277, 291–2.

that shows little interest in their activities and expects almost nothing to come from their work." Few could "distinguish between chemists and glass-blowers" and most thought chemists were either mere pharmacists or "technicians dabbling in secret processes."[19]

Like Boyle, Venel thought chemistry's reputation was problematic. It was still being overly associated with 'lowly' practical arts like pharmacy and metallurgy, from which it had emerged during earlier times. But while Boyle felt that the "Un-Philosophical" chemists would benefit from adopting the physical tenets of the mechanical philosophy, Venel believed that the reason chemistry was so "scarcely cultivated" and so widely misunderstood was precisely because it has "accepted the prevalent physical systems." By submitting chemical phenomena to the duality of matter and motion, Boyle aimed to reformulate chemical discourse along what he had aptly designated as "physico-chymical" or "chymico-physical" principles, terms which he used interchangeably.[20] Venel thought that as chemistry "has become successively Cartesian, corpuscular, Newtonian, academic or experimental" it had given up its independence. This loss of autonomy was at the core of its midcentury crisis. "As long as chemists and physicists each philosophize in their own manner on the objects of their study … things will be fine," he continued, but "if one treats physical things as *more chimico*, and the other treats chemical things as *more physico* … then things are not well at all."[21] In the early eighteenth century Boerhaave still echoed some of Boyle's doubts but his views on fire and heat confirmed Venel's concerns about the unwarranted blurring of ontological and epistemological boundaries.

In his *New Method of Chemistry* (Fig. 2.1) Boerhaave wrote of physical corpuscles as the ultimate immutable elementary constituents of bodies. Regarding chemical "operations" and analysis—with particular reference to distinctly chemical processes like "calcination, fixation, vitrification, sublimation, fermentation, putrefaction, digestion, [and] purification"—he stated[22]:

> a separation of parts thus effected does not shew that those parts have pre-existed in the body, such as they now appear; since those very operations whereby the parts become separated and disengaged from the rest, may make great alterations in them: so that it is a false conclusion of the generality of chemists, that their elements were really contained in the compounds … it is by no means so clear as chemists imagine, that their art gives the true first elements of things; and that we may judge of the compounds by the principles into which they are chemically resolvable.[23]

[19] Venel "Chymie ou Chimie," 408.

[20] The first phrase is taken from the title of Boyle's "Essay on Nitre: A Physico-Chymical Essay … Redintegration of Salt-Petre"; the second from the subtitle to 1661 *The Sceptical Chymist*, which reads "Chymico-Physical Doubts & Paradoxes, Touching the Spagyrist's Principles […]."

[21] Venel "Chymie ou Chimie," 416 (italics in original).

[22] Herman Boerhaave, *A New Method of Chemistry; Including the History, Theory, and Practice of the Art: Translated from the Original Latin of Dr. Boerhaave's Elementa Chemiae, as Published by Himself. To which are added notes; and an appendix, shewing the necessity and utility of enlarging the bounds of chemistry*, 2nd edition, trans. P. Shaw (London: T. Longman, 1741), vol. 1, 157.

[23] Ibid., 158.

Even though Boerhaave was suspicious of the relation between chemical analysis and the elementary constituents of bodies, he avoided a mechanistic-reductive interpretation of chemical phenomena. In fact, he admitted the existence of several chemical elements.[24] "It must indeed be allowed that fire, air, water, earth, alcohol, mercury, the presiding spirit in each body [the "*Spiritus rector*"], and several others, do, when absolutely simple, appear to be fine, permanent elements." Boerhaave defined the "*Spiritus rector*" as a highly subtle, penetrable, and virtually insensible agent, peculiar to each body "whereby it is accurately distinguished from all others." Reminiscent of the traditional aerial niter, "when pure and single [this spirit] flies off, by its great mobility mixes with the air, and returns into the common chaos of volatiles ... it falls down with snow, hail, rain, or dew ... By such revolution it passes into new bodies, whose mass it animates and directs."[25] Unlike Boyle before him, Boerhaave did not rule out the possibility of a certain number of elements although "whether such bodies can by any contrivance be procured and exhibited perfectly pure, has not yet been demonstrated ... [for] nothing of this simplicity is found in the common operations of the chemists."[26] Boerhaave's list of elements is intriguing. Despite being discussed in close relation to chemical analysis and practice, the inclusion of the "presiding spirit" suggests that Boerhaave, influenced more by Newton than by Boyle, did not perceive matter as thoroughly passive and inert.

Even more striking is the fact that this list of elements might seem at first glance like a throwback to the very systems that came under Boyle's forceful attack. Among Boerhaave's elements we find the four Aristotelian elements "fire, air, water, earth" alongside mercury, which was one of the Paracelsian *tria prima*. Subtle and spirituous, mercury, which was sometimes called spirit, has been conventionally associated with chemical reactivity. Why did Boerhaave invoke—as late as the 1730s–1740s and in a chemical textbook that linked chemical phenomena with corpuscular explanations—the elementary status of such traditional systems?[27] Why did he include spirits alongside material elements? He clarified, after all, that these elements have

[24] On Boerhaave's view of instruments "as tools within the practice and philosophy of chemistry and as principles of organization within the structure of Boerhaave's chemical courses" see John C. Powers, "Chemistry Without Principles: Herman Boerhaave on Instruments and Elements," in Lawrence M. Principe, ed., *New Narratives in Eighteenth-Century Chemistry* (Dordrecht: Springer, 2007), 45–62 (on 47).

[25] Boerhaave, *New Method*, p. 168. Although he assumed this spirit to be largely insensible, Boerhaave suggested, at times, that it manifested itself through smell, being "so subtile as only to be perceivable by its smell or taste." (Ibid.). For historical background see Allen G. Debus, "The Paracelsian Aerial Niter," *Isis* **55** (1964), 43–61.

[26] Boerhaave, *New Method*, 167–68.

[27] It should be noted that Boerhaave was not unique in this regard. In fact, the 'elementary' status of the four elements remained uncertain throughout most of the century. But the later debates were conducted within a very different epistemological context, mostly related to greater questions about the relationship between matter, metaphysics, language, and chemical method. Cases in point are the debate on the composition of water between Priestley, Cavendish, Kirwan, and Lavoisier; the variety of phlogistic debates concerning the nature of fire; or the distinction between air as a gaseous state or as chemical species as seen in the increasing number of newly discovered gases.

not yet been "procured and exhibited perfectly pure" in the "common operations of the chemists." Boerhaave's elements encompassed material and immaterial, permanent and changing, theoretical and demonstrable dimensions. But how did they figure in his conception of matter and material change? Or, put differently, what was their chemical, and correspondingly physical, status?

The *New Method* consisted of three parts. The second part, which will concern us here, included Boerhaave's views on the goals of chemistry, observations on metals, salts, acids, and minerals, alongside substantial chapters on the operations and instruments of chemistry, where Boerhaave discussed the four elements, with particular focus on fire. These were the accounts Fourcroy referred to as "masterpieces." Boerhaave followed up the chapter entitled "Of the Operations of Chemistry" with a discussion of "The Effects, or Productions of Chemistry," especially in relation to natural philosophy, medicine, and the mechanical arts. Whereas the first and second subjects occupied less than three and four pages of the entire book, respectively, the third spanned 27 pages. The chapter on the mechanical arts included sections on painting, enameling, art of glass, metallurgy, natural magic, art of wine, brewing, and alchemy, in which Boerhaave professed his intention "to add a few, but candid and ingenious considerations, on the great use of chemistry to alchemy."[28] He explained:

> Those arts are vulgarly call'd mechanical, which are performed by the labour of the hands; being very different from that of mechanics, which lays down the powers of bodies, arising from the common properties of all bodies, which geometricians explain. Chemistry is of no service to this latter; but is of singular advantage to the former, which are employ'd in working upon, and making changes in bodies.[29]

Despite upholding a general corpuscular conception of matter—and despite Venel's lumping together of "Boyle, Newton, Keill, Freind, [and] Boerhaave" under the heading of "true physicists"—Boerhaave was not a reductive mechanist like Boyle had been early on, especially in the 1660s. The difference could be understood as one of emphasis. Like other mechanists, Boerhaave also appealed to matter and motion in explaining chemical phenomena. But whereas a mechanist like Boyle laid more emphasis on *matter*—on the sizes, shapes, and arrangements (to which he referred as textures) of particles—Boerhaave, finding this type of reasoning too speculative, stressed the generalized and universal action of *motion* within and upon matter, which could be described in a corpuscular way. "Chemistry is employed in changing the bodies contained in the three classes above specify'd [vegetal, animal, mineral], and the change it produces in them is effected by means of motion alone."[30] This tied in with Boerhaave's view of chemistry as being closer to a mechanical art "performed by the labour of the hands" than to "mechanics," which dealt with generalized abstractions and was the domain of "geometricians." The former was experimental and particular, the latter abstract and universal.

---

[28] Boerhaave, *New Method*, vol. 1, 200 (italics mine).

[29] Ibid., 178.

[30] Ibid., 155.

Further addressing the relationship between chemistry and the mechanical arts, Boerhaave proclaimed, "I shall begin with the instruments which are always required when any thing is to be done by art."[31] Boerhaave adopted and deployed "instruments"—a key feature of the mechanical arts and of contemporary experimental physics (also indicated above by Venel)—as a central metaphor in his chemical outlook. His detailed examinations of these instruments, comprising the main bulk of the first of two volumes of the *New Method*, was what related the discussion back to his conception of elements. We begin with his definition of chemical instruments, followed by their identification:

> In all arts which direct bodies to be chang'd, the name instrument is given to certain things capable of a particular motion; which being thence apply'd to the body intended to be chang'd, produces the requisite change therein: thus, in the present art [chemistry], there are certain bodies by whose means the requisite actions are produced. These, with the best chemists, we usually reduce to six principal ones; fire, water, air, earth, menstruums, and utensils.[32]

As we have seen, fire, water, air, and earth were regarded as "fine, permanent elements." But they also figured as instruments—for occasioning "requisite change[s]" within bodies—alongside "menstruums, and utensils." The list of utensils included "the furniture of a laboratory"; "crucibles and melting pots"; "evaporating vessels"; and lutes and furnaces.[33] This peculiar alignment implied a view of an instrument as a changeless agent of change in other bodies through motion. If an element or a menstruum was akin in this sense to a furnace or a retort it should facilitate a chemical process without engaging in it, like the slicing of bread with a knife or the boiling of water in a vessel. This view assumed not only the immutability but also the materiality of these elements-instruments.

This twofold nature of Boerhaave's elements (immutable and material) reflected his understanding of the relation between chemistry and physics. In view of the irreducible abundance of chemical particularities, he related chemical phenomena not to mechanistic agendas but to experimental physics. "Certain bodies have peculiar powers," he noted,

> from whence effects arise, that do not come within the compass of any general theorem; but depend on a certain constitution, peculiar perhaps to some one body. In forming such theory [in chemistry], a direct use may indeed be made of the demonstrations in physics, as particularly in mechanics, hydrostatics, and hydraulics; since the properties common to all bodies, and what further affections certainly flow therefrom, have their place in chemistry.[34]

In examining "the properties common to all bodies" Boerhaave avoided hypothesizing about the ultimate nature of matter. The limit was set at a mechanical-instrumental level, as it were, where properties "peculiar perhaps to some one body" should be

[31] Ibid., 205 (italics mine).

[32] Ibid., 205.

[33] Ibid., 580–93; titles of the subsections in the chapter "Of the Chemical Apparatus, and Vessels."

[34] Ibid., 3.

considered by taking into account "demonstrations in physics" rather than speculations about its "certain constitution."

In a Newtonian vein, Boerhaave sought to carefully draw by way of induction general laws from single instances, while keeping theorizing at bay: "We allow of no other *theory* in chemistry, except what is built on general laws; which must originally have been deduced from a multitude of common incontestable facts."[35] Even though he accepted a controlled degree of theoretical generalization, his frustration with the persistent idiosyncrasies of experimental chemistry was palpable:

> each operation is considered an example of one of the single cases, from which the general theorem was first deduced. By which method the perfection of chemistry may be arrived at, without incumbring [sic] ourselves, more than needs, in a pursuit, which of itself is sufficiently tiresome. For want of this, the endless labours and processes of chemists are of little moment, and only serve to keep them in a continual chase, without furthering them a jot; whence instead of a reward at last, they are often left considerable losers.[36]

Again, we find in Boerhaave echoes of Boyle's old complaint about the "Un-Philosophical" ways of chemists, who lacked a coherent theoretical framework and were therefore consigned to a vain "continual [experimental] chase." Boerhaave thus urged chemists to avail themselves of physical knowledge about "the properties common to all bodies." After all, both sciences studied the same bodies, which were comprised of the same matter. Their ways of conceptualizing this matter and their respective methods might be different, as Venel pointed out, but they were not mutually exclusive. Boerhaave's methodological stance, at least in the context of "general theorem[s]," was more in line with the metaphor of the physical and mechanical action of an instrument than the hypothetical features and behavior of unobservable particles.

Out of the six instruments, the discussion on fire was by far the most extensive and intricate.[37] Boerhaave set fire apart as the only element-instrument that could "perhaps be exhibited pure and elementary … [whereas] no human art can produce the least drop of pure water, and much less of any of the rest, as air and earth, or the like."[38] Furthermore, "no chemical operation ever was, or can be hereafter performed, to which fire does not contribute; and this cannot be so extensively said of the rest."[39] Air, he thought, "whose action and concurrence are used in most of the operations, both of nature and art … is more compound than fire itself."[40] Water, despite its ubiquity, is even more difficult to pin down, for it is "one of the most difficult subject in all natural philosophy to be acquainted with," primarily because it

---

[35] Ibid., 2.

[36] Ibid., 5.

[37] Water and menstruums are also examined at relative length. The chapters on air and earth are the shortest. Boerhaave's exploration of fire has later gained popularity among French chemists and was commonly referred to as *Traité du feu*.

[38] Ibid., 167.

[39] Ibid., 205.

[40] Ibid., 379.

"continually mixes itself with chemical subjects, and especially with air."[41] Earth was a "principle, or element, in the composition of bodies, entering them either as an ingredient, or giving them a power of performing various operations in nature and art." Strictly speaking, earth was a simple, hard, and brittle body that did not melt in either fire or water; it was expanded, impenetrable, and possessed of gravity.[42] Boerhaave's definitions of the elements-instruments suggested that he was considering them as vehicles of chemical and physical change on the one hand, and as components of bodies on the other. The phrase "action and concurrence," which he here used in relation to air, was emblematic of this approach. The issue of composition, however, which encompassed both chemical and physical transformations, was the most complex and vexing.[43]

Boerhaave first tried to ascertain the "signs" of fire or the "mark[s], whereby fire is known to be present in whatever place." In line with his definition of an instrument—an entity capable of occasioning "a particular motion" in other bodies—his exploration of fire encompassed a wide range of physical and chemical processes and phenomena, including flame, heat, light, color, combustion, expansion, among many other.[44] Given this synoptic approach, and given his aim to "rigidly abstain from speculations," the ensuing account was ambitious and complex.[45] It consisted of an impressively rich blend of experimental observations alongside theoretical speculations, transcending well beyond the confines of a traditional Baconian natural history of fire. This encyclopedic scope inspired Fourcroy (and Macquer) to dub it a masterpiece "containing all that was at that time known" on the subject.

A detailed analysis of this account is beyond my present purpose. In the next sections I focus on one of the most challenging and vexing features of Boerhaave's account of fire, at the confluence of its role as a physical effect, chemical element, and instrument of material change. Exploring Boerhaave's approach to the question of the materiality and ponderabilty of fire, with particular attention to the issue of weight gain in the calcination of metals, provides insights into competing styles of thinking about and experimenting with fire, heat, and light The use of "burning instruments," especially mirrors and lenses, was of particular significance in this context.

[41] Ibid., 436.

[42] Ibid., 473.

[43] See Robert Siegfried and Betty J. T. Dobbs, "Composition, A Neglected Aspect of the Chemical Revolution," *Annals of Science* **24** (1968), 275–293.

[44] Ibid., 209–10.

[45] Ibid., 207.

2.3 "Fire, Whether Void of Gravity"

Attempting to determine the true "effects of fire, in hopes that among them, we may find some one, which will enable us to discover and estimate the presence and quantity of this active element," Boerhaave observed:

> on a careful inquiry I do not find any body, to which we may not apply that which all men call fire, whether solar, culinary, or subterraneous; and all such bodies to which fire is thus applied, without one exception, are hereby render'd bigger, swell, and rarify, yet without any observable difference in their weight.[46]

In spite of the emphasis on physical expansion and rarefaction, Boerhaave considered fire in a broad and general sense, as suggested by his reference to "solar, culinary, or subterraneous" fire. Solar fire, sometimes known as celestial fire, emanated from the sun. Subterraneous fire "manifests itself in fiery eruptions of the earth, volcanoes, or burning mountains." Boerhaave's mention of "culinary fire," conventionally perceived as "that [which is] employed in all chemical operations,"[47] is most notable, especially since his rejection of weight change in "bodies to which fire is thus applied" related directly to the problem of chemical combination and composition.

The effect of weight gain in combustion and calcination has puzzled chemists for a long time. Boyle tackled this phenomenon in his 1673 *Essays of Effluviums*. Why did substances undergoing combustion gain weight, and in particular why did metallic calxes weigh more than the original metals? How did fire and associated phenomena like heat and light relate to these transformations? Did fire combine with matter, and if so, how? What was the nature of the relationship between its action and the elementary constituents of bodies?

Fire, according to Boerhaave, "diffused thro' all spaces and bodies," constantly moved and imparted motion to matter. It thus "immediately begins to expand all bodies, and remove their parts from their natural closeness … hinder[ing] the proper cohesion of the elements." Due to the propensity to "expand itself more equably," fire "continually produced certain operations inseparable from it; all which principally aim at this, *viz.* to remove the elements from one another." This tendency offset the natural inclination of "bodies which are continually striving against the separation of their elements," resulting in a mechanism of "perpetual action and re-action" between the natural bodies and fire.[48] True to his instrumental conception of both fire and elements, Boerhaave's account of the separation of the latter by the action of the former retained a considerable measure of ambiguity about their ultimate natures and, perhaps more importantly, the relationship between them.

---

46 Ibid., 212–13.

47 "Fire," in *The Complete Dictionary of Arts and Sciences, in which the whole circle of human learning is explained [...]*, eds., Temple Henry Croker, Thomas Williams, Samuel Clark (London: for the authors, 1765), vol. 2.

48 Boerhaave, *New Method*, vol. 1, 246.

Boerhaave indeed treated "fire as mass or aggregate" and not as mixt, which is why Venel branded the Dutchman's research as physical rather than chemical. Yet Boerhaave did not reduce his "elements" to the universal material homogeneity characteristic of aggregates. Unlike Venel, he did not rely on the Stahlian framework based on the distinction between homogenous masses—understood as chunks of inert and universal matter—and heterogeneous mixts—viewed as intimate combinations of dissimilar elementary constituents (resembling, in a sense, the modern notion of a chemical molecule). Nor did his account map onto such matter theoretical notions. Despite his emphasis on material instrumentality and mechanical-like motion, Boerhaave recognized the difference between chemical and physical changes, even if his reasoning was physically oriented. In discussing the "operations of chemistry" he linked the complexities of the chemical realm not to corpuscular configurations but solely to the various manifestations of motion at the particulate level:

> Suppose, for instance, a single corporeal mass wholly at rest; that is, all its particles mutually quiescent among themselves, such as it was at the beginning, must it not remain the same and unchanged in all future times? Tho' the whole power of chemistry were applied to it, yet if it raise no motion in any of its parts, it must remain as it did; or suppose a motion communicated to the whole mass, whereby it is removed to another place, without making any further change in the constituent parts of the body, the idea of such body will still remain the same … But if motion be raised among the parts, we may conceive an inexhaustible variety of effects and changes producible therefrom. Chemistry then, is totally employ'd, either in uniting, or in separating; there being no third operation in nature: so that to these are all its multitude of operations reducible, without one exception.[49]

Even without distinguishing between physical universal attraction and chemical elective affinity, Boerhaave made clear that all chemical "effects and changes," bar none, arose not from motion at the level of the "whole mass" but at the level of "the constituent parts of the body."

In an observation related to menstruums (see Sect. 2.5) Boerhaave came close to depicting chemical affinity—as a particular, unquantifiable, and relational "power" inherent in bodies. It "appears in the history of menstruums," he claimed, "that multitudes of bodies have certain relative powers, which only shew themselves upon placing them near one another." Following this claim he considered once more the connection between the action of the elements-instruments and the essentially irreducible variety of chemical appearances, this time employing a more chemical-experimental language centered on notions of composition and mixture: "by separating compounds into their simples, and mixing simples with simples, an infinite number of things before unknown may be produced." This statement, however, signaled the extent to which Boerhaave was willing to rely on what could be seen as uniquely chemical accounts. For he immediately thereafter linked this statement on elements, or "simples," with one on "instruments," against the previously established background of chemistry as a mechanical art. "If we only consider a single body retaining the same bulk, but only changed in its figure, that is in the disposition

[49] Ibid., 156.

of its surface, yet we shall find it capable of exerting very different powers." "The same piece of steel," he continued, "according to the different figure which is given it, will produce instruments of very different powers; as a wedge, knife, poniard, lancet, sphere, cube, cylinder, prism, pyramid, cone, &c."[50] Such underlying tensions reached their boldest expressions in explanations that were regarded as uniquely chemical.

In the second section of the chapter "Of Fire"—titled "Of Fewel, or the Pabulum of Fire"—Boerhaave set out to "examine those bodies wherein fire may be preserved the longest time." Following a traditional style of natural historical exposition, he surveyed experiments and observations conducted on vegetal, animal, and mineral matters (to which he referred here as the "fossil class").[51] Examining the effects of fire and heat on inorganic substances forced him to address the 'weight gain paradox'. This turned out to be one of the thorniest questions in matter theory and chemistry, a challenge that within only a few decades would emerge as a key issue in the Chemical Revolution. If according to the traditional paradigm combustion meant loss of phlogiston, why did metals exposed to fire gain weight?

Boerhaave dedicated an entire subsection to the question of "fire, whether void of gravity" but his attempt to answer it was evasive and inconsistent. Curiously, his way of approaching this ostensibly *physical* query was by way of an explicit and near-complete deferral to the experimental conclusions of *chemists*, whom he (rightly) identified as authorities on the subject:

> *Du Clos* shew'd before the Academy of Sciences, that a quantity of antimony, exposed to the burning-glass, encreased [sic] a sixteenth part in weight … *Homberg*, carrying his enquiries still further into these matters, by experiments made with *Tschirhause*'s glass … seems to have proved still more clearly the possibility of the union of true fire with bodies, its consequent concretion, and forming therewith a new and very different body, considerably encreased in weight … That such the event of these, and some experiments of Mr. *Boyle*, I do not doubt; being equally convinced of the ability of those great men for making experiments, and their candour in relating them.[52]

This overt deferral was clearly due in part to Boerhaave's goal to stay close to experimental reality, or what he had called the "multitude of common incontestable facts." But the inconsistencies ran deeper. Even though his account was virtually given over to a summary of chemists' findings, he ended up rejecting their conclusions, while offering little in their stead. Staying with his twofold definition of fire as a material element and immutable instrument of physical and chemical change, Boerhaave's thoughts on whether and how fire combined with matter remained inconclusive. In light of empirical evidence, the question was unavoidable, forcing Boerhaave to take a firm stand in face of conflicting physical and chemical conceptions of matter and material change. From a chemical perspective, best exemplified by Venel's assessment, Boerhaave's notion of fire was not only incompatible with experiments—which, on Boerhaave's own account, had been convincingly adduced

[50] Ibid., 157.

[51] Ibid., 300, 336.

[52] Ibid., 339–40.

by chemists—but more importantly, it resulted in a conceptual impasse. Boerhaave's incongruent and unsatisfactory solution was an instance of the undesired confusion implied by Venel's cautionary note to avoid treating "physical things as *more chimico* … [and] chemical things as *more physico*".[53]

In particular, by invoking the authority of these chemists, Boerhaave tapped into a particular experimental lineage, related to Boyle's treatise on "New Experiments, To Make the Parts of Fire and Flame Stable & Ponderable," one of his *Essays of Effluviums*.[54] Although he leaned toward the view that fire was "void of gravity," Boerhaave underscored the lasting influence of Boyle's arguments to the contrary, especially those presented in what he termed Boyle's "treatise of *the ponderability of the flame*."[55] In that treatise, Boyle postulated that if "Flame may act upon some Bodies as a Menstruum [solvent], it seems no way incredible, that, as almost all other Menstruums, so Flame should have some of its own Particles united with those of the Bodies expos'd to its action." The flame thus acted through a "Coalition of its own particles with those of the fretted body, and thereby permanently adding Substance and Weight to them."[56]

The chemical-experimental lineage Boerhaave relied on in this context included, alongside Boyle, the "solid arguments" of Samuel Duclos and Wilhelm Homberg, whose research on fire was unique in that they also used burning mirrors, or speculums, and burning glasses to focus the rays of the sun and deploy them experimentally.[57] Such instruments had been in use since antiquity but it was only in the late seventeenth and eighteenth centuries that they were used more systematically in chemical experiments proper.[58] Lavoisier, for instance, used a large lens to obtain carbon dioxide by burning diamond. Joseph Priestley used a smaller lens, made by William Parker (Fig. 2.6), to obtain mercuric oxide by heating mercury.[59]

Long regarded by natural philosophers as the quintessential carriers of the matter of pure fire, the sunrays were concentrated through reflection or refraction by these instruments. Although he distinguished between the empirical manifestations of fire

[53] Venel "Chymie ou Chimie," 416 (italics in original).

[54] Robert Boyle, "New Experiments, To Make the Parts of Fire and Flame Stable & Ponderable, (*Essays of Effluviums*, 1673)," in Michael Hunter and Edward B. Davis, eds., *The Works of Robert Boyle* (London: Pickering and Chatto, 1999–2000), vol. 7, 227–336.

[55] Boerhaave, *New Method*, vol. 1, 339.

[56] Boyle, "New Experiments," 306, 330.

[57] For a recent study, revealing the influence of this lineage on leading French chemists like Étienne François Geoffroy and especially Pierre-Joseph Macquer (particularly about their own work with burning mirrors and lenses to establish the nature of metals), see Christine Lehman, "Alchemy Revisited by the Mid-Eighteenth Century Chemists in France: An Unpublished Manuscript by Pierre-Joseph Macquer," *Nuncius* **28** (2013), 156–216. For the cultural context see Gregory Lynall, "'Bundling up the Sun-Beams': Burning Mirrors in Eighteenth-Century Knowledge and Culture," *Journal for Eighteenth-Century Studies* **36** (2013), 477–90.

[58] Knowles Middleton, "Archimedes, Kircher, Buffon, and the Burning-Mirrors," *Isis* **52** (1961), 533–43.

[59] W. A. Smeaton, "Some Large Burning Lenses and their Use by Eighteenth-Century French and British Chemists," *Annals of Science* **44** (1987), 265–76.

and of light, since "the most powerful fire may be without the least appearance of light, and the strongest light without producing the least heat," Boerhaave referred to the entire range of phenomena related to such instruments as "physical effects."[60] He found burning mirrors and lenses to be particularly effective in collecting and reproducing highly intense fire. "The sun does not emit any fiery matter," he stated, "this luminary body only has a power of directing the fire already existing in any place, into parallel right lines, from whence the same quantity of fire being again collected out of its parallelism by reflection or refraction, and united still closer … is enabled to produce all its effects."[61] François Villette's famous speculum, for instance, "produce[d] a fire in its focus, almost as vehement, as if the rays had been received immediately from the sun itself."[62]

Boerhaave recognized the findings of these chemists as proving that "elementary fire is capable of concreting instantaneously with bodies, and considerably augmenting their weight."[63] He thus accepted the empirical findings of Boyle, Duclos, and Homberg but rejected their theoretical explanations, especially the inference that the increase in metals' weight implied the ponderability of fire. Reminiscent of Boyle's tendency to embrace the experimental work of other chemists while rejecting their interpretations, Boerhaave finally concluded:

> Elementary fire must be truly corporeal … immutable in itself, and the great instrument or means of changing all other things. But whether fire has also that further property, which some of the greatest men of the present age hold inseparable from all bodies, *viz.* weight or gravity … does not so certainly appear from the consideration of the whole history of fire.[64]

For Boerhaave fire was a material, immutable, and weightless instrument of physical and chemical change: it could change bodies while remaining unchanged. Whether it united or separated bodies, composed or decomposed them, it did not alter their weight.

Boerhaave's rejection of the ponderability of fire was at odds with experimental records. But from a physical theoretical standpoint the denial that fire had "weight or gravity" rendered his claim unproblematic. The inapplicability of similar assumptions within a chemical context, however, in which composition and material constitution were paramount, led to considerable difficulties:

> Such an increase in weight, however, does not happen in all bodies thus calcined, but only in those abounding with a rapacious sulfur, as antimony, lead, tin, iron, and orpiment; whence this addition is, perhaps, owing to the corrosion, attrition, and mixture of the particles of the sulfur with those of the other bodies, separated by fusion in such operation. But the increase of weight, in bodies exposed in glass vessels, from the supposed insinuation of fire into them, is so small, that it may, perhaps, rather be ascribed to something communicated to them from the glass itself. Experiments of this kind should be purposely made, but with great caution, as we may be easily led into error.[65]

[60] Boerhaave, *New Method*, vol. 1, 212, 268.

[61] Ibid., 281.

[62] Ibid., 268.

[63] Ibid., 339.

[64] Ibid., 359.

[65] Ibid., 340.

Boerhaave's hesitations were evident. He rhetorically minimized the quantities involved and played up the likelihood of being "led into error." Most notably, he went as far as to ascribe the weight increase "to something communicated to [the bodies] from the glass itself" although, as we have previously noted, he considered "utensils" as instruments on par with the four elements and the menstruums. His stab at a chemical explanation was even more tentative, ascribing the weight gain to "perhaps ... the corrosion, attrition, and mixture of the particles of the sulfur." Ironically, if unsurprisingly, the chemists whose conclusions he rejected provided a much better sense of what Boerhaave might have meant by this obscure depiction of the "addition" of sulfurous particles—a substance that has indeed been traditionally associated with inflammability. Then again, Boerhaave wrapped up the section in which he discussed these chemical experiments by claiming to have "explain'd the chief *physical* ways hitherto known of collecting and preserving fire."[66]

2.4 Instruments of Fire

Samuel Cottereau Duclos (1598–1685) was one of the founding members of the French Royal Academy of Sciences and the establisher of its chemical laboratory. Wilhelm Homberg (1652–1715) was the most prominent chemist in the Academy around the turn of the eighteenth century and its most important chemist since Duclos (and during the post-1699 reform era).[67] While Homberg was relatively well known in his time, Duclos remains an obscure figure in both the chronicles of the Academy and of chemistry in general. As a protestant and alchemist with Neoplatonic tendencies, he was seen as a dissenter and some of his writings have been banished from publication by the Academy.[68]

In April 1667, less than half a year after the establishment of the Academy, Duclos presented before the assembly three weekly memoirs (dated 16th, 23rd and 30th) concerning "experiments on the augmentation of weight of certain substances in their calcination by the heat of the sun, or by ordinary fire." Applying a "burning mirror" (Fig. 2.3) to a sample of antimony, Duclos found that after "one hour" the metal "had been reduced to ashes" and its weight had increased by one tenth of the metal's original weight. Having repeated and diversified the experimental setup—using regulus of antimony[69] made with iron and saltpeter, as well as by adding

[66] Ibid., 342 (italics mine).

[67] For general information about Homberg see Kim, *Affinity*, ch. 2.

[68] For background on Duclos and the Academy see Victor D. Boantza, *Matter and Method in the Long Chemical Revolution: Laws of Another Order* (Burlington: Ashgate, 2013), 17–26.

[69] "Antimony metal can be made to form a visibly crystalline structure by slowly cooling the molten antimony beneath a thick layer of slag. The antimony is reduced from stibnite (Sb_2S_3) by heating it with iron and saltpeter until fusion ensues." Chymical Products: Star regulus of antimony, The Chymistry of Isaac Newton, W.R. Newman, ed., December 2013. http://webapp1.dlib.indiana.edu/newton/reference/chemProd.do

alcohol—he concluded that "a strong combustion of this material resulted in a great augmentation of weight."[70] Duclos further established that whether submitted to the concentrated rays of the sun through a burning mirror, or baked in an intensely hot oven, metals gained weight in a similar way. While Boerhaave considered the weight increase as marginal, Duclos deemed it considerable.

Duclos repeated the antimony experiment using other metals like copper, iron, lead, tin, and mixtures thereof, meticulously tracing the changes in their weights as well as changes in their consistency, color, and other sensible qualities. He attributed the observed augmentation of weight to three main causes. First, "the sulfurous quality of the substances"; second, their inflammability in open air; and third, "the motion of the air, agitated by the heat of these burning materials." By "sulfurous" Duclos meant "not only inflammable substances but also those of a similar nature (that is to say, in which the quality of fire predominates and renders the [combustible] qualities stronger and more apparent)."[71] He thus considered the metals he had subjected to the heat of the burning mirror to be sulfurous in nature; and the more sulfur they contained, the more weight they gained during calcination. The qualities of combustibility and sulfurousness were closely bound, as "everything sulfurous is susceptible to fire depending on its degree of sulfurousness [sulphureité] and according to the quality and condition of the other ingredients in its mixture [composition]."[72]

Duclos supplemented the qualitative argument with a materialistic claim: "all augmentation of weight results from the addition of parts." The addition of these parts, which he considered "earthy and sulfurous," was facilitated by air during combustion or calcination. The "airy sulfurs are exhaled during inflammation," Duclos explained, but the "earthy sulfurs are set on fire without inflammation and brought into combination with the earthy parts [of a body] by a salt. These are the parts we identify as the augmenters of weight."[73] These earthy sulfurous parts, then, were the primary material cause for the augmentation of weight. Duclos further explained that the air itself did not actively participate in the process but merely served as a mechanical and *passive* agent. By contrast, the process of chemical combination was governed by an *active* mechanism of affinities:

> the addition of the earthy sulfurous parts … occurs by means of the air, which is drawn by the fire during combustion. As the heat excites the motion of the air surrounding the burning matter, the air moves, and since no void can be left in its place it is followed successively by rushes of air, and as long as the motion continues, this succession of air persists … the air, unable to combine with fire, adds nothing of its own but carries other substances that are mixed with it, such as aqueous vapors or igneous exhalations. The former have no affinity to sulfurous matters and cannot combine with them easily. It must therefore be the latter, contained in the air, which by the affinity [symbole] of their qualities join the earthy sulfurs and thus augment their weight.[74]

[70] Procès-Verbal de séance de l'Académie Royale des Sciences, t. 1, fols 39–41 (hereafter: AdS, PV, followed by tome and page numbers).

[71] Ibid., fols 45–7.

[72] Ibid., fol. 48.

[73] Ibid., fol. 48.

[74] Ibid., fol. 49.

Duclos' notion of affinity implied a qualitative kinship—symbolicity or reciprocity—that was here centered on the quality of sulfur. The "igneous exhalations" were possessed of this quality and could therefore combine with the "earthy sulfurs." What was most interesting about this explanation was the way fire, the apparently immediate source of change, was rendered a secondary cause: it merely imparted motion to the air, which was in turn a passive vehicle, a medium carrying material parts. Duclos espoused an instrumental view of these traditional elements (Fig. 2.2).

In the late 1660s Duclos laid the experimental foundations at the Academy for harnessing the sunrays in chemical experiments. A quarter of a century later, as Boerhaave indicated, Homberg revived these interests and carried them further. In his review for the year 1702, the Academy's secretary, Bernard de Fontenelle, noted the unique qualities of fire, the "universal solvent," as applied to the "decomposition of bodies." Fontenelle enthusiastically described the installation of a new burning glass at the Jardin du Palais-Royal, consisting of a great improvement over the older burning mirrors employed by Duclos, whose burning mirrors were concave and focused the sunrays by reflection (Figs. 2.3, 2.4, and 2.6). The new burning glass, as

Fig. 2.2 "The solar calcination of antimony." Nicaise Le Febvre, *Tracité de la Chymie* (Paris, 1660), v. 1, p 899. Le Febvre had worked and studied under Duclos sometime in the first half of the seventeenth century. The "artist" in the illustration is said to be using a "mirror" but the positioning of the sample in relation to the instrument and the source of light suggest burning by refraction (lens) rather than reflection (mirror); *miroir ardent* and *verre ardent*, or *brûlant*, where sometimes used interchangeably

Fig. 2.3 "Catoptrique. Utilisation d'un miroir ardent." Demonstration with a burning mirror (possibly at Versailles or at the Paris Observatory). Jacques Cassini, *Illustrations de Theses Mathematicae de Optica* (Paris, 1691)

Fontenelle remarked, "3 or 4 feet in diameter, which is an extraordinary size" was convex and focused the rays by refraction (Figs. 2.2, 2.5, and 2.6). These remarkable lenses, which had been ordered by the Duc d'Orléans, were produced by the German mathematician and entrepreneur Ehrenfried Walther von Tschirnhaus (1651–1708). As the Duke's personal physician, Homberg was among the first to put them to experimental use, paying particular attention to metals, especially gold.[75] Boerhaave extoled in this context the "divers experiments" using "the Duke of *Orelans*'s burning-glasses." They were performed "in the king's garden in *Paris* … to discover the nature of fire: which will be thought necessary to be here historically related, both as they give a considerable insight into the doctrine of fire, and as their effect was the greatest that has been observed of the kind."[76]

In his inaugural report on the subject Homberg declared how the burning glass "may not only afford great progress in establishing the principles [elements] of chemistry but it may well open a gate to a new Physics, like the Microscope and the Air-Pump had done in their time."[77] Half a century later, recall, Venel would com-

[75] Bernard de Fontenelle, "Sur des experiences faites a un miroir ardent convexe" *Histoire de l'Académie Royale des Sciences. Avec les mémoires de mathématique & de physique, pour la même année* (1702), 34–5. (Hereafter *Histoire* or *Mémoires*, followed by year.)

[76] Boerhaave, *New Method*, 277.

[77] Wilhelm Homberg, "Observations Faites par le moyen du Verre Ardent," *Mémoires* (1702), 149.

Fig. 2.4 A large burning mirror is prominently featured in this depiction of Louis XIV's visit to the Academy (c. 1671). In E. C. Watson "The Early Days of the Académie des Sciences as Portrayed in the Engravings of Sébastien Le Clerc." *Osiris* **7** (1939), 576

DIOPTRIQUE.

EFFETS DES VERRES BRULANS

DE TROIS OU QUATRE PIEDS DE DIAMETRE.

Monſieur Tſchirnhaus a taillé des verres brûlans, dont les effets ſont au-deſſus de tout ce qu'on a encore vû. Voici ce qu'il nous en a communiqué.

1. Il faut placer le verre *a*, *b*, en ſorte qu'il reçoive les raïons du Soleil à plomb, ce que l'on connoîtra, lorſque le foyer ou l'image du Soleil *E*. paroîtra parfaitement ronde ſur la planche, *c*, *d*, qui doit être parallele au verre, *a*. *b*, & ſi cette figure étoit ovale, il faudroit incliner & tourner le verre juſqu'à ce qu'elle fût arrivée à cette parfaite rondeur, & alors le bois s'enflamera dans un moment, le plomb ſe fondra, les ardoiſes ſe vitrifieront, &c. & cela dans une diſtance de douze pieds, le foyer étant du diametre d'un pouce & demi.

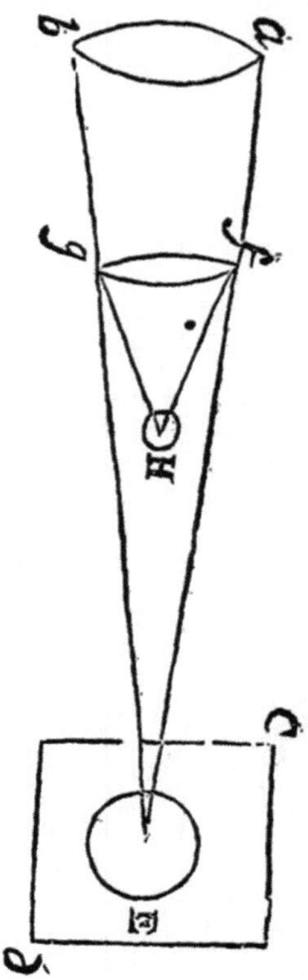

2. Pour rendre ce foyer plus vif, il faut le retrecir par le moyen d'une ſeconde lentille qui doit être placée en *f*, *g*, parallele au premier verre *a*, *b*, & alors le foyer qui

Fig. 2.5 Homberg's schematic illustration of the burning glass setup. By using one lens (ab) light is focused at E; much greater temperatures are achievable by adding a second lens (fg), the focus being at H. In Bernard de Fontenelle, "Effects of Burning Glasses of Three or Four Feet in Diameter," *Histoire de l'Académie royale des sciences* (Paris, 1699), 90

a

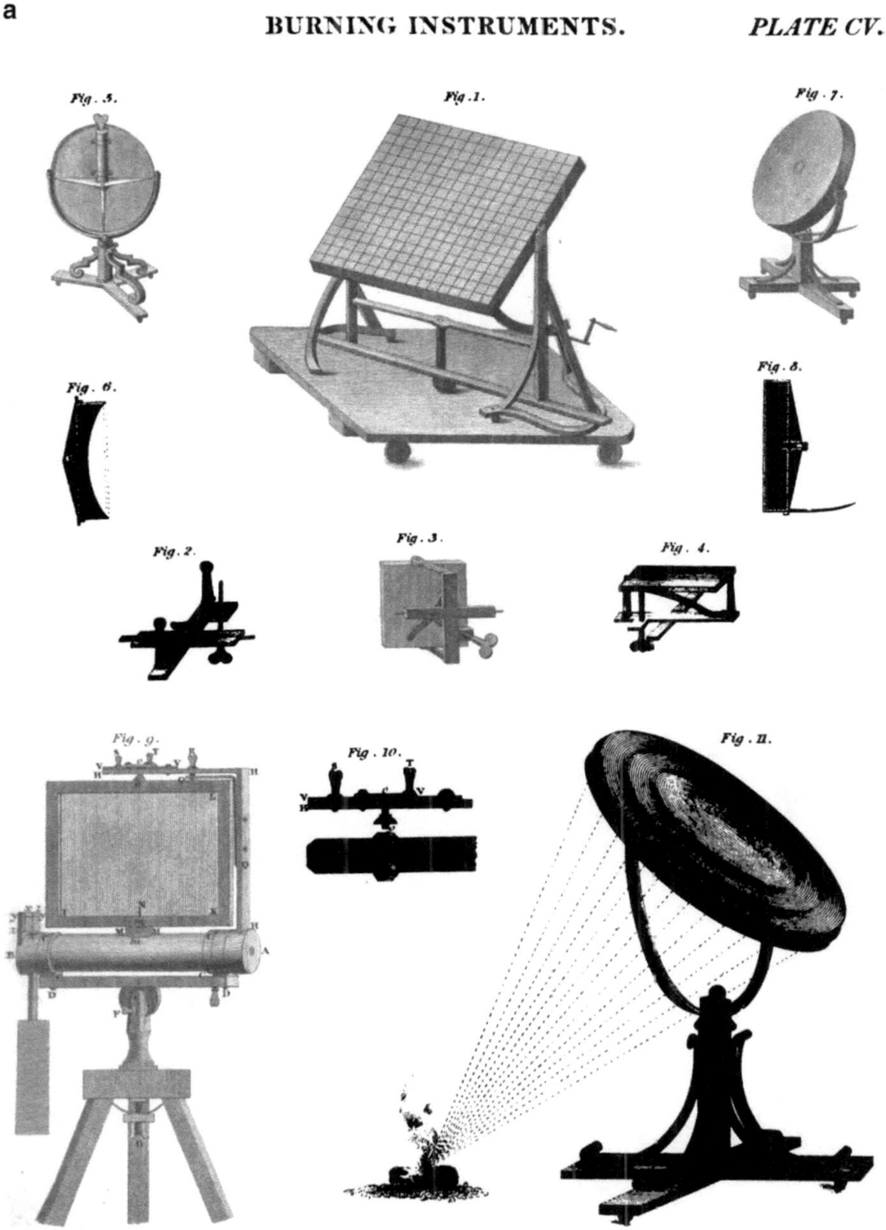

Fig. 2.6 Eighteenth-century burning instruments: burning mirrors used by Buffon ca. 1747 (plate CV above—top row and bottom right); double burning lens by William Parker, similar to the ones used at the Academy on a larger scale (plate CVI below). *The Edinburgh Encyclopaedia*, conducted by David Brewster (Edinburgh, 1830), vol. 5

b

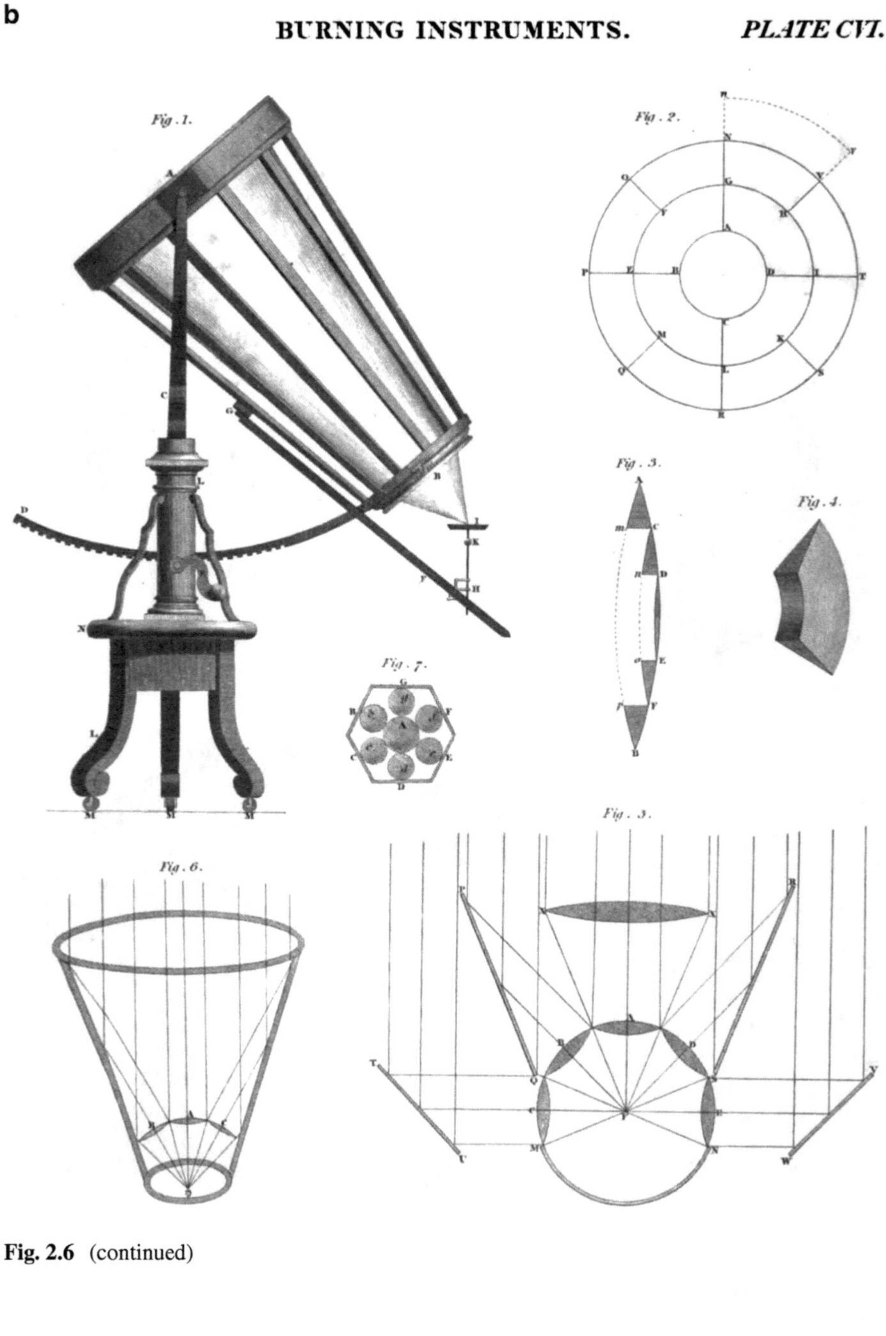

Fig. 2.6 (continued)

plain that "we cannot do for chemistry what elegant machines, optics, and electricity have done for physics." Such instruments embodied contemporary preoccupations with the disciplinary boundary between chemistry and physics. In the *Essays of Effluviums* Boyle reported similar attempts to use "good Burning-glasses … upon divers Bodies, as Lead, Quicksilver, Antimony, &c." but his experiments were hampered by the "frequent interposition of Clouds and Mists." Using a crucible containing kindled sulfur, Boyle shifted his attention from the action of sunlight to whether "Flame it-self may be as 'twere incorporated with close and solid Bodies so as to increase their bulk and weight."[78] Duclos and Homberg were thinking along such lines too.

In 1700, 2 years before the arrival of the big burning glass, Homberg wrote:

> we have incontestable examples where the matter of fire enters into certain bodies, remains there for a long time and augments their weight, as we see with the regulus of antimony calcinated by the burning mirror. We cannot attribute the augmentation in the weight of the antimony to volatile salts or to the oil of coal, because the metal does not come into contact with the coal in the fire … [we are] therefore obliged to admit here an the introduction of particles of fire into the antimony as a cause for its weight gain.[79]

Although Homberg did not acknowledge Duclos by name, the reference is unmistakable. Even more interesting is the way Homberg reinterpreted Duclos' experiment. For Duclos, fire was merely a passive agent that put other material parts in motion. For Homberg, fire combined chemically, lodging itself within material bodies. By using a burning instrument to heat up the metal instead of exposing it to an open fire, Homberg could rule out the possibility of the transference of any "volatile salts" or "oil of coal" from the fire to the metal. Pure light was being transformed into intense heat and applied directly to the metal.

Several years later Homberg revisited this experiment and acknowledged his predecessor: "Mr. Duclos has performed this operation [calcination of antimony] at another time using one of the mirrors at the Observatory." This time, however, Homberg regarded the experiment as a confirmation that "the matter of light alone,"

> and without the involvement or mixture of any combustible matter, can enter into bodies, remain in them, rendering them more fixed [less volatile] and augmenting their weight considerably; this is the calcination of regulus of antimony by the rays of the Sun [performed] by the burning mirror.[80]

What had previously been attributed to the particles of "the matter of fire" was now attributed to "the matter of light." And his reference to the fixity or volatility of the bodies bore out the chemical context. Homberg's explanation was part of a broader cosmological vision that revolved around sulfur as the only truly active chemical principle, which he also identified with the matter of light. This notion stemmed in part from his experimental realization that the sulfur principle—the empirically

[78] Boyle, "New Experiments," 302–3.

[79] Wilhelm Homberg "Observations sur la Quantite d'Acides absorbées par les Alcalis Terreaux," *Mémoires* (1700), 69.

[80] Wilhelm Homberg, "Suite des Essays de Chimie. Article Troisieme. Du Souphre Principe," *Mémoires* (1705), 94.

demonstrable "oily or fatty matter"—could no longer be considered elementary since it could be further decomposed into simpler matters.[81] Homberg subsequently distinguished between what he called "sulfurs or sulfurous matters"—obtainable through distillation and chemical analysis—and "the true sulfur principle … sensibly detectable only when attached to other matters that serve as a vehicle."[82] In this sense, as Kim has suggested, Homberg's principle of sulfur resembled phlogiston, eluding detection yet chemically activating other bodies. This is not the place to explore Homberg's chemical philosophy; suffice it to observe that he considered light (and fire) as a chemical kind that combined with matter *chemically*. Although inaccessible by chemical analysis, Homberg surmised, several operations indicated that "it is the matter of light which is our sulfur principle, and the sole active principle of all mixts."[83]

Considered abstractly, both Homberg and Duclos linked the influence of solar rays to the weight gain of metals during calcination. But the different interpretations of the mechanism of action represented two different approaches to chemical analysis, composition, and the role of fire in chemical change. To understand Duclos' conception of fire we need to turn to his broader view of chemical analysis, which he had initially envisioned as an important research tool within one of the early Academy's most ambitious collective projects—the Natural History of Plants.[84]

In 1668 Duclos presented a memoir "concerning the method of the history of plants."[85] Over the following 2 weeks, in two consecutive memoirs (16 and 23 June)—one on distillation and one on solution analysis—he pitted distillation against solution analytical chemistry and physical mechanism against non-mechanistic chemistry in the context of chemical analysis and particulate theories. Duclos' decision to dedicate two sessions to the theoretical aspects of chemical practice was revealing. The informal reason behind Duclos' decision to discuss

[81] Kim, *Affinity*, 93; Homberg's chemical theory and its relation to chemical practice and analysis is discussed by Kim, *Affinity*, 66–110; Alice Stroup, "Wilhelm Homberg and the Search for the Constituents of Plants at the seventeenth-Century Académie Royale des Sciences," *Ambix* **26** (1979), 184–201; and Mi Gyung Kim, "Chemical Analysis and the Domains of Reality: Wilhelm Homberg's Essays de chimie, 1702–1709," *Studies in History and Philosophy of Science* **31** (2000), 37–69. For Homberg's views on the relations between chemistry and physics see Rémi Franckowiak and Luc Peterschmitt, "La Chimie de Homberg: Une vérité certaine dans une physique contestable," *Early Science and Medicine* **10** (2005), 65–90.

[82] Homberg, "Suite des Essays," 88–9.

[83] Homberg, "Suite des Essays," 89; Kim, *Affinity*, 96. For Homberg's "chemistry of light" see Lawrence M. Principe, "Wilhelm Homberg: Chymical Corpuscularianism and Chrysopoeia in the Early Eighteenth Century" in Christoph Lüthy, John E. Murdoch, William R. Newman, eds., *Late Medieval and Early Modern Corpuscular Matter Theories* (Leiden: Brill, 2001), 535–56; Lawrence M. Principe, "Wilhelm Homberg et la chimie de la lumière," *Methodos* **8** (2008). Retrieved 26 January 2014, from http://methodos.revues.org/1223.

[84] See Alice Stroup, *A Company of Scientists: Botany, Patronage, and Community at the Seventeenth-Century Parisian Royal Academy of Sciences* (Berkeley: University of California Press, 1990); Boantza, *Matter and Method*, 67–82.

[85] AdS, PV, t. 4, fols 48r–55v.

these topics was clearly linked to the Natural History of Plants project, which by 1668, given the growing importance of chemical analysis, Duclos presided over.

Duclos asserted that "the principal means of analysis are fire, air, and dissolutive liquors," thus setting the background for the ensuing discussion.[86] According to Duclos, fire decomposed or separated the parts of a mixt either through the action of heat alone or by way of inflammable combustion.[87] This is how fire analysis worked:

> The heat of the fire excites a motion in the mobile parts [of the mixt] according to their degree of mobility so those that share the same degree of mobility cannot separate at the same degree of heat, which in agitating them equally makes them rise together, and [hence] they separate only from the [relatively] less mobile parts. Those that are unequally mobile separate from each other because the most agile, being most rapidly rarified by the heat, rise first and detach from the less mobile, which might follow when chased by a stronger fire.[88]

This explained how distillation worked to separate the constituents of mixts and signaled the limitations of fire as an analytical tool, the underlying principle being that two constituents could be different—in essence, nature, or constitution—yet by sharing the same degree of mobility would "rise together" and not separate. By inference, the chemical essence of a substance was unrelated to motion, which was a physical attribute.

Duclos underscored the relation between heat and motion by challenging "those who claim that heat brings together things of the same nature and separates those of different natures." Again, whereas "nature" was associated with chemical essence, "heat" was viewed as a physical phenomenon. In fact, it was not heat that actively either separated or united the constituents of bodies, since its sole action was to impart motion. The resulting separations or unions were consequences of the generally imparted motion—based on the constituents' propensities to move—and not of heat itself. Duclos' reference to potential unions suggested yet another sense in which he was critical of the analytical capacity of fire, since during distillation some constituents "merely liquefy and attach to those which are fixed, producing a new composition of parts."[89] Significantly, in treating the action of fire, Duclos avoided interpreting such unions by recourse to affinities, resemblances, or correspondences among entities bearing essential similarities, as exemplified by his use of "symbole," which he reserved for chemical transformations. He allowed only for the physico-corpuscular principle of motion in explaining the separation and creation of new compounds.

Like Boyle, Duclos maintained that fire not only failed to decompose mixts into their elementary constituents but created new combinations.[90] Closely acquainted

---

[86] Ibid., fol. 58r.

[87] Ibid., fols 58r–v.

[88] Ibid., fol. 58v.

[89] Ibid., fols. 59r–v.

[90] For controversies on fire analysis in the early modern period see Allen Debus, "Fire Analysis and the Elements in the Sixteenth and the Seventeenth Centuries," *Annals of Science* **23** (1967), 127–47; Frederic L. Holmes, "Analysis by Fire and Solvent Extractions: The Metamorphosis of a Tradition," *Isis* **62** (1971), 128–48. See also Mi Gyung Kim, "The Analytical Ideal of Chemical Elements: Robert Boyle and French Didactic Tradition of Chemistry," *Science in Context* **14** (2001), 361–95.

with chemical experimental reality, he admitted that "fire can occasion, by the power of its heat, not only separation, but also union," since it "separates the volatile parts from the fixed ones by the movement occasioned by its heat," thus uniting "into a new compound the parts which are less fixed with those that are more fixed."[91] Duclos' association of fire with the properties of volatility and fixity was significant, as he proceeded to assert that fire could resolve "mixts composed of volatile or combustible parts." Yet for the analysis of the most fixed materials or the least mobile and volatile—usually earthy and dry since fixity was related to lack of humidity—Duclos prescribed the use of "resolutive menstruums" or solvents.[92]

Furthermore, Duclos' distinction between 'superficial' (mechanical) and 'radical' (essentially transformative) types of analysis represented his understanding of the physical–chemical divide. A similar line of reasoning was discernable in his interpretation of the augmentation of weight in metals exposed to the fire of a burning mirror. In both cases fire only imparted motion—from which followed either distillatory fractions according to their degrees of mobility—or, in the case of the burning mirror—the circulatory motion of air around the inflammable matter that in turn *enabled* the union of the "igneous parts" to the metal.[93] The actual union was explained in terms of affinity ("symbole"), an account reserved for distinctly chemical and non-mechanical types of material change. This mechanism of affinity was *causally* independent of and only *circumstantially* related to heat or fire. Likewise, unions resulting from distillation occurred on two levels: the superficial, which could be undone by heat; and the radical, which could not be decomposed by heat and thus called for an entirely different means of analysis—"menstruums."

As we have seen, Boerhaave invoked traces of this ontological demarcation. He distinguished between changes produced by a "motion communicated to the whole [of a corporeal] mass" from those arising from a "motion … raised among the parts." The latter produced "an inexhaustible variety of effects and changes." As to the former, even if "the whole power of chemistry were applied to it … if it raise no motion in any of its parts, it must remain as it did," that is, unchanged. Boerhaave's distinction was general, abstract, and heuristic. It implicitly turned on the difference between motion at the (physical) level of homogenous "mass" and motion at the (chemical) level of the heterogeneous "parts." We have also noted Boerhaave's attempt to reconcile this type of reasoning with his notions of *instruments* of analysis and material change, on the one hand, and *elements* as instruments and material constituents, on the other. The fact that this attempt arose in the context of an observation on "menstruums" was not incidental.

---

91 AdS, PV, t. 4, fol. 60r.

92 Ibid., fols 60r–61v.

93 The subject of the inflammable particles is further linked in Duclos' cosmology to what he termed the igneous spirit, a universal agent of change that was a material–immaterial hybrid.

2.5 Menstruums, Chemical Causality, and the "Appetite of Union"

In his account of solution analysis and "dissolutive liquors" Duclos provided a crucial clue regarding their action: "since most mixts that do not have a strong compaction between their parts, are resolved independently through putrefaction in their own humidity, the chemists conducted the resolution of less humid substances by way of putrefaction, by the addition of a regulative liquor."[94] Such "regulative liquors" would "facilitate putrefaction" or fermentation, processes that Duclos linked to analysis and decomposition. Duclos identified three kinds of menstruums: corrosive, extractive, and resolutive. The corrosive, when applied to solid mixts, brought about the discontinuation of their mass and broke them down into "integrant, highly subtle, particles." The extractive acted for the most part by precipitation to separate a certain part of the mixt. Only the resolutive solvent, however, could bring about a "radical resolution"—of a mixt into its elementary constituents.[95] Duclos referred to the latter in his research proposal for plant analysis as the "extreme analysis of plants, which provided the most accurate knowledge of the constitution of a subject."[96] Since it rarified the parts of the mixt and enhanced their mobility, the corrosive resolution could be used as a preparatory step to the extreme one.

All these menstruums, Duclos proclaimed, consisted of "salts, resolved and spiritualized," namely "reduced into highly penetrating liquors." These salts were either mercurial, sulfurous, or mixt.[97] The mercurial liquors were acidic and merely corrosive (like aqua fortis) while the liquors of sulfurous salts were purely extractive (like alcohol). Only the mixed spirits, which contained both mercurial and sulfurous salts, could cause a radical resolution. "These mixed menstruums are the true solvents intended for true [radical] chemical analyses," Duclos stressed, and were therefore "useful in the study of the principles of natural mixts as well as in performing analytical observations that facilitate our knowledge of the nature and qualities of mixts."[98] Radical solution analysis, then, was a distinctly chemical and non-mechanical process. Its activity generated by the combination of mercurial and sulfurous salts, it was analogous to irreversible and essentially transformative processes like putrefaction and fermentation. As we have noted, Duclos thought that salts played a crucial role in composition too, for he argued that the "earthy sulfurs," which he regarded as the true "augmenters of weight" in calcined metals, were "brought into combination with the earthy parts [of the metal] by a salt." Chemical composition and decomposition, or unions and separations, might be affected by

---

[94] Ibid., fols. 63v–64r.

[95] Ibid., fols. 64r–v.

[96] Ibid., fol. 53v.

[97] Duclos' reference is evocative of the medieval and Renaissance dyadic view of composition (especially of metals), consisting of the two principles of sulfur and mercury.

[98] Ibid., fols. 64v–65v.

heat but were not directly caused by it. Heat operated at the physical level of homogeneous aggregate; salts, affinities, and some menstruums, to name a few, acted at the 'deeper' chemical level of heterogeneous mixt.

Duclos' view of menstruums exemplified his delineation of the physical and chemical spheres. Based on a quest for a universal explanation, Boerhaave's account of fire embedded an attempt to collapse such divisions. For Boerhaave, fire was an element and instrument of both physical and chemical change. As an element it was material and as an instrument it was immutable; and it acted by way of its perpetual motion and tendency to expand equally in space. Like Duclos' heat, it functioned by imparting motion, but it did not do so by the sheer agitation of the parts of bodies; it actually—that is *physically*—penetrated into bodies. This inner presence rarified bodies but did not augment their weight. Homberg's fire combined directly with matter through the sulfur principle. Whereas Duclos' immaterial fire (and air) enabled the physical addition of material parts, the actual chemical combination occurred at a level that was governed by affinities and elective attractions. Ontologically, the latter were separate from heat and its main effect—motion.

Boerhaave recognized the chemical singularity of menstruums, his "fifth [instrument], which is reckoned almost peculiar to chemistry; being that wherein chemists place their chiefest excellence; and whereto they ascribe the greatest effects of their art."[99] Reminiscent of Duclos' consideration of menstruums in non-mechanical terms, Boerhaave set this "instrument" apart from the other five (i.e. the four elements and lab utensils):

> It is therefore the property of a menstruum, to be itself equally dissolved, at the time it dissolves the solvend. And this property obtains in all solutions performed by the means of menstruums … and hence it appears, that this action of menstruums *differs from all mechanical separations*, where the dividing instrument remains entire and whole, both in the act of separation, and afterwards; as we evidently see in the case of a knife, sword, saw, piercer, &c. for all these instruments, whilst they divide, are not themselves divided but remain almost the same as before.[100]

Out of Boerhaave's six instruments only menstruums and utensils were not also considered as elements. Since they were perceived as 'instrumental' rather than 'elementary' they did not have to be immutable. This might seem paradoxical at first glance: the shifting of emphasis toward the instrumentality of a menstruum lifted the ontological restrictions entailed by the very definition of an instrument—its capacity to remain unchanged while effectuating material change—like "in the case of a knife, sword, saw, piercer, &c." The way around this paradox, however, was precisely what defined the uniqueness of menstruums as non-mechanical. Since only mechanical instruments were bounded by the immutability requirement, this principle at once described and prescribed the nature of solvents, whose action "differs from all mechanical separations."

Once more, the separation (or interaction) between the physical and chemical, and in particular between the ontological levels at which material change took place, was

[99] Boerhaave, *New Method*, 489.

[100] Ibid., 489 (italics mine).

not an easy one in Boerhaave's system. Despite his seemingly non-mechanical view of menstruums, he hastened to qualify his stance, for "there may be room to suspect, that the single particles of a menstruum act, in solution, like the above-mentioned mechanical instruments [i.e. knife, sword … etc.], in which case, each particle of a solvent would have its proper size, figure, hardness and gravity, and act by these properties, which may be called mechanical."[101] Whereas fire had to be imponderable, the particles of solvents had gravity. As a whole, a solvent would change during chemical dissolution, but it might still remain unchanged at the particulate level. This seemed to suggest a 'whole' (menstruum) that was 'greater than the sum of its parts' (corpuscles), a notion that boded well with non-mechanical perceptions. The constituent parts, or corpuscles, of a solvent could be viewed mechanically, but their overall chemical action was irreducible to mechanistic principles. In light of Boerhaave's overall notion of matter as a homogenous aggregate, which entailed a certain view of analysis, this micro–macro distinction was not superfluous:

> tho' we go on dividing and abstracting to infinity, [a] body will still consist of other lesser ones; which … *will be perfectly like the greater*; and those again consist of other similar lesser ones … certain corpuscles unite into little masses, so firmly cohering as that *no power implanted in nature, or to be excited by art, can divide them into lesser parcels*; and consequently they must remain the same, what violence soever be applied.[102]

We have already encountered Boerhaave's depiction of fire's general way of interacting with matter. Found in constant motion, it insinuated itself within the interstices of bodies and "immediately begins to expand all bodies, and remove their parts from their natural closeness … hinder[ing] the proper cohesion of the elements." Endeavoring to expand equally in space, fire "continually … remove[s] the elements from one another." The bodies were in turn "continually striving against the separation of their elements," resulting in a mechanism of "perpetual action and re-action" between the fire and material bodies. In a Cartesian vein, Boerhaave asserted that the "particles of fire … must necessarily be exceedingly subtile, as they readily penetrate all, even the densest bodies." Although its particles were the "smallest of all bodies yet known," fire did not combine with bodies.[103] Toward the end of the chapter on fire, underscoring its elementary-material and instrumental-immutable nature, Boerhaave clarified:

> Fire, therefore, according to this doctrine, must be all corporeal, immutable, incapable of any alteration of figure, and *unable* either *to join* and grow to itself, or to *other bodies*; and in the mean while must have the *greatest power of dividing* all other things, as it can always enter the pores of bodies to be divided, and exert its power therein, so as to break and dissolve their parts and branches.[104]

Both fire and menstruums were instruments of dissolution, yet it was the former's inability to unite chemically with bodies that distinguishes it from the latter.

---

[101] Ibid., 489–90.

[102] Ibid., 222.

[103] Ibid., 359.

[104] Ibid., 361 (italics mine).

Unlike fire, which comprised a physical and mechanical means of separation based on properties pertaining to aggregates—like mass, extension, impenetrability, and expansion (self-repulsion)—menstruums acted on a different ontological level; or, more accurately, since the physical and the chemical were not mutually exclusive, on a different level *as well*. Duclos, for instance, ranked menstruums according to their levels of activity. Although no menstruum was entirely mechanical, the corrosive type, whose main action was to loosen up the mass and volatilized the particles, could be employed in preparation for the resolutive one. More generally, Venel noted that "the principles [elements] of mixtures … are connected by a link different from the one governing the formation of aggregates or relationships of mass. The first can be broken down by mechanical as well as by chemical means. The second can be separated by chemical means alone."[105] Boerhaave, for whom motion was so central in all processes of material change, observed that fire could "promote the action" of solvents. Yet dissolution "seems to be performed by a certain power with which the parts of the menstruum endeavour to attract the dissolved parts, rather than to repel them. We are not therefore to imagine this is a mechanical action … but rather an *appetite of union*." This point is crucial to understanding Boerhaave's perception of the demarcation between the physical and the chemical. Like Duclos, Venel, and other chemists, Boerhaave recognized the key difference between "the *common* causes of motion" and "a *particular* cause, not common to all bodies, exerted between the solvent and solvend," the importance of which he acknowledged in no uncertain terms, "for he, who can learn the power of menstruums, will understand the whole nature of chemistry."[106]

2.6 Conclusion

Boerhaave's view of menstruums embodied a clear recognition of the ontological and methodological differences between physical and chemical categories, and more importantly, an admission of the implausibility of reducing the latter to the tenets of the former. His account of fire, on the other hand, straddled the disciplinary boundaries. *Pace* Venel's warning, it creatively blended notions and categories from the two domains. Flawed as it was, Boerhaave's effort to understand and characterize fire, or his "history of fire," was impressive in scope, breadth, and detail. Equally evident, however, was just how difficult it was "to *distinguish* the very action of fire from that of other concurring *causes*," as his words in the opening epigraph suggested. The second quote in the epigraph, taken from Ephraim Chambers' entry on fire in his *Cyclopædia*, captured some of the key tensions underlying this challenge. Fundamentally, fire was defined as a corporeal body in motion, whose "Motion is argued from its expanding the Air … And its Corporeity is proven by Experiment." This perspective was partly in agreement with Boerhaave's view, according to

[105] Venel "Chymie ou Chimie," 413.

[106] Ibid., 493, 495 (italics mine).

which bodies subjected to fire would universally be "render'd bigger, swell, and rarify." Likewise, he maintained that "elementary fire must be truly corporeal ... immutable in itself, and the great instrument or means of changing all other things"—by occasioning motion within them. For Boerhaave, fire's predominant attributes were its corporeality and motion. Chambers was a writer and encyclopedist, not a learned natural philosopher or a seasoned chemist. But his alignment, pairing motion with expansion, and corporeality with experiment, was instructive, especially in light of our examination of Boerhaave's (and others') theories of fire. The associations presented by Chambers implied that while fire's motion was deduced from the physical expansion of air (as a medium), its corporeality or materiality was deduced from chemical experimentation. He mentioned in this context the prolonged heating of mercury, which ended up being "reduced into a Solid; and its Weight considerably increased." According to Boerhaave, bodies expanded "yet without any observable difference in their weight."

Venel dismissed the relevance of the work of "true physicists"—including Boyle, Newton, and Boerhaave, among others—to chemistry because their works were "marked by lack of experimentation." Nonetheless, Boerhaave's study of fire relied extensively on experimental data, as he sought to deduce his principles "from a multitude of common incontestable facts." This rendered his staunch belief in the weightlessness of fire, despite its corporeality, and especially his denial of the weight gain observed in metallic calxes, all the more striking. Such inconsistencies and conflicts were largely absent from the accounts of the chemists we have examined (following Boerhaave's own cue). Boyle and Homberg thought that the increase in weight was due to the addition of fire particles. The two differed in that Homberg identified those particles with an actual chemical substance, "the principle of sulfur." The differences between chemical and physical perceptions of materiality and causality were significant. For a chemist like Homberg, Boyle's mechanistic interpretation was merely heuristic and non-causal since the general recourse to particles required in itself a causal ontological explanation. The chemist worked with and manipulated palpable substances, bearing particular qualities that could not be generalized or accounted for outside their patterns of chemical behavior and interaction in relation to other substances. Physical matter was universally homogenous and inert; chemical matter was fundamentally heterogeneous, specific, and (electively) active. Homberg considered the sulfur principle as "the matter of light ... and the sole active principle of all mixts." Boerhaave related the weight augmentation to the "corrosion, attrition, and mixture of the particles of the sulfur with those of the other bodies" yet he failed to incorporate this conjecture into a coherent chemical explanatory scheme. Ultimately, by communicating motion to matter, fire occasioned physical and chemical change, the causes of which could not be usefully distinguished from one another. In this sense, Venel was right to observe that Boerhaave "treated [fire] physically ... as mass or aggregate."

Duclos' views on the subject throw these perspectives into sharper relief. Unlike Boerhaave, Duclos did not regard fire as corporeal; unlike Homberg, he held that it did not combine with matter. Like Boerhaave's fire, by imparting motion to matter and exciting its parts, it facilitated separations and unions. While the former depended on the physical tendencies of material parts to move, the latter were directed by an order of chemical affinities. Hence fire was both an instrument of analysis and a facilitator of composition, but it neither penetrated into matter nor combined with it, for "all augmentation of weight results from the addition of parts," by which Duclos meant other substances. Duclos' account of fire analysis and chemical composition, and especially his interpretation of the weight increase in metallic calxes, illustrated the importance he attributed to the distinction between physical and chemical material transformations and the divergent mechanisms underlying them. Overshadowed by the alluring universal simplicity of the Newtonian promise, this kind of keen ontological and methodological awareness, so crucial to the practicing chemist, gradually faded away. By the middle of the century, as Venel noted, true chemists were "not very numerous, having their own language, laws, mysteries, and living rather isolated." These shifts, reformations, and disciplinary crises made it increasingly difficult to determine whether the significance of Boerhaave's work on fire belonged among the ranks of chemical "masterpieces" or physical "axiom[s]."

Chapter 3
At the Medical Edge or, The Beddoes Effect

Larry Stewart

In the spring of 1794, James Watt was desperate. The chronic illness of his daughter Jessy had made him search out every possible remedy in correspondence with the chemist Joseph Black in Edinburgh, and with physicians like William Withering, Erasmus Darwin, and Thomas Beddoes. These reports of her symptoms make uneasy reading:

> Jessy…has been for sometime much molested by Hysterical fits which were at one time extremely violent, but have become less so either from the influence of the season, or from the use of the common palliatives, of which tincture of castor & laudanum mixed, & the tincture of fuliginis have been the most useful. She still has a cough & expectoration, the latter of the common kind free from all symptoms of pus, and at sometimes when the expectoration is difficult she has pains in her chest & some tendency to inflammation, the remedy for which has been generally, an alkaline mixture I believe the volatile saline Julep with oyster shells Magnesia & some Sal. Tartr. this seem to operate by allaying the febrile symptoms & attenuating the mucus. The Myrrh Medicine, was only given when most free from febrile symptoms, to restrain the ever copious expectoration, which it has generally affected, but has seldom been continued so long as 10 days together & always laid aside on any appearance of pain in the side. It is the only tonic medicine which has agreed with her, all the others have speedily induced fever. At present her feverish fits, especially after being exhausted by the Hysterics, they commonly go off of themselves. Her great disease is weakness & want of appetite, which I think owing to a lurking fever, proceeding from or causing the affection of the lungs. We have very lately put an issue in her arm from its having had very great effects in some similar Cases, But we have the greatest hopes from air & exercise, & as soon as she can bear the travel one Stage a day, Mrs. W. sets out with her for the coast of Hampshire, where Dr. Withering is to meet them & take the direction of her health.[1]

[1] Birmingham Central Library [BCL], James Watt Papers 4/12/28. Watt to Joseph Black, May 15, 1794. Reproduced in Robert G.W. Anderson and Jean Jones, eds., *The Correspondence of Joseph Black*, **2** (Farnham: Ashgate, 2012), 1236–1237.

L. Stewart (✉)
College of Arts and Science, University of Saskatchewan, Saskatoon, Saskatchewan, Canada
e-mail: l.stewart@usask.ca

J. Buchwald, L. Stewart (eds.), *The Romance of Science: Essays in Honour of Trevor H. Levere*, Archimedes 52, DOI 10.1007/978-3-319-58436-2_3

The young Jessy had been through many a difficult episode of fevers, coughs, and what Watt described as hysterical fits. By June, the battle was lost. For Jessy, all efforts failed. While surely tragic, the moment was unfortunately not unique. It revealed the all too common desperation and a babel of unproven answers. As with so much disease, and no effective remedy at hand—just the province of quacks remained. Watt's own technocratic resilience allowed him to think a solution might yet be found even when ubiquitous consumption conquered all. Often mysteriously ill himself, Watt turned his many talents to the secrets of Joseph Priestley's newly-discovered airs in pulmonary complaints. Amongst the chemically adept, Watt was hardly alone. Control of nature by chemical cures was a fertile field for generating hope and for the exploitation of the desperate.

Here was a messianic vision tied to the chemical achievements of the late Enlightenment. Indeed, the growth of experimental knowledge seemed as though it reflected a progress that was ultimately providential.[2] Watt's own controversial friends like the Unitarian Priestley and radical Thomas Beddoes were ardent proponents of both republican politics and the benefits arising from chemical experiment. If there were limits to reform in politics and in medicine, there were also opportunities—seen by Beddoes and Watt in the very airs Priestley, and so many others across the Channel, had made it their business to explore. It was for this very reason that Beddoes came to assert the social consequence of chemistry. When Beddoes set up in the Bristol suburb of Clifton, after his politics had denied him the Chair in chemistry at Oxford, he still thought it possible his public chemical lectures might be of some benefit.[3] In the aftermath of Priestley's landing in America, and amid the collapse of a French democratic impulse, Beddoes declared of his Bristol lectures that "the effect of a number of people receiving agreeable ideas together may be to soften animosity & that there will be thus a chance of preventing some acts of barbarity in the times that I fear are coming."[4] This was prescient, amidst the obstinate resistance of the medical and political establishment. Much then rested on his chemistry—in the hope that utility might ameliorate social or political differences, not to speak of aiding those caught in the crisis of debilitating, and frequently fatal, ailments (Fig. 3.1).

Chemical prescriptions were widespread. Beddoes' vast network of correspondence revealed the urgent task of attacking disease with the aerial discoveries. The most well-known of these exchanges perhaps lies in that between Beddoes and Watt especially in the aftermath of Jessy's death at the age of 15. Aside from smallpox, consumption was among the most serious and lethal ailments Europeans faced. Even so, two things can be gleaned from Beddoes' network: first, he clearly was not alone

[2] Cf. Steve Fuller, *Science* (Durham: Acumen, 2010), chapter 8.

[3] Trevor H. Levere, "Dr. Thomas Beddoes at Oxford: Radical politics in 1788–1793 and the fate of the Regius Chair in Chemistry," *Ambix* **28** (1981), 61–69, reprinted in Levere, *Chemists and Chemistry in Nature and Society 1770–1878* (Aldershot: Variorum, 1994).

[4] BCL, James Watt Papers [JWP], W/9/7. Beddoes to James Watt, April 21, [1796?]. He expressed the identical sentiment to Davies-Gilbert in 1798. See Cornwall Record Office, Davies-Gilbert Correspondence, DG 42/2. Beddoes to Giddy, April 14, 1798; DG 42. Beddoes to Giddy, June 1, 1798.

Fig. 3.1 Thomas Beddoes, M.D. (By permission of the Wellcome library, London)

in the assumption that pneumatic chemistry could create new remedies and, secondly, the notion of aerial cures was not confined to consumption but was widened to attacking a vast range of infections. This network of aerial apostles and experimentalists included physicians, surgeons, chemists and even engineers all of whom, with hopeful patients, thought the pneumatic chemistry might lead to profitable results.

The chemistry of airs was hardly then of recent concern. But since Joseph Priestley's identification of the 'most respirable portion of air' there was a great deal of interest in the manner in which new airs might be identified and tested.[5] Yet, could such knowledge be credibly put to use? This surely occurred to Priestley as he particularly sought out Matthew Boulton in 1777. He had dragooned Boulton "to procure air as it is actually breathed by the different manufacturers in this kingdom and hope you will be so obliging as to procure me the proper samples from Birmingham." Identifying an easy means of collecting airs, Priestley went on to say, "I only wish you would do this in those places where you expect the air to be the worst, on account of bad fumes or a number of people working together &, and not at your place in the country, but in the middle of the town. I should be glad also to have the air of some of your closest streets, and likewise the best in your neighbourhood, noting the state of the weather."[6] Place was important especially with an

[5] Victor D. Boantza, "Collecting airs and ideas: Priestley's style of experimental reasoning," *Studies in History and Philosophy of Science* **38** (2007), 506–522; and Boantza, "The rise and fall of nitrous air eudiometry: Enlightenment ideals, embodied skills, and the conflicts of experimental philosophy," *History of Science* **51** (December, 2013), 377–412. (I wish to thank Dr. Boantza for providing me with an early draft of his essay).

[6] Robert E. Schofield, ed., *A Scientific Autobiography of Joseph Priestley, 1733–1804: Selected Scientific Correspondence, with Commentary* (Cambridge, Mass.: MIT Press, 1966), 161–162. Priestley to Boulton, before November, 1777.

emerging, largely amateur, experimentalism. But, by the next decade, the explicit objective was to address the increasingly obvious relationship of manufacturing towns with pulmonary afflictions. Hence, the Manchester physician and chemist, Thomas Percival sought out Watt—writing about a conversation he had with Priestley regarding Watt's apparent claim of a "method of destroying smoke, which issues from fire engines, furnaces, & other works." Percival believed that this was "of great importance to the inhabitants of Manchester, who appear to be peculiarly incident to pulmonary afflictions, & [as he was] apprehensive will become more & more so from the rapid increase of the cotton manufactory."[7] As it turned out, Percival's alarm obviously reached rather beyond the boundaries of Manchester.

Industrial air, either as cause or aggravation of pneumatic disease, was little understood but widely noticed. But there were also many ailments, urban or not, that might succumb to chemical knowledge. In an age when diagnosis was uncertain, and treatment elusive, any wisp of chemical airs induced novel experiments and trials, by medics of many kinds. Testimony was essential. Thomas Beddoes set out to gather as much information as possible from anyone who manufactured tales of hope in the chemist's laboratory. The range of his contacts was immense and, as Jessy's case would show, his programme urgent. It was extremely likely, therefore, that Watt and Beddoes would fall in together, not only because of their common connection to the Edinburgh chemical professor, Joseph Black. Among Beddoes' correspondents, far too numerous to mention here, we find many who were also joined to Watt's own network such as the illustrious Black, the physicians Erasmus Darwin, James Lind, and John Ewart, brother of an engineer who worked with Watt, all of whom explored treatment with airs.[8] Of course, besides the French chemists of Paris and Dijon, like Antoine Lavoisier or Louis-Bernard Guyton de Morveau, whom Beddoes knew well, there were also those like the Birmingham surgeon John Barr who worked alongside Watt in the treatment of patients. Barr, for example, reported a case of hydrothorax [an excess of pleural fluids, possibly from heart or liver disease] where the patient, even having consulted "some of the ablest Physicians in England" including Darwin, could not sleep without being propped up. After the first application of oxygen he reportedly slept "as flat as a flounder." Seemingly cured, Barr sent him off with an air holder filled with oxygen.[9] There were many

[7] BCL, JWP 4/48/7. Percival to Watt, September 16, 1786. On Percival, see Trevor Levere and Gerard L'E Turner, *Discussing Chemistry and Steam. The Minutes of a Coffee House Philosophical Society 1780–1787* (Oxford: Oxford University Press, 2002), and Trevor Levere, Larry Stewart, and Hugh Torrens, *The Enlightenment of Thomas Beddoes. Science, medicine and reform* (London and New York: Routledge, 2017).

[8] On John Ewart see Larry Stewart, "His Majesty's Subjects: From Laboratory to Human Experiment in Pneumatic Chemistry," *Notes & Records of the Royal Society*, **63** (September, 2009), 231–245, esp. 231–232; and Thomas Beddoes, M.D. and James Watt, Engineer, *Considerations of the Medicinal Use and Production of Factitious Airs.* Part III. Second edition, corrected, and enlarged (London: Printed for J. Johnson, St. Paul's Church-Yard, 1796), 23, 24, 101, 155. On Lind see BCL, JWP 4/23/19. Beddoes to Lind, Sept. 23, [1794?]; 4/23/20. Beddoes to Watt, Sept. 18, [1794?].

[9] BCL, JWP 4/23/13. Barr to Watt, September 20, 1795.

such claims, and as many doubts, not just in Birmingham, or London, but across the width of empire. Seduced by airs were Dr. William Saunders of Guy's Hospital, the London surgeon Jonathan Wathen, and the chemist and physician (and probable consumptive) Adair Crawford, Beddoes even receiving similar communications on the curative properties of nitric acid from as far afield as Dr. Helenus Scott in Bombay.[10] From Birmingham to Bombay, promise followed airs. These were more than flickers of a world-wide hope for chemical release from unconquered afflictions, some obviously the ready consequence of venerian sports, or others like consumption their transmission not so well understood but unfortunately far too common.

3.1 The Breathing Campaign

New airs might offer hope to some, but access remained a problem. Even the most respirable gasses were not ready to hand for any medical application. Watt and Beddoes set their minds to expand the ready availability of machines both "for the chemists as well as the practitioners of pneumatic medicine."[11] Watt's newly-designed breathing apparatus soon attracted an overwhelming demand, from the afflicted and from amateurs, so much so that Watt decided he could not supply them directly. Commercial distribution was through Chippendale, in London's Fleet Street.[12] But this raised other issues, especially as growing numbers of practitioners could readily be accused of quackery even if the futility of their pneumatology was not immediately certain.[13] Quite the contrary (Fig. 3.2).

[10] On Scott see Mark Harrison, "Medical experimentation in British India. The case of Helenus Scott,"in Hormoz Ebrhaminejad, ed., *The Development of Modern Medicine in Non-Western Countries* (London and New York: Routledge, 2009), 23–41; also Guelph University Library, XS1 MS A164/ 94, James Dinwiddie Correspondence, Helenus Scott to James Dinwiddie, August 23, 1796; On Saunders see Jan Golinski, *Science as Public Culture. Chemistry and Enlightenment, 1760–1820* (Cambridge and London: Cambridge University Press, 1992), 61; BCL, JWP 4/23/28. Beddoes to Watt, July 7, [1794]; See also [Robert Thornton], *The Philosophy of Medicine: Or, Medical Extracts on the Nature of Health and Disease, Including the Laws of the Animal Oeconomy, and the Doctrines of Pneumatic Medicine.* **5** (London: Printed by C. Whittingham, Dean-street, Fetter-Lane, 1800), 388–391.

[11] BCL, JWP 4/23/25. Beddoes to Watt, August 21, 1794.

[12] Thomas Beddoes and James Watt, *Considerations on the Medicinal Use, and on the Production of Factitious Airs.* Part I. By Thomas Beddoes, M.D. Part II. By James Watt, Engineer. Edition the Third. Corrected and Enlarged. (Bristol: Printed by Bulgin and Rosser, For J. Johnson, in St. Paul's Church-Yard, London, 1796), 222. Chippendale's brother apparently tried the breathing apparatus of Robert Thornton in London but complained the vital air "had done him harm." BCL, JWP 4/23/15. Robert Thornton to Beddoes?, August 27, 1795.

[13] Cf. Michael Neve, "Orthodoxy and Fringe: Medicine in Late Georgian Bristol," in W.F. Bynum and Roy Porter, eds., *Medical Fringe & Medical Orthodoxy 1750–1850* (London and Sydney: Croom Helm, 1987), 40–55, esp. 48ff.

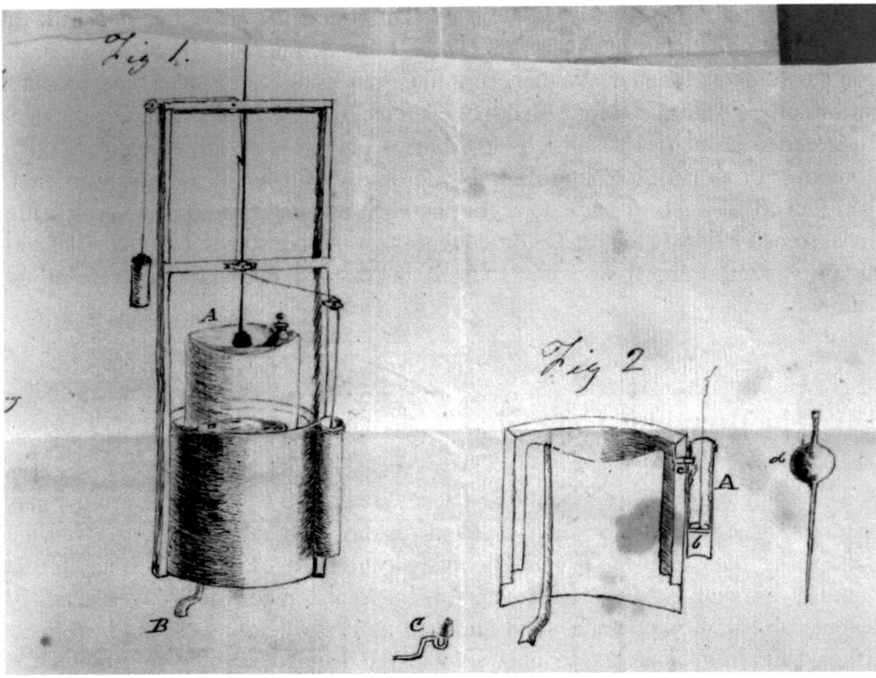

Fig. 3.2 Draft Mercurial Apparatus, Thomas Beddoes to James Watt, January 8, 1795. MS. 3219/4/27/9 (By permission, The Library of Birmingham)

There were many supportive medics and their patients who swore that airs achieved results. This project was not ultimately confined to enthusiasts like Watt or Beddoes. For his part, Beddoes seemed determined to challenge the label of quack and his published reports of trials, many of apparent success, were not quite as questionable as we may now assume. But Beddoes' reputation, sullied by his radical politics, was rather more favourable than pneumatic disappointment would suggest.[14]

Pneumatic chemistry had received great attention amongst medical practitioners and various chemical entrepreneurs willing to attempt new remedies. One such was the botanist-physician, and later lecturer, Robert John Thornton in London who became known as a promoter of the Linnean system.[15] In his early career as

[14] On the hopes of Watt and Beddoes and concerns over the damage Beddoes' republicanism did to their project see David Philip Miller and Trevor H. Levere, "'Inhale it and See?' The Collaboration between Thomas Beddoes and James Watt in Pneumatic Medicine," *Ambix* **55** (March, 2008), 1–24; Michael Neve, "Orthodoxy and Fringe," 48.

[15] For his lectures before the Spitalfields Mathematical Society, see Dalhousie University Library, MS 2–726, B75. Dinwiddie fonds, Journal, May 19, 1808; Jan Golinski, "From Calcutta to London: James Dinwiddie's Galvanic Circuits," in Bernard Lightman, Gordon McOuat and Larry Stewart, eds. *The Circulation of Knowledge Between Britain, Indian and China. The Early-Modern World to the Twentieth Century* (Leiden and Boston: Brill, 2013), 75–94, at 89, n. 61.

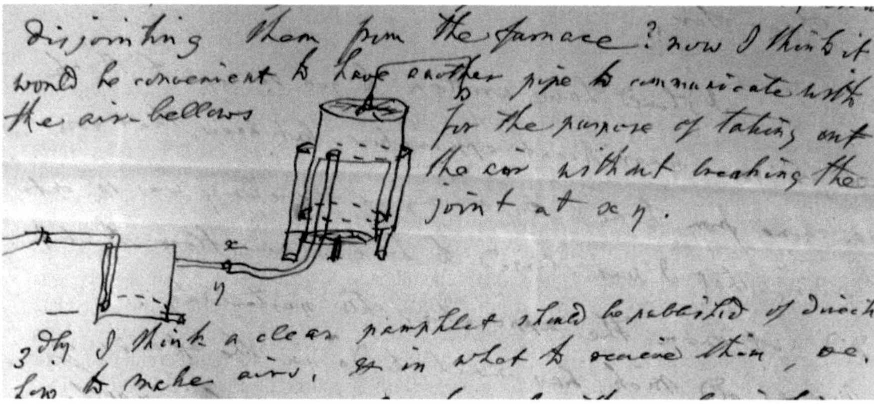
disjointing them from the furnace? now I think it
would be convenient to have another pipe to communicate with
the air-bellows
for the purpose of taking out
the air without breaking the
joint at x y.
3dly I think a clear pamphlet should be published of directions
how to make airs, & in what to receive them, &c.

Fig. 3.3 Erasmus Darwin to James Watt, August 17, 1794. MS.3219/4/104 (By permission, The Library of Birmingham)

physician, Thornton quickly developed an extensive pneumatic practise as well as designing his own breathing machine. In the winter of 1794–5, Thornton claimed that he had "upon an average given the air to 15–20 patients in a day… when not one besides myself in London have [practised] it."[16] But many others were interested. Erasmus Darwin likewise urged the infirmaries in Nottingham and Shrewsbury purchase the breathing apparatus[17] (Fig. 3.3).

There were two critical problems: no one knew how the airs actually worked, as it appeared they did and, even so, there was such a demand for devices from both practitioners and patients that it could not be met. These were revealed in a letter to Watt in December, 1794. Beddoes had also discussed pneumatic treatments with Ewart and with a Dr. Robertson who, seemingly, had come to the view that carbonic acid gas [formed from water over heated iron] "stimulates the absorbents–whence hopes in tubercular phthisis." Beddoes reported that he also had "a case of this kind, where the lungs have been 20[?] weeks ulcerated, in which the pneumatic treatment is distinctly advantageous, & yet I cannot prevail upon them to have an apparatus, though very rich."[18] There is little doubt that the apparatus was expensive. This was a clear obstacle for some. Even the instrument used unsuccessfully by Lord Daer later demanded 14 guineas. Promises came with a hefty price.

Even so, medical enthusiasm remained as with Thornton. But Beddoes worried that Thornton's successes might eclipse their own. He copied to Watt portions of Thornton's letters which "might bear the interpretation of a wish to monopolize the pneumatic practice in London" even if trials were widely attempted, especially by amateurs. There remained resistance to the expense of the device, as from a Mr. Gladwell

[16] BCL, JWP 4/65/2. Thornton to James Watt, July 22, 1795.

[17] BCL, JWP 4/65/6. Darwin to Watt, April 29, 1795.

[18] BCL, JWP 4/65/7. Beddoes to Watt, December 12, 1794.

in Clifton, "rich but excessively avaricious," who had demurred over the price to obtain Daer's apparatus. Beddoes believed the cost of

> This unexpected affair hardly leaves room for Dr. Thornton–who says as follows: the first intimation of the oxygen air for the removal of gutta serena was in a lady, who was a patient of mine for a nervous head-ach. By accident, she took up at church a prayer-book, & was surprized to find her eye-sight so strengthened that she cd read in it, when before, so small a print was perfect confusion. Having mentioned this, I was anxious for the trial in the worst case of defective energy of ye optic nerve. The 1st patient, Mr. Hill (a surgeon) tried the air with, was the sister of Ld. Walpole. She early lost one eye, though the defect cd not be discovered by a bye-stander–& the sight of the other had become very feeble. Electy. was conjoined with the vital air. She cd not discern, as the told Mr. Wathen, the dial-plate belonging to the Horse Guards clock, but before she left him she cd readily perceive both the hour & minute hands. I gave you the acct of benefit received in gutta serena by a poor patient, who contented with a moderate sight, left off attending … a lad was 7 weeks in St. Thomas's Hospital & turned out incurable. Mr. Wathen, having no success in such a desperate case, sent him to inhale the vital air. In 4 days he…had a glimmering of sight & in 15 discerned objects.[19]

Beddoes was obviously conflicted over London's enthusiasm. It was precisely the spread of these trials and devices that had the potential to provide credible evidence for pneumatic medicine. There could be no surrender to disease or profession. His publication of the many claims of cures was the most effective means of attracting further patients or volunteers and reinforcing hope.

Beddoes' early view had entirely misread Thornton. Indeed, it was Thornton who explicitly addressed the fundamental issue of access to the new airs. In the summer of 1795, Thornton sought to encourage Watt's designs by telling him, "I am anxious now to multiply practitioners & for this purpose I wish you would turn your mind to the contrivance of a graduated glass apparatus, & tin jappanned reservoir, & having these in abundance, & of a moderate price, neat, simple, & elegant, they would encrease the sale of the pneumatic apparatus, & extend the benefit of your discoveries as well as those of Dr. Beddoes…"[20] Indeed, Thornton clearly acknowledged the energy and contribution of Beddoes and Watt. In his own practice, Thornton feared he too would be "branded as innovator", in some circles as nasty an opprobrium as 'projector' or 'calculator'. If this was an age of innovation, much of it political, many challenged orthodoxy and power, and were thus not welcomed (Fig. 3.4).

Neither Beddoes, nor Watt, nor Thornton could be easily dismissed as dangerous quacks. There were too many experiments for that. Thornton was well aware of various continental trials, by Felice Fontana in Florence or Christoph Girtanner in Gottingen, as of "our learned friend Beddoes" that affected the vascular system, whether these be the fault of venom or of experimental injections such as laudanum.[21]

---

[19] BCL, JWP 4/65/7. Beddoes to Watt, Nov. 15, 1794. Eye disease particularly seemed to have been one of the problems attracting new methods of treatment, however controversial. See Neve, "Orthodoxy and Fringe," 45.

[20] BCL. JWP 4/65/5. Thornton to Watt, July 13, 1795.

[21] On Fontana see Simon Schaffer, "Measuring virtue: eudiometry, enlightenment and pneumatic medicine," in Andrew Cunningham and Roger French, eds. *The medical enlightenment of the eighteenth century* (Cambridge and New York: Cambridge University Press, 1990), 281–318.

Fig. 3.4 Robert John Thornton, by F. Bartolozzi (By permission of the Wellcome Library, London)

Thornton's plan, however, focussed on patients. He was to create a room in Bennet Street, St. James's, in which to conduct trials of different atmospheres. Here he followed Watt's suggestion that the "conveying to the sores of ulcerated lungs different ingredients may be carried on with care in such an apartment." He was, however, alarmed about the risks, especially among those privately adventurous where dangers were real: Thus, he wrote, "I depend most on the hydrocarbonate for producing a permanent benefit. Your observations on the fumes of vinegar are I believe very just, a patient in the country poured warm vinegar on chalk, thinking to produce Fixed air, and inhaled the fumes. Soon after he felt most violent pains in his chest, inflammation augmented, and he became a speedy martyr to the trial." He revealed "from my own personal experience, there is always uneasiness, when I have inhaled the different factitious airs at forced inspirations."[22] But Watt also reminded Beddoes that there were many untried airs "likely to be useful in pulmonic complaints"—and listed some.[23]

Of course, the practical issue was how airs might be applied. Self-medication was a commonplace, even more so among the chemical physicians.[24] Beddoes' particular concern with Lord Daer related to Daer's unsuccessful attempt to attack consumption both by factitious airs and ultimately in the warmth of Lisbon. Daer was, like Beddoes, a pupil of Black in Edinburgh and an associate of Dr. Adair Crawford, of St. Thomas's Hospital and of the Woolwich Military Academy. A noted chemist, Crawford also suffered from pulmonary complaints. Thornton told Crawford that he attributed his own apparent recovery from Phthisis in part to his "breathing an atmosphere with a diminished proportion of oxygene", to exercise on horseback, and by "the vertiginous motion on shipboard, and now & then a nauseating emetic." Newly identified were the atmospheric airs. Yet he noted "freedom from complaints in my chest, while in the moist air of Cambridge, and that having not long back gone to Oxford to see some friends at the university...was almost immediately affected with pain under the sternum, and… other marks of pulmonary affection."[25] Airs, evidently, divided Oxbridge.

Practise could not depend on well-defined chemical principles, even amidst the controversial revolution wrought by disciples of Lavoisier. Indeed, Thornton's account of Crawford's trials with hydrocarbonate relied primarily on "the theory first suggested by [his] ingenious friend Dr. Beddoes." Crawford had come to believe the hydrocarbonate "produced a far greater effect on the animal aeconomy, than either inflammable or azotic gases..." He "inhaled after dark nearly a quart of hydrocarbonate diluted with 500 pints of common [and] declared, 'that it transfused over his body at the time a soothing tranquillity, such as opium is known sometimes to produce, but with slight vertigo.' The following day, he mentioned…the circumstance of the correction of the offensiveness of his expectoration, which he said 'he wholly attributed to the air'." Crawford's was a desperate case: "He inhaled

[22] BCL, JWP 4/23/14. Thornton to Watt, September 27, 1795?; JWP 4/23/8. Beddoes to Watt, on Mrs. Keir. October 30, 1794? On Girtanner, see David M. Knight, *Dictionary of Scientific Biography*, **5**, 411; Robert John Thornton, *Medical extracts. On the nature of health, and the laws of the nervous and fibrous systems* (London, 1798), **1**, 73.

[23] BCL, JWP, Copy Books, Watt to Beddoes, October 20, 1797.

[24] See, for example, BCL, JWP Copy Books, Watt to James Watt, jun., November 16, 1797.

[25] On Crawford see Trevor Levere and Gerard L'E Turner, *Discussing Chemistry and Steam. The Minutes of a Coffee House Philosophical Society* (Oxford: Oxford University Press, 2002); Claire L. Nutt, 'Crawford, Adair (1748–1795)', *Oxford Dictionary of National Biography*, Oxford University Press, 2004 [http://www.oxforddnb.com/view/article/6637, accessed 25 May 2013]; Larry Stewart, "Pneumatic Chemistry, Self-Experimentation and the Burden of Revolution, 1780–1805," in Erika Dyck and Larry Stewart, eds., *The Uses of Humans in Experiment. Perspectives from the 17th to the twentieth Century. Clio Medica* ***95*** (Leiden and Boston; Brill, Rodolpi, 2017), 139–169, esp. 136–137; Stewart, "A Jacobin cloven foot," in Trevor Levere, Larry Stewart and Hugh Torrens, *The Enlightenment of Thomas Beddoes. Science, medicine, and reform* (London and New York: Routledge, 2016), 143–144.

a second & a third time the modified air, when he went to Lymington 'with a fever' as he told [Thornton] 'on his spirits' and then in a few months this great good man paid the debt to Nature." Despite such prominent failures, Thornton's own trials with factitious airs included the important optical instrument maker, George Adams, and the use of super-oxygenated air on the elevated, like the Duke of York. To report on these could hardly escape the desired attention. Thornton, nevertheless, ultimately confessed he was unable to give a "full trial of the aerial practice... which the *new chemistry has introduced to minds nurtured in Philosophy*... "[26] He had turned his attention to botany and the Linnean system. Innovation in medical practice had, however, followed discovery in the laboratory. It was the purpose of publicity to expand aerial trials.

Despite growing criticism, it was partly through Thornton that Beddoes's reputation continued to be tied to pneumatic innovation. In a 5 volume *Philosophy of Medicine* (1799–1800), Thornton elevated Beddoes to the company of Darwin and Ewart in England, and of Lavoisier in France. Had it not been for Beddoes, Thornton argued, "aerial remedies had probably met with the same fate as transfusion" in the seventeenth century.[27] Of course, Beddoes and Watt were at the centre of a network of correspondence they openly published. It is interesting, for example, that the notion of acid remedies for venereal disease in the place of mercury, notably from Helenus Scott in India and William Blair of London's Lock Hospital, attracted much attention because of Beddoes' promotions. At Woolwich, where were found both Crawford and William Cruickshank, surgeon and chemical lecturer, attention was to be paid to hydrocarbonates and to the treatment of lues venera. Here, to compare results to the older mercury treatments, detailed statistical tables were kept. These involved a wide range treatments including nitrous acid, oxygenated muriat of potash [chlorate of potash] and of manganese, lemon juice, and nitrous acid in combination with oxygenated muriat of potash and manganese.[28] In the laboratories of pneumatic chemists, new acids were significant by-products inducing more tests of their medical effects, especially in sites like the Royal Hospital in Plymouth. Thus, the Plymouth surgeon Francis Geach wrote to Beddoes in 1797 on the success of nitrous acid in secondary venereal affections. This, of course, Beddoes published

---

[26] BCL, JWP 4/23/15 (3). Thornton to Beddoes, August 20, 1795 encl. in Beddoes to Watt, August 27, 1795. My italics.

[27] [Robert Thornton], *The Philosophy of Medicine: Or, Medical Extracts on the Nature of Health and Disease, Including the Laws of the Animal Oeconomy, and the Doctrines of Pneumatic Medicine.* Vol. I (London: Printed by C. Whittingham, Dean-street, Fetter-Lane, 1799), 427.

[28] Harrison, "Medical experimentation in British India," 34–35. On Cruickshank see William Blair, *Essays on the Venereal Disease* (London: J. Johnson, 1798), 35, 50ff; [Robert Thornton], *The Philosophy of Medicine*, Lance Day and Ian McNeil, eds., *Biographical Dictionary of the History of Technology* (London: Routledge, 1996), 181–2; *A Journal of Natural Philosophy, Chemistry & the Arts*, ed. William Nicholson, V (September, 1801), 201–211; A. Coutts, "William Cruickshank of Woolwich", *Annals of Science* **15** (1959), 121–133; K. D. Watson, 'Cruickshank, William (*d.* 1810/11)', *Oxford Dictionary of National Biography*, Oxford University Press, 2004; online edn, Jan 2008 [http://www.oxforddnb.com/view/article/57592, accessed 25 May 2013]; cf. Thomas Garnett, *Popular Lecture on Zoonomia, or the Laws of Animal Life* (London, 1804); William Henry, *An Epitome of Chemistry* (London, 1801).

and was also reproduced shortly thereafter by Thornton in his *Philosophy of Medicine*.[29] Beddoes, typically, also produced a circular letter requesting accounts of cases from any medic who may have knowledge of such chemical results.[30] Beddoes found in his net many a promising case.

If Thornton had reason to fear the label of innovation, Beddoes embraced it. Early on, before Beddoes and Priestley challenged the graft of old corruption, hope in many forms emerged from flasks and furnaces. Birmingham manufactured promises and glazes, alkalis and soap bubbled into fortunes. Apart such from industrial uses, whatever Beddoes' political aims they were surely not much different from Priestley or from James Keir. Yet, in promoting medical progress, Beddoes admitted it was possible his democratic sympathies had at least put one more obstacle in the way of attracting the support of the cautious. He told Watt that he knew well that his "politics have been very injurious to the airs" but "it requires great self-denial to look on patiently & silently when such great interests are at stake." Medicine mixed with his social and political sensibility. Beddoes then was the kind of innovator who attracted hostility from power, even when his intention was "to repress vengeance & calm violence."[31] But, how indeed, was it possible to make a mission out of their airs so much at variance with the views of the medical establishment and to the great alarm of Edmund Burke? The broader benefit, Beddoes proposed, was through scientific lectures, in geology and chemistry, possibly even in botany as Thornton also attempted, "attended by persons of opposite parties."[32] Despite hostility amidst fears of revolution, a picture emerges of many willing to put their science in the service of transforming society.

3.2 The Galvanic Turn

Beddoes sought every advantage the new philosophy might bring. Past practice was clearly no surety either for the ill or the powerful. As with the case of Watt or Priestley, a claim to secular providence might convey ideals of improvement. In Beddoes' early enthusiasm for the airs and Watt's breathing apparatus, family members were among the first subjects. He tried hydrocarbonate on Mrs. Keir, although it made her giddy. In her case, once described by Beddoes as desperate, he maintained the gas could be used safely as he "never had a disagreeable much less a

[29] Thomas Beddoes, M.D.[published by], *Reports principally concerning the Effects of the Nitrous Acid in the Venereal Disease, By the Surgeons of the Royal Hospital at Plymouth, and By Other Practitioners* (Bristol, Printed by N. Biggs, For J. Johnson, St. Paul's Church-yard, London, 1797).

[30] An original of the printed circular is in Cornwall Record Office, Davies-Gilbert MSS. 42/3, September 5, 1797.

[31] BCL, JWP 4/23/15. Beddoes to Watt, December 26, 1795. Watt thought Beddoes' "incorrigble." See JWP 4/12/18. Watt to Joseph Black, October 9, 1796.

[32] CRO, DG 42/2. Beddoes to Giddy, April 14, 1798. On Beddoes's science and his politics see Levere, Stewart and Torrens, *The Enlightenment of Thomas Beddoes* (2017).

dangerous occurrence…"[33] Likewise, despite some reservations, he reported Anna Beddoes had tried inhalations of nitrous oxide. The effect was such that she "could walk much better up Clifton Hill [and] frequently seemed to be ascending like a balloon."[34] But we know a convergence of remedies had long attracted Beddoes. As early as 1792 Beddoes, even before his alliance with Watt, had become an enthusiast of the news from Italy regarding Volta's discoveries.[35] Of course, medical electricity had by then a devoted following.[36] Indeed, he reported on experiments undertaken in James Keir's home in West Bromwich on the taste brought on by the bi-metallic effect of lead and silver placed on the tongue. Here he also convinced Miss Keir and 'the ladies', probably Anna Beddoes as well, to attest to the taste and smell. Even for incidental experimenters, Beddoes saw a broader consequence. He was immediately convinced that "the fact of medicine will be changed; & every body will study it as the most curious & interesting of all the sciences." But more significantly, he enthused, "There will soon be an end of physicians & apothecaries; or at least a gradual elimination of a body of men, who do & have done entirely more mischief than good, will begin to take place."[37] This fit nicely with his general assault on the entire harvest of doctors and quacks.[38] Indeed, the obvious failures and abuses of medicine is precisely what made any new measure appear seductive. Beddoes held hard to hope in airs and electric shocks.

By the end of the century, there were many prepared to enter the fray. One such was Tiberius Cavallo whose *Essay on the medicinal properties of factitious airs* appeared in 1798. This elicited a long and friendly letter from Watt in which Watt provided a detailed list of the properties of factitious airs and the effects which he had experienced in the laboratory. He thanked Cavallo for the compliments bestowed upon the apparatus, "the Child of my old age." Even though there were many

---

[33] BCL, James Watt Papers, 4/23/5. Beddoes to Watt, January 1, 1795; 4/23/8. Beddoes to Watt, October 30, 1795.

[34] Quoted in Golinski, *Science as Public Culture*, 168. According to Watt, Mrs. Beddoes was thought to have "an incipient consumption, of which she was cured by Digitalis." BCL, JWP Copy Books, Watt to Joseph Black, December 8, 1799, reprinted in Anderson and Jean Jones, eds., *The Correspondence of Joseph Black*, **2**, 1386.

[35] See Guiliano Pancaldi, "On Hybrid Objects and their Trajectories: Beddoes, Davy and the Battery," *Notes & Records of the Royal Society*, **63** (September, 2009), 247–262; also Jan Golinski, "From Calcutta to London: James Dinwiddie's Galvanic Circuits," in B. Lightman, G. McOuat, and L. Stewart, eds., *Circulating Knowledge, East and West* (Leyden: Brill, 2013), 74–94.

[36] See Paola Bertucci, "Therapeutic Attractions: Early Applications of Electricity to the Art of Healing," in Henry Whittaker, C.U.M. Smith and Stanley Finger, eds., *Mind, Brain and Medicine. Essays in Eighteenth-Century Neuroscience* (New York: Springer, 2007), 271–283.

[37] CRO, DG 41/19. Beddoes to Giddy, September 12, 1792.

[38] See Roy Porter, "Taking histories, medical lives: Thomas Beddoes and biography," in Michael Shortland and Richard Yeo, eds., *Telling Lives in Science. Essays on Scientific Biography* (Cambridge and New York: Cambridge University Press, 1996), 215–242, esp. 227–229; Roy Porter, *Quacks. Fakers & Charlatans in English Medicine* (Stroud and Charleston, SC.: Tempus Publishing, 2000), 168–169, 176–177.

differences in experience of the airs, Watt acknowledged, "the subjects is difficult and requires more experience, every one shd however contribute his mite."[39]

Tiberius Cavallo was one of the more shrewd promoters of aerial innovation following Priestley. As Simon Schaffer has shown, as early as 1781, Cavallo was exploring the benefits of eudiometry partly through his connections with the Tuscan toxicologist Felice Fontana who had appeared in London in 1778. Even tests of the salubrity of airs, however, were not uniformly regarded as useful either in manufacturing, agriculture, or indeed in medicine. Cautionary tales abounded even for Cavallo who was seeking to gain a foothold in the market for eudiometers.[40] By the end of the century he had published an *Account of the Medicinal Properties of Factitious Airs* in which he took note of the various gases that showed promise after Priestley's claim for the benefits of experimentally produced oxygen. In his later contemplations Cavallo, in 1798, repeated Beddoes' own reservations about the harmful properties of breathing raw oxygen commonly prepared from the heating of manganese.[41] Cavallo's work was a revealing reflection on the much-contested practice of pneumatic medicine. But like Beddoes, Watt, and Thornton, Cavallo also provided a list of cases reported by several physicians, surgeons, and chemists. These especially included those from his friend the Windsor physician James Lind, as well as Carmichael and Barr in Birmingham, along with numerous claims about the effects of airs.[42]

Like others, Cavallo various schemes in the scientific marketplace also turned to the uses of electrical stimulation. This was not a new notion at all, but became increasingly significant with reports of Alessandro Volta's experiments and Luigi Galvani's claim that "muscular contraction is really an electrical phenomenon."[43] For Cavallo this was another opportunity, especially during the development of various portable electrical machines. All the while, Cavallo was developing his own 'travelling philosophical box' which was presumably intended for the use of gentleman philosophers in the field. He had his eye on the instrument trade for those interested in electricity and magnetism. In his extensive correspondence with Lind, Cavallo focussed primarily on reports of electrical discoveries in Italy, discussions at the Royal Society, and the design of apparatus.[44] He knew of many medical trials, read many accounts, but also made his own attempts at using oxygen in 1797 on Mrs. Partington, quite likely the wife of the medical electrician Miles Partington.

---

[39] BCL, JWP Copy Books, Watt to Cavallo, February 9, 1798.

[40] Schaffer, "Measuring virtue," 309, 313.

[41] Tiberius Cavallo, *An Essay on the Medicinal Properties of Factitious Airs. With an Appendix on the Nature of Blood* (London: Printed for the Author, And sold by C. Dilly, in the Poultry; P. Emsley and D. Bremner, in the Strand, 1798), 33–34.

[42] Cavallo, *An Essay on the Medicinal Properties of Factitious Airs*, 149ff.

[43] BL., Add Mss. 22,897, 'Letters of Tiberius Cavallo', Cavallo to Lind, June 10, 1792.

[44] On which, see Tiberius Cavallo, *The Elements of Natural or Experimental Philosophy*, vol. II, ed. F.X. Brosius (Philadelphia: Dobson, 1813), 190–199. https://babel.hathitrust.org; and Giuliano Pancaldi, *Volta. Science and Culture in the Age of Enlightenment* (Princeton and Oxford: Princeton University Press, 2003).

The link between Cavallo and Partington is an interesting one. We know that Partington was sceptical of pneumatic claims, but when oxygen was tried on Mrs. Partington by means of one of Watt's portable devices she showed much improvement. While primarily an electrical practitioner Partington proposed that, with Watt's advice, he would engage in pneumatic therapies "towards diffusing a scientifical [sic] knowledge of it among all Medical Gentlemen who are inclined to purse the same plan of Operation with myself."[45] This has strong echoes of the Priestley-Beddoes doctrine of democratic diffusion.

At this same time, Cavallo was frustrated in his own attempt to secure parts for the pneumatic apparatus. He complained about Chippendale's lack of co-operation, finding him "as Kind as the Keeper of Hell. He would sell nothing but the whole apparatus."[46] Cavallo was then finishing his *Essay on the medicinal properties of aerial fluids*, and had recommended Watt's device for those interested in further trials. The continually well-informed Cavallo still reported to Lind on the use of nitrous airs and dilute nitrous acid. Much of what he ultimately published on medical cases he seems to have obtained from Lind.[47] But Lind was nervous about the connections to innovation of a different kind. While Lind was very curious, he had strong reservations. He had nevertheless already told James Watt in 1795 that the links between French politics and British chemists were to him unnerving. As a Windsor physician, it is not terribly surprising that he would declare "a detestation of having any connexion with such a set of miscreants notwithstanding I am induced to believe from several reasons that in many Diseases, the practice will be of real use."[48] While Cavallo was pursuing some trials of airs, he remained primarily concerned with developments in electrical apparatus especially of the redesigns of Galvanic apparatus.[49] Cavallo clearly grasped that, just like the marketing of pneumatic breathing devices, medical electricity might be made more available in the doctor's surgery. Here too, wherever knowledge was spread, there were many complaints of quackery.[50]

It is clear that many promoters of pneumatic chemistry often became enamoured with Galvanic practise.[51] Perhaps this simply followed from a common laboratory experience. But in therapy against undefeated disease there was no apparent boundary between the chemical or electrical practices. To take but one little-known

[45] BCL, JWP, W/9/9. Partington to Watt, January 24 [likely 1798].

[46] BCL, JWP W/9/11. Cavallo to Lind, January 15, 1798.

[47] BL., Add Mss. 22,897, 'Letters of Tiberius Cavallo', Cavallo to Lind, July 2, 1796; February 22, 1797; March? 1797; June 17, 1797.

[48] BCL, JWP 4/65/21. Lind to Watt, February 20, 1795.

[49] BL., Add Mss. 22,897, 'Letters of Tiberius Cavallo', Cavallo to Lind, June 13, 1800.

[50] See Paul Elliott, "'More Subtle than the Electrical Aura': Georgian Medical Electricity, the Spirit of Animation and the Development of Erasmus Darwin's Psychophysiology," *Medical History*, **52** (2008), 195–220, esp. 213–214.

[51] Cf. [anon], "Observations and Experiments on the Galvanic Power," in William Nicholson, ed., *A Journal of Natural Philosophy, Chemistry and The Arts*. V (May, 1801), 41–45, esp. 44; "Electric Shock by means of Galvanism," 80.

example of those who followed the pneumatic trials, Michael La Beaume became a leading Galvanic inventor and practitioner in the early nineteenth century. Self-styled 'medical galvanist', he set out to attack many of the very ailments which Beddoes had attempted to address.[52] Like Cavallo, La Beaume sought a niche in the marketplace—this more evidently medical. That his promotions explicitly owed a great deal to the pneumatic physicians and to gas chemistry is evident from his use of galvanism alongside "the nitro-muriatic-acid bath, or sponging and chlorine.[53] It is of no small interest that La Beaume, in a letter to Giovanni Aldini, Galvani's famous nephew, wrote of the virtues of medical electricity in diseases which Beddoes and his correspondents repeatedly encountered ultimately without success.[54] Using Aldini's pile, in his residence in Russell Square, La Beaume tried shocks on himself in order to get an appreciation of the stimulus that would be required to affect any sensation. He certainly thought a "very sensible or beneficial excitement" would arise in the galvanized patient. But too many competitors gave galvanism a dubious character. Just as had been the case with pneumatic practitioners, La Beaume complained of "ignorant and unprincipled adventurers, or automaton operators, who, without skill or conscience, may hazard its application in improper cases, but put the credulous patient into torture and peril."[55] Hence, he claimed to have tried various remedies since about 1800 in "secondary diseases only, which indeed I have repeatedly done myself with advantage, as in *Gutta Serena, Deafness, &c.* But in some of these kinds of cases, it must be acknowledged that the hopes of the patient as well as the practitioner have been disappointed, not only from the efficacy, but from the unskilful application of the remedy."[56] This sounds all too familiar.

By the 1820s La Beaume confronted his own doubts. He had tried the factitious airs and, like others, found in them no panacea:

> I have indeed both heard and read of many cures which have been ascribed to the Medical Gasses, but on making the *strictest* enquiry, I am now warranted in asserting, that the most *efficient medicines* were administered to the patients in conjunction with the Vital Air—and

[52] On La Beaume see Alan Rauch, *Useful Knowledge. The Victorians, Morality, and the March of Intellect* (Durham and London: Duke University Press, 2001), 107–108. He was, from 1843, apparently running the Galvanic and Electric Institution in Argyle Street, Regent Street. It appears La Beaume was declared bankrupt by 1847. See *The Jurist*, **11** (2), (London, 1848), 217.

[53] See La Beaume, *On Galvanism with Observations on Its Chymical Properties and Medical Efficacy* (London, 1826).

[54] On Aldini see Rob Iliffe, "Galvanic Humans," in Erika Dyck and Larry Stewart, eds., *The Uses of Humans in Experiment. Perspectives from the 17th to the twentieth Century* (Leiden and Boston: Brill/Rodopi, 2016), 52–79.

[55] M. La Beaume, 'Communication' in D. Uwins, S. Palmer and Mr. Gray, eds., *The London Medical Repository, Monthly Journal, and Review*, **12** (1819), 256–260.

[56] M. LA BEAUME, Medical and Surgical Electrician, F.L.S. &c, *Observations on the Properties of the Air-Pump Vapour-Bath, in the Cure of Gout, Rheumatism, Palsy, &c with Occasional Remarks on the Efficacy of Galvanism, in Disorders of the Stomach, Liver, & Bowels, with some New and Remarkable Cases.* Second edition, Greatly Enlarged (London: F. Warr, 1819), 265–266.

thus, the *merit* of cures performed by the *skill* of the Physician with the aid of *ordinary* remedies, has been *solely attributed* to a mere doubtful auxiliary.

Under these impressions, it is impossible, that *I could honestly recommend* the Factitious Airs as an *efficient* remedy. If, however, any of my patients have still a predilection in their favour, I am willing to make further trials of them, as an auxiliary to aid the means I employ,—they may procure these Gasses in their purest state at a reasonable price, from the *Medical Hall*, Piccadilly:—those who cannot afford to purchase them, may receive instructions from me for preparing them by a simple process, with very little trouble, and at a trifling expense.[57]

La Beaume's views of opponents of innovation strikingly echoed those of Beddoes: "The *venal* age in which we live, and the *mercenary trading spirit* which pervades the whole circle of society, and which is but too *glaringly* conspicuous even in the more *elevated* rank of professional pursuits—have degraded the practice of medicine into a *'a Guinea Trade,'* and rendered it subservient to the interests of a race of *petty venders* of *drugs* and *chymicals*."[58] Of course La Beaume, like Beddoes, may well have had ample reasons to expect interference from proper physicians and their allies.[59] Then La Beaume took an instrumental turn. He tried the route of designing improved batteries of 2–300 plates by which he developed his own reputation.[60] By the 1820s, he was sufficiently successful he had moved to the fashionable area off Regent Street, and clearly had found connections that would serve him. Possibly as early as July 1831, and certainly by 1839, he was among the lists of medics in the Royal Household.[61] But his career was checkered and competition fierce. There are several references in notices of his bankruptcy in the early nineteenth century before his death in 1856.[62] Such was the price of his enthusiasms for a practice that was often deemed dubious, despite his apparent patrons and royal connections.

57 M. La Beaume, *Remarks on the History and Philosophy, but Particularly on the Medical Efficacy of Electricity, in the Cure of Nervous and Chronic Disorders: And in Various Local Affections, As Blindness, Deafness, &c. Illustrated with many New and Striking Cases: Together with Observations on Galvanism, as an Efficient Substitute for Mercurial Remedies, In Bilious and Stomach Complaints. Dedicated by Permission, To His Royal Highness Prince Leopold of Saxe Coburg, &c.* Second Edition, Greatly Enlarged. (London: F. Warr, Red Lion 1820), 365–367. The reference is to Richard Reece's Medical Hall, 168–171 Piccadilly, near Bond Street. Cf. Marieke M.A. Hendrikson, "Consumer Culture, Self-Prescription, and Status: Nineteenth-Century Medicine Chests in the Royal Navy," *Journal of Victorian Culture*, **20** (2015), 147–167.

58 La Beaume, *Remarks on the History and Philosophy, but Particularly on the Medical Efficacy of Electricity*, 341–342.

59 On the orchestrated abuse directed at Beddoes see Jan Golinski, *The Experimental Self. Humphry Davy and the Making of a Man of Science* (Chicago and London: University of Chicago Press, 2016), 39–41.

60 'M. La Beaume's New Galvanic Batteries', in *Mechanics Magazine, Museum, Register, Journal, and Gazetter*, **175** (Saturday, December 30, 1826), 549–551.

61 *Medical Times and Gazette* **12** (1856); *Gentleman's Magazine* (1831); A.M. Cooke, "Queen Victoria's Medical Household," *Medical History* **26** (1982), 307–320, esp. 319.

62 *London Gazette* (1847), Court of Bankruptcy, where he petitioned for protection from the bankruptcy process and was described as a medical electrician residing in Penzance and Cornwall.

3.3 Conclusion

It is a mistake to label such innovators simply 'quacks', as Beddoes and Watt and Thornton had often feared would happen. The difficulty of promoting pneumatic or galvanic remedies inevitably attracted the label of innovator, or worse. Success as a medical practitioner meant acquiring, as Beddoes had fully understood, both testimony and patrons. In Beddoes' case, despite support from midland's industrialists and liberal aristocrats, he also attracted the hostility of the well-connected Joseph Banks, tied to the Board of Trade and President of the Royal Society. La Beaume tried another route—instrumental invention and even sought to establish connections through the Linnean Society.

In the midst of their pneumatic campaigns, aerial practitioners had known full well what was coming. Immediately upon publishing his *Medical Observations* (1793), Beddoes remarked in a work he addressed to Darwin: "... although it is evident that none but beneficial consequences can result to the public from the prosecution of my design, yet it requires very little knowledge of the world to perceive the danger to which I am exposing my reputation. It is impossible to engage in a new and arduous undertaking without incurring the ridicule and obloquy: Of course I must expect to be decried at home as a silly projector, and by others as a rapacious empiric. Secrecy, as you well know, and promises unrestrained by sense or shame, constitute the essential character of empiricism: Now I trust that the [*Medical Observations*]... will bear ample testimony that on medical subjects I have no disposition to secrecy."[63] If openness was an answer, it was hardly a sufficient one. For his part, Watt remained defiant. In 1798, he told Professor John Robison in Edinburgh that he was often styled a quack "by which appellation I am honoured by the enemies of innovation."[64] Beddoes had enough negative press not aided, of course, by his political interventions. But Beddoes, and those who inherited his hopes of chemical and electrical achievement, nevertheless had quite a following that should not now relegate them to the margins of medical utility.

---

[63] Thomas Beddoes, M.D., *A Letter to Erasmus Darwin, M.D. ON A New Method of Treating Pulmonary Consumption, and Some other Diseases hitherto found Incurable*. (Bristol: Printed by Bulgin and Rosser, Broad-Street; Sold by J. Murray, No. 32, Fleet-street, and J. Johnson, No. 72, St. Paul's Church-yard, London; also by Bulgin and Sheppard, J. Norton, J. Cottle, W. Browne and T. Mills, Booksellers, Bristol). [1793], 3–5; See also Thomas Beddoes, M.D., *Observations on the Nature and Cure of Calculus, Sea Scurvy, Consumption, Catarrh, and Fever: Together with Conjectures Upon Several Other Subjects of Physiology and Pathology* (London, Printed for J. Murray, no. 32 Fleet Street, 1793).

[64] BCL, JWP Copy books, Watt to Robison, January 30, 1798.

Chapter 4
"Men of Letters" and "Men of Press Copies": The Cultures of James Watt's Copying Machine

David Philip Miller

For men of letters and men of business in the late eighteenth century the proliferation of correspondence and other manuscript documents presented an escalating problem of record. The information in the documents that they generated had to be retained for reference and other purposes. The traditional way of making copies of manuscripts was to transcribe them by hand. Clerks were primarily copyists. As business and correspondence networks expanded, the volume of material requiring copying, as well as issues of confidentiality, created problems in scribal culture. Interest in other ways of generating copies, which had been a recurring theme from the late seventeenth century, became an obsession in the late eighteenth and nineteenth centuries. James Watt's copying machine, patented in 1780, was one of the key inventions addressing these issues. I consider here aspects of its origins, marketing, development and diffusion in business, and, particularly, in the British Civil Service with emphasis upon understanding the cultural springs, and the ongoing cultural relations and significance, of this information technology. Beyond its functional value, the copying machine was in its early days a collectable device exhibiting and symbolizing ingenuity. Such appeal probably explains some of the eagerness of men of learning to acquire it. In the nineteenth century, as its use became commonplace in business offices and government bureaucracies, the copying machine became a more firmly entrenched symbol of merely mechanical work to be contrasted with the creative work of the man of letters.

D.P. Miller (✉)
University of New South Wales, Sydney, NSW, Australia
e-mail: dp.miller@unsw.edu.au

J. Buchwald, L. Stewart (eds.), *The Romance of Science: Essays in Honour of Trevor H. Levere*, Archimedes 52, DOI 10.1007/978-3-319-58436-2_4

4.1 Introduction

The copying machine invented by James Watt (1736-1819) is probably among the better-known "obscure" inventions.[1] Early biographers discussed it as a testament to Watt's versatility and fecundity as an inventor, and later chroniclers of his life have done much the same.[2] Whilst the genius of Watt, and of his partners in the copier project, Matthew Boulton and James Keir, must be given its due, as an episode in the histories of information the invention must also be seen as *one* of the responses in the eighteenth and early nineteenth centuries to a growing demand for mechanised reproduction. The copying machine was a response to the business, and to a lesser extent personal, need for a more efficient way of making copies of correspondence and other manuscript materials. As such it has a place in the history of scribal and office practice extending into the early twentieth century. Its success, as I will argue, was as part of a technological system of writing practice.[3]

However, the copying machine was also more than this. Boulton and Watt engaged at various times in efforts to mechanically reproduce paintings and sculpture as well as manuscript writings and engineering drawings. This suggests that the copying machine might be seen also as a cultural product of a larger Enlightenment project that sought to mechanise reproduction. These were early moves in the process by which art was industrialised and transformed in the way famously, if obtusely, discussed by Walter Benjamin.[4] The copying machine was one of a number

[1] René Schils, *How James Watt Invented the Copier: Forgotten Inventions of Our Great Scientists* (New York: Springer, 2012), 37–42 begins: "James Watt is best known as the inventor of the steam engine and driver of the Industrial Revolution. This reputation is not entirely deserved, as his invention was actually an improvement on a steam engine invented half a century earlier. Watt was, however, the real inventor of the copying machine." (37)

[2] The story of Watt and the copying machine can be traced through: François Arago, *Historical Eloge of James Watt ....Translated from the French with Additional Notes and an Appendix by James Patrick Muirhead.* (London: John Murray, 1839); James Patrick Muirhead, *The Origin and Progress of the Mechanical Inventions of James Watt.* 3 vols. (London: John Murray, 1854); Samuel Smiles, *Lives of Boulton and Watt. Principally from the Original Soho MSS.* (London: John Murray, 1865), 265–68; H.W. Dickinson, *The Garret Workshop of James Watt* (London: HMSO, 1929); Richard L. Hills, "James Watt and his Copying Machine," *Studies in British Paper History, Vol 1, The Oxford Papers*, ed. Peter Bowers (Kidlington: British Association of Paper Historians, 1996), 81–88; Richard L. Hills, *James Watt. Volume 2: The Years of Toil, 1775–1785* (Ashbourne: Landmark Publishing, 2005).

[3] The place of Watt's copying machine in the *longue durée* of copying practices is indicated by Ian Batterham, *The Office Copying Revolution* (Canberra: National Archives of Australia, 2008) and also by Barbara Rhodes and William Streeter, *Before Photocopying: The Art and History of Mechanical Copying, 1780–1938* (Delaware: Oak Knell Press, 1999).

[4] Walter Benjamin, "L'oeuvre d'art à l'époque de sa reproduction mécanisée," *Zeitschrift für Sozialforschung* **5** (1936), 40–68. Benjamin's is a much cited and much critiqued essay. If Clive James's critique of Benjamin's claims concerning the impact of mechanization on the "aura" of artworks is sound (and I suspect that it is) then the most important consequences of changes in the technology of reproduction lie, perhaps not surprisingly, in the area of reproduction and availability rather than in meaning and significance. See Clive James, *Cultural Amnesia: Notes in the Margin of My Time* (London: Picador, 2007), 47–56.

of projects through which Boulton & Watt, and others, sought to exhibit and market ingenuity as a commercial product, as Maxine Berg has argued.[5] Beyond its functional value, then, the copying machine was in its early days a collectable device exhibiting and symbolizing ingenuity. Such appeal probably explains some of the eagerness of men of learning to acquire it. In the nineteenth century, as its use became commonplace in business offices and government bureaucracies, the copying machine became more firmly a symbol of merely mechanical work to be contrasted with the creative work of the man of letters and to substitute systematically for the class of hand copyists.

There was also a moral dimension to the action of copying and hence to the copying machine. In the early days, the machines were sometimes given as gifts. If the recipient was a creative writer then the gift was often interpreted as promoting a moral good by saving the writer's precious time. Such was the thought, for example, when Thomas Wedgwood presented a Boulton & Watt copying machine to William Godwin in 1795.[6] In a world where copying machines were available, devoting time to hand copying could be a moral lapse, even exhibiting sloth or desultoriness. This was a message of Jefferys Taylor's instructive novel for youth *Harry's Holiday* in which Harry Pemberton's week-long escape from the regulated life imposed by his parents saw him engage in a hive of miscellaneous and misdirected activity including a rather scrappy hand copying of Joseph Priestley's *New Chart of History*. The enumeration of Harry's errors in doing this included his misconstrual of "the value of labor in an economy of mechanical reproduction".[7] As we will see in considering the copying machine in the workings of the British Civil Service, hand copying and machine copying, and the question of who should be doing what, were moral and class issues in Victorian Britain. They divided "Men of Letters" and what we might call "Men of Press Copies".

In what follows, I first explore the origins and early development of Watt's copying machine as both spurred by functional demand and as Enlightenment project. I then consider the broader development of the technology and culture of the copying machine in the Victorian period, and especially the debates about its use within the British Civil Service where the machine helped to realize a distinction between creative and mechanical work and the social hierarchies associated with it.

[5] See Maxine Berg, "From Imitation to Invention: Creating Commodities in Eighteenth Century Britain," *The Economic History Review* New Series **55** (2002), 1–30 and Barbara Fogarty, "The Mechanical Paintings of Matthew Boulton and Francis Eginton," in *Matthew Boulton: Enterprising Industrialist of the Enlightenment*, ed. Kenneth Quickenden, Sally Baggott and Malcolm Dick (Farnham: Ashgate Publishing, 2013), 111–126.

[6] When Godwin used the occasion to question the giving of gifts and presents, Wedgwood argued that bestowing the gift on Godwin would advance knowledge and so public welfare. (See Pamela Clemit, "William Godwin and James Watt's Copying Machine: Wet-Transfer Copies in the Abinger Papers," *Bodleian Library Record* **18** (2005), 532–60.

[7] Taylor's novel is interpreted thus in Daniel Rosenberg, "The Young and the Restless," *Cabinet*, Issue 29 (2008). I am grateful to Professor Rosenberg for drawing the novel to my attention.

4.2 Origins of the Watt Copying Machine

The copying machine that Watt invented and developed between 1778 and 1780 was by no means the first effort to develop a machine to make copies of manuscript documents, though it was unusual in employing, as Watt described it, a "chemical" process.[8] While Watt's invention was about constituting chemically suitable inks and treated papers to allow successful wet press copies of a document to be made, prior schemes were mechanical in nature, involving as they did devices –"polygraphs"– that enabled the writer to produce a copy or copies at the same time as the original document.

Both Sir Christopher Wren and Sir William Petty were granted very early patents (in 1631 and 1648 respectively) for polygraph devices, but neither came into wide use. In the late eighteenth and early nineteenth century Erasmus Darwin, Marc Isambard Brunel, John Isaac Hawkins, and Ralph Wedgwood were among those of Watt's contemporaries who laid claim to polygraph machines. Although Thomas Jefferson was a user,[9] polygraphs had limited success among private individuals, government offices and businesses. A number of issues probably contributed to this lack of up-take. First, polygraphs were typically complex, large devices, clumsy to use in an office and virtually impossible to transport without disassembly and reassembly which involved tortuous fine-tuning. Second, and more interestingly, they united rather than divided labour in that they made the creator of a document also responsible for producing the copies of it.[10] The creative and the mundane aspects of producing a manuscript and copying it had been, for a long time, at least ideally divorced from each other. The letter copy press was to offer this possibility in a way that polygraphs did not.

Carbon copying was also developed, and Ralph Wedgwood was an important innovator here too. This technology did have the advantage of more readily producing multiple copies, but standard quills and metal nibs could not be employed easily and the need for special styluses was an obstacle for many. The technology found niche markets (notably for the production of receipts) but made less headway than press copying where documents were concerned, at least until the typewriter came along in the later nineteenth century. Even then, the press copying of typewritten documents was widespread practice for some time before typewritten carbon copies became standard practice. The press copy technology thus shared with these other less successful approaches impetus from a burgeoning demand for some method of producing copies other than by transcription. Some of its advantages derived from

---

[8] The copy machine thus provides further evidence for the claim that thinking about chemical processes was central to much of Watt's invention, including his improvement of the steam engine, in a way that was systematically obscured by nineteenth century historiography. See David Philip Miller, *James Watt, Chemist: Understanding the Origins of the Steam Age* (London: Pickering & Chatto, 2009).

[9] Silvio Bedini, *Thomas Jefferson and his Copying Machines* (Charlottesville: University of Virginia Press, 1984).

[10] See Rhodes and Streeter, *Before Photocopying* (ref. 3), 16–17.

a more ready assimilation to existing, and developing, clerical hierarchies and the ergonomics of office practice.

The proximate reasons why Watt decided to develop a manuscript copying technology were two: the promptings provided by Erasmus Darwin's development of his "polygrapher" or "bigrapher", and Watt's immediate need to relieve his own burden of work.

Darwin discussed his polygrapher in letters to his friend Charles Greville in late 1778 and early 1779. The first of these letters itself exists in two copies, it being unclear which is the original, and has been claimed as the first perfect mechanical copy.[11] It appears that Darwin was unwilling to patent the device and hoped that he might sell it to a London instrument maker who would exploit it. He also apparently lent the device to Watt in the hopes that Boulton & Watt might take it on as a commercial proposition. Darwin also reportedly discussed the mechanical copier at a Lunar Society meeting early in 1779. But rather than follow Darwin's lead, Watt decided to develop his own entirely different process.[12]

When he turned his mind to copying processes Watt had been travelling in Cornwall on steam engine business and was engaged in an extensive and elaborate correspondence in connection with it. The exigencies of keeping copies, particularly of his business letters, proved very difficult. He was wary about using clerical help because of the commercial sensitivity of so much of the correspondence, and the piracy and dirty-dealings that characterised the tough business environment in which he was working. Watt was also tired of having to repeatedly write out instructions regarding the steam engines that he was selling and erecting. The creation of multiple copies of engineering drawings was also a problem to be solved.[13]

The process that Watt developed involved pressing a dampened piece of thin, unsized paper against an original that had been written in a specially developed ink. The press copy reproduced the original in reverse facsimile and could be read through the other side of the paper. This basic concept of the copying process, and an early working version of it were established sometime in March or April 1779, since by May that year Watt was telling Erasmus Darwin and Gilbert Hamilton about the technique.[14] He was already using it to copy his own letters. However, for

[11] Erasmus Darwin to Charles F. Greville, 12 December 1778, 16 May 1779 and 7 June 1779, in *The Collected Letters of Erasmus Darwin* ed. Desmond King-Hele (Cambridge: Cambridge University Press, 2007), 157–158, 165–168.

[12] Arago, *Historical Eloge* (ref. 2), 93–4. Arago's story that Watt immediately saw a better approach to copying and invented the copying press overnight must be taken with a pinch of salt, given the puffery generally engaged in by Arago, with Watt's son James Watt Jr. at his shoulder, in discussing the exploits of the "Great Steamer". See David Philip Miller, *Discovering Water: James Watt, Henry Cavendish and the Nineteenth-Century Water Controversy* (Aldershot: Ashgate Publishing, 2004), 111–122.

[13] See J.H. Andrew, "The Copying of Engineering Drawings and Documents," *Transactions of the Newcomen Society* **53** (1981–82), 1–15 at 2–4.

[14] Gilbert Hamilton to Watt, 9 May 1779, quoted in Michael Cook, "Towards a History of Recording Technologies: The Damp-press Copying Process," *Journal of the Society of Archivists* **32** (2011), 35–49 at 37, and Watt to Erasmus Darwin, 12 May 1779 in Muirhead, *The Origin and* Progress

Watt a rudimentary system was one thing, a patentable process that he was satisfied with quite another. He experimented long and hard in 1779 to develop inks that would not "run" but would give good copies, to develop treated paper that would dampen well, and take good copies, and to design a roller press and a screw press that would produce a faithful copy without damaging or smudging the original. The patent that was finally enrolled on 31 May involved all these elements. However, development did not end with the patent; Watt was still experimenting with paper into late 1780 and with inks as late as 1784.[15]

Watt held a half share in the patent and Matthew Boulton and James Keir held one quarter each. Boulton financed the patent, Keir was to manage the business. To conduct that business these three formed a partnership under the name "James Watt & Co" to produce, market and sell the copying systems. The presses were produced at the Boulton & Watt Soho works in Birmingham, where a special production line was established and attempts were made to estimate production costs. These business methods were themselves remarkably innovative. Rather than sell the machines and related supplies directly, the partnership appointed the London stationer James Woodmason as their sales agent. Sourcing and preparing paper and ink were taken over by Woodmason quite early in the piece, and by 1781 their provision was further devolved: it was decided that Watt would simply make available his recipe for the ink and his instructions for preparing the paper and allow the users to supply those elements of the system.

The business plan, in place by November 1779, was to sell machines by subscription, aiming at 1000 subscriptions at five and a half guineas for which subscribers would receive "the secret", a press, and proper paper and materials. The original idea was that only when the subscription had been filled would manufacture begin and the machines be supplied. But this plan was overambitious and readily abandoned. By June 1780 only 400 subscribers had signed up. Presses were being made anyway and the first batch was sent to Woodmason in August 1780. Meanwhile Boulton had embarked on a marketing campaign that involved demonstrations at the Houses of Parliament, at the Royal Society, and in London coffee houses. Deluxe versions of the machines were sold to the Kings of England and Prussia, the Duke of Buccleuch, Lord Mansfield and other nobles. Courting the elite was Boulton's *modus operandi* in all his business ventures. In this case, as in others, Watt considered that this was misdirected effort and that merchants, whose money was just as good, should be targeted.

Sales remained steady through the 1780s and in 1794 the business was turned over to James Watt Jr. and Mathew Robinson Boulton, who continued to develop it, especially through a new, portable copy press. By this time, however, the patent had expired, and exploitation of the letter press system and the production of copying

(ref. 2), Vol. 2, 115. Muirhead dates the letter only "1779", the date is identified as 12 May by Hills, *James Watt, Volume 2* (ref. 2), 191.

[15] This and the following paragraphs rely on the excellent accounts of Watt's technique and of the business arrangements of James Watt & Co given in Hills, "James Watt and his Copying Machine" (ref. 2) and Hills, *James Watt, Volume 2* (ref. 2), 190–211.

machines for sale was open to others. It was through them that the copying machine became ubiquitous in business, mercantile and legal offices in the first half of the nineteenth century.

Apart from selling machines, Boulton and Watt also gave them to friends, and they spread through the Republic of Letters becoming an important instrument in the commerce of information. While the giving of machines as gifts was in part a marketing technique it was also recognition of their potential importance and value in the advancement of learning. Tellingly, on the occasion of his demonstration of the copying machine to a meeting of the Royal Society of London in May 1780, Boulton told Watt not only that it "afforded much satisfaction to a crowded audience" but also that he "did not show the list of subscribers and the proposals, nor dishonour philosophy by trade in that room…".[16] In learned circles the copying machine was treated as an example of philosophical genius applied to reproduction of information, and an instrument for the propagation and support of enlightenment.

4.3 The Copying Machine in the Broader Culture of Mechanized Reproduction

Although the primary purposes of Watt's copying machine venture were to develop a commercial business around it, and to bring greater efficiency and facility to Watt's and the firm's own administrative and engineering work by copying letters and engineering drawings, it must be understood also as part of a wider culture of mechanized reproduction which manifested itself in a variety of ways in the work and business of Boulton and Watt. The connections between these projects are also evident in some commonalities and similarities of technical means and devices to accomplish them.

At the very time that Watt was conjuring the letter press process, his partner Boulton was involved, in partnership with Francis Eginton, in an attempt to mechanize the reproduction of paintings, so as to produce multiple copies that had the look and feel of original oil paintings.[17] Eginton, who was an employee and partner of Boulton in various ventures, was the inventor of the process. He adapted existing processes for transfer from paper to solid surfaces and also the aquatint process, in order to produce mechanical paintings. The image was first etched onto a copper plate, then printed onto paper, transferred to canvas giving an impression that was then overpainted in oils. Clearly this was not a total mechanization of painting but rather a process approximating to "painting by numbers" that involved human skills. Young workers (referred to as "the boys") executed the first overpainting of the canvas. The images were then retouched and finished by professional artists. As

[16] Boulton to Watt, 14 May 1780, quoted in Smiles, *Lives of Boulton and Watt* (ref. 2), 267.

[17] Fogarty, "The Mechanical Paintings" (ref. 5).

Barbara Fogarty notes, the fact that Boulton on occasion sold what were entirely hand-painted pictures as mechanical ones suggests that what he was selling was the *idea* of such ingenuity in mechanical reproduction. On a visit to Soho in 1780, Princess Dashkova was apparently moved to buy several mechanical pictures and a Watt copying machine by much the same admiration for their ingenuity as examples of the fruits of technology and science.[18]

The mechanical painting venture was not, in the end, a success, and Boulton abandoned involvement in 1781. Nevertheless, the episode is important in illustrating both the Enlightenment fascination with mechanizing ingenuity, and the business concern to trade upon it. In addition it seems likely that the stage of the process involving transfer to canvas directly inspired Watt's approach to the copying machine via special inks to produce an offset copy, since they are very similar processes.

Thanks to the preservation of the garret workshop where Watt worked during his later years, with its prominently featured sculpture copying machines, we have long been aware of this preoccupation of his old age.[19] Perhaps understandably, Watt's concern with copying sculpture has often been considered a foible when compared with his heroic efforts in steam technology and their transformative influences. That this interest was not merely a foible becomes apparent, however, when we view it in the broader context of the technology and business of "creative copying".[20] At the time of his death, Watt was drawing up patent specifications for his sculpture copying machines and had in view their commercial exploitation to feed the burgeoning bourgeois neoclassical taste for busts of the famous and the worthy, not least of himself! There were thus strong continuities here with many of Boulton's ventures, and indeed Wedgwood's, in the commercial reproduction of art and design on pottery and metal goods. Further reinforcing the sense that mechanical reproduction was a central theme of Watt's inventive life are the continuities between his sculpture copying work and his early involvement with the Delftfield pottery, which had

[18] Fogarty, "The Mechanical Paintings" (ref. 5), 123–4; Olga Baird, "Benjamin Franklin, Catherine Dashkova and James Watt's 'Art of Copying'," *Benjamin Franklin and Russia: The Philosophical Age. Almanac 31.* (St Petersburg, 2006), 121–129.

[19] The garret workshop was at Heathfield House and had been preserved untouched by Watt Jr. until his death in 1848. Later occupants of Heathfield House also preserved the workshop. It was opened to select visitors and also, notably, on the occasion of the centenary of Watt's death in 1919. Before Heathfield House was demolished the workshop was removed in 1924 to the Science Museum. See Smiles, *Lives of Boulton and* Watt (ref 2.), 493–5, 512–14; Dickinson, *The Garrett Workshop* (ref. 2); Christine MacLeod, *Heroes of Invention: Technology, Liberalism and British Identity 1750–1914.* (Cambridge: Cambridge University Press, 2007), 261, 388; Christine MacLeod and Jennifer Tann, "From Engineer to Scientist: Re-inventing Invention in the Watt and Faraday Centenaries, 1919–1931," *The British Journal for the History of Science* **40** (2007), 389–411 and most recently Ben Russell, *James Watt. Making the World Anew* (London: Reaktion Books, 2014), 7–8, 203–6, 224–33.

[20] This is an insight perhaps first had by Uglow, see Jenny Uglow, *The Lunar Men. The Friends who Made the Future* (London: Faber & Faber, 2002), 304–8.

initiated a lifelong interest in copying techniques and in materials such as artificial stone and marble.[21]

When the copying machine is seen in this context we can appreciate that as an object embodying ingenuity it was a desirable possession in the same sense, and for the same reasons, that an early mechanically reproduced painting or piece of sculpture was. In both cases, of course, the effect faded as the ingenuity of the mechanical reproduction of ingenuity lost its novelty.

4.4 Copying Machines and the Civil Service in the Nineteenth Century

Watt's copying machine, as modified and improved, and as integrated into a system of practice, became an actor in the development of bureaucracy and the mechanization of administration in the nineteenth century. Thus its story and its culture intersect with that of the "government machine".[22] Over the longer haul it was the next generation of technologies of document production, reproduction, storage and retrieval—notably the typewriter, the stencil and the vertical filing system—that most clearly brought revolutionary change to the production and handling of information in the late nineteenth and early twentieth centuries. That revolution has been seen as primarily rooted in the United States corporate sphere.[23] Nevertheless, the adoption of the copying machine in the British Civil Service is an interesting and instructive episode, taking us to the heart of the writing practices of the "graphosphere", in which, if our Foucauldian brethren are to be believed, the power and authority of the bureaucracy, and ultimately of the State, was grounded.[24]

By the 1850s copying machines were almost universally deployed in mercantile houses, law firms and large organizations such as shipping and railway companies.[25]

---

[21] See Jane Insley, "James Watt and the Reproduction of Sculpture," *Sculpture Journal* **22** (2013), 37–66.

[22] Agar insightfully roots twentieth-century developments in computing in processes of government and bureaucracy, beginning with Civil Service reforms in the nineteenth century. See Jon Agar, *The Government Machine: A Revolutionary History of the Computer* (Cambridge, MA: The MIT Press, 2003).

[23] See JoAnne Yates, *Control through Communication: The Rise of System in American Management* (Baltimore, MD: Johns Hopkins University Press, 1989).

[24] Studies of the role of writing practices in the constitution of power have been extensive in the last decade. For a sample, see Patrick Joyce, *The Rule of Freedom: Liberalism and the Modern City* (London: Verso, 2003), 98–143; Ilana Feldman, *Governing Gaza: Bureaucracy, Authority and the Work of Rule, 1917–1967* (Durham, NC: Duke University Press, 2008); Miles Ogborn, *Indian Ink: Script and Print in the Making of the East India Company*. (Chicago: The University of Chicago Press, 2007); Patrick Joyce, "Filing the Raj: Political Technologies of the Imperial British State," in *Material Powers: Cultural Studies, History and the Material Turn*, ed. Tony Bennett and Patrick Joyce (London: Routledge, 2013), 102–123. Ben Kafka, "Paperwork: The State of the Discipline," *Book History* **12** (2009), 340–53 is a very useful survey.

[25] Rhodes and Streeter, *Before Photocopying* (ref. 3), 8.

By contrast, even though it seems that the earliest press copy produced for government business dates to 1786[26] their use by government departments was patchy even in the mid-nineteenth century. While some departments, notably the Admiralty, made extensive use of copying machines, copying by hand and the employment of tribes of clerical copyists was deeply entrenched in the Civil Service, particularly so in Treasury.[27]

A fascinating insight into the debates within the Civil Service about the respective value of hand copying and press copying is provided by the proceedings of the Commission on Civil Services Expenditure in the early 1870s.[28] The Commission was charged with finding economies in the Civil Services establishments in the face of their rapid expansion and escalating cost. The issue of the substitution of copying machines for the labour of large numbers of low-level clerks was one of four main issues that the Commissioners focussed upon in pursuing their brief. It would be fair to say that the Commissioners' questioning betrayed an advocacy of the technology. As a result, virtually all the main witnesses before the Commission were questioned about the copying machine issue. Their responses were mixed.

Officers of the Treasury, which did not use copying machines, presented the strongest reasons for maintaining the *status quo*. Reginald Welby, then Principal Clerk for Financial Business in the Treasury,[29] made the Treasury case against copying machines clear. The main point was that press copies were not durable enough—with frequent reference to them over decades the thin copy paper on which press copies were taken was inadequate—and of poor quality since press copies were often blurred and spoiled. Welby's remarks to this effect were challenged by his questioners, who also sought to impress upon him the successful and almost universal use of press copying in the business world, the legal evidential value of press copies, their veracity as facsimiles of the original, and the possibilities of enormous savings of labour and time. In the face of this barrage, Welby's relatively junior status perhaps caused him to back off and defer to contrary opinions. Other Treasury

[26] Agar, *The Government Machine* (ref. 22), 60, quoting Yates.

[27] See Barbara L. Craig, "The Introduction of Copying Devices into the British Civil Service, 1877–1889," in *The Archival Imagination. Essays in Honour of Hugh A. Taylor*, ed. Barbara L. Craig (Ottawa: Association of Canadian Archivists, 1992), 105–33 and also Emmeline W. Cohen, *The Growth of the British Civil Service* (London: Frank Cass, 1965, originally published 1941), 125. For instructive comparison see the account of clerical culture within the Excise in the eighteenth century in John Brewer, *The Sinews of Power: War, Money and the English State, 1688–1783.* Reprint edition (Cambridge, MA: Harvard UP, 1990), 64–87.

[28] *Third Report from the Select Committee on Civil Services Expenditure; Together with the Proceedings of the Committee, Minutes of Evidence, Appendix and Index* (1873). Consulted via House of Commons Parliamentary Papers Online. Craig, "The Introduction" (ref. 27) gives an overview of this report and of others dealing with the subject. See also Cohen, *The Growth* (ref. 27), 125–34.

[29] *Third Report* (ref. 28), 1–38 for Welby's evidence. Reginald Earle Welby, 1st Baron Welby (1832–1915), was educated at Eton and Trinity College Cambridge, and had entered the Civil Service as a Treasury Clerk in 1856. He was later to become Assistant Financial Secretary (1880) and Permanent Secretary (1885–1894), and was raised to the peerage in 1894 on his retirement. (*Oxford New Dictionary of National Biography*, hereafter *ODNB*).

witnesses were more senior and more forceful. Ralph Lingen,[30] Permanent Secretary of the Treasury, affirmed the frailty of the paper and the frequent poor quality of copies as major drawbacks, to which he added the claim that press copying produced a greater bulk of papers than hand copying because neither large sheets nor both sides of the paper could be used. He could imagine circumstances in which press copying might be useful, indeed he had seen it used in his prior position in the Education Office:

> But in an office like the Treasury we constantly, where the questions are enduring ones, have to refer to books and papers of 20, 30, and even more years ago; those books have to be sent about in different divisions of the office, and they have constantly to be marked for extracts, copying, reference, and for matters of that kind. It appears to me that the introduction of pressed copies or not, depends on the nature of the business of an office.[31]

Once again the Commissioners challenged the witness's views, but he held firm, conceding only that in some Departments other than Treasury the issue of copying machines might be worth further investigation. When the Chancellor of the Exchequer[32] himself appeared, he endorsed the views of his officials as to the unsuitability of press copying for the work of Treasury even though in the abstract the machine seemed to offer great economy and accuracy.

A dissenting voice with experience in Treasury came in the form of the Rt Hon William Baxter MP, who had been Secretary to the Admiralty under Gladstone (1868-71) where he was responsible for replacing hand copying by press copying. Baxter was Financial Secretary to the Treasury from 1871 to 1873, a position that he left shortly after his evidence to the Commission. He stressed in that evidence[33] the immense saving of time and labour that press copying made possible as well as the accuracy of the record it provided. Baxter had read the previous evidence given and addressed the contrary arguments in detail.

The dominant Treasury position was supported by the witness from the Home Office, its Under Secretary Adolphus F.O. Liddell. He too was concerned with matters of durability but on a grander scale. There were, he said, books now in the Warrant Office from the time of Charles II, which were "just as fresh and good as they were when they were written". In a long-standing tradition entries had always

---

[30] Ralph Robert Wheeler Lingen, 1st Baron Lingen (1819–1905), of an old Birmingham family was educated at Trinity College Oxford, and became a Fellow of Balliol. He was called to the bar in 1847 but accepted a position in the Education Office instead, where, in 1849, he succeeded Sir James Kay-Shuttleworth as Secretary. He held that position until 1869 when he moved to Treasury (*ODNB*).

[31] *Third Report* (ref. 28), 146.

[32] *Third Report* (ref. 28), 228. The Chancellor of the Exchequer at this time was the Rt Hon Robert Lowe, 1st Viscount Sherbrooke (1811–1892).

[33] *Third Report,* (ref. 28), 246–253. William Edward Baxter (1825–1890) had a significantly different background from most of the other witnesses. The son of a Scottish businessman, he was educated at Dundee High School and University of Edinburgh and was a partner in his father's firm before becoming liberal MP for Montrose Burghs from 1855–1881. (*ODNB*). His unqualified enthusiasm for the copying machine is no doubt partly explained by his business experience and outlook.

been made in the same kind of large book with good thick paper. Copying machines were all well and good, "the only question is, whether, as some of the Home Office papers became historically interesting in other times, that thin paper would last as well as the books that we use now".[34] The preference of the great houses in the City for the copying machine did not impress Liddell who noted that such businesses were interested only in preserving records for the duration of a partnership. Nor was he convinced that such machines produced identical copies since only if they were well done were they readable. Faced with the claim that the hand copies made in the Home Office were liable to mistakes that the machines would not make, he protested, perhaps losing sight of the central issue of economy, that all the Home Office copies were read over!

Officials from some other departments were much more positive. Mr. Robert Herbert, Permanent Secretary of the Colonial Office (and, incidentally, formerly the first Premier of Queensland!) considered that the use of copying machines was far preferable to hand copying. The latter produced documents that could not necessarily be relied upon to be accurate whereas the press copy "is like a photograph, a fac-simile of the thing copied". For Herbert this was a major consideration and providing that press copies were carefully taken their legibility and durability were entirely satisfactory.[35] Press copying had been universal in his time at the Board of Trade but, ironically, at the Colonial Office they relied upon keeping drafts as a record of the fair copies sent out. He did not indicate whether he planned to change this.

Herbert's former superior at the Board of Trade, its Permanent Secretary, Thomas Farrer[36] was another strong advocate of the copying machine. He recounted how the machine had been introduced first into the Marine Department from about 1850 to deal with new work and from then was progressively introduced in all departments until "everything that can be done by a machine is done by a machine".[37] This had produced substantial economies— Farrer estimated that there were perhaps 10 per cent fewer clerks as a result of the use of copying machines. The figure, he felt, might prove to be higher elsewhere. He rejected arguments about poor quality and durability of press copies, considering that this was readily obviated by proper training and procedure.

---

[34] *Third Report* (ref. 28), 221.

[35] *Third Report* (ref. 28), 164–165. Sir Robert George Wyndham Herbert (1831–1905) was educated at Eton and Balliol and elected a Fellow of All Souls, Oxford. He was secretary to Gladstone when the latter was Chancellor of the Exchequer but then migrated to Australia in 1855, where he became Colonial Secretary of Queensland in 1859, and then its first Premier until 1866. On his return to England he was Assistant Secretary to the Board of Trade from 1868–1870 and then became Permanent Secretary of the Colonial Office (*ODNB*).

[36] Thomas Henry Farrer (1819–1899), educated at Eton and Balliol College Oxford, was called to the Bar at Lincoln's Inn in 1844 and entered the public service in 1850 as Secretary to the Naval Department (later the Marine Department) of the Board of Trade. He became Permanent Secretary of the Board in 1867. Created Baronet in 1883 and raise to the peerage as 1st Baron Farrer in 1893, he was among the strictest advocates of free trade (*ODNB*).

[37] *Third Report* (ref 28), 189.

Those who resisted the use of copying machines eventually lost the argument and the Commission firmly recommended that their use be made "universal" within the Service. Even before this, however, there were shifts taking place in other parts of the system of work practices that favoured the machines. One particularly interesting example is handwriting. Guidance given on handwriting to young men preparing for Civil Service examinations included the following: "*Handwriting*—Having regard to the use of copying machines in some of the departments, the handwriting is required to be of that even stroke which allows legible copies to be taken by pressing". It was also noted, however, that the Commissioners had found it impossible strictly to enforce this handwriting rule without detriment to the Service.[38] As we have seen, the legibility of press copies was one of the issues that critics focussed upon. As is typical in technological system change,[39] moves were afoot in other parts of the system to ensure better legibility by adapting handwriting to the requirements of the machine, but they proved difficult to implement in a totalising fashion because of other considerations, presumably the impossibility of rejecting otherwise strong candidates on this specific aspect of the quality of their handwriting alone. As press copying became standard procedure in the public offices there would have been numerous other ways in which work practices and customs were modified in order that it be accommodated and to ensure that problems were ironed out.

Another issue that was important to the adoption of press copying, and also became a factor in its longevity, was the legal status of press copied documents. The fact that press copies were facsimiles was significant. In the early days of the invention Boulton had been criticised by those concerned about the possible use of the technique in forgery, but these concerns faded when it became clear that press copies could never be more than relatively crude "flimsy" documents, incapable of reproducing detail of the sort required by the forger.[40] The copies were nevertheless facsimiles and so a good case could be made for them as faithful reproductions of the content of an original, derived from that original by a purely physical process. An 1848 judgment by Lord John Campbell made the ruling that press copies were authentic copies of originals and were therefore valid legal evidence, without the need for corroborating witness in the absence of the original document. This legal status of press copies, another fine example of how the ramifications of technological

---

[38] *Guide to Employment in the Civil Service: Being a Complete Epitome of the Examinations for the Various Departments of the Public Service, organized according to the Recommendations of the Civil Service Commissioners*. (London: Cassell & Co, 1867), 15. Cassell's *Hand-book of letter-writing* (Cassell & Co., 1861) includes samples of the copying machine-friendly handwriting that was recommended, and candidates were advised to use model copy books, also published by Cassell, to develop suitable handwriting by tracing over given forms.

[39] Thomas P. Hughes, "The Evolution of Large Technological Systems," in *The Social Construction of Technological Systems* ed. Wiebe E Bijker, Thomas P. Hughes and Trevor Pinch, (Cambridge, MA: The MIT Press, 1987), 51–82 is an early and influential account of technological system change.

[40] Smiles, *Lives of Boulton and* Watt (ref. 2), 266.

systems give them inertia, assisted the technique's survival in some court and legal offices into the 1950s.[41]

There is another important dimension to the debates about press copying within the British Civil Service. This is their connection with the larger question of the types of staff recruited to the Service. In the reforms of the early 1850s, which included the institution of selection by competitive examination, recruits to the Service were university educated. Part of the rationale for this was that these civil servants would have to mix, and work effectively, with the ruling elite so that a shared educational background was advantageous. We see, for example, in the biographies of those called to testify to the Expenditure Committee the regularity with which a public school education (with Eton most prominent) followed by Oxford or Cambridge provided that background. But reforms also created another type of Civil Servant of a much lower grade and perceived standard, of whom only a general commercial education was expected. The function of this group (variously referred to as the second class of clerks, or assistant clerks) was to undertake the routine work, including copying, that was wasteful of the time and capacities of the higher civil servants. The work of the second class of clerks was explicitly referred to as "mechanical" in character, in contrast to the creative work of the higher class.[42]

These issues thread through the evidence of the various witnesses to the Commission on Civil Services Expenditure and undoubtedly account for some of the differences of opinion about copying machines. For example, where confidentiality was important, letters would have to be handled by a higher clerk whose education and gentlemanly status was taken to ensure the personal qualities required to maintain such trust. In such situations the letters composed would have to be hand copied since the higher clerk could not be required to perform the mechanical work of press copying.[43] Baxter, who as we saw was a fervent advocate of the copying machine, thought that it might reduce the number of clerks by about 2000 out of 16,000, but would also have other effects:

> I believe that there are now men pretty high up in the various departments who have spent the best portion of their lives in doing nothing but copying letters. That is not only an extraordinary waste of time, but it has a distinct tendency to prevent a person from exercising his intellect, or developing his mental powers.... Many excellent servants have been spoilt by having been kept grinding for years doing nothing but copying letters.[44]

The decision that copying machines should be standard in the Civil Service enshrined a symbolic division between the ruling elite and the higher echelons of that Service on the one hand, and the rest, the "mechanicals". The former made creative judgments and composed the documents of the Service, the latter merely copied them and moved them around. Mechanical work could in principle, and sometimes in practice, be done by machines. The copying machine could thus save

---

[41] Rhodes and Streeter, *Before Photocopying* (ref. 2), 8. On the inertia of technological systems see Hughes, "The Evolution", (ref. 39).

[42] Craig, "The Introduction" (ref. 27), 109 and Cohen, *The Growth* (ref. 27), 125–34.

[43] See Mr. Melly questioning Mr. Welby, *Third Report*, (ref. 28), 16–17.

[44] Evidence of William Baxter, *Third Report*, (ref 28), 250.

expenditure on lower level clerks. It could also, in the wake of the Trevelyan reforms, police the boundary, in the sense of being used, like the lower clerks themselves, to ensure that the talents of the higher reaches of the Service were not wasted in mundane activity.

4.5 Conclusion

James Watt was a creature of the Enlightenment and a participant in the Republic of Letters. Yet that same Enlightenment outlook ushered in a world of mechanized reproduction not only of goods but of information and of design. In the histories of information there is a direct line from the Enlightenment project through the records' production, preservation and retrieval practices of business and the Civil Service, to the creation of the "government machine" of the digital age. As developed and systematised by many other hands in the course of the nineteenth century, Watt's copying machine played a significant role in this process.

As in all technological change, however, the functional possibilities of the copying machine could only be realized by virtue of complex adjustments and evolving skills that adapted it to existing work practices. In the case of the British Civil Service this took time and was not universally possible so that in some situations old work practices involving hand copying persisted. Nevertheless, at a symbolic level, the copy press came to represent, and in some ways enforce, distinctions between creative workers and the mechanicals in the realm of paperwork, between "Men of Letters" and "Men of Press Copies". What began as part of an Enlightenment vision of the industrialization of originality through mechanical reproduction became a component and symbol of the class society of high Victorian Britain. The copying machine and its culture of use resonated with, and to a significant extent reproduced, distinctions within that social order.[45]

[45] An important caveat must be entered here. Although a decisive distinction between creative and mechanical work served well both political aspiration and social symbolism, no decisive or uniform separation was achieved in practice in the British Civil Service. See Craig, "The Introduction" (ref. 27), 117.

Chapter 5
Poetry, Chemistry, and Wisdom

David Knight

> *Oh, most magnificent and noble nature!*
> *Have I not worshipped the with such a love*
> *As never mortal man before displayed?*
> *Adored thee in thy majesty of visible creation,*
> *And searched into thy hidden and mysterious ways*
> *As Poet, as Philosopher, as Sage? (John Davy, Fragmentary Remains, Literary and Scientific, of Sir Humphry Davy (London: Churchill, 1858), 14.)*

So, at an uncertain date, wrote Humphry Davy, friend of Samuel Taylor Coleridge and the preeminent chemical discoverer and lecturer of his day. Chemical and natural philosophers of the nineteenth century were willy-nilly the heirs of the Romantic movement as well as the Enlightenment represented by Antoine Lavoisier; and in Britain (as in the German lands) this meant thinking dynamically, conceiving forces rather than matter to be fundamental. Those who aspired like Davy to combine poetry and dynamical natural philosophy would move beyond Understanding to Reason, achieving true wisdom and becoming sages.

I have known Trevor Levere since we were both graduate students, I working on Davy and he on Coleridge. He was pursuing the then-surprising notion that Coleridge (and his circle) were interested in and enthusiastic about science, even knowledgeable about it, provided that it was dynamic, based upon forces and powers, rather than materialistic, based upon particles, mechanics, and algebra: Natura naturans rather than Natura naturata. I was seeking to understand what made Davy tick. Sir Harold Hartley, eminent engineer and a valuable patron to me, was publishing his excellent biography of Davy the chemist[1]: but he confessed to me that he had been able to make little or nothing out of Davy's last and most popular works, the dialogues *Salmonia: or Days of Fly Fishing* (1828) and particularly the posthumously-published *Consolations in Travel; or the Last Days of a Philosopher* (1830). Davy, a sick man facing death after some kind of stroke, avoiding company and often sozzled with laudanum, dictated this book to John James Tobin, the

[1] Harold Hartley, *Humphry Davy* (London: Nelson, 1966).

D. Knight (✉)
Durham University, Durham, UK
e-mail: d.m.knight@durham.ac.uk

J. Buchwald, L. Stewart (eds.), *The Romance of Science: Essays in Honour of Trevor H. Levere*, Archimedes 52, DOI 10.1007/978-3-319-58436-2_5

medical student who was his companion and amanuensis in 1828–1829 as they moved around Italy and Carniola (now Slovenia). The dialogues were perhaps a substitute for real conversations. Davy saw the book as a kind of testament or legacy; as well as a good deal of geology and chemistry, it is saturated with a dualistic pantheism and an urge to move beyond understanding to reason and wisdom which were typically Romantic. In their young days in Bristol, Coleridge and Davy were (like their patron and friend, Thomas Beddoes) political radicals. Both abandoned temporarily the established Church of England in which they were brought up, Coleridge for Unitarianism, a 'feather-bed to catch a falling Christian' as it was called in the Darwin and Wedgwood family. Coleridge nearly became (like Joseph Priestley) a Minister in this liberal sect, until the Wedgwoods came to his rescue with an annuity. Later, Davy and Coleridge both turned politically conservative; but whereas Coleridge returned to firm and influential allegiance to the Church, Davy seems to have been a nominal Anglican but never joined a particular congregation. Rather surprisingly for a British man of science in that century, but not un-Romantically, he rejoiced at the very end of his life in Roman Catholic Emancipation (1828): unfazed by superstition, a story-teller brought up in Cornwall on the 'Celtic fringe' of Britain, he was sympathetic to what he saw of that church when in Ireland, Italy and Carniola. With one of us interested in Coleridge's science, and the other in Davy's poetry and philosophy, the congenial Trevor and I could not but get on well; and his friendship, support and helpful criticism over the years have been a delight.

Chemistry has never been the easiest science from which to launch into natural theology, the usual vehicle for wonder and wisdom.[2] Unlike astronomy or natural history that evoke wonder at the starry heavens and the adaptations and workmanship evident in animals, chemistry does not lend itself to contemplation. It is the science of the secondary qualities, tastes and smells, colours and textures: a hands-on activity, a craft as well as a science, dedicated not only to understanding but also to improving the world. Anyone who has studied it will be familiar with burnt fingers, stinks acrid and aromatic, amazing changes of colour, the unctuous or gritty texture of solids, the viscous and viscid qualities of fluids, the taste of solutions inadvertently sucked up a pipette, and triumphs like preparing crystals that melt at the right temperature or making a wonky T-joint from glass tubing, that go with acquiring the necessary skills. To chemists Creation cannot but be a work in progress, needing human assistance; as Davy wrote, the chemist exerts

> on a scale infinitely small a power seeming a sort of shadow or reflection of a creative energy and which entitles him to the distinction of being made in the image of God and animated by a spark of the divine mind. Whilst chemical pursuits exalt the understanding, they do not depress the imagination or weaken genuine feeling.[3]

[2] Russell Re Manning, ed., *The Oxford Handbook of Natural Theology,* (Oxford: University Press, 2013); my 'Chemical Sciences and Natural Theology', from which parts of this essay are expanded, is at 434–48.

[3] Humphry Davy, *Consolations in Travel, or the Last Days of a Philosopher* (London: Murray, 1830), 245.

Creative was an adjective not normally applied to humans at that date, but to the Creator.[4] Awarding the Copley Medal to William Buckland for his paper on the Kirkdale Cavern that had seemed to justify belief in Noah's Flood, Davy commented

> The more we study nature, the more we obtain proofs of divine power and beneficence; but the laws of nature and the principles of science were to be discovered by labour and industry, and have not been revealed to man; who with respect to philosophy, has been left to exert these god-like faculties, by which reason ultimately approaches, in its results, to inspiration.[5]

Coleridge likewise had little patience with the search for 'evidences' prominent in William Paley's *Natural Theology*,[6] writing:

> To understand God by the World, instead of the World by God ... is an attempt to see the Sun by Moonlight.
>
> Assume the existence of God, - and then the harmony and fitness of the physical creation may be shown to correspond with and support such an assumption; - but to set about *proving* the existence of God by such means is a mere circle, a delusion.[7]

He would not have been impressed by devotees of Intelligent Design.

From a quite different position to Coleridge's, the prominent evangelical Thomas Gisborne refused to subscribe to Paleyan natural theology: he saw a world which had been very good in God's eyes, but was fallen. Human sin had desecrated it. It was now a sphere of probation, or Botany Bay, full of pain and predation, toil and death: and inferences to God's goodness and wisdom ceased to be possible when our ancestors were expelled from paradise.[8] Mankind was naturally evil, but within reach of mercy; pain, disease and warfare proved the penal character of life here on Earth; technical advances were accompanied by drawbacks; and unbelief was sinful. This was an unusually pessimistic vision among evangelicals, but not uncharacteristically stern: attempts to argue for, or prove, God's existence were futile, and penitence and faith were required for salvation. None of the chemists we shall be looking at believed that. They saw the world as essentially the product of a benevolent Creator.

So despite criticisms the tradition of Paley refused to die, indeed flourished, and in 1836 Buckland published a *Bridgewater Treatise* on the power, wisdom and goodness of God manifested in Geology and Mineralogy: he owned and lightly annotated a copy of *Consolations,* now in the library of the Geological Society, where he will have found a more subtle and less orthodox theology, impatient of the

4 See Samuel Johnson, *A Dictionary of the English Language,* 1(1773); and William Enfield, *A Pronouncing Dictionary* (London: Crosby, 1815), s.v.'creative'.

5 Humphry Davy, *Collected Works*, ed. John Davy (London: Smith Elder, 1839–1840), vol. 7, 41.

6 William Paley, *Natural Theology,*[1802], ed. Matthew D. Eddy and David Knight (Oxford: University Press, 2006).

7 Kathleen Coburn, ed., *Inquiring Spirit: a new Presentation of Coleridge from his Published and Unpublished Prose Writings* (London: Routledge, 1951), 120, 381.

8 Thomas Gisborne, *The Testimony of Natural Theology to Christianity,* (London: Cadell & Davies, 1818).

argument from Design. It was not until 1833, at the Cambridge meeting of the infant British Association for the Advancement of Science, that Coleridge told the assembled savants that they should not be calling themselves philosophers *tout court,* and William Whewell coined the word 'scientist' (which did not catch on for another generation). Davy, and his protégé Michael Faraday, saw themselves as chemical philosophers (a subset of natural philosophers); and Davy in Coleridgean vein wrote about such a person:

> He should resemble the modern geometricians in the greatness of his views and the profoundness of his researches, and the ancient alchemists in industry and piety. I do not mean that he should affix written prayers and inscriptions of recommendations of his processes to Providence, as was the custom of Peter Wolfe and who was alive in my early days, but that his mind should always be awake to devotional feeling, and in contemplating the variety and beauty of the natural world, and developing its scientific wonders, he will always refer to that infinite wisdom, through whose beneficence he is permitted to enjoy knowledge; and in becoming wiser, he will become better, and will rise at once in the scale of intellectual and moral existence, his increased sagacity will be subservient to a more exalted faith, and in proportion as the veil becomes thinner through which he sees the causes of things, he will admire more the brightness of the divine light by which they are rendered visible.[9]

Given the uncertainties of apparatus and reagents in the early nineteenth century, we may sympathise with Woulfe, the inventor of a widely-used wash-bottle for gases.

Consolations begins with a reverie in the Colosæum in Rome, where a 'Genius' reveals the progressive past of the human race, who "owe their improvements to a few superior minds"; and unveils the future, when human souls liberated from their material bodies will migrate in more spiritual form to another planet.[10] In Davy's day, the word genius was ceasing to mean djinni and taking on its modern sense of a quality in an unusual person, perhaps inspired by such a spirit. Coleridge indeed defined it:

> as the faculty which *adds* to the existing stock of power and knowledge by new views, new combinations; (by discoveries not accidental but anticipated, or resulting from anticipation). In short, I define GENIUS, as originality in intellectual construction: the moral accompaniment, and actuating principle of which consists, perhaps, in the carrying on of the freshness and feelings of childhood into the powers of manhood.[11]

Newton, Wordsworth's 'Voyager through strange seas of Thought, alone', would certainly count: he had found in universal gravitation the dynamics and harmony, the order and beauty that he sought, which lay behind Kepler's laws of planetary motion.[12] Gravity was universal, but chemistry was based upon elective affinities and a second Newton was required to account for its laws. Priestley had looked forward to that day:

---

[9] Humphry Davy, *Consolations in Travel*, 254-5. On *Consolations* see James Secord, *Visions of Science: Books and Readers at the Dawn of the Victorian Age* (Oxford, University Press, 2014).

[10] Humphry Davy, *Consolations in Travel*, 20.

[11] Coburn, ed., *Inquiring Spirit*, 340.

[12] David Knight, *Voyaging in Strange Seas: the Great Scientific Revolution,* (New Haven: Yale University Press, 2014).

> Hitherto philosophy has been chiefly conversant about the more sensible properties of bodies; electricity, together with chymistry, and the doctrine of light and colours, seems to be giving us an inlet into their internal structure, on which all their sensible properties depend. By pursuing this new light, therefore, the bounds of natural science may possibly be extended, beyond what we can now form an idea of. New worlds may open to our view, and the glory of the great Sir Isaac Newton himself, and all his contemporaries, be eclipsed by a new set of philosophers, in quite a new field of speculation.[13]

Davy, demonstrating experimentally in 1806 that electricity and chemical affinity were manifestations of one power, and the following year isolating potassium electrically, had sailed into this new world: perhaps he was the Newton of Chemistry.

Davy's social ascent to the Presidency of the Royal Society in 1820 (visible in the swagger portrait that now hangs in its rooms, by Sir Thomas Lawrence, PRA, who had risen equally far) dismayed his old friends, particularly Coleridge who thought also that he had turned into an atomist and materialist. Though he had joined the Establishment, that claim was not the case, as Davy declared in *Consolations:*

> The doctrine of the *materialists* was always, even in my youth, a cold, heavy, dull and insupportable doctrine to me, and necessarily tending to atheism. When I had heard with disgust, in the dissecting rooms, the plan of the physiologist, of the gradual accretion of matter and its becoming endowed with irritability, ripening into sensibility and acquiring such organs as were necessary, by its own inherent forces, and at last rising into intellectual existence, a walk into the green fields or woods by the banks of rivers brought back my feelings from nature to God; I saw in all the powers of matter the instruments of the deity … I saw *love* as the creative principle in the material world, and this love only as a divine attribute.[14]

The physiologist he has in mind will have been enthused by the French, probably through the radical surgeon William Lawrence whose book was condemned by the Lord Chancellor as blasphemy: meaning that Lawrence lost his copyright in it. The effect, the law being an ass, was that he could not prohibit pirated editions, and the book helped make medical schools hotbeds of medical materialism.[15] Davy may also have been thinking of Mary Shelley's *Frankenstein,* 1818, which was largely inspired by his lectures.

Sir Harold warned me against turning science into an intellectual game remote from practical experience. Davy was ready with stories, uplifting rhetoric, hypotheses and doctrines: but his fame depended upon his experimental skills in electrochemistry, in working on chlorine and iodine, and on the understanding of flame that led to his safety-lamp. Later he worked in Italy, under the aegis of the sculptor Antonio Canova, on the pigments in paintings found in Pompeii. Davy's earliest researches at the Royal Institution (dedicated to applying science to the common

---

[13] Joseph Priestley, *The History and Present State of Electricity*, 3rd ed. (London: Bathurst & Lowndes, 1775), vol. 1, xv.

[14] Humphry Davy, *Consolations in Travel*, 219.

[15] William Lawrence, *Lectures on Physiology, Zoology, and the Natural History of Man* (London: Benbow, 1822); Adrian Desmond, *The Politics of Evolution: Morphology, Medicine, and Reform in Radical London* (Chicago: University of Chicago Press, 1989).

purposes of life) were on tanning and agriculture, and it was for these that he received the Copley Medal. In his famous Inaugural Lecture of 1802 he foresaw a bright day just dawning when applied science would transform society, bringing prosperity and avoiding the political convulsions so unhappily evident across the Channel in France as Revolution led to military despotism: it proved to be so. Well-aware like Thomas Beddoes of the history of chemistry, not unsympathetic to honest alchemists, Davy rejoiced in the improvements that his science had brought about. Ores were refined into useful metals (new ones like Platinum being added to the list), ceramics and glass were made, spirits were distilled, drugs, dyes and pigments, fertilisers, oils and perfumes, and gunpowder (for engineering as well as firearms) were prepared, making lives longer, easier and arguably more enjoyable. But all this was the fruit of human ingenuity, guided by God-given reason but not directly by providence like the instincts and adaptations bestowed upon animals, the influence of weather, and the light from Sun, Moon and stars.

Actual voyagers of discovery took with them a large crew, and perhaps some specialists: the solitary and towering genius is what the public expects but the historian rarely uncovers. Newton admitted that he stood on the shoulders of giants, and Davy drew inspiration from his reading of Lavoisier and Priestley; contact with his patrons Davies Giddy, Thomas Beddoes, and Joseph Banks; his friendships with Thomas Poole the tanner, and with Coleridge; and in London from his widening circle of friends and acquaintances. Friends and colleagues helped him decide that his mysterious 'potagen' was a metal, anomalous as that seemed, that could therefore be called potassium; that oxygenated muriatic acid was really an element, chlorine; and that he should give the eager bookbinder's apprentice Michael Faraday a job. But although Davy trained his cousin Edmund, his brother John, and then Faraday in a kind of informal system of apprenticeship, he never founded a school, and indeed given his situation at the Royal Institution he could not have done so. A team of his friends taught chemistry to medical students at Guy's Hospital[16]; William Thomas Brande, his immediate successor at the RI, laid on courses for such students in the wake of the Apothecaries Act of 1815 that made them compulsory; but they did not happen in Davy's time as Professor. Later as PRS Davy stood aloof from plebeian scientific groups; and his version of a theology of nature, less argumentative than Paley's, remained idiosyncratic. Two things perhaps made it unusual: it pointed to the dangers of experimental chemistry (thus as adventurous as other kinds of exploring), and was written in sickness and exile, with death in prospect: praise for the benevolence and wisdom of God comes more readily from those in more comfortable circumstances. Davy had made his reputation with self-experiments on nitrous oxide, laughing gas, in association with Coleridge: he not only analysed it and other oxides of nitrogen, but saw its potential as a recreational drug and possible treatment. Next he recklessly breathed nitric oxide which badly burnt his throat and mouth, and then 'carburetted hydrogen', containing carbon

[16]William Babington, Alexander Marcet and William Allen, *A Syllabus of a Course of Chemical Lectures read at Guy's Hospital* (London: Phillips, 1816).

monoxide: his friends managed to bring him round, but in the words of Joseph Cottle, his friend and the publisher of *Lyrical Ballads*:

> Such are some of the appalling hazards encountered by Mr. Davy in his intrepid investigation of the gases. These destructive experiments, during his residence at Bristol, probably, produced those affections of the chest to which he was subject throughout life, and which, beyond all question, shortened his life. Nothing at this moment so excites my surprise, as that Mr. Davy's life should have been protracted, with all his unparalleled indifference concerning it, to the vast age, for him, of fifty years.[17]

Although not comparable in scale to what happened with astronomy and natural history, where popularisation usually took that form, there was within chemistry a tradition of theology of nature, and even of the more rigorously-demonstrative natural theology, going back to Paracelsus and prominent in Robert Boyle.[18] Contemplating God's role in the world, Davy wrote that:

> on so high and incomprehensible a subject, … it is wiser to confess our entire ignorance, and to bow down in humble adoration to the one incomprehensible Cause of all being. … perhaps there is no more pleasurable state of the human mind than when, with intense belief, it looks forward to another and better state of existence, or is absorbed in the adoration of the supreme and eternal Intelligence.[19]

Not everyone was so persuaded of the inscrutability of God, and the danger of patting him on the back for a particularly ingenious adaptation, as Paley sometimes seemed to do. Five chemists prominent in Davy's and the next generation, Samuel Parkes, James Finlay Weir Johnston, William Prout, George Fownes, and George Wilson, in rather different ways did what Faraday would refuse to do, that is, mingling science and religion.

Samuel Parkes was a Unitarian and a manufacturing chemist, based at the Haggerstone Chemical Works in Hackney, east London, not far from the Olympic Games site in 2012. Hackney, where Priestley had taught at a Dissenting Academy after his house in Birmingham was sacked, and where a 'Phalanx' of high-churchmen gathered, was a more salubrious spot than it subsequently became as noisome industry like Parkes' proliferated and brought rows of cheap housing. Parkes' very successful *Chemical Catechism,* which included a product placement, was addressed (through their parents) to boys in the upper and middling ranks of society, to generate enthusiasm for the science and appreciation of how useful it was in managing an estate or entering a profession; and moreover

> To exhibit, in a popular form, a body of incontrovertible evidence of the wisdom and benevolence of the Deity, in the establishment and modification of those laws of matter which are so infinitely and beautifully varied, and whose operation is too minute to be the object of general notice.[20]

---

[17] Joseph Cottle, *Reminiscences of Samuel Taylor Coleridge and Robert Southey* (London: Houlston and Stoneman, 1847), 269.

[18] Michael Hunter, *Boyle: Between God and Science* (New Haven: Yale University Press, 2009).

[19] Humphry Davy, *Salmonia*, 3rd ed. (London: Murray, 1832), 173-4.

[20] Samuel Parkes, *The Chemical Catechism: with Notes, Illustrations, and Experiments*, 4th ed. (London: Lackington Allen, 1810), ix; subsequent quotations are from 510, 463, 169, 407, 22.

Though aimed at boys, it was originally composed for his daughter. For some of the 256 experiments described at the back of the book, he was 'indebted to the polite communication of Professor Davy', to whom an encomium comes at the end of the notes, hailing him indeed as chemistry's Newton:

> In sciences kindred to chemistry, the knowledge of the nature of the alkalies, and the analogies arising in consequence … will certainly form a new æra in the history of Chemistry, and render the name of Davy ever dear to those who delight in the investigation of nature.

> Immortal Newton thus with eye sublime
> Mark'd the bright periods of revolving time;
> Explored in Nature's scenes the effect and cause,
> And, charm'd, unravell'd all her latent laws.

The idea that the laws of chemistry are too minute to be generally noticed seems new with him. The book is set out with a main text, and copious notes in smaller type which frequently dwarf it. Parkes explained that the tyro should ignore the notes at the first reading, and come back to them later: the catechism was the traditional form of church teaching, and the book can remind us of Thomas Kuhn's paper on the role of dogma in science.[21] It is the notes that contain the effusions of theology of nature, and often verse as well, for Parkes was not in fact concerned with incontrovertible proofs but with praise and wonder. Most of the notes do indeed amplify the text, and refer to publications; but referring to the alkalis (pp. 168–9) he noted how 'infinite Power and Wisdom, conjoined with consummate Beneficence, can effect the most *important* changes, by the most *unlikely* agents', adding (a little oddly) from Erasmus Darwin a stanza on cyclic changes, in Priestleyan materialistic mode:

> Organic forms with chemic changes strive,
> Live but to die, and die but to revive;
> Immortal matter braves the transient storm,
> Mounts from the wreck, unchanging but in form.

Elsewhere, on page 407, he quoted Bishop Richard Watson whose chemical lectures in Cambridge had been a huge success in the previous century: 'every secondary cause we discover, is but a new proof of the necessity we are under of ultimately recurring to him [God] as the one primary cause of everything'. Parkes himself wrote (p. 45) that 'perhaps nothing tends to give so satisfactory an assurance of the goodness of the ALMIGHTY, as the consideration of the various *uses* to which different substances in nature may be applied'. Marvelling at cyclic processes and the chemistry of life, a boon which was lent to us, he ended the book with the Biblical text 'God saw everything that he had made, and behold it was very good' with a supporting note, a last word, from Paley.

Poetry and religious feeling thus coexist in the notes with reports of analyses; for after all, Parkes was seeking to make converts to chemistry, not to a church. Parkes' frontispiece in some editions was a chemical curiosity, etched on glass using fluoric acid; in others, the text boasts a delightful illustrated title-page, where in what was

[21] Thomas Kuhn, "The Function of Dogma in Scientific Research," in Alistair Crombie ed., *Scientific Change* (London: Heinemann, 1963), 347–69.

then a new and high-tech process coal gas was being generated and chandeliers lit, all rather dangerously near draped curtains. Though he was no Gradgrind, his rhetoric to parents concluded with a passage in praise of facts:

> If 'knowledge is power', surely the *love* of knowledge, and a taste for accurate investigation, is the most likely way for conducting to opulence, respectability, and rational enjoyment. Moreover, it is a necessary consequence of attention to this science, that it gives the habit of *investigation,* and lays the foundation of an ardent and inquiring mind. If a youth has been taught to receive nothing as true, but what is the result of *experiment,* he will be in little danger of being led away by the insidious arts of sophistry, or of having his mind bewildered by fanaticism or superstition. The knowledge of *facts* is what he has been taught to esteem; and no reasoning, however specious, will ever induce him to receive as true what appears incongruous, or cannot be recommended by demonstration or analogy.

Despite the Romantic flavour of his notes, Parkes was, like Davy, firmly rooted also in the Enlightenment, devoted to useful knowledge, and aware that it was power.

With the same intention of drawing people into chemistry, but with a more austere background as a Scottish Presbyterian (later Free Kirk member), Johnston, a well-known chemist and lecturer working in the infant University of Durham, published at the end of his life a popular work, *The Chemistry of Common Life* (1855). It became widely-known and admired, both in English and in translation. The chapters cover topics like 'The Water we drink', 'The Bread we eat', 'The Liquors we ferment' and 'The Narcotics we indulge in'. It is like Parkes' book pervaded with a theology of nature, though Johnston was very conscious of the insufficiency of natural theology without revelation, which told us more of God's purposes and character than could ever be inferred from chemical (or other scientific) discoveries. His conclusion was at once both effusive and sobering:

> The Deity willed that this corner of His vast work should be the theatre of new displays of wisdom, of consummate contrivance, of a wonderful fitting-in of means to the accomplishment of beneficent ends, and at last the seat of an intellectual being, with capacity to study and comprehend and admire His works – to praise, and love, and serve Him.[22]

But all this could be stopped by God at any time – and the crowning lesson from science must be the insignificance of human beings in the great scheme of things.

In all these works theology of nature was unavoidably present, but they were not formal natural theologies, arguing directly for God's existence and His wisdom and benevolence in creation. But by the time Davy died some such works were starting to appear that were explicitly concerned with chemistry and natural theology. In 1829 the Earl of Bridgewater had died, bequeathing £8000 to the Royal Society to commission works on science as demonstrating the goodness and wisdom of God. Prout, physician and chemist, took on the last one, a rag-bag of chemistry, meteorology, and the function of digestion.[23] That chemistry only featured in the last book of

---

[22] James F. W. Johnston, *The Chemistry of Common Life* (Edinburgh: Blackwood, 1855), vol. 2, 448.

[23] Jon R. Topham, "Beyond the Common Context: the Production and Reading of the Bridgewater Treatises," *Isis* **89** (1998), 233–62.

the series of eight, and had to share with such curious bedfellows, indicates that it was seen as much less apt than astronomy, geology, and even the human hand, each of which got a whole book to themselves. Prout, the Royal Society's Copley Medallist of 1827, was well-known for his identification of hydrochloric acid in the stomach juices. He is remembered in the history of chemistry for his tentative hypothesis of 1815, at first anonymous but then acknowledged, that the chemical elements were polymers of hydrogen (or perhaps something even simpler). This idea, in effect making hydrogen Boyle's fundamental 'corpuscle', owed something to Davy, who could not accept Dalton's postulate that all the elements had essentially different and indivisible atoms. Prout gave the hypothesis a quantitative basis: the atomic weights of the elements, relative to hydrogen, would be whole numbers. At first, analyses by Thomas Thomson and his students in Glasgow seemed to confirm this, and Thomson gave his book the grand title *An Attempt to Establish the First Principles of Chemistry by Experiment*.[24] Subsequent work, notably by Jacob Berzelius, showed that they diverged, often not much but sometimes considerably, from integers.[25] Nevertheless, various elements, potassium and sodium for example, were extremely similar to each other; and they also behaved very like the ammonium radical, NH4, known to be compound: it was not implausible that Dalton's many different kinds of atoms were stable configurations of one, or very few, ultimate particles. Prout's Hypothesis, though experimentally falsified, therefore survived to tantalise chemists with a vision of a simpler world, notably Josiah Cooke (1827-94) founder of the chemistry department at Harvard University.

Prout admired Paley, and often quoted or summarised *Natural Theology* in his Bridgewater Treatise. He remarked that:

> The argument of design, assures [man] that, insignificant as he is, while he investigates and approves of the order and harmony around him, he is exerting faculties truly god-like.[26]

Our reason differed from God's only in not being infinite. Chemistry, a science founded in Prout's opinion (echoing Thomson's title) 'solely on experience', was particularly apt for demonstrating design more subtle than mechanical contrivances. Having created the chemist's atoms, polarized 'molecules' that are clearly 'manufactured articles' (as John Herschel had famously declared), God had gone on to make a 'harmonious, connected series' of compounds that fuel a cumulative argument for His wisdom and benevolence. Thus water, liquid at just the right range of temperatures for conditions on Earth, illustrates the fine adjustments that prove design, which is even more evident in the complex chemical processes that go on in living 'organised beings'. The way in which the properties of compounds are so different from those of their components is a source of wonder and fascination in

---

[24] Thomas Thomson, *An Attempt to Establish the First Principles of Chemistry by Experiment* (London: Baldwin, Craddock & Joy, 1825).

[25] William H. Brock, *The Fontana History of Chemistry* (London: Fontana, 1992), 160-2; Brock, *From Protyle to Proton: William Prout and the Nature of Matter* (Bristol: Adam Hilger, 1985).

[26] William Prout, *Chemistry, Meteorology, and the Function of Digestion considered with reference to Natural Theology*, 2nd ed. (London: Pickering, 1834), 7.

what Prout clearly found a delightful science, though in the preface to the second edition he had to deny that any argument for "design had been founded on the supposed molecular arrangements he had given."[27] For the chemist of his generation, distinguishing hypothesis from all-important facts was crucial, though to the modern reader it is Prout's theory of matter that makes his book interesting rather than run-of-the-mill. The Bridgewater Treatises were a surprising success from the publishing point of view, being widely discussed and selling well; but Prout's was never one of the popular or particularly controversial ones. He refrained from working out how his theory of matter might throw light upon intelligent design. If he had stuck to his guns, and been less pusillanimous about hypotheses and imagination, he would have been a more interesting chemical philosopher - even perhaps a sage.

In 1844, only 10 years later but in a different world, George Fownes published his essay, focussed unlike Prout's entirely upon chemistry, which had won the prize of a hundred guineas established at the Royal Institution by one Hannah Acton.[28] He referred to the (relatively) small number and useful distribution of elementary substances that constituted all the variety that we see in the world, and to the laws of chemical composition, as pointing to a wise lawgiver. Thus the contiguity, for example, of iron ore and vast forests in otherwise-barren parts of Sweden were indicators of a divine plan. The labour required to extract useful materials was not a curse, but a blessing: it went with health and happiness, whereas idleness was the real curse for humanity. Fownes was well-aware of the new developments in chemistry, and his book has 'flow charts' indicating the course of reactions, diagrams indicating the play of affinities, tables, symbols for the elements, and even equations. More important than these then-hypothetical constructions was the dramatic rise of organic chemistry, particularly associated with Liebig, with whom Fownes had studied in 1839, and his new methods of analysis. That was Fownes' field of interest and research. Animals and plants had been found to consist almost entirely of carbon, hydrogen, oxygen and nitrogen, in great series of compounds where they were differently arranged: sometimes indeed the same elements in exactly the same proportions constituted two or more different substances, in the phenomenon of isomerism. At last animal and vegetable chemistry was being understood.

Thus the body was a low-temperature furnace, in which carbon and nitrogen were gently combusted: chemical reactions could evidently go on under very different circumstances, and at different speeds. Chemical equilibria, reversible reactions, and cyclic processes, were hugely important in the processes of nature, and chemists who were beginning to understand them could not but reflect that:

> It is difficult to avoid the conclusion, that these exquisitely beautiful laws and relations have been framed and adjusted to each other by an intelligent mind.[29]

---

[27] Ibid., xiii; for chemistry's basis in experience, 19.

[28] George Fownes, *Chemistry, as Exemplifying the Wisdom and Beneficence of God* (London: Churchill, 1844).

[29] Ibid., 184.

They indicated enlightened, active benevolence. Natural theology was therefore for Fownes the ultimate object of science; and chemistry an excellent subject for such religious meditation. Certainly, as an accessible account of where chemistry stood in this time of rapid change, it was excellent; and natural theology had often sugared the bitter pill of scientific instruction.

Fownes' textbook, *A Manual of Elementary Chemistry, Theoretical and Practical,* published the following year, established itself as a classic, in print in revised editions for over 40 years. His career looked promising, and he was appointed Director of the Birkbeck Laboratory at University College, London – the first purpose-built academic laboratory in Britain. But in that year he was struck down by lung disease, and died in 1849. Similarly, Prout by the time he wrote his *Bridgewater Treatise* was suffering from severe deafness which seriously impaired his social and scientific life. Nevertheless, there seems to be some truth in the idea that dilating upon the wisdom and goodness of God was is easiest for those like Paley generally in good health and comfortable circumstances; and Prout and Fownes also did well financially out of it, as some crabby critics noted. There is a facile aspect to what they wrote, as George Wilson pointed out in his writings, interesting and profound reflections upon God and chemistry by a lifelong invalid, who had in 1843 suffered the amputation of his left foot without anaesthetics (for which he subsequently became a fervent advocate). Wilson was in 1855 appointed the first (and perhaps last) Professor of Technology in the University of Edinburgh, and spent much time explaining what that meant; and he was also director of the museum, which in due course became the Royal Museum of Scotland and preserves elements of his vision. In 1858 he was urged to apply for the better-endowed chair of chemistry, but his failing health made that impossible. After his death his sister Jessie wrote his biography, including his poems and his horrifying subjective account of his amputation; and also edited articles and papers into a volume to which she gave the title *Religio Chemici.* He had intended if spared to write a book with that title, on the model of Sir Thomas Browne's *Religio Medici* (1642/1643); what we have instead is a stimulating collection of essays, including brief and perceptive biographies of Boyle, Dalton, and William Hyde Wollaston as well as the religious reflections that would presumably have filled the projected volume. He delighted in the history of science, and had written both a full-scale biography of the reclusive Henry Cavendish, and a book on colour-blindness and its implications for engine drivers and sailors, to which Clerk Maxwell wrote an appendix.

The allegorical title-page of *Religio Chemici* shows an angel, the winged Hermes, and chemical apparatus, with a Biblical quotation (*John,* 1, 3–4); and the book contains, as well as the biographies, an essay on 'The Chemistry of the Stars' that raises the then-popular topic of extra-terrestrial life. William Whewell (against) and David Brewster (for) had turned this into a controversial theme involving questions about creation, evolution and redemption; Wilson was judicious as usual. Three deal more explicitly with our subject: 'Thoughts on the Resurrection', 'Chemical Final Causes', and 'Chemistry and Natural Theology'. He was a man of firm evangelical faith, with somewhat variable denominational loyalties: initially Baptist, when in England Anglican, and then later Congregational. On the resurrection of the dead,

Wilson went for bodily resurrection, from a few of our actual particles, with personal identity conserved. He did not, as he might have done having suffered amputation, ponder at what age and in what state we might be resurrected. On final causes, he looked like Fownes particularly at the components of living organisms: "no creature is a fortuitous concourse of atoms."[30] There is an endless flux of particles in a stable organism, "'a temple always complete, and yet always under repair'" in an amazing concatenation of processes that maintain life. While he declared that we go wrong if we make teleology, the search for final causes, the chief object of scientific inquiry, we cannot seriously doubt that the study of the chemistry of living organisms directs us to God. And thus the contemplation of final causes gives us a foretaste of heaven.

Much more original and interesting was his examination of natural theology, which took the form of a kind of essay-review of the books by Prout and Fownes. He began in the traditional manner, in this case looking at the atmosphere, the mix of active oxygen and inert nitrogen being just right for living beings: if the Earth is our mother, the atmosphere could be called our foster-mother. This theme was to be taken up by Cooke, in lectures published in 1864.[31] But then Wilson asked whether we do in fact see unthwarted benevolence.[32] The atmosphere goes also with lung-diseases, and with the miasmas that bring plagues; while other chemical elements and compounds, all too common, go with poisons and death. Prout and Fownes, like other writers on natural theology, have too often ignored "the *dark* side." They blandly, vaguely and unsatisfactorily allude to the greater good. Men, and women even more, need something better. Women, Wilson noted, scarcely read natural theology, because it was full of "intricate and unfeminine science" and failed to lift "the dark veil." They especially could not avoid or turn away from pain, suffering and death, from disease and bereavement, and from natural disasters that befall humans, and that both now and in the remote past revealed in the fossil record, have always overtaken animals. Moreover, though we are constantly under repair, we do not remain as good as new: we age, and wonder why. Looking briefly outside chemistry, Wilson noted that carnivores were "made to destroy": "God has been very kind to the shark." Agony, pain and suffering are no "*transient,* incidental, occasional thing", but must be part of our world-view.

Chemistry, he added, cannot weigh pain against happiness in some kind of utilitarian calculation. We may even perhaps be made to wonder whether an evil as well as a good being is at work in the world; but in reality there need be no apprehension of that as a scientific conclusion. But it does all add up to an insoluble problem, an enigma: "Chemistry can prove that God is light, but not that in him is no darkness at all." The mystery of pain will haunt us through our whole lives; but we can still reasonably hope that while the curtain is indeed thick, the light shines through, and

[30] George Wilson, *Religio Chemici: Essays,* ed. J. Wilson (London: Macmillan. 1862), 114, 121, 164.

[31] J. B. Cooke, *Religion and Chemistry, or Proofs of God's Plan in the Atmosphere and its Elements* (New York: Scribner's, 1864).

[32] Wilson, *Religio Chemici,* 26, 31, 43, 50.

that death will be swallowed up in victory. That is, the promises of revealed religion in the Christian dispensation will console us in a world where benevolence cannot be clearly discerned. We might say that for Wilson the subjective experience of pain, and the miseries he saw around him in the lives of people and animals, revealed the remoteness and thinness of natural theology: the First Cause was not the same as God the Father, who loves and consoles us, and his Son Jesus, the man of sorrows and acquainted with grief.

We have come a long way from Davy's pantheism to more orthodox religion, from those years when "Bliss was it in that dawn to be alive/But to be young was very heaven!" of Wordsworth's, Coleridge's and Davy's youth on to sober mid-Victorian times, and thoughts of suffering and death. The intervening years were also a time in which the promises of Baconian science were being fulfilled in the technology Wilson displayed in his museum. Verse, which Erasmus Darwin had used to disseminate his science, and in which Davy had expressed his deepest thoughts, was now used by 'Red Lions' at the B.A.A.S. and by Clerk Maxwell for comic effect. But in 1870 John Tyndall, the successor and heir to Davy and Faraday at the Royal Institution, gave an address to the B.A.A.S. on the use of the imagination in science, where, controlled by experiment, he believed it to be essential. Science was not just technical problem-solving using Baconian induction: imagination was not only for lovers, lunatics and poets. Coleridge can have the last word; he had written (quoting Milton) of

> the strong hold which in all ages chemistry has had on the imagination. If in SHAKSPEARE we find nature idealized into poetry, through the creative power of a profound yet observant meditation, so through the meditative observation of a DAVY, a WOOLLASTON, or a HATCHETT;
>
> By some connatural force
> Powerful at greatest distance to unite.
> With secret amity things of like kind,
>
> we find poetry, as it were substantiated and realized in nature: yea, nature itself disclosed to us ... as at once the poet and the poem.[33]

The chemist might indeed still hope to be poet, philosopher and sage.

---

[33] Samuel T. Coleridge, *The Friend*, [1818], ed. Barbara E. Rooke (London: Routledge, 1969), vol. 1, 471.

Chapter 6
Facts or Fantasies in the Chemistry Lecture Theatre?

Robert G.W. Anderson

Abstract Towards the end of the eighteenth, and into the nineteenth century, subscription courses of chemistry lectures were established in Great Britain, not only for those intending to become doctors or chemists, but for a more general public. Indeed, such courses became fashionable pastimes. It is remarkable how many contemporary references there are to women not only attending these courses in Great Britain but commenting on the personal attributes of the lecturers. There are also a number of poems and cartoons which support the idea that these kinds of feelings existed.

This paper considers the comments which were recorded and speculates to what extent the lecturers could be considered to be true celebrities of their time within certain female groups.

The current level of involvement of women in science, as students and as practitioners, has been only slowly achieved.[1] Before the twentieth century, those who were involved in any form of scientific activity were few. Indeed, the small numbers then educated beyond a fairly basic level of competence were unlikely to have the sciences included in their curricula, other than possibly a study of natural history. Women were not admitted to universities in Great Britain until the late 1800s, and even then they were often taught separately from men. They were scarcely ever admitted to scientific academies.[2] It is possible to detect certain changes starting to take place around 1800, one of these being that women were able to attend some of

[1] Londa Schiebinger *The Mind Has No Sex? Women in the Origins of Modern Science* (Cambridge Mass: Harvard University Press, 1989). There were exceptions to the general rule that women were not involved in science, for example, a few wrote pedagogic scientific works which became highly popular. See also Gerald Dennis Meyer *The Scientific Lady in England* (Berkeley and Los Angeles: University of California Press. 1955) and Patricia Phillips *The Scientific Lady: A Social History of Women's Scientific Interests 1520–1918* (London: Weidenfeld and Nicolson, 1990).

[2] In the case of the Royal Society of London, women could not become fellows until 1945. See Joan Mason, "The Women Fellows' Jubilee," *Notes and Records of the Royal Society of London* 49 (1995), 125–40.

R.G.W. Anderson (✉)
Chemical Heritage Foundation, 315 Chestnut Street, Philadelphia, PA 19106, USA
e-mail: rgwa2@hermes.cam.ac.uk

J. Buchwald, L. Stewart (eds.), *The Romance of Science: Essays in Honour of Trevor H. Levere*, Archimedes 52, DOI 10.1007/978-3-319-58436-2_6

the subscription courses in science. This was not restricted to the middle classes and above. The rise of mechanics institutes in the 1820s could provide opportunities for women of poorer classes to come into contact with science, but the picture is mixed one. The earliest, the Edinburgh School of Arts, was attended by men alone until 1869, while the London Mechanic's Institution, founded in 1823, admitted women members from 1830. Higher up the social scale, attending science lectures became a fashionable pastime for the wealthier classes of both sexes by the end of the eighteenth and in early nineteenth century.[3] Some of their teachers even became well-known public figures, their followers flocking to hear them. A number of courses were devised and taught by university teachers (probably to enhance their incomes, though they may have admitted only to the philanthropic aspect of their extra-mural teaching). It would have been unlikely that such a cultural change would occur without repercussions. This paper looks at various aspects of women learning science in public contexts in Britain, and concentrates largely on graphic representations and contemporary commentaries. These have to be treated with considerable care – but neither should they be ignored.

A number of prurient stories developed, implying that certain male chemistry teachers were considered to be strongly attractive to the women in their audience. Of one (in fact, rather pompous) Scottish professor, Thomas Charles Hope, it was said: "the Ladies are ready to break their necks after him" and other lecturers were said to be pursued just as enthusiastically.[4] The veracity of these stories needs to be judged with care.

From early in eighteenth-century Britain, informal courses of natural philosophy and chemistry thrived. Lecturers sometimes often moved round from place to place with their apparatus, with which, for a subscription fee, they performed their demonstrations. In larger towns and cities lecture theatres might be available, and courses tended to be given on a more regular basis. Women must have attended some of these. In 1741 the lecturer in natural philosophy, J. T. Desaguliers, pointed out "Note, Ladies attend the lectures as well as gentlemen.[5] Such demographics of audiences were sometimes catered for: a 1796 course on the philosophy of natural history, held in Newcastle in the north of England, exacted one guinea per gentleman, but ladies were admitted at half-price.[6] (This charging regime is not so very different from some night clubs of the present day!). There were wide variations in

[3] For the emergence of science into the public domain, see Jan Golinski *Science as Public Culture: Chemistry and Enlightenment in Britain, 1760–1820* (Cambridge: Cambridge University Press, 1992) and (for a rather earlier period) Larry Stewart *The Rise of Public Science* (Cambridge: Cambridge University Press, 1992).

[4] Hope was professor of medicine and chemistry at the University of Edinburgh from 1795 (initially jointly with Joseph Black) then sole professor until his resignation in 1843.

[5] Alan Q. Morton and Jane A.Wess *Public & Private Science. The King George III Collection* (London: Science Museum, 1993), 64.

[6] Robert Spence Watson, *The History of the Literary and Philosophical Society of Newcastle-Upon-Tyne (1793–1896)* (London: Walter Scott, 1897) p.204. The course was offered by Henry Moyes.

the quality of teaching being offered. Many teachers concentrated on spectacular demonstrations. Some made serious contributions to their subjects while a few can be considered as being little more than charlatans. The courses given by the serious scholars often mirrored, at a somewhat simpler level, the lectures which they gave to their regular students at universities. Not infrequently, there were concerns that the understanding of mathematics needed would be too great for those without a grounding in the subject. The Kelso School of Arts (Kelso is a small border town in the south of Scotland) advertised in 1824 a course of "Chemical Lectures not requiring mathematical knowledge to make [the chemistry] intelligible" which were given by the local Church of Scotland minister, the Reverend George Gray.[7] Clearly these were popular, but they were only successful up to a point, as the annual report reads: "Mr Gray's introductory Lecture was delivered to an overflowing audience, consisting of ladies, gentlemen, mechanics and tradesmen… Mr. Gray did not deem it expedient to institute any formal examinations, as the difficulty of fully understanding many of the technical terms of Chemistry… presents unavoidable obstacles."[8]

Mr. Gray was clearly a conscientious teacher, but there were some shady characters within the teaching fraternity. One of these was 'Dr Gustavus Ketterfelto' who by 1798: "has been Travelling these 34 Years past [sic] through most parts of Europe, and has lectured in LONDON for many Years, with great applause where he has been honoured with some of the Royal family." Katterfelto's printed syllabus demonstrates his marketing skills.[9] His apparatus included Two Mathematical Scales "by which he will discover how to make a guinea appear six or seven grains heavier in an instant", Six Different Phosphorus "Whereby he explains many curious and surprising experiments", and A Model of a House "Whereby he explains how a thief may be catched or killed when robbing a house." Subscriptions were one guinea for a course of 12 lectures, but servants were admitted to the "Back Seats" for one shilling per lecture, which was presumably preferable to standing outside in the rain with the sedan chair until their masters and mistresses emerged, edified, even excited, by their philosophical experiences.

A significantly more respectable teacher than Katterfelto was the Edinburgh Professor of Chemistry, Thomas Charles Hope. Hope was an extremely popular university lecturer, whose formal classes attracted hundreds of registered students every year.[10] One of his most famous pupils was the young Charles Darwin, who as a medical student in 1825 recorded: "The instruction at Edinburgh was altogether by lectures, and these were intolerably dull, with the exception of those on

[7] Robert G. W. Anderson, 'Chemistry Beyond the Academy: Diversity in Scotland in the Early Nineteenth Century' *Ambix* **57** (2010), 84–103 (at pp. 94, 95).

[8] *First Report of the Committee of the Kelso School of Arts* (Kelso: Kelso Mail Office, 1825), 14.

[9] Watson, *Literary and Philosophical Society of Newcastle upon Tyne* (n.6); see facsimile syllabus opposite p.206.

[10] As many as 554 persons subscribed for Hope's chemistry course in 1823–124, see Jack Morrell, 'Practical Chemistry in the University of Edinburgh, 1799–1843' *Ambix* **16** (1969), 66–80 (on p.76, n. 84).

chemistry by Hope."[11] Like his predecessor, Joseph Black, he gave extra-mural classes, the earliest of his on record being in 1801. In 1826 (repeated in 1828), Hope decided to diversify his teaching still further and he started a session course of Popular Lectures on Chemistry for the public, charging two guineas per person.[12] This meant that for the first time women could attend a chemistry class taught within the confines of the University. There is a wickedly amusing, but probably highly exaggerated, description of the enterprise by the prominent Edinburgh lawyer, Henry Cockburn, in a letter of 1826 to a fellow lawyer, Thomas Kennedy MP:

> The fashionable place here is now the College, where Dr. Thomas Charles Hope lectures to ladies on Chemistry. He receives 300 of them by a back window, which he has converted into a door. Each of them brings a beau, and the ladies declare that there was nothing so delightful as these chemical flirtations. The Doctor is in absolute extacy with his audience of veils and feathers, and can't leave the Affinities. The only thing that inwardly corrodes him, is that in an evil moment, when he did not expect to draw £200, he published that he was to give the fees to found a Chemical prize, and that he now can't retract, though the said fees amount to about £700. Horrible --------. I wish some of his experiments would blow him up. Each female student would get a bit of him.[13]

In fact, the popular course which Hope gave was rather rigorous, comprising 22 lectures, each of one and a half hours duration, covering many of the more important aspects of heat, pneumatic chemistry, acids, salts and electricity. The seriousness of his intent is indicated by the way he signed-off the courses with the following comment: "You are now aware that Chemistry is an Experimental Science, & will readily be convinced be pursued in the closet alone – it is necessary to engage in the labors of the Laboratory – Experiments must be performed & processes conducted".[14] Whether these comments might have been intended for his female as well as his male audience is unclear. William Thomas Brande of the London Institution said of women "It is not our intention to invite them to assist in our Laboratories" and Jane Marcet said that it was inappropriate for them to enter "into the minute details of practical chemistry."[15] Sometimes chemical experiments could be pursued by women – but at home. George O'Brien Wyndham, 3rd Earl of Egremont (1751–1837) provided facilities for his (from 1784) live-in mistress, Elizabeth Ilive (1769–1822).[16] He was an acknowledged philanderer, and this was the main reason that his

---

[11] Paul H. Barnett and R.B. Freeman (eds.) *The Works of Charles Darwin Volume 29*: Nora Barlow, ed., *The Autobiography of Charles Darwin* (New York: New York University Press, 1989), 93.

[12] R.G.W. Anderson *The Playfair Collection and the Teaching of Chemistry at Edinburgh 1713–1858* (Royal Scottish Museum: Edinburgh, 1978), 42.

[13] Letter, Henry Cockburn to Thomas Francis Kennedy, 27 February 1826, see *Letters Chiefly Connected with the Affairs of Scotland, from Henry Cockburn to Thomas Francis Kennedy* (London 1874) pp.137–38; also, Jack Morrell, "Science and Scottish University Reform: Edinburgh in 1826," *British Journal for the History of Science* **6** (1972), 39–56 (at p.55).

[14] Edinburgh University Library, Centre for Research Collections, MS Gen 271, Bundle 124: in envelope marked: 'Progress Popular Lect 1826/&1828'.

[15] See Golinski *Science as Public Culture* (ref. 3), 261.

[16] A. McCann, "A Private Laboratory at Petworth House, Sussex, in the Late Eighteenth Century," *Annals of Science* **40** (1983), 635–655.

eventual marriage in 1801 to Ilive lasted only 2 years. Wyndham could not quite be described as a chemist, though he was known as an agriculturalist, and he took a deep interest in both, as he was in the process of developing his vast estates. He was elected a Fellow of the Royal Society and consulted the chemist Charles Hatchett and mineralogist Charles Greville concerning the viability of exploiting his mineral deposits. To support Ilive's interest he had a chemistry laboratory established at Petworth House in Sussex (of which a large quantity of glassware and ceramics from it still survives in the house).[17] Did she feel an attraction to his scientific leanings, or was it was more his wealth? This is a question which cannot easily be answered, though she departed rather rapidly when, finally, she could not tolerate the continuing existence of "my Lord's Seraglio", and they divorced.

Henry Cockburn was not the only person to suggest that female members of chemistry lecture audiences felt an attraction to their lecturers. John Rotheram had enjoyed this reputation somewhat earlier.[18] When the great Edinburgh chemist Joseph Black started to feel his age and teaching was taking too much out of him, he appointed the chemist Rotheram in 1791 to help him out. Rotheram himself had delusions that he would take over from Black, but it was not to be: Black instead chose Hope as his successor, and that meant that Rotheram had to find another job, which he managed in 1795, becoming Professor of Natural Philosophy at St Andrew's University. Rotheram was not to know that among his students there was a young disaffected poet, John Strachan, who was struggling to make his way through university and who was not at all impressed by the teaching he was receiving.[19] Strachan emigrated to Canada in 1799, and in 1802 wrote 'A Dialogue', a vicious verse in which he was harshly critical of the teachers who had taught him at St Andrews:

> "Learning's pursuits in these Northern climes,
> Deserve a keener pen to mark their crimes.
> Dull Heron first, but confident, and vain,
> A critic great with "clouda" in a train,
> Next Stewart with boldness muds his master's stream,
> And gloomy Erskine chocks the sacred flame.
> Thomas gapes and writes a dedication.
> Mansfield yawns and gives the presentation;
> Tho' nonsense all, the fool must have some food,
> And Mansfield swears it's for the public good.
> "By Jove!" cries Rothram squinting at his legs,
> "Fresh snow in puddings far excels good eggs".[20]

---

[17] Mary Holbrook, *Science Preserved* (London: HMSO, 1992), 195–96.

[18] John Rotheram (c.1750–1804), graduated MD at Uppsala in 1775.

[19] John Strachan (1778–1867) was a significant figure in the Church and education in the history of nineteenth-century Canada. See his biography by G. M. Craig in *Dictionary of Canadian Biography* IX (Toronto: University of Toronto Press, 1976), 751–766.

[20] Wanda Campbell, ed., *Poetry by John Strachan* (London, Ontario: Canadian Poetry Press, 1996), 22 and 33–4.

Fig. 6.1 Henry Moyes, etching by John Kay, 1796

This macaronic, unpublished at the time, is fairly obscure - but Strachan added a lengthy and crushing note to his couplet, making reference to Rotheram's extra-mural lectures (and extra-mural admirers):

> ...Let us no longer be told of the diligence of the old Philosophers and their address in communicating instruction, for they were only able to draw on the attention of young men. But John Rotheram M.D., Surgeon, Professor of Natural Philosophy, Chemist, Pharmacopeist [sic], Apothecary, Editor of 29 publications (few of which he understood), and elder of the Kirk of Scotland, has given several different lectures on Natural Philosophy in an assembly room to the grand edification of the Ladies who forsook the dance to catch the words of the Dr. distilling from his lips as honey from the comb.

One heart-throb of a chemical lecturer was truly an itinerant - which is surprising, because he had been blind from the age of three. Henry Moyse was born in 1749 near Kirkcaldy, in Fife, Scotland, and started his lecturing career in earnest in Manchester, Leeds and Birmingham when he was 33 years old[21] (Fig. 6.1). In May 1784 he turned up in Boston, and he commenced a remarkable two-year long tour down the east coast of America. He became hugely popular with his audiences. His first course in Boston was so crowded that he was implored to offer a second, but his plans had been made, and he travelled down to New York, and then to Philadelphia to offer further courses. Again, his teaching was received with adulation, especially

[21] John Anthony Harrison, "Blind Henry Moyes, 'An Excellent Lecturer in Philosophy'," *Annals of Science* **13** (1957), 109–125.

by his female followers. The event was recorded in a letter of March 1785, sent by the author and lawyer, Francis Hopkinson, to Thomas Jefferson:

> I have mentioned Dr Moyes. I shall now tell you who he is. He is a Scotchman by birth and a Philosopher by Profession. He arrived I believe a year ago at Boston and has come thence to this City, giving lectures in natural philosophy all the way. He spent the beginning of this Winter at New York, where he became very popular and a great favourite of the Ladies in particular, who crowded to his Lectures, and happy was she who [could] get him to dine or drink Tea at her House. Having gone thro' his Course there and reaped no small Honour and Profit, he is now performing with us... He exhibits three Evenings a week in the College Hall, [and] he has already given 10 or a Dozen Lectures to an Audience of not less than 1000 and most commonly 1200 Persons. The Ladies are ready to break their necks after him. They throng to the Hall at 5 o'clock for places, altho' his Lectures do not begin till 7.[22]

Moyes formed a friendship with fellow chemist Benjamin Rush - in fact, both had studied in Edinburgh under Joseph Black. From a letter between the two men, it seems that Moyse may have attempted a romantic association while in Philadelphia. When he arrived at Charleston on the way to the ship which would return him to England, Moyse wrote to Rush, on 24 May 1786, to say: "Remind me kindly to Mrs. Rush, & to Miss Polly Stockton. Tell the latter that I can scarcely forgive her for so cruelly disappointing me at Philadelphia."[23] It is not entirely certain who Polly Stockton was, but "the endowments of her heart" had been described by another admirer, John Pierce, 2 years earlier, as being "goodness and sensibility".[24] Moyes sailed back shortly afterwards. He remained peripatetic (and unmarried) for the rest of his life, teaching in Scotland, Ireland and England, and dying in Doncaster in December 1807. A sentimental poem purporting to have been written in 1807 and having the approval of 300 Ladies, 'The Edinburgh Ladies' Petition to Dr. Moyes', was published in *The New Monthly Magazine* some 28 years later. It is followed by 'Lord Byron's Reply'. The authorship of the latter is extremely doubtful; it is a mystery why these pieces should have been published as long as 28 years after Moyes' death.[25]

Humphry Davy, with his smouldering good-looks, was one chemist who clearly had a following of women admirers.[26] Davy started his chemistry career at the Pneumatic Institution in Bristol, a therapeutic center which had been set up by Thomas Beddoes, the purpose of which was the curing of disease, usually tuberculosis, through the inhalation of gases. It was here that Davy conducted experiments on the analgesic effects of laughing gas (nitrous oxide), which were published by

---

[22] Julian P. Boyd, ed., *The Papers of Thomas Jefferson...* volume 8 (Princeton: Princeton University Press, 1953), 50–51.

[23] Eva V. Armstrong and Claude K. Deischer, 'Dr. Henry Moyes, Scotch Chemist' *Journal of Chemical Education* **24** (1947), 169–74 (on p.173).

[24] John Pierce to Sarah Pierce, 9th April 1784, in Elizabeth C. Barney Buel, ed., *Chronicles of a Pioneer School from 1792 to 1833*, compiled by Emily Noyes Vanderpoel (Cambridge, Mass: [Harvard] University Press, 1903), 346–348.

[25] 'The Edinburgh Ladies' Petition to Dr. Moyes, with Lord Byron's Reply' *The New Monthly Magazine* **44** (1835), 420–24. See Appendix to this paper for the Ladies' poem in full.

[26] Jan Golinski, "Humphry Davy's Sexual Chemistry," *Configurations* **7** (1999), 15–41.

him in 1800. Davy moved to the recently established Royal Institution in Albemarle Street, London in 1801, and started teaching in April. The lecture theatre there was rapidly becoming one of the most fashionable venues in which to be seen by London society. The lawyer, Francis Horner, wrote in his diary on 31 March 1802:

> The audience is assembled by the influence of fashion merely: and fashion and chemistry form a very incongruous union. At the same time it is a trophy to the sciences; one great advance is made towards the association of female with masculine minds in the pursuit of knowledge, and another domain of pleasing and liberal enquiry is included within the range of polished conversation[27]

The lectures attracted an enthusiastic audience of both men and women, the latter constituting about one-third of the annual subscribers in the first year.[28] Women were said to be more attentive: a Spanish visitor, Don Manuel Alvarez Espriella, wrote: "Part of the men were taking snuff to keep their eyes open, others more honestly asleep, while the ladies were all upon the watch, and some score of them had their tablet and pencils, busily noting down what they heard, as topics for the next conversation party."[29] Humphry Davy started teaching at the Royal Institution in 1802 and became an instant hit with his audience, especially, it was said to women, some of whom may have had things other than conversation parties on their minds. A friend of Davy's, Thomas Poole, recorded that: "when [Davy] first lectured at the Royal Institution, the ladies said, "Those eyes were made for something besides poring over crucibles.""[30]

This suggested attraction women felt for Davy is obliquely indicated in a well-known 1802 caricature of James Gillray, 'Scientific Researches! – New Discoveries in PNEUMATICKS! – or – an Experimental Lecture on the Powers of Air – ' (Fig. 6.2). A chemical demonstration at the Royal Institution is shown. The youthful, coiffed, Humphry Davy is shown with a smile on his face as he waits to pass a gas enema to the demonstrator, Thomas Young, who is making Sir John Coxe Hippisley inhale a gas from another machine. The result is that Hippisley has produced a monstrous fart; in fact, the whole of his backside has exploded, ripping his clothes. What is of particular interest is the way in which Gillray has depicted the audience. There are six women, 14 men, and a girl and a boy, seated on benches. A woman and the girl are taking notes (as Don Manuel said they did), and one man appears to be taking snuff; none is actually shown asleep. The fashionably be-hatted women are clearly captivated by what they are experiencing, though whether it is because of Davy's beauty, or Hippisley's unfortunate fate, is difficult to judge.

---

[27] Richard D Altick, *The Shows of London* (Cambridge, Mass: Belknap Press, 1978), 367.

[28] For a discussion of women at the Royal Institution in its early years, see Patricia Phillips *The Scientific Lady: A Social History of Women's Scientific Interests 1520–1918* (London: Weidenfeld and Nicolson, 1990), 193–99.

[29] Robert Southey, *Letters from England by Don Manuel Alvarez Espriella*, volume 3 (London: Longman, Hurst, Rees and Orme, 1808), 284, 285.

[30] David Knight, *Humphry Davy: Science and Power* (Oxford: Blackwell, 1992), 50.

Fig. 6.2 'Pneumatic Researches' by James Gillray, 1802

This is not the only chemistry lecture of the time to be depicted: another, by Thomas Rowlandson and Augustus Pugin, shows the chemist Frederick Accum lecturing at the Surrey Institution on Blackfriars Road in London.[31] This is less of a caricature than Gillray's portrayal of a chemical lecture (Fig. 6.3). What is of particular interest here is that the large crowd of spectators is dominated by smartly-dressed women, though they are depicted as being far more decorous than Gillray's blousy group. A related drawing by Rowlandson in the London Museum, and an engraving of it, also show Accum lecturing, with a significant proportion of the audience being women (though the engraver has changed the sex of a bulky male in the foreground of the drawing to an equally bulky woman).[32]

Yet another cartoon, engraved by Isaac Robert Cruikshank, comes from one of the volumes describing tours of th fictional Doctor Syntax, published in 1820 (Fig. 6.4). The location is the London Institution in Finsbury Circus, which operated from 1815 to 1912. The lecturer talking to an assistant is possibly intended to be Michael Faraday; Dr. Syntax looks on in his quizzical way. The audience, which comprises nearly half of those present, is not entirely well-behaved: a young man is

[31] Plate in W. H. Pyne and William Combe *The Microcosm of London* (London: R. Ackermann, 1808–11).

[32] R. Burgess, "Humphry Davy or Friedrich Accum: A Question of Identification," *Medical History* **16** (1972), 290–93.

Fig. 6.3 Frederick Accum teaching at Surrey Institution, 1808–1811

Fig. 6.4 Dr. Syntax (?Faraday teaching), 1820

flirting with a girl and shows no interest in the chemistry being demonstrated. The text, which accompanies the image, reads:

> The visit to our Doctor next,
> Still sticking to his prying text,
> Intent through everything to pore,
> And gather rare and useful lore,
> Was to the London Institution,
> Where learning lives by contribution.
> Though oft, too oft, our man so holy
> Was doomed to hear from tongue of Dolly
> Full many a lecture, we may tell,
> Keen, and delivered very well,
> Yet still he wished some learned oration
> To hear from those who fill the station
> Of grave instructors, doctors wise.
> Who speak as if all learning lies
> Within the compass of their skull
> They seem so confident and full,
> And who perform the offices
> Of high priests to the sciences.
>
> Thus with a mind bent to explore
> And art and science to adore,
> Syntax expressed his resolution
> Of visiting the Institution,
> To spouse, who lov'd to shew her face
> At every gay and public place,
> And who like other ladies fair,
> Made it her most especial care
> To see and hear with pleased attention
> Whate'er surpassed her comprehension,
> For ladies now will e'en command
> Things that they do not understand.
> No wonder then the fair ones learn
> To lecture warmly in their turn,
> When taught to beat the horrid cross drum
> By the Gamaliels[33] of the rostrum.
> Pleased was our lady to attend
> The doctor thus their ears to lend
> To the renown'd, enlightened oracle
> Who oft descants so oratorical
> On chemistry, and all the arts,
> That shew you the component parts
> Of whatso'er in earth and air
> Employs the mind's enquiring care,
> And how ingredients are compounded,
> And likewise how they are confounded.[34]

[33] The Rabbi Gamaliel was one of the greatest teachers in Judaism.

[34] *The Tour of Doctor Syntax Through London, Or, The Pleasures and Miseries of the Metropolis. A Poem* (London: J Johnston, 1820), 205–206.

Fig. 6.5a Andrew Ure teaching at Anderson's College, 1825

Another lecturer at the London Institution, the chemist William Thomas Brande (1788–1866), positively encouraged women to "… partake of that healthy and refined amusement which results from a personal perception of the variety, order and harmony, existing in all the kingdoms of nature; and to encourage the study of those more elegant departments of science, which at once tend to exalt the understanding and purify the heart.".[35]

A somewhat different intention was served by two contrasting coloured woodblock prints by William Heath, "Rival Lectures", which appeared in *The Northern Looking Glass*, in 1825.[36] One shows the Andersonian Institution, which had been founded in 1796 in Glasgow for those unable to follow the routine of traditional academic study (Fig. 6.5a). A chemistry lecture is being given, probably by Andrew Ure, who taught there from 1803 to 1830. Three rows of the male-female audience are indicated, and at least six ladies wearing large hats are shown. One of the men and a woman are shown canoodling in the front row. The scene is lavishly lit by eight gas flares. The companion print indicates a rougher scene and is almost certainly

[35] Golinski, *Science as Public Culture* (ref. 3), 241.

[36] *The Northern Looking Glass* **1** (14 November 1825), 34. This periodical publication started life as *The Glasgow Looking Glass*. It stopped publication in 1826 after only 2 years; it is considered a contender to be the first periodical publication in comic strip form.

Fig. 6.5b John Steel teaching at Glasgow Mechanics Institution, 1825

meant to be the Glasgow Mechanics' Institution in Shuttle Street, founded in 1823 (Fig. 6.5b). The lecturer is probably intended to be John Steel (dates unknown), who taught chemistry and mechanics from its earliest days. There are no women present here, only lines of rowdy men, one of whom clearly has a black eye. Only two flares light the scene. The caption accompanying the cartoons emphasizes the intended moral: "The social equilibrium cannot be maintained unless the higher ranks keep in intellectual advance of the lower". Lecture-going was able to emphasize admission of women to the former but not the latter stratum of society.

Over the period considered in this paper, from the later eighteenth century to the middle of the nineteenth century, women did make significant advances in a subject which, to that point, had been essentially a men-only preserve. Although she did not attend Peter Shaw's lectures in London, Marie Thiroux d'Arcanville (1720–1805) was asked by Pierre-Joseph Macquer to translate his *Chemical Lectures* of 1731, which was published as *Leçons de Chymie Propres á Perfectionner la Physique, le Commerce et les Arts* (Paris 1759). Two British women chemists of the late eighteenth century were married to chemists and may have been influenced to take up the subject by them. It is unclear whether Elizabeth Fulhame attended chemistry lectures or whether she was entirely self-taught. Practically nothing is known about her, save that she was the wife of Thomas Fulhame MD, and that she wrote *An Essay on Combustion with a View to a New Art of Dying and Painting* (London 1794). Rather more is known about Jane Marcet (1769–1858), author of the many editions of *Conversations on Chemistry* first published in London in 1805 or 1806.

It was Humphry Davy's Royal Institution lectures which inflamed her interest, and she wrote in the Preface of a later 1806 edition of her work:

> On attending, for the first time, experimental lectures, the author found it almost impossible to derive any clear or satisfactory information from the rapid demonstrations.
>
> Which are usually, and perhaps necessarily, crowded into popular courses of this kind. But frequent opportunities having afterwards occurred of conversing with a friend on the subject of chemistry, and of repeating a variety of experiments, she became better acquainted with the principles of that Science, and began to feel highly interested in its pursuit. It was then she perceived, in attending the excellent lectures delivered at the Royal Institution, by the present Professor of Chemistry, the great advantage which her previous knowledge of the subject, slight as it was, gave her over others who had not enjoyed the same means of private instruction[37]

The question arises as to where Jane Marcet first attended chemistry lectures which initially she found to be of so little value. The private instruction might have been available from a number of London-based chemists. Her husband, Alexander Marcet was a physician at Guy's Hospital, where chemistry had been taught from 1781 by William Babington, though it is extremely unlikely that she would have been admitted to the lecture course for male trainee-doctors.

A further issue is whether the commentators and illustrators were providing descriptions and representations which bore any close relationship to real situations. In the case of Henry Cockburn's ladies, who are said to have clambered through a window to hear Hope's lectures, this would seem unlikely. The truth of the matter is that the Old College was built slowly and in phases, so that it is likely that temporary admission arrangements had to be made for everyone.[38] Cockburn's biographer in the Oxford DNB refers to the "inspired exaggerations" in his writings. There was probably no love lost between the two men, Cockburn being a liberal Whig while Hope was instinctively a Tory. They could scarcely have avoided each other in the Edinburgh social and intellectual hothouse of the early 1800s.[39] Cockburn would eulogize Joseph Black in his old age ("a striking and beautiful person, tall, very thin and cadaverously pale… he glided like a spirit through our rather mischievous sportiveness.").[40] For Cockburn, the much less charismatic, lumpen Hope was no match for this vision of past greatness.

Equally, some illustrations referred to could be considered caricatures. Certainly the point of Gillray's 'Discoveries in Pneumaticks' is intended primarily to ridicule

---

[37] Most easily found in the reprint, Jane Marcet *Conversations on Chemistry* volume1, Aileen Fyfe, ed., (London: Thoemmes Continuum and Edition Synapse, 2004), v, vi.

[38] The major problems in constructing the new University of Edinburgh buildings was that the original architect, Robert Adam, died soon after building work had commenced in the 1780s and the war with France was protracted, The appointment of Adam's successor, William Playfair, and redesigning and recommencing the project was a stop-go affair. See Andrew G. Fraser *The Building of Old College* (Edinburgh: Edinburgh University Press, 1989), particularly chapter 8, 'The Playfair College: Phase Two, 1823–1840', 243–78.

[39] Both men had studied at the Royal High School, followed by the University of Edinburgh, both were directors of an official state body to erect a National Monument in Scotland (the unfinished reproduction of the Parthenon which remains on Calton Hill, Edinburgh, today), and so on.

[40] Henry Cockburn *Memorials of His Time* (Edinburgh: Adam and Charles Black, 1856), 50–1.

Fig. 6.6 Richard Owen teaching at Museum of Practical Geology, 1857

the victim of the experiment, not provide an historically accurate representation of the mixed audience in the Royal Institution lecture theatre. The presence of women at lecture courses may have been deliberately exaggerated by illustrators. In many they are shown in prominent positions at the ends of rows, and this treatment continued through the century. To take three examples: in the woodcut 'Professor Owen Lecturing at the Museum of Practical Geology' (Fig. 6.6), five bonneted ladies are shown very close to the viewpoint of the artist.[41] The Royal Polytechnic Institution in Regent Street, London, provided audiences with scientific spectacle, and one of the most successful lecturers was 'Professor Pepper' (John Henry Pepper (1821–1900)); a print of 1869 shows him demonstrating the 'Great Induction Coil' to a large, mixed audience (Fig. 6.7). What is immediately noticeable is that the artist has placed most of the women in a group in the foreground of the woodcut, whilst another five women are shown in profile in a special enclosure.[42] In the third example, the French chemist Michel Eugène Chevreul (1783–1886), at the age of 100, is shown demonstrating a chemical experiment. In the left foreground, a prominently placed young woman has a notepad on her knees (Fig. 6.8). Chevreul is focusing in on her in what can only be called a flirtatious gaze.[43] A print of an imaginary lecture

[41] *Illustrated Times* (18 April 1857), 252, see Nicolaas Rupke *Richard Owen: Biology without Darwin* (University of Chicago Press: Chicago 2009), 30. At this time, Richard Owen (1804–92) was Superintendent of the Natural History Department at the British Museum while it was still in Bloomsbury.

[42] *Illustrated London News* (17 April 1869), 401.

[43] "Chevreul in his Laboratory," unknown periodical, volume 21 (1886), 337.

Fig. 6.7 'Professor' Pepper teaching at the Royal Polytechnic Institution, 1859

Fig. 6.8 Michel Chevreul teaching in Paris, 1886

Fig. 6.9 Richard Doyle, 'Manners and Customs of ye Englyshe', 1850

class was produced by Richard Doyle as one of the new series of his 'Manners and Customs of ye Englyshe' series drawn for *Punch, or the London Charivari* in 1850[44] (Fig. 6.9). This makes gentle fun of a lecturer teaching animal anatomy to a large expectant audience, comprising both men and women. The English in general are being satirised in the series (Doyle was an Irishman), the artist bringing out a wide range of expression, from intense concentration, to boredom, to flirtation (a recurrent theme for artists portraying lecture audiences).

Evidence concerning women attending the British Association for the Advancement of Science meetings (which started in 1831) has been analysed in some depth.[45] They were excluded from the scientific meetings in the early days, though at the 1837 meeting they were allowed into certain sections and by 1839 they could attend all of them. The BAAS meetings were rather a different matter from the lecture courses already discussed. For most, they were away from home and residential, and their intent was not so much pedagogic as a means for scientific men to meet and discuss developments in their disciplines. Most women came for the social aspects and to support their husbands. William Buckland emphasized this when, in 1832, he wrote, "Everybody whom I spoke to on the subject agreed that if

[44] *Punch, or the London Charivari ....* (1850), 'Manners and Customs of ye Englyshe. New Series No.10. A Scientific Institution. During ye Lecture of an Eminent 'Savan''.

[45] Rebekah Higgitt and Charles W. J. Withers, "Science and Sociability: Women as Audience at the British Association for the Advancement of Science, 1831,"-1901' *Isis* **99** (2008) 1–27.

the Meeting is to be of scientific utility, ladies ought not to attend the reading of papers – especially in a place like Oxford – as it would overturn the thing into a sort of Albemarle dilettanti meeting instead of a serious philosophical union of working men."[46] In an 1872 print, 'Science and Pleasure at Brighton' which appeared in *The Graphic* (Fig. 6.10), the damage appears to have been done. It shows a languid scene, away from the intellectual ferment of the lecture theatre. Its description of the most prominent figure, an attractive young woman lost in a reverie, reads: "The lady in the foreground, anxious to preserve a momento of the visit, is sketching the picturesque old building, and her attendant swain looks on in evident admiration of her handiwork.".[47]

In spite of the early presence of women at courses of chemistry lectures it was a long time before academic institutions were prepared to allow females to attend lectures on equal terms with men. At University College, London, separate classes for women in physics and chemistry were arranged for the session starting in 1869 and men and women were not taught together until 1878.[48] The University of Cambridge was not so advanced. A women's college, Newnham, founded in 1871, built its own chemistry laboratory in its grounds in 1879 as young ladies were excluded from the University's main chemistry laboratory.[49] The women were taught at Newnham by a particularly dynamic and competent female chemist, Ida Freund (1863–1914). However, she shared donnish characteristics with her male counterparts, one girl writing home to say: "Miss Freund is the presiding genius; a jolly stout German, whose clothes are all falling in rags off her back." Though numbers were small to begin with, it was becoming less unusual for women to be studying science and mathematics[50] and it seems doubtful whether serious female chemistry students in late nineteenth-century Cambridge would have formed emotional attachments to male teachers of their subject, as appears to have happened several decades earlier in public lecture courses.

One explanation for why stories developed about women having romantic feelings for their chemistry teachers is that the accounts and illustrations were invariably written, drawn and published by men. For some writers and artists,

---

[46] Letter, William Buckland to Roderick Impey Murchison, 27 March 1832, quoted in Jack Morrell and Arnold Thackray, *Gentleman of Science. The Early Years of the British Association for the Advancement of Science* (Oxford: Oxford University Press, 1981), 150.

[47] *The Graphic* (31 August 1872), 186. The "picturesque old building" is simply described as "an old ivy-clad ruin, rich in archaeological associations".

[48] [W.P. Ker, ed.] *Notes and Materials for the History of University College, London* (London: H K Lewis, 1898), 14, 15.

[49] Liba Taub and Ruth Horry, "At the Sign of the 'Laboratory of Physical Chemistry': Preserving and Presenting Histories of Chemistry in Cambridge," in Marta C Lourenço and Ann Carneiro, eds., *Spaces and Collections in the History of Science: The Laboratorio Chimico Overture* (Lisbon: Museum of Science of the University of Lisbon, 2009), 195–205.

[50] Women were first examined in Cambridge in 1882. In 1890, Philippa Fawcett was ranked first in the mathematical tripos. She could not be termed 'Senior Wrangler', a title which up to that time only men had held. The University solved the problem by officially deeming her to be "above the Senior Wrangler". It was not until 1948 that Cambridge admitted females as graduands, full members of the University.

Fig. 6.10 'Science, and Pleasure', BAAS Brighton, 1882

there was no compunction in distorting what they wrote and drew: it was their stock-in-trade. But alternatively, the response can be explained by men feeling threatened by the presence of women participating in what had until then been an exclusively male preserve. The result was to ridicule the motives and desires of the opposite sex.

Appendix

The Edinburgh Ladies' Petition to Dr. Moyes, with Lord Byron's Reply

Dear Doctor let it not transpire
How much your lectures we admire
How at your eloquence we wonder,
When you explain the cause of thunder;
Of light'ning and electricity,
With so much plainness and simplicity;
The origin of rocks and mountains,
The seas and rivers, lakes and fountains,
Of rain and hail, of frost and snow,
And all the winds and storms that blow;
Besides an hundred wonders more,
Of which we never heard before.
But now, dear Doctor, not to flatter,
There is a most important matter,
A matter which you never touch on,
A matter which our thoughts run much on,
A subject if we right conjecture,
Which well deserves a long, long lecture,
Which all the ladies would approve __
The Natural History of Love.
Oh! List to our unified voice,
Deny us not, dear Doctor Moyes;
Tell us why our poor tender hearts
So willingly admit Love's darts?
Teach us the marks of love's beginning,
What is it makes a beau so winning?
What is it makes a coxcomb witty,
A dotard wise, a red coat pretty?
Why we believe such horrid lies,
That we are angels from the skies,
Our teeth are pearl, our cheeks are roses,
Our eyes are stars – such charming noses!
Explain our dreams waking and sleeping,
Explain our laughing and our weeping,
Explain our hoping and our doubting,
Our blushing, simpering and pouting.
Teach us all the enchanted arts
Of winning and of keeping hearts.

Teach us, dear Doctor, if you can,
To humble that proud creature man;
To turn the wise ones into fools,
The proud and insolent to tools;
To make them all run helter-skelter
Their necks into the marriage-halter;
Then leave us to ourselves with these,
We'll rule and turn them as we please.
Dear Doctor, if you grant our wishes,
We promise you five-hundred kisses;
And rather than the affair be blunder'd,
We'll give you *six score to the hundred.*
Approved by 300 Ladies, 1807.

Chapter 7
Poetry in War and War in Nature. From Vauban to *Naturphilosophie* to Clausewitz

Janis Langins

Abstract Carl von Clausewitz (1779–1831) knew the practice of war and was one of the great theoreticians of war. In the eighteenth century there was a flurry of writings on warfare that proposed systems of conduct of warfare but Clausewitz radically departed from them. This essay will suggest possible intersections between his ideas and themes that were popular with the so-called *Naturphilosophes*, who were influential in promoting what historians of science have called Romantic Science at the beginning of the nineteenth century.

My favourite book by Trevor Levere is his *Poetry realized in nature: Samuel Taylor Coleridge and early nineteenth-century science*.[1] In this paper I discuss a person whose name never appears in this book and whom the omnivorous reader Coleridge probably never read – the Prussian military theorist Carl von Clausewitz (1779–1831). I will attempt a speculative sketch of a thinker who was often obscure like the *Naturphilosophes* but with whom both he and Coleridge were familiar and I will argue that both came under their influence to a greater or lesser degree. Further, I will argue that Clausewitz tried to develop in his unfinished book *On War* a theory of war that, although he would not have called it a science, had some features of the science that was close to that of the *Naturphilosophes* and were not those of the Newtonian science dominant in England and France. Both Coleridge and Clausewitz reacted strongly and negatively to the materialist strand in the Newtonianism of their day and rejected the attempts by some to expand its applicability beyond mechanics. For Clausewitz, the attempt to do so for the terrible human drama of warfare was unjustified, futile, and harmful.

Science and war have mutually influenced each other in the early modern era. Research in ballistics goes back to Galileo. Although his results – correct in ideal theoretical cases – were of little use in practice, such research continued in the

[1] Trevor H. Levere, *Poetry realized in nature: Samuel Taylor Coleridge and early nineteenth-century science* (Cambridge and New York: Cambridge University Press, 1981).

J. Langins (✉)
University of Toronto, Toronto, ON, Canada
e-mail: j.langins@utoronto.ca

J. Buchwald, L. Stewart (eds.), *The Romance of Science: Essays in Honour of Trevor H. Levere*, Archimedes 52, DOI 10.1007/978-3-319-58436-2_7

eighteenth century into the era of the digital computer.[2] For their part, there were military writers from the Renaissance onwards who dreamed of a science of war that imitated a formal mathematical model with axioms, maxims, and general conclusions. The military historian Azar Gat, on whose more sophisticated studies of military thought I have relied heavily, has also argued that science was not an unchanging mode of thought and corpus of knowledge.[3] When one asserts that science influenced thinking on warfare, he argues, one must ask what *kind* of science? The science of the renowned Hapsburg general Raimondo Montecuccoli (1609–1680) was not the science that influenced French military writers of the Enlightenment and this was a science that only partially influenced Clausewitz.[4] Thanks to scholars like Levere, the *Naturphilosophie brand of* science formerly dismissed as fanciful elucubrations of German philosophers letting their imaginations run riot under the influence of *Sturm und Drang,* has received more respectful attention and its impact on scientific activity, philosophy, and areas beyond, even if ephemeral, has been evaluated more carefully, if not sympathetically.[5] By presenting some parallels between the thought of Clausewitz and the *Naturphilosphes* I hope to show that the spirit of the times in which he lived influenced his thinking as much as the new kinds of wars in which he fought.

I will begin my discussion of Clausewitz's theories by looking at how the career and military exploits Sébastien le Prestre de Vauban (1633–1707) were used to support the erroneous claim that he had provided a model for a science of warfare in France. I will follow with a glance at French Newtonian science and proceed to the challenges to that kind of science in Germany. This will lead me to Clausewitz and I will present some parallels between his thinking and that of the *Naturphilosophes* and Romantic thinkers close to them. I will contend that Clausewitz was both a poet and a philosopher of war.

Vauban, Louis XIV's great fortifier and taker of fortresses, enjoyed a great reputation in his lifetime for his military achievements and character, and that reputation has persisted to a remarkable extent up to the present day as can be seen by numerous

---

[2] Brett Steele, "Military 'Progress' and Newtonian Science in the Age of the Enlightenment" in B.D. Steele and Tamera Dorland, eds., T*he Heirs of Archimedes: Science and the Art of War through the Age of Enlightenment.* Dibner Institute Studies in the History of Science and Technology (Cambridge, Mass.: MIT Press, 2005).

[3] Azar Gat, *The Origins of Military Thought from the Enlightenment to Clausewitz* and *The Development of Military Thought: The Nineteenth Century* (Oxford: Clarendon Press, 1989) *and and his The Development of Military Thought: The Nineteenth Century* (Oxford: Clarendon Press, 1992*).*

[4] T.M. Barker, *The Military Intellectual and Battle: Raimondo Montecuccoli and the Thirty Years War* (Albany: State University of New York Press, 1975).

[5] See for example Hans Eichner, "The Rise of Modern Science and the Genesis of Romanticism." *PMLA* **97**:1(January, 1982), 8–30; Barry Gower, "Speculations in Physics: The History and Practice of Naturphilosophie." *Studies in the History and Philosophy of Science* **3** (1973), 301–356; and Robert J. Richards, *The romantic conception of life: science and philosophy in the age of Goethe* (Chicago: University of Chicago Press, 2002).

biographies, coffee table books of his fortresses, and interest groups.[6] A considerable part of Vauban's fame is associated with the exact sciences of his time. Although Vauban was a member of the *Académie royale des sciences*, he did not contribute to any scientific discoveries and his membership was that of a *membre honoraire* – someone who had an interest in science but was elected as much because of his social and administrative status and interest in science than for any distinct contribution to it. The science that was associated with Vauban's name falls into the rather broad category of Newtonian science – a science that had not attained a dominant position in France during his lifetime, which overlapped with Newton's.

Newton's *Principia* was known to Vauban's colleagues in the *Académie royale des sciences* but the dominant scientific world view in France was that of Descartes. The defenders of Descartes remained strong in the *Académie* and its *secrétaire perpetuel* Bernard le Bovier de Fontenelle (1657–1757) remained a Cartesian throughout his long life.[7] The elegant cosmological system of Descartes had a great attraction for Frenchmen of the classical age well into the Enlightenment and Newtonianism, which rejected Cartesian mechanics, had some difficulty in making headway in France. By the 1740s, however, in some measure due to the influence of the anglophile Voltaire and his mistress Madame du Châtelet, Newtonianism had effectively triumphed among French philosophers and scientists.[8] Newton, along with Bacon, had become one of the great heroes of the Enlightenment whose manifesto, the *Encyclopédie,* had commenced publication a few years before Fontenelle's death. The spectacular and universal results of Newtonian mechanics in the domain of navigation, celestial mechanics, measurements of the earth, and dynamics had left a strong impact. on mentalities and it had extended beyond these domains into other areas. Newtonianism's success, coupled with Baconian dreams for the application of science to improving man's estate, led to what today would be called scientism – the belief that the methods of the natural sciences could be applied to all dimensions of the social sphere. Its main proponent was another *sécretaire perpetuel* of the *Académie*, Marie-Jean-Antoine-Nicolas de Caritat, marquis de Condorcet (1743–1794).[9] Condorcet, the protégé of Anne-Robert-Jacques Turgot (1727–1781)

---

[6] A recent excellent one based on access to family archives is Michèle Virol, *Vauban: de la gloire du roi au service d'état*. (Seyssel: Champ Vallon, 2003). See also Anne Blanchard, *Vauban* (Paris: Fayard, 1971) and English speaking writers like Reginald Blomfield, *Sébastien le Prestre de Vauban, 1633–1707* (1938; New York: Barnes & Noble, reprint ed., 1971) and F. J. Hebbert and G. A. Rothrock, *Soldier of France: Sebastien Le Prestre de Vauban, 1633–1707*. (New York: Peter Lang, 1989).

[7] See Fontenelle's *Eloge de Monsieur le chevalier Neuton* (Paris: 1728) and the commentary on it in A. Koyré, *Newtonian studies* (Cambridge, Mass.: Harvard University Press, 1965), ch. III.

[8] It has been argued that by that time, Descartes' physics and cosmology was on the defensive. See Pierre Brunet, *L'introduction des théories de Newton en France au XVIIIe siècle* (Genève: Slatkine Reprints, 1970).

[9] On Condorcet, see Keith Baker, *Condorcet, from natural philosophy to social mathematics* (Chicago: University of Chicago Press, 1975) and also his article on the legitimation of progressive politics by science, "Scientism at the End of the Old Regime: Reflections on a Theme of Professor Charles Gillispie", *Minerva* **2** (1987), 21–34.

and Jean le Rond d'Alembert (1717–1783), was a leader of the party of the *philosophes,* a liberal aristocrat who went on to become an enthusiastic supporter and eventually victim of the French Revolution. His *Esquisse d'un Tableau historiques des Progrès de l'Esprit humain* published in 1795 also inspired a number of nineteenth century epigones like Auguste Comte (1798–1859) and the ideology of progress that has only come to be questioned seriously in our own time. Vauban, who had always had an interest in collecting economic and social data that he felt would be of use for taxation and military purposes and before him Sir William Petty (1623–1687) had already taken a quantitative approach to demography and economics.[10] Condorcet went farther with the concept of "Social Arithmetic" and dreamt of using science to construct voting systems that would lead to the most accurate and fair representation of the wishes of electorates. For this he probably deserves a place among the precursors, if not the creators, of political science.

From one perspective, these attempts that may strike us as naive reveal a certain residual Cartesianism. Although Newton had a system of the world, believed that it was part of God's plan, and that God was an active and ongoing presence in Nature, he had never aspired to create an all-encompassing system deduced from first principles alone and was perfectly consistent and self-sufficient. Nor was his methodology an almost exclusively deductive one. In spite to his famous disclaimer about not feigning hypotheses, they were working tools rather than unassailable rational entities. Experimentation played a major part in Newton's methodology and opened up not just means of potential verification but new research agendas. In short, one crude categorization of Newtonianism and Cartesianism could be that the former is open-ended and the later is closed – the latter strives for an elegant and universal system and the former provides an equally elegant and more verifiable basic structure of mathematical laws with a potential for application.

In his preliminary discourse on the *Encylopédie* d'Alembert, one of its two initial editors, contrasted the *esprit systématique* with the *esprit de système*. He was highly critical of the latter and he saw it as typical not only of the scholastic philosophers who still dominated theological faculties and universities in France but also of French Cartesians.[11] It was at once a source of specious intellectual comfort and a trap that prevented the development of knowledge and often perpetuated error. True philosophers were warned to avoid slipping into this mode of thought. However, turning one's back on the spirit of system did not mean that the world was to be perceived as a chaotic mass of unrelated facts. Using reason as a guide one could approach the world with a systematic spirit and discover patterns and truths that were coherent and useful. They would not be fences and barriers but

[10] A recent collection of articles sees him as an "architect of modernity." See Thierry Martin and Michèle Virol, eds., *Vauban, architecte de la modernité?* ([Besançon]: Presses universitaires de Franche-Comté, 2008.)

[11] Jean le Rond D'Alembert, *Discours préliminaire de l'Encyclopédie publié intégralement d'après l'édition de 1763, avec les avertissements de 1759 et 1763, la dédicace de 1751, des variantes, des notes, une analyse et une introduction*, F. Picavet ed., (1763, reprint ed., Paris: A. Colin, 1894), 30 and 116–117.

provide a base that was solid – even more solid than the putative systems beloved by system builders and at the same time provide richer practical knowledge. This is analogous to the simple schematic distinction between Cartesian and Newtonian approaches proposed above. But it also poses some problems. Both modes of thought agree that reason is important and paramount. The builders of systems are not uninterested in reality and the results of thinking systematically often lead to systems of explanation. Did not Newton talk of a System of the World? Did not Descartes look at rainbows? The distinction is not entirely clear cut. D'Alembert himself had a great respect for Descartes for undermining the systems of scholastics and this was not entirely a result of national affinity. Hankins has argued that d'Alembert was in some ways a crypto-Cartesian[12] His Newtonianism is that of the *Principia* and his interests lie solidly in the area of rational mechanics where one departs from a limited number of axiomatic principles to deduce laws of mechanics. It was this French brand of Newtonianism that was dominant in mid-century and which was also invoked as a justification and basis for thinking about social phenomena, including war.

There is little need to warn historians of science of the Protean nature of Newtonianism. Added to Newton's characteristic caution in advancing views on matters where he had legitimate and well-founded doubts are his different fields of interest. Even if we ignore Newton the Theologian – not at all easy to do – the Newton of the *Principia* and the Newton of the *Opticks* are almost two different Newtons and his disciples and admirers, even when they were capable of seizing the nuances of Newton's pronouncements, had free rein to develop several brands of Newtonianism. In his matter theory, Newton provided support for two streams of thought that Levere has called "mechanism" and "dynamism".[13] In the latter one can detect a further subdivision between the force atoms of John Rowning (1701?-1771) and the point atoms of the Jesuit scientist Roger Joseph Boscovitch (1711–1787). These are a far cry from the materialistic views on imponderable fluids that were probably the majority view among researchers of the phenomena of electricity, magnetism, and chemistry in the second half of the eighteenth century, especially in England and France, where Lavoisier's new chemistry gave added support to imponderable fluids for heat and light.[14] The dynamical view of matter introduced forces into science. This, along with ideas of attraction at a distance, was uncongenial with French scientists, especially those for whom Cartesian repugnance for such entities still exerted a strong influence.

In England, Newtonianism had penetrated most forcefully into the world of business and technology. There were many "projectors" of commercial and industrial schemes who hoped to exploit Newtonian ideas – or the reputation enjoyed by those

[12]Thomas L. Hankins, *Science and the Enlightenment* (Cambridge and New York, Cambridge University Press, 1985), 3–7.

[13]Trevor H. Levere, *Affinity and Matter: Elements of Chemical Philosophy 1800–1865* (Oxford: Oxford University Press, 1971), 8–22.

[14]Antoine-Laurent. Lavoisier, *Traité élémentaire de chimie: présenté dans un ordre nouveau et d'après les découvertes modernes: avec figures*. (Paris: Chez Cuchet, libraire, 1789).

ideas – popularized in numerous public lectures and coffee shops.[15] In France, it appears that these ideas were important in the development of scientism with a focus on social and political concerns. They also coloured military thought. Soldiers were not philosophers but they were increasingly aware of the usefulness of science in military affairs. This was not entirely a new phenomenon in the eighteenth century. Artillerists and fortifiers were the first to pay attention to the new discoveries in dynamics. From the Renaissance onwards, the importance of mathematics had been acknowledged and even exaggerated among builders of fortresses. Some of them came close to seeing mathematics as a tool to establish impregnability of fortifications. And it was Vauban, the greatest practitioner of siege warfare of his day, who in spite of himself, provided a justification of military thinking that consciously claimed to be scientific. The Newtonian conquest of French salons and academies had made science popular and even fashionable. It would be strange that soldiers would have been immune to the effects of this powerful intellectual current.

In the introduction to his book *Art de la Guerre, par Principes et par Règles,* published posthumously in 1748, Jacques de Chastenet, Marquis of Puységur (1665–1743), one of Louis XIV's generals who eventually attained the rank of Marshal of France, wrote:

> I am undertaking … to show that without war, without troops, without armies, and without being obliged to leave one's house, by study alone, with a bit of geometry and geography, one can learn all the theory of field warfare from its most detailed to its most general parts. This [can be done] in the same way that Marshal Vauban, by the theory contained in the book he has left us and the practice he established in conformity [with that theory], teaches us the art of fortifying, attacking, and defending fortresses. [This is] something which is taught daily by people who have never been to war nor have supervised work on fortifications.[16]

Puységur was one of a growing number of influential French writers who wrote on the art and science of war during the Enlightenment. Yet his argument that one could learn a science of war from books was something that in later years would attract the ridicule of Carl von Clausewitz who was always insistent that one could definitely not learn anything practically useful about the conduct of war "without being obliged to leave one's house, by study alone with a bit of geometry and geography." Writing on the art of war has a long history that stretches back to Antiquity but it was during the Enlightenment that there was a growing belief that the art of war could be patterned on the modern science that had triumphed in France in its Newtonian form about the same time Puységur's book was published.[17] Science was to be conscripted by war on the one hand to provide a methodology and approach to rationalize the increasingly complex elements of war such as logistics, transport, and the increasingly technical arms such as artillery and fortification and on the

[15] Larry Stewart, *The rise of public science: rhetoric, technology, and natural philosophy in Newtonian Britain, 1660–1750*. (Cambridge: Cambridge University Press, 1992).

[16] Jacques.-François de Châstenet de Puységur, *Art de la Guerre, par Principes et par Règles* (Paris, Charles-Antoine Jombert, 1748), .2.

[17] Robert S. Quimby, *The Background of Napoleonic Warfare: The Theory of Military Tactics in Eighteenth-Century France* (New York: Columbia University Press, 1957).

other hand to discover the underlying laws of war analogous to the laws of nature. There was a fourfold increase between 1748 and 1756 in the rate of publication of military texts over the rate from the beginning of the seventeenth century until that time. And between 1756 and the Revolution there was further fourfold increase to a level that would be maintained throughout the nineteeth century.[18]

There are three concepts that were the basis for Puységur's thinking. First, there was the belief that it is possible to create a true science of war – presumably a science that would generate a recipe for victory in any given case. It is highly plausible that this stems from the triumph of Newtonianism, the spectacle of its practical applications, and the temptation to expand its methods and results into domains beyond the exact sciences. The second idea was that modern science could generate a system – that one can deduce an unequivocal and coherent matrix of maxims and principles that have general applicability to military field operations. The messy business of war could be grasped intellectually without having to smell powder and accumulate actual military experience. Here was the love of systems as a residue of Cartesianism. D'Alembert himself was not entirely immune from this in spite of his awareness of its dangers. It is not surprising that lesser spirits were not and that popular French Newtonianism was an amalgam of Newtonian and Cartesian thought. Third, Vauban was considered to have provided a concrete example of the correctness of these views. Puységur explicitly justified his claims of having discovered a science of war by invoking the example of the great fortifier, who was alleged to have created a perfect system for siegecraft and fortifications.

In this Puységur was undoubtedly mistaken. It is a curious aspect of Vauban's thought that in spite of what many have written about him since his own time up to our own, Vauban was far from being an author of systems. Vauban had said that he had no intention of publishing a book on the science of fortification. The book to which Puységur was referring, published in a pirated edition in Holland about the same time as he was writing his own, was probably *De l'attaque et de la défense des places* (1737), a manuscript originally presented to the King in 1704. This book gave technical details of siegecraft and fortifications as well as some analysis of mistakes made in sieges familiar to Vauban. It did not give any overarching recipe for a perfect siege and a perfect defence. Years after Vauban's death the mathematics professor, writer on military subjects and contributor of the article "Fortification" to the *Encyclopédie,* Guillaume Le Blond (1704–1781), had even attributed three "systems" of fortification to Vauban, something that has been repeated again and again.[19] Vauban himself never had such a tripartite division of "systems" nor did he ever claim to have created a system in the sense of Puységur.[20] Vauban himself constantly stressed the

[18] Gat, *The Origins of Military Thought*, 25 citing bibliographic work by Pöhler as a source.

[19] P. Lazard, *Vauban 1633–1707* (Paris: Alcan, 1934).

[20] J. Langins, "L'ingénieur systématique contre les systèmes. La théorisation de la pensée de Vauban et la publication de ses écrits" in *Vauban, bâtisseur du Roi-Soleil,* I. Warmoes, V. Sanger, eds., Musée national des monuments français (France), Musée des plans-reliefs (France) and Cité de l'architecture et du patrimoine (Paris France). (Paris, Somogy: Cité de l'architecture et du patrimoine; Musée des Plans-reliefs, 2007), 330–339.

importance of conforming to the actual location, nature of the ground, existing fortifications, and available resources. In spite of Louis XIV's largesse for military operations, there were fiscal limits that could not be ignored. Continuous compromise and improvisation when building fortifications was necessary at all times. Pragmatism rather than systems was the prominent feature of Vauban's thinking.

It is undeniable, however, that Vauban's successes in his long and illustrious military career and the consequent reputation he enjoyed with friend and foe alike had resulted in the impression that he had a formula for invincibility that could be mechanically applied, as evidenced by the number of books referring to his methods by foreigners and Frenchmen alike.[21] Among the latter there was a large representation of people who were mathematics masters rather than soldiers. Mathematics exercises in schools sometimes took the form of fortification projects.[22] A veritable cult of Vauban existed among French military engineers but since they were familiar with the vicissitudes and unpredictability of warfare, paying a high price in blood for their experience, there were few who indulged in the system building of those to whom they contemptuously referred as the "professors of the square and the compass."[23] However, they were quite content to leave the impression that they did have a secret of invincibility based on science and mathematics left by Vauban. The advanced education in mathematics and science they received at the Royal Military Engineering School, founded at mid-century at Mézières, only enhanced the reputation of the scientific soldier.

But for Puységur and others, Vauban's siege operations that had worked like clockwork with the inevitable surrender of enemy fortresses – Vauban had participated in 50 sieges and had an unblemished record of success in all of them – were worthy of admiration and emulation. This suggested that one could construct systems of rules that would transform the art of war into something radically different from ways in which the chaotic and unpredictable wars of the past were conducted. Azar Gat rightly believes that there was a new context and mindset permeated with the prestige of Newtonian science (if not always with an accurate knowledge of it) and its spectacular predictive and deterministic success that worked not only in astronomy but fortification as well. It was certainly a factor in the upsurge of military writing and the belief that war could lend itself to intellectual analysis and the notion that military operations could be conducted with great predictability. Puységur's fixing on the example of Vauban and siege warfare suggests that this was the ideal example of the possibilities of his project to create an "art" according to "principles and rules."

---

[21] For example, Abbé Du Fay, *Manière de fortifier selon la Methode de Monsieur de Vauban, avec un Traité Préliminiaire des Principes de Géométrie* (Paris: J.-B. Coignard, 1707). An earlier edition had appeared in 1692.

[22] *Thèses de Mathématique, de Géométrie, de Trigonométrie rectiligne, et de Fortifications, qui seront soutenues au Collège de Louis le Grand, le Vendredy 25 Juin 1751, à trois heures de l'après midi* (Paris: Thibout, Impimeur du Roi, 1751).

[23] [Charles.-René Fourcroy de Ramecourt], *Mémoires sur la fortification perpendiculaire, par plusieurs officiers du génie* (Paris: Nyon, 1786), 112. Fourcroy was chief of the French military engineering corps at the end of the eighteenth century.

Neither Puységur nor the French were the only ones who had such ideas on warfare. At the end of the century, the Prussian Dietrich von Bülow (1757–1807) was also one of several military authors in Germany who attracted widespread attention with his theories of warfare.[24] Von Bülow's geometrical constructions for the conduct of military operations popularized ideas like the base of operations as well as interior and exterior lines that were to find a place in the curricula of military academies. Yet the attempts to establish an impeccable theory of war did not lead to the kind of unanimity that was the hallmark of Newtonian mechanics. Especially after the disasters of the Seven Years' War, French military theorists clashed about military tactics as well as the more strategic considerations regarding the conduct of operations. Jacques-Antoine-Hippolyte, Comte de Guibert (1743–1790), author of the *Essai général de tactique* (1773), marked an extension from the narrow considerations of what would today be called tactics, and even from the broader considerations of the conduct of operations, to look at international politics and the national temper of peoples.[25] In some ways he was a herald of the new ways of warfare that appeared in the wars of the French Revolution.

But the idea of an almost mechanical science of warfare was still strong at the end of the Old Regime. The elaborate geometrical designs for marshaling troops and schemes for flanking movements presented in the works of both François-Jacgues de Mesnil-Durand (1736–1799), the advocate of the *ordre profond,* and the defenders of the *ordre mince*, in spite of their differences, could fit in well with Puységur's attempts to create a scientific art of war. An exaggerated example of such attempts is Jacques-Antoine Réveroni, baron de Saint-Cyr (1767–1829), military engineer, briefly a teacher at the École Polytechnique during the Revolution, and writer of mediocre novels. He believed that war has mechanisms identical to those of Newtonian mechanics as can be seen from the title of his 1808 *Essai sur le mécanisme de la guerre*. He changed the title of his revised edition in 1826 *Statique de la guerre.* He had already turned his attention to art in his 1803 *Essai sur le perfectionnement des Beaux arts, Par les Sciences Exactes, ou Calculs et Hypothèses sur la poésie, la peinture et musique*. In the preface to this he wrote "It is beyond doubt: the great writers, the great artists would have been good applied mathematicians because true genius is universal."[26] Announcing his aim of mathematizing all the fine arts and linking them together may have put him at the extreme of cultivated opinion but did not put him outside it.

In this context it is not surprising that there were efforts to scientize the study of war in the wake of the dominant Newtonian paradigm for the exact and natural sciences. But towards the end of an illustrious military career and much meditation on the subject – meditation that spawned an extensive body of uncompleted work, including the monumental *On War* just before his death – Clausewitz turned his back on much of this tradition and suggested a new way

[24] Dietrich von Bülow, *The spirit of the modern system of war* (London: T. Egerton, 1806).

[25] Jean-Paul Charnay, *Guibert ou le Soldat philosophe* (Paris: Centre d'Etudes et de Recherches sur les Stratégies et les Conflits de l'Université de Paris-Sorbonne, 1981).

[26] Jacques-Antoine Réveroni, *Statique de la guerre ...*, Vol. 1, 3.

of looking at war that remains influential to this day. Like Puységur, Clausewitz also pointed to the developments in siege warfare in the eighteenth century (without explicitly mentioning Vauban) as an inspiration for an intellectual approach to war. Siege warfare had indeed taught soldiers that one could *think* about war.[27] However, he clearly did not believe that this was applicable to thinking about war in general. Even worse was the illusion that one could create a tidy system – characterized by abstract descriptions and general maxims into which one could fit all there was to know about war. In his preface, using an example from another critic of the penchant for tidiness and formal descriptions, Clausewitz quoted the Göttingen physics professor and satirist Georg Christoph Lichtenberg (1742–1799) mocking system builders with his extract from Prussian fire-fighting regulations:

> If a house is on fire, one must above all seek to save the right wall of the house on the left, and on the other hand the left wall of the house on the right. For if, for example, one were to try and protect the left wall of the house on the left, one must remember that the right wall of the house is on the right of its left wall, and thus, since the fire is also to the right of this wall and of the right wall (for we have assumed that the house is on the left of the fire) the right wall is closer to the fire than the left, and the right wall of the house could burn down if it were not protected before the fire would reach the left, protected wall; consequently something could burn down that was not protected, and sooner than something else, even if this something else was also unprotected; consequently the latter must be left alone and the former must be protected. To fix the point firmly in mind, one need only remember: if the house is to the right of the fire, it is the left wall that matters; and if the house is on the left, it is the right wall.[28]

He continued "The author does not want to put off the intelligent reader with such trite wisdom …".

Clausewitz did not reject the Enlightenment's faith in rationality, its search for order, and its faith in a scientific approach. This was illustrated not only in his remarks on siege warfare but in his general tone. Although he rejected most of the ideas of his predecessors, he conceded that

> Efforts were … made to equip the conduct of war with principles, rules, or even systems. This did present a positive goal, but people failed to take adequate account of the endless complexities involved.[29]

Clausewitz was nothing if not a thinking soldier, who had faith in reason to uncover the fundamental laws of Nature but also to help uncover those of warfare. However, he went beyond this basic attitude in two ways. Firstly, he rejected what may be called the residual aspects of Cartesianism in French Newtonian thought.

---

[27] Carl von Clausewitz, *On War* (1832; Reprint with Commentaries by Peter Paret, Michael Howard, and Bernard Brodie, Princeton,: Princeton University Press, 1989), 133. It is perhaps significant that Raymond Aron, one of the foremost students of Clausewitz, who was as much interested in Clausewitz's ways of thinking as in his specifically military ideas entitled his 1976 *Penser la Guerre, Clausewitz.*

[28] Ibid., 61–62 quoted from an unpublished manuscript written between 1816–18.

[29] Ibid., 134.

> … absolute, so-called mathematical, factors never find a firm basis in military calculations. From the very start there is an interplay of possibilities, probabilities, good luck and bad that weaves its way throughout the length and breadth of the tapestry.[30]

Newton's search for general laws was always closely tied to his experimental method and was severely disciplined in avoiding Descartes' aim of creating a philosophical system whose intellectual coherence and elegance pushed experiment and empirical results into the background. Puységur had displayed this in his overweening and extravagant ambition to create a complete system of military thought. Clausewitz, however, felt that, although thinking about war to develop a theory of war was necessary, only parts of this theory could be tidy and abstract. The elements of struggle and violence between human beings, as well as the presence of the goddess Chance on the battlefield, prevented this.

Secondly, even if the model of classical Newtonian science could not be universally applied, science was not impotent. There were lessons from another kind of science that were useful in thinking about war. Clausewitz goes beyond the Newtonian paradigm in embracing a new kind of inchoate paradigm that was popular in Germany during his lifetime.

In his work on the military thought of the Enlightenment and its transformation at the end of the eighteenth century, Gat has given this "climate of ideas" the general label of "The German Movement."[31] He stresses its heterogeneity and its sometimes contradictory elements, such as a continuation of some of the ideas of the Enlightenment but also the reaction against it so eloquently described in Isaiah Berlin's works on the Counter Enlightenment.[32] Gat sees four themes within it. Firstly, while the Enlightenment was solidly anchored to Cartesian rationalism, the model of Newtonian science, British empiricism, and neoclassical conceptions in the arts, a new generation of German thinkers felt that the world was too complex to fit into what they felt was a constricting and jejune structure. Such a structure was at best inadequate and at worst misleading and false. Nature was complex and diverse in its manifestations and it did not fit into tidy abstract systems. It was also dynamic: simple forces interacting in multifarious ways produced the richness visible to humans. Secondly, while admiring the achievements of British empiricists in their construction of human consciousness, they rejected the materialistic implications and the belief in the uniformity of the human mind that started out as a blank slate and had no active powers. They believed instead in the creative force of the human mind and the importance of subjective impressions. Humans were not mechanisms and did not behave mechanically. Thirdly, this emphasis on the creative, the personal, and the subjective led to a reaction against neoclassicism in the arts. Instead of the cultures of ordinary folk and primitive peoples being dismissed as barbaric and ugly because they did not conform to the rational norms of underlying beauty and the

[30] Ibid., 86.

[31] Gat, *The Origins of Military Thought*, 142ff.

[32] Isaiah Berlin, Isaiah and H. Hardy, *Three critics of the Enlightenment: Vico, Hamann, Herder* (London: Pimlico, 2000) and I. Berlin and H. Hardy, *The roots of romanticism* (Princeton: Princeton University Press, 2013).

canons of neoclassicism they were admired instead. Their vitality and spontaneity were favourably contrasted with what were seen as the desiccated and effete productions of Enlightenment artists. Finally, there was a historicist view of the evolution of mankind. Every age had its distinct essence and needed to be understood on its own merits. Present day codes of conduct and life could not be used to judge the past nor could the putative lessons of the past be applied to the present.

Although the centre of gravity of these ideas was Germany, there were hints of similar attitudes in other countries and even in France, at the epicentre of the Enlightenment there were people like Denis Diderot (1713–1784), the other editor of the *Encyclopédie,* who rejected d'Alembert's focus on mathematics and mechanics.[33] Diderot favoured the life sciences and has elements of pre-Romanticism in his thought. Romanticism did indeed have indigenous roots in other countries like England but the powerful impulse of Gat's German movement and Berlin's Counter-Enlightenment eventually made it more broadly European in the nineteenth century.

Although a soldier and not a philosopher, Clausewitz was exposed to all these streams of thought. Always a voracious reader, he and his wife attended the salons of Berlin and he was familiar with the ideas of Kant and undoubtedly other more or less rebellious epigones who defended the claims of the new kind of science – *Naturphilosophie* – in contrast to the regnant Newtonian science of the time. Azar Gat, Peter Paret, and Bernard Brodie all agree that Clausewitz was influenced by Kant.[34] Gat plausibly claims that Clausewitz was familiar with Kantian ideas through his exposure to their popularization by Johann Gottfried Karl Christian Kiesewetter (1766–1819), a student and friend of Kant who taught philosophy at the Military Academy in Berlin.[35] Grasping the richness, complexity, and subtlety of Kant's thoughts had eluded many of his best known followers and admirers Clausewitz would have been in excellent company if his Kantianism was primitive and distorted. But the combination of Kant's faith in Enlightenment science, along with the revision of the basis for this faith, and delineation of its boundaries of effectiveness is a visible and a major part of Clausewitz's writing. There is no sign that he adopted the more extravagant notions of *Naturphilosophie*. I feel, however, that an examination of his *On War* reveals that he had accepted a number of themes that were congenial to it.

It is perhaps fitting that a major figure in German Romantic thought, a friend of many *Naturphilosophes* and the author of the *Farbenlehre,* who was keenly interested in science but was critical of some of Newton's followers' pretentions to universality, provides a symbolic link between new ideas in science and the radically

[33] The classic text of Diderot in this connection is *De l'interprétation de la Nature* in D.Diderot, *Oeuvres Philosophiques*, ed. Paul Vernière (Paris: Garnier, 1956).

[34] Peter Paret, Peter, "The Genesis of *On War,*" 15 and B. Brodie, "The Continuing Relevance of *On War*", p.48 in the English translation of *On War* (Princeton: Princeton University Press, 1976). Against this, another great student of Clausewitz, Raymond Aron, t. 1, 436–437, rejects any influence by Kant.

[35] Gat, *The Origins of Military Thought*, 176 and 206.

changing nature of war. Goethe was present at a key moment in the history of the French Revolution. His words at the battle of Valmy in 1792, where the French repelled the invading Prussian army, have often been quoted: "From this place, and from this day forth begins a new era in the history of the world ...". The wars that pitted Revolutionary and Napoleonic France against the rest of Europe for almost a quarter of a century saw a new type of warfare that also saw the development of new theories of which those of Clausewitz proved to be the most original although not initially the most influential.[36] Clausewitz had joined the Prussian army at the age of 12 and had participated in the military campaigns of the 1790s, the battles of Jena, Borodino, and finally Waterloo. He was also a disciple and protégé of Gerhard von Scharnhorst (1755–1813), himself a prominent military thinker and reformer who had pondered the failures of the vaunted Prussian army against French. The "new era in the history of the world", which Goethe saw beginning in warfare and politics, had also begun in literature, art, and science and there were interactions between them. Clausewitz was pivotal to these interactions. Although he did not claim to be proposing a science of war, his theories of war display the imaginative insight that the *Naturphilosophes* felt was the indispensable tool for creating a Romantic science.

Here again a possible comparison between Coleridge and Clausewitz suggests itself. Both left a large volume of unfinished and occasionally contradictory work that has often frustrated or delighted their exegetes. For both, this was not merely the result of intellectual fastidiousness or intellectual humility but the result of struggling to understand the relations between abstract thought and ambient reality. By his own admission only the first of the eight books in Clausewitz's defining work *On War* is complete. He had worked on it for a lifetime and it is often sibylline and difficult to read. Coleridge, too, left many of his works unfinished and many of his most profound insights are found in the marginalia of his extensive readings. He wanted to create bridges between Man, Nature, and God in order to create a harmony among them that would vivify a sterile and mechanistic science, open an avenue for imagination and fulfilment for humans, and ultimately link these to the Divine. To encompass these aspects, considered isolated by many of his contemporaries, he wanted to ascend to a new level of thought that Levere, following M.H. Abrams, calls "metascience."[37] The best interpreter of this enterprise would be the poet who would be able to see the poetry in nature. Clausewitz, too, saw a poetry of a grimmer sort in war, which for him was a central and inescapable feature of society, but instead of the poet he sees the enlightened and philosophical soldier as the central figure in his vision. He grappled with the idea of war in theory and practice. His theory was not a system to provide certainty, but a method to guide judgement of the soldier when confronted with the uncertain and fluid circumstances of war both on the battlefield and in policy decisions.

[36] That man was Antoine-Henri de Jomini (1779–1869), who outlived von Clausewitz for many decades. On Jomini, see John Shy, "Jomini" in *Makers of modern strategy: from Machiavelli to the nuclear age*, ed. P. Paret (Princeton, N.J.: Princeton University Press, 1986), 186–216.

[37] Levere, *Poetry realized in nature*, 3–5.

For him, there was clearly no adoration of a science of frozen systems; his "theory" – he did not like to call it a science – was closely linked to art and was never definitive and complete. Its function was not dogma but heuristics and its main function seemed to be to provide a method – what in Coleridge's felicitous phrase was "a distinct science, the immediate offspring of philosophy, and the link or *mordant* by which philosophy becomes scientific and the sciences philosophical."[38] For Clausewitz, the purpose of thinking about war was to produce a theory that dealt with passion and violence, chance and probability, as well as rational considerations of politics and policy. A conventional scientific approach to some aspects of war – the administrative, logistical, and technical aspects – was necessary but by no means sufficient, whether one called it a science or a theory. This thinking would take into account the fact of human passions such as violence between humans and it would require humans, specifically soldiers in the form of great captains, to intuit the reality of war with the help of theory. As with poets who are the privileged interpreters of Nature for Coleridge, so for Clausewitz there was an ongoing and active communion between the great captains and the idea of war.

These views are highly compatible with Romantic Science evident in the Clausewitz's ideas on war as polarity, dynamism, and holism. The importance of polarity had already begun to appear in the increasingly important eighteenth-century science of electricity and magnetism with speculation on different kinds of electrical charges and magnetic forces that had an attractive or repulsive effect on each other. Charges interacted with forces between them. Newton had been typically cautious about forces and, while conceding they existed, he had restricted himself to descriptions of their effects without speculating on causes or their ultimate natures. Romantic scientists were less inhibited and assigned a metaphysical existence to forces and their dual nature. One result of this was the primacy of force over matter and the rejection of the materialist world view that was so often associated with Newtonianism. Boscovitch had already advanced this with his dynamical atomism and force fields that were alternatively repulsive and attractive, which gave matter its appearance of impenetrability, and provided a potential explanation for affinity and the apparent diversity of matter. This view also related the notion of polarity to dynamism, the second aspect of Romantic science, which rejected the idea of an inert matter. The world appeared as the theatre of the clash between opposing forces, simple in themselves but leading to the diversity and vitality of Nature. Moreover, Nature being alive, all its elements interacted with each other in a tightly meshed whole. A soulless mechanical determinism that shocked thinkers like Coleridge was replaced with a dynamic determinism governed by powers.[39] The Romantic holistic view saw Nature as a living, perpetually changing entity whose parts were continually interacting, not only with themselves but with the active mind of the observer who communed with Nature rather than viewed it as something detached. Clausewitz's Nature was not imbued with the gentle and often religious romanticism of a Coleridge. War was, regrettably but undeniably, a part of Nature

[38] Coleridge, cited by Levere, Ibid., 83.

[39] Levere, *Poetry realized in nature*, 3.

and even a major part of it. His Nature was not the benign Nature of the Romantics. He was no idolater of blood and iron, as some have claimed, but his Nature was more a Hobbesian Nature red in tooth and claw. This may be regrettable but he saw it as a stubborn empirical fact with which one must deal with a clear head.

There was also no sign of a religious sentiment in Clausewitz. There was no mention of God or gods in Clausewitz, not even Mars. There were no gods for him but there was a goddess in war and her name was Fortune or Chance. Luck as a prominent feature of war undermined all mathematical formulas attempting to ensnare it. However, this does not imply the surrender of the military intellectual to chaos and mindlessness. For example, probability permits a limited but real tool to understand and plan for war. The harmony between the powers and forms of the mind and the powers and forms of nature allowed one to grapple with war, never with total success but never fruitlessly. Just as the thinking of philosophers and scientists should be active and imaginative, the soldier also needed these qualities to orient himself in the fog of war.

For Clausewitz, the intellect was necessary to achieve success in war but it should not be reduced to classical Newtonian science. The abstract and analytical categories of that science were useless and misleading in understanding war – war needs a holistic approach. "It would obviously be a mistake, he wrote, to determine the form of an ear of wheat by analyzing the chemical elements of its kernel, since all one needs to do is to go to a wheat field to see the grown ears."[40] Indeed, "...in war more than in any other subject we must begin by looking at the nature of the whole; for here more than elsewhere the part and the whole must always be thought of together."[41] These words strongly resonated with Goethe and the *Naturphilosophes*. And here he joined the Romantic scientists who saw Nature as multiform, omnipresent, and stretching beyond the narrow limits of an analytical framework. The "...conduct of war branches out in almost all directions and has no definite limits; while any system, any model, has the finite nature of a synthesis."[42]

Some writers on military affairs had not gone further because of their belief that what Maurice de Saxe (1696–1750) called the "sublime" aspect of war was a matter of genius – only the intuitive powers of superior minds might occasionally reach this sphere. Yet Clausewitz believed that his work would enable probabilistic insights into the underlying nature war and how to wage it. What for him was certain, however, was that the systems and mathematical laws of earlier writers on war were useless. In ways that closely aligned with Romantic philosophers' views on polarity and action, he also thought that "War ... is not the action of a living force upon a lifeless mass ... but always the collision of two living forces. The ultimate aim of waging war ... must be taken as applying to both sides. Once again, there is interaction."[43]

---

[40] Clausewitz, *On War*, 61.

[41] Ibid., 71.

[42] Ibid. 134.

[43] Ibid. 77.

To arrive at the underlying truth about war, Clausewitz stressed two – at first sight contradictory – methods. On the one hand, he did not abandon empiricism. One cannot stay at home as Puységur proposed and learn the art of war from books containing descriptions of systems. One must base one's thinking solidly on the historical facts and experience of warfare and also on the facts of politics and national resources including human ones. A detailed knowledge of these factors was necessary and one must pay attention to the quality and reliability of data. Hence, he distrusted the accounts of ancient authors on warfare because of the lack of detail and omissions in their accounts. Indeed, he felt that only the period from 1740 to 1815 could provide reliable data because of its proximity to his own time and the more extensive accounts of warfare during this period.[44]

On the other hand, he spoke of the *coup d'oeil*, a popular concept in the writings on war of the Enlightenment. The *coup d'oeil*, which could be translated as "glance" and which Clausewitz leaves in the original French in his text, was the experienced, instantaneous, and inspired grasp of the topographical situation and the positions of contending armies on the battlefield. It was something characteristic of the most successful generals and needed both innate genius and cultivation to acquire. Here we see another example of Romantic thought in the Enlightenment that does not conform to its more prominent characteristics. It was something more than the mere physical processing of topographical facts: it "refers [he writes] not alone to the physical but, more commonly, to the inward eye."[45] His "inward eye" suggested the idea, also popular among the Romantic scientists, that the active interpenetration of the subject and the object was required to find scientific truths. Sight alone is a passive sense that merely channeled physical perception towards the active mind where the powers of imagination transformed it into something higher than a mere mechanical impact. In its caricatural form, this process implied a German professor sunk in deep thought in his darkened study gaining access to the secrets of nature by flashes of insight.[46] The contradiction between the stress on empiricism and the "inward eye" did not pose a problem for Clausewitz or the *Naturphilosophes* since contradiction, too, was a clash of opposites and science did not have to be tidy to be useful.

Clausewitz's "science", although some its elements can be traced to the dynamical strand of Newtonianism was primarily a reaction against it. The soldier, the poet, and the natural philosopher went beyond the simple sensual knowledge of Locke and Condillac impinging on an inert *tabula rasa.* By virtue of their engagement with and harmony with Nature they actively transformed perception into a deeper understanding of Nature of which they were a part. The interaction and clash of forces in Nature and on the battle field could never be reduced to simple and static forms but took on the creative features of poesy. For the natural philosopher, the ascent from simple sense to understanding could lead to underlying ideas that were accessible to the human mind that was a product of Nature. No uniform and predictable practical results could be expected but rather a range of possibili-

[44] Ibid. 173–4.

[45] Ibid.102.

[46] I owe this image to Trevor Levere.

ties that opened yet more insights. For the soldier, the talented and intuitive eye scrutinizing the chaos of war and battle could grasp no recipes for guaranteed victory. But it provided a heuristic that gave principles and rules of conduct that enabled one to navigate better through the shoals of chance and the fallibility of human character. In both cases this was possible because Man and Nature were alike and the latter was thus accessible to the former.[47]

I have tried to draw attention to similar aspects of *Naturphilosophie* and the ideas of *On War*. However, Clausewitz was a highly original and sometimes opaque thinker – another link between him and *Naturphilosophie*. He would never have claimed that he was creating a science of war, although science was useful in understanding war. He wrote that both the ideas of a science of war and an art of war were inapplicable to his work although the latter term was more suitable than the former although an analogy to commerce might be the best of all.[48] Not belonging to either the arts or the sciences it was part of "man's social existence".

It is perhaps here that we can see one of the places where Clausewitz parts ways with Coleridge and the *Naturphilosophes*. For both of them, Man was the focus of interest as possessor of an intellect that could commune with Nature and approach its secrets through imagination. Man was important for Clausewitz as well, as displayed in his remarks on Genius and the importance of the great commander in war.[49] But War was a central element of society and of "man's social existence" and society was his focus as much as the individual. For Clausewitz science and philosophy – whether of the Newtonian sort or that of *Naturphilosophie* – served as an aid to the critical analysis of war and as justification and exaltation of the great military leader who must perforce be a military philosopher. However, war was a social phenomenon like commerce and deserves similar cold blooded analysis although both teem with human emotion, danger, daring, and vicissitude.

By putting war into the domain of "man's social existence" Clausewitz by no means renounces the possibility of thinking fruitfully about war. But this thinking was not to discover a system of war analogous to Vauban's alleged system of siegecraft. It was to obtain probabilistic guides for the conduct of war that individual human genius, thoroughly imbued with practical knowledge, experience, and the context of the particular battlefield, could grasp the dynamic polarity of war to achieve victory. Nineteenth-century scientists, like the German chemist Justus von Liebig (1803–1873). derided *Naturphilosophie* for its obscurity, pretentiousness, and the sterility of the Romantic Science associated with it. But the accusation of sterility for Romantic Science was, however, only partly justified because scientists tlike the pioneer of electromagnetism Hans-Christian Ørsted

[47] Levere, *Poetry realized in nature*, 63, who attributes this thought to the German chemist Ritter.

[48] Clausewitz, *On War*, 148–9. Michael Howard, "The Influence of Clausewitz" in Clausewitz, *On War*, 43–44 notes that this aspect of Clausewitz's thought elicited favourable comment from both Engels and Marx.

[49] "On Military Genius" in Clausewitz, *On War*, 100ff.

(1771–1851) were strongly influenced by *Naturphilosophie* and were friends of its prominent leaders. Moreover, the example of Clausewitz suggested that it had an impact not just on science, philosophy, and art but also on the bloody phenomenon of war. Ørsted demonstrated the interaction between magnetism and electricity in 1821 when Clausewitz very likely heard of this sensational discovery. It could have provided a metaphor congenial to the *Naturphilosophes* that Clausewitz used to describe his theory of war.

> War is more than a true chameleon that slightly adapts its characteristics to the given case. As a total phenomenon, its dominant tendencies always make war a paradoxical trinity – composed of primitive violence, hatred, and enmity, which are to be regarded as a blind natural force; of the play of chance and probability within which the creative spirit is free to roam; and of its element of subordination, as an instrument of policy, which makes it subject to reason alone. [...] Our task therefore is to develop a theory that maintains a balance between these three tendencies, like an object suspended between three magnets.[50]

[50] Clausewitz, *On War*, 89.

Chapter 8
John Herschel's Geology: The Cape of Good Hope in the 1830s

Gregory A. Good

The theme which unites this article came to me as a charge – a charge I brought upon myself, but a charge nevertheless. An international group of historians of geology (INHIGEO) was to meet in Cape Town in 2016. Our group had no active members in South Africa, which presented a problem for the conference organizers: How could we have a history of geology field trip? I volunteered that John Herschel had lived at the Cape in the 1830s and that there must be a story and a field trip to be found there. I did lead a field trip "On the Trail of Charles Darwin and John Herschel: The Cape in the 1830s." I also discovered an unanticipated depth of archival material, which I will be exploring for some time, and which informs this story of the encounter between John Herschel, the astronomer, and the geology of Cape Town, South Africa, in in the 1830s.

I frame this story in the space of interaction between Herschel and Charles Lyell, back in England. First, a remarkable letter from Herschel to Lyell, and read in part by most geologists in England, emphasizes how ardent a participant Herschel was in geological field work and in theoretical debate. This addresses Herschel's place in the geological community at the time. It helps us understand that it was Herschel the experienced geologist who Darwin was to visit in 1836, as much as it was Herschel the astronomer.

The second frame of this article is the field work undertaken by Herschel in the Cape region. Herschel began his field work in January 1834, and in the 2½ years before Darwin visited, Herschel explored surprisingly far afield in search of rock types, hot springs, minerals, and strata. What Herschel learned about the geology of the Cape region was within the frameworks of debate in the geological community, but as 'new ground' the field observations both inspired and tested geological theories. This is a story of field work, but it is also a story of grand visions of the scale

G.A. Good (✉)
American Institute of Physics, College Park, MD, USA
e-mail: ggood@aip.org

J. Buchwald, L. Stewart (eds.), *The Romance of Science: Essays in Honour of Trevor H. Levere*, Archimedes 52, DOI 10.1007/978-3-319-58436-2_8

and origin of geological change. Herschel acted as a remote observer and discussant for scientists back in Britain and Europe interested in understanding Earth's dynamism and history.

8.1 The Mountain Setting of Herschel's Observatory at Feldhausen, Cape Colony

John Herschel arrived off the Cape of Good Hope on 15 January 1834, along with his wife, children, a nanny, and a mechanic-assistant. On the 16th, from an anchorage in Cape Town's harbor, Herschel described the scene (Fig. 8.1):

> The situation is most remarkable, hemmed in on the sides by steep promontories & backed by the Mural precipice of the Table Mountain which rises sharp & sudden behind it.[1]

The family quickly settled on a property about 10 km out of Cape Town, in a grove and garden, with a house and several out-buildings. Here John Herschel erected his 20-foot reflector telescope.[2] With a mirror 18-inches in diameter, it was the largest functioning telescope anywhere in the world at the time. From 1834 to 1838 he swept the skies, searching out and painstakingly mapping over 10,000 double stars, nebulae, and star clusters.[3] Herschel's stay at the Cape is best remembered for this astronomical work.

During the day, Table Mountain and Devil's Mountain shaped Herschel's reflections on the Earth and its place in the Cosmos. Herschel investigated the rocks, their crystals and minerals, and he thought about the large forces continuously reshaping Earth's surface. He also measured solar radiation and organized meteorological and

---

[1] David S. Evans, Terence J. Deeming, Betty Hall Evans, and Stephen Goldfarb, eds., *Herschel at the Cape: Diaries and Correspondence of Sir John Herschel, 1834–1838.* (Austin and London: University of Texas Press, 1969), 37. Evans selected passages, in particular, from John Herschel's Travel Journal XIII, 1829–1837. This is in the Herschel Family Papers, Container 22.14 (W0063a on the old numbering system), University of Texas, Harry Ransom Center. Note that although Evans calls these "diary" entries, in the Herschel Family Papers, this series of books is termed "Travel Journals," the series "Diaries" being entirely different. Brian Warner includes many of Herschel's *camera lucida* sketches in *Cape Landscapes: Sir John Herschel's Sketches 1834–1838.*(Cape Town: University of Cape Town Press, 2006). Warner's book, with Herschel's *camera lucida* sketches, illustrates well Herschel's fascination with landscape and its connection to geology.

[2] There are a number of secondary works, reflecting on and analyzing the activities of Herschel at the Cape. Donald Fernie's *The Whisper and the Vision: The Voyages of the Astronomers* (Toronto/Vancouver: Clarke, Irwin & Company Limited, 1976), 57–106, includes a chapter on Herschel's years at the Cape. A more recent scholarly account sets Herschel's time at the Cape in a broader cultural and political context: Steven Ruskin, *John Herschel's Cape Voyage: Private Science, Public Imagination and the Ambitions of Empire* (Aldershot, Hants, England and Burlington, Vermont, USA: Ashgate, 2004).

[3] Omar W. Nasim, *Observing by Hand: Sketching the Nebulae in the Nineteenth Century,* (Chicago: University of Chicago Press, 2013).

Fig. 8.1 John Herschel's free-hand sketch on shipboard, from the Lion's Head and Rump on the *left* to one of the Buttresses at the south end of Table Mountain (By permission of National Library of South Africa, Cape Town, Herschel Album I, INIL 9088)

tidal collaborations. He carefully measured the atmosphere. This period of intense astronomical observing was also critical in the development of Herschel's ideas about the Earth and about terrestrial phenomena.[4]

When John Herschel arrived in Cape Town in 1834, he was already famous as an astronomer and natural philosopher. He had nearly been elected president of the Royal Society of London in 1830. He had applied immense energy to discovering and mapping double stars, star clusters, and nebulae in the English skies in the 1820s. He was anxious to carry this work through personally in the southern hemisphere, not simply out of respect to his father, William Herschel, who had started these investigations, but to assure uniformity and completeness. Herschel came to Cape Town mainly to observe deep-sky objects. But he still had the daylight hours and a span of days around full moons when he could not observe deep space. He went afield at every chance he got, sometimes for geology, sometimes botany, and sometimes just for the joy of drawing landscapes.

[4] Gregory A. Good, "A Shift of View: Meteorology in John Herschel's Terrestrial Physics." In *Intimate Universality: Local and Global Themes in the History of Weather and Climate,* eds. J.R. Fleming, V. Jankovic, and D.R. Coen (New York: Science History Publications, 2006), 35–67.

Because Herschel is usually thought of as an astronomer, his involvement in geology may come as a surprise. But he started studying minerals and crystals in his 20s,[5] and became a member of the Geological Society in London in 1824. In the 1820s Herschel undertook what might be called Grand Tours of Europe, except for his expeditionary level of organization for fieldwork. He travelled to the Italian volcanoes with the volcanologist Charles Daubeny. He briefed Charles Lyell and Roderick Murchison on geological sites to visit in the Alps and other European ranges. And he collected crystals and rock samples. Geologists trusted Herschel as a source of samples and geological descriptions, but they also valued his firm grasp of chemical and physical science and his application of these to thinking about the Earth and geo-processes. Herschel understood the geologist's commitment to field work and description, but he added in skill sets other geologists lacked.

8.2 John Herschel's Geological Fieldwork at the Cape

When not at the telescope, Herschel was a keen observer of landscapes, strata, and rocks. He visited Table Mountain, Paarl Rock, Franschoek, and beyond, making geological observations like those he had made in the Alps in the 1820s. Herschel frequently made forays to nearby hilltops, passes, and bays. He occasionally ranged further, but even his furthest exploration (to Genadendal) was only 300 km round trip from his observatory. (Horse cart got him there in 3 days.) Herschel's travels to several well-known geological sites around the Cape colony set the stage for Darwin's arrival. This section provides an overview of what Herschel had learned about Cape geology before Darwin's arrival in May 1836.[6]

Herschel's first geological experience after landing at Cape Town in January 1834 was a visit to a local slate quarry. Brian Warner tells of Herschel's landing: "Almost the first scientific comment in his diary after reaching Cape Town was to note the absence of fossils in the local slate quarry. The immediate vicinity of Cape Town is largely devoid of fossils, which Herschel cleverly found out by consulting a well-digger…".[7] On 19 January 1834 he noted:

> In the Evening walked out to a quarry on the side of Lion Hill just out of Town. It is Slate (or at least blue hard stone) in nearly vertical and very well defined Strata, with a cross stratification extremely distinct, & breaks into very regular blocks almost exactly like the rhomboid of Carbonate of Lime. Saw no traces of organic remains in any part of this quarry.[8]

[5] Gregory A. Good, "John Herschel's Optical Researches and the Development of his Ideas on Method and Causality," *Studies in the History and Philosophy of Science,* **18** (1987), 1–41. Herschel used polarized light to probe crystalline structure.

[6] Brian Warner provides a useful summary of Herschel's travels around the Cape in *Cape Landscapes,* the chapter "Journeys into the Interior," at 131–161. Warner reproduces 21 of Herschel's *camera lucida* drawings.

[7] Brian Warner, "The Years at the Cape of Good Hope," in D.G. King-Hele, ed., *John Herschel 1792–1871: A Bicentennial Commemoration* (London: The Royal Society, 1992), 51–66, quote at 51.

[8] Evans, *Herschel at the Cape,* 37–38, Herschel's journal entry for 19 January 1834.

Lion's Head is at E and Lion's Rump at e in Fig. 8.2. Herschel often walked or rode his horse around the area, exploring the geology and mineralogy. Two weeks after visiting the slate quarry he rode up to the kloof or pass between Table Mountain and Devil's Peak (the latter marked A in Fig. 8.2).[9]

In February 1834, only a month after arriving at the Cape, Herschel made his first foray near Table Mountain.

> In the afternoon Rode out with M. [his wife, Margaret] and D.S. [Duncan Stewart, Margaret's brother] towards Hout's Bay and from the Hill which flanks the pass leading into it got a most beautiful view.[10]

In late July that first year, Herschel wrote: "Rode with Mr Stewart [Steuart] (High Sheriff) to Hout's Pass, and climbed Constantia Berg, whence took a semi-Panorama." From this pass he saw the view to Hout Bay in the west and more mountains beyond the flats to the east. He could see the climb to the summit of Table Mountain. He saved it for another day and returned home that night.

On an April Sunday, Herschel rode over to Wynberg Church (in the village, marked F in Fig. 8.2), but found there was no service. So to put the morning to good use, he examined a local outcropping:

> Returning, examined the great granite block in Mr Hare's grounds. It is a single stone without a flaw 16 paces long & 10 broad of an irregular oval form and rounded surface projecting about 6 feet above the soil, & nearly smooth. The granite is chiefly White Felspar (sic) in large crystals with some black mica and quartz in small quantity. Very hard & little appearance of weathering. It is accompanied by much other granite in loosely scattered masses similarly projecting & is doubtless only a projection of a great outcropping Granite Rock. Taking a course out of the Road towards the Table Hill I observed the outcrop of Granite masses (of similar nature) in several other parts so that doubtless the substratum of Wynberg Hill is Granite.[11]

He had first examined this granite block after church three Sundays earlier and already considered it part of the "substratum" of Table Mountain, "laid bare."[12] Early in May, Herschel returned to Wynberg Hill to examine other granite outcrops, which may be indicated in Fig. 8.2 by the short, thick lines above F. On the property of a Major Rogers "on the high side of Wynberg Hill" he described two "great knolls of granite which project from its summit…"[13] Again he saw white granite crowded with feldspar and black mica, but with little quartz. He convinced himself that Wynberg Hill's granite continued to the west under the Table Mountain rocks. Rocks and minerals crowded Herschel's daytime gaze as nebulae did his telescope at night. (Today these outcrops are largely gone or built over in suburban Claremont and Wynberg, although a few may remain with picturesque names like the Rifle Butts and Hen and Chickens Rock.)

---

[9] Evans, et al., *Herschel at the Cape,* 44, Herschel's journal entry for 3 February 1834.

[10] Evans, et al., *Herschel at the Cape,* 45, Herschel's journal entry for 3 February 1834.

[11] Evans, et al., *Herschel at the Cape,* 63, Herschel's journal entry for 20 April 1834.

[12] Evans, et al., *Herschel at the Cape,* 58, Herschel's journal entry for 30 March 1834.

[13] Evans, et al., *Herschel at the Cape,* 65, Herschel's journal entry for 4 May 1834.

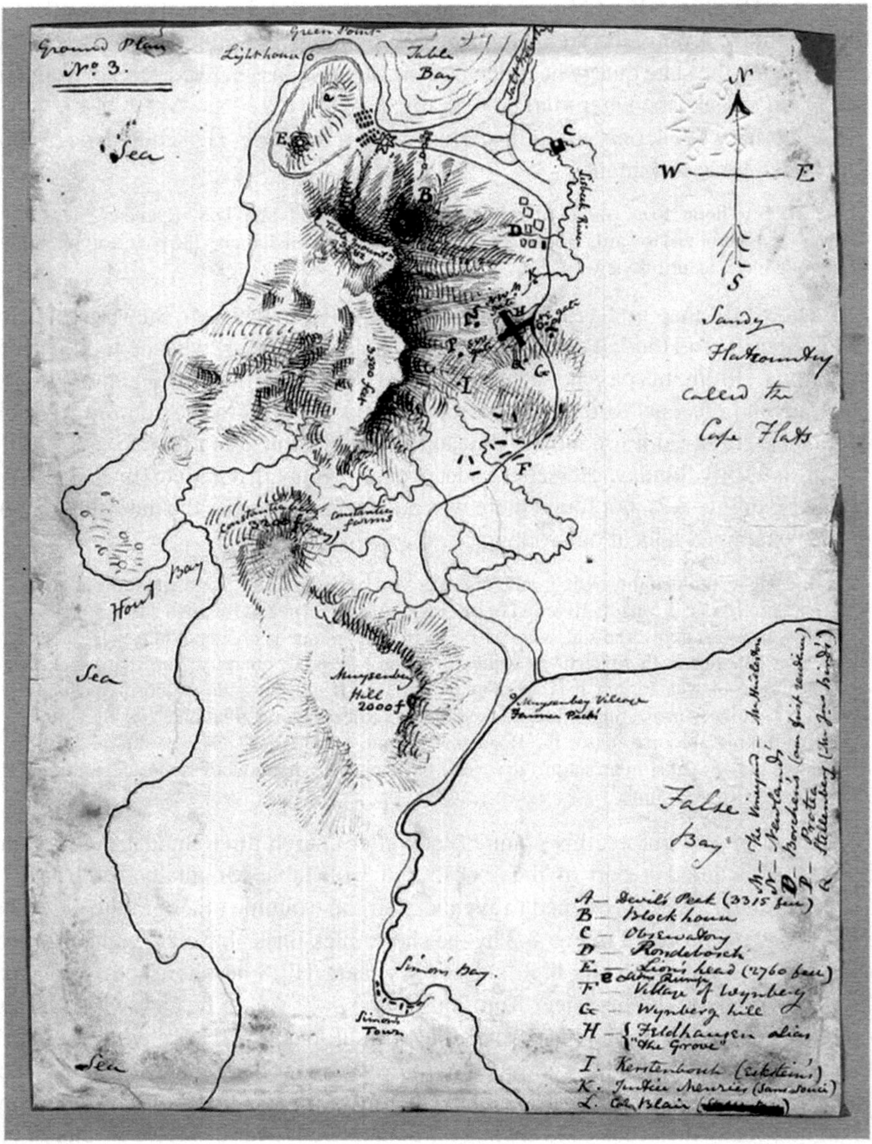

Fig. 8.2 Herschel's "Ground Plan No. 3" includes a prominent X (the intersecting lanes on his property, Feldhausen). This map ranges from the Atlantic Ocean on the west to False Bay on the east, from Table Bay in the north to Simon's Town in the south. This is printed in Evans et al., Herschel at the Cape, p. 121. This copy is a scan of the manuscript, by permission of National Library of South Africa, Cape Town, MS B 258 Ground Plan No. 3

From these minor heights, Herschel observed not only the dominating mass of Table Mountain to the north and west, but also he saw Constantiaberg and Muysenberg to the south and the Hottentots Holland Mountains far to the east. He started with the local rocks and gradually was pulled further afield. He and his wife Margaret often explored nearby on daytrips. He also developed a circle of field work companions at the Cape. The most prominent were Thomas Maclear (Cape Astronomer), Dr. Andrew Smith (an explorer), Sir John Bell (son in law of the First Earl of Malmesbury), and Robert Wauchope (a British naval officer). Wauchope also acted as field intelligence for Herschel, sending him reports from places Herschel had not visited. John Steuart, the high sheriff of the area, accompanied Herschel on his first climb to the summit of Table Mountain and on other nearby trips.

In June of that first year at the Cape, Herschel explored a route up one of the kloofs (perhaps Skeleton Gorge in Kirstenbosch Botanical Gardens), but was repulsed after clambering up the talus and several rock sections, where he found his way blocked by "mural precipices."

> Good heavens! What a view from this point over Simons Bay & the flats with those noble hills beyond![14]

Both Table Mountain and the Hottentots Holland Mountains – "those noble hills" – occupied Herschel's thoughts and later drew him on.

In July, Herschel returned to Constantiaberg and Hout Pass where he could see the Atlantic Ocean and Hout Bay to the west. He visited again, this time with Maclear, to take bearings, compare the mountains with published accounts, and to draw several landscapes using the *camera lucida,* the drawing aid which he had mastered in the 1820s.[15] Herschel clearly delineated the strata in his sketch of the buttresses of Constantiaberg, overlooking the east side of Hout Bay.[16] Herschel was reconnoitring, as this would be the start of his route to the summit of Table Mountain. Note that one of these two visits coincided with a full Moon (20 July 1834), when Herschel could not work at the telescope on dim stellar objects. So he explored the Earth instead.

A few days after the next full Moon (19 August 1834), Herschel returned to Hout Pass on a Saturday with Steuart and climbed Constantiaberg. Herschel's map in Fig. 8.2 shows this mountain with an altitude of 3200 feet, but with the word "query" beside it.[17] On this occasion, Herschel made a small panorama, with three segments on one sheet, showing Sir Lowry's Pass, with many individual peaks identified east of the Cape Flats.[18] The recon continued.

---

[14] Evans, et al., *Herschel at the Cape,* 70, Herschel's journal entry for 4 June 1834.

[15] Evans, et al., *Herschel at the Cape,* 82 and 85, Herschel's diary entries for 21 July 1834, and for 31 July 1834.

[16] Warner, *Cape Landscapes,* 128, Figure 6.13. Warner dates Figure 6.13 to May 1836, hence not the drawing mentioned here. His landscape (126, Figure 6.12) eastward in 1834 was much more romantic, the mountains far away, rendered in an India ink wash.

[17] Constantiaberg is now reckoned to be 3041 feet (927 meters) high.

[18] This panorama is not reproduced in Warner's *Cape Landscapes*. The original is in the National Library of South Africa, Cape Town, Herschel Portfolio II, INIL 9214.

Herschel passed through Hout Pass during his first ascent of Table Mountain on 10 November 1834, another Saturday. Along with Maclear, Steuart, and a Mr. Agar, Herschel ascended gradually northward through wind-eroded sandstone outcroppings, hummocks, and vegetation. They reached the broad summit – the "great flat top of the hill" – late in the morning and explored from (what is now) Maclear's Beacon to the beginning of the descent through Platteklip Gorge to Cape Town: according to Herschel, "a vast and deep Cleft of wh took sketch."[19] Herschel ranged about the top to get views in all directions, from the Devil Berg around to Hout Bay and the Cape to the south. The party then descended toward Hout Pass and arrived home in the evening.

The ascent of Table Mountain was just the start of a week-long excursion. After taking Sunday off, Herschel started on 10 November for the mountains across the Cape Flats, on the eastern horizon. He and Margaret took 2-month-old Louisa and a nanny, a coachman, a groom, and a friend, Dr. George Withers, on a trip to Paarl Rock, with furthest points reached at Wagenmakers Valley (now Wellington) in the north and then Franschhoek in the southeast.[20] Herschel was extending his geological descriptions and landscape drawing.

Herschel's travel journal during the first day's journey to Stellenbosch notes tersely that the Helderberg mountain's "tall & mossy down contrasts strangely with sharp rugged pinnacles." He paused during the day to draw several views of the Groot Drakenstein, Jonkershoekberge, Stellenboschberg, and Helderberg. The next day the party continued north, and Herschel noted in his journal

> A most superb view of the Amphitheatre formed by Simonds berg French Hook Donkersberg and Elderberg embracing Stellenbosch…. Klamus [Klapmuts]… The hills are mostly covered very deep with Alluvium of sand & clay.[21]

The names of the peaks are slightly different today, but recognisable: Simonsberg and Klapmuts, for example.

In the village of Paarl the next morning, Herschel and Rev. George Withers set out to summit Paarl Hill (Paarlberg), a granite dome overlooking the village. Herschel wrote in his journal:

> On Paarl. A most remarkable Dike cuts across it like the top of a brick wall. General Direction from ESE to WNW ($\theta = 287°$, $\theta' = 108°$) breadth 12 in – 14 in in broadest part but thins off & subdivides & reunites…[22]

This dike ran, he said, 200 paces. Herschel stayed a half hour on the summit, where he measured the bearings of other peaks and sketched them roughly

---

[19] Evans, et al., *Herschel at the Cape,* 100. The extended diary entry is not reproduced in this book. The manuscript (see footnote 1) contains much more, including sketches. If Herschel used the *camera lucida* on this first ascent, no drawings made with it are in the archives.

[20] Evans includes much of the relevant portion of Travel Journal XIII in *Herschel at the Cape,* pp. 100–106. He again omits rough sketches in the journal. Note that several peaks bear variations of the name Drachenstein. Warner discusses this first trip to the eastern mountains in *Cape Landscapes,* pp. 131–133.

[21] Evans, et al., *Herschel at the Cape,* 101–102, Herschel's journal entry for 11 November 1834.

[22] Evans, et al., *Herschel at the Cape,* 102–103, Herschel's journal entry for 12 November 1834.

Fig. 8.3 Groot Drakenstein Range. Herschel's notes on front, none on back: "Altar Mountain. Donker-berg the Southern Buttress," a camera lucida drawing by Herschel (By permission of National Library of South Africa, Cape Town, Herschel Album I, INIL 9066)

in the journal. He would have used his barometer to measure the altitude of Paarlberg, if the jarring wagon ride had not loosened its screws allowing all the mercury to escape. Herschel would have to return to complete his study of Paarlberg. By 10 am, "after a Desperate broil and scramble," Herschel and Withers had rejoined their party in the village and were on their way north to Wagenmakers Valley.

The exploration continued the next day southeastward up the valley toward Franschhoek Pass. Herschel described the "great Buttress Hill at the Corner & the Simonds berg beyond it," the Drachenstein sandstone, and the change of the dip of the strata (Fig. 8.3).

> The Sandstone here in the M^{n} [Mountain] to right of the Pass dips at 70 or 80° thus making a saddle while further down the valley the beds are horizontal.[23]

Approaching the top of the pass, Herschel described the rock in more detail:

> A great bed of Pisolitic Iron *very blue* & like a conglomerate of peas & Beans with their pods with dirt between. It looks so blue that [illegible] Phosphate. Took specimens. The general mass of the rock is a hard siliceous stone [containing quartz or silicon dioxide crystals]. I found one block of regular siliceous Pudding.

[23] Evans, et al., *Herschel at the Cape,* 103–1-4, Herschel's journal entry for 13 November 1834.

Further on he noted more feldspar and quartz, no mica, and the dip of the strata to the south southeast at 50°. Herschel's keen interest in the geology spanned from its mineralogy to the larger structure of the strata.

Herschel ended the November 1834 excursion just past the highest point of Franschhoek Pass, turned about, and rode back toward Stellenbosch. He described the mountains and geology and made a few *camera lucida* drawings. He noted one peak he called "the great buttress shoulder of the Drachenstein" and said: "…so after all I suppose we must call the huge abrupt guardian mass which forms the corner (SW) of the French Hook valley."[24] He encountered a thick bed of pisolitic iron oxide on the slopes at the base of the mountains, deep ravines eroded in beds of clay, and noted that the Berg River had cut a deep channel revealing the "huge confused mass of water rolled stones large & small, from half a ton to sand – all smoothly rounded, as if by a debacle or long succession of Torrents."[25] As the journey ended, the Moon reached its full phase again.

The year 1834 was a busy first year of exploration. During1835 Herschel cast his daytime eye on meteorology, tides, and solar radiation, in addition to geology. He read accounts of fossil fish discovered in remote areas of South Africa he would not visit and his friend Wauchope collected rock specimens for him on Ascension Island. Herschel was (according to his wife) "running over the Heavens & the Earth" with John Bell, who was chief secretary to the Cape governor.[26] He wrote to Roderick Murchison, William Whewell, Francis Baily, and Charles Lyell in England about the geology he had observed and about the need for long-term observations of tides and sea-level.[27] Coincident with another full Moon, Herschel made his first visit to Sir Lowry's Pass in September 1835 to collect flower bulbs (another passion) and examine rocks.[28] During the November full Moon, Herschel travelled to Simon's Town and to the Cape of Good Hope itself, broadening his geological observations yet further.[29] He examined mineral deposits on granite boulders in Simon's Bay and a basalt dike.

In January 1836, 2 years after arriving at the Cape, Herschel returned to Paarl for a second, more dedicated scientific visit to the granite dome overlooking the town. His sole companion this time was the Cape astronomer, Maclear. In his travel journal he described the hills, rocks, and soil along the way. This time he carried his barometer more carefully and brought his *camera lucida* and drawing board to the summit, while Maclear carried a theodolite for measuring the bearings of distant peaks.

---

[24] Warner, *Cape Landscapes,* 141, identifies this as today's Drakensteinpiek.

[25] Evans, et al., *Herschel at the Cape,* 105–106, Herschel's journal entries for 14 and 15 November 1834.

[26] Margaret Herschel to Duncan Stewart (her brother), 6 June 1835, printed in Brian Warner, ed., *Lady Herschel's Letters from the Cape, 1834–1838*(Cape Town: Friends of the South African Library, 1991), 73–79.

[27] Herschel's connection to Murchison, going back 10 years, was strengthened by the presence of Murchison's brother at the Cape: John Herschel to Roderick Impey Murchison, 8 March 1835, Geological Society (London).

[28] Warner, *Cape Landscapes,* 132. Warner quotes Herschel's journal for 3 September 1835.

[29] Evans, et al., *Herschel at the Cape,* 195–196.

Starting early in the morning, in 2 h they were atop Paarl Rock, having removed socks and shoes for better grip. Herschel observed a stream two thirds of the way up, and described the soil as "where not granite, very hard baked pot clay…" He merely noted the dike he saw on the first visit, and then he and Maclear turned their attention to barometric readings (for altitude), temperature, Beaufort scale estimates of wind speed, and precise sightings for a geodetic survey connection of Paarl Rock to Wynberg Hill and the Royal Observatory. Herschel drew a panorama and observed the barometer for altitude, but he and Maclear had to concede defeat on one part of the project. They had arranged to have flares launched from the Cape Observatory and at Wynberg for two more links in the geodetic survey of the colony. But a haze obscured the view and Herschel decided that staying out overnight atop Paarlberg was a bad idea.

They climbed down via Diamond Rock and ate dinner in its cave, and indulged in "long draughts & a plentiful ablution of feet & hands" in the brook. Herschel described a granite flake 28 paces long and 7 broad, and another "remarkable specimen of granite exfoliation in a block of vast size almost exactly spherical on a highly inclined slope. Took a sketch of it. Situation most romantic." Maclear and Herschel returned home the next day, a Sunday, taking the time to climb to the top of Klapmuts Peak, with the help of three Africans to carry their instruments.[30]

8.3 Herschel Among the Geologists

I stop the tale of Herschel's geological field work at the Cape in early 1836 because of two events that year. The first event was the writing of a letter to Charles Lyell, in which Herschel both described his field work to date and in which he speculated quite freely on the geological causes of the contemporary landscape. The second event was the famous visit of Charles Darwin to the Cape on the return voyage of the *Beagle*, during which he met Herschel several times and visited geological sites Herschel knew intimately. First, the letter to Lyell.

On 20 February 1836, after 2 years of residence at Feldhausen and 3 months before Darwin arrived at the Cape, John Herschel wrote an 11-page letter to Charles Lyell, thanking him for the new edition of his *Geology*, in fact, Lyell's fourth edition of the *Principles of Geology*.[31] This remarkable, densely packed letter reveals the

[30] Evans, et al., *Herschel at the Cape,* 211–213. Warner points out that Maclear and Herschel timed this trip to avoid attending a dinner in honor of the Cape governor whose policy toward indigenous peoples they strongly opposed.

[31] John Herschel to Charles Lyell, 20 February 1836, American Philosophical Society Archives, B.D25.L.1 Another copy is at the Royal Society, RS:HS 25.4.20. The APS manuscript was transcribed and published with commentary by Walter F. Cannon, "The Impact of Uniformitarianism: Two Letters from John Herschel to Charles Lyell, 1836–1837," *Proceedings of the American Philosophical Society,* **105**(1961), 301–314. Herschel referred to passages in Lyell's fourth edition: *Principles of Geology,* 4th ed. (London: John Murray, 1835). Lyell's version appeared as "Extracts from a Letter from Sir John F.W. Herschel to C. Lyell…," *Proceedings of the Geological Society of London,* **2** (1833–1838),548–552, 596–598.

depth of Herschel's geological researches and the degree of his engagement with Lyell's arguments and evidence in *The Principles of Geology.* This letter is like an outcropping of a stratum normally deep beneath the landscape. It follows years of fieldwork by Herschel, and it is embedded in ongoing debates not only among geologists, but also among a broader range of scientists. Herschel had published several detailed discussions of geology and mineralogy between 1830 and 1833 and had an ongoing correspondence not only with Lyell and Murchison, but also with Adam Sedgwick, William Buckland, and other geologists.

Herschel's 1836 letter is best known for the passage that includes the words: "… that mystery of mysteries the replacement of extinct species by others." The phrase "mystery of mysteries" was highlighted by Darwin in *The Origin of Species,* and this has effectively joined Darwin to Herschel via evolution. For example, Michael Ruse and Sandra Herbert highlight this passage in their discussions of Lyell's early positions on species.[32] The phrase is common in the literature, so common its origin is often not realized.

Herschel did in fact devote attention to the botany of the Cape region, describing plants and speculating about the rise and fall of species. He did write that the Creator, acting through intermediate natural causes, originates "fresh species." And Darwin and Lyell both were deeply interested in his views of how new species appear, as is clear in their letters and notebooks.

But Herschel's letter is mostly geological. Historians have noted this. Ruse related Herschel's geological theorizing in the letter to his 1830 discussion of his *vera causa* principle.[33] Walter F. Cannon placed this letter in the context of the uniformitarian-catastrophist debate, and also provided a mostly reliable transcription of most of Herschel's letter.[34] Leonard Wilson echoed Cannon, with more attention to Lyell's reaction to Herschel's ideas about the physical causes of geological processes.[35] Martin Rudwick instanced that Lyell turned to Herschel for guidance on the relation of the Earth's spheroidal shape to its internal make-up.[36] Herbert focused on Darwin's reaction to Herschel's theory on the roles of sedimentary deposition and isotherms in raising continents.[37]

These analyses, however, do not exhaust the geological contents of Herschel's 1836 letter. Two aspects of this letter are complementary: the importance to Herschel

---

[32] Michael Ruse, *The Darwinian Revolution: Science Red in Tooth and Claw* (Chicago: University of Chicago Press, 1979), 84–85; Sandra Herbert, *Charles Darwin, Geologist.* (Ithaca: Cornell University Press, 2005), 317–318.

[33] Ruse, *Darwinian Revolution,* 56–61. This issue requires extensive re-examination since new sources and perspectives have appeared.

[34] Cannon, "Impact of Uniformitarianism," 302–303. Cannon discussed the species question, but asserted that the rest of the letter was "still of greater importance to the historian." (p. 302).

[35] Leonard Wilson, *Charles Lyell, the Years to 1841: The Revolution in Geology.*(New Haven: Yale University Press, 1972), 286 ff.

[36] Martin J.S. Rudwick, *Worlds Before Adam: The Reconstruction of Geohistory in the Age of Reform* (Chicago: University of Chicago Press, 2008), 479–481.

[37] Herbert, *Charles Darwin, Geologist,* 209–212. Herbert's case is based on Darwin's post-Beagle notebooks and demonstrate his familiarity with Herschel's letter and ideas.

of Lyell's dynamic vision of geology, climate, and even living things; and second, Herschel's commitment to test these ideas against field work. How do Herschel's long descriptive passages, built on the 2 years of geological field work related in the previous section, relate to Earth's deeper processes? Herschel's letter is an integral part of his developing sense of terrestrial sciences beyond geology.

Herschel was thoroughly intrigued by Lyell's ideas long before 1836. He had been reading volume 3 in particular of Lyell's *Principles of Geology* during the 1834 voyage to Cape Town. He said he had transformed his views on geology.[38] Herschel's eye raced across the face of the Earth as it is now, a view of a single, frozen moment, but which he saw as swept along by geological processes over long periods of time. Indeed, before Herschel even approached the Cape, he already anticipated observing signs of either uplift of the land or of sea-level change. In the context of tidal research, but also with sea-level in mind, Herschel asked Whewell: "Do tell me of things of a local nature requiring precise numerical measure. I propose to fix two or 3 coast levels very exactly as points of geological reference." Herschel shared Lyell's vision of a dynamic Earth in which continents and oceans rose and fell, but he brought with him a commitment to measurement of that change.

Herschel's 1836 letter to Lyell sounded this theme again. By now, Herschel said he had read Lyell's newer edition of the *Principles*, for the third time! He pronounced it: "one of those productions which work a complete revolution in their subject by altering entirely the point of view in which it must thenceforward be contemplated." He found it awe-inspiring.

A few points from Lyell's *Principles* especially struck Herschel. The first was the way Lyell linked climate change to geological change. The second was the roles of ocean erosion and ocean sedimentation in long-term geological change.

The heart of Herschel's letter is a long passage regarding deep-Earth processes, which leads off with the quotable line, "Now for a bit of theory." Herschel outlined consequences of Lyell's geological processes and then entertained various scenarios in which knowledge of physics and properties of matter might offer assistance.

He described an Earth on which continents rest atop an outer crust, afloat on a fluid or semi-fluid matter. But this is not a uniform lava, he said, but partly solid, partly fluid, partly gas or steam. What would happen, Herschel asked, if as Lyell proposed, waves eroded most of the rock that forms new strata on the ocean bottom? What effect would this have over time in the heat and pressure profiles at depth beneath the ocean floor? Herschel asked how the accumulation of matter would affect the semi-fluid matter beneath the crust. One result would be a general explanation of volcanos, as occurring on the edge of areas experiencing a large transfer of matter. Herschel wrote Lyell, "It has always been my greatest difficulty in geology to find a *primum mobile* for the Volcano, taken as a general, not a local phænomenon."[39] Any rupture at a structural weakness would reproduce volcanic

[38] Herschel to Whewell, 28 January 1834, Trinity College Archives, Cambridge, UK, Add. Ms.a.207[23].

[39] Herschel to Lyell, 20 February 1834, paragraphs 8 through 14 concern physical theory of Earth processes in Cannnon, "Impact of Uniformitarianism," 306–307.

phenomena in all their variety: off-gassing, dust clouds, froth, pumice, and lava. (Most of this discussion was omitted from Lyell's extracts in the *Proceedings of the Geological Society.*)

Herschel continued with important questions related to heat, outward heat flow, and the behaviour of materials at high temperatures and pressures. Heat would continue to flow outward from Earth's center, "that great desideratum of the Huttonian theory." Here Herschel revealed himself both a whimsical and a serious gentleman. He denied the ideas that "great currents of melted matter are circulating in the nucleus of the globe," and that "great waves of Caloric" are racing about. He said, simple deposition of new strata alters the thermal equilibrium. The new strata effectively draw heat to themselves.

Herschel's pages of climatological and geological description far surpass the theoretical paragraphs in bulk and imagery.[40] Herschel sought out accounts of rivers drying up and of changes in winds over generations across South Africa and the globe. He reported that the river at Kuruman in the Northern Cape once was "a great and rapid river up which abundance of Hippopotami came." It no longer flowed. Herschel told Lyell of his own geological field work, starting with his second summiting of Paarl Rock "a big, rounded, weathered mass of granite of beautiful uniformity of texture & enormous dimension." He proceeded to describe the great, bifurcated dike, which he told Lyell was the same granite as the general mass, but infinitely smaller grained & much harder." He linked this to a black basalt dike near Simon's Bay, which he detailed. This basalt was dark and magnetic, in laminated columns, with thoroughly embedded "hornstone" (like flint). He did not bother Lyell with azimuths and dip angles, although he had measured them.

Herschel did, however, regale Lyell with narratives of older basaltic rocks, including columns, being in-filled or encrusted in travertine or other calcium carbonate-rich materials in Lancerote, Simon's Bay, and on the Cyclopoean Islands, which Herschel had visited and collected specimens from 10 years earlier.

To contextualize the Simon's Bay basaltic dike, Herschel casually remarked that the dike cuts through the granite "wh[ich] forms the base of all the Table Range as far as Cape Point." This, of course, is a conclusion Herschel mentioned as early as March and April 1834, during his ambles around Wynberg's granite outcroppings. He was extending this granite stratum to underlie all of the Table Range to the Cape of Good Hope.

Herschel then turned to relating general writings about the deposition of soft strata below oceans to his field work across the "Great Valley" or Cape Flats to the mountains on the other side. He described a landscape of islands, flats filled with detritus, hillocks, peaks, and receding oceans. The surrounding mountains, he wrote, all consisted of granite, slate, and sandstone. The Cape Flats and the lower slopes were covered by "enormous deposits" of iron-bearing clay, derived from these mountain materials. The thickness of these deposits was revealed by deep

---

[40] Herschel to Lyell, 20 February 1834, paragraphs 17 through 28 concern climate and geological field work, Cannon, "Impact of Uniformitarianism," 308–311.

erosional gullies. No fossils occurred here, "both the slate & the super-incumbent but unconformable sandstone being entirely destitute of organic remains".[41]

Herschel had thought since before landing in Cape Town in January 1834 that he would find evidence in South Africa that the oceans had once been higher and have been receding. Here it was. Herschel showed in a sketch within the text the heights of different peaks he had climbed, measured, drawn, and sampled in his excursions of 1834 to early 1836: Table Mountain, Lion Head and Rump, Tigerberg, Klapmuts 'cone', the Saxenberg hills, and the Hottentots Hollands. Herschel asserted that the "mural cliffs" above and talus slopes below, along with the deposits on the Flats, showed clearly "that the sea once stood at a much higher level here." He concluded: "Nothing can I think prove more clearly a former sea level at x *gradually* subsiding to the present level gg – but a very few feet below the existing level of the great flat plain of sand which extends from Simons to False Bay.".[42]

Herschel's letter to Lyell mingled historical and physical geology together. Clearly, Herschel had given careful consideration to every element of geological theory he included in this letter; he wanted these ideas to be broadly distributed and discussed. Herschel intended for his friends in England to publish extracts of the February 1836 letter.[43] Lyell published a synopsis of parts of the letter, liberally sprinkled with unacknowledged quotes and paraphrases, in the *Proceedings of the Geological Society of London,* and Charles Babbage published longer extracts as an appendix in his *Ninth Bridgewater Treatise* (1838).[44] Rudwick has noted that Lyell only reluctantly gave Babbage permission.[45] Notably, both Lyell and Babbage omitted some of Herschel's geological theorizing, and both omitted almost all of his field work.

Herschel spelled out his vision in a section not published until Cannon in the 1960s:

> Thus the circuit is kept up – The primum mobile is the degrading power of the sea & rains (both originating from the suns action) above and the inexhaustible supply of heat from the enormous reservoir below always escaping at the surface, unless when repressed by an addition of fresh clothing, at any particular part. In this view of the subject, the *tendency is outwards.* Every continent deposited has a propensity to rise again; and the destructive principle is continually counterbalanced by a reorganizing principle from beneath.[46]

---

[41] Herschel to Lyell, 20 February 1834, paragraph 21, in Cannon, "Impact of Uniformitarianism," 309.

[42] Herschel to Lyell, 20 February 1834, paragraph 23, in Cannon, "Impact of Uniformitarianism," 309–310.

[43] John F.W. Herschel, "Extracts from a letter from Sir John F.W. Herschel to C. Lyell, Esq., dated Fredhausen (sic), Cape of Good Hope, 20th February, 1836," *Proceedings of the Geological Society of London* **2**(1833–1838)548–552 and 596–598. Charles Babbage, *The Ninth Bridgewater Treatise, A Fragment* (London: John Murray, 1838), 225–236. Babbage also prints part of Herschel's letter to Roderick Murchison, 15 November 1836, at 237–241 and an extract of a later one to Lyell on 12 June 1837, at 241–247.

[44] The extracts of Herschel's February 1836 letter to Lyell both leave out large, and different, passages of Herschel's manuscript versions at the Royal Society of London and the American Philosophical Society.

[45] Rudwick, *Worlds Before Adam,* 548.

[46] Herschel to Lyell, 20 February 1836. The passage is in Cannon, "Impact of Uniformitarianism," 308–309.

This is the Sir John Herschel who Charles Darwin and Captain Robert Fitzroy came to visit in May and June of 1836. Certainly Herschel was an astronomer, but he was also deeply committed, too, to investigating the history and physics of the Earth.

8.4 Conclusion: Where Does This Story of Herschel's Geological Work in South Africa Lead?

On the most limited scale, this was not the end of the story of John Herschel's travels and geological researches around the Cape area. More of Herschel's field work remains to be integrated, and also Herschel's drawings of mountains have yet to be examined fully. Herschel's geological interactions with Lyell and Murchison have not yet been fully assessed. But much more importantly, given Herschel's easy acceptance by scientists and gentlemen of many nationalities, what other communities of researchers did Herschel join to pursue his own goals beyond the small community of British geologists? This trace will carry us deeper into the varieties of practitioners with whom Herschel maintained contact, but not in this article. His transition from natural philosopher interested in geology to one of the main organizers of British geophysical science is a story yet to be told. His geological investigations in South Africa, along with his coincident collaboration in tidal and meteorological research, set the stage for John Herschel's pivotal role in organizing Britain's geomagnetic and broader geophysical projects in future decades.

Chapter 9
More Food for Thought: Mill, Coleridge and the Dismal Science of Economics

Margaret Schabas

> *On the whole, there is more food for thought—and the best kind of thought—in Coleridge than in all other contemporary writers (Mill, letter to John P. Nichol, 1834; Mill 1963, 12:221)*

Abstract Mill and Coleridge make for odd bedfellows. Not only were they at opposites ends of the political spectrum, Mill a liberal if not a socialist, and Coleridge a staunch conservative after a jejeune brush with the Jacobites, but they also drank from two different philosophical streams, Mill from British empiricism and Coleridge from German idealism. Mill endorsed the forward march of Newtonian physics, and argued that economics was a science in the same manner. Coleridge, a Romantic, found the reductionist appeal to atomism and the deductive theory of Ricardo almost repulsive. Mill was respectful of Christian belief but kept his economics entirely secular. Coleridge was deeply religious and built his economics on a duty to care for the poor and destitute. For some forty years, Coleridge wrote extensively on the political and economic debates of his day, both as a journalist and as an essayist. While there is nothing that approximates the theoretical depth offered by Mill's economics, I will argue that there is nonetheless evidence of Coleridge's imprint, for example the fostering of individual freedom, strong dislike of the "commercial spirit" and romantic appeals to the end of economic growth.

More than any other nineteenth-century philosopher, John Stuart Mill (1806–1873) was responsible for sustaining the tradition of British empiricism and passing the baton on to his godson Bertrand Russell. Mill famously contributed to almost every branch of philosophy: logic, language, mind, epistemology, feminism, history and philosophy of science, ethics, and political philosophy. A positivist at heart, he steered most away from metaphysics, although lengthy passages in his later

M. Schabas (✉)
University of British Columbia, Vancouver, BC, Canada
e-mail: mschabas@mail.ubc.ca

J. Buchwald, L. Stewart (eds.), *The Romance of Science: Essays in Honour of Trevor H. Levere*, Archimedes 52, DOI 10.1007/978-3-319-53436-2_9

writings on August Comte and Sir William Hamilton, contain prolonged reflections on the nature of causation, identity, and free will.[1]

In the final book of his *System of Logic* (1843), Mill offered lengthy arguments to advance the axiomatic-deductive pursuit of economics.[2] His vision of an edifice of general laws rendered more fine-grained by secondary and tertiary laws to account for the disturbing causes found purchase in his massive text, his *Principles of Political Economy* (1848).[3] This multi-volume work offers a much richer and detailed account of human nature, historical forces, and institutions than any work since Adam Smith's *Wealth of Nations* (1776). His *Principles* is a masterpiece, not only for its breadth and vision but also for its ability to advance the analytical core and to resolve many outstanding debates. David Ricardo may have been deemed, by Benjamin Jowett, as the greatest mind since Plato or, by John Maynard Keynes, to have "conquered England as thoroughly as the Inquisition conquered Spain," but it was Mill who insured that the main tenets of the Ricardian system bore grandchildren, and thus endured for a second or even a third generation. The Ricardo-Mill system was challenged at its very foundations in the 1870s, by the work of William Stanley Jevons, Francis Ysidro Edgeworth and Léon Walras most notably, but it took another decade or two before the pillars were fully toppled and Alfred Marshall's *Principles of Economics* (1890) became the dominant text.[4]

Although Mill wrote his massive two-volume text in less than 2 years, he had first cut his teeth on Ricardo's economics at the age of 13, and had been raised to be a radical voice in the many political and economic debates that were overhauling post-war Britain of the 1820s and 1830s.[5] Before the age of 20, Mill began to write essays on economics that canvassed a wide range of topics, some theoretical and some practical (I will henceforth just use economics in place of political economy). These 30-odd essays displayed remarkable erudition and analytical skill, on the core questions of value and distribution, on currency and banking, public finance, trade and the well being of the working class. They established Mill as a technical and philosophical economist of the highest order.

The most enduring essay, entitled "On the Definition of Political Economy; and on the Method of 'Investigation Proper to It'", was first drafted in 1831, but only published 5 years later, in the *London and Westminster Review* (October 1836). Mill revised it again and it was reissued as the last of a collection, in 1844, entitled "Essays on Some Unsettled Questions of Political Economy". It was the first extensive foray into the epistemological standing of economics, at least in English, since the writings of David Hume and Adam Smith. Both of the Scottish philosophers had promoted the view that propositions and theories in the moral sciences were less likely to be outlandishly wrong than those in the natural sciences where, for example, Cartesian vortices had held a firm grip over many savants for almost a century. The reason is

---

[1] See Skorupski 1989, 1998.

[2] Mill 1973(1872).

[3] Mill 1965(1871).

[4] See De Marchi 1974.

[5] See Ryan 1979; Capaldi 2004. Also see Mill 1981(1873).

that in the moral realm we have an added resource, namely our introspection, that tells us whether or not a given proposition is even remotely plausible.

Mill was more inclined to establish an epistemological equality between physics and economics. Both use the axiomatic-deductive method and build toward a system of laws and broader theories. Like physics, economics must commence with abstract and hence false assumptions about human behaviour, the single-minded pursuit of wealth for example. But Mill also remarks that we know readily that this is a false proposition: "Not that any political economist was ever so absurd as to suppose that mankind are really thus constituted, but because this is the mode in which science must necessarily proceed."[6] In that sense, introspection and commonplace observation serve as invaluable aids to remedy what might be mistaken for truth.

Furthermore, if economics was to secure a set of laws on production and distribution—and Mill believed that it had done so in part already–it must assume perfect competition, even though, Mill concedes, this is an idealized condition that has never and cannot ever exist. There are numerous customs and institutional constraints that are woven into all markets. Even the prior conditions of property rights insure that perfect competition is an idealization. Mill emphasizes that mechanics also posits laws using idealized conditions; it is well understood that bodies are not perfectly elastic or inclined planes free of friction. When it comes to practical mechanics, the laws are tailored to the specific problem at hand with the full acknowledgment of the imperfections in the material objects. Similarly, in economics, Mill advocated that one must learn about customs and institutions that intrude into the otherwise competitive conditions, and one must develop a specific wisdom about the complexities of human nature by reading history and literature.[7]

Mill is strongly wedded to the uniformity of nature (including human nature) such that any of these frictions or what he calls "disturbing causes" is also subject to laws. The problem, especially in the moral realm, is that we rarely know these secondary or tertiary laws. Nevertheless, Mill insists, everything has a cause. There cannot be exceptions to laws; any given deviation is a signal that yet another law is operating in tandem with the known law. As part of the search for the laws that govern the disturbing causes, the economist must look beyond the subject matter at hand. As Mill cautioned, "the mere political economist, he who has studied no science but Political Economy, if he attempt to apply his science to practice, will fail."[8] Mill proposed a new science, which he named "ethology", to study character formation, and he hoped that this microanalysis would form the building blocks for a more general social science. This project, alas, had no significant followers.

Mill's core tenets in political economy can be placed quite firmly on the trajectory that reaches back to Adam Smith, Thomas Robert Malthus, Jean-Baptiste Say, and Ricardo. Mill sought to settle many of the debates sparked by Ricardo, and to address the burning reforms of the day on banking, trade, and poor relief. The received view deems Mill as so successful in settling the disputes between, for

[6] Mill 1967 (1836), 4:322.

[7] See Hausman 1981.

[8] Mill 1967 (1836), 4:331.

example, Nassau Senior and James Ramsey McCulloch, that he is considered the last major classical economist. Of the few who wrote on the core theory after Mill, John Elliott Cairnes or Henry Fawcett for example, there was little need to don boxing gloves and or blow whistles as had Mill. Indeed, if we can take Mill at his word, he decided to write the massive tome to make peace between warring camps, but in doing so, he clearly broke new ground, however modest his proclamations.

The primary emphases of the classical school were on long-term growth and capital accumulation, with much concern paid to the problem of overpopulation and the falling rate of profit. A core commitment was the labour or cost theory of value and, with Ricardo, most attention was directed to the distribution of the annual product among the three classes. Landowners were generally deemed of little to no economic value, while the capitalists brought vision and momentum to the entire system. The classical economists were also remarkably sympathetic to the working class and promoted high wages and increased education, particularly as the means to combat population growth and the dreaded stationary state.[9]

In my earlier work, I have argued that Mill stood less at the summit of classical economics and more in the valley that rose up toward the neoclassical school.[10] His *Principles* and essays on economics were pivotal in furthering the understanding of economic phenomena as mental rather than material. Mill foregrounded human agency and hence psychology to a degree without precedent, and thus also gave much momentum to the role of the individual actor in the determination of the key economic phenomena. Ricardo, by contrast, had left little scope for individual deliberation. A producer of wool might respond to a higher profit in the silk industry, but everything happened mechanistically. There was, in short, little to no room for free will. The Ricardian system was as deterministic as the Laplacian universe and unfolded with as much predictability. The so-called "iron law of wages," for example, is a clear instance of the strict determinism imbedded in the Ricardian account. Mill was a determinist, but believed there was a complex web of laws in operation at any moment. He gave more scope not only for group action but also for individual action, and was also inclined to see the entire system in evolutionary terms.[11] As individuals themselves shift their moral anchors over time, the world would move beyond the commercial (capitalist) era. Mill waxed poetic about a future time when capital accumulation had ground to a halt. The stationary state, as it had been known since Smith, would bring not despair and suffering, but a greater flourishing of humanity, a world in which we would fulfil our individual predilections without "elbowing and shoving," and where we would find genuine personal development by communing with nature.

Mill's *Autobiography* made famous his unorthodox upbringing, the strict pedagogical diet provided by his father, James Mill, as part of a plan to form a pure Benthamite being. At the age of 20, however, Mill experienced a nervous breakdown, followed by severe depression. He reacted with hostility toward his father

[9] See Hollander 1986; Winch 1996.

[10] See Schabas 2005.

[11] See Schabas 2015.

and Bentham and what he considered their trivialization of the human condition. He only found his way back to equanimity by reading William Wordsworth, and thus came to value intuitionism and transcendental thinking for the first time. For unknown reasons, Mill did not read Samuel Taylor Coleridge (1772–1834) as part of his cure in 1826, even though Coleridge and Wordsworth were conjoined as the Lake Poets and had co-authored a collection of poems in 1795. But 2 years later, in 1828, Mill became acquainted with two of Coleridge's disciples, F.D. Maurice and John Sterling, at the London Debating Club. Mill then commenced reading Coleridge, and made his acquaintance by venturing up to Highgate where Coleridge held forth in salon-like gatherings. We know very little about these encounters, except that it seems they conversed more than once. But more importantly, Mill took the trouble to read many of Coleridge's political tracts, both published and unpublished, and clearly considered them much "food for thought" as the opening quote suggests.[12] There is reason to believe that Mill had quite a firm understanding of the poet's broader social and philosophical vision in the years leading up to Coleridge's death in 1834 and that Mill sustained these interests for at least another decade while a close friend of Sterling who died in 1844. Mill would remain receptive to other idealist influences, notably Thomas Carlyle and T.B. Macaulay. Moreover, because Mill brandished swords with William Whewell and William Hamilton on logic and epistemology, he was inclined to revisit and reflect on the merits and demerits of German idealism and Kantianism for the rest of his active writing career.[13]

For some 40 years, Coleridge had written extensively on the political and economic debates of his day, both as a journalist and as an essayist.[14] Some of his key works were shepherded posthumously into print by his children, who made an effort to insure that his social philosophy was promoted well into the 1860s. After some early tussles with Edmund Burke and the import of the French Revolution, Coleridge shed his sympathies with the republicans and by around 1800 focused on domestic debates, for example religious toleration or suffrage. His mature and best known publications on political and economic thought are *The Statesman's Manual* (1816), *A Lay Sermon* (1817), and *On the Constitution of the Church and State* (1830), all cited and admired by Mill. We also learn, from Mill's letter to Nichol (partly cited in the epigraph), that he discerned in Coleridge's political writings a "most distinct thread of connection." That thread remains unspecified, but from my estimation, Coleridge's devout Christianity would be a reasonable guess. His political writings are encrusted with explicit appeals to the Bible. One, *The Stateman's Manual*, is subtitled: "The Bible the Best Guide to Political Skill and Foresight". Coleridge's sense of Christian stewardship permeates his entire sensibility and positions him squarely in the broader movement known as Christian political economy that

---

12 See Turk 1988.

13 See Snyder 2006.

14 See Link 1948; White 1938 [1970, reprint]; and Morrow 1990.

commenced with Malthus, and was furthered by dozens more, including Edward Copleston, Thomas Chalmers and Charles Kingsley.[15]

Mill and Coleridge make for odd bedfellows. Not only were they at polar opposites of the political spectrum, Mill a liberal if not a socialist, and Coleridge a staunch conservative after a jejeune brush with the Jacobites. They also drank from two different philosophical streams, Mill from British empiricism and Coleridge from German idealism. Mill was mechanistic while Coleridge found the reductionist appeal to atomism almost repulsive. Mill was respectful of Christian belief but remained entirely secular in his philosophical work, including his political economy. Yet, Mill proclaims in his private letter of 1834 to Nichol: "Few persons have exercised more influence over my thoughts and character than Coleridge has; not much by personal knowledge of him, though I have seen and conversed with him several times, but by his works, pieced together by what I have otherwise learned of his opinions".[16]

As for public displays of indebtedness to Coleridge, Mill took the trouble to compose and publish a lengthy essay in the *London and Westminster Review* entitled "On Coleridge" (1840), and to pair it with an earlier essay "On Bentham" (1838). There is nothing surprising in Mill writing a tribute to Bentham given his upbringing, but the decision to juxtapose Bentham with Coleridge was motivated, it seems, from a meta-judgment that they are the two most influential thinkers of the age, even though they were "inhabitants of different worlds."[17] Coleridge, Mill avers, "has been the great awakener in this country of the spirit of philosophy" and "Bentham excepted, no Englishman of recent date has left his impress so deeply in the opinions and mental tendencies of those among us who attempt to enlighten their practice by philosophical meditation."[18] Mill acknowledges that only in the field of poetry is Coleridge unequivocally among the canon; his philosophy is still contested and its longevity and influence yet to be measured, but Mill is of the view that the "time has not yet come for Coleridge" and his assessment is thus part and parcel of this effort to adjudicate.[19]

Notwithstanding the adage not to speak ill of the dead, Mill spares almost nothing in criticizing or exposing the shortcomings of both Bentham and Coleridge. Bentham is deemed a mere "boy" for having lived a one-dimensional life, for having never suffered pain or heartbreak, and for having projected his own simplistic emotional make-up onto others. Bentham truly could not see why pushpin was

---

[15] Coleridge, however, had strong antipathy to Malthus for his pessimistic predictions. On the early stages of Christian economics, see Waterman 1991. Kennedy also claims that Coleridge can fit into a movement he calls humanistic economics, to which Carlyle, Dickens, Ruskin, Morris, Shaw and H.G. Wells belong as well as some of the Christians like Kingsley. Literature and economic thought were often conjoined in this period. See Ryan 1981; Connell 2001; Gallagher 2006; and Bronk 2009. Kennedy positioned Coleridge in a movement he calls humanistic. See Kennedy 1978, 43–48.

[16] Mill 1963, 12:221.

[17] Mill 1969 (1840). See Priestley 1969 10:120.

[18] Mill 1969 (1840), 10:120.

[19] Mill 1969 (1840), 10:122.

inferior to poetry as long as individual and collective utility was maximized. Any wisdom regarding eudemonia or the problems of weakness of the will, drawn from the ancient Greek philosophers, was entirely lost on Bentham.

Mill says far less about Coleridge's personal life, although given his later rebuke in *On Liberty* regarding the immoral standing of bearing children one cannot support, it would seem that Coleridge would have been a prime candidate for harsh words. His failed marriage, prolonged inability to provide adequately for his children, and his prolonged opium addiction, can only give pause when tested against Mill's harm principle. For all the simplicity and innocence of Bentham's life, at least as Mill knew it, Coleridge's was one of anguish and upheaval, misguided spiritualism (at least by Mill's reckoning), and political folly. But the two shared an appreciation for the ancient reverence for a self-examined life, for struggles to be endured in the achievement of artistic heights, and for what Mill would later call the "Art of Living".

In his letter to Nichol, Mill claims that Coleridge's views on politics are "principled" and "systematic." But in his published essay of 1840, Mill judged Coleridge more critically, noting both the distortionary tendencies of his religious devotion and the absence of any system. The following captures well the entire tenor of the essay: "In the details of Coleridge's political opinions there is much good, and much that is questionable, or worse. In political economy especially he writes like an arrant driveller, and it would have been well for his reputation, had he never meddled with the subject." The footnote to this somewhat mutes the harsh judgment: "Yet even on this subject he has occasionally a just thought."[20] Mill had the *Lay Sermon* (1817) in mind with this remark, and Coleridge's quite penetrating account of financial crises. It might be worth noting that Thomas De Quincey was also very critical of Coleridge's economics, whose ignorance of the subject meant he committed "vulgar errors long ago refuted even by Necker".[21]

Perhaps what is more telling is the proprietary stance of Mill or De Quincey in claiming expertise in economics. At first glance, neither had any claim to higher authority. Coleridge, with his theology degree from Cambridge, had identical credentials to Malthus, the first Englishman to hold the title of professor of political economy, or to Richard Jones, Malthus's successor at the East India College. Coleridge trumped Ricardo or Mill, who lacked academic credentials altogether (Ricardo had never been to university; Mill had spent a couple of years studying at the University of Montpellier but held no degree). It was precisely in the late 1820s and 1830s that the British universities established professorships in political economy, at Oxford, Cambridge, Trinity College in Dublin, King's and University Colleges in London. But as was often true of appointments in the natural sciences, the professors themselves normally had no formal training in the subject. Arguably, the first example of a professor of economics with an advanced degree in the subject is Jevons, who took a Master's degree in political economy at the University College, London in 1862, and eventually became the professor there, in 1876. Mill's smug

[20] Mill 1969 (1840), 10:155.

[21] Quoted in Kennedy 1978, 24.

remarks about Coleridge's command of economics can only be based on a careful reading, not on the lack of credentials as might be the case today were a world-renowned poet to be so bold as to weigh in on economic theory or policy.

The fact that Mill or De Quincey took the time to comment on Coleridge's economics suggests that there were others who took the poet's ideas seriously. As a journalist of some renown, Coleridge had engaged in a number of economic debates, notably the restoration of a convertible currency, and on the Corn Laws that protected the landed gentry. As a teenager, he had read James Steuart's *Principles of Political Oeconomy* (1767) and retained an interest in the agrarian well being of the Scottish Highlands in mature writings. Steuart was the leading contributor to British economic thought between the more influential contributions of David Hume (1752) and Adam Smith (1776), but Coleridge may have first been drawn to Steuart for his Jacobite allegiances. Coleridge later read Smith, since he voices some support for his recognition that middle men serve a valuable role that, contrary to popular belief, do not in effect raise prices. Coleridge's praise for Smith went against the judgment of his close friend William Wordsworth, who found Adam Smith highly distasteful.[22] Surprisingly, Coleridge from time to time spoke in favour of free trade, albeit with some qualifications.[23] As for other eighteenth-century contributors to the subject, Coleridge remarked en passant of the physiocrats (and perhaps Turgot), "at the commencement of the French revolution, in the remotest villages every tongue was employed in echoing and enforcing the almost geometrical abstractions of the physiocratic politicians and economists."[24] He believed that economists were inclined to scientific pretentions that were unjustified and, as a result, had a direct harm on the lives and well being of ordinary people. He wrote scathingly about Earl of Lauderdale's use of economic theory to justify unregulated factory conditions for children, and celebrated Robert Peel's measures that in principle reduced their hours and introduced inspections.

From the essay "On Coleridge," there are many signs that Mill appreciated Coleridge's heuristics for the pursuit of knowledge. Mill came to think more organically, to appreciate the virtues of intuitionism and, above all, to recognize that truths normally come in parts and not wholes. Mill developed an interesting account of how robust and enduring truths emerge from a struggle with falsehood and this became a central motivation for his wider appeal to liberalism, particularly the freedom of expression and belief. Most propositions are only partly true and hence must be subject to severe scrutiny and juxtaposed forcefully with other propositions that might be equally false. Much inquiry might be construed as a complex dialectic of these ill-resolved propositions. Coleridge also gave Mill much to ruminate about on "the Art of Living" in a way that Bentham simply did not. Finally, there is reason to think Coleridge left an imprint on Mill's elitist views on education and governance. For a socialist, Mill's peculiar decision to grant educated citizens more weight in elections is always difficult to understand. For Coleridge, it would make perfect

[22] See Henderson 2013.

[23] According to Kennedy (18), "Coleridge, in the main, agreed with the theorist of *laissez faire* on the given free scope to self-interest in commerce." Mill, however, took Coleridge to be an opponent.

[24] Coleridge 1972, 15–16.

sense since he believed firmly in the elite of society guiding the underclass and under-educated. Both shared a deeply rooted fear of conformity or what Mill, drawing on Alexis de Tocquevilles study of American culture, aptly called the "tyranny of the majority."[25]

Coleridge had many more views about economic policies. He opposed the move to a convertible currency (the gold standard) that started to surface in the early 1810s and took full hold by 1820.[26] He advocated high taxes as a remedy to the post-Napoleonic depression of the 1815–1820. Above all, he sought to protect the landed interest because they were essential in sustaining and expanding the agrarian sector. The production of food, for Coleridge, was sacred and could not be analyzed with the same logic of supply and demand as the commercial sector. Self-interest and the pursuit of profits were subordinate to religious obligations. God had given the land and its bounty to insure that all of his children would be fed and hence those who directed the agrarian sector had a Christian duty of care to provide subsistence. A shortfall of grain would need to be met one way or another.

As a result, Coleridge devised a complex array of views on the Corn Laws. In part, he welcomed their protectionist fetters and went further to argue for agrarian autarky. But he also saw them as entirely too self-serving for his liking, and prone to keeping the price of food too high for the common good While the initial laws introduced in the early 1800s, as a result of the war with France, met his approval, Coleridge came to see these measures as too restrictive. After a series of post-war riots due to poor harvests and general hardship, Coleridge was more inclined to wish them dismantled and the unrestricted import of grain reinstated. What he feared, however, was that the commercial spirit would insinuate itself into the agrarian and, in that sense, he was not in favour of free trade per se, but trade that upheld the overriding ethic of care.

Although Coleridge died before Mill's *Principles* appeared, it is unlikely he would have been persuaded by Mill's more surgical measures of severe death duties or fiscal reform. Coleridge advocated sustained high taxes during the post-war years for 1815 to 1821, but he wanted taxes that would not deplete the primary leadership role of the landed gentry, because they embodied the proper conservative ethos and promoted the more dignified form of labour. Coleridge had a romantic view of farming, even though his own experiment at husbandry, as a young man, proved unsuccessful.

Trevor Levere's book, *Poetry Realized in Nature* (1981), promotes an image of Coleridge as curious of virtually every field of science. Coleridge made strong pronouncements about Kepler, Lavoisier, Cuvier, Oken, and so forth, even though he had not become an adept astronomer, chemist, or anatomist. His close friendships with Humphrey Davy and Thomas Beddoes gave him an appreciation for chemistry, and his lifelong affiliation with physicians, including his older brother, fed him more empirical material on physiology, but it is fair to say that Coleridge was more dilettante than savant. What comes through is an image of a man of immense audacity

[25] See Collini et al. 1983.

[26] Alexander Dick (2013) offers a thorough account of Coleridge on currency and monetary reform.

or, to return to De Quincey's critique, a man "too proud to learn from anyone else."[27] Levere only refers once to Coleridge's efforts at economics, in his remark about the excessive geometry of physiocracy, because his eye was directed to the natural sciences and the theme of Coleridge's organic imagery and transcendental modes of thinking.[28] Coleridge had embraced *Naturphilosophie* in 1815 and this sustained him till the end of his life. Coleridge sought a system that would unify and comprehend the sense in which Nature is alive and intelligible, full of powers that are life-affirming. Science must be part of a broader Christian mission to link our free will or soul to nature. Because nature was developmental, it had a history and thus revealed a purpose. This teleological stance meant that history was itself part of science. It was essential in scientific inquiry to single out prominent phenomena but not to remain at the superficial level of observation. To delve more deeply, one must treat nature as a set of powers, not of things. Matter was itself the result of polar powers, "the phantom of the Absolute Will", and to understand it was to enter, transcendentally, into a world of symbols.

Coleridge also believed that the mind as the seat of free will is inextricably conjoined with human action and that this is part and parcel of his transcendental mode of thinking. It is this theme that I will propose also seeped into Mill's conception of economics, and that this in turn strengthens the view that Coleridge left a significant imprint on Mill's economics. As part of Mill's insights into the very nature of economic phenomena, there was a greater sphere for the will, for what has come to be enshrined in mainstream economics as "rational choice." It may have been in part his appreciation of Coleridge on this inner essence of the human spirit, the potential for individual choices that were not mere manifestations of the mob that helped steer Mill in this direction. Coleridge would not have favored the emphasis on consistency that contemporary rational choice theory makes paramount, but he would have appreciated the sense in which individuals would be free to choose and rank their preferences in whatever fashion they desire. Coleridge could never be accused of succumbing to the "tyranny of the majority." He played to his own drum for his entire life, a character trait that Mill, who had been groomed to be like Bentham, much appreciated.

The classical economists, Smith, Malthus and Ricardo, had broached the law of the falling rate of profit. More worrisome was the fact that once the profit rate fell to the point that there was no further incentive to accumulate capital (Ricardo estimated a rate of 6% or 7%), growth would halt and human population saturate the land. This was called the Stationary State, but Mill, rather than fearing its onset, urged that it arrive more quickly. He saw this future state as one to be embraced, or rather, to be anticipated such that the population was below saturation point but the

[27] Quoted in Kennedy 1978, 24.

[28] Economics had remarkably high standing at the time, at least in Britain. The leading scientists, William Whewell, John Herschel, and Charles Babbage, all recognized economics a science, and in prominent publications, including Herschel's *Preliminary Discourse on a Study of Natural Philosophy* (1831). They also helped to highlight its representation in Section F of the BAAS (see Snyder 2006).

capitalist engine still petering out. In this respect, he shared Coleridge's deeply-rooted aversion to the "commercial spirit." For Mill, the Stationary State meant that the "elbowing and shoving" of the commercial error would decline if not disappear, and that people would become ends and no longer means. It meant that some wilderness would still be preserved such that true solitude was possible. This, Mill believed, was the only cradle for genuine individual development. Mill placed much value on self-reflection and the inner life, and in that respect he owed a considerable debt to Coleridge.

Mill also positioned the human mind as the proximate cause of economic phenomena. In that respect, his ability to move beyond a strictly physical interpretation of the laws of value and distribution could also reflect indebtedness to Coleridge. Mill grants human agency and the will more specifically a much greater scope than his classical predecessors. There is, in fact, very little room for individual deliberation in the core accounts of Adam Smith and David Ricardo. Each economic group or class follows its particular goals, to raise profits, rents or wages, or vice versa, depending on overriding interests. Within each group there is no reason to differentiate specific types of economic agents or actors. But with Mill, we start to see a more fine-grained account of both the working class and the capitalist class. Some are trade unionists, and others cottiers or basic unskilled workers; among the capitalists, some are idle rentiers collecting dividends on annuities, while others are enterprising and ambitious investors. More to the point, Mill promotes the view that each of us must take on the burden of enhancing our individualism, of experimenting with our lives, and with trying hard not to take anything for granted simply because it is bequeathed to us by the status quo. The need to have a reflective and examined life is paramount in Mill, and there is reason to maintain that he arrived at this and sustained these beliefs in part because of his exposure to the writings of Coleridge.

Acknowledgement I would like to acknowledge the contributions of Alexander Dick, Larry Stewart, C. Tyler DesRoches, and Christopher Mole in preparing this essay, as well as funding from SSHRC. I also benefitted from presenting this paper to the UKHET conference in Shanghai (September 2016).

References

Bronk, Richard. 2009. *The romantic economist: Imagination in economics*. Cambridge: Cambridge University Press.

Capaldi, Nicholas. 2004. *John Stuart Mill, A biography*. Cambridge: Cambridge University Press.

Coleridge, Samuel Taylor. 1969–2002 [1772–1834]. *The collected works of Samuel Taylor Coleridge,* multiple editors, 11 vols. Princeton: Princeton University Press.

———. 1972 [1817]. *The Lay Sermons*. In *The collected works of Samuel Taylor Coleridge,* ed. R.J. White, vol. 6. Princeton: Princeton University Press.

Collini, Stefan, Donald Winch, and John Burrow. 1983. *That noble science of politics: A study of nineteenth-century intellectual history*. Cambridge: Cambridge University Press.

Connell, Philip. 2001. *Romanticism, economics and the question of 'Culture'*. Oxford: Oxford University Press.

De Marchi, Neil B. 1974. The success of Mill's principles. *History of Political Economy* 6: 119–157.

Dick, Alexander. 2013. *Romanticism and the Gold Standard: Money, literature, and economic debate in Britain 1790–1830*. Basingstoke: Palgrave and Macmillan.
Gallagher, Catherine. 2006. *The body economic: Life, death, and sensation in political economy and the Victorian novel*. Princeton: Princeton University Press.
Hausman, Daniel. 1981. John Stuart Mill's philosophy of economics. *Philosophy of Science* 48: 363–385.
Henderson, James P. 2013. Beggars: Jeremy Bentham versus William Wordsworth. *History of Political Economy* 45 (3): 415–442.
Hollander, Samuel. 1986. *The economics of John Stuart Mill*. Toronto: University of Toronto Press.
Kennedy, William F. 1978 [reprint of 1958]. *Humanist versus economist, The economic thought of Samuel Taylor Coleridge*. Westport: Greenwood Press.
Levere, Trevor H. 1981. *Poetry realized in nature: Samuel Taylor Coleridge and early nineteenth-century science*. Cambridge: Cambridge University Press.
Link, Arthur S. 1948. Samuel Taylor Coleridge and the economic and political crisis in Great Britain, 1816–1820. *Journal of the History of Ideas* 9: 323–338.
Mill, John Stuart. 1963. *The earlier letters of John Stuart Mill, 1812–1848*. In *Collected works of John Stuart Mill*, ed. F.E. Mineka, vol. 12. Toronto/Buffalo: University of Toronto Press. [abbreviated as CWJSM].
———. 1965 (1871). *Principles of political economy*. In *Collected works of John Stuart Mill*, ed. John. M. Robson, vols. 2–3, 7th ed. Toronto/Buffalo: University of Toronto Press.
———. 1967 (1836). On the definition of political economy and on the method of investigation proper to it. In *Essays on economics and society. Collected works of John Stuart Mill*, ed. J.M. Robson, vol. 4. Toronto/Buffalo: University of Toronto Press.
———. 1969 (1838). Bentham. In *Essays on ethics, religion and society. Collected works of John Stuart Mill*, ed. J.M. Robson, vol. 10. Toronto/Buffalo: University of Toronto Press.
———. 1969 (1840). Coleridge. In *Essays on ethics, religion and society. Collected works of John Stuart Mill*, ed. J.M. Robson, vol. 10. Toronto/Buffalo: University of Toronto Press.
———. 1973 (1872). *A system of logic: Ratiocinative and Inductive. Collected works of John Stuart Mill*, ed. J.M. Robson, vols. 7–8. Toronto/Buffalo: University of Toronto Press).
———. 1981 (1873). *Autobiography and literary essays. Collected works of John Stuart Mill*, ed. J.M. Robson and J. Stillinger, vol. 1. Toronto/Buffalo: University of Toronto Press.
Morrow, John. 1990. *Coleridge's political thought: Property, morality and the limits of traditional discourse*. New York: St. Martin's Press.
Priestley, F.E.L. 1969. "Introduction" to John Stuart Mill, *Bentham (1838)* and *Coleridge (1840)*. In *Essays on ethics, religion and society. Collected works of John Stuart Mill*, ed. J.M. Robson, vol. 10. Toronto/Buffalo: University of Toronto Press).
Ryan, Alan. 1974. *J. S. Mill*. London: Routledge and Kegan Paul.
Ryan, Cheyney C. 1981. The friends of commerce: Romantic and Marxist criticisms of classical political economy. *History of Political Economy* 13 (1): 80–94.
Schabas, Margaret. 2005. *Natural origins of economics*. Chicago: University of Chicago Press.
———. 2015. John Stuart Mill: Evolutionary economics and liberalism. *Journal of Bioeconomics* 17 (1): 97–111.
Skorupski, John. 1989. *John Stuart Mill*. London: Routledge.
———, ed. 1998. *The Cambridge companion to Mill*. Cambridge: Cambridge University Press.
Snyder, Laura J. 2006. *Reforming philosophy: A Victorian debate on science and society*. Chicago: University of Chicago Press.
Turk, Christopher. 1988. *Coleridge and Mill*. Aldershot: Avebury.
Waterman, A.M.C. 1991. *Revolution, economics and religion: Christian political economy, 1798–1833*. Cambridge: Cambridge University Press.
White, R.J. 1938 [1970, reprint]. *The political thought of Samuel Taylor Coleridge*. London: Jonathan Cape.
Winch, Donald. 1996. *Riches and poverty: An intellectual history of political economy in Britain, 1750–1834*. Cambridge: Cambridge University Press.

Chapter 10
"These Can Not All Have an Interest for England": Symmetry, Beauty and the Trouble with Romanticism in Britain

Gordon McOuat

In mid-nineteenth century Britain, it was advised not to begin a work on natural philosophy with a discussion of beauty and the sublime – in any case, not in a serious scientific work. Certain circumstances blocked that romantic route. This paper is about one or two of those circumstances.

10.1 Erring on the Side of Beauty

On the 6th of January 1848, a besieged and beseeching 71-year-old Hans Christian Oersted dispatched a letter to his old friend, Edward (soon to be Sir Edward) Sabine. Sabine was part of the experimental, electromagnetic, old guard, indebted to Oersted for dipping needles, magnetism and mapping, which made Sabine's grand Humboldtian magnetic survey of the world possible[1]:

> I desire by your intervention and perhaps that of some others of my English friends to find a publisher for a translation of a book of mine. I intend to make a collection of my popular treatises and other papers published in several Danish journals and transactions. These can

[1] Indeed, it was under Humboldt's suggestion, in a letter to the President of the Royal Society, 1836, that magnetic stations were established throughout the British Empire. Cf. "Sabine" *DNB*. On Humboldtian science, see Susan Faye Cannon, *Science in Culture: The Early Victorian Period* (New York: Science History Publication, 1978), chapter 4. As Michael Dettelbach points out, Sabine's Humboldtianism was a translated version. See his fine discussion in Michael Dettelbach, "Humboldtian Science", in Nicholas Jardine, James Secord and Emma Spary, eds., *Cultures of Natural History* (Cambridge: Cambridge University Press, 1996), 287–304.

G. McOuat (✉)
History of Science and Technology Programme, University of King's College/Dalhousie University, 6350 Coburg Road, Halifax, Nova Scotia, Canada
e-mail: gmcouat@dal.ca

J. Buchwald, L. Stewart (eds.), *The Romance of Science: Essays in Honour of Trevor H. Levere*, Archimedes 52, DOI 10.1007/978-3-319-58436-2_10

> not all have an interest for England, but I think that a choise [sic] of them should be acceptable.[2]

Oersted looked to England, buoyed by its enlightened patronage of his experimental advances. He had visited there on a number of occasions. The Royal Society had honoured him with the prestigious Copley Medal in 1820, and he was much appreciated in English experimental circles as a founding father. Yet few of his papers – experimental or otherwise – had been translated into English and it was mainly through his network of correspondence and visits that he was publicly well known.

"I should at all times wish to sue for approbation of your enlightened countrymen," Oersted began. He outlined his "considerable pecuniary" interest in the matter of the proposed book. The continental political situation was dark and tense. Denmark was entangled in the Schleswig-Holstein revolt, and liberals of Oersted's circle were pressing for a new constitution. Publishing in Denmark was anything but lucrative, liberal England offered advantages, both in speed and prestige (both scientific and liberal). Oersted presented a strategy: an English translation of his book would be provided to the publisher "three months before the Danish edition is brought to sale", and thus benefit from the English connection before going Danish.

The projected "little volume" – Oersted hoped it would be enlarged by engravings (it wasn't) – was to be engaged not with the "experimental researches" by which Oersted was so well known in England, "but", as he put it, with "the view of a natural philosopher upon subjects appertaining to the intellectual world." Oersted noted that it was not uncommon amongst English scientists to pronounce on philosophical and methodological issues, William Whewell and John Herschel being the most well known. Yet, he expected no competition with what he called that "excellent book" of his friend, Herschel: the *Preliminary Discourse on the Study of Natural Philosophy.*[3] Herschel's book fast became the mainstay of discussions of scientific method while Whewell's, to the author's chagrin, was much under-appreciated.[4] Yet, while Herschel opened with utilitarianism and the conquest of nature, physics, progress and the inductive method, Oersted promised his English audience a much more heady romantic brew. According to his plan, Oersted's text would *begin* with beauty and the sublime, using that as a foundation to discuss morality and spirit, along with genius, reason and religion, and science:

---

[2] Oersted to Sabine, October 6, 1848. In M.C. Harding, ed., *Correspondence de H.C. Örsted avec divers savants* (Copenhagen: H. Aschenhoug, 1920), vol. 2, 508–10. Sabine was General Secretary of the British Association for the Advancement of Science. Oersted was elected "corresponding member" in 1834.

[3] John Herschel, *Preliminary Discourse on the Study of Natural Philosophy* (London: Longman, Rees, Orme, Brown and Green, 1830). In the *Preliminary Discourse*, Herschel admits that "it required the eye of a philosopher such as Oersted to seize the indication, refer to its origin, and thereby connect two great branches of science." § 127. He later compares Oersted's philosophical doggedness in the face of the disunity of electricity and magnetism to Columbus and his notion of the necessary existence of the New World. § 376.

[4] See W. Whewell, "Review of J. Herschel's *Preliminary Discourse on the Study of Natural Philosophy (1830),*" *Quarterly Review* **90** (1831), 374–407. Whewell wished that his *History* and his *Philosophy of Inductive Sciences* had made a moral and political impact.

> A considerable part of these researches, of which only a part is printed, have for object the physical principles of beauty and their connection with the general laws of nature. I have even expatiated from the first view to the relation of these principles to the moral precepts and to the truth in general.
>
> Several [of the other treatises planned for the book] concern the genius of natural science, connected with a view of the genius of sciences in general.[5]

Sabine obliged Oersted's request – perhaps reluctantly – and put him in contact with Longmans, the publisher of Sabine's magnetic surveys and Herschel's own *Preliminary Discourse*.[6] Longmans solicited a sample of the work and Oersted provided his own translation of the opening piece, "The Fountain" – a short dialogue on the relations between beauty and the sublime, the inner and outer world, which would set out the theme of the rest of the book.[7] The dialogue was distinctly post-Kantian, much closer to Schelling without any acknowledgement of that source. Where Kant, following Burke, had commenced with the separation of beauty and the sublime, giving over to the totality and transcendental "terror" of the latter in an unresolved dynamic tension between reason and the sensible object, Oersted united the two in the inner life of Reason at one with the object. Without such an inner life, where the Sublime gives meaning to the Beautiful, "that which might otherwise be called beautiful would be dead."[8] By turning his attention to a "fountain", and away from the usual candidates for the sublime – huge mountains, frightening storms, and an imposing God – Oersted likewise encountered the sublime and inner beauty in artificial and experimental things. Here was Oersted's philosophy of nature. He thought, at least, that the practical British might concur.

Alas, the reception could not have been less rewarding. Longmans, ill satisfied with the intended direction of the work, rejected the book out of hand. Undaunted, Oersted turned to his old friend Herschel for advice.[9] Oersted translated a *second*

---

[5] Oersted to Sabine, loc cit. On genius in romanticism see Simon Schaffer, "Genius in Romantic natural philosophy," in Nicholas Jardine and Andrew Cunningham, eds., *Romanticism and the Sciences* (Cambridge: Cambridge University Press, 1990), 82–98.

[6] Edward Sabine, ed. *Observations made at the Magnetical and Meteorological Observatory at Toronto in Canada*, 3 vols. Printed by order of Her Majesty's government. (London: Longman, Brown, Green, and Longmans, 1845–1857).

[7] Oersted, "The Fountain," in H.C. Oersted, *The Soul in Nature*. Leorna and Joanna B. Horner, trans., (London: H.G. Bohn, 1852), 28–40. For a discussion of Oersted's *The Soul in Nature*, see David Knight, "The Spiritual in the Material," in *Hans Christian Ørsted and the Romantic Legacy in Science*, eds. R. Brain, R.M. Cohen and O. Knudson (Dordrecht: Springer, 2007), 417–422. On the sublime in romantic philosophy and art, see Thomas Weiskel, *The Romantic Sublime: Studies in the Structure and Psychology of Transcendence* (Baltimore: Johns Hopkins University Press, 1976), and Immanuel Kant, *Critique of Judgment*, (1790) trans. Werner Pluhar, (Indianapolis: Hackett, 1987).

[8] One of Oersted's interlocutors, "Frank" in the English dialogue, echoes the Kantian perspective. "I well remember that once, in a mountainous country, when an enormous precipice rose before me, I was overcome by a sense of desertion and death." Oersted (1852), 38. Oersted, through "Frank", counters with a much more unified picture of beauty and the sublime.

[9] Herschel's relations with Oersted had been close. Yet, we may wonder what Herschel thought of the proposed book in light of his original panegyric on Oersted at the 1835 meeting in Southampton: "How often did [Herschel] wish to heaven that he could trample down, and strike for ever to the

dialogue for Herschel – the "Spiritual in the Material" – in the hope that *this* might better appeal to a more practical, and what now seemed like a less aesthetic, British audience. Herschel, however, was not impressed and, after reading the newly translated piece, advised against publication, observing that "much of it would not be palatable to the English readers." Oersted was now acutely aware of the difficulty: "Perhaps I have erred in beginning my communications to the English public by the philosophy of beauty?"[10] Indeed.

Oersted hunted, in vain, for another publisher. By the time that the work was finally released in English, as part of Bohn's Philosophical Library series – (Bohn was noted for publishing such translations) – Oersted was dead, a Danish edition was already in print, and the order of the dialogues was reversed. "Beauty", indeed, would come second, after "spirit and matter."[11] The reordering didn't work. Charles Darwin summed up the general reception as "dreadful."[12]

What was it that blocked this British reading, this (re)introduction of a fully-blown Romanticism, beauty and the sublime into England at mid century? British Baconianism, or lack of aesthetic sense, might be one suspect. But the British had already had their fare share of experience with continental imports, and indigenous romantic encounters, not all of it negative.[13] Encounters with the "sublime" began in England, and surely could have returned.[14] Moreover, the British, even if they were Baconian, were flexible. Indeed, it has been claimed that in the first half of the nineteenth century, Britain was in a veritably awash in Romanticism. Yet, by the mid century, this was no longer the case. The romantic road was blocked, at least in the natural sciences. Oersted attempted his publication at the wrong time.

---

earth, the hasty generalization which marked the present age, and bring up another and a more safe system of investigation." Quoted in Oersted (1852), introduction, xviii. This might shed a slightly different light on Herschel's claim that Oersted should be regarded "as a fertilizing shower descending from heaven, which brought forth a new crop, delighted to the eye and pleasing to the heart." Ibid.

[10] Loc. cit.

[11] These negotiations over opening dialogues help explain the rather peculiar ordering of the eventual English translation. The 1852 English translation begins with "Here we are all assembled again this evening, which is so fine that it seems to invite us to pursue our conversation of last night." (p.1), which is really a continuation of the end of the second dialogue in the book.

[12] "dreadful", Darwin's reading notes, May 11, 1852. DAR128:3, Cambridge University Library Darwin Collection.

[13] For British transcendentalism, see Phillip R. Sloan, "Transcendentalism in Britain." In Philippe Huneman, ed. *Understanding Purpose. Kant and the Philosophy of Biology* (New York: University of Rochester Press, 2007), 149–189.

[14] Indeed, it was Wordsworth who criticised the separation of beauty and the sublime in Burke and wished to find the sublime in everyday things. The classic discussion of Wordsworth and the sublime is found in Albert O. Wlecke, *Wordsworth and the Sublime* (Berkeley: University of California Press, 1973).

10.2 Defining Romanticism in an English Context

In recent years, after so many years of neglect, the history of British romanticism and science has emerged as a growth industry. (Indeed, one is at pains to say anything new here.) Trevor Levere and his colleague David Knight were there at the beginning: Trevor Levere with his detailed and sympathetic reading of Coleridge's philosophy of nature, David Knight with his numerous studies of the life and ideas of British scientists during the so-called Romantic era.[15] It is no coincidence that both David and Trevor began their careers as historians of chemistry and as students of Alistair Crombie. They have shown that understanding chemistry in the nineteenth century is impossible without understanding romanticism, elective affinities, dynamism and force.

Prompted by Levere, historians of science have since rediscovered the era, following Humboldt in his travels, unfolding transformationism,[16] resurrecting archetypes, recognising genius,[17] and knocking Darwin, Tyndall and the positivists down a peg or two.[18] Trevor, David and subsequent historians have examined Romanticism mostly in its heyday, and have discovered its vestiges in later works of British scientists. For some, the "dynamic" system of science was never lost on the British, even in their most materialist and positivist stages – so much so that now we face some difficulty in deciding who might be a Romantic. Perhaps everyone was. Where the positivists once saw discontinuity with Romanticism, now we see (perhaps too much) continuity.[19] However, this chapter aims to be a bit more precise in its encounter with Romanticism, not so much in understanding what it meant to be a British romantic, but in how it became very hard (and in one place, rather impossible) to be one around mid-century. It hopes to discover how a small pocket of resistance could, with some chance, and much beautiful display, come to block any further imports, any more romantic explorations of the kind Oersted imagined. And it could do so

---

[15] Trevor Levere, *Poetry Realised in Nature: Samuel Taylor Coleridge and Early Nineteenth Century Science* (Cambridge: Cambridge University Press, 1981); David Knight, *Science in the Romantic Era* (Aldershot: Ashgate, 1998).

[16] Evelleen Richards, "A Question of Property Rights: Richard Owen's Evolutionism Reassessed," *British Journal for the History of Science* **20** (1987), 129–171; Robert J. Richards, *The Meaning of Evolution: The Morphological Construction and Ideological Reconstruction of Darwin's Theory* (Chicago: University of Chicago Press, 1992); Nicholas Rupke, *Richards Owen: Victorian Naturalist* (New Haven: Yale University Press, 1994).

[17] Schaffer, "Genius in Romantic natural philosophy", loc cit.

[18] Some go as far as claiming Darwin as a "Romantic", surely pressing the issue too far. See the debate around Robert Richards, *The Romantic Conception of Life* (Chicago: University of Chicago Press, 2002).

[19] See, for example, Nicholas Jardine, "*Naturphilosophie* and the Kingdoms of Nature," In Nicholas Jardine, James Secord and Emma Spary, eds., *Cultures of Natural History* (Cambridge: Cambridge University Press, 1996), 230–245, especially on 244: "for all their denunciations of the 'harebrained aesthetic blathering' (Du Bois-Reymond) of the *Naturphilosophen*, the experimental practices of the new sciences owed as much to German Romantic *Eigentümlichkeit* as to French analysis."

without the political and philosophical confabulations so associated with the mid-century revolutions on the continent. It did so, in one sense, by bringing one of the concerns of romantic science to bear on another, namely a more Humboldtian encounter with "being there", standing against an apriori notion of beauty, symmetry and truth displayed in some of the more transcendental texts.[20] I concentrate on natural history, and less so on magnetism, simply because Romanticism in England, by the mid century at least, lived more in the life sciences than amongst the physicists and the mathematicians. Debates about the system of life were more conducive to transcendental ideas, more systematic in their enrolment. At bottom lay natural history. Natural history gives us our "Systemae Naturae" and our "natural kinds", our grounds for the other sciences. If we find (and block) romanticism there, we can find (and block) it everywhere.

What do we mean by Romanticism, even in such a barren context as Britain? Here we should resist the dulling temptation to define. Definitions, as William Whewell constantly needed to remind us, at best come at the end after all the work is done.[21] Rather, perhaps we should catch the *mood* of Romanticism, and, mainly, the players. One of the very best *very* short works on Romanticism and the sciences remains Ernst Hamm's recent small survey in our *Reader's Guide to the History of Science*. Hamm, a fine Canadian historian of Goethe's geology and a student of Levere, notices at least two contemporary nineteenth century schools exploring the history and meaning of Romanticism (albeit Romanticism mainly as a *literary* movement).[22] Henrich Heine (1797–1856), once a self-confessed member of the "school", stressed the reaction against Napoleonic occupation, the rejection of things modern: democracy, Enlightenment, cosmopolitanism and anything French.[23] Rudolf Haym (1821–1901), Heine's more detached near-contemporary, saw Romanticism as originating and unfolding around Schelling in particular, and his union of Goethe's vision of nature with an objectified Fichtean idealism. It is with Schelling's absorption of Goethe, that we have a vision of nature as at one with spirit, evolving (to use the original sense of the term) via polarized forces giving the very *construction* of matter, of action, of being itself.[24]

Kant's attempt to lay out a constructivist foundation to the Newtonian world view simultaneously gave us the *construction* of matter out of apriori polar forces of

[20] Perhaps this is nothing new. In a certain sense, Darwin's worry about the fecundity found in a Humboldtian world played havoc with any vestige of transcendentalism left in his naturalism. Here is where we must dissent from Robert Richards reading of the material.

[21] William Whewell, "On the Use of Definitions," *Philological Museum* **2** (1833), 263–272. Whewell's attack on definitions was part of a concerted effort to keep Benthamism at bay. See Gordon McOuat, "The Origins of 'Natural Kinds': Keeping Essentalism at Bay in the Age of Reform," *Intellectual History Review* **19** (2009), 211–230.

[22] Ernst Hamm, "Romanticism," in A. Hessenbruck, ed., *Readers Guide to the History of Science* (London: Fitzroy, 2000), 656–658.

[23] Heinrich Heine, "The Romantic School," in Helen Mustard and Max Knight, eds., *Heinrich Heine: Selected Works* (New York: Random House, 1973).

[24] Rudolf Haym, *Die romantische Schule: Ein Beitrag zur Geschichte des deutschen Geistes* (Berlin: Rudolph Gaertner, 1870).

possible experience according to law (in his *Metaphysical Foundations of Natural Science* and the *internal* (as opposed to utilitarian-external) aesthetic teleology of the *Critique of Judgment* – two seemingly radically different visions uniting as one Schelling's romantic philosophy of nature.[25] Romanticism and *naturphilosophie*, in some sense, stem from there, even though we can also find hints in earlier, often German, resistances to Newtonianism and materialism.[26] But, if this direct reading of Kant and the post-Kantians is what we mean by Romanticism, then there were indeed very few British Romantics, at least amongst the natural philosophers: Coleridge, perhaps, and Humphrey Davy at times. Kant's *Metaphysical Foundations of Natural Science*, for example, first found its English translation very late in the century (1883), by a Marxist and staunch anti-feminist.[27] Kant's greatest importer, William Whewell, could hardly be called a Romantic. His early and sustained hatred for all things Benthamite and many things analytic and French, set him on one side of Mill's Coleridge/Bentham divide.[28] But even Whewell was flabbergasted and repelled by the more Schellingly imports. He would write on "polarity", but warned about the "rash with which men of great name in Europe (for example, Schelling or Hegel) have built up systems of arbitrary and baseless polar antitheses from their own imagination."[29] Whewell's consilience and method was meant as something unique.

Hamm, Heine and Haym are careful to point out that, for all the "mood", Romanticism is located in a time and place, and should be thought of as having certain causal connections to that place.[30] This should serve as a caution to those still interested in the morphological state of *Zeitgeists* (a more sophisticated version of "in the air", or what was once called a "history of ideas"). For it was not ever an appeal to dynamism, nor a notice of a gestalt unity in nature that make a Romantic. If so, Linnaeus and Leibniz would be Romantics. So too would most of the British,

---

[25] I. Kant *Metaphysische Anfangsgründe der Naturwissenschaft* (1786); *Kritik der Urteilskraft* (1790).

[26] Of course, this potted history of Romanticism does no justice to the philosophical and cultural conditions which gave rise to German romanticism. See R. Richards, *The Romantic Conception of Life,* loc cit.

[27] Ernest Belfort Bax, ed., Kant's *Prolegomena and Metaphysical Foundations of Natural Science* (London: Bohn's Philosophical Library, 1883). Bax (1854–1926), a son of a successful English businessman, studied music and Hegel in Germany, turning to Marx and joining the Social Democratic Federation in 1882. In 1884, he left the Federation to form the more decidedly Marxist, Socialist League. He was later known for his rabid anti-feminist writings, e.g. *The Fraud of Feminism* (1913). Earlier, Schelling's *Introduction* to his *Outline* had been translated by Tom Davidson in, significantly, in the first year of *The Journal of Speculative Philosophy*. F.W.J. Schelling [1799], "Introduction to the Outlines of a System of Natural Philosophy," *Journal of Speculative Philosophy* **1** (1867), 193–220.

[28] Cf. J.S. Mill, "Bentham," *Westminster Review* **29** (1839), 467–506. J.S. Mill, "Coleridge, "*Westminster Review* **33** (1840), 257–302.

[29] W. Whewell, "On the Idea of Polarity. Talk to the Royal Institution," *Athenaeum* **1110** (1849), 119.

[30] Roy Porter and Mikulas Teich, eds., *Romanticism in National Context* (Cambridge: Cambridge University Press, 1988).

including even the most staunch Baconians. This morphological of history view forgets the breaks, the tensions and the real contingent developments, which do not *merely* unfold from the *idea* of Romanticism. Rather, we should go local.

10.3 Transcendentalism and Scottish Brains

If we look for routes of transcendentalism into England, at least in the life sciences, and we pass by the first wave of Coleridgean imports, we would have to go to Scotland. For it is there, early in the century, in the less Anglicanized university medical school we may now see the growth of a grand philosophical, transcendental, morphology, standing opposed to the functionalist, Paleyian, natural history of the Oxbridge network.[31] Robert Knox, W.B. Carpenter, J.H. Green, Robert Grant, borrowed freely from continental sources. And it is mainly via Edinburgh that transcendentalism came to England. "Why Scotland" has yet to be *fully* explored in the literature. The very lack of Paleyite natural theology so attached to the Oxbridge network is one important and well-explored part of local conditions that gave rise to a trancendentalism north of the borderlands.[32] The deep interest in an extra-curricular comparative anatomy is another. The great and quirky botanist, Hewett C. Watson (1804–1881), thought that he had found a third reason: in the peculiar nature of Scottish heads. In his extensive statistical phrenological study of botanists' brains, Watson had discovered the right bumps for "Theory" and for mere "Collection". The Scottish botanists had the former in excess.[33]

Whatever the case, with the Scottish diaspora of the early nineteenth century, transcendentalism drifted down to England, to be accepted and propagated in radical (and not so radical) London.[34] One of the main targets for its bayonets was the notion of *external* final cause so entrenched in English natural history of the natural theology school. In the Paleyian natural theological tradition of the Oxbridge network, God was indeed found "in nature", but not immanently. The study of nature,

[31] As described by Philip Rehbock, *The Philosophical Naturalists: Themes in Early Nineteenth Century Biology* (Madison: University of Wisconsin Press, 1983); and Adrian Desmond, *The Politics of Evolution: Morphology, Medicine and Reform in Radical London* (Chicago: University of Chicago Press, 1989).

[32] Although one might be cautious about certain "functionalist" explanation given by historians. See, for example, N. Rupke, *Richard Owen*, 64ff: "Whereas Cuvierian functionalism fitted the career ambitions and conditions of Oxbridge's clergymen-naturalists, idealism was tailor-made for those of Owen and his metropolitan *confreres*." Only in functionalist accounts are views "tailor made."

[33] H.C. Watson, "On the Relationship Between Cerebral Development and the Tendency to Pursuits, – and on the Heads of Botanists," *Phrenological Journal* **8** (1833), 97–108.

[34] See Adrian Desmond, *The Politics of Evolution* (Chicago: University of Chicago Press, 1989). Cf. Philip Rehbock, *The Philosophical Naturalists*. In his concentration on radical London, Desmond forgets the provinces. For some provincialism, see G. McOuat, "Rules, Meanings and Species: The Politics of Language and the End of Definitions in Nineteenth Century Biology," *Studies in History and Philosophy of Science* (1996) **27**, 473–519.

in the Paleyite tradition, was a search for *functional* perfection, for the planned final cause that was always "external" to the material processes that made up the object of scientific study. God, beauty and morality were there, but once removed. For the Paleyites, the transcendentalists had mixed up matter and final cause (such as Oersted's "spirit and matter"), edging dangerously towards a pantheism where God was "nothing but" nature.[35]

10.4 Fives All the Way Down

Much of the historiographical concentration on transcendentalism has focused on two aspects of the tradition arising from a search for an *inner*, immanent, principle of teleology: the "archetype" and "transformationism". Both were strongly linked to Owen's comparative anatomy at the Royal College of Surgeons and its outposts. In natural history, however (as opposed to comparative anatomy[36]) they were found in the peculiar attempt at building a natural system built on recurring types, circles and symbols, namely the now-obscure "quinarian" system.

In part, the quinarian system was a strong reaction against the Paleyite natural system too. It found its greatest strength amongst the Young Turks attempting a radical reform of natural history away form the received Linnaean system: in the Zoological Society of London, the British Museum, and in the provincial natural history societies.[37] In quinarianism, affinities and analogies between organisms were immanently *symbolic* relationships between numerically specific sets of beings, replicating their ordered relationships by analogy from level to level, nested sets upon nested sets. Affinities between organisms set themselves out in circles, and circles touching circles at "inosculating" points. On every level the members

---

[35] The best version of this attack on transcendentalism is to be found in the lengthy "Preliminary Discourse" of work by the Oxbridge doyen, Adam Sedgwick, *A Discourse on the Studies of the University of Cambridge*. 5th edition (Cambridge: Cambridge University Press, 1850). Differences between the "British" style of natural theology and that of the Germans is already found in the famous correspondence between Leibniz and Samuel Clarke, the latter representing the Newtonian picture.

[36] It must be remembered that Richard Owen's program for comparative anatomy was but one centre of possible authority in natural history. The British Museum, decidedly Cuverian, was much more successful at claiming to be the centre of naturalist authority. It did so in the face of Owen's morphology. See G. McOuat, "Cataloguing Power: Delineating Competent Naturalists and the Meaning of Species in the British Museum," *British Journal for the History of Science* **34** (2001), 1–21.

[37] For the Zoological Society, see Adrian Desmond, "The Making of Institutional Zoology in London: 1822–1836," *History of Science* **23** (1985), 158–185, 223–250. For the British Museum, see G. McOuat, "Cataloguing Power: Delineating Competent Naturalists and the Meaning of Species in the British Museum," *British Journal for the History of Science* **34** (2001), 1–28. For the provincial radicals, see G. McOuat, "Species Rules and Meaning," *Studies in History and Philosophy of Science* **27** (1996), 473–519.

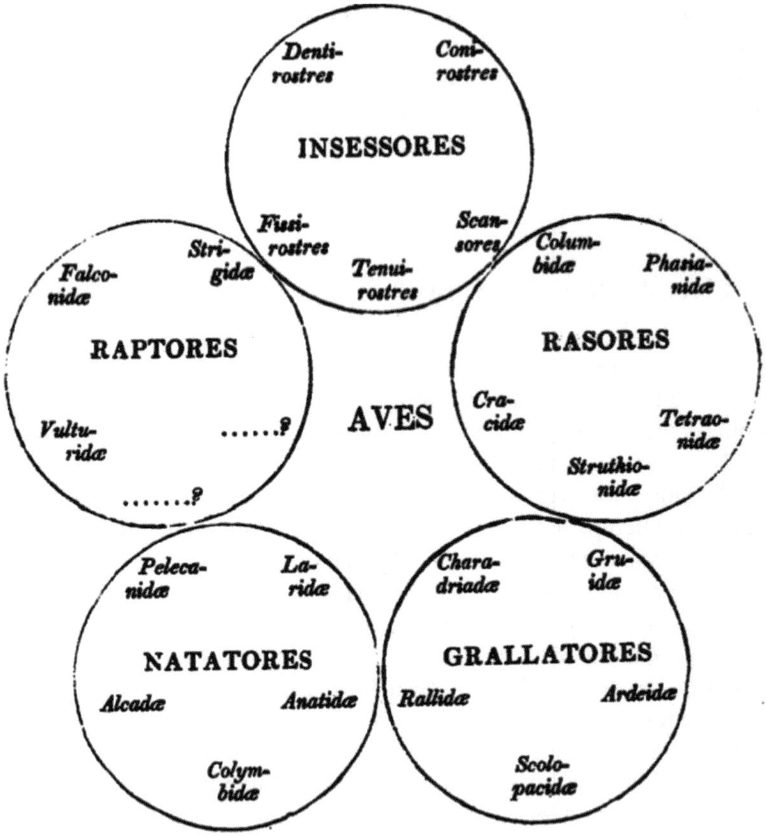

Fig. 10.1 A version of the quinarian classification of birds, N.A. Vigors, "Observations on the Natural Affinities that Connect the Orders and Families of Birds," *Transactions of the Linnean Society of London* **14** (1824), 395–517, at 509

turned in upon themselves in an ordered and perfect loop.[38] On each successive level, groups paralleled each other, repeating the pattern of affinity and analogy from level to level. Here, teleology was not a striving towards an externalist functional relationship, but one of internal and eternal recurrence of the same plan, over and over again at different levels (Fig. 10.1).

Now, there is some question as to how much quinarianism was truly "romantic' or transcendental. Quinarianism was very much a home-grown phenomenon. William Sharp Macleay (1792–1865), by all accounts the father of the system, considered it a particularly English synthesis (or, we might say, anachronistically, a "sublation") of the two competing French systems: the unilinear progressive system of Lamarck and the branching system of Cuvier. Synthesise a branch and a line and

[38] If there was a particular "type" missing in the circle, quinarians inserted a "?" placeholder, knowing full well that nature would provide the type eventually.

you get a circle, so thought Macleay. Progression happened, not by universal unilinear striving towards complexity, but an eternal unfolding of morphological structure at each level of being, repeating that which came before (albeit, not in *time*).[39]

David Knight has shown, convincingly, that William Swainson (1789–1855), the great propagandist for the new system, opened his book *Preliminary Discourse in Natural History* (published in the same series as Herschel's) on a most romantic theme. By fronticepiecing William Jones of Nayland (1726), anti-Newtonian and purveyor of a world of symbols, instead of Herschel's Francis Bacon, Swainson had nailed his colours to the post of anti-enlightenment world of symbols and immanent repetition.[40] The emphasis was on beauty, plan and symbol in a most *anti*-Enlightenment passion.

> Nature seems to delight in showing us glimpses of that beautiful and consistent plan upon which she has worked, by giving us a few instances of symbolical or analogical representations, so striking and unanswerable in themselves, that they are perceived and acknowledged by all. What for instance could be more perfect than the analogy between the Bengal tiger and the African zebra?[41]

Here words, things and signs were intimately intertwined. It was decidedly anti-Lockean, to be sure.

Whether a home grown or imported Romanticism, the quinarian system indeed had *very* close affinities with the premier *naturphilosophie* system of the great German romantic biologist, Lorenz Oken (1770–1851), especially to that elaborated in his grand synthesis of the complete transcendental system, his *Lehrbuch der Naturphilosophie* of 1809.[42] Oken, naturalist, political activist, quintessential *Naturphilosoph,* took Kant's notion of teleology seriously, taking it to its full metaphysical conclusion.[43] Oken, like the quinarians, also thought deeply in fives. Historians of science have emphasised his transcendentalism and his debt to Goethe and the *Naturphilosophie* of Schelling and Fichte. The concentration on the notion of "evolve" and developmentalism has often, even amongst the cautious, led to conceptualising Oken's *archetype* transformationally in some modern sense. Rather, Oken thought of the archetype (and transformationism) symbolically, aesthetically.

Oken had his own loathing for the lack of symmetry of Cuvier and the functionalists,[44] preferring his own version of rolling sets of fives. The (in)famous

---

[39] W.S. Macleay, "Remarks on the Identity of certain General Laws which have been lately observed to Regulate the Natural Distribution of Insects and Fungi," *Philosophical Magazine* **62** (1823), 192–200. W.S. Macleay *Horae Entomologicae* (London: Bagster, 1819).

[40] David Knight, "William Swainson, Types, Circles and Affinities," in J.D. North and J.J. Roche, eds., *The Light of Nature* (Dordrecht: Nijhoff, 1985), 88. See William Swainson, *Preliminary Discourse on the Study of Natural History* (London: Longmans, 1834).

[41] William Swainson, *Preliminary Discourse*, 91.

[42] Lorenz Oken, *Lehrbuch der naturphilosophie* (Jena: Fromann, 1810).

[43] For Oken, see Olaf Breidbach, Hans-Joachim Fliedner and Klaus Ries, eds., *Lorenz Oken (1779–1851): Ein politischer Naturphilosoph* (Weimar: Hermann Böhlaus, 2001). Olaf Briedbach and Michael Ghiselin, "Lorenz Oken and *Naturphilosophie* in Jena, Paris and London," *History and Philosophy of the Life Sciences* **24** (2002), 219–247.

[44] See Levere, *Poetry Realized in Nature*, 211–212.

Lehrbuch begins with "zero" (the nothingness), creating being out of the nothingness by positing its own being and limits. But it quickly turns to five. In Oken, archetypes of five repeat themselves throughout nature, from mineralogy to chemistry, to natural history, morality and aesthetics. The anthropologically symbolic identity was with the five senses. As Oken's English bulldog, Richard Owen, put it:

> the animal classes are virtually nothing else than a representation of the sense organs, and that they must be arranged in accordance with them:

1. Dermatozoa – Invertabrata (skin)
2. Glossazoa – Fishes (tongue)
3. Rhinozoa – Reptiles (nose)
4. Otozoa – Birds (ear)
5. Opthalmozoa – mammals (sight)

Each class takes as its starting point from below and consequently all of them pass parallel to each other, from mineralogy to plants and animals.[45]

Oken's races of man are analogous (actually, symbolically identical), with the rolling fives of the classification of nature:

> The skin-man is the *Black*, African
> The Tongue-man is the *Brown*, Australian-Malayan
> The Nose-Man is the *Red*, American
> The Ear-Man is the *Yellow*, Asiatic-Mongolian
> The Eye-Man is the *White*, European

And the Eye-Man begins the relationship of five once again at the aesthetic level, evolving to the final unity of being-for-itself of Nature.

For Oken, man is the "panaesthetic" animal – the comparison and unity of all the world's symbols, and this *free* comparison is *Reason*. Man is the ultimate goal or internal-aesthetic purpose of Nature's design (*Lehrbuch* §3620). The terminus or goal of nature is, in man, to revert again into herself at a self-conscious level. (This is no more evident in Oken's discussion of the famous facial angle, so important to romanticist aesthetics. "The face", for Oken, "is beautiful, whose nose runs parallel to the spine (*Lehrbuch* §3632). In Greek aesthetic art, the facial angle of the philosopher is represented as 100%, he correctly notes.) This reflecting back upon nature at the aesthetic level is not *mere* repetition or reflection. Rather, it is the site of creative genius, the moment of the romantic sublime, the identification of all of Nature in one moment, one particular. Thus, for Oken, philosophy is the second highest of the sciences. The highest science is, remarkably, the "Art of War", where freedom, aesthetic and the act are at one: "The Art of War is the highest, most exalted art, the art of freedom and of right, of the blessed condition of Man and of humanity – the Principle of Peace." (*Lehrbuch* §3652).

One can imagine the incredulity such statements of ecstatic, even incomprehensible, romantic sublime would meet amongst an English audience. Yet, in 1847, Oken's *Lehrbuch* appeared in an English translation, surprisingly under the

---

[45] Richard Owen, "Lorenz Oken," *Encyclopedia Britannica* (8th edition) **16** (1858), 498–503.

imprimatur of the prestigious, and rather conservative, Ray Society.[46] Its publication in the Ray Society series marked the *Lehrbuch* as a serious work and sparked a storm of controversy in British naturalist circles, especially amongst the Paleyians – a controversy almost as vociferous as that which whirled around the anonymous and transcendental *Vestiges of the Natural History of Creation* in the years before.[47] This time, rather than an anonymous tract published for the lay reading public (*Vestiges*), a wholly transcendental work (*Lehrbuch*) received what seemed to be a public approval by one of the most prestigious of naturalist societies. Yet, the *Lehrbuch's* publication was not uncontroversial, and threatened the very existence of that Society, creating a deep split in the governing council. The reaction by the dominant Paleyites was ferocious, with the Cambridge geologist Adam Sedgwick leading the attack, mocking, accusing, and quoting (out of context didn't seem to matter) some of the wild ecstasies of Oken's deductions:

> I have read his work, and I have striven to perceive some glimmerings of steady light among the mists of his first sixty or seventy pages, and have seen but an *ignis fatuus* playing: here and there, on a darkness that is palpable and impenetrable.[48]

For Sedgwick, the problem with both the *Lehrbuch* and the *Vestiges* rests on their confusion about final causes: "'tis the crown and glory of organic science that *does*, through final cause, link material to moral; and yet *it does not* allow us to mingle them in our first conception of laws and our classification of such laws, whether we consider one side of nature or the other."[49] Here Sedgwick reiterated the classic

[46] Lorenz Oken, *Elements of Physiophilosophy*, Alfred Turk, trans. (London: Ray Society, 1847). The publication of the Tulk translation of Oken's *Lehrbuch* is an interesting story in itself. Much of the related correspondence is kept in the Ray Society papers in the Natural History Museum, London, and in the H.E. Strickland papers, Zoology Department, Cambridge University. The Ray Society was formed in 1844 in part as a reaction to the lowering attention to "natural history" in the burgeoning British Association for the Advancement of Science.

[47] [Robert Chambers], *Vestiges of the Natural History of Creation* (London: Churchill, 1844). On the *Vestiges*, see James Secord, *Victorian Controversy: the Extraordinary Publication, Reception, and Secret Authorship of Vestiges of the Natural History of Creation* (Chicago: University of Chicago Press, 2000). The first editions of the *Vestiges* were decidedly quinarian in focus. Later editions downplayed it.

[48] Adam Sedgwick, *A Discourse on the Studies of the University of Cambridge* (Cambridge: Cambridge University Press, 1850),. cciii. Sedgwick included a 25 page appendix on Oken, quoting, without comment, some of the more outrageous claims made in the Elements. "Supplement to the Appendix, No. viii, A series of extracts from Oken's 'Physiophilosophy'", 222–246. Sedgwick follows his comments and appendix on Oken with a sustained attack on the *Vestiges*. It is noteworthy that Sedgwick couches his critique in a form of inductive utilitarianism: "A general would be mad, were he to mount his artillery on an air-balloon. A miller would be mad, were he to set his machinery in movement, and pretend to grind flour for the food of man, without a supply of corn for the operation of the mill. A manufacturer would be mad, who would pretend to make some fabric for our use by the mere rotation of his mechanism, and without supplying it with any of the raw material on which his mechanism is to do its work. An not one jot less mad is the transcendental Idealist, who pretends to work out the general truths of physical science without an experimental appeal to nature, and without first resting on its evidence." Ibid, cci.

[49] Adam Sedgwick to Charles Darwin, December 24, 1859, in reaction to what Sedgwick thought was Darwin's similar error! DAR 98: B17–18, Cambridge University Library.

mechanistic account of aesthetics/ethics against the eminent notions of *Naturphilosophie* and Romanticism.

10.5 Spectacle and the Beauty of Refutation

At the centre of the British Oken-storm was Hugh Edwin Strickland (1811–1853), major player behind the establishment of the Ray Society, editor and translator of many of its works, and the prime combatant against *any* transcendental system. Four years before the publication of Oken's *Lehrbuch*, Strickland thought that he had effectively blocked any attempt to import a "sublime" system into natural history with his spectacular display of a non-Romantic system of nature at the BAAS meeting in Cork.

Strickland, graduate of Oxford and son of a Baronet, had become a major force in his vociferous attacks on reformist systems of natural history, beginning with local provincial radical Chartists/naturalists of his hometown in the midlands, and then moving on to his full frontal confrontation with the metropolitan "Procrusteans" (as he liked to call the followers of quinarian and Romantic natural systems).[50] Philosophically, Strickland was an avowed Lockean – this was no more evident than in the decidedly Lockean, gentlemanly and enormously influential "Rules of Zoological Nomenclature" (1842) of which Strickland was the primary author.[51] The "Rules", the first of their kind in any science, literally decided who could and who could *not* be a "competent naturalist" permitted to designate and dub new living forms. Strickland was also notable in his transporting of such terms as "analogy", "homology" and "affinity" from transcendental morphology, re-defining them in terms more conducive to contingent and local Paleyian-Oxbridge systems of natural history.[52] He was also most instrumental for his spectacular and unrelenting attack on the "procrustean"-aesthetic systems in natural history – an attack successful enough to make being a procrustean rather difficult in English circles.

Strickland received his education under the geologist William Buckland in the Oxbridge system, but expanded it in a Humboldtian way, by examining the lay of the geological land in a wider context, with a more general overview of the "set of the land." For him, the "natural system" in classification was more like a geological map – contingent, sloppy, detailed, and in no way mathematically neat or apriori.[53]

[50] For a description of Strickland's fights and career, see G. McOuat, "Species, Rules and Meaning," *loc cit.*

[51] H.E. Strickland, et al., "Report of a committee appointed 'to consider the rules by which the Nomenclature of Zoology may be established on a uniform and permanent basis.'" *BAAS Report* (1842), 105–121. The committee included the range of establishment, conservative, natural historians: Henslow, Jenyns, Richardson, Ogilby, Phillips, Westwood, Waterhouse, Yarrell, and Darwin.

[52] Mario Di Gregario, "Hugh Edwin Strickland (1811–53) on Affinities and Analogies: or, The case of the Missing Key," *Ideas and Production: A Journal in the History of Ideas* **7** (1987), 35–50.

[53] H.E. Strickland, "Outline of Whewell's *History of the Inductive Sciences*, Vol. 3 Re. Affinity of Animals," (1838), in H.E. Strickland papers, University of Cambridge, Zoology Department.

It was the geologist in him who could pen the following words in his notebook after reading a passage from William Whewell on "the natural system": "The affinities of animals may possibly be neither rectilinear nor circular, but irregular like the distribution of water and land, and like that, capable of being laid on a map – not however by any *apriori* theory, but by long observation like a geographical survey."[54] And it is to just such a map-metaphor that Strickland turned for his final blow: the "spectacular" refutation of all "Procrustean"/aesthetic systems, where he would turn beauty *against* the sublime, introduce contingency in nature, and make it impossible for anyone like Oersted to even begin there.

Early in his career, Strickland had already taken on the quinarians (and Lamarckians and "bifurcationists") in print. Strickland found repugnant the notion that natural history should be tied down to any symmetry given by natural law. "There is no department of the creation which is tied down to mathematical laws."[55] Granted, there are mathematical laws in physics and chemistry, but these in no way present a form of "symmetry" demanded by the transcendentalists. For example, Strickland said, the relationship between the relative distances of planets do not conform to any numerical law, the fixed stars give no arrangements (except conventional ones), and there is no symmetry in islands and continents. This last point was the most germane.

Importantly, Strickland was also one of the gentlemanly Oxbridge natural philosophers (now termed "scientists") behind the establishment and success of the great "theatre of science," the British Association for the Advancement of Science.[56] And it is to that "theatre" that Strickland turned to present his spectacular end to transcendental/aesthetic systems.

In 1840, at the Glasgow meeting of the *British Association*, Strickland outlined his spectacle of refutation. In front of an audience prickly with representatives of rival "natural systems", including many transcendentalists, Strickland threw down the gauntlet.[57] He unveiled plans for a map of the "natural system of birds." It resonated with a geological "style." Yet, the sketch presented at Glasgow was but a promise of a system, a hint of what was to come. Three years later, Strickland made good his promise to map out the birds. At the Cork meeting of the BAAS in 1843, Strickland unveiled his well-prepared, large, beautiful wall chart on "The Natural Affinities of the Insessorial Birds."[58] The fragments reproduced in the *BAAS Report*,

---

On the importance of Humboldtian maps for early nineteenth century British geologists, see James Secord, *Controversy in Victorian Geology: The Cambridge-Silurian Debate*. (Princeton: Princeton University Press, 1986); Martin Rudwick, "The Emergence of a Visual Language for Geological Science," *History of Science* (1976) **14**, 149–195.

[54] Strickland, *"Outline of Whewell's History of the Inductive Sciences"* Strickland Papers.

[55] H.E. Strickland, "On the True Method of Discovering the Natural System in Zoology and Botany," *Annals and Magazine of Natural History* **5** (1841), 184–195.

[56] The BAAS was, in the words of Jack Morrell and Arnold Thackray, a "theatre of science run by missionaries intent on converting their audiences to their views." J. Morrell and A. Thackray *Gentlemen of Science. Early Years of the BAAS* (Oxford: Clarendon, 1981).

[57] For a verbatim account of Strickland's report and the conciliatory response by the quinarians, see *Athenaeum* **677** (17 October, 1840), 825.

[58] H.E. Strickland, "Description of a Chart of Natural Affinities of the Insessorial Order of Birds." *BAAS report* 1853, Part 2, 69.

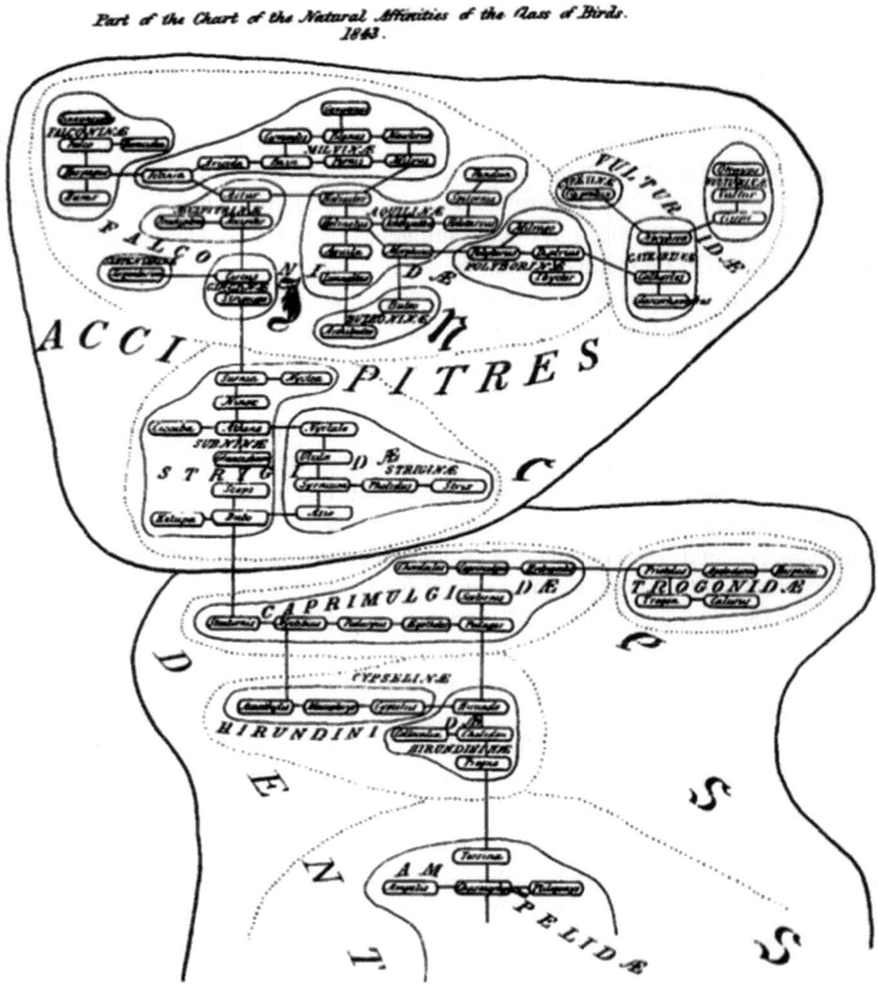

Fig. 10.2 Extract of H.E. Strickland, "Chart of the Insessorial Birds." from William Jardine, ed. *Memoirs of Hugh Edwin Strickland.* (London: John Van Woorst, 1858), facing page cciv

the *Annals of Natural History*, and in Jardine's *Memoirs of Hugh Strickland* hardly do it justice. It is a truly striking piece[59] (Fig. 10.2).

Yet, its beauty does *not* lay in its symmetry or anything *apriori*. It lies in the wealth of geological metaphors.

[59]The original Strickland map was long thought destroyed. It has since been rediscovered and beautifully restored at the Zoology Museum, Cambridge University, thanks originally to Rae Symonds and Adrian Friday of the Museum.

> The natural system is an accumulation of facts which are to be arrived at only by a slow inductive process, similar to that which a country is geographically surveyed. If this be true, it is evident how erroneous must be all these methods which commence by assuming an a-priori system, and then attempt to classify all created organism in conformity with that system.[60]

Here we should note that the classification itself is not specifically "geographical". (The British were not quite ready for that move.[61]) Classes are not laid out according to geographical centres of creation or anything along those lines. Rather, the map is *informed* by a geographical/geological style of thinking – a knowledge that sees the diversity and multitudinousness of relations between things as wholly natural, and, more importantly, drawing upon the notion of "being there" in all the encounter with the accidental, local fecundity of nature, like "the appearance of a cluster of islands depicted on a map."[62] Hence, nothing a-priori.

And, exactly at this point Strickland makes his direct appeal to the separation of beauty and the sublime, art and nature. For Strickland, analogies found in nature exhibit the utmost variety and irregularity. Variety is "the beautiful law of nature and what distinguishes her productions from those of art, and it is that which man often exerts his highest effort in vane to imitate."[63] Now, in the mapping of natural systems, contingency and variety were the main moral and aesthetic factors at play. He would drive this home in his later inaugural lecture as geology professor at Oxford:

> Many of our clergymen too, are destined to pass their lives in remote parishes, far from educated neighbours, and where their parochial duties are too light to occupy their whole time. The monotony of such a position will often, in spite of the best principles, react upon the nerves, and render a person listless, if not discontented. To a mind thus diseased, I can prescribe no better remedy than this: Make a geological map of your parish. Form a collection of all its animal, vegetable, and mineral productions. When this is done, extend your researches to the neighbouring parishes, or to the whole country.[64]

In addition to his conservative supporters, present at the Cork meeting were a host of transcendentalists, quinarians, and a few "dichotomists", tired and almost broken from their unending disputes.[65] The effect of his imagery must have been profound.

---

[60] H.E. Strickland, "On the True Method", loc cit. The report in the *Athenaeum* of September 16, 1843, goes into greater detail: "Being a purely inductive process, the details of any branch of natural history may be in this way worked out and depicted without recourse to any theoretical misconceptions... the true affinities of organic structures branch out irregularly in all directions, and that no symmetrical arrangement or numerical uniformity is discoverable in the system of nature when studied independently of preconceived theory." at 851.

[61] The British were rather cold to the idea of classification by geographical distribution until well into the century.

[62] *Athenaeum*, Sept 14, 1843, 851.

[63] Strickland, loc cit.

[64] H.E. Strickland, "On geology, in Relations to the Studies of the University of Oxford," in William Jardine, ed., *Memoirs of Hugh Edwin Strickland* (London: John Van Voorst, 1858), 217–219. See Jean-Marc Drouin and Bernadette Bensuade-Vincent, "Nature for the People," in Nicholas Jardine, James Secord and Emma Spary, eds., *Cultures of Natural History* (Cambridge: Cambridge University Press, 1996), 409–425, on 420.

[65] For a contemporary account of the event, see the *Athenaeum* (16 September, 1843), 851.

It is certain that Strickland's spectacle had some effect, at least in the *style* of British transcendentalism after 1843. Dov Ospovat in his remarkable *Development of Darwin's Theory*, describes the Cork meeting as delivering "the mortal blow" against quinarianism and transcendentalism in natural systems (although Ospovat fails to mention Strickland's map!).[66] Indeed, after 1843 one sees a remarkable decline in symbolic archetypical systems of nature, of discussions of the sublime in natural historical works. Even Richard Owen, who was in attendance at Strickland's talk, found himself waxing Baconian after 1843.[67] It can safely be said that an encounter with the "sublime" was no long a part of a system of nature. Here Oersted's passage to England was blocked and his appeals fell on deaf ears.[68]

10.6 Grand Tours

But, how did Oken's *Lehrbuch* get published so late in the day, and by Strickland's Ray Society, no less? Although nominally governed by a council of naturalists, with the aim to publish British natural history monographs that would "not normally see the light of day," the Ray Society was a rather loose coalition. Strickland was in charge of publications. Yet, when the translation by Alfred Tulk (no relation to the Swedenborgian, alas) was rushed through and put to press, Strickland was absent, out of the country.[69] He had just married Catherine, the daughter of the prominent naturalist, William Jardine, publisher of the decidedly conservative *Naturalist Library*.[70] When Tulk's translation was being brought to press, the newlywed couple were on their own romantic "Grand Tour" (Strickland's second tour) of Holland, Bremen, Hamburg, Copenhagen, Malmo, Lund, Stalsund, Berlin, Leipzig, Dresden, Switzerland, Frankfurt and Brussels – mapping all the way.

---

[66] Dov Ospovat *The Development of Darwin's Theory: Natural History, Natural Theology, and Natural Selection, 1838–1859* (Cambridge: Cambridge University Press, 1981).

[67] For a discussion of changes in Owen's style and system, see N. Rupke, *Richard Owen*, loc cit.

[68] This is not to imply that no-one was a quinarian or transcendentalist after 1843. Quinarianism retained some interest, mainly on the periphery. Transcendentalism was transformed. See Jennifer Coggan "Quinarianism after Darwin's *Origin:* The Circular System of William Hincks," *Journal of the History of Biology* **35** (2002), 4–42.

[69] Tulk went on to translate H.M. Chalybäus, *Historical Development of Speculative Philosophy from Kant to Hegel* (Edinburgh: T&T Clark, 1853).

[70] It is worth noting Strickland's reaction to the publication. In a letter to Charles Bell, author of the Bridgewater treatise, *The Hand: Its Mechanism and Vital Endowments as Envincing Design* (1833), Strickland hoped that the further success of the Ray Society would "soon efface the recollection of this one injudicious step." But, he found the book interesting both *psychologically* ("to observe the kind of nonsense which a fearful mind produces when it... deserts the Baconian, or inductive method of reasoning") and *ethnologically* (where it "shows how very different Germans are from Englishmen"). Interestingly, Strickland thought that some of Oken's views of the Deity are "sublime and commendable". To Prof. Bell, November 13, 1847, in W. Jardine *Memoirs of Hugh Edwin Strickland*, ccxix–ccxx.

Chapter 11
Science Born of Poison, Fire and Smoke: Chemical Warfare and the Origins of Big Science

Andrew Ede

Big Science has transformed the practice of science. It has also changed our expectations of what topics science will address, and how scientists are trained and employed. Today, many scientists routinely expect to work in teams and in collaborations of teams rather than as individual researchers. Multinational projects such as the Human Genome Project and the International Space Station receive funding in the millions and billions of dollars. Funding agencies and the public have become accustomed to science that makes headlines, not just advancements that add to the body of scholarly knowledge. The importance of Big Science makes it a target for historical investigation, and such investigations can help us understand the impetus for the creation of this phenomenon and why it has taken the form that it currently exhibits. The historical problem is that most examinations of Big Science have presented physics programs during the Cold War as the foundational moment for Big Science. In fact, Big Science originated 40 years before the Manhattan Project and came from chemistry, not physics.

It is easy to understand why physics became so closely associated with Big Science. Physics has been dominated by big projects and has received massive popular coverage since the end of World War II. In 1961 the subject was discussed by Alvin M. Weinberg, director of the Oak Ridge National Laboratory. In an opinion piece for *Science*, Weinberg warned that "…we must not allow ourselves, by short-sighted seeking after fragile monuments of Big Science, to be diverted from our real purpose, which is the enriching and broadening of human life."[1] Weinberg was not

[1] Alvin M. Weinberg, "Impact of Large-Scale Science on the United States," *Science* **134**, n. 3473, (1961) 164.

A. Ede (✉)
Department of History and Classics, University of Alberta, Edmonton, AB, Canada
e-mail: ede@ualberta.ca

J. Buchwald, L. Stewart (eds.), *The Romance of Science: Essays in Honour of Trevor H. Levere*, Archimedes 52, DOI 10.1007/978-3-319-58436-2_11

against Big Science *per se* (he was running one of the largest nuclear research facilities in the world at the time), but against funding scientific projects that were simply aimed at making political points against the Soviet bloc or designed to appeal to the public. Science for Weinberg was still conceptually controlled by individual scientists or small, often informal, teams. For Weinberg, the "Big" part of Big Science was primarily a statement about equipment and facilities (and thus the money) that might be required to carry out the investigations.

An important investigation of the transformation of science appeared in 1992 when Peter Galison and Bruce Hevly published *Big Science: The Growth of Large-Scale Research.*[2] Galison, Hevly *et alia* focused largely on physics and argued that Big Science did more than bring scientists together with money to produce grand devices and take on big projects. The way problems are conceived, the training of new scientists, the necessary inclusion of non-scientists such as administrators, lawyers, technicians and communications specialists in the teams and networks of teams change both the process and the narrative of scientific research. The size and conception of scientific projects is clearly illustrated by the paper "The ATLAS Experiment at the CERN Large Hadron Collider that appeared in the *Journal of Instrumentation* with a list of 2926 "authors" from 169 research facilities.[3] At one level, the authorships simply acknowledges the practical contributions of the people who worked on the project, but it also says that the conceptualization and practice of the scientific issues was a collective effort.

What Weinberg, Galison and Hevly and many others share is that they primarily locate Big Science in the world of physics. While physics (especially if astronomy and space exploration are grouped with physics) commands much attention and is dominated by Big Science, the fascination with modern physics obscures the historical origin of Big Science. Big Science came from chemistry, not physics. It was not the Manhattan Project and the Cold War that created the framework for Big Science, it was chlorine and mustard gas. Furthermore, if Big Science includes some measure of social impact, chemistry has transformed society more than physics. The modern world could not exist without the infrastructure and products that are produced by the chemical industries. Even in terms of numbers of practitioners, in America there are about three chemists for every physicist. The American Physical Society had 51,523 members in 2014 compared to 158,000 members of the American Chemical Society.[4]

---

[2] Peter Galison and Bruce Hevly, *Big Science: The Growth of Large-Scale Research* (Stanford: Stanford University Press, 1992).

[3] Martin Aleksa and the Atlas Collaboration, "The ATLAS Experiment at the CERN Large Hadron Collider," *Journal of Instrumentation* **3** (2008) *JINST* S08003.

[4] APS "Membership 2014," aps.org/about/governance/annual-reports/upload/annrep2014.pdf; ACS, "ACS By the Numbers 2014," acswebcontent.acs.org/annualreport/highlights.html.

The history of Big Science can be linked to what seems like an unlikely event to start a revolution in science: The Berlin Conference of 1884–1885. Called for by Portugal, and hosted by German Chancellor Otto von Bismarck, the conference focused on dividing up the last unclaimed colonial territory in Africa. Germany, recently unified, hoped to gain international status and much needed natural resources by laying claim to African land. Germany's colonial efforts were largely thwarted. France held the most land (although much of it was inhospitable), while Britain claimed vast territories, including the productive and strategic areas of Egypt and South Africa. Even lesser powers Portugal and Belgium controlled Angola and the Belgian Congo. What Germany got was undeveloped, strategically unimportant and not especially rich in natural resources. Exploiting the German territories in African would not be easy.

Compared to its neighbours, Germany lacked the arable land of France, the vast natural resources of Russia and the global market controlled by Britain. Without easily exploited colonial holdings, Germany had two potential courses of action: territorial expansion or increased productivity. Although elements in the German government and military pushed for expansion, eventually contributing to the terrible consequences of 1914 (and returning as *Lebensraum* in Nazi ideology), the more constructive answer was to apply science and technology to make up the gap in resources. What resulted was a broad collaboration between the universities, industry and government. Informal and personal connections between researchers and industry often led to new research projects and the transfer of knowledge from lab to factory.[5]

Germany took a major step towards Big Science by establishing the Kaiser Wilhelm Gesellschaft in 1911. The first two Kaiser Wilhelm Institutes (KWI) were the KWI for Chemistry, and the KWI for Physical Chemistry and Electrochemistry. These were joined in 1912 by the Coal Research Institute, the KWI for Biology, the KWI for Biochemistry and the KWI for the Physiology of Effort. Physics did not get an institute until 1917.[6]

Chemistry was a key national science. Germany needed nitrates for intensive agriculture and industry, but had very limited access to natural sources of nitrates. The largest commercial source in the period was caliche (a mix of sodium nitrate and potassium nitrate), an ore mined primarily from the Atacama desert in northern Chile. Although Chile nationalized nitrate production, British and French companies controlled the market. Britain also had sources of potassium nitrate from the Bengal region in India. To make matters worse for Germany, the Atacama supply

---

[5] For an overview of German physics in this period, see Christa Jungnickel and Russell McCormmach, *The Intellectual Master of Nature; Theoretical Physics from Ohm to Einstein* (Chicago: University of Chicago Press, 1986).

[6] For a list institutes, see en.wikipedia.org/wiki/Kaiser_Wilhelm_Society.

was expected to run out in less than 30 years, so long-term plans could not be based on such a limited resource.[7]

Fritz Haber's work on fixing atmospheric nitrogen is the best known project that illustrates German effort to use science to deal with limited resources. The story of nitrogen has several of the elements of Big Science. It was a strategic project that involved researchers, industry and the government, and it had national and international implications. It cost a great deal of money, largely paid for by Badische Anilin- und Soda-Fabrik (BASF), one of Germany's biggest companies. Although the size of the initial team working on the project was small (primarily Haber and Robert Le Rossignol, with some students and a laboratory technician), by the time it reached industrial scale many more scientists and engineers, including metallurgists at Krupps, were involved.

Haber, working with Robert Le Rossignol, demonstrated the process of turning atmospheric nitrogen into ammonia in 1909. Ammonia was a key ingredient in fertilizer needed for intensive agriculture in order to feed Germany's growing population. Ammonia was (and is) used by the textile industry and as a cleaning agent. It was also the base stock used to manufacture high explosives because it is easily transformed into nitric acid. If you have the technical capability to produce ammonia, you have more than enough technology to produce nitric acid.

Haber process (Ammonia from atmospheric nitrogen)	
$N_2 + 3\ H_2 \rightarrow 2\ NH_3$	150–245 atmospheres
	300–355 °C
	4 catalysts with iron and K_2O, CaO, SiO_2 and Al_2O_3
Ostwald process (Nitric acid from ammonia)	
$NH_3 + O_2 \rightarrow NHO_3 + H_2O$	4–10 atmospheres
	217 °C
	Platinum catalyst

It was at BASF that Haber met Carl Bosch, who oversaw the industrial development of the process. The Haber-Bosch process, as the industrial system came to be known, began serious production in 1913.[8] This was called "Brot aus Luft," or "bread out of air."[9] Bosch would go on to help found and lead the chemical cartel IG

[7] For a brief economic history, see Michael Monteón, "The British in the Atacama Desert: The Cultural Bases of Economic Imperialism," *The Journal of Economic History* **35** (March, 1975), 117–33.

[8] Dietrich Stoltzenberg, *Fritz Haber: Chemist, Nobel Laureate, German, Jew* (Chemical Heritage Foundation, 2004), 96.

[9] The phrase "Brot aus Luft" appears frequently in association with Haber's work, but it is not clear who said it first or when it was first said. The most likely candidate was Max von Laue. Max von Laue, "My Development as a Physicist," in P.P. Ewald, ed., *50 Years of X-Ray Diffraction* (Utrecht:

Farben while Haber became the first director of the KWI for Physical Chemistry and Electrochemistry, serving from 1911 until 1933.[10]

Although ammonia for fertilizer would not overtake cyanimide until after the war, by early 1916 BASF was producing 24,000 tonnes of fixed nitrogen annually with plans to expand to 36,000 tonnes. In 1917–1918, BASF produced 105,000 tonnes of fixed nitrogen.[11] Originally intended for the commercial market, the entire supply was purchased by the Ministry of War.[12] A number of historians have argued that without Haber's work, Germany could not have fought the war after 1916 or early 1917. The American chemist Jerome Alexander went further, saying "No nitrates no explosives, and without the Haber process it is doubtful if Germany would have started the war, for which she carefully prepared.".[13]

By the beginning of World War I, the features of Big Science existed, at least in nascent form, in the German scientific-industrial nexus. The practice of science was still relatively local, with individuals or small teams of scientists and technicians working on projects, but commercialization, multi-team collaboration, national funding of key scientific disciplines and the increasing number of working scientists was changing the conceptual landscape of scientific practice. Increasingly, German scientists expected to work with other scientists, engineers, and technicians, as well as negotiating for support from business and government. When war came, the landscape would change further as scientists put national interest ahead of 'curiosity-driven research.'

The Germans thought the war would be short, just as the Franco-Prussian war had been. Germany's military was better trained and equipped than most of her opponents at the beginning of the war, but the gap proved insufficient to secure immediate victory. Within a few months, Germany's technical advantage was lost and the Triple Entente's larger populations and access to resources slowly started to change the nature of the war. Germany was faced with a war of attrition. German production of conventional weapons and munitions was at maximum capacity and adding new manufacturing capacity would take time. Haber recognized that the chemical industries were already producing materials that could be used as weapons without reducing the resources for conventional munitions. This was the strategic

International Union of Crystallography, 1962), 307. Modern authors such as Daniel Charles and Thomas Hager use the phrase without attribution. Haber, in his Nobel lecture said "...nitrogen fertilization of the soil brings new nutritive riches to mankind and that the chemical industry comes to the aid of the farmer, who, in the good earth, changes stones into bread." Fritz Haber, "The Synthesis of Ammonia from it Elements," *Nobel Lecture*, June 2, 1920, 339.

[10] See Stoltzenberg, *Fritz Haber,* ch. six.

[11] Stoltzenberg, *Fritz Haber,* 102.

[12] Stoltzenberg, *Fritz Haber*, 102.

[13] Jerome Alexander, "Nobel Award to Haber," Letter to the Editor, *New York Times*, Feb. 3, 1920, 14.

imperative for chemical warfare. Trench warfare, which placed soldiers in confined spaces at close proximity to the enemy, was the tactical impetus.

There is some confusion in historical accounts about the early use of chemical agents. German, French, British and American sources may agree on when chemicals were used at the beginning of the war, but not what agents were used or exactly how they were used. What is universally true of these early events is that they had no effect on the enemy. For example, Germany claimed that French forces used 26-mm. rifle grenades containing ethyl bromacetate (a form of tear gas) in August, 1914.[14] The French denied the charge, but it was known that ethyl bromacetate had been developed by the French prior to the war.[15] If the French did use tear gas, the effect went unnoticed by the soldiers. The claim, however, that France used gas became the justification for the use of chemicals by Germany. The Germans likely tested chlorsulfate of ortho-dianisidine (a lung irritant) in October, 1914, but it isn't clear if they used it by itself, or added it to shrapnel shells. Germany then used xylyl bromide-filled artillery shells against the Russian at Bolimow in January, 1915, but with a melting point of 18 ° C and sub-zero temperatures on the battlefield, it was not very effective.[16] A similar attack on the English at Nieuport in March 1915 was equally impotent.[17]

Haber, who had been appointed the head of the Chemistry Section in the Ministry of War, oversaw the transition of the Germany's powerful chemical industries from peace to wartime production. He also helped transform the Kaiser Wilhelm Institutes (particularly his own) into research centres looking at military questions. He had attempted to join the military, but was rejected for active service because he was too old at 46. At some time in December, 1914, Haber suggested the use of chlorine to the German High Command. Haber was given permission to begin work on the offensive use of chemicals in February, 1915. He began to organize the people and material for a chlorine gas cloud attack. Ypres, in Belgium, was chosen, not because it was the best location, but because it was the only section of the western front whose commander had not objected to the use of gas.[18]

---

[14] Augustin Mitchell Prentiss, *Chemicals in war: A treatise on chemical warfare* (London: McGraw-Hill, 1937), 132. A number of sources incorrectly say the chemical was xylyl bromide, which was used extensively by Germany and known by its code name "T-Stoff."

[15] Prentiss, *Chemicals in War*, 132.

[16] www.chemicaldictionary.org/dic/X/Xylyl-bromide_2072.html; Prentiss, *Chemicals in War* 134.

[17] Prentiss, *Chemicals in War*, 134.

[18] Daniel Charles, *Mastermind: The Rise and Fall of Fritz Haber* (New York: HarperCollins, 2005), 160.

On April 22, 1915 at about 5:30 pm, the modern era of chemical warfare began. With the wind finally blowing in the right direction, Haber ordered his gas soldiers to open the valves on the 5730 gas cylinders and release the 168 tons of chlorine. It blew across no-man's land and onto the Allied line. The brunt of the attack fell on French Colonial troops, as well as some British soldiers. The front line collapsed, leaving a 7 km opening. German forces walked slowly behind the gas cloud and took the front line almost without opposition. Three things stopped the German forces from breaking through the lines completely. First, some Allied forces, particularly the 1st Canadian Division, behind the front lines on the eastern flank of the attack put up a strong fight despite the gas. Second, night was approaching and German forces were reluctant to march into the darkness. Third, the German command, who had seen the attack as at best a test and at worst a stunt, had not committed sufficient troops to continue the attack. The tired German soldiers occupying the abandoned trenches and began to dig in for a counter attack rather than continue on the offensive.[19]

Most historians follow Augustin Prentiss' *Chemicals in War* when discussing the casualties from gas attacks, and he sets the number for Ypres at 15000 injuries and 5000 deaths.[20] In reality, we have no clear numbers for the attack. British aid stations listed hundreds of injuries, but only about 150 deaths. The French have few records and none from aid stations. Many of the dead were left where they fell, never to be counted.

What followed was an arms race among the combatants for more powerful chemical agents and better defenses. Despite new toxins appearing on the battlefield, in general the effectiveness of chemical warfare declined after the early attacks. Chemical weapons shifted from a tactic role to a strategic one where chemicals were used for three main objectives: denying territory by contaminating it, increasing the psychological and physical stress of front line combat, and attempting to overwhelm the support system such as aid stations and hospitals behind the lines.

The success of Germany's chemical weapons made governments around the world aware that they did not know how many scientists were available or where those scientists might be. Britain and France were forced to play scientific catch up. Of the two nations, British science was more affected. The British government was forced to formalize (and pay for) the relation between science and the state, while France had a much longer history of state support to build on during the war. The long-standing, but often unclear, link between the British government and the Royal Society of London did serve some of the science needs of the state, but the Royal

[19] For an overview of the German experience, see L.F. Haber, *The Poisonous Cloud: Chemical Warfare in the First World War* (Oxford: OUP, 1986); for the Canadian perspective, see George H. Cassar, *Hell in Flanders Fields. Canadians at the Second Battle of Ypres* (Toronto: Dundurn Press, 2012).

[20] Prentiss, *Chemicals in War*, 663.

Society was not set up to organize or fund scientific research. In 1915, the Department of Scientific and Industrial Research was created as a branch of government to promote, co-ordinate and help fund scientific work. Its budget was significant at £1 million in the 1 year.[21] Rather than establishing its own laboratories, the DSIR distributed most of the money to university researchers.

The war also led other Commonwealth nations to take science more seriously. Both Canada and Australia established federal research organizations in 1916. In Canada, the Honorary Advisory Council for Scientific and Industrial Research undertook an inventory of scientists and offered science fellowships at Canadian universities, but it did not build its own laboratory until 1928.[22] In Australia, the Advisory Council of Science and Industry was formed, but it did no serious work during the war, strangled by government infighting over autonomy and funding.[23]

The greatest transformation was in the United States. Although very late to enter the war, the war affected science in America from almost the first days of the conflict. In particular, science education was affected. A significant portion of American science students pursuing graduate work went to Europe, and especially Germany. Cutting off this connection led American colleges and universities to start or increase graduate teaching in the sciences, with chemistry being by far the largest area of study.[24]

Although education was immediately affected, the war seemed distant to most Americans. The War Department was aware that the military was not well prepared to engage in a major conflict. In 1914 the US Army consisted of 4572 officers and 88,444 enlisted men, significantly below the organizational cap of 100,000.[25] In addition, more than a quarter of the Army's forces were in the Philippines, raising concern in the Department of War that the Army was even less prepared for a serious conflict in Europe than appeared on paper. When news of the chlorine attack at Ypres reached America, it added to concern among some senior military commanders such as General John Pershing that the Army was falling behind European military capability. Chemical warfare did not fit US military doctrine, which was based largely on British and Prussian models adopted in the late nineteenth century. These models emphasized the use of cavalry and light infantry armed with rifles and bayo-

---

[21] H.F. Heath and A.L. Heatherington, *Industrial Research and Development in the United Kingdom* (London: Faber & Faber, 1946), 250–251.

[22] "The National Research Council," www.thecanadianencyclopedia.com/en/article/national-research-council-of-canada/

[23] C.B. Schedvin, *Shaping Science and Industry: A History of Australia's Council for Scientific and Industrial Research 1926–49* (North Sydney: Allen & Unwin, 1987), 14–15.

[24] See Arnold Thackray et al., *Chemistry in America, 1876–1976* (Dordrecht: D. Reidel, 1985), ch.2 and 3.

[25] "Costs Less to Arm than Ransom City," *New York Times* (November 7, 1915), 17.

nets. Most of the artillery was antiquated and not integrated with the other services. The use of machine guns was not part of infantry doctrine. The different branches of the ground services trained separately and many of the tactical methods of the Army were the same as those used during the American Civil War.

Not only did chemical warfare not fit with the prevailing doctrine, it was controlled by a group of men that the Army in particular had shunned—scientists. The primary technical support unit for the Army, the US Army Corps of Engineers, was heavily oriented towards civil engineering and had more experience with canals, railroads and dams than chemicals. The Navy employed some scientists, in part because they had supported oceanographic work by the U.S. Coast and Geodetic Survey since its founding in 1807, but they were not important to the infrastructure of naval activity. The other technical groups were the Ordnance Department, the Signal Corps, the Sanitation Corps and the Surgeon General's Office, each with some scientific personnel. The sudden demand for scientific knowledge and laboratory expertise far outstripped the capabilities of the units as they were constituted at the beginning of the war.

On the national level, the relation of the scientific community and the federal government was in many ways more tenuous than what existed in Britain. The National Academy of Sciences (NAS) had been founded in 1863, somewhat ironically, as a response to the Civil War. The act of incorporation stated that "The Academy shall, investigate, examine, experiment, and report upon any subject of science or art."[26] Its raison d'être was to offer the best scientific advice to the government, but the government rarely took up this offer. In terms of funding for science more generally, the federal government paid for some science through various departments, such as the Bureau of Mines, but the government was legally prevented from expending money it did not directly oversee, and so the NAS operated more often as a little-used advisory board than a scientific institution.

The NAS under President William Henry Welch was reluctant to do anything about science and war. In July, 1915 the astronomer George Ellery Hale urged Welch to offer the NAS's services to President Wilson, but Welch did nothing. Even the news of the use of chemical weapons did not stir the Academy to action. Only after a Germany U-boat attacked the British passenger ferry SS *Sussex* in March, 1916 endangering American lives did Hale persuade Welch to act. On April 26, 1916, representatives from the Academy met with President Wilson. The delegation included Welch, Hale, Charles Walcott, Edwin G. Conklin and Robert S. Woodward. President Wilson suggested that the Academy set up a committee to look at war work, but he had some reservations about the project, asking that his approval not be publicized.[27]

---

[26] NAS, *A History of the First Half-Century of the National Academy of Sciences 1863–1913* (Washington, DC: NAS, 1913), 13.

[27] NAS, *A History of the First Half-Century*, 209.

The president's guarded interest prompted the NAS to create the Committee on the Organization of the Scientific Resources of the Country for National Service. The verbose title was generally reduced to the War Committee. It was chaired by Hale with biologist Edward Conklin, physician/pathologist Simon Flexner, physicist Robert Millikan, and chemist Arthur A. Noyes. They proposed the formation of a National Research Council (NRC) that would function as a super-committee overseeing war work. The NRC was approved by the NAS on June 19, 1916, but President Wilson continued to be reluctant to publicly support the organization. Wilson would not officially sanction the NRC by presidential decree until 1918. The NRC's first job was to create an inventory of scientific resources (people, equipment, production facilities and laboratories), identify needed projects, and then promote inter-organizational co-operation to solve the identified problems.

Hale got the Academy to present a resolution offering its services. On July 16, 1916 the resolution said:

> ...the President of the Academy be requested to inform the President of the United States that, in the event of a break in diplomatic relations with any other country, the Academy desires to place itself at the disposal of the Government for any services within its scope.[28]

The wording was careful because of President Wilson's reservations. Wilson's political maneuvers were complex. If he publicly accepted the NAS offer of services, he feared it would look like he was preparing for war when a major plank in his re-election campaign was to keep America out of the war. At the same time, by the summer of 1916 he was fairly sure that the US would become involved in the war in some way, so he was quietly making plans that he hoped would not be needed. To complicate the NAS offer, the President had also authorized a Council of National Defence (CND) with representatives from the Secretary of War, the Navy, Interior, Agriculture, Commerce and Labor. Its driving force was Hollis Godfrey, President of Drexel Institute and an engineer.[29] Although it had started as an intergovernmental advisory group, Godfrey pushed for a more active role in directing and overseeing industrial and scientific work. On a political level this made sense since it could work without much public attention and it avoided the problem of funding non-governmental work. Getting the mostly academic scientific community to work on war problems was a more difficult proposition. The funding issue would not be resolved until after the war, although a kind of *entente cordiale* developed that left civilian research primarily in the hands of the NAS and the NRC, and industrial development and government-run research under the CND which funneled much of its funding through the military.

One of the first things that the NAS did after receiving Presidential approval was to send a delegation to Europe to try to assess the situation. Hale and Welch con-

[28] NAS, *A History of the First Half-Century*, 208.

[29] NAS, *A History of the First Half-Century*, 213–4.

sulted with British and French scientists and physicians, and much of their enquiries were about chemical warfare.[30] Hale and Welch came back more concerned than ever that the United States needed to prepare for modern war. Their concern was not limited to chemical warfare since they had also witnessed the devastating effect of the machine gun and modern artillery. It is not surprising that one of the first issues addressed by the NAS was the supply of nitric acid for munitions. The well-known chemist Arthur A. Noyes headed the NAS Committee on Nitric Acid Supply, and the committee concluded that although it was desperately needed, the government could not build a synthetic nitric acid plant (based on the Haber-Bosch process) in less than 18 months.[31] The War Department began to stockpile nitrates and ordered the construction of four nitric acid plants, but as Noyes predicted, they were not completed until after the war.

Despite the efforts of the NAS and the CND, America was inching, not racing, towards change in science. If anything, the war caused more confusion and the proliferation of committees rather than actual work. Not counting administrative committees such as the one on membership, the National Academy of Sciences established 132 committees and subcommittees during 1917–1918, but only undertook 27 projects including the Location of Submarines with Seagulls and Seals and the somewhat more significant Effect of Altitude on Pilots.[32] Some committees, such as the Committee on Noxious Gases and Noyes' Committee on Nitric Acid Supply, would contribute to serious work, but most did little more than lament the lack of American science and recommend spending money to improve the situation. Although the number of people involved in war science was growing, it could not be said that this was Big Science in a financial sense. The total expenditure by the NAS between 1916 and 1918 was somewhere between $347,500 and $540,000.[33] Compare this to Haber's allotment of 6 million marks (about US $4.3 million in 1918 dollars) just for the KWI for Physical Chemistry and Electrochemistry.

The birth of Big Science in America came out of the chaos of war, and chemical warfare in particular. At the entry of the United States into the war on April 6, 1917 there were more than a dozen committees and several laboratories working on various aspects of chemical warfare. These included the Office of the Surgeon General, the Sanitation Corps, the Army Corps of Engineers, the Signal Corps, the Navy, the Department of Ordnance, plus committees run by the CND, NAS, NRC, the AAAS and the American Chemical Society. The group that was doing the most actual

[30] NAS, *A History of the First Half-Century*, 213.

[31] NAS, *A History of the First Half-Century*, 211.

[32] Alice M. Quinlan, "NAS-NRC Central File 1914–1918," (Washington, D.C.: NAS, 1977).

[33] Figures differ because the NRC received money that was not part of the NAS funding, but was still technically part of NAS accounting. NAS, *A History of the First Half-Century*, 229 and 229 n.70.

work, in terms of research and development, was the Bureau of Mines under the Department of the Interior. The Bureau of Mines had been looking at developing gas masks for miners since 1910. Its director Van H. Manning had hoped to apply this knowledge to chemical warfare, and he offered the Bureau's expertise to the Department of War. The Department of War did not take up the offer, in part because it came from a civilian branch of the government, but there were other reasons for the neglect. The Army was struggling to adjust to the new forms of warfare at the same time as it was moving to a war footing. If that wasn't enough, what Manning probably did not know at the time was that the Surgeon General had been made responsible for developing an American gas mask in 1916. The Surgeon General's people lacked the technical expertise to design or manufacture such a device.[34] The Surgeon General's office asked the Quartermaster Corps to put the mask out to tender, but the Quartermaster Corps could not look for suppliers for a device that had not even been designed. The Quartermaster Corps said they would try to find a supplier when they got detailed specifications. Two years went by without any progress.[35]

The impetus for rationalization came from General Pershing, the commander of the AEF. Pershing had created a Gas Service Section, and its first operational unit was the 30th Engineer Regiment (Gas and Flame), activated on August 15, 1917.[36] The regiment had very little equipment and depended on the French and British for supplies and information. In a letter to the War Department and the General Staff, Pershing recommended (it was in reality more of a demand) that a single organization be created to deal with chemical warfare: "It is recommended that all Chemical or Gas Service in the U.S. be united under one head who shall be an officer of the U.S. Army with suitable rank and staff.".[37]

Pershing's recommendation was strongly supported by George August Hulett, who was the Professor of Physical Chemistry at Princeton University. He had a strong connection with the Bureau of Mines and worked as a consultant for the Bureau. It was probably from this connection that he was asked by the NRC to join the Scientific Mission to France and England.[38] This was led by Joseph Sweetman Ames, a physicist at Johns Hopkins University, along with George K. Burgess, a physicist from the Bureau of Standards; Harry F. Reid, a geophysicist from Johns

[34] Charles Lynch, Frank W. Weed and Loy McAfee, *The Medical Department of the United States Army in the World War* (Washington, D.C.: Government Printing Office, 1923), 504

[35] Lynch, Weed and McAfee, *The Medical Department*, 506.

[36] Leo P. Brophy and George J.B. Fisher, *The Chemical Warfare Service: Organizing for War* (Washington, D.C.: Government Printing Office, 1989), 6.

[37] Maj. Gen. John Pershing to Newton D. Baker, Secretary of War, in E.C. Sullivan, "George August Hulett," (Washington DC: National Academy of Sciences, 1960), 92–3.

[38] Sullivan, "George August Hulett," 91.

Hopkins University; Richard P. Strong a physician specializing in tropical diseases, recruited from Harvard; and Linsly R. Williams, physician and deputy health commissioner for the state of New York.[39]

Hulett urged the War Department and the NRC to consolidate work on chemical warfare, and in turn the new NAS President Charles Walcott advised President Wilson to authorize the creation of a single organization to oversee research and production for chemical warfare. This idea was not without opposition, as Manning strongly objected to transferring the Bureau of Mine's research and scientists into military hands. He argued that the scientific work should remain free of military control. The Secretary of War thanked the Bureau for its good work, but presented the Department of War's position that the consolidation would lead to better co-ordination and greater efficiency. This point was reiterated by Sibert, but Manning objected, saying that the transfer would "… convey the impression to the country at large that the Bureau of Mines had fallen down on the job."[40] Secretary Baker said there was nothing but praise for the Bureau's work, while Sibert restated the necessity for efficient work. Manning replied "… that more efficient work could be done by a research organization by leaving it out of military organization, because the technical and scientific men would work better by request than by order.".[41]

Manning had good reason to challenge the centralization of control in the hands of the military. The Military Committee of the NRC determined that the first priority of any government effort should be gas masks for the Army. Manning was asked to begin the development of an American mask, something that his scientists had already been working on. He assembled new researchers and began to talk with industrial producers. By May, 1917, 21 university labs, three government labs and three industrial companies were assigned specific tasks for mask creation and production.[42] He had accomplished far more in a few months than the Army Surgeon General and the Quartermaster Corps had been able to do in 2 years. Consolidations did not destroy Manning's work. By December, 1917 the Bureau of Mines was overseeing 277 civilians (mostly chemists), 34 officers and 200 enlisted men, and had been given a budget of $762,000.[43]

[39] Henry Crew, "Joseph Sweetman Ames," (Washington DC: National Academy of Sciences, 1944), 191.

[40] Memorandum Regarding Conference Held in the Office of the Secretary of War, NARA RG 70, Special Projects, Boxes 111, 3.

[41] Memorandum Regarding Conference, 3.

[42] Memorandum Regarding Conference, 3.

[43] Memorandum, NRC to the Secretary of War, June 9, 1917.

Although there was more political fighting about who would control chemical war research, on June 25, 1918 President Wilson signed Executive Order 2894 creating the Chemical Warfare Service (CWS) and consolidated both civilian and military work in a single body. At the recommendation of Pershing, Major General William L. Sibert was made the first Director. Although Sibert did not have a background in chemistry, he had a great deal of experience with big projects. He had graduated from West Point in the top 10% of the class and been commissioned in the Army Corps of Engineers. He worked on a variety of projects, including the Panama Canal and was briefly the commander of the 1st Infantry Division.[44] He knew how to build things and how to negotiate the upper levels of military politics. He set about recruiting scientists into the CWS, first from the military and then from universities and colleges. He was so aggressive that he was eventually ordered to stop raiding the other branches of the military.

It was quickly determined that there was little capacity for new chemical production in the private sector, as the majority of chemical works were either unsuitable or already committed to conventional supplies. The CWS had to plan and build its own facilities. When the CWS was created it was given part of the military property knowns as the Gunpowder Neck Reservation. The Reservation would become the Aberdeen Proving Grounds, near Baltimore on Chesapeake Bay. The CWS facility was renamed the Edgewood Arsenal. The building of the Edgewood Arsenal and Gas Defense Plant was complex and often confusing. Some parts were well planned, while other sections were *ad hoc* additions. It was certainly the case that what was actually needed, what was planned, and what got built were not strictly in accord. It combined scientific facilities, a military base and the largest chemical weapons production plant in the world. At its peak, it had the capability to produce 675 tons of offensive chemical agents a week and fill 1 million artillery shells a month.[45]

Colonel William H. Walker of the Ordnance Department was ordered to take the overall command of the construction at Edgewood. He was an analytical chemist from the Massachusetts Institute of Technology who had been commissioned a lieutenant colonel and made chief of the Chemical Service Section. He was given a budget of $19,040,000 for "…construction of an initial operation of filling plants, chemical plants, chlorine plants and additional power installation, as well as the necessary cantonments, hospital, storehouses, etc., and a gas shell proving ground with laboratory and animal farm."[46] In all, 665 separate structures ranging from sheds to complex chemical production plants were constructed, along with miles of

---

[44] Chemical Corps Regimental Association, www.chemical-corps.org/honors/sibertbio.htm

[45] Leo P. Brophy & George J.B. Fisher, *The Chemical Warfare Service: Organizing for War* (Washington, D.C.: Government Printing Office, 1970), 12.

[46] C.B. Wheeler, Orders for Colonel William H. Walker, Apr 2, 1918. NARA CWS 175. Box 4.

roads, a narrow gauge railway, and both docks and mainline rail spurs for shipping were built in less than 2 years.[47]

The CWS undertook a massive research and development program. It built laboratories at the Edgewood Arsenal and at the American University Experimental Station (AUES) in Washington, D.C. Scientists were conducting their research while workmen constructed the buildings around them. For most of the projects, the scientists were playing catch-up, trying to replicate the materials and understand the chemical behaviour of agents already being used. When Germany introduced mustard gas (not really a gas, but an oily liquid $(ClCh_2Ch_2)_2S$), the Americans, British and French chemists reverse engineered the compound and then figured out how to produce their own versions.

The major original American research effort led to the creation of lewisite, which was believed to be the most toxic agent to be mass produced during the war. Although it was put into production, it was never used on the battlefield, having been developed too late in the war. Lewisite was named after the chemist and lead investigator Winford Lee Lewis. Lewis had earned his Ph.D. in chemistry at the University of Chicago in 1909, and was teaching at Northwestern University in Illinois. When the US entered the war, he went to Washington to offer his services and was initially asked to serve in the Food Division of the Sanitary Corps, but opted to volunteer for the Gas Service where he was commissioned as a captain.[48]

The discovery of the toxic potential of lewisite had a serendipitous start. The compound $ClCHCHAsCl_2$ (2-chloroethenylarsonous dichloride) was first noted by Julius A. Nieuwland in his doctoral dissertation "Some Reactions of Acetylene."[49] Nieuwland, who was ordained as a Catholic priest in 1903, completed his dissertation under the supervision of Father John Griffin at the Catholic University of America in Washington D.C. What was not included in the dissertation was any mention that during his work Nieuwland had been hospitalized for several days because of exposure to the by-product of mixing acetylene and arsenic trichloride.[50] When Lewis was directed to find a faster acting chemical agent than mustard gas (depending on exposure, the reaction to mustard can take up to 3 h to appear), he was working at the Experimental Station at the American University. He felt the conditions were not safe enough for his research team and moved them to the Catholic University to use the chemical laboratories at Maloney Hall. This was where Lewis met Father Griffin, who related the story of Nieuwland's illness during

[47] R.C. Marshall, Jr. and Edward B. Ellicott, *Completion Report of Construction of the Edgewood Plant of the Edgewood Arsenal* (Washington, D.C.: [Government Printing Office], 1919), 16.

[48] Joel A. Vilensky, *Dew of Death. The Story of Lewisite, America's World War I Weapon of Mass Destruction* (Bloomington, IN.: Indiana University Press, 2005), 8.

[49] Vilensky, *Dew of Death*, 4.

[50] Vilensky, *Dew of Death*, 4.

his dissertation research.[51] This led Lewis and his team of 34 chemists to investigate and eventually synthesize the toxic compound.

Lewisite process[52]	
$C_2H_2 + AsCl_3 \rightarrow ClCHCHAsCl_2$	2–4 atmospheres
	160 °C
	$AlCl_3$ (aluminum chloride) as a catalyst

The CWS decided to develop lewisite and the job of industrializing the process and overseeing the establishment of a production facility was given to James Bryant Conant. Conant received his Ph.D. from Harvard in 1916 and that same year he joined Harvard to teach organic chemistry.[53] Like Lewis, when war was declared, he offered his services to the military and was commissioned as a second lieutenant in the CWS, but was quickly promoted, leaving the Service with the rank of major. Conant chose the town of Willoughby, Indiana as the site for setting up the lewisite factory.[54] It was near Cleveland, where the CWS already had established part of its gas mask production. The Ben Hur plant, as the production facility was called, cost about $5 million to build and it was kept under serious security.[55] The 542 enlisted men and 22 officers who ran the plant were not allowed to leave the plant site and mail was delivered to a post box in Cleveland.[56] Sibert wanted 3000 tons of lewisite for the expected spring campaign and Conant and his team did their best to reach this goal.[57] Although the Ben Hur plant never came close to the 3000 ton target, at the end of the war, the plant was probably producing 10 tons of lewisite a day. The daily production and total production, however, are not really known.

Although lewisite was not used on the battlefield, the use of chemical weapons contributed to the appalling level of casualties during the war. About 124,000 tons of offensive chemicals were used by all sides during the course of the war.[58] The

[51] Vilensky, *Dew of Death,* 19–21.

[52] W. Lee Lewis and G.A. Perkins, “the Beta-Chlorovinyl Chloroarsines,” *Industrial and Engineering Chemistry* **15** (Mar. 1923),290–5.

[53] Paul D. Bartlett, “James Bryant Conant,” (Washington, D.C.: NAS, 1983), 93.

[54] Vilensky, *Dew of Death,* 39.

[55] Vilensky, *Dew of Death,* 43.

[56] Vilensky, *Dew of Death,* 43–8.

[57] Vilensky, *Dew of Death,* 39.

[58] Prentiss, *Chemicals in War*, 652.

Russians, poorly equipped, badly trained and disastrously led, suffered the most casualties from gas as well as conventional weapons, while the United States lost the highest percentage of soldiers to chemical agents. This high rate of casualties demonstrated a lack of experience with chemical attacks. American forces entered the war facing the most toxic agents and had not gained experience from exposure to the less toxic or more easily defended against chemical agents used earlier in the war.

Casualties from chemical weapons and total casualties during the war[59]

Country	Chemical wounded	Chemical deaths	Total chemical casualties	Total war casualties (Deaths and wounded)	Ratio of chemical casualties to total casualties %
Russia	419,340	56,000	475,340	9.1 million	7.5
Germany	191,000	9000	200,00	7.1 million	3.5
British Empire	180,597	8109	188,706	3.0 million	7.3
France	182,000	8000	190,000	6.2 million	3.5
USA	71,345	1462	72,807	340,698	26.8

Chemical warfare was Big Science. Haber at the height of his war-time authority oversaw a large scientific organization that was centred in his WKI for Physical Chemistry and Electrochemistry. It included about 1300 people, including 150 scientists.[60] Among Haber's gas troops were a number of notable scientists including James Franck, Otto Hahn and Gustav Hertz who would go on to do important work after the war. Although there is no clear accounting of Germany's expenditure on chemical warfare, it was in the tens of millions of marks. Even at the end of the war, with Germany on the verge of defeat, Haber would ask and get from the Ministry of War a grant of 6 million marks (about US $1.05 million in 1918) to continue the work at the Institute.[61]

In response to the use of chemical warfare, the United States began a crash program of building and research and in the process established the roots of Big Science in America. For the first time, large numbers of scientists were brought together and worked in big teams and networks of teams. In the end, Sibert and his second in command Amos Fries tried to recruit 1749 chemists (including 165 men with chemical experience, but no academic credentials) from across the military and from the civilian chemical community. By September 1918, they had recruited 1294 scien-

[59] Prentiss, *Chemicals in War,* 651–3.

[60] Stoltzenberg, *Fritz Haber,* 140.

[61] Stoltzenberg, *Fritz Haber*, 149–50.

tific and technically trained men into the CWS.[62] It was by far the largest assembly of scientists in a single research organization in history. The CWS built modern laboratories that rivalled the best available in Europe, particularly at the Edgewood Arsenal and the American University Experiment Station.[63] Where the federal government had previously had little interest in science and very limited knowledge of the technical industries operating in America, by the end of the war government was acutely aware of the strategic importance of science and industry and had endorsed the creation of the NRC and the CND. Important ties were created between the academic researchers and industry. In the short time the CWS had existed, it had spent more than US $83 million (US $1.1 billion today). Most of the expenditure was spent on facilities, but research and development was funded at a level not to be seen again until the next war.[64] This figure seriously underestimates the total spent on chemical research because it does not include the value of work done prior to the consolidation, especially by the Bureau of Mines, nor does it account for the work done by the universities at their own expense.

Some of the conditions that led to the creation of the CWS faded after the war. The CWS had only existed for 18 months, so it did not have the time to establish much of a foundation or institution presence. It was disliked by the other branches of the military, particularly the Army Corps of Engineers, who campaigned actively for the dismantling of the CWS and the transfer of its duties to the Engineers. Congress tried to place the CWS under the Ordnance Department of the Army (largely because the CWS filled artillery shells), even though Ordnance did not want it.[65] Many of the chemists and other scientists who had worked for the CWS in uniform or as civilian researchers were happy to return to more peaceful work. In addition, the public was strongly against chemical warfare, so there was little political will to preserve the research capability.[66]

What did not fade away were the lessons learned by the scientists themselves about how much research could be done with large teams, good facilities and big budgets. Many of the members of the CWS returned to civilian life with new skills, connections and a desire to see America as a scientific leader. Just a brief sample of

[62] "Chemical Warfare Service," *Journal of Industrial and Engineering Chemistry* **10** (9) (Sept., 1918), 675–684.

[63] G.A. Burrell, "Report of Work Done at Bureau of Mines Experiment Station, American University," (Washington, D.C.), May, 1919, NARA CWS 175. Boxes17, 1–30.

[64] Leonard P. Ayres, *The War With Germany. A Statistical Summary* (Washington, D.C.: Government Printing Office, 1919), 133.

[65] Editorial, "Chemical Warfare Service Endangered," *Journal of Industrial and Engineering Chemistry* **12** (January,1920), 2–3.

[66] See for example, "The Chemical Warfare Service," Editorials, *Industrial and Engineering Chemistry* **12** (April,1920), 314.

the researchers who worked for the CWS suggests that they took the lessons they learned during the war with them. In particular, James Conant, with his experience overseeing the production of lewisite, was well prepared for his role in the Manhattan Project.

A brief sample of CWS alumni[67]

Name	CWS rank	Post-war activity
Frederick E. Breithut	Major	Chemical Trade Commissioner, US Embassy, Berlin.
James Conant	Major	President of Harvard. Member and Chair of the National Defense Research Committee, overseeing (among other projects) the Manhattan Project. Member of the Interim Committee that advised President Truman to use the atomic bomb on Japan.
Frederick G. Keyes	Captain	Created the MIT Research Laboratory of Organic Chemistry in 1925 and the Research Laboratory of Inorganic Chemistry in 1930. Helped establish Eastman Laboratories.
Bryon Goss	Captain	President, Erie Chemical Company.
Joel Henry Hildebrand	Major	Taught at Berkley. Won Priestley Medal. NAS Council and member of the Citizens Advisory Committee on Education. The Joel Henry Hildebrand Award is given in his honour by the ACS.
William Henry McAdams	Captain	Taught chemical engineering at MIT and Harvard. Member of the National Research Council. Worked on the Manhattan Project. Winner of the William H. Walker Award.
William McPherson	Lt. Colonel	Head of the Department of Chemistry and Interim President, Ohio State University.
Samuel Parson Mulliken	Major	Worked on the plutonium project (part of the Manhattan Project). Nobel Prize 1966, Priestley Medal 1983. Taught at the University of Chicago in both the chemistry and physics departments.

Even the post-war activity of the CWS was emulated by the post-war nuclear program. After World War I, Amos Fries attempted to utilize war research for peaceful purposes. There were a number of projects such as an investigation into the use of chlorine to treat pneumonia and mustard gas to treat tuberculosis. Chlorine was a complete failure while mustard gas was discovered to be anti-bacterial, but it was neither specific enough nor safe enough for medical use.[68] Arsenical compounds related to lewisite were successfully used to treat syphilis. A significant research effort was undertaken to use war gases on the boll weevil which attacked cotton plants. The problem of persistence and danger to humans eventually ended the use

[67] See James McKeen Cattell, *American Men of Science* (New York: Science Press, 1927, 1933).

[68] See for example, H.J. Corper and O.B. Rensch, "The Effect of Mustard Gas (Dichlorethylsulphide) on Experimental Tuberculosis," *Journal of Infectious Disease* **28** (March, 1921), 286–93.

of war chemicals as insecticides, but not the idea.[69] In the nuclear age there was nuclear power and Operation Plowshare. Operation Plowshare was an ill-fated collaboration among the Atomic Energy Commission, the US Geological Survey and the Bureau of Mines that started in 1958 and ended in 1975. It attempted to use nuclear explosives for peaceful purposes such as mining, fracking to release natural gas and instant harbour creation. After 27 tests, the program was stopped because of a combination of public concerns about radioactivity from nuclear tests and the lack of support in Congress for the commercialization of nuclear explosives.[70]

The primary characteristics of Big Science included large teams, big budgets and interconnections with what today would be called "stakeholders" such as government, higher education, industry and the military. All were present in the work of Fritz Haber and even more so in the work done by the CWS. The NRC and the CND, founded to deal with war science, formed the institutional backbone of national research after the war. The research organization built by the CWS did not last much past the end of the war, but it served as a template used during the next war, particularly by the Manhattan Project. One of the often neglected or misunderstood aspects of Big Science is the perception of need. While scientists might dream of massive research projects, those projects do not become reality unless the larger community (primarily, but not exclusively, government) sees the need for investing in the endeavour. Whether Big Science is good science is often a matter of research objectives, but its existence started with poison, fire and smoke during World War I.

[69] See H.W. Walker and J.E. Mills, "Chemical Warfare Service Boll Weevil Investigation. Progress Report," *Industrial and Engineering Chemistry* **19** (June, 1927), 703–11. See also Edmund Russell, *War and Nature. Fighting Humans and Insects with Chemicals from World War I to* Silent Spring (Cambridge: Cambridge University Press, 2001), 64–6.

[70] See US Department of Energy, "Plowshare Program. Executive Summary," Office of Scientific and Technical Information, plowshare.pdf.

Chapter 12
Politics, Morality, Innovation, and Misrepresentation in Physical Science and Technology

Jed Buchwald

The pressures of politics, the desire to be first in innovation, moral convictions, and the potential dangers of error are all factors that have long been at work in the history of science and technology. And every so often, the need to reach a result may require leaving out a few steps here and there. Historians think and argue best through stories, so what follows are several tales, each of which exemplifies one or more of these aspects, though some reach back nearly 200 years. The first concerns the depletion of the ozone layer; the second involves the discovery of electric waves by Heinrich Hertz in 1888; the third concerns the controlled production of electromagnetic radiation by Guglielmo Marconi and John Ambrose Fleming in the early 1900s; the fourth portrays the circumstances surrounding Joseph von Fraunhofer's discovery and use of the spectral lines in the 1810s; our final case involves a bitter controversy between the physicist Hermann von Helmholtz and the astronomer Friedrich Zöllner in the 1890s.[1]

[1] Each of these examples derives from the following publications, where full details can be found. On the ozone layer: Jed Z. Buchwald and George Smith, "*The Ozone Layer* (review)," *American Scientist* **89** (2001), 546–49. On Hertz: Jed Z. Buchwald, *The Creation of Scientific Effects: Heinrich Hertz and Electric Waves* (Chicago: The University of Chicago Press, 1994) and Manuel G. Doncel, "On the Process of Hertz's Conversion to Hertzian Waves," *Archive for History of Exact Sciences* **43** (1991), 1–27. On Fraunhofer: Miles W. Jackson, *Spectrum of Belief: Joseph von Fraunhofer and the Craft of Precision Optics* (Cambridge, MA: The MIT Press, 2000). On Marconi and Fleming: Sungook Hong, *Wireless: From Marconi's Black-Box to the Audion* (Cambridge, MA: The MIT Press, 2001). On Helmholtz: Jed Z. Buchwald, "Helmholtz's Electrodynamics in Context: Object States, Laboratory Practice and Anti-Idealism," in *Hermann von Helmholtz and the Foundations of Nineteenth-Century Science*, ed. D. Cahan (Berkeley: University of California Press, 1993), 334–73.

J. Buchwald (✉)
California Institute of Technology, Mail Code 101–40, Pasadena, CA 91125-4000, USA
e-mail: Buchwald@caltech.edu

J. Buchwald, L. Stewart (eds.), *The Romance of Science: Essays in Honour of Trevor H. Levere*, Archimedes 52, DOI 10.1007/978-3-319-58436-2_12

12.1 The Ozone Layer

The first tale is also the most recent. In 1995, an MIT chemist by the name of Mario Molina shared the Nobel Prize in chemistry for his work on the atmospheric reactions that produce the depletion of the ozone layer. Since they were first handed out at the beginning of the last century, the Nobel Prizes have been the gold standard by which scientists have judged success. When you win a Nobel, your work has been canonized by the priestly guardians of science, all of whom live in Sweden near the Valhalla of the old Nordic gods. With a few exceptions, one would not want to argue with the appropriateness of the awards handed out over the last century, though the losers might want to do so. And one would certainly not gainsay the Prize awarded for the superb chemistry developed by Molina. But the research that led to the award bears discussion because it holds an interesting and timely lesson.

The background to this work goes back quite far, to 1879 in fact, when ozone in the upper atmosphere was first recognized. Soon thereafter, it was realized that ozone shields the surface of the Earth from ultraviolet radiation. When it was also discovered that ozone is substantially present only in an altitude range of 15–50 km, and that it exhibits patterns of annual variation, scientists began to investigate what sort of chemical processes might be involved. Then, in the 1930s, a new industrial development took place when it was found that chlorofluorocarbons were an ideal refrigerant—these CFCs, as they were soon widely called, were non-toxic, and non-reactive and so seemed to be perfectly suited for widespread use in the rapidly growing world of commercial refrigeration. Although scientists noted quite early that CFCs could easily accumulate in the stratosphere, at the time this raised no alarms. In fact, the first public discussion of the possibility that human activity might compromise the ozone layer occurred only in the 1960s and then did not involve the effects of CFCs, but rather of the high altitude flights of the proposed supersonic transport (SST). This was one element (though certainly not the only one) in the debates of the day that led the US Congress to stop funding for the SST.

This set the stage for the three events that led to Molina's Nobel, which he shared with Paul Crutzen and Sherwood Rowland: research in the early 1970s identified a sequence of chemical reactions by which CFCs could gradually deplete the ozone layer. Then, surprisingly late, in the mid-1980s the worsening development each spring since the mid-1970s of an ozone "hole" over the Antarctic was announced; this was followed by a highly organized effort to gain data that revealed, among other things, that the size of the Antarctic "hole" varied over the course of a year.

And here we come to a surprising and significant point. The chemistry that had been proposed for ozone depletion, and for which the Nobel was granted, requires ultraviolet radiation, and the various models that were built using this chemistry had predicted about a 5% general depletion of the layer near the equator by the 1980s. But the ozone hole occurred at the Antarctic. Not only is ultraviolet radiation comparatively small at the poles, the depletion effect there was far greater than the meager 5% predicted for the equator. Clearly something was up. Competing speculations were proposed by different groups of scientists. Then, as often happens, new experimental data altered the discussion. In the late 1980s, the Antarctic Airborne Ozone Experiment produced what amounted to "smoking gun" evidence that chlorofluorocarbons were the culprit, after which efforts focused on the detailed chemistry involved. In 1987, the Montreal Protocol for eliminating the use of CFCs was signed, but only in 1994 was a widely accepted chemical process that could explain the full depletion cycle evolved. The process of elaborating the chemistry continued thereafter.

Despite the failures of the original chemical system that had linked high atmospheric concentrations of CFCs to low concentrations of ozone, no chemist today thinks that the first theory was actually false; it was instead only incomplete, for the principal reactions singled out did remain part of the story. Moreover, the original chemistry had been instrumental in focusing the research community's attention on CFCs. This holds a lesson for scientific research that goes beyond the overly simple notion that a theory just makes predictions that either fit the data or do not, and that, if the fit is bad, then the theory must be tossed out. On the contrary, here we see that the fit between theory and experiment was not only bad, it was pretty much the reverse of what it should have been. Nevertheless, research scientists had no workable alternative but to continue pushing the original chemistry until it had been sufficiently fine-tuned to accommodate observation.

There are at least two lessons in this story, one scientific and the other political. The scientific lesson is this: good researchers are stubborn; they do not simply cave in when faced with discrepant data. The data may be complex, the connections between model and data even more so, so that when the foundations of a theory are otherwise firm it is not a good idea to discard it too quickly. This is all the more true when dealing with something like the atmosphere, for here scientists encounter the difficult question of whether large scales require different ways of thinking than do the small scales that had been dealt with until the middle of the twentieth century. After all, the atmosphere cannot be put into a tabletop device. But it can be modeled on a computer, and a very great deal of recent science, as well as engineering, is done with computer models either because of computational complexity, scale, or both. This trend raises important questions about just what the relation between a model and a natural system might be, and how discrepancies between data and computational outcomes should be dealt with, particularly when the system in question cannot actually be manipulated. Not only can we not yet play around with the atmosphere, one hopes that we would not want to do so even if we could, given the possibility of disastrous consequences, which brings us to the moral and political lesson of the ozone episode.

Morality and politics are not words that easily go together, and not because politicians are inherently untrustworthy, though most Americans think with good reason that they probably are. Politics is and always has been the art of persuading others to do what the politician wants done, and that for a complicated variety of reasons. Some of those reasons may even be moral, however unlikely that may seem. But most of them are, and must be, pragmatic, which does not mean that they are necessarily immoral. Pragmatism comes in at least two forms: the pragmatics of the moment, and the pragmatics of vision. What might be expedient and perhaps even useful in the short term may have unfortunate consequences as the years go by.

The ozone layer discussions that took place in the US House and Senate exhibit these characteristics. There was originally considerable talk that the strength of evidence supporting the connection between the ozone hole and CFCs was much too weak to support drastic action, which was probably true in the first few years. But data and theory were pushed further, and soon the connection between them became quite close—though certainly not of the same order as, say, the statement that if you let go of an apple it will fall to earth. Precisely because the shrinking ozone layer could have drastic ecological consequences, the connection did not need to be so tight as that in order to mandate a policy with vision. Because policy must be developed, decisions must be made in the light of a balance of risks against potential gains. Science can shed light on the likelihood of risk and gain, but not always with the unbreakable certainty that policymakers might wish because many questions call for decisions that cannot wait until the finished scientific scheme is in place. Standards of evidence when risk is central can and must be very different from standards of evidence in either comparatively finished science or science in the earliest stages.

These differing standards can be a serious source of continuing confusion in disputes over such matters as ozone depletion. Indeed, they clearly have been in respect to global warming, which, unlike the ozone debates, has not prompted sufficient action by a recalcitrant and scientifically ignorant US Congress, despite an overwhelming consensus among scientists—a consensus that is no less strong than the one that prompted action against CFCs. Those whose vested interests may be adversely affected by policy decisions made on the basis of this evidence can always invoke the standard of a perfectly finished science to argue for delay. All the while, the policy question is best viewed as a balance of risks against gains, given all currently available information.

12.2 Hertz's Experiments on Electric Waves

Let us turn now to something rather different, something that will take us into the heart of the research enterprise as it was practiced over a century ago. Our story here has two dimensions: one concerns the character and meaning of what appear to be misrepresentation in science; the other concerns the thrill that comes with a new discovery. The two topics are, we shall see, not unconnected. Claims concerning scientific misrepresentation have, in the last two decades, become increasingly common and increasingly shrill, more often than not because government money—and thus politics—is involved. Depending on the circumstances, such a thing might even be considered fraud.

What, though, is fraud? Definitions can become rapidly, and legalistically, complicated, but an assertion that some claim is fraudulent presupposes at the least that a deliberate attempt has been made to lead others to believe something the defrauder knows to be untrue. But more than mere deception is necessary for a claim to be truly and properly fraudulent: the purpose of the deception must be venal; its goal must be to produce a significant benefit to the defrauder or to those for whom he is acting. Without this element of personal benefit, we are not dealing with true fraud but rather with something quite different, namely misrepresentation, and there can be many motives for that.

Fraud and its sibling, willful misrepresentation, have existed in every area of human endeavor that involves persuasion. Science is certainly no exception, since claims made by its practitioners are crafted to convince others. Those who persuade many Americans, apparently by the millions, that they will soon by transported rapturously to heaven in their family cars certainly do no less, this being a major theme of several very successful recent books sold widely in airport kiosks.[2] But there is a difference. The scientist knows that any attempt to persuade may come to shipwreck on the shoals of a future observation, experiment, or calculation, and that this does not lie altogether under his or her powers of persuasion. Whereas if the Rapture does not happen tomorrow, then the true believer will just wait for another day. Belief is utterly and essentially impervious to evidence. Therein lies a critical difference between those who foolishly attempt scientific fraud and those who merely misrepresent, because the power of evidence can eventually destroy a scientific fraud, whereas a misrepresentation may be designed to convey just how powerful the evidence for a particular claim may be.

Consider the case of the German theoretician and experimentalist, Heinrich Hertz (Fig. 12.1). Born in 1857, Hertz undertook the first series of experiments in the late 1880s to demonstrate the existence of electric waves. He began with wires. In those early years of telegraphy and telephony the highest frequency that had yet

Fig. 12.1 Heinrich Hertz (Credit: https://commons.wikimFigure)

[2] To wit, the egregious, apocalyptic *Left Behind* series by the Christian dispensationalist and John Birch Society member Tim LaHaye, co-authored with Jerry B. Jenkins.

Fig. 12.2 James Clerk Maxwell (Credit: http://www.clerkmaxwellfoundation.org/html/maxwell_21.html)

been produced was about 15 kilohertz—of course the very word for frequency, hertz, refers to our discoverer. He was the first to produce waves in wires in the megahertz range, and to show how to detect and to control them. Along the way, he also discovered how to design a spark-switched oscillator, which was a device with an air gap that behaved just like a closed wire when the air broke down at sufficiently high potential. This discovery enabled Hertz to generate waves in air and to investigate all of their properties.

We will turn in a moment to how Hertz felt as he worked this golden vein of discovery in his laboratory, but first let us jump ahead to see how he presented his work to the scientific world, for here he faced a knotty problem. At that time—the 1880s—there were several competing theories for electromagnetism and only one of them, the British physicist James Clerk Maxwell's (Fig. 12.2), predicted the existence of radiation. But even that one, many leading British physicists of the day felt, did not lead to the possibility of artificial electromagnetic radiation, since most people thought that only processes at the molecular level would do so, and that they would generate only the ultrahigh frequency oscillations that constitute light. This was because no one, even in Britain among Maxwell's followers, had worked out the details of what would later become antenna theory—which in fact became one of Hertz's major theoretical contributions. So Hertz was faced with the very difficult problem of convincing essentially everyone, even the British, that he had actually managed to produce and to control something that none of them (though for different reasons) thought possible.

That conundrum led him into misrepresentation. In the months after his discovery, Hertz was asked by the editor of the premier physics journal of the day, the *Annalen der Physik*, to write an account of what he had done. As he thought about how best to present a convincing narrative, Hertz decided to divide the story into three distinct parts, each one being principally concerned with a separate part

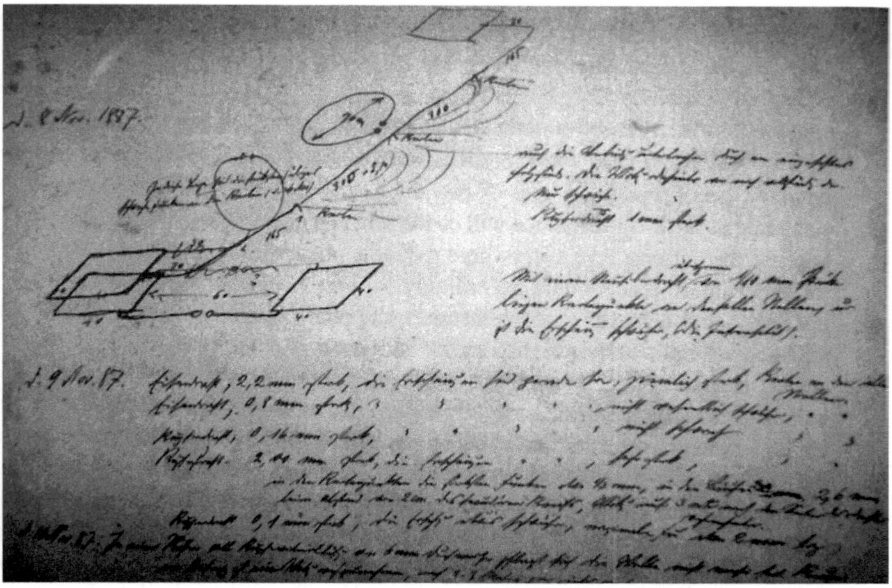

Fig. 12.3 A page from Hertz's laboratory notebook (Credit: By permission of H. G. Hertz)

of the apparatus. In the printed trilogy that resulted, Hertz claimed that he was relating the sequence of experiments and thoughts that resulted in his discovery.[3]

Until fairly recently, we had only this printed work to go on. But some time ago Hertz's actual laboratory notebook was found in the possession of one of his descendants in Germany (Fig. 12.3). That notebook tells a very different story from the public account. It shows unequivocally that Hertz had considerably altered the true course of events in ways that made his path to discovery seem to be much more logical and linear than it was. Hertz had to work very hard to produce electric waves, and there were many more stumbling blocks along the way than he explained.

Were Hertz's actions fraudulent in any truly meaningful sense? Hardly. Hertz did not intend to mislead his readers in order to create in them a false sense of his experimental and logical abilities. That would indeed be fraud. But such was not Hertz's purpose, for he knew very well that his results were difficult to understand. In writing his trilogy, he deliberately chose to lead his readers step by step through unfamiliar territory, to guide them by the hand. This is not fraud; it is good pedagogy. No doubt Hertz might have remarked in a note that he was not literally recounting the paths he had followed, but that would only have raised questions about his results, deflecting attention from them to their production.

So here we have a situation in which the public misrepresentation of the actual course of events served the important purpose of getting results into scientists' hands rapidly and effectively. The tactic certainly did work, for laboratories in Britain, France, Germany, Italy, and elsewhere rapidly began producing and probing

[3] H. Hertz, *Electric Waves* (trans. D. E. Jones, 1962 reprint of the 1893 edition), New York: Dover Publications, pp. 95–136.

the nature of electromagnetic radiation. Indeed, within scarcely a decade, Guglielmo Marconi in Italy, with the help and advice of the English engineer-scientist John Ambrose Fleming, was transmitting long-wave radiation across dozens of miles. Today, Hertz would perhaps have been more explicit about the actual sequence of events. But he might very well have written his trilogy in very much the same way, just warning the reader that this was not quite how the discovery happened. For that, he would have waited until his Nobel Prize address (unfortunately he died very young, in 1894, and so just missed the first awards, which came in 1901.) We will return in a moment to this instance of rapid progression from a fundamental discovery to a new technology, but first let us rejoin Hertz in his laboratory, where we can learn a second lesson.

Fortunately, Hertz kept a diary and wrote his parents frequently about his life and work.[4] In the early winter of 1887, his experiments were going well, and he wrote home that he had had "good luck with my experiments, and though there were some mishaps … I have never before been on such fertile soil, prospects are opening right and left for new, interesting experiments."[5] But a few days later, things had begun to turn sour, because instead of finding proof that electric waves exist, he seemed to be discovering that they do not. And so, he later explained, "disheartened, I gave up experimenting."[6] But he did not stop for long. He returned to the bench, not as enthusiastic as before, but nevertheless determined to go ahead, because he now felt that disproving the existence of something would itself be very important. Not, of course, as exciting as finding something new that no one had ever seen before.

But, as Hertz continued to work, he played around with his device to make new sorts of measurements, trying to be utterly certain about his negative results. And when he did this, he suddenly obtained indications that waves do indeed exist. The old measurements, he decided, had been flawed by disturbing effects. Over the next weeks and months, he successfully tracked electric waves throughout his laboratory; he measured their polarization; he refracted them; in short, he became the first person to produce and to manipulate artificial electromagnetic radiation. He was mining a vein of pure gold and he wrote an arresting letter to his teacher and mentor, the great German polymath Hermann von Helmholtz at the University of Berlin, in the spring. This letter says much about what happens when a truly creative scientist moves beyond the pressures of competition, beyond the immediate cares and concerns of daily affairs, to glimpse something that no human ever had before. He wrote: "I now have the … feeling that I am … on my own ground and territory and almost certainly not competing in an anxious race and that I shall not suddenly read in the literature that someone else had done it all long ago. It is really at this point that the pleasure of research begins, when one is … alone with nature and no longer worries about human opinions, views and demands.".[7]

---

[4] Johanna Hertz, ed., *Heinrich Hertz: Memoirs, Letters, Diaries*, 2nd enl. ed., trans. L. Brinner, M. Hertz, and C. Susskind (San Francisco: San Francisco Press, 1977).

[5] Ibid., 237.

[6] Heinrich Hertz, *Electric Waves, being Researches on the Propagation of Electric Action with Finite Velocity through Space*, trans. D. E. Jones (1893; New York, NY: Dover, 1963), 8.

[7] Hertz, *Memoirs* (ref. 3), 255.

The human world, with its anxiety-producing demands and pressures, disappeared for Hertz as he wandered at will through this unexplored land. Wonder and delight at the discovery of the utterly new have always been a hallmark of the finest creative science and engineering, and they remain so to this day. In 1995, two groups, one at the University of Colorado NIST-JLA laboratory, the other soon after at MIT, also produced an object that had never before existed, called a Bose-Einstein condensate. A graduate student at MIT named Marc-Olivier Mewes reflected on the moment when he first saw the condensate, that it was "one of those rare times in physics when you discover a really new effect. It makes you feel kind of strange," he continued, because "you're seeing something that nobody else has ever seen before" (MIT News, Jan. 29, 1997). That had been Hertz's own sentiment over a century before.

What lessons might be drawn from this? One is certainly this: however pressing the desire to beat the other fellow may be—and one would not wish at all to underplay the power of competition—nevertheless, when the truly creative scientist or engineer at last reaches deep into discovery, then the mundane fades into the background and it is the power and wonder of the unknown that carries him or her forward.

There is of course something rather inhuman, or perhaps it would be better to say, unhuman, about this. In most other areas of great creativity, the human world forms an essential part of the enterprise. Literature, after all, concerns human experience; artistic creation may have some of the elements that so gripped Hertz and Mewes, but the world that it produces is, if not unconstrained by nature, nevertheless not so directly entangled in an unknown reality. One does not think of a writer or of an artist as being "alone with nature" in quite the same way. Physically alone he or she may very well be, but the human world can never be altogether far from mind.

In that sense, the scientist really may live in an unhuman world, at least for a time, and this is perhaps one of the things that makes scientific work seem to be so very strange to many people. Even as citizens of technologically advanced countries live in a world increasingly designed by science and engineering, a world that is filled with devices and effects that have never existed before and that in ever more powerful ways mold daily life, belief in the irrational remains as widespread as it has ever been, perhaps even more so in the United States, where millions read their daily astrological charts and believe the earth to have been miraculously created some 6000 years ago. Why is this so? It is not solely because of the sorry state of science education in this country, though that is certainly a factor, and a very difficult one to overcome. The reasons run deeper, and are much more ancient, than mere educational failure.

The universe of magical belief governed the earliest civilizations in Egypt and Mesopotamia, where people believed unhesitatingly that mysterious, conscious forces governed human destiny, forces that required appropriate propitiation to avoid disaster. Yet these antique powers were not unhuman, inimical though they may have been to human destiny, precisely because they could be propitiated; they could be appealed to in ways that were not altogether different from human supplication. The world of modern science since the seventeenth century offers no such hope. We cannot hope to appease the gods because natural law offers no such mechanism. Nature no more listens to human desires than a hurricane attends to a grass hut. Nature is altogether, utterly indifferent to our existence; it is quite literally, and to many, terrifyingly unhuman. Our world is more than ever fabricated out of this apparently indifferent material reality; nature may not care whether humanity exists or not, but people do.

Fabrication is what humans really do best, for if anything is a hallmark of the human species it is the ability to engineer new worlds out of the natural environment, which brings us to another aspect of scientific discovery, its close link to technology. The connection between the two realms is inevitable, profound—and yet strained by a potential tension. Since its true beginnings in the seventeenth century, laboratory science has always used the craftsman's workshop, and eventually large-scale industry, for its instruments. The first air pump was produced by skilled English journeymen, whose names are lost to history; many of Galileo's devices were forged by Tuscan guildsmen; and Hertz's galvanometers and induction coils were built by German instrument makers. In a quite direct sense, laboratory science has always been closely linked to the artisanal world, and, moreover, artisans in the Renaissance and later often thought of themselves as engaged in work that revealed the secrets of nature. However, when we think today of the connection between science and technology, we envision something that does not evoke the guild or the craftsman, but rather applied science, whose usual image is this: a new effect or process is discovered in the laboratory as a result of investigations whose goals had little to do with the world of practical application; afterwards industry capitalizes on the discovery. On this view, science today produces, and industry consumes.

This has never been true, certainly not in such a bald form, and it is not true today. There are instances here and there that do look something like the cliché, but even there the interactions are much more complicated than it might seem. Moreover, there are nearly as many instances in which research driven by essentially practical motivations has led to significant scientific discoveries as the reverse. Let us take a brief look at two examples of past connections between science and technology.

12.3 Marconi's and Fleming's Wireless Demonstration

The first takes us back to electromagnetic waves. By the early 1900s, Marconi and Fleming (Fig. 12.4) were working to produce a useable system that could transmit signals without wires. Here they faced a number of problems. First of all, though the idea of wireless communication arose quite rapidly after Hertz's discovery, two issues cropped up nearly at once. The first was distance: it seemed to everyone who knew anything that reaching farther than line-of-sight to the horizon would be no more possible than seeing around the curvature of the earth. This seemed to limit wireless usefulness to ship-to-shore or ship-to-ship communication, where each could see the other. In other words, wireless might at best replace the old system of signaling by line-of-sight semaphore stations. But even this was problematic.

Fig. 12.4 Guglielmo Marconi (*left*) and John Ambrose Fleming. Credit: Marconi (*left*) from Library of Congress; Fleming (*right*) from Electrical World (1890), vol XVI, pg. 467

Semaphore signaling had been used by militaries since the days of Napoleon, and it worked extremely well, albeit with the critical disadvantage for navies that it worked only over land. Still, where they could be used, semaphore signals did not interfere with one another; you just had to look in the right direction. Here, wireless was at a considerable disadvantage, because devices based on Hertz's original oscillator were heavily damped. As a result they had what would later be called a very wide bandwidth—which meant, though this was not well-understood at the time, that you could not separate one signal from another.

The Hertz oscillator, so useful for scientific discovery, was clearly useless for practical communication. Yet all but one of the competing systems in the early 1900s were based directly on Hertz's device. The sole exception was Marconi's. He had developed a method to narrow the bandwidth, in effect to tune the oscillator. This meant that messages sent at different frequencies did not swamp one another, and (of equal importance at the time) the ability to choose a specific frequency held out the hope of ensuring military secrecy and of communicating at sea. Only those who knew the right frequency could hear the message (at least until simple methods of receiver tuning were developed).

Marconi and Fleming held a public demonstration in London to exhibit the virtues of their system. It nearly failed because of one of the earliest instances of industrial sabotage. One of their rivals, knowing the time and location of the demonstration, swamped the delicate tuning of the Marconi system by transmitting a wide-bandwidth signal sent from a typical Hertzian oscillator. The transmission ceased just moments before the true message was sent from another Marconi device, which prevented public embarrassment, but only because the saboteur's timing was off. Now one might say, how unfair. But fairness was hardly the issue. The failed sabotage actually demonstrated something quite important. Namely, that Marconi's new system was a good one only in a world where Marconi-like systems excluded all others. If the world also had noisy, wide-bandwidth radiators, then Marconi's device would be utterly useless. What eventually happened was that the noisy devices were everywhere legislated out of existence, and only narrow-bandwidth oscillators were permitted. Since even in wartime the benefits of wireless communication were great for each side, no one thereafter attempted to block transmissions by swamping *all* the airwaves.

Here, we see just how complex the path from discovery to technology can be, even when it seems that the original discovery was quite complete, as Hertz's did indeed seem to be at the time. First of all, the very nature of Hertz's device blocked its exploitation for over a decade. But second, the eventual success that Marconi achieved was not simply fixing a problem with the original discovery. Not at all. Marconi had literally to forge an entirely new technological world in which only devices of his particular kind were allowed to exist. Far from just developing a new technology to fill an uncertain market niche, Marconi had to persuade governments, and through them industry, to build an exclusive realm uniquely controlled and defined by Marconi-like devices. Since in the first decade or so only Marconi knew how to build such things, his company had an extraordinary monopoly, one that only the First World War and rapidly growing competition eventually destroyed.

Even in this case, where it seemed that we had a nearly pure instance of the application of a new scientific discovery to industry, the situation was vastly more complicated, involving as it did major departures from the original science in direct connection with economics, society, and government. Let us turn next to a different example, from much earlier in the nineteenth century, where the situation seems rather the reverse, one in which new science emerged from new industry, to see what lessons we can find there.

12.4 Fraunhofer's Glass

In the years after the final defeat of Napoleon in 1815, Britain thoroughly secured its dominance over the rapidly evolving industrial world. Its factories produced textiles, munitions, and soon other products, some of which had never existed before. British imperial dominance ensured markets for its products, with the notable exception of the former American colonies, which, in New England, were themselves rapidly industrializing, while the American South was mired in the horrors of slavery, with the inevitable consequence that the South remained a producer principally of raw products for the New England and British mills. Germany was not yet a unified country, consisting of 300 different principalities linked to one another in a common customs union, prefiguring the European Common Market. Industrial plants were beginning to develop in some of these principalities, which however faced the competitive colossus of Great Britain.

In circumstances like these, the best tactics for success are not to challenge the colossus on well-established grounds unless one has truly significant competitive advantages in quality or price. Since labor in Germany was if anything more costly than in Britain, price competition seemed unlikely to succeed. Quality was another matter. Here a man named Joseph Fraunhofer decided to challenge Britain in an area of comparatively minor economic, but great strategic and scientific importance: the manufacture of the finest optical glass, in which Britain had for a century been predominant.

Glassmaking had always been much more of an art than a science; it depended on the skilled craftsmen's knowledge of the right temperatures, the right times, and ways to mix, what proportions to add in and when, and the appropriate cooling procedure. The chemical properties of glass mixes were almost entirely mysterious. Large-scale industrial production depended to a high degree on methods for controlling and replicating craft knowledge, but even so, in Britain glass foundries were beginning to look more like factories than like sites of traditional craftwork.

The young Joseph Fraunhofer, who had the support of the Bavarian government through personal connections, decided to challenge England. To do so, he produced a unique combination of craftwork, industrial production, and scientific acumen. Fraunhofer took advantage of existing craft skills in Bavaria by building his foundry in a secularized monastery, which had been expropriated under Napoleon. The generous size of the location, the monastic tradition of dedicated craft labor, and the very layout of the monastery, which was congenial to the keeping of trade secrets, suited Fraunhofer's plans perfectly. The foundry he constructed bore little resemblance to contemporary British ones: it was much smaller, more conducive to craftsman-like control of the processes (Fig. 12.5).

In this unusual environment, Fraunhofer and his workers rapidly developed new methods for producing extremely high-quality optical glass. But there was a problem: how could you convince buyers that this new glass was vastly better than any other—especially English—glass? There was no accurate, easily repeatable way to test optical quality at the time. The usual method was to examine the refraction of

Fig. 12.5 Fraunhofer (Credit: http://www.wikiwand.com/de/Joseph_von_Fraunhofer)

the glass, but one long-standing problem here was that all glass is dispersive, that is it produces a rainbow of colors from white light. Consequently, in order to specify quality, you had to decide just what part of the spectrum to measure, and this meant very high inaccuracies.

It is exactly here that scientific discovery merged with new technology. In looking hard for ways to market his glass by examining its refraction at a very high level of detail, Fraunhofer discovered what were soon thereafter named the "Fraunhofer lines" in the solar spectrum. These were to become major sources of scientific interest during succeeding decades as physicists linked them to atomic processes. But this was not what Fraunhofer used them for, though he was pleased eventually to be accorded the status of a research scientist as well as a producer of very good glass. His goal had been to find ways to market the new glass, and the lines were exactly what he needed. What he did was to use the lines as markers in the spectrum, and to measure the refraction of the glass at each of them. That way he had a precise method for specifying the quality of his glass—and a way to force his English competitors to compete on his own grounds, for now they had to be able to produce and to measure the spectral lines just as Fraunhofer had. That alone was not an altogether easy thing to do until Fraunhofer himself invented the diffraction grating. Before then, anyone wanting to compete with Fraunhofer had to master the comparatively intricate technique that he himself described in print. If you could not do it properly, then you could not compete. Much like Marconi many decades later, Fraunhofer created his own technical world by forcing everyone else to use the very procedures he had himself invented in order to compete.

And compete they did, or at least they tried. English scientists visited Fraunhofer's factory, as much of it as he allowed them to see. They took back to England specimens of his optical glass, and then they tried to reverse-engineer it by chemically analyzing its composition. The finest English chemist of the day, who was soon to make fundamental discoveries in electromagnetism, Michael Faraday, tried hard to understand Fraunhofer's glass, and he was advised by England's greatest optical scientist, John Herschel (Fig. 12.6). Unfortunately for the English, they failed utterly. Fraunhofer's clever use of craftsmen's knowledge, which he refused to divulge, preserved his control of optical glass, which remained for a century and a half a German specialty and near monopoly.

The discovery of the spectral lines, then, originated in Fraunhofer's desire to compete in a market controlled by the English. Yet as Fraunhofer continued his work, he became profoundly involved in new scientific research that did not have clear industrial application; he was among the first to probe the implications of the new wave theory of light for instruments, for this was how he came to invent the diffraction grating. In Fraunhofer's world, the links between new science and new technology were so tight that it is probably pointless to distinguish between the two areas.

We have looked at several past technologies and sciences, at their mutual connections, and at some of the motivations and work that went into producing them. We will close with a different kind of story, one of science and moral virtue. This tale takes us back to Germany in the third quarter of the nineteenth century, shortly after the region's many principalities had been unified. Berlin had become the

Fig. 12.6 Michael Faraday (*left*) and John Herschel (Credit: Faraday (*left*) by Thomas Phillips from https://en.wikipedia.org/wiki/Michael_Faraday#/media/File:M_Faraday_Th_Phillips_oil_1842.jpg; Herschel (*right*): by Alfred Edward Chalon from https://upload.wikimedia.org/wikipedia/commons/7/74/John_Herschel00.jpg)

directing center of the new country, its university the most powerful and productive one of all. German science had by this time begun to surpass its British and French competitors at nearly every level; German training methods, based on the apprenticeship of the graduate student to a doctoral supervisor, had become an engine producing highly competitive, intensely committed researchers. In fact, by the end of the century this German system became the model for university training in the United States and remains with us to this day.

12.5 Helmholtz's Cosmopolitanism

The new rector of the University of Berlin was Hermann von Helmholtz (Fig. 12.7), under whom the young Hertz would shortly apprentice. Helmholtz was one of the discoverers of the principle of energy conservation; he had created the first mathematics for fluid vorticity, had produced the trichromatic theory of color vision, and had completely renovated the study of sound. At just about the time that he became rector at Berlin, he had also developed an entirely new form of electrodynamics, which he taught to Hertz. Any one of these achievements would likely have won him a Nobel Prize today, though, like Hertz, he died several years before the prize was created.

Decidedly Germanic in many ways, Helmholtz was thoroughly cosmopolitan when it came to science. He was an internationalist and, more than that, had a pro-

Fig. 12.7 Hermann von Helmholtz (Credit: http://www.wissen.de/lexikon/helmholtz-hermann-ludwig-ferdinand-von)

found belief in the virtues of free investigation, unconstrained by ideologies or religious beliefs. Several years before his rectorship, Helmholtz had translated into German an extraordinary textbook of the day, the *Treatise on Natural Philosophy*, which had been co-authored by one of the most famous Scottish scientists of the time, his close friend William Thomson, who became Lord Kelvin, and Peter Guthrie Tait. This admiration for a foreign product did not sit well with a number of Helmholtz's German colleagues, who were increasingly steeped in the poisonous atmosphere of xenophobia that would eventually send Germany, and very nearly the world, to utter destruction.

Among these colleagues was a scientist by the name of Friedrich Zöllner, who had not long before invented the first photometer that could be used to produce reliable values for stellar magnitudes. Zöllner accused Helmholtz of propagating un-Germanic science because of his dealings with and friendship for the British, in particular Thomson, whom Zöllner thought to be addicted to the crudest of materialistic beliefs, whereas he, Zöllner (and he was hardly alone in this), was certain that the world was guided by a uniquely Germanic spirit. "Judging from what [Zöllner] aims at as his ultimate object," Helmholtz scornfully remarked,

> it comes to the same thing as [the philosopher Arthur] Schopenhauer's Metaphysics. The stars are to "love and hate one another, feel pleasure and displease, and to try to move in a way corresponding to their feelings." Indeed, in blurred imitation of the principle of Least Action, Schopenhauer's Pessimism, which declares the world to be indeed the best of possible worlds, but worse than none at all, is formulated as an ostensibly generally

> applicable principle of the smallest amount of discomfort, and this is proclaimed as the highest law of the world, living as well as lifeless.[8]

So much for Zöllner's metaphysically based science. But that was not the only target of Helmholtz's disdain. From the point of view of someone like Zöllner, no one should pursue science except according to Germanic ideological principles. He was a vocal and influential enemy of academic freedom, on whose remarks the Nazis would draw decades later. This disgusted and dismayed Helmholtz. He replied in his inaugural address as rector at Berlin on that very topic. Helmholtz praised the great freedom of the German university, where "the most extreme consequences of materialistic metaphysics, the boldest speculations upon the basis of Darwin's theory of evolution, may be taught with as little restraint as the most extreme" pursuit of religious belief. Where, he continued, though "it is forbidden to suspect motives or indulge in abuse of the personal qualities of our opponents, nevertheless there is no obstacle to the discussion of a scientific question in a scientific spirit."

The free pursuit of scientific research was for Helmholtz a model for intellectual freedom, and a model as well for a tolerant and a moral society. Germany abandoned that in the 1930s, when biology was deformed by racism and when it became expedient to think that there was such a thing as "German physics." The statue honouring Heinrich Hertz at the University of Karlsruhe, where he had first produced electric waves, was thrown out because he had Jewish ancestry, while Hertz's own assistant at Bonn, Philip Lenard, who had won the Nobel Prize in 1905 for his work on cathode rays, denounced his mentor and extolled the virtues of a purified Germanic physics.

The United States welcomed many of those who fled Nazi Germany, though it was not welcoming enough. Science and scholarship in the United States achieved their heights as a direct result of this forced European flight, heights that might never have otherwise been reached. In the decades since World War II, science and technology in this country have benefited in extraordinary ways because so many from around the world have come to study and often to stay here. Let us hope that the present climate of wariness, fear, intolerance and, not least, Congressional disdain for scientific and humanistic truth dissipates and that the freedom which Helmholtz so admired will in time be embraced even by those who would silence their opponents.

---

[8] H. Helmholtz, "Helmholtz on the Use and Abuse of the Deductive Method in Physical Science," trans. Crum Brown, *Nature* **11** (1874), 149–51.

Chapter 13
Fishing an Extreme Environment: Science, Sovereignty and Hudson Bay

Jennifer Hubbard

Was a deep-sea commercial fishery possible in Hudson Bay? This question brought the Dominion Government of Canada, through the Department of Marine and Fisheries, to sponsor several mostly-forgotten expeditions beginning in the 1880s to gauge its fishery potential. Logically, fisheries should have been possible, given the importance of many fisheries in far northern waters–the groundfish fishery in Barents Sea north of Norway, for example, and the whale fishery in northern Hudson Bay. Sea fisheries had apparently not developed yet in Hudson Bay due to its inaccessibility, the use of only small craft for near-shore fishing, and the climate's severity. Or did the lack of a fishery reflect a real absence of fish? The Canadian government's queries were perplexing, however, given the challenges in fishing Hudson Bay's ice-shrouded waters. Its interest in the region had been so scanty that parts of northern Hudson Bay had not yet been charted. Also, the prolific east-coast fishery already harvested too many fish for domestic markets, and faced depressed prices in international markets. Toronto and other inland markets were well-supplied by Great Lake fish. Simply put, Canada did not require more fish.

Other interests, then, drove the Department of Marine and Fisheries to fund the scientific surveys of Hudson Strait and Bay. Indeed, Canadian bureaucrats and politicians were responding to potential Arctic territorial claims from Arctic science expeditions sponsored by other countries. Scientific research had become fundamental to the economic justification of sovereignty claims. As will be seen, support for scientific expeditions waxed and waned according to fluctuating sovereignty and economic interests. Another factor was the new technology that eased attempts to close the northern frontier. Prior to 1913 Hudson Bay was navigated only by sail-powered ships, some with auxiliary engines. By 1930, ships with powerful engines were equipped with echo-sounders, and rugged air planes were used for scientific surveys.

J. Hubbard (✉)
Ryerson University, Toronto, ON, Canada
e-mail: jhubbard@arts.ryerson.ca

J. Buchwald, L. Stewart (eds.), *The Romance of Science: Essays in Honour of Trevor H. Levere*, Archimedes 52, DOI 10.1007/978-3-319-58436-2_13

Several expeditions between 1880 and 1914 had elements of fisheries research, but a real understanding of Hudson Bay's unique biological profile did not emerge until scientists of the Biological Board of Canada joined the quest to survey Hudson Bay's fish resources. Lacking Arctic experts, in 1920 the Board turned to Danish biologist Frits Johansen, a veteran of the Canadian Arctic Expedition of 1913–1916, who shared explorer Vilhjammer Stefansson's belief that Nordic peoples had special arctic expertise. Johansen's three Hudson Bay excursions, however, were limited in scope and usefulness. The Hudson Bay and Strait Fishery Expedition of 1930, completely directed by Biological Board scientists, finally generated answers by combining fishing trials with oceanographic surveys. Oceanographer Harry B. Hachey and fisheries biologist A.G. Huntsman used European oceanographic and fisheries research methods to analyse Hudson Bay's unique oceanography and its influence on fish production. These forever debunked analogy-driven beliefs–based on extensive fisheries in other Arctic oceans–about Hudson Bay. Canadian sovereignty, in turn, was assisted by this increased scientific understanding.

13.1 Background Issues: Environmental History of Hudson Bay and International Law

Hudson Bay was formed by glacial depression of Precambrian Shield volcanic rock during several glaciations in the past million years. Early in the Holocene warm period, the Laurentide Ice Sheet dammed the northeast, creating two vast glacial lakes: Lake Agassiz and Lake Ojibway. The ice dam collapsed around 8500 years ago; freshwater flooded through Hudson Strait. Salty water flowing back through Hudson Strait created the Tyrrell Sea 7800 years ago. Isostatic rebound drained the lowlands, leaving Hudson and James Bays.[1]

The Hudson Bay ecosystem is shaped by Atlantic waters entering through Hudson Strait, setting up cyclonic circulation within the bay, and by fresh-water from its watershed. A huge watershed reaching to the Alberta-British Columbia continental divide, Baffin Island in the north and North Dakota and Minnesota in the south, drains into Hudson Bay. Not surprisingly, its floor is formed of deep alluvial silt. The water inflow creates a thick layer of brackish water that overlies denser, saline Atlantic waters. Hudson Bay's stable vertical salinity profile is one of its most salient features. Sodden peat bogs surround James Bay and southwestern Hudson Bay. Further north, muskeg (poorly-drained land overlying permafrost) stunts tree growth. The tree line, which roughly denotes the Arctic,[2] runs near the Manitoba-Nunavut border on the west coast, but much further south to the northern edge of

[1] D.B. Stewart and W.L. Lockhart, "An Overview of the Hudson Bay Marine Ecosystem', Chapter 2 'Ecological Overview," Canadian Technical Report of Fisheries and Aquatic Sciences No. **2586** (2005), 2, 7–8, http://www.dfo-mpo.gc.ca/libraries-bibliotheques/toc-tdm/314704-eng.htm

[2] Trevor H. Levere, *Science and the Canadian Arctic: A Century of Exploration 1818–1918* (Cambridge: Cambridge University Press, 1993), 12.

James Bay in the east. Dense boreal forests only emerge far inland, away from the bays' chilling influence.

Hudson's Bay Company officers were the first to collect and in some cases describe specimens of Hudson Bay's wildlife. James Isham's unpublished "Observations on Hudson's Bay" (1743) offered the region's first natural history.[3] The Hudson's Bay Company in the 1800s supported "scientific collecting and observing", especially helping the Smithsonian Institution.[4] Company officers sent specimens to the Smithsonian, and offered trading post facilities, transportation, and other logistical support to its naturalists.[5] The region's varied wildlife includes 118 species of migratory seabirds and raptors, which feed and breed along the tall cliffs on the eastern shores or in the bountiful salt marshes, eel grass meadows and tidal flats in the south and southwest.[6]

Naturalists found familiar northern molluscs, barnacles, sea anemones, amphipods and sea squirts in the intertidal zone. Beyond the ice scour zone they dredged up benthic (bottom dwelling) molluscs, sea urchins, starfish, annelid worms, sea spiders, crabs, shrimps, hydroids, and sea sponges. Pelagic species included comb jellies, copepods, sea butterflies, and about 60 species of fish, including Arctic cod, plaice, capelin, whitefish, and walleye.[7] In 1771, however, when the Hudson Bay Company spent £20,000 (equal to over £3.3 million in 2016) to bring in Spitsbergen fishermen to establish a blue water fishery, only three fish were caught.[8]

The Thule (1000–1500 AD) were the first bowhead whale hunters.[9] A defining feature of Hudson Bay is the "southerly presence of Arctic marine mammals", including the migratory beluga whales often found near James Bay, Arctic bowhead and narwhal, and occasional minke, northern bottlenose and commercially-important sperm whales. Killer whales are drawn by walruses and native and migrating seals.[10] Whaling became the only commercial fishery when American and Scottish whalers began hunting bowhead whales near Southampton Island in the

[3] Ted Binnema, *Enlightened Zeal: The Hudson's Bay Company and Scientific Networks, 1670–1870* (Toronto: University of Toronto Press, 2014), 66.

[4] Levere, *Science and the Canadian Arctic*, 100, 344–45.

[5] Ibid., 345–55; Binnema, *Enlightened Zeal*, 238–75.

[6] 'Canadian Biodiversity: Hudson Plains', http://canadianbiodiversity.mcgill.ca/english/ ecozones/hudsonplains/hudsonplains.htm

[7] D.B. Stewart and W.L. Lockhart, "Summary of the Hudson Bay Marine Ecosystem Overview: Prepared by Arctic Biological Consultants, Winnipeg, for Canada Department of Fisheries and Oceans, Winnipeg, MB" (2004), Draft vi + 66,14-16, http://www.dfo-mpo.gc.ca/Library/280328.pdf

[8] A.G. Huntsman, 'Fishery prospects for Hudson and James Bay', *Canadian Fisherman* 5:8 (1918): 896–8; Letter from W.A. Found to B.M. Stitt, 27 May 1932. Library and Archives Canada, RG 23 Vol. 1206 File 726-4-5 [3].

[9] Stewart and Lockhart, "Summary of the Hudson Bay Marine Ecosystem Overview," 26.

[10] Ibid., 26, 17.

1850s.[11] Along with Norwegian whalers they completed over 150 whaling voyages by 1915.[12]

Hudson Bay was only sparsely settled. American whalers' small settlements on Southampton Island and at Chesterfield Inlet were inhabited by the Inuit who worked for them. "Major" centres grew around Hudson's Bay Company trading posts–Churchill, York Factory, Fort Severn, Fort Albany and Fort George–with populations ranging from a few dozen to above one hundred people. They relied on mail packets sent via dog sleds in the winter, or boats by summer (along the Hayes River south through Lake Winnipeg to the Red River Settlement), to communicate with the outside world.[13] Visiting ships also offered sporadic mail service.

Inaccessibility and a dearth of economic interests minimized Canada's interest in regions still dominated by London-based Hudson's Bay Company. The British government, having claimed the Arctic Archipelago, lost interest when the Northwest Passage failed to materialize, especially after the Franklin Expedition of 1845. Only possible competing claims by the United States or other nations spurred their attempts to consolidate sovereignty over the region.

How was Arctic sovereignty to be recognized? International interest in claiming polar territories was very recent. With no relevant case law (the earliest cases settled by international courts began in the 1920s) sovereignty was difficult to validate. Established international law was a series of conventions triggered by sixteenth-century European expansion. Legal scholars justified conquest based not on "what nations ought ideally to do, but rather what they were able to do without arousing active opposition from other states."[14] Europeans justified appropriating land from aboriginal peoples of North America, Australia and large parts of Africa by arguing their land was not effectively occupied since it was not in continuous possession or use. Farming constituted occupation; therefore "uncultivated land could legitimately be taken by those who would farm it."[15] While native North Americans planted maize, squash and beans together, this was not seen as farming since the small plots were set in unploughed vaguely cleared areas, and lacked fences–a sign of ownership to Europeans.[16] Ironically, we now know their "insufficiently intensive"

---

[11] Randall R. Reeves and Susan E. Cosens, "Historical Population Characteristics of Bowhead Whales (*Balaena mysticetus*) in Hudson Bay," *ARCTIC* **56** (2003), 283–292; Shirley Tagalik, "The History of Whaling," accessed March 8, 2016, http://www.inuitcontact.ca/index.php?/vignettes_en/history_whaling

[12] Keith J. Crowe, "Whaling and Eskimos: Hudson Bay 1860–1951. By W. Gillies Ross," *ARCTIC* **29** (1976), 124–25.

[13] Lorraine Brandson, "Precious Cargo! Mail packets in the Hudson Bay region," Parks Canada Wapusk News **5** (2012). Accessed at http://www.pc.gc.ca/eng/pn-np/mb/wapusk/ne/ne1/ne1_2012/ne1_ 2012_6.aspx.

[14] Janice Cavell and Jeff Noakes, *Acts of Occupation: Canada and Arctic Sovereignty, 1918–25* (Toronto: University of British Columbia Press, 2010), 63.

[15] Ibid., 65–6.

[16] Charles C. Mann, *1493: Uncovering the New World Columbus Created* (New York: Alfred A Knopf, 2011), 46–47.

practices were probably more productive per acre than European farming at the time.[17]

European scholars argued that territory could be acquired "by the first discoverer in the name of his country."[18] In practice European nations claimed new territory through discovery and symbolic acts of possession, but refused to acknowledge other nations' sovereignty claims unless they actually settled and used the land. By 1900 recognition of sovereignty required "acts of political administration" which involved maintaining local official establishments and authorities to uphold order. This posed definite challenges when applied to empty or sparsely populated polar territories. Ungoverned territories, like much of the Arctic Archipelago, were now legally termed *terra nullius*. Under international law discoverers of a new land were given "inchoate title" with limited rights.[19] If the discovering country made no later actions, after many years (an undefined period) other nations would be free to occupy and settle the land.

Another problem existed for Canada. Nineteenth-century customs regarding freedom of the seas recognized nations' sovereignty as extending to three miles beyond the coastline. Later, the 1958 United Nations Convention on the Law of the Sea would allow nations to claim waters within a bay where a straight baseline between the headlands did not exceed 15 miles.[20] Thus in 1900 Canada could not claim even the 40-mile wide Bay of Fundy. This did not prevent the Dominion government from declaring sovereignty over Hudson Bay in the 1914 Fisheries Act,[21] but such proclamations would hold no water in the face of international objections.

With no immediate means to establish sufficient centres of local authority, the Dominion government turned to scientific expeditions as a proxy. Expedition leaders were enjoined to proclaim Canadian sovereignty, while conducting hydrographic, geological, mineralogical and fisheries surveys. Such research, besides constituting a Canadian presence, established Canadian expertise concerning the Arctic Archipelago and signalled the intent to use these regions.

[17] Jane Mt. Pleasant, "The Paradox of Plows and Productivity: An Agronomic Comparison of Cereal Grain Production under Iroquois Hoe Culture and European Plow Culture in the Seventeenth and Eighteenth Centuries," *Agricultural History* **85** (Fall, 2011), 460–92.

[18] Cavell and Noakes, *Acts of Occupation*, 64.

[19] Cavell and Noakes, *Acts of Occupation*, 65–6.

[20] United Nations, 'Articles concerning the Law of the Sea with commentaries', *Yearbook of the International Law Commission, 1956*, Vol. II., 268. Accessed at http://legal.un.org/ilc/texts/instruments/english/commentaries/8_1_8_2_1956.pdf.

[21] "The Value of Hudson Bay," *The Globe*, Toronto, 23 October 1930.

13.2 Scientific Expeditions and Arctic Sovereignty

While the British government had turned away from the Canadian Arctic after the doomed Franklin Expedition, internationally scientific and economic interest in the Arctic grew considerably around 1870. Their respective governments strongly supported German North Polar and Greenland expeditions, the Austro-Hungarian expedition that discovered Franz Joseph Land in 1872–1874, and A.E. Nordenskiöld's Swedish geological expeditions to Spitsbergen and his navigation of the Northeast Passage in 1878–1879. British interest only reawakened in 1874 when William A. Mintzer of the U.S. Navy Corps of Engineers requested a mining land-grant on Baffin Island. Assisting in the search for missing 1871 US *Polaris* expedition crew, he had noted rich mica deposits. Heat-resistant mica was popular as a glass-substitute in kerosene lamps.[22]

The British government officially ignored Mintzer, but the British Admiralty internally confessed a "very imperfect" knowledge of the region which hampered the official response.[23] Mintzer's request set in motion efforts to confirm British ownership, even though Britain had no desire to administer the Arctic Archipelago, which had no known value.[24] The practical solution was to transfer ownership to Canada, while hiding from Mintzer and the United States the British government's internal confusion. However, the clerk of the Queen's Privy Council insisted that Canada's Parliament give its sanction, since "the acquisition of this additional territory…will entail a charge upon the revenue of the country."[25]

The other response, according to Trevor Levere, was that Mintzer's formal request piqued the government into supporting Arctic scientific exploration after stonily ignoring years of lobbying by the British scientific community. These promoters included Professor William Carpenter (1813–1885), President of the British Association for the Advancement of Science.[26] Carpenter had organized and participated in the 1868 *Lightning* Expedition to dredge for marine life in the ocean depths, and organized the then-ongoing *Challenger* Oceanographic Expedition of 1872–1876, the first large-scale expeditions focused on oceanography..[27]

The British Arctic Expedition of 1875–1876 to northern Greenland and Ellesmere Island therefore included marine science in its broad scope. Commanded by Sir George Nares (1831–1915), *Alert*, a refitted sloop strengthened for ice-breaking, and *Discovery*, a former whaling ship, were used to survey and chart coastlines,

---

[22] S.L. Osborne, *In the Shadow of the Pole: An Early History of Arctic Expeditions, 1871–1912* (Toronto: Dundurn, 2013), 32.

[23] Osborne, *In the Shadow of the Pole*, 32–33.

[24] Osborne, *In the Shadow of the Pole*, 35.

[25] Levere, *Science and the Canadian Arctic*, 263; Osborne, *In the Shadow of the Pole*, 35, 41–2, 37.

[26] Levere, *Science and the Canadian Arctic*, 263.

[27] See Margaret Deacon, *Scientists and the Sea, 1650–1900* (Aldershot: Ashgate, 1971, 1997), 306–365; Helen Razwadowski, Introduction to "Focus: Knowing the Ocean: A Role for the History of Science," *Isis* **105** (2014), 335–337; and Michael S. Reidy and Helen M. Rozwadowski, "The Spaces In Between: Science, Ocean, Empire," *Isis* **105** (2014), 338–351.

measure tides, ocean currents, temperatures and icebergs, and sample marine life. Ornithologist Henry Wemyss Feilden and ship's surgeon and botanist Edward Lawton Moss collected plant, bird and fish specimens, and dredged up molluscs, "corals and other unlikely species".[28] Feilden witnessed myriad birds feeding on abundant marine life in Smith Sound and Melville Bay. Further north, fewer birds and plants were seen, but marine life proliferated: many molluscs, star-fishes, and crustaceans were dredged up; mid-water sampling collected sea butterflies and crustaceans. Above Baffin Island they collected ten fish species also found in the Barents and Bering Seas including *Salmo naresii* (named after Nares)[29]–in fact a local variation of *Salmo salar*. *Discovery* and *Alert*, overwintering on Ellesmere's Lady Franklin Bay and northeast tip respectively, established a British presence.[30]

This expedition did little to advance British Arctic sovereignty, however. In 1876 Mintzer once again ran mining operations at 'Mintzerville' on Baffin Island without permission, thus making an unofficial claim to the land. The Dominion government ignored Mintzer despite British dismay. When the US Army sponsored the Howgate Polar Colonization Preliminary Arctic Expedition to Lady Franklin Bay in 1877, some Ottawa politicians rejected British demands that Canada accept sovereignty, seeing no benefit, only responsibility.[31] Not until September 1, 1880 did Canada take control of all remaining British North American possessions and territories through an Order in Council.[32]

Canada's economic interest in Hudson Bay came later–sparked by problems arising from the 1879 federal National Policy programme, which had resulted in a railway spanning Canada. The Canadian Pacific Railway used its monopoly to charge exorbitant freight rates for transporting grains and other goods through the Prairie Provinces to Montreal. Frustrated western farmers in 1884 called for a new railway from The Pas, Manitoba to Hudson Bay to provide an alternative, cheaper–and at about 365 miles shorter, also faster–shipping route to England through Hudson Strait. Prime Minister John A. Macdonald supported the idea. A railway to Hudson Bay would expand Canada's dominion to its northern coast and improve western access to European markets. Furthermore, he argued, opening up this route "will be exceedingly valuable on account of the known mineral wealth and the wealth of the fisheries."[33]

Macdonald's allusion to the 'wealth of the fisheries' could not have meant the small existing estuarine or blue-water whale fisheries. It must instead have expressed the hope that exceedingly productive fisheries, like those of the Barents Sea and around Iceland, would develop. Just a year earlier, in 1883, a huge sockeye salmon fishery had opened up in Bristol Bay, Alaska. The potential also existed, as the

---

[28] Levere, *Science and the Canadian Arctic*, 266–71, 273, 278.

[29] Ibid., 303–4.

[30] During the Cold War the Canadian forces established Alert, a permanent scientific and military station, near location where HMS *Alert* overwintered.

[31] Osborne, *In the Shadow of the Pole*, 41–2.

[32] Ibid., 47.

[33] Ibid., 77, 53–4.

British Arctic Expedition had discovered, in Davis Strait and Baffin Bay. A Hudson Bay commercial sea fishery therefore seemed reasonable.

Macdonald's commercial visions were backed by Dr. Robert Bell (1841–1917), assistant director of the Geological Survey of Canada. In February 1884 he recommended this "easy and cheap route to the markets of the world" to a House of Commons "Navigation of Hudson Bay" committee. While most sailors reported only two to three months of clear sailing in Hudson Strait, Bell's own experience suggested a long ice-free season. The committee later favourably reported that, since 1610, around 800 commercial, military and passenger vessels had crossed Hudson Bay safely "without marine charts, without an accurate knowledge of these waters, without a system of telegraphic communication, and without the aid of steam."[34]

13.3 Canadian Hudson Bay and Strait Expeditions 1884–1904

The "Navigation of Hudson Bay" report convinced the House of Commons to sponsor a series of sea-going expeditions, all having scientific aspects, under the aegis of the Department of Marine and Fisheries. The first, carried out from 1884 to 1886, had a meteorological and geological focus. Later expeditions also assessed Hudson Bay's potential for fishing, mining and as a transportation corridor, and shared a mandate to affirm Canadian sovereignty. Trevor Levere has noted that "science and colonialism, imperialism and nationalism provide recurring themes" in polar exploration, as do "the interplay of science and internationalism."[35] Canada's early Arctic efforts were largely national and nationalistic. Other nations' similarly nationalistic expeditions included the Norwegian *Fram* expedition (1893–1896) headed by Fridtjof Nansen (1861–1930), who locked *Fram* in polar ice in an attempt to drift to the North Pole. Nansen and other leading Arctic explorers won lasting fame because of the insights gained (Nansen predicted the Polar ocean), but even more because of risks they undertook for national glory. Canada's Hudson Bay expeditions were also risky: several participants died during overwinter investigations, since northern Hudson Bay remained as remote as Ellesmere Island. Their aims, however, were economic and prosaic, hence their relative obscurity. Despite this economic focus, however, fishery investigations remained cursory at best. Science related to navigation had a higher priority, but another problem was that nobody in Canada knew how to survey of deep-sea fish populations. This only changed in 1914–1915, when Johan Hjort (1869–1948), the eminent Norwegian fisheries biologist, led the

[34] Ibid., 55; Binnema, *Enlightened Zeal,* 358; and *Report of the Committee of the House of Commons to Enquire into the Question of the Navigation of Hudson's Bay*, 1884, quoted in Osborne, *In the Shadow of the Pole,* 55.

[35] Levere, *Science and the Canadian Arctic*, 8.

Canadian Fisheries Expedition and trained Canadians in methods he and other European scientists were pioneering (see below).[36]

The Hudson Bay and Strait Expeditions of 1884, 1885 and 1886, to determine the navigation season, were commanded by meteorologist and hydrographer Lieutenant Andrew Robertson Gordon (1851–1893). Robert Bell acted as medical officer and geologist. He also collected over 50 specimens of insects, mammals, birds, and fish, and dredged molluscs from Hudson Bay.[37] The 1884 expedition was to construct six meteorological stations on opposite sides of Hudson Strait, at the Atlantic entrance (ice fields blocked one site), the Hudson Bay entrance, and at the mid-point.[38] Three young meteorologists manned each station in overwinter shifts, recording daily tides, temperatures, weather and ice conditions. Stupart Bay's extra observer, Robert Stupart, took continuous magnetic observations. The Ashe Inlet station, left intact, was reused by the Canadian Geological Survey during an 1897 expedition (see below).[39]

In 1884 the department chartered the three-masted sealing ship *SS Neptune* from Job Bros. of Newfoundland (supplier of many ships for later expeditions). The British Navy's much-less-malodorous HMS *Alert* was borrowed for the 1885 and 1886 expeditions. As with every later expedition, they voyaged from Halifax, Nova Scotia, the nearest large Canadian port.[40] In 1884 *Neptune* arrived in early August, the usual time Hudson's Bay Company ships entered the Hudson Strait. To find the earliest date for safe passage, in 1885, *Alert* attempted entry on June 15th, but suffered considerable ice damage. After repairs she returned on August 4th, encountering two Company ships struggling through ice. In 1886 *Alert* entered the strait on July 9th, but took 11 days to ram through field ice within its 200 mile length, and needed repairs before traversing Hudson Bay in a mere four days.[41]

Gordon's *Report on the Hudson's Bay Expedition* (1886) made several recommendations. Hudson's Bay Company rendered blubber to make oil for British markets, and Canada should also. While the Company paid Canada to fish in Hudson

[36] Jennifer Hubbard, "Johan Hjort: The Canadian Fisheries Expedition, International Scientific Networks, and the Challenge of Modernization" *ICES Journal of Marine Science* **71** (2014), 2000–2007.

[37] This expedition, like many that followed, also was required to make ethnographic observations and collect Inuit artifacts, but this dimension of their work is not discussed here.

[38] Osborne, *In the Shadow of the Pole*, 81.

[39] N.B. McLean, *Report of the Hudson Strait Expedition 1927–28* (Ottawa: F.A. Acland, 1929), 7–9. Port De Boucherville became the location of a Canadian tide station; Port Burwell in the 1920s became the site of a Royal Canadian Mounted Police detachment, and the settlement of Killiniq grew around it, but the residents were resettled in 1978.

[40] Halifax was the closest port to Hudson Strait prior to Newfoundland confederating with Canada in 1949.

[41] Osborne, *In the Shadow of the Pole*, 57–62, 70–1, 75–7. The Company got its first 'fully steam-powered ship for travel to Hudson Bay built to specification–the two-masted *Nascopie*' in 1912. See N. Hall, "Constructed Contexts: Ships as Workplaces," doingcanadianhistory N.0: Northern Arc: the Significance of Seafaring to Western Canadian History. Accessed at https://hallnjean.wordpress.com/ships/constructed-contexts-ships-as-workplaces/.

Bay, American whalers did not, and Gordon was concerned by lost taxes and tariffs. Of the two locations being considered for a railway terminus, he recommended Churchill: its basin was well-suited for deepwater port facilities; but at York Factory, the Nelson River's shallow estuary would need considerable and ongoing dredging. (*Alert* was forced to anchor 12 miles from shore.) Extensive surveys and charting were needed; Gordon had made the first hydrographic surveys around the two proposed ports. Treacherous shoals required beacons. He warned, moreover, that Hudson Strait was only navigable 4 months a year, and magnetic fluctuations and the north magnetic pole's proximity made compasses unreliable in western Hudson Bay.[42]

Gordon's report, together with declining wheat prices, scuttled the Hudson Bay railway project. It languished until 1896, when Wilfrid Laurier's new government supported a railway shipping route to strengthen federal ties and Dominion governance in the north. In 1897 his government provided the Department of Marine and Fisheries with $35,000 to study Hudson Bay's transportation, mining, and commercial fisheries' potential. The Minister, Sir Louis Davies, also ordered it "to take as formal possession of the country as possible, to plant the flag," and to admonish "natives and foreigners" to observe the laws, "particularly the customs laws of Canada." Davis wanted the expedition's ship to carry a small cannon to enable "firm" treatment of American whalers on eastern Baffin Island. (In fact the whalers here were mostly Scottish.) No evidence suggests, however, that *Diana*, a rented sealer, was so equipped.[43]

Davis put Dr. William Wakeham (1844–1915) in command. Born in Québec City, Wakeham had a medical degree from McGill, and had worked as a doctor in the Gaspé region and Québec City before seeking a career with the Department of Marine and Fisheries. Promoted to Inspector of Canadian Fisheries for Canada's east cost in 1879, he had commanded the fisheries patrol steamer *La Canadienne*, visiting remote coastal hamlets dispensing medical services, issuing fishing permits and inspecting fishing operations. He had served on the 1891–1896 "Joint Commission with Great Britain Relative to the Preservation of the Fisheries in Waters Contiguous to the United States and Canada.",[44] examining, with his American counterpart, Richard Rathbun (1852–1918), fisheries and pollution problems in boundary waters. The Boundary Waters Treaty of 1909 used their recommendations.[45]

---

[42] Osborne, *In the Shadow of the Pole*, 62–72, 86–89; Eric L. Mills, *The Fluid Envelope of Our Planet: How the Study of Ocean Currents became a Science* (Toronto: University of Toronto Press, 2009), 240.

[43] Osborne, *In the Shadow of the Pole*, 94–95,103–4.

[44] Ibid.,103.

[45] Joseph Gough, *Managing Canada's Fisheries: From Early Days to the Year 2000* (Sillery, Québec: Septentrion, 2006), 101–02. Rathbun was Curator of Marine Invertebrates for the Smithsonian Institution, but had earlier worked for the US Fish Commission. See Amy Ballard, "Richard Rathbun: The Smithsonian's Renaissance Man," http://nmnh.typepad.com/100years/2011/06/richard-rathbun-the-smithsonians-renaissance-man-part-1.html#sthash.wg1o9jSJ.dpuf.

Wakeham's fishery expertise and his crew of experienced whalers, sealers and fishermen would lead one to expect the 1897 expedition would do a fishing prospect in Hudson Bay. This was not the case. The scientific mission was secondary to proclaiming Canadian sovereignty, for which Wakeham had to leave Hudson Bay and sail *Diana* to Kekerten Island in Cumberland Sound. The site of a Scottish whaling station since 1857, it was mainly populated by Inuit, who processed the roughly 125 whales caught annually in return for supplies of tea, biscuits, tobacco and other commodities. On August 17, Wakeham assembled the settlement's approximately 140 men, women and children, and formally raised the Union Jack to demonstrate for the first time Canada's exclusive authority over Baffin Island and adjacent territories. Wakeham also declared Dominion rule at Uumanaqjuaq, on Blacklead Island, another Scottish whaling station.[46]

Wakeham had dropped off three Geological Survey members at sites on Hudson Strait and Bay: Robert Bell, Albert Peter Low, and George Albert Young. Low and his party conducted geological surveys at the Ungava Peninsula inlet that Wakeham named Douglas Harbour. Bell's party used the intact 1884–6 expedition base at Ashe Inlet. Both groups also collected fish specimens, but not systematically. They sent these to Professor Edward E. Prince, the Dominion Commissioner of Fisheries, for identification and analysis.[47] Near Churchill, on September 2nd, Wakeham used *Diana* to make the first official trawl-fishing attempt on western Hudson Bay, but fierce waves quickly ended the experiment.[48]*Diana* had to leave next day to pick up the geologists and convey them to St. Johns by October 1st, where they could telegraph the Geological Survey to prevent it from sending out relief expeditions.

About navigation conditions Wakeham had nothing good to say. At the end of June *Diana* entered Hudson Strait and became surrounded by pan ice. The crew used dynamite to keep her free, but on June 30th she was trapped; ice pressure squeezed her out of the water on the 31st. "Only Nansen's *Fram* was said to be more ice-worthy than *Diana*", a rebuilt former Hudson's Bay Company transport ship, but the crew nevertheless prepared to abandon ship. Fortunately, the ice shifted and released her. Her propeller now twisted and broken, it took two weeks to reach Hudson Bay; repairs took five more days. The Strait was clear from late July to mid-September, but in late September *Diana* battled thick snow storms to transport the geologists to St. Johns. Blizzards and heavy seas drove Wakeham away when he returned on October 16th to determine the end of the safe navigation season. His report that Hudson Strait navigation dates extended from, at best, mid-July to October 20th vanquished support for the Hudson Bay shipping route.[49]

[46] Ibid., 113–4.

[47] E.E. Prince, "Description of specimens of Sea-trout, Caplin, and Sturgeon from Hudson Bay," Report of the Sixty Seventh Meeting of the British Association for the Advancement of Science Held at Toronto in August 1897 (London: John Murray, 1898), 687–88; manuscript in Huntsman Collection, University of Toronto Archives, B1978-0010 Box 150 File 2.

[48] Osborne, *In the Shadow of the Pole*, 114–15.

[49] Ibid., 116–19, 103–4, 120–1.

Wakeham's flag-raising activities got almost no Canadian or international attention. Arctic sovereignty remained unresolved. The North Pole itself was an unknown quantity: was it land or an ice-sheathed ocean? Sixteenth-century Dutch blue-water sailors had told tall tales of sailing to the North Pole that probably gained credence because of polynias. These openings in otherwise thick Polar ice (the Bering Sea often has huge ones) are caused by upwelling warm water or by wind driving the ice pack away from the coast. Alexander Dallas Bache (1806–1867) of the US Coast Survey, Matthew Maury of the US Naval Observatory (1806–1873), and Louis Agassiz (1807–1873) of Harvard University all believed an open Arctic ocean was warmed by the Gulf Stream. A late adherent was George Washington DeLong, who discovered and claimed for America three islands north of Siberia during the 1879–1981 expedition that took his life.[50]

Lured by the unknown, explorers used scientific goals to draw official funding. The doomed Russian Polar Expedition (1900–1902) conducted hydrographic and geological research on New Siberian Islands (Novosibirskiye Ostrova) before disappearing in search of the mythical Sannikov Land to the north. In 1898, a *Fram* expedition veteran, Otto Sverdrup (1854–1930) sailed the refitted *Fram* to survey Greenland's northern coast. Blocked by ice, he turned west, and between 1898 and 1902 discovered the last large islands, Amund Ringnes, Ellef Ringnes, and Axel Heiberg, all the while making botanical, zoological and geological collections and observations.[51] In 1903, Roald Amundsen (1872–1928) re-mapped the shifting north magnetic pole to scientifically justify his attempts to find the Northwest Passage. In 1906 he succeeded in the latter, albeit using some sea-passages too shallow for commercial shipping.

While Amundsen eschewed claiming land for Norway, Sverdrup tried to secure his newly discovered islands. King Oscar II "wasn't interested in valueless, barren land in the High Eastern Arctic",[52] but in international law, nevertheless, Norway now had precedence. The Norwegian government later raised the claim in 1928, using it to bargain with the United Kingdom over the status of Jan Mayen in the Arctic and Bouvet Island in the Antarctic.[53]

The nationalistic activities of Sverdrup, and those of Lieutenant Robert Edwin Perry's 1898–1902 expedition, seeking American honour in the race to reach the North Pole, raised Canadian sensitivities. They had not asked the Dominion government for permission to cross territories Canada regarded as its own. Exacerbating the situation, in 1903 British diplomats settled the Alaska-Yukon boundary dispute in favour of US demands. Canada thereby lost many islands and a long narrow slice of mountain territory in northern British Columbia–the "Alaskan Panhandle".

[50] Michael F. Robinson, "Reconsidering the Theory of the Open Polar Sea," in *Extremes: Oceanography's Adventures at the Poles*, eds., Keith R. Benson and Helen Rozwadowski (Sagamore Beach, Mass.: Science History Publications, 2007), 16–17, 22, 9–20.

[51] Levere, *Science and the Canadian Arctic*, 365–7.

[52] Osborne, *In the Shadow of the Pole*, 132.

[53] On 11 November 1930, Norway recognized Canadian sovereignty over the Sverdrup Islands. Pierre Berton, *The Arctic Grail: The Quest for the North West Passage and the North Pole*. (Toronto: Random House of Canada Ltd., 1988), 629.

Furthermore, Americans in 1903 established a new whaling colony on Herschel Island in the western Canadian Arctic, and were still whaling in the eastern Canadian Arctic and mining on Baffin Island. In response, according to Trevor Levere, the Dominion government prioritized scientific Arctic exploration through the "coast guard, the Department of Mines and the GSC [Geological Survey of Canada], and other government agencies..."[54]

In 1903 the Departments of Marine and Fisheries and of the Interior, and the Geological Survey of Canada funded the Eastern Arctic Expedition to set up a Royal North West Mounted Police post while overwintering at the 50-year old American whaling base at Fullerton Harbour in northwestern Hudson Bay. America took note. The *New York Times* reported that "Americans have been making a very good thing out of the fisheries for many years past. The result of the expedition will be to drive the Americans out."[55] Geologist Albert Peter Low (1861–1942) commanded the expedition, which included "a naturalist, a botanist, and a topographer and meteorologist." Science, once more, was used "to reinforce claims to sovereignty, with scientists as advance guards of government."[56] The next summer Low sailed *Neptune* north through Smith Sound, taking possession "of islands not previously explicitly claimed for Canada", while also prospecting for minerals. Low later reported that whaling and mining offered strong economic motivations for Canadian ownership, but the Hudson Bay whale fishery would fail if not regulated.[57] Indeed, although a small fishery was revived in the 1940s, the American whale fishery ended in the early 1900s due to overfishing.

13.4 Hudson Bay Fisheries and Arctic Expeditions 1913–1914

The Dominion government in 1913–1914 began to sponsor expeditions specifically to investigate fish of the Hudson Bay and Atlantic Canada. Norwegian Johan Hjort, the world's leading fisheries biologist, was hired by the Biological Board of Canada, under the Department of Marine and Fisheries, to head the Canadian Fisheries Expedition of 1914–1915. This expedition succeeded in its not-incidental goal of introducing Canadian scientists to new European methods in the emerging discipline of fisheries biology. Hjort notably trained Archibald Gowanlock Huntsman (1883–1973), who a few years later became the first full-time director of the Atlantic Biological Station in Saint Andrews, New Brunswick. The two may be regarded as joint founders of Canadian fisheries science.[58] The Canadian Fisheries Expedition

[54] Levere, *Science and the Canadian Arctic*, 376.

[55] Osborne, *In the Shadow of the Pole*, 143.

[56] Levere, *Science and the Canadian Arctic*, 372.

[57] Levere, *Science and the Canadian Arctic*, 374–5.

[58] Jennifer Hubbard, *A Science on the Scales: Canadian Fisheries Biology 1898–1939* (Toronto. University of Toronto Press, 2006), Chapter 2; Vera Schwach and Jennifer M. Hubbard, "Johan Hjort and the Birth of Fisheries Biology: The construction and transfer of knowledge, approaches

was centred on the Gulf of St. Lawrence, and did not encounter the extreme ice conditions, fogs, and fierce storms that characterized Hudson Bay and Strait. These required heroic fortitude, best illustrated by the experiences of the Hydrographic Survey of Canada (later the Canadian Hydrographic Service) during annual surveys which ran from 1910 to 1914.[59] In the 1911 Hudson Bay Expedition, three sail-powered ships with auxiliary steam engines suffered severe ice damage; in 1912 the schooner *Chrissie Thomey* was destroyed after being crushed by ice. It is not surprising, then, that the Canadian Fisheries Expedition's success contrasts starkly with the fiasco of the Hudson Bay fishing surveys of 1913 and 1914. Their failure, furthermore, also highlights the need to let scientists, not politicians or bureaucrats, organize and run scientific investigations.

Archibald Gowanlock Huntsman (1883–1973) working on water samples, in an undated image likely taken in the mid-1920s. He instigated fisheries biological and fisheries oceanographic research in Hudson Bay for the Biological Board of Canada and the Department of Marine and Fisheries in the 1920s.

---

and attitudes, Norway and Canada, 1890–1920," *Studia Atlantica* **13** (2010), 20–39; and Jennifer Hubbard, "Johan Hjort: The Canadian Fisheries Expedition, International Scientific Networks, and the Challenge of Modernization" *International Council for the Exploration of the Sea, Journal of Marine Science* **71**(2014), 2000–2007.

[59] "Sketch of Hydrographic Field Service – Mr. R.J. Fraser – 1907–1926," Accessed at http://www.canfoh.org/Vignettes/Fraser/mr__rj_fraser.htm.

The Hudson Bay fishing surveys were sponsored by the Department of Railways, since a rail line was finally under construction[60] and the Department of Marine and Fisheries, since J.D. Hazen, its Minister, had received requests for fishing permits. If a Hudson Bay fishery opened, fish could be transported quickly and in good condition to Toronto and other markets.[61]

Professor J.B. McCarthy of King's College, Windsor, Nova Scotia,[62] attempted a preliminary survey in 1913. The Department of Railways rented the sealer *S.S. Beothic*, which ran aground, and had to return to Halifax for repairs. Arriving at Port Nelson on September 11, McCarthy had no time for survey fishing on Hudson Bay. At anchor in the Nelson River estuary, *Beothic's* hull was damaged by a gale on the 19th that swept everything from the deck, including McCarthy's collections. McCarthy saw lots of white fish, trout and a few sturgeon in Nelson River and noted trappers fished there when short of food supplies. *Beothic* was required to leave on September 30th and make three stops in Hudson Strait–at each end and in the middle. McCarthy's queries revealed no cod had ever been taken at Erik Cove just inside Hudson Strait during many fishing attempts. However the region's rivers featured arctic herring and arctic salmon, and fish abounded in Hudson Strait. His own cod jigging on the Atlantic end yielded abundant cod, but time was too short at Ashe inlet–the mid-point–to fish. McCarthy concluded that "the prospects of finding abundance of fish in the Bay itself are very bright indeed".[63]

In 1914 the Dominion government amended the Fisheries Act to assert Canadian sovereignty over Hudson Bay. Superintendent of Fisheries William Ambrose Found (1873–1940) argued "the Government should gain as exact knowledge of the conditions in the Bay and tributary waters, as possible," though he himself believed "the fisheries there are not of the importance that the public are disposed to regard them."[64] Three concurrent fishing survey expeditions–suggested by Commander Wakeham–were jointly funded by the Department of Marine and Fisheries and the Department of Naval Services.

For the first survey, two graduate students travelled to James Bay. Wakeham commented: "There is no difficulty whatever in getting from Ottawa to James Bay in canoes. School boys have made the trip in and out during their summer hollidays (sic). Now that we can go more than half way by rail, the trip is a short and easy one."[65] The two students were C.D. Melville, a graduate student in Ottawa, and

[60] The Dominion government made the mistake of choosing Port Nelson near York Factor as the terminus, despite Gordon's earlier–and correct–advice about the near-impossible challenges of dredging this river for this location. The need to correct this mistake would delay the completion of the project until 1929.

[61] Memo from J.D. Hazen to His Royal Highness, the Governor General in Council, 17 March 1914, Library and Archives Canada RG 23 Vol. 1205 File 726-4-2 [1].

[62] Kings College moved to Halifax in 1922.

[63] Letter from J.B. McCarthy to the Deputy Minister of Marine and Fisheries, 1 November 1913, Library and Archives Canada RG 23 Vol. 1206 File 726-4-2 [2].

[64] W. A. Found, "Memorandum Re Fisheries Survey of Hudson Bay," 2 March 1914, Library and Archives Canada RG 23 Vol. 1206 File 726-4-2 [2].

[65] W. Wakeham, memorandum "RE Sending an Expedition to Hudson Bay to enquire into the fishing possibilities of the Bay and its tributary rivers," 16 March 1914, Library and Archives Canada RG 23 Vol. 1205 File 726-4-2 [1].

Arthur R.M. Lower (1889–1988), best known as a Canadian economic historian, but then a Ph.D. candidate at the University of Toronto. Melville and Lower, laden with Department-supplied tents, rifles, nets and lines, made their way to James Bay, where they hired Métis fishermen as assistants.[66] Their thorough work merely confirmed the presence of 16 freshwater species in rivers that drain into James Bay. Trout, pike, sturgeon and white fish supported fisheries amounting to over 13,000 pounds per year on some rivers. However, they found it difficult to set nets in the fast currents of the seaweed and filth-choked James Bay.[67]

For the second expedition Wakeham recommended Fisheries Service officer Napoleon (Nap) Comeau, of Godbout Québec.[68] Comeau was ostensibly put in charge of the expedition to experimentally fish Hudson Bay and its river tributaries and collect specimens for the Fisheries Museum (1884–1918) in Ottawa.[69] Government bureaucrats, however, did not see the need to give him or other experts a say in the mission's organization, as will be seen. When the Biological Board tried to include a University of Toronto graduate student, E. Horne-Craigie– who had experienced a Scottish North Sea fishery and hydrographic survey[70]–the vessel master of the rented schooner *Burleigh* argued space was too short, and George J. Desbarats (1861–1944), Deputy Minister of the Naval Service, backed him.

Comeau had a terrible time of it. He was supposed to conduct fishing surveys as soon as *Burleigh* entered the bay, but the gasoline 40-horsepower auxiliary engine gave her a top speed of two and a half miles per hour, too slow for trawling. Her hull was too deep for near-shore work or travel into estuaries, and when under engine power any adverse breeze or current pushed her backwards. Her resulting reliance on sails meant she needed a lot of sea room.[71] The first mate was old, possibly suffering from vertigo, and never worked. The second mate was a good hand, but Comeau counted only three reliable sailors on the crew. One supposed sailor did not know port from starboard, and another could not read a compass; the ship "chased to all points of the compass" one foggy night, colliding with pieces of heavy ice. Often the watches were not manned; on several occasions she was nearly driven onto reefs or old ship wrecks. One night, high winds and ice collisions knocked 12

[66] W.A. Found wrote 'halfbreed' rather than Métis. Letters from W.A. Found to A.R.M. Lower, 2 June 1914; and from W.A. Found to C.D. Melville, 2 June 1914, Library and Archives Canada RG 23 Vol. 1206 File 726-4-2 [3].

[67] A.R.M. Lower, "A Report on the Fish and Fisheries of the West Coast of James Bay," Library and Archives Canada RG 23 Vol. 1206 File 726-4-2 [5].

[68] W. Wakeham, "Sending an Expedition to Hudson Bay."

[69] Letter from W.A. Found, Superintendant of Fisheries, to Napoleon Comeau, 30 March 1914. Library and Archives Canada RG 23 Vol. 1206 File 726-4-2 [1]. For a history of the Fisheries Museum, see William Knight's recent dissertation "Modeling Authority at the Canadian Fisheries Museum, 1884–1918"(Carleton University, 2014), which can be accessed at: https://curve.carleton.ca/system/files/theses/31615.pdf.

[70] Letter from A.B. Macallum to A. Johnston, 22 June 1914. Library and Archives Canada RG 23 Vol. 1206 File 726-4-2 [3].

[71] Letter from Napoleon Comeau to W.A. Found, 7 September 1914 at Port Nelson. Library and Archives Canada RG 23 Vol. 1206 File 726-4-2 [4].

poorly-stowed barrels of gasoline loose; Captain Bartlett, exhausted from retrieving these, nearly ran *Burleigh* into an obstacle the next day. Bartlett did not bother to set night watches or use running or anchor lights, leading to a near-collision with the Port Nelson "tug-and-tow" one night. He never used the lead line. Comeau used his own hand line when he suspected *Burleigh* was running toward shallows and reefs (the captain ignored him until Comeau's soundings proved *Burleigh* was in imminent danger). Captain Bartlett told him he took his orders from Ottawa, not Comeau. Unfortunately, Bartlett's personal agenda was to break the speed record for reaching Port Nelson from Halifax.[72] He did not allow *Burleigh* to stop for fishing trials at any point. He did allow Comeau to shoot a bear, but was already chugging away as Comeau and his men struggled to bring the carcass back by canoe (the promised power dory had not materialised and the launch was too unwieldy).[73]

In the end Bartlett ran *Burleigh* aground near the wreck of the *Caerense* in the Nelson River estuary. Comeau jumped ship at Port Nelson on September 6th, refusing to risk life and limb on the return voyage, as Captain Bartlett expressed interest in breaking the speed record for the return journey.[74] A railway terminus and port facility construction had begun that year. The intended Marconi wireless, however, was not installed, since the ship carrying it, *Caerense*, had also gone aground 15 miles downstream just before *Burleigh*'s arrival, and had not yet been salvaged. Comeau wrote to W.A. Found of his predicament on September 7th. The mail service, run along a tote road by MacMillan Brothers of Winnipeg,[75] was good enough for G.J. Desbarats, by September 14th, to have arranged a berth for Comeau on another ship returning to Halifax.[76]

This communication also may have been facilitated by new technology. For example, CGS *Acadia* had entered service in the Hydrographic Survey in 1913 and spent the summers of 1913 and 1914 in Hudson Bay. A steam-powered ship, with small masts for signal flags, not sails, she continued to assist hydrographic and oceanographic research until 1969. She rescued 28 people after the SS *Alette* was crushed by ice in 1913; for the first time events in Hudson Bay were nearly instantly communicated to Halifax via *Acadia*'s wireless telegraph.[77]

[72] There is no hint in the records as to this captain's first name. He is certainly not to be confused with the famous and competent Captain Bob Bartlett, captain of the *Karluk*.

[73] Letter from Nap. Comeau to W.A. Found, 26 September 1914 at Port Nelson. Library and Archives Canada RG 23 Vol. 1206 File 726-4-2 [4].

[74] Letter from Nap. Comeau to W.A. Found, 7 September 1914 at Port Nelson.

[75] David Malaher, "Port Nelson and the Hudson Bay Railway," *Manitoba History* **8** (1984). Accessed at http://www.mhs.mb.ca/docs/mb_history/08/hudsonbayrailway.shtml.

[76] Letter from G.J. Desbarats to the Chief Engineer, Department of Railways and Canals, 14 September 1914. Library and Archives Canada RG 23 Vol. 1206 File 726-4-2 [4].

[77] In 1915 *Acadia* was loaned for the Canadian Fisheries Expedition and shortly thereafter was seconded to the navy; she was only restored to the Hydrographic Survey in 1921. *Acadia* also saw military service during the Second World War. Today the restored CSS *Acadia* is moored in Halifax at the Maritime Museum of the Atlantic. CSS replaced CGS as the prefix and stands for Canadian scientific ship or Canadian survey ship. See Kelly Shiers, "CSS Acadia, survivor of the seas, turns 100," *Chronicle Herald*, 26 October 2013, http://thechronicleherald.ca/thenovascotian/1162703-

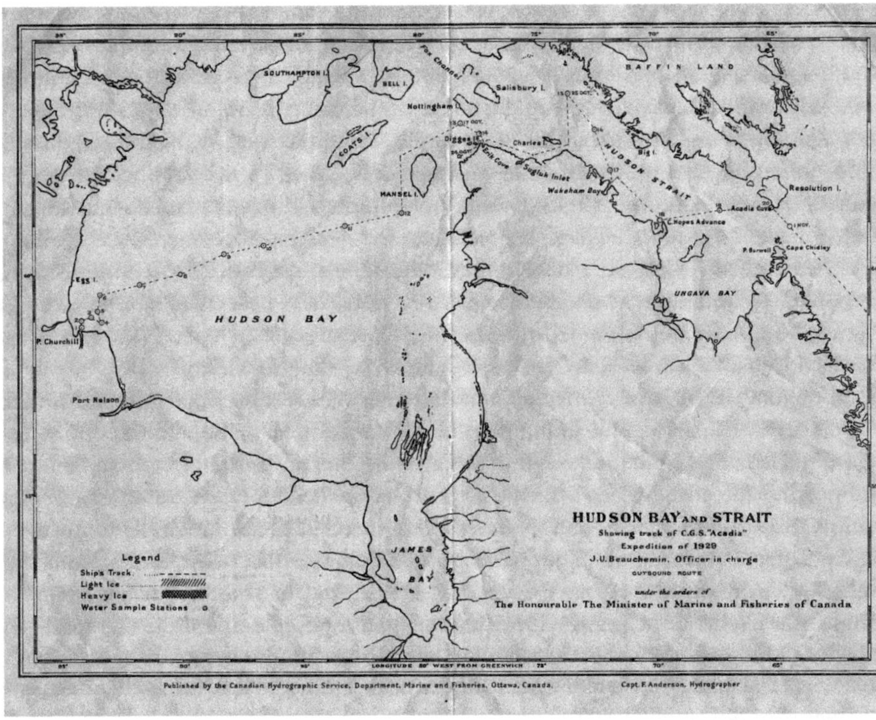

Hudson Bay and Strait stations taken by CGS *Acadia* during 1928 portion of the Hudson Strait Expedition of 1927–1928. *Acadia* was used for auxiliary hydrographic work for this expedition.

Comeau's report about Hudson Bay fisheries was considerably briefer than his letters of complaint about Captain Bartlett's conduct, since he had been given no time to do trial fishing. Comeau was nevertheless able to conclude that should any fisheries develop, nature had already "enacted" efficient protective "fishery regulations", since ice conditions precluded fishing in the Bay before June 15th or after September 15th. He did recommend, shockingly to modern sensibilities, that: "Everything possible should be done to encourage the destruction of the white whale (beluga) in these waters–they devour more fish in one day than has been taken by all the natives in 10 years."[78]

Wakeham, in charge of the third expedition (an Arctic patrol), did manage experimental fishing in late summer. He saw abundant salmon, trout and whitefish in the

css-acadia-survivor-of-the-seas-turns-100; and O.H. Meehan, "The Canadian Hydrographic Service From the time of its inception in 1883 to the end of the Second World War: Chapter 2: The Hydrographic Survey of Canada from its Formation to the First World War 1904–1914," in *The Northern Mariner* **14** (2004),92, http://www.cnrs-scrn.org/northern_mariner/vol14/tnm_14_1_43-103.pdf.

[78] Napoleon Comeau, "Protection of the fisheries," Library and Archives Canada RG 23 Vol. 1206 File 726-4-2 [5].

rivers, but caught nothing but sculpins in the Bay itself.[79] Officially the Department of Naval Services remained optimistic, Desbarats telling one enquirer the studies "were not sufficiently exhaustive to warrant the preparation of fishery charts" but that "the information obtained leaves little room to offer doubt that codfish – Greenland cod – abound in the Eastern side of the Bay."[80] In fact, neither Comeau nor Wakeham had found any cod. Wakeham's role was now ended. Commissioned to command CGS *Princess*'s scientific cruises for the Canadian Fisheries Expedition in 1915, he died days after his ill-health cut short the first cruise in May.

After spending $7500 on Comeau's fiasco (over CDN $150,000 in 2014 dollars), and another $5000 for the overland fisheries expeditions,[81] the Dominion government set aside fishing issues until the completed Hudson Bay Railway could bring in coal to power the steam trawlers used for test-fishing. The First World War, a subsequent crisis in railway finances and the realization that the government had chosen the wrong port meant that freight trains did not reach the port of Churchill until 1929.

13.5 Frits Johansen and the Biological Board's Hudson Bay Fisheries Expeditions 1920–1929

While the Hudson fishing survey expeditions were ongoing, a more famous expedition had begun that also linked scientific exploration with Canadian Arctic sovereignty. The continued possibility that other nations' expeditions might reach and claim undiscovered Arctic islands inspired various Canadian government departments to sponsor jointly the Canadian Arctic Expedition of 1913–1918, led by Vilhjalmur Stefansson.

[79] Ibid.

[80] Letter from G.J. Desbarats to Wallace Russell, 21 January 1915. Library and Archives Canada RG 23 Vol. 1206 File 726-4-2 [5].

[81] Letter from G.J. Desbarats to A.K. Maclean, 11 March 1915. Library and Archives Canada RG 23 Vol. 1206 File 726-4-2 [5].

Fritz Johansen at work surrounded by his sampling gear on Hudson Bay's shores. The image likely dates from 1920 as he is wearing a rather battered boater hat. Boaters were popular in the early 1920s.

The Canadian-born, ethnically Icelandic Stefansson had grown up in the United States and studied anthropology at Harvard. Norwegian Arctic expeditions inspired him to join the 1906 Anglo-American Polar Expedition. While it failed to discover new territory north of Alaska, Stefansson ended up living among the Inuit of the Mackenzie delta, adopting their food and customs. Together with zoologist Rudolph Anderson, Stefansson then mounted an anthropological expedition in 1908–1910, sponsored by the American Museum of Natural History in New York city. Stefannson discovered the "Copper Eskimos", whose blond hair, fair skin and use of copper implements inspired him to claim this Inuit tribe was descended from Norsemen. While this claim destroyed his scientific reputation, he still made valuable contributions by collecting artifacts and recording their lifestyle and customs when still untouched by western society.[82]

Stefansson and Anderson's next expedition was originally sponsored by the American Museum of Natural History and the National Geographic Society. Stefansson asked Canada for further funding. As one of his goals was to discover new lands, the Canadian government decided to finance the entire expedition through the Department of the Naval Service, under Deputy Minister G.J. Desbarats.

---

[82] Richard Duibaldo, *Stefansson and the Canadian Arctic* (Montreal: McGill-Queens University Press, 1978), 49–54.

Its 15 scientific members were divided into two groups. Stefansson's northern party would explore for new lands north of the Beaufort Sea, while the southern party would do scientific work along the Northwest Territories's coasts.[83]

The entire expedition was riven by rivalries and charges of incompetence, starting when Stefansson hastily chose the nearly derelict *Kurluk* as the expedition's vessel.[84] As most of the expedition had predicted, it was crushed in the Arctic ice in the first season. Sixteen scientists and crew died; starving survivors were rescued from Wrangle Island. Stefansson had left to hunt before the disaster and carried on. He became the last man to discover significant new land masses, including Brock, Borden, Meighen, and Lougheed Islands, claiming these for Canada. His party also mapped and took depth soundings of huge portions of the Beaufort sea.[85]

The Southern Party, led by Anderson, included Frits Johansen, a Danish expert on Arctic fishes, in charge of marine biology and entomology. Dredges and other marine sampling equipment were seconded by Desbarats from the Biological Board,[86] but Johansen was rarely allowed on the schooner *North Star* due to ice conditions, ship damage, or the low priority given to his work by Anderson. Frustrated, he was forced to focus on collecting and studying insects.[87]

Both Johansen and Stefannson self-identified as Nordic men, who like the Inuit were uniquely qualified to thrive in the Arctic.[88] Unlike Canadians, who orient their economic interests southward, the Nordic people had developed and articulated an identity around their northern and Arctic settings. This focus was partly economic: Sweden, Norway and Iceland were (and are) singularly dependent upon natural resources, especially migratory fish. Starvation and hardship followed sharp declines in the herring fisheries in the 1860s and 1870s;[89] the government of Sweden and Norway began funding scientists to investigate these fluctuations. Zoologist Georg Ossian Sars (1837–1927) and meteorologist Henrik Mohn (1835–1916) wanted to investigate polar ocean currents to explain the coastal climates of Scandinavia, the Faeroes, and Shetland Islands, and to "throw light ... above all, upon the biological characteristics of our migratory fishes." Together they organized the *Vøringen*

---

[83] Stuart E. Jenness, *Stefansson, Dr. Anderson and the Canadian Arctic Expedition, 1913–1918: A Story of Exploration, Science and Sovereignty* (Gatineau: Canadian Museum of Civilization, 2011), 6–9.

[84] Jenness, *Stefansson*, 18–31; Duibaldo, *Stefansson and the Canadian Arctic*, 70–72; 76–82.

[85] Duibaldo, *Stefansson and the Canadian Arctic*, 2.

[86] G.J. Desbarats, Memorandum for E.E. Prince Re Stefansson Expedition, 14 April 1913. Library and Archives Canada RG 23 Vol. 404 File 4232 Reel T 3392.

[87] Stuart E. Jenness, *The Making of an Explorer: George Hubert Wilkins and the Canadian Arctic Expedition, 1913–1916* (Montreal: McGill-Queen's University Press, 2004), 273, 274, 288.

[88] For a discussion of Stefansson's embracing of the north and northern identity, see Tom Henighan, *Vilhjalmar Stefansson, Arctic Adventurer* (Toronto: Dundurn Press, 2009), 139–158. Jenness comments that Johansen was the only member of the Southern Party who liked and even admired Stafansson, perhaps because he, too, hoped one day to be famous. See Jenness, *The Making of an Explorer*, 326.

[89] Eric L. Mills, *Biological Oceanography: An Early History 1870–1960* (Toronto: University of Toronto Press, 2012), 77.

Expedition, in imitation of the British *Challenger* Expedition. In the summers of 1876–1878 the expedition investigated the movements of the Gulf Stream and polar water masses to as far away as Greenland (the Greenland Sea). The *Vøringen* made transverses across the Norwegian Sea to the Faeroes and Shetland Islands, and ventured northward to Jan Mayen and Spitsbergen (Svalbard), collecting specimens, measuring water currents, temperatures and depths, and mapping deep-water channels in the eastern Arctic continental shelf. The expedition also visited Bear Island and Svalbard, and mapped Jan Mayen, collecting zoological and botanical materials. After this expedition, Norwegians increasingly embraced "the notion of Norway as a polar nation…For since then, Norwegians have considered the polar areas as a broad northward extension of economic and strategic concern of the state." Norwegian interest in sovereignty over Jan Mayen and Svalbard dates from this period. Although Mohn and Sars tried to keep their expedition free from geopolitics, Scandinavian scientific expeditions were implicitly linked to sovereignty issues. Indeed, a large aspect of fisheries science is also political and related to sovereignty.[90] Nansen's later famous *Fram* expedition (1893–1896), according to Norwegian marine science historian Vera Schwach, "marked Norway as a polar nation, claiming national and strategic interests in arctic areas."[91]

Denmark's similarly increasing scientific interest in the north ultimately led Fritz Johansen (1882–1956) to Canada in 1911. Born in Copenhagen, the son of artist and professor Viggo Johansen, he studied under fisheries biologist Harald Blegvad, director of the Danish Biological Station. His obligatory military service in 1904 was spent as a biologist on the naval vessel *Beskytteren* investigating Faeroe Island and Icelandic fisheries. His copious accounts in the Danish Biological Station's *Fisheriberetningen (Report of the Fisheries)* led to his being appointed as a zoologist to the 1906–1908 Danmarkekspeditionen (The Denmark Expedition) to North-East Greenland. Responsible for all zoological collections aside from mammals and birds, he made an outstanding collection of insects. He worked up parts of the collection upon his return, and wrote the introduction to I.C. Nielsen's book on Greenland's insects, describing climate conditions and various species' distributions. He also published observations of seal and walrus life histories, a pamphlet on 'Freshwater Life in North-East Greenland', and a 1912 paper on Greenland's north-eastern fishes which he helped to illustrate.[92]

---

[90] See Mills, *Biological Oceanography*, 75; Vera Schwach, "Faded Glory: The Norwegian *Vøringen* Expedition, 1876–1878′, in *Extremes: Oceanography's Adventures at the Poles*, eds. Keith R. Benson and Helen Rozwadowski (Sagamore Beach, Mass.: Science History Publications, 2007), 35, 36–7, 31, 34; Jennifer Hubbard "In the Wake of Politics: The Political and Economic Construction of Fisheries Science 1860–1970,"*Isis* **105** (2014), 364–78; and Jennifer Hubbard, "Changing Regimes: Governments, Scientists and Fishermen and the Construction of Fisheries Policies in the North Atlantic 1850–2010," in *A History of the North Atlantic Fisheries: Volume 2, From the Mid-Nineteenth Century to the Present,* eds. David J. Starkey, Jon Th. Thór, and Ingo Heidbrink (Bremerhaven: Deutsches Schiffahrtsmuseum, 2012), 129–176.

[91] Schwach, "Faded Glory", 56.

[92] R. Spärck, "Frits Johansen 17 Juli 1882–24 April 1957," *Videnskabelige Meddelelser Dansk Naturhistorisk Forening* **119** (1957), ix–xii. I am grateful to Dr. Torben Wolff of the Zoological Museum, University of Copenhagen, for supplying me with Frits Johansen's scientific obituary and images reproduced in this article, and to Dr. Vera Schwach for her translation.

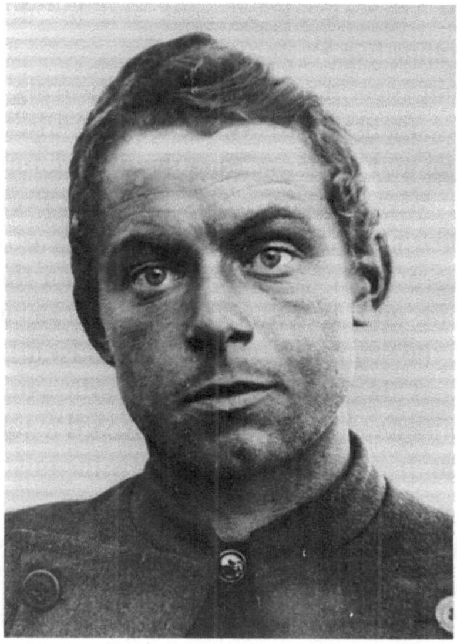

Fritz Johansen (1882–1957)

Despite this promising beginning and the support of his mentor, botanist and ecologist Eugenius Warming, he failed to obtain an academic degree. He could not collect his thoughts during examinations. Thus in 1911 he sought his fortune in the New World, working briefly for the US National Museum before transferring to the Victoria Memorial Museum (National Museum of Natural Sciences) in Ottawa.[93] There Johansen met Stefansson and signed on to the Canadian Arctic Expedition in 1913. He refused to join the Northern Party, however, fearing Stefansson's offhand approach to safety and survival.[94] After his return to Ottawa in 1916, the Department of Naval Services employed him to work up his collections and write on Arctic insect life, vegetation and crustaceans. His unpublished manuscript on Arctic fishes was later used by the Canadian Museum of Nature for an important work on the fishes of Arctic Canada.

Johansen took a break in the summer of 1917 to assist A.G. Huntsman, curator of the Atlantic Biological Station, in a biological survey in the Gulf of St. Lawrence, modeled after Hjort's Canadian Fisheries Expedition. Johansen collected specimens

[93] Spärck, "Frits Johansen."

[94] A single *Colias* specimen was collected by Johansen 'in 1916 on a small hill near Bernard Harbour. He identified it as *Colias meadii*, a species known only from the Rocky Mountains between Alberta and Colorado. Many years of further collecting in northern Canada failed to turn up any additional examples of this species, so the specimen was assumed to be mislabelled or misidentified. In 1988, Jim Troubridge and Kenelm Philip used Frits Johansen's diary to relocate the hill and rediscover a thriving colony of this butterfly, which was described as a new species and named *Colias johanseni* after its original discoverer'. See Canadian Biodiversity Information Facility, "The History of Butterfly Study in Canada," accessed March 8, 2016, http://www.cbif.gc.ca/eng/species-bank/butterflies-of-canada/history-of-butterfly-study-in-canada/?id=1370403265516.

for the Ottawa Museum of Natural Sciences and for the Station. He was also assigned to study the cunner fish.[95]

There is no doubt that meeting Johansen, and hearing of his Canadian Arctic experiences, piqued Huntsman's curiosity. He asked Andrew Halkett, a naturalist in the Department of Marine and Fisheries (and the now-defunct Fisheries Museum in Ottawa),[96] and Reverend W.G. Walton, a missionary at Fort George on James Bay, for specimens. In 1919 Huntsman drew up a summary of the fauna in James and Hudson's bay, which were entirely Arctic species even though, Huntsman noted, Hudson Bay is sub-arctic and does not freeze over. He also outlined its known hydrography, lamenting the scant information on Hudson Bay's currents and salinity profiles, and speculated about how the copious inflowing freshwater mixed with Atlantic salt water. Huntsman, the Biological Board's most influential scientist, wanted more information. It was needed, as seen by his mistaken understanding that the Bay did not freeze over in winter.[97]

Huntsman was receptive, then, when Johansen asked the Biological Board and the Hudson's Bay Company to co-sponsor him to survey Hudson Bay's fauna in 1920. The Board agreed to pay expenses up to 500 dollars in return for a report. Johansen collected widely around southern and eastern James Bay as far north as Richmond Gulf in Hudson Bay. His report, however, was never completed. Specimens were ultimately identified by Dr. William Healey Dall (1845–1927), the American malacologist and expert in northern biota (the Dall's porpoise of the North Pacific was named after him) who had worked for the US Coast and Geodetic Survey, the Smithsonian, and the US Geological Survey.[98]

Johansen had not kept his part of the bargain. Worse, he appears to have copied Stefansson's modus operandi of running up expenses without permission.[99] Hudson's Bay Company refused to cover his exorbitant expenditures. The Department of Naval Services, under which federal fisheries services remained until 1921, also refused, arguing the Biological Board had hired him. The Board was

[95] Margaret S. Rigby and A.G. Huntsman, "Materials Relating to the History of the Fisheries Research Board of Canada (Formerly the Biological Board of Canada) for the Period 1898–1924," Fisheries Research Board of Canada Manuscript Report Series (Biological) **660** (1958), 151, 153.

[96] See William Knight's doctoral dissertation, "Modeling Authority at the Canadian Fisheries Museum, 1884–1918." (Carleton University, 2014). Available at https://curve.carleton.ca/system/files/theses/31615.pdf.

[97] See correspondence between Huntsman and Andrew Halkett, and Huntsman and W.G. Walton, 1919, and Huntsman's handwritten manuscript, "The Fauna of James and Hudson Bays" in Huntsman Collection, University of Toronto Archives, B1978-0010, Box 150 File 2.

[98] "Marine Mollusca collected in James and Hudson Bays 1920, F. Johansen (Identified by Dr. W.H. Dall)," Huntsman Collection, University of Toronto Archives, B1978-0010 Box 150 File 3.

[99] The Biological Board of Canada and all federal fisheries services remained under the Department of Naval Services until 1922. A.P. Knight discussed Johansen's expenses in a letter to J.P. McMurrich, 24 June 1922. Huntsman Collection, University of Toronto Archives, B1928-0010 Box 15 File 8.

forced to pay out by setting aside $200 annually; the Department of Marine and Fisheries paid the balance in 1923,[100] and terminated his contract in 1926.[101]

In 1927 Johansen pleaded with Biological Board chairman James Playfair McMurrich (1859–1939) to get him a berth as an experienced Arctic scientist on the Hydrographic Survey's ship *Beothic* in return for published reports and specimens he would collect from Hudson Bay.[102] McMurrich rebuffed him: "the Board had a rather unpleasant experience in connection with a former trip that you made…, and I would hesitate greatly to expose the Board to a repetition of that experience."[103] In spite of this, Huntsman's interest led a reluctant Biological Board to hire him again, but for different venture than the one outlined by Johansen.

Huntsman took advantage of the upcoming 1927–1928 Hudson Strait Expedition sponsored by the Department of Railways and Canals, assisted by Naval Defence and Marine and Fisheries, to open up the trans-Atlantic commercial route. The *Halifax Herald* described it as the "largest, most comprehensive, and most scientific expedition the Canadian government has ever despatched" with aerial surveys "on a scale never hitherto attempted in such northern latitudes."[104] Minister of Railways and Canals C.A. Dunning boasted this "romantic undertaking" was the first to mass "scientific aid…in the development of a new navigation route."[105] Canada's first aerial Arctic expedition set up three large bases on Hudson Strait: on Nottingham Island; Wakeham Bay in the middle; and Port Burwell at the Atlantic end. Each had an officers' dwelling, prefabricated quarters, a wireless radio communications building, storehouse, blubber house, two hangars, and two Fokker airplanes equipped with floats, skis and wheels for aerial surveys of ice flows.[106] In 1928 *Acadia*, newly furnished with a Sperry gyrocompass to neutralize the region's severe magnetic fluctuations, undertook hydrographic surveys for the expedition, returning in 1929 endowed with a British Admiralty echo sounder.[107]

---

[100] Biological Board of Canada, 'Minutes of the Annual Meeting Held in the Library of the Department of Marine and Fisheries, Hunter Building, Ottawa, at 2:30 P.M., Monday, May 21st, 1923', Huntsman Collection, University of Toronto Archives, B1978-0010 Box 12, File 6.

[101] Letter from A.P. Knight to J.P. McMurrich, 29 June 1922; Letter from F. Johansen to J.P. McMurrich, 15 June 1927. Huntsman Collection, University of Toronto Archives, B1978-010 B1978-010 Box 15 File 8; and Box 12 File 13; Spärck, "Frits Johansen."

[102] Letter from F. Johansen to J.P. McMurrich, 15 June 1927. Huntsman Collection, University of Toronto Archives, B1978-010 Box 12 File 13.

[103] Letter from J.P. McMurrich to F. Johansen, 21 June 1927. Huntsman Collection, University of Toronto Archives, B1978-010 Box 12 File 13.

[104] "Expedition will Sail on Saturday Next," *Halifax Herald*, 14 July 1927. Huntsman Collection, University of Toronto Archives, B1978-0010 File 150 Box 8.

[105] "Aviators to leave Ottawa this week upon new, risky kind of exploration work," *Ottawa Citizen*, Wednesday July 6, 1927. Huntsman Collection, University of Toronto Archives, B1978-0010 File 150 Box 8.

[106] Letter from F. Johansen to A.G. Huntsman 28 August 1927, sent from Wakeham Bay. Huntsman Collection, University of Toronto Archives, B1978-0010 File 150 Box 8.

[107] Maritime Museum of the Atlantic, "*CSS Acadia*", accessed March 8, 2016, https://maritimemuseum.novascotia.ca/what-see-do/exhibits/css-acadia

The Hudson Strait Expedition was expected to help open up Arctic commerce. On these grounds, Deputy Minister of Fisheries Alex Johnston telegraphed Huntsman to ask about adding fish investigations. "It would seem possible," Huntsman replied on June 21st, "to...settle the question of the suitability of…the strait and at the northern end of Hudson bay for fishes."[108] That day Huntsman also wrote to McMurrich recommending Johansen,[109] and apprised Johansen of a possible job. He made it impossible for McMurrich to refuse:

> The Deputy suddenly expressed a desire to have fisheries work done on the Hudson Strait Expedition, which starts from Halifax on July 1st...I have been arranging for gear and supplies and trying to get a man. Our people are all fully engaged for the summer [...] But there is Johansen. You know the Deputy's attitude toward him, as well as our own experience. Yet he is the man who is keen to do just that thing, ready to rough it, and pertinacious in carrying anything through that he likes, as well as having had the necessary experience...I have said so much to Johansen about his failure to fit into the scheme of things...I have hopes of his changing his attitude and getting along better…[110]

Huntsman especially wanted "planktonic and hydrographic sections to determine... whether there is a real basis for getting big fisheries there ever…"[111] McMurrich reluctantly hired Johansen. He warned Huntsman to insist that Johansen heed his budget: 'It would never do to let Johansen go on such an expedition with unlimited credit guaranteed by the Board.".[112]

While the Hudson Strait Expedition's icebreaker CGS *Stanley* carried most of the expedition staff, Johansen travelled on the supply ship SS *Larch*. Although directed by Huntsman to lay out several series of drift bottles,[113] Johansen only did so during the approaches to Hudson Strait, since expedition commander Major N.B. McLean refused to let the ships change course– as required to set out drift bottles–within the strait itself.[114] Johansen prepared a journal and reports on the fishes and fresh water crustaceans he collected. His collections of fossils, insects, fish and marine life, and plants were split between the Natural History Museum in Ottawa and the Biological Board.[115] His water samples–all from the surface since no stations were allowed to make hydrographic sections–were sent to Henry B. Hachey

---

[108] Letter from A.G. Huntsman to A. Johnston, 21 June 1927. Huntsman Collection, University of Toronto Archives, B1978-0010 Box 150 File 4.

[109] Letter from A.G. Huntsman to J.P. McMurrich, 13 June 1927. Huntsman Collection, University of Toronto Archives, B1978-0010 Box 150 File 4.

[110] Letter from A.G. Huntsman to J.P. McMurrich, 28 June 1927. Huntsman Collection, University of Toronto Archives, B1978-0010 Box 12 File 13.

[111] Ibid.

[112] Letter from J.P. McMurrich to A.G. Huntsman, 7 June 1927. Huntsman Collection, University of Toronto Archives, B1978-0010 Box 12 File 13.

[113] To determine the direction of currents. 'Drift Bottles', Huntsman Collection, University of Toronto Archives, B1978-0010 Box 150 File 4.

[114] Letter from F. Johansen to A.G. Huntsman, 27 July 1927, sent from Port Burwell. Huntsman Collection, University of Toronto Archives, B1978-0010 Box 150 File 4.

[115] Letter from F. Johansen to A.G. Huntsman 28 August 1927, sent from Wakeham Bay. Huntsman Collection, University of Toronto Archives, B1978-0010 File 150 Box 8.

(1901–1985), then at the University of New Brunswick Department of Physics, but soon to become the Biological Board's first professional oceanographer (see below).[116] In 1928 Johansen stayed in Denmark with his ailing father, so B.W. Taylor of McGill's zoology department took his place. Major McLean again prohibited most of the oceanographic work Huntsman wanted done.[117]

Johansen was rehired in 1929. Huntsman had convinced his superiors that the Hudson Bay Railway's imminent completion offered a unique opportunity for a thorough survey of hydrographic conditions and aquatic life-forms near Fort Churchill (which now boasted a movie theatre) before shipping-contamination occurred.[118] Johansen arrived at the end of July with an impressive array of fishing and plankton nets, hooks, lines, dredges, a sounding line, drift bottles, a surface thermometer, specimen bottles and boxes.[119] He pitched a tent on a boulder-strewn beach near the Churchill River. There, he wrote, "I take my Viking-dip in the brackish water every day", having "the advantage of being in the primitive and (comparatively) unspoiled nature, and of associating with people (missionaries, traders, gov't Depts etc.) who have a real interest and business in the country, and not with people looking for jobs or silly tourists." Unauthorized, he hired an Icelandic fisherman and carpenter from Lake Gimli, K. Einarsson, as an assistant; he dredged, made plankton tows, and took water temperatures and samples during "excursions by day or by night"' in a small boat repaired by Einarsson. A Dr. Walton living in Fort Churchill sold him formaldehyde and alcohol for preserving specimens. Johansen took only small lots of alcohol, telling Huntsman "it is not safe to have too much of it in my tent owing to the failings of mankind."[120] Presumably he was worried about his hired help and visitors.

Johansen collected as many fish species at different life stages as possible to determine their abundance, food, parasites, and life histories. He told Huntsman: "I do not spare myself as to my work; if low tide occurs in the middle of the night I am down at the beach then."[121] J.B. Skaptason, Supervisor of Fisheries, visited and reported to William Found, now Deputy Minister of Fisheries, that Johansen's work

> seems to consist more of a general, than that of a specially [*sic*] fishery research work...His operation seemed to consist largely of staking out of a net at low tide, leaving it in during high water, and visiting it for results when the tide ebbs. He had taken in this way, two cod

[116] Letter from A.G. Huntsman to H.B. Hachey, 1 November 1927. Huntsman Collection, University of Toronto Archives, B1978-0010 Box 150 File 4.

[117] Letter from A.G. Huntsman to F. Johansen, 25 April 1929. Huntsman Collection, University of Toronto Archives, B1978-0010 Box 153 File 9; and letter from B.W. Taylor, 22 October 1928, Huntsman Collection, University of Toronto Archives, B1978-0010 Box 150 File 6.

[118] Letter from A.G. Huntsman to J.J. Cowie, 3 July 1929. Huntsman Collection, University of Toronto Archives, B1978-0010 Box 153 File 9.

[119] Huntsman's ambitions, as revealed by this equipment were to be largely disappointed by Johansen. 'Articles shipped to Fort Churchill to be used by Mr. Johansen. Shipped by C.P. Exp. Jul6/29'. Huntsman Collection, University of Toronto Archives, B1978-0010 Box 150 File 10.

[120] Letter from F. Johansen to A.G. Huntsman, 10 August 1929. Huntsman Collection, University of Toronto Archives, B1978-0010 Box 153 File 9.

[121] Ibid.

> fingerlings and quite a collection of capelin, he had caught one cod about two pound, this I did not see as it was preserved and packed away. He told me it was Greenland cod.[122]

Johansen also captured many cisco, "Arctic" trout, and smaller quantities of whitefish.[123]

Although Huntsman was pleased, an unauthorized delay of over a week meant Johansen had been paid for doing nothing. Money was once again an issue. Johansen's unauthorized hire of his assistant and additional expenses were noted with disfavour.[124] J.J. Cowie, Secretary-Treasurer of the Board and Chief Inspector of Fisheries, refused his request for a raise and an extended (paid) report deadline.

Johansen left suddenly for Denmark. His mother had died and his father was unwell, but he promised a formal report. Cowie was sceptical, and seized the collections Johansen stored in his museum office in Ottawa. In 1930, Johansen returned to discover that Huntsman and Biological Board member R.H. McGonigle, on Cowie's orders, had ransacked his rooms.

> From these quarters of mine, these two gentlemen have then removed, without even asking my permission, all the specimens and papers and most of the scientific books found there, comprising not only the minor part… belonging to the Biolog. Board, but thousands of Dollars worth of my private property…[125]

including materials from before Johansen's arrival in Canada in 1911. Using a list from Johansen, Huntsman and McGonigle then had to itemize and recover his property– dispersed to the University of Toronto and various biological stations–to ship to Denmark.[126]

This unfortunate conclusion belies the importance of Johansen's contributions to Canadian Arctic entomology and aquatic zoology. He made the first systematic collection of aquatic fauna in the Churchill River region, and his expertise in taxonomy and natural history gave competent guidance to understanding species and their distributions. A large bay on the south coast of Victoria Island was named in his honour. Unfortunately, his inability to organize his ideas left him unable to prepare larger works. Johansen spent several years sitting at the library in the Zoological Museum in Copenhagen trying to write up his Canadian material. Sadly his growing disorganization and eccentricity left him unable to fulfil his obligations.[127]

---

[122] Letter from J.B. Skaptason to W.A. Found, 19 October 1929. Library and Archives Canada, RG 23 Vol. 1504 File 769-29-2 [1].

[123] Ibid.

[124] Letter from F. Johansen to the Biological Board of Canada, 5 July 1929; and letter from A.G. Huntsman to F. Johansen, 21 October 1929. Huntsman Collection, University of Toronto Archives, B1978-0010 Box 150 File 7; and Box 153 File 11.

[125] Letter from F. Johansen to J.J. Cowie, 30 June 1930. Library and Archives Canada, RG 23 Vol. 1504 File 769-29-2 [1]. In defence of the Board, it had been paying rent for Johansen's office since 1927.

[126] R.H. M'Gonicle, "Report of the Frits Johansen Collections," 17 December 1930. Huntsman Collection, University of Toronto Archives, B1978-0010 Box 150 File 3; Letter from F. Johansen to J.J. Cowie, 30 June 1923. Library and Archives Canada, RG 23 Vol. 1504 File 769-29-2 [1].

[127] Spärck, "Frits Johansen."

13.6 Hudson Bay and Strait Fishery Expedition of 1930

Fishery Supervisor Skaptason's 1929 report had concluded that "nothing is authentically known regarding deep sea fisheries of the Hudson's Bay" and recommended "bringing in a properly equipped vessel from the outside, manned by experienced fishermen." He added "there is actually some evidence of deep sea fish, which … makes it possible there may be some valuable fishing in the deeper area's [*sic*] of the bay."[128] Fishing corporations, enthused by the new railway, also demanded a government-sponsored deep-sea fishing expedition. The newly-formed Department of Fisheries acquiesced. Hudson Bay remained sparsely settled,[129] and a fishery would help remedy this. Meanwhile, scientists, although not oblivious to these economic interests, simply wanted to understand the biological productivity of Hudson Bay and Strait.[130]

The 1930 Hudson Bay and Strait expedition had two components. S.J. Walker, the central region's Supervisor of Fish Culture, went to Churchill to oversee two experienced fishermen doing experimental gill net and drift net fishing. The Biological Board's recently-hired oceanographer, Harry B. Hachey, led the seagoing expedition. Hachey had first visited the Atlantic Biological Station in 1927, where American oceanographer Henry B. Bigelow, a frequent visitor, convinced him of the importance of oceanography.[131] That summer he studied the circulation of Passamaquoddy Bay (between New Brunswick and Maine) for Huntsman, who was eager to avert an American proposal to enclose this large bay with a tidal power dam. In 1928 Huntsman hired Hachey away from the University of New Brunswick physics department.

That the 1930 Hudson Bay and Strait Expedition generated any conclusive information was owing to Huntsman's persistence. He ensured it became Canada's (and the world's) first quasi-oceanographic survey of Hudson Bay and Strait by insisting on, among other things, plankton productivity profiles to give context for conditions affecting fish populations. He wrangled with Cowie about putting a scientist in charge and no doubt put forward Hachey's name. When the Department originally planned an adjunct role for Hachey, Cowie telling Huntsman not to be concerned,[132] Huntsman retorted that any intelligent crew member could take samples. He insisted Hachey be put in charge, thus making science the primary focus, with fishing trials

[128] Letter from J.B. Skaptason to W.A. Found, 19 October 1929.

[129] As it does to this day. The current population around all of Hudson Bay is about 32,000. See Environmental Management of Enclosed Coastal Seas (EMECS), 'Hudson Bay'. Accessed at http://www.emecs.or.jp/guidebook/eng/pdf/13hudson.pdf.

[130] Huntsman, for example, promoted the possibilities of a Hudson Bay fishery in the popular press in 1918. See A.G. Huntsman, "Fishery prospects for Hudson and James Bay," *Canadian Fisherman* **5** (1918), 896–8.

[131] Mills, *The Fluid Envelope of Our Planet*, 240.

[132] Letter from J.J. Cowie to A.G. Huntsman, 10 July 1930. Huntsman Collection, University of Toronto Archives, B1978-0010 Box 150 File 11.

an important secondary activity. Otherwise, the Biological Board would avoid involvement in another "abortive" investigation.[133]

Huntsman prevailed; Hachey commanded the expedition. The Department chartered the steam-powered beam trawler *Loubyrne*, captained by Sigurd Sorensen, which was also equipped for experimental dory fishing.[134] Sorensen had 24 years' experience in the North and White Seas and Gulf of St. Lawrence fisheries. His officers and crew were also experienced trawler operators; if there were fish to be taken, these men could get them.[135]

Conditions in Hudson Strait were too icy for experimental trawling, but during the inward journey across Hudson Bay, throughout the operations thereafter, and during the return through the Strait, Hachey made plankton tows, released drift bottles, and at set stations (some revisited) took water samples and readings to obtain a temperature and salinity profile of the bay.[136]

Loubyrne began trawling operations just before reaching Churchill on August 8th.[137] Walker's fishing operations near Churchill had returned a few trout, cisco, and whitefish–not even enough for bait for *Loubyrne*'s fishing efforts.[138] On August 10th *Loubyrne* left to fish in northwestern Hudson Bay. Birds abounded but the Bay's waters yielded few signs of life except for one walrus and several beluga near Churchill. Only jellyfish, sagitta, shrimp and copepods were netted in the surface waters. An hour-long bottom trawl 50 miles from Churchill yielded mud, "100 sea peaches, 100 basket stars, various shrimp…sea urchins, star fish (few)", and one six inch sculpin "that looked as if it hadn't eaten for years."[139] Between resupply visits to Churchill, the expedition crew, lacking bait, resorted to trawling, dory and seine fishing; deeply disappointed by each empty haul, they became fed-up and homesick.[140] Hachey continued oceanographic sampling and measurements.[141] Huntsman

---

[133] Letter from A.G. Huntsman to J.J. Cowie, 4 July 1930. Huntsman Collection, University of Toronto Archives, B1978-0010 Box 150 File 11; Library and Archives Canada RG 23 Vol. 1206 File 726-4-5 [2].

[134] "Loubyrne's Expedition to Investigate Commercial Fishery in Hudson Bay: Beam Trawler to Sail from here Today to End Old Controversy," *The Halifax Herald,* 19 July 1930. Huntsman Collection, University of Toronto Archives, B1978-0010 Box 150 File 11.

[135] Preliminary Report [of H.B. Hachey] on the Hudson Bay and Strait Fisheries Expedition of 1930. Library and Archives Canada RG 23 Vol. 1206 File 726-4-7 [1].

[136] Preliminary Report [of H.B. Hachey].

[137] W. A. Found, "Memorandum re Hudson Bay Fishery Expedition," 5 September 1930. Library and Archives Canada RG 23 Vol. 1206 File 726-4-5 [2].

[138] Walker and his colleagues took 'only 352 edible fish...in netting operations which were carried on almost daily from July 16 to September 9.' Letter from H.B. Hachey to A.G. Huntsman, 9 August 1930; and "Fishing Prospects Poor in Hudson Bay," *The Fundy Fisherman* 26 December 1930. Huntsman Collection, University of Toronto Archives, B1978-0010 Box 150 File 11.

[139] Letter from H.B. Hachey to A.G. Huntsman, 9 August 1930. Huntsman Collection University of Toronto Archives, B1978-0010 Box 150 File 11.

[140] Letter from W.A. Found to A.G. Huntsman, 21 August 1930. Huntsman Collection University of Toronto Archives, B1978-0010 Box 150 File 11.

[141] Letter from A.G. Huntsman to H.B. Hachey, 18 August 1930. Huntsman Collection University of Toronto Archives, B1978-0010 Box 150 File 11.

had also insisted upon deep tows to show if any life forms, let alone commercial fishes, existed in the depths.[142]

On September 9th, after a month of operations, *Loubyrne* departed for Hudson Strait.[143] Despite the crew's best attempts no trawling was possible here. The waters were too deep and the currents too swift, even close to shore. *Loubyrne* occupied three stations to enable scientific observations, then sailed for Halifax. Hachey estimated the expedition had fished and sampled along 3000 of a total of 8000 miles travelled, used hand lines for over 7 h, drift nets for 12 h, long lines for 3 h, and trawled nearly 60 h.[144]

Hachey's verdict was stark. In "the whole of Hudson Bay, not one commercial fish was taken" in 22 days of commercial fishing; even commercially valueless fish were rare.[145] Near Nelson, one small, cod-like fish, two kinds of lump fish, an eel pout, sand launce, and capelin were caught, the largest 7 inches long at best, and of no commercial interest.[146] Anadromous river fish–salmon, trout, white fish, capelin, sand launce and so on–did not stray far into the Bay. In Hudson Strait fishing was impossible.

The Toronto Globe commented that the results "will be a disappointment to many Canadians", and would reintroduce issues of Canadian sovereignty over Hudson Bay, but "the railway to Churchill, meantime, will serve another valuable purpose in opening a region rich in mining resources."[147] The Montreal *Gazette* reported the expedition's failure to catch a single fish of commercial value in 22 days of fishing, and lamented the dashed "visions…of trainloads of fish being carried southward and westward over these lines to the great inland markets of Canada and the United States."[148] The *Manitoba Free Press*, unconvinced, cited one Department of the Interior worker who had seen abundant salmon trout and other fish at Churchill. It criticized the scant time actually spent fishing and the expedition's preconceived notions that Hudson Bay would not form "a base for commercial fishing operations."[149]

Since Lake Winnipeg's valuable fisheries were being ravaged by parasites, Manitobans were especially disappointed. After analysing Hachey's report line by line, the National Resources Committee of the Industrial Development Board of

---

142 Ibid.

143 Preliminary Report [of H.B. Hachey].

144 Ibid.

145 To be profitable, Hachey noted, a commercial trawler like *Loubryne* had to catch in the vicinity of 1000 pounds of fish per week. See *ibid.*

146 Letter from H.B. Hachey to A.G. Huntsman, 26 August 1930. Huntsman Collection University of Toronto Archives, B1978-0010 Box 150 File 11.

147 "The Value of Hudson Bay," *The Globe*, Toronto, 23 October 1930. Library and Archives Canada RG 23 Vol. 1207 File 726-4-7 [1].

148 "Hudson Bay of no use to Fishermen," *The Gazette*, Montreal, 23 October 1930. Library and Archives Canada RG 23 Vol. 1207 File 726-4-7 [1].

149 "Fish in the Bay," *Manitoba Free Press*, 11 December 1930. Huntsman Collection, University of Toronto Archives, B1978-0010 Box 150 File 8.

A flashy two-page article in the Manitoba Free Press on the incipient Hudson Bay Fisheries Expedition of 1930 indicates the importance officials in Winnipeg attached to developing new commercial fisheries in Hudson Bay.

Manitoba concluded the fishing time had been too limited. Eighty hours of fishing along 8000 nautical miles was inadequate to profile Hudson Bay's 576,000 square miles. The expedition's need to refuel and reach various stations on time had prevented it from fishing in bays and inlets (possible havens for fish).[150] The Committee pointed out that the White Sea in northeastern Russia was similar to Hudson Bay, yet supported a huge autumn cod fishery. When Deputy Minister of Fisheries Found forwarded these complaints to Huntsman, he countered that they were "a natural reaction to an unfavourable report, when a favourable one is keenly desired."[151] A

[150] National Resources Committee, Industrial Development Board, Manitoba, 'Analysis of Dominion Government Report on Hudson Bay and Strait Fisheries Expedition of 1930', 1 February, 1932. Huntsman Collection, University of Toronto Archives, B1978-0010 Box 152 File 15.

[151] Letter from A.G. Huntsman to W.A. Found, 11 April 1932. Huntsman Collection, University of Toronto Archives, B1978-0010 Box 152 File 15.

final criticism came from B.M. Stitt, MP for The Pas, Manitoba. Found responded: "The suggestion...that an explanation of the negative results of the investigation might possibly be the use of liquor on board the trawler...in your remarks in the House of Commons, was the first of its kind to reach this Department…I think it well to attach for your information a copy of [Mr. Hachey's] reply. As it speaks for itself, it is unnecessary to comment on it here."[152] Unfortunately, Hachey's response is missing.

13.7 Conclusion: Hudson Bay Sovereignty Through Science, Not Fish

Still unanswered was the question of why Hudson Bay had so few fish. Huntsman sent Hachey's collections to the Smithsonian Institution (to which duplicate specimens were gifted), the British Museum of Natural History, and experts across North America for identification and analysis.[153] Copepods and zooplankton, for example, were sent to a veteran of the Canadian Fisheries Expedition, Dr. Arthur Willey at McGill. Phytoplankton was sent to Dr. Viola Davidson, who had been investigating phytoplankton during summers at the Atlantic Biological Station since 1924.[154] Huntsman asked for any hypothesis that might link the assemblage of species with poor fishing conditions.

The experts were mystified, although C. McLean Fraser at the University of British Columbia commented, regarding the depauperate collection of northern hydroid species, that "in my experience every good fishing area...is a good hydroid

[152] Letter from W.A. Found to B.M. Stitt, 27 May 1932. Library and Archives Canada RG 23 Vol. 1206 File 726-4-5 [3].

[153] Dr. Hugh P. Bell at Dalhousie University worked up the sea-weeds; Dr. C. McLean Fraser, Director of the Pacific Biological Station and a University of British Columbia professor, identified the hydroids; Dr. Philip Cox of the University of New Brunswick examined the sticklebacks and lumpfishes; and Vladimir Vladykov, a European expert on fish systematics, dealt with the rest of the fish. R.C. Osborn of Ohio State College examined the Bryozoa; and Dr. W.T. Calman of the British Museum of Natural History identified several specimens of *Cumacea*. Three scientists from the Smithsonian Institution Division of Marine Invertebrates in Washington DC were enlisted: Dr. C.R. Shoemaker identified amphipods and other invertibrates; echinoderms were examined by Dr. A.H. Clark; and decapods by Waldo L. Schmidt. See correspondence between A.G. Huntsman and H. P. Bell, A.H. Clark, P. Cox, C. McLean Fraser, R.C. Osborn, W. L Schmidt, C.R. Shoemaker, A. Willey. Huntsman Collection, University of Toronto Archives, B1978-0010 Box 150 Files 11 and 13, and Box 152 File 14; and letters from A.G. Huntsman to J.J. Cowie, 18 October and 24 December 1930, Huntsman Collection, University of Toronto Archives, B1978-0010 Box 150 File 11; and Libraries and Archives Canada RG 23 Vol. 1207 File 726-4-7 [1].

[154] In 2010 the Canadian Coast Guard Specialty vessel CCGS *Viola Davidson*, named in Dr. Davidson's honour, was launched to assist fisheries research in the Canadian Atlantic. Davidson is credited with being the first published female biologist at the Saint Andrews Biological Station. Huntsman repeatedly attempted to convince his Department of Biology colleagues at the University of Toronto to hire Dr. Davidson, who had been his graduate student. His representations were rejected by male colleagues saw no place for a woman within their faculty.

area."[155] Waldo L. Schmidt, writing for his colleagues in the Smithsonian Museum's Division of Invertebrates, commented "it appears that you certainly have a wonderful lot of fish food available over the ground examined. It is Cape Cod and Grand Banks material if ever there was any, and if you have no abundance of commercial food fishes feeding on this large supply there must be something else that causes [this]..." such as low temperatures. He added that biologist "Will" Thompson,[156] who had been visiting, "said it looks much as though your problem was on the order of the Sea of Okhotsk which had vast quantities of crustaceans but few fish." He wondered the expedition had lacked "practical fishermen."[157] No answers, then, were forthcoming from the Smithsonian.

Hachey in his preliminary report had proffered an unsatisfactory suggestion that Hudson Bay's mud floors, uncongenial to cod or many other fish, caused the fish deficit.[158] A better explanation did not unfold until Hachey analyzed his oceanographic samples and measurements:

> Intense stratification in the upper 25 m, decreasing as the waters of the open ocean are approached, gives Hudson bay [*sic*] the character of a large estuary. Below 50 m the waters are for all purposes dynamically dead, thus resulting in a cold saline body of water which probably undergoes very little change from season to season.[159]

Hachey determined that despite fresh water from James and Hudson Bays circulating out through Hudson Strait, the thick overlay of brackish surface water (from massive river runoff) never mixes thoroughly with deeper saline water in the bays. The vertical stability of the heavily stratified waters in the water column traps nutrients in the deep layers. The lack of nutrients available to plankton in the surface waters limits planktonic productivity–and hence food for fish.

Eric L. Mills has pointed that Hachey's efforts during and following the Hudson Bay investigations introduced dynamic oceanographic science to Canada. Dynamic oceanography, a method for calculating currents and providing a dynamic understanding of currents in large bodies of water, had already in a sense visited Canada once before, during the Canadian Fisheries Expedition of 1914–1915. Following the expedition, Hjort had put Swedish oceanographer Johan W. Sandström in charge of analysing the hydrographic data. Sandström and his colleague Bjørn Helland-Hansen had over the previous decade developed mathematical models for calculating the direction and speed of currents, thus pioneering dynamic physical oceanography. Sandström's 'The Hydrodynamics of Canadian Atlantic Waters' was published in the expedition's report: *Canadian Fisheries Expedition, 1914–1915:*

155 Letter from C. McLean Fraser to A.G. Huntsman. Huntsman Collection, University of Toronto Archives, B1978-0010 Box 150 File 11.

156 William F. Thompson (1888–1965), the pre-eminent west coast fisheries biologist, was at that time the director of the International Fisheries Commission headquartered in Seattle.

157 Letter from W.L. Schmitt to A.G. Huntsman 25 February 1931. Huntsman Collection, University of Toronto Archives, B1978-0010 Box 152 File 14.

158 Preliminary Report [of H.B. Hachey].

159 H.B. Hachey, "The general hydrography and hydrodynamics of the waters of the Hudson Bay region," *Contributions to Canadian Biology and Fisheries* **7**(1931), 93–118.

Investigations in the Gulf of St. Lawrence and Atlantic Waters of Canada (1919). Not only 'the first full-scale application of the mathematical analysis of ocean currents', it was also intended to serve as a training manual for future Canadian oceanographers.[160] Outside of Scandinavia, however, few scientists were prepared to understand the new approach. Now, 11 years later, "with some panache", Hachey used the data generated from 17 oceanographic stations made in Hudson Bay to teach himself dynamic oceanography, using 'techniques learned from Sandström's monograph of 1919 and E.H. Smith's "A Practical Method for Determining Ocean Currents".[161] Hachey's expeditionary work strengthened Canadian sovereignty in terms of scientific capability, if not in terms of fish.

Hachey's oceanographic data was too temporally and geographically limited to reveal Hudson Bay's seasonal dynamics. Scientists still believed it never froze over: it was too far south. During the Second World War, however, RCAF pilots flying across Hudson Bay in winter reported it to be ice-covered. Indeed "except near the Belcher Islands, along the coasts, and in Hudson Strait" it is completely ice covered from January to April.[162] It "is unique among the world's arctic oceans in that it is nearly covered by ice in winter and is free of ice in summer".[163] Ice formation extrudes salt water, keeping lighter fresh water at the top. The strength of the pycnocline (the steep density gradient marking the border between brackish and saline waters) in summer is reinforced by the influx of fresh water. The lighter brackish layer, warmed by the sun, remains separated from the underlying dense, highly saline deeper water layers,[164] which remain at −2.2 degrees Celsius, prevented from freezing by their high salinity. This temperature is too cold for fish reproduction in many species.[165] Higher productivity occurs in bays and estuaries, and where upwelling of nutrient-rich waters occurs near islands. The large population of bowhead whales present before whalers arrived indicates a likely "area of high productivity and therefore vertical mixing in northwestern Hudson Bay", but no such area has been discovered. Harsh winter conditions, and lack of "a commercially exploitable offshore fishery resource" have continued to limit the research needed to understand the structure and dynamics of Hudson Bay's marine ecosystem and its marine species's life histories.[166]

Despite its shortcomings, the 1930 Hudson Bay and Strait Fisheries Expedition, thanks to Huntsman's insistence on the primacy of scientific research over

---

160 Ibid., 111, 134–136.

161 Mills, *Fluid Envelope of our Planet*, 242.

162 Eric W. Danielson, Jr., "Hudson Bay Ice Conditions," *ARCTIC* **24**:(1971), 90, 94.

163 D.B. Stewart and W.L. Lockhart, "An Overview of the Hudson Bay Marine Ecosystem. Chapter 5. Oceanography," Canadian Technical Report of Fisheries and Aquatic Sciences no. **2586** (2005), 29, 40. Accessed at http://www.dfo-mpo.gc.ca/libraries-bibliotheques /toc-tdm /314704-eng.htm.

164 Stewart and Lockhart, "Overview...Chapter 5. Oceanography," 29, 40.

165 Natural Resources Committee Industrial Development Board of Manitoba, "Analysis of Dominion Government Report on Hudson Bay and Strait Fisheries Expedition of 1932." Huntsman Collection, University of Toronto Archives, B1978-0010 Box 152 File 15.

166 Stewart and Lockhart, "Overview...Chapter 5. Oceanography," 29, 40–41.

experimental fishing trials, finally offered an explanation for the grim and mystifying findings of earlier expeditions. Before this it was impossible to erase the widespread conviction that commercially exploitable fish stocks lurked in Hudson Bay's icy depths. The Hudson's Bay Company's efforts to establish a fishery in 1771 using experienced Spitsbergen fishermen could only, at best, have covered a portion of the Bay. When hopes were revived following the Dominion of Canada's acceptance of Arctic sovereignty, Wakeham's half-hearted fishing attempts, Comeau's thwarted efforts, and Johansen's shore-based surveys did nothing to dispel these. Experts could still theorize that given its historically important whale fishery, Hudson Bay must also be able to support other fisheries. Hachey's analysis finally dashed these hopes. Yet the same conditions that have limited Hudson Bay's fish populations as well as thorough scientific studies from Wakeham's 1897 expedition onwards, have in a sense helped to secure Canada's sovereignty. The port of Churchill has never been an economic success. The extended season of ice cover and difficult navigation conditions have minimized international (and even national) economic interest in the area–aside from burgeoning arctic tourism. Rather than settlement or economic development, Canadian scientific expeditions continue to be important extensions of state sovereignty, just as they were in the past.

Chapter 14
Collectors, Displays and Replicas in Context: What We Can Learn from Provenance Research in Science Museums

David Pantalony

Abstract In this chapter, I outline three prominent themes that have emerged in my provenance research in the collections of the Canada Science and Technology Museum (CSTM). First, I explore the role collectors have had on the kinds of histories we collect, preserve, research and display, and the means through which their activities build historical value for a body of preserved materials. Through the history of collectors (many are scientists), we depart from traditional narratives and learn more about their own time, culture, preoccupations and conceptions of science. Second, I look at the historical function of display in the history of museum objects, and their surprisingly long history of being on display, in many cases much longer than their technical use. These display histories tell us much about the nature of scientific communication, culture and identity. Third, I look at the history of scientific replicas and recreations, and their function in making and shaping the history and culture of science. I treat replicas as artifacts from their time of construction, and not just as representations of a previous historical moment.

At the Conservatoire national des arts et métiers (CNAM) in Paris one can view Antoine Lavoisier's original instruments in a permanent exhibition about the chemical revolution, a reminder of their established place in French and science history. The instruments are a valuable resource for the history of chemistry as well as a material record of Lavoisier's experimental efforts to redefine chemistry in the eighteenth century. In an essay on language, instruments and the chemical revolution, Trevor Levere placed these instruments at the center of this revolution by arguing that Lavoisier's "approach to chemistry was conditioned in part by the skills of his instrument makers."[1] Nicholas Fortin's work on the balance, for example,

[1] Trevor Levere, "Lavoisier: Language, Instruments and the Chemical Revolution "in *Nature, Experiment, and the Sciences: Essays on Galileo and the History of Science*, ed. Stillman Drake, Trevor Harvey Levere, and William R. Shea (Dordrecht; Boston: Kluwer Academic Publishers, 1990), 207–23, at 209–10.

D. Pantalony (✉)
Curator of Physical Sciences and Medicine at the Canada Science and Technology Museum, Ottawa, ON, Canada
e-mail: dpantalony@techno-science.ca

J. Buchwald, L. Stewart (eds.), *The Romance of Science: Essays in Honour of Trevor H. Levere*, Archimedes 52, DOI 10.1007/978-3-319-58436-2_14

shaped Lavoisier's expectations for precision.[2] For Levere, the chemical revolution resulted in part from a dynamic interaction between instrument makers, instruments and users.[3]

Whereas Levere described how these instruments shaped Lavoisier's experimental practice, Marco Beretta has studied their history of use and ownership beyond the laboratory, and their odyssey from scientific instruments to cultural artifacts that have shaped history, identity and culture.[4] Following Lavoisier's death, the authorities performed an inventory of the entire collection; they dispersed some of the possessions, but allowed Mme Lavoisier to retain the majority of her husband's apparatus. She kept them in good condition and even encouraged scientists to use them. After her death in 1836, the collection went to her relatives, the Chazelle family of Champaissant, Sarthe. In 1935 a surviving member of the family organized a small, local museum for the collection. In 1943, the instruments resurfaced dramatically in Paris during Nazi occupation in an exhibition at the Palais de la découverte on the bicentennial of Lavoisier's birth. In the early 1950s, Pierre Samuel Dupont purchased and donated them to the French government and CNAM. For Dupont, the head of the American Dupont chemical empire, this was a grand gesture to return something that should be rightfully in French hands. He was descended from Eleuthére Irénée Du Pont, the founder of the Du Pont company who had apprenticed under Lavoisier, before moving to America in 1800. The instruments have been an iconic attraction at the CNAM for over 60 years.[5]

---

[2] Trevor Levere, *Chemists and Chemistry in Nature and Society, 1770–1878*, ed. Trevor Levere (Aldershot: Variorum, 1994), 313–32.

[3] Under the supervision of Levere, I adopted this as a central theme in my PhD dissertation on Rudolph Koenig and the instruments he constructed related to the acoustical research of Hermann von Helmholtz. David Pantalony, *Altered Sensations: Rudolph Koenig's Acoustical Workshop in Nineteenth-Century Paris* (Dordrecht; New York: Springer, 2009).

[4] A. Truman Schwartz, "Instruments of the Revolution: Lavoisier's Apparatus," *Bulletin for the History of Chemistry* **5** (1989), 31–34; Marco Beretta, "Lavoisier's Collection of Instruments: A Checkered History / Marco Beretta.," in *Musa Musaei: Studies on Scientific Instruments and Collections in Honour of Mara Miniati.*, ed. Paolo Galluzzi, Marco Beretta, Cario Triaric (Firenze: Leo S. Olschki, 2003), 313–34.

[5] The iconic 1788 Jacques-Louis David portrait of Lavoisier, his wife and instruments has had an equally complex ownership history. Metropolitan Museum of New York, Inventory No. 177.10. Purchase, Mr. and Mrs. Charles Wrightsman Gift, in honor of Everett Fahy, 1977. Full provenance for MMA 1977.10: Antoine-Laurent Lavoisier, Paris (1788–d. 1794); Mme Antoine-Laurent Lavoisier, later Countess Rumford, Paris (1794–d. 1836); her great-niece, comtesse Pierre-Léon Bérard de Chazelles (Jeanne-Marie-Laure-Hélène-Gabrielle Ramey de Sugny), Paris, and later the Auvergne (1836–1876 [his death] or 1888 [her death]); her son, comte Étienne Bérard de Chazelles, Paris, and château de la Canière, near Aigueperse (by 1888–d. 1923; his estate, 1923–24; sold by his heirs to Wildenstein); [Wildenstein, Paris and New York, 1924–25; sold to Rockefeller]; John D. Rockefeller Jr., New York (1925–27); Rockefeller Institute for Medical Research, later Rockefeller University, New York (1927–77; sold to MMA)

14.1 Provenance Research and Scientific Instruments

The Lavoisier artifacts show us how much we can learn about science and culture by following historical artifacts throughout their history from creation to present. In this chapter, I apply the methods of museum provenance research, or the full history of ownership of an artifact as a window into uncovering multiple histories of scientific instruments beyond their original scientific place and function. Following the work of Igor Kopytoff, this has been referred to as the biographical approach to provenance studies.[6]

Here I outline three prominent themes that have emerged in my provenance research in the collections of the Canada Science and Technology Museum (CSTM). First, perhaps an obvious theme, I explore the role collectors have had on the kinds of histories we collect, preserve, research and display, and the means through which their activities build historical value for a body of preserved materials. Through the history of collectors (many are scientists), we depart from traditional narratives and learn more about their own time, culture, preoccupations and conceptions of science.[7] Second, I look at the historical function of display in the history of museum objects, and their surprisingly long history of being on display, in many cases much longer than their technical use. These display histories tell us much about the nature of scientific communication, culture and identity.[8] Third, I look at the history of scientific replicas and recreations, and their function in making and shaping the

---

[6] Igor Kopytoff, "The Cultural Biography of Things: Commoditization as Process," in *The Social Life of Things: Commodities in Cultural Perspective*, ed. Arjun Appadurai (Cambridge and New York: Cambridge University Press, 1986), 64–91.

[7] There is a wide variety of histories of collectors and collections that shed light on the history of science museums: William J.H. Andrewes, "The Life and Work of David Pingree Wheatland (1898–1993)," *Journal of the History of Collections* **7** (1995), 261–68; Ken Arnold, *Cabinets for the Curious: Looking Back at Early English Museums*, Perspectives on Collecting (Aldershot, England; Burlington, VT: Ashgate, 2006); Edmund De Waal, *The Hare with Amber Eyes: A Family's Century of Art and Loss*, 1st American ed. (New York: Farrar, Straus and Giroux, 2010); Paula Findlen, *Possessing Nature: Museums, Collecting, and Scientific Culture in Early Modern Italy*, Studies on the History of Society and Culture (Berkeley: University of California Press, 1994); O. R. Impey and Arthur MacGregor, *The Origins of Museums: The Cabinet of Curiosities in Sixteenth and Seventeenth-Century Europe* (Oxford Oxfordshire, New York: Clarendon Press; Oxford University, 1985); Richard Kremer, "A Time to Keep and a Time to Cast Away: Thoughts on Acquisitions for University Instrument Collections," *Rittenhouse* **22** (2008), 188–210; Marta C. Lourenço, *Between Two Worlds; the Distinct Nature and Contemporary Significance of University Museums and Collections in Europe* (PhD Dissertation, Conservatoire national des arts et metiers, 2005); David Pantalony, Richard L. Kremer, and Francis J. Manasek, *Study, Measure, Experiment: Stories of Scientific Instruments at Dartmouth College*, 1st ed. (Norwich, Vt. Lebanon, NH: Terra Nova Press; Distributed by University Press of New England, 2005); Simon Schaffer, "Object Lessons," in *Museums of Modern Science: Nobel Symposium 112*, ed. Svante Lindqvist, Marika Hedin, and Ulf Larsson (Canton, MA: Science History Publications/USA, 2000), 62–76.

[8] Silke Ackermann, Richard L. Kremer and Mara Miniati, *Scientific Instruments on Display*, History of Science and Medicine Library (Leiden; Boston: Brill, 2014).

history and culture of science.[9] I treat replicas as artifacts from their time of construction, and not just as representations of a previous historical moment.

14.1.1 Collectors in Context: Building the Canadian Surveying Landmark Collection

Scientific instruments and objects often have not been saved and preserved by historians of science, nor curators of science. But scientists, practitioners and technicians have saved them. The story of who collected them, how and why is quite telling of their own historical period.

On the shelves of the warehouse at the CSTM, there are 138 historic survey markers from across Canada. They are made of wood, iron and stone, and span dates 1762–1939. Canadian land surveyors collected them from every region of Canada.

Upon first sight, the markers appear to be fairly simple and similar to each other. Upon closer examination their details, wear, varying size and style, materials and construction evoke distinct histories of exploration, surveying, practice, resources and place. One broken wooden piece clearly shows "Dawson Tow Site" [Townsite][10] (Fig. 14.1). In 1901, Dominion Land Surveyor (D.L.S.), A. J. McPherson placed this post at the southeast corner of the town in the Yukon. He secured it in a stone mound, common for that time.[11] In 1960 R.T. McCurdy, D.L.S., discovered this piece while updating the survey; he replaced it with a "standard post, 4 pits and a mound."[12] Surveying and collecting was a dangerous business. On August 11, a day before finding the Dawson marker, McCurdy recorded in his field book that they "saw one bear in AM [and] grizzly BEAR following them down the line – very hasty retreat uphill [and] returned to camp 4:00 PM.".[13]

All the historic survey markers came to the National Museum of Science and Technology Museum (now CSTM) in 1973 from the Surveys and Mapping Branch of the federal department of Energy, Mines & Resources (now Natural Resources Canada or NRCan). It was part of a collection of "Canadian Landmarks" gathered over twenty years by the Dominion Land Surveyor (DLS) network across Canada. Ralph W. Clark, D.L.S. started this collection in 1952 in the post-World War II period of rapid expansion, aerial photography, surveying and updates.[14]

---

[9] K. B. Staubermann, *Reconstructions: Recreating Science and Technology of the Past* (Edinburgh: NMS Enterprises, 2011).

[10] The apparent misspelling appears as "Townsite" on another marker in the collection, "Klondike Townsite." Canada Science and Technology Museum (CSTM) Artifact Number 1973.0762.

[11] Canada Lands Survey Records, FB37158 CLSR YT and Plan No. 9060 CLSR YT

[12] CSTM Artifact Number 1973.0761

[13] R. T. McCurdy 1960, FB.34390, Canada Lands Survey Records.

[14] L.M. Sebert, "One Hundred and Sixty-Three Landmarks," *The Canadian Surveyor* XIX, no. 2 (1965): 220–23; J.H. Webb, "Ralph William Clark (1907–1970)," *SLSA Corner Post: Quarterly Newsletter of the Saskatchewan Land Surveyors' Association* **29**, no. 4 (2008), 124–27.

Fig. 14.1 A wooden survey post, broken in two, deteriorated but still showing its original marking. Canada Science and Technology Museum, Artifact Number 1973.0761

"Because today's survey monuments have changed to more standardized type of iron post, the items represented in this collection will soon become extinct,"[15] Clark wrote. The post-World War II surveyors saw the markers as representing the efforts by their predecessors to mark and survey the early Dominion.

Rarely does the public see these kinds of survey markers out of context. Only registered surveyors can remove the markers, and if they do, they must dispose of them so they do not resurface later and cause legal complications. Clark and his colleagues, however, started gathering the markers from across Canada and preserving them in Ottawa with official sanction from a Landmark Committee at the Department of Energy Mines and Resources. They preserved delicate materials, restored broken pieces and mounted them for display.[16]

Clark started his collection in 1952. He was performing a resurvey of one block of a subdivision in Banff, Alberta when he and his survey party became puzzled by a missing marker. Closer examination of the area "with a sweep of the hand among the surface debris" revealed two metal discs (top one inscribed "27/IV") and a galvanized pipe that was roughly in the right location, but did not fit the original survey description. The party went on to finish their job, but Clark kept the historic remnants.

> The more the surveyor [Clark] at the Banff site thought about the strange marker the more interesting he perceived it to be, and eventually he decided to postpone final disposal and to take the pipe with its discs back to Ottawa for the information of his colleagues on the staff of the Surveyor General of Canada Lands.
>
> And that is how Ralph Clark, D.L.S., began his collection of Canadian landmarks, in the summer of 1952.[17]

In 1914 Dominion Land Surveyor C.M. Walker had placed this post in the "North East corner of Lot 27, Block 4, Villa Lot Section."[18] In 1952 Clark brought it to a

---

[15] Documentation of Canadian Landmarks, Dept. of Energy Mines & Resources, Surveys and Mapping Branch, Supplementary Information for CSTM Artifact Numbers 1973.0683–1973.0817.

[16] Webb, "Ralph William Clark (1907–1970)."

[17] Sebert, "One Hundred and Sixty-Three Landmarks," p. 221. Information about this specific marker can be found in Supplementary Information CSTM Artifact Number 1973.0708.

[18] CSTM Artifact Number 1973.0708

Fig. 14.2 Limestone Marker from 1762. Photo: Canada Science and Technology Museum, Artifact Number 1973.0716

new home in Ottawa. For Clark, the discs and pipe represented the complexities of past practice in surveying.

The Landmark collectors obtained the oldest stone marker from the Kahnawá:ke Mohawk Reserve on the south shore of the St. Lawrence near Montreal. (Fig. 14.2). In 1765 J. Raimont and L. Guyer placed the limestone post on the aboriginal lands following the order from a military court of field officers in Montreal. The marker is inscribed it "G.R. III" (King George the Third) and "Honi soit qui mal y pense" [Evil be to him who evil thinks].[19] It also has the inscription "1762," but it was posted following Royal Proclamation in 1763.[20]

Wooden markers are also prominent in the collection. Surveyors "blazed" tree sections when they were directly on a surveying line. Arthur Noakes, B.C.L.S. marked such a "bearing tree" in 1915 on the Cowichan Reserve in British Columbia. He inscribed the Douglas Fir with "B.T. 13" and the tree eventually grew over and hid the original blaze, but the inscription was found transferred in reverse to the resin on the new growth as a scar.[21]

Several of the historic markers are inscribed with "IR" for "Indian Reserve." In 1873, J. McLatchie, one of the more a well-travelled Dominion Land Surveyors

[19] CSTM Artifact Number 1973.0716

[20] Don W. Thomson, *Men and Meridians: The History of Surveying and Mapping in Canada* (Ottawa: R. Duhamel, Queen's Printer, 1967), vol. 2, 277–8.

[21] CSTM Artifact Number 1973.0782

of his day,[22] placed an iron post at the east boundary of the Rolling River Reserve, Manitoba. Clark replaced it in 1953.[23] McLatchie's survey took place amidst much strife between the Red River (1869) and North-West (1885) Rebellions. He had even been part of the Dominion Land Surveyors Intelligence Corps of 1885.[24] Many of the IR markers, in fact, come from a similar context of discord. An unknown surveyor used a shotgun barrel to mark the division between Kingsclear Reserve and St. Ann's Catholic Church in New Brunswick. The prolific D.L.S. surveyor R.O. Semper found this marker in 1963 and brought it back to Ottawa.[25]

14.1.1.1 Mapping the Surveyors

Historic survey markers are therefore ideal for detailed historical and geographical provenance studies. We can trace the use of surveying instruments and markers in precise geographical detail through the surveyors's field books, plans and maps. In 1894 W.A. Austin documented the 1762 stone marker at "Caughnawaga" Reservation [now Kahnawá:ke] and drew it in his field book.[26] (Fig. 14.3). It had been knocked off the stone foundation. He does not record if he put it back together, but by 1957 Quebec Land Surveyor, Gaston Bolduc had come across the same limestone marker and recorded that he had "Found top of tablet under nearby cottage – top bears inscription G.R. III 1762.".[27]

The above story high-lights a tension that emerges in detailed, geographical provenance studies—the closer one gets to the exact "find spot," (a term used by archeologists) the more one encounters various uncertainties. Historian Daniel Rueck has written about the history of surveying at Kahnawá:ke, and the accounts of locals moving or destroying markers. For the local aboriginal community, surveying represented an unwanted, outside authority, as well as carrying different notions of property.[28] The story behind the broken, relocated stone was not a trivial part of the provenance, but essential to understanding its local context.

The preservation of these monuments, above all, represented growing nostalgia within a tight community during profound transformation in post-World War II

[22] James Grierson MacGregor, *Vision of an Ordered Land: The Story of the Dominion Land Survey* (Saskatoon: Western Producer Prairie Books, 1981), 12, 44, 52, 61, 111.

[23] CSTM Artifact Number 1973.0694

[24] The CSTM has several instruments, notebooks, books, manuals and photos from McLatchie in the artifact file nos. 1978.0715–1978.0753. One photo album contains several photos from his service in the Intelligence Corps in 1885.

[25] CSTM Artifact Number 1973.0786.

[26] W.A. Austin Sept. 1894, FB347 CLR, Canada Lands Survey Records.

[27] Gaston Bolduc 1958, Plan No. 50311 CLSR, and Gaston Bolduc 1957 FB.1279 CLSR, Canada Lands Survey Records.

[28] Daniel Rueck, "I Do Not Know the Boundaries of This Land, but I Know the Land Which I Worked': Using HGIS in the Study of Indigenous Environmental History," in *Historical GIS in Canada*, ed. Jennifer Bonnell and Marcel Fortin (Calgary: University of Calgary Press, 2014), 129–52, on 138

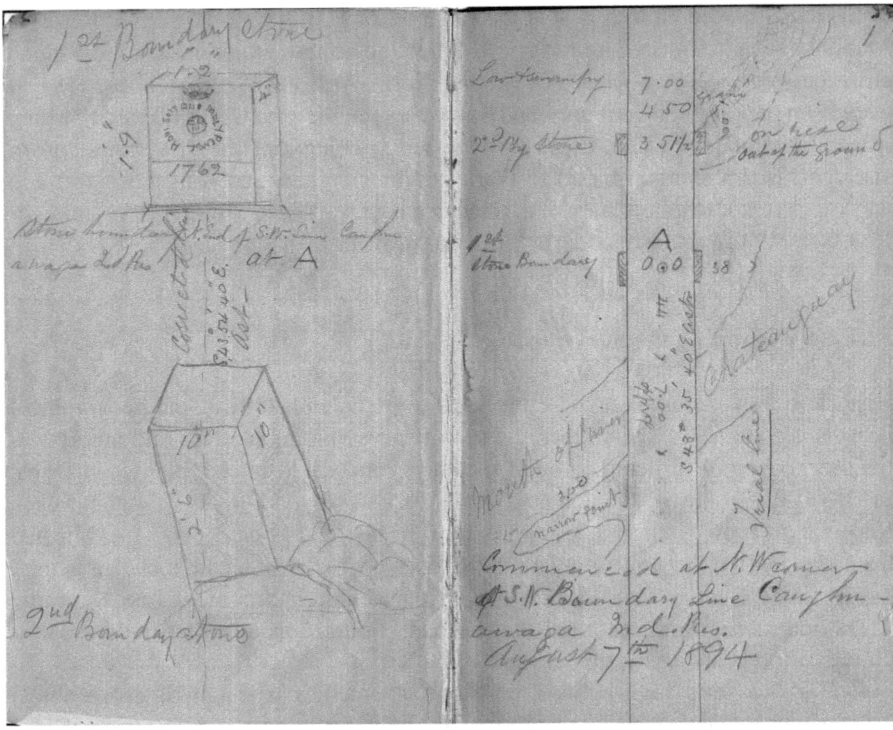

Fig. 14.3 Field notes with sketch of 1762 marker by W.A. Austin Sept. 1894, FB347 CLR, Canada Lands Survey Records

Canadian surveying.[29] When the marker collection came to the national museum in 1973, R.O. Semper, the chair of the Historic Survey Monument Committee at Energy Mines and Resources (EMR) remarked that the "posts have become all the more valuable as specimens of an era gone by."[30] In the 1950s and '60s, several staff at EMR demonstrated a collective historical consciousness. Don Thomson began researching and documenting the history surveying and mapping in Canada, which formed the basis for his three-volume classic *Men and Meridians*.[31] There were a number of other historical accounts of surveying released around the time of Canada's centennial.[32] Following the lead of Clark, Semper and Thomson, Louis Sebert, a retired mapper and surveyor at EMR, provided guidance to the National

[29] *Mapping a Northern Land: The Survey of Canada, 1947–1994*, ed. Gerald McGrath and L. M. Sebert (Montreal: McGill-Queen's University Press, 1999).

[30] Supplementary Information CSTM Artifact Numbers 1973.0683–1973.0817

[31] Thomson, *Men and Meridians: The History of Surveying and Mapping in Canada.*

[32] A. J. Birrell, *Into the Silent Land: Survey Photography in the Canadian West, 1858–1900: A Public Archives of Canada Travelling Exhibition*, ed. Public Archives of Canada (Ottawa: Public Archives Canada, 1975); Courtney C. J. Bond, *Surveyors of Canada, 1867–1967*, ed. Surveying Canadian Institute of (Ottawa: Canadian Institute of Surveying, 1966); Thomson, *Men and Meridians: The History of Surveying and Mapping in Canada.*

Museum of Science and Technology for collecting the modern instruments used for mapping and surveying. He had an active interest in the history of exploration and surveying in Canadian history, especially in the post-World War II period.[33]

Sebert had trained with the classical survey instruments, but he had also used devices such as the tellurometer that employed radio signals for measuring distance. Amidst these dramatic post-war advances, Sebert found meaning in preserving simple objects such as a 1960s range pole. Surveyors often viewed these technologies as common and routine, and thus discarded them. Sebert, however, viewed the pole as an important representative of the surveyor's practice. It is a simple surveying tool but it journeyed to many places throughout Canada.[34]

The surveyor-collectors were also nostalgic for instruments connected to specific projects. In 1922, realizing the significance of a nineteenth-century Steinheil theodolite found in Topographical Survey Branch, W.W. Cory, Deputy Minister of the Department of the Interior personally transferred it to the Dominion Archivist. Dominion Land Surveyor J.J. McArthur's had used the instrument to survey the Rocky Mountains between 1887 and 1892. From 1893 to 1894, surveyors used it to establish the boundary between Alaska and British Columbia. It rested in the National Archives until 1937 when archivists sent it to the National Research Council (NRC) where scientists had created a small survey museum. The NRC had to close the museum due to needed research space during World War II, so the objects went into boxes and storage. In 1967 and 1984 several of these instruments were transferred to the newly formed National Museum of Science and Technology (Fig. 14.4).

14.1.2 Display as History: Mathematical Models in Context

In several instances of researching the provenance of scientific instruments, one comes across long histories of display and storage, sometimes much longer than the original scientific use of the instrument. The display of scientific objects (in museums and/or scientific institutions) has a core function within science.[35] To illustrate this theme, I focus on a group of nineteenth-century string mathematical models (28 in total) at the CSTM that have a history of ownership, function and institutional context that enriches and complicates their traditional history within mathematics.

Their general history relates to a field of mathematics called descriptive geometry, a branch of geometry where one performs geometric operations in three-dimensions.

[33] L. M. Sebert, *The Evolution of Surveying and Navigational Instruments and Their Use in Mapping* (Ottawa: National Museum of Science and Technology, 1990); L. M. Sebert, *Mapping a Northern Land: The Survey of Canada, 1947–1994*; *Surveying in Canada: A Proposal for an Exhibit* (Ottawa: National Museum of Science and Technology, 1989).

[34] Supplementary Information CSTM Artifact Number 1993.0200

[35] David Pantalony, "Biography of an Artifact: The Theratron Junior and Canada's Atomic Age," *Scientia Canadensis* **34** (2011), 51–63; Ackermann, Kremer, and Miniati, *Scientific Instruments on Display*.

Fig. 14.4 Steinheil Theodolite c. 1870. Signed "Steinheil in Munchen No. 17861" Inscribed "T.S. 45" (Topographical Survey) Canada Science and Technology Museum, Artifact Number 1966.0527

This field derived from the work of French mathematician Gaspard Monge (1746–1818) in the context of engineering drawing and aids to visualization in the form of textbooks and three-dimensional string models for teaching. A student of Monge, the mathematician, Théodore Olivier (1793–1853), improved, produced and popularized the string models. In the 1840s and 50s, the Paris instrument firm Pixii made and sold them to schools in Europe and several emerging institutions in the United States such as Harvard, Princeton, West Point, and Columbia. Union College obtained original Olivier models in 1869.[36] The Conservatoire national des arts et métiers (CNAM) had original models.[37] The University of Coimbra in

[36] V. Ennis Pilcher, *Early Science and the First Century of Physics at Union College, 1795–1895* (Schenectady, N.Y.: Union College, 1994).

[37] Peggy Aldrich Kidwell, Amy Ackerberg-Hastings, and David Lindsay Roberts, *Tools of American Mathematics Teaching, 1800–2000* (Baltimore; Washington, D.C.: Johns Hopkins University Press; Smithsonian Institution, 2008); Amy Shell-Gellasch, "The Olivier String Models at West Point," *Rittenhouse* **17**, no. 2 (2003), 71–84.

Portugal bought several Olivier models along with a comprehensive collection of French scientific instruments.[38]

At West Point, the models were part of French traditions in technical education and training engineers. However, the professors used them as demonstrations, and not for allowing students to explore descriptive geometry with their own hands. Profess Arthur Hardy commented:

> In descriptive geometry, the Academy had a magnificent collection of models, but they were shown us *after* the study was finished – in other words, mental discipline was the object – practical helps and ends were secondary.[39]

14.1.2.1 Museum Display as Function and History

The models in Ottawa have a history related to museums and display with a different context of use. Upon closer examination of the "Intersection of Two Cylinders" model, we see the original museum number 1872–125 (Fig. 14.5). This number derives from the South Kensington Museum in London, the predecessor of the Science Museum. In 1872 the South Kensington commissioned the Parisian instrument maker, Fabre De LaGrange to make forty-five Olivier models for their education collection.

The South Kensington collected the models as part of the education collection, but they were also part of a broader nineteenth-century museum context. First, they fit into the encyclopedic ideal of the museum as an attempt to collect every type of object from a wide classification of human activities and knowledge. Second, the museum encouraged scientists, teachers and technicians to study the models to learn about construction and use. Third, the models were part of a new kind of national display deriving from world exhibitions.[40]

The 1872 catalogue described the principle purpose of the models for illustration in the Industrial Arts:

> This collection illustrates the principal types of the class of surfaces which can be traced out in space by the motion of a straight line. These surfaces, on account of the facility with which they can be constructed and represented and of the ease with which their intersections can be determined, are of more consequence than any others in the geometry of the Industrial Arts.[41]

---

[38] Carlota Simões and Carlos Tenreiro, "O Gabinete de Geometria da Faculdade de Ciências e a sua colecção de modelos para o ensino," in *História da ciência na Universidade de Coimbra: 1772–1933*, ed. Carlos Fiolhais, Carlota Simões, Décio Martins (Coimbra: Universidade de Coimbra, 2013), 193–207.

[39] Shell-Gellasch, "The Olivier String Models at West Point," 78.

[40] R.M. Brain, *Going to the Fair: Readings in the Culture of Nineteenth-Century Exhibitions* (Cambridge: Whipple Museum, 1993).

[41] C. W. Merrifield, *A Catalogue of a Collection of Models of Ruled Surfaces, Constructed by M. Fabre De Lagrange; with an Appendix, Containing an Account of the Application of Analysis to Their Investigation and Classification* (London: South Kensington Museum, 1872), 3.

Fig. 14.5 Olivier String model. A preserved label from the Science Museum reads: "Intersection of Two Cylinders, Made by De LaGrange, Paris. In this model one of the cylinders pierces the other giving two separate loops of intersection. Inv. 1872.125" Photo: Canada Science and Technology Museum, Artifact Number 1973.0241

The museum acted as a meeting place to study the models. Students of geometry, we read in the catalogue, used the models to study changing geometric surfaces:

> The models are constructed with especial reference to the possibility of changing their shape, by moving some of the supports of the strings, by altering the lengths or positions of certain parts, or by converting upright forms into oblique. This possibility of deformation, as the process is technically called, greatly enhances the value of the models, by allowing them to represent a much greater variety of surfaces than if they were fixed.[42]

However, the author noted that students did not handle the models directly:

> They are, however, too delicate to be much pulled about, and, unless they are very cautiously handled, the strings are apt to become entangled or break. They should never be used except by a person who understands them, and they should not be shifted without some good reason.[43]

These models have existed in an evolving museum context throughout their history. In the 1870s, the South Kensington Museum combined collections from decorative arts and science and industry; this framework had emerged from the Great Exhibition of 1851. The objects co-existed with others from around the world

---

[42] Ibid, 4.

[43] Ibid, 4.

and the visitors would have seen many similar items with wood, silk thread, brass displaying Victorian engineering aesthetics. The models were also usable in theatrical, education and demonstration aspects of science and mathematics. The "1872" inventory number was most likely added later when the museum split into the Victoria and Albert Museum and the Science Museum. It was also marked in pencil as "E.35. – 1872", referring to the education department of the South Kensington Museum. Throughout the twentieth century, curators at the Science Museum transformed the use of the models from those of education display to artifacts of history and culture.[44]

14.1.2.2 Broken: The Recent History of the Models

By mid-twentieth century, storage and disrepair came to define their function, history and context. In February 1952, H. Saunders of the Science Museum re-strung one of the models and recorded this on the bottom of the wooden base.[45] Several of them had fallen into disrepair, were no longer displayed and in 1973 were eventually sold. At the CSTM, we have kept them mostly in storage with occasional display as beautiful mathematical objects.[46] In 2010, the ones in disrepair caught the attention of our artist-in-residence Robert Bean who focused on their entanglement and disintegration as the subject of a broader photographic portrait of obsolescence; Bean showed these photographs in exhibitions in Toronto, Nova Scotia and Paris.[47]

The "broken" models are also worthy of in-depth material scrutiny.[48] When the conservation staff at the CSTM restored several of them based on their original

[44] In 2011 the Science Museum exhibited their mathematical instruments and models in "From Order to Obsession: A View of Mathematics." Curated by Jane Wess, the exhibition highlighted four themes in the history of mathematics – beauty, power, necessity and play. The catalogue placed the models in "Necessity" highlighting their practical functions. "Most people engage in Mathematics not because they consider them beautiful, find it powerful or enjoy it, but because they simply have to." Science Museum, *From Order to Obsession: A View of Mathematics* (London: Science Museum, 2011), section "Necessity."

[45] CSTM Artifact Number 1973.0241

[46] In 1989, CSTM showed one in an exhibition called *Exploring the Collection*. In 1990, CSTM displayed a model in "A Way with Math" mostly devoted to contemporary mathematical education and teaching. In 1998 curators placed "Intersection of Two Cylinders" (Artifact Number 1973.0241) in the entrance lobby as part of an open-collection display.

[47] Robert Bean's Exhibition List displaying this instrument: "Obsolescence and Inscription: Robert Bean and Ilan Sandler". Saint Mary's University Art Gallery, Halifax, Nova Scotia. October 19 – December 1, 2013; "273 (brushing information against information)." Circuit Gallery, Toronto. November 2011; "Illuminated Manuscripts" and "Polyphony," Centre culturel canadien, Paris, France. September 26 – November 16, 2011; "Illuminated Manuscripts," McLuhan Program in Culture and Technology (The Coach House), the University of Toronto, a site-specific installation commissioned by the CONTACT Photography Festival and the McLuhan Program in Culture and Technology, May – August, 2011

[48] Edward Jones Imhotep and I co-curated an exhibit related to this theme – "Broken: An Interactive Exhibit" at the *Materiality: Objects and Idioms in Historical Studies of Science and Technology*, York University, May 2–3, 2013.

design, use and materials, we gained more insight into the amount of effort, skill and choice that went into their original productions. Skilled artisans intricately and precisely designed these models based on Olivier's mathematics and previous models; skilled people harvested, produced, dyed and sold the silk; skilled instrument makers constructed the precision brass parts and wooden frames; the instrument maker and his workers designed, assembled and strung these parts into a complex mathematical machine. The maker arranged, displayed and marketed similar ones at fairs, salons and exhibitions. The condition of artifacts in museum storage is perhaps one of the best windows into the complexities and efforts inherent in the making, demonstration and display of these objects, but also of their evolving value (or lack of) as cultural entities.

14.1.3 Replicas in Context: Babbage, Guatelli and Cohen

In 2010 a donor brought to our museum what appeared to be nineteenth-century Babbage difference engine. It appeared to be an intricate brass calculating machine with layers of Escher-like gears, numerical dials, and classical clock-like pillars. Close examination showed that it was certainly not brass, nor nineteenth-century - some of the gearing and finish were from the mid to late twentieth century. It was extremely heavy (Fig. 14.6).

The machine was in fact a replica from the 1970s of an original trial portion of a Difference Engine from 1833. It had a rich history involving a colourful artisan and scholar named Roberto Guatelli, IBM sponsorship, and a network of influential historians of science and designers. An identical model had been part of the 1972 exhibition *The Computer Perspective* produced by IBM and the California designers, Ray and Charles Eames. Harvard historian of science, I.B Cohen, who had already consulted for IBM to develop their historic collection, wrote the introduction to the catalogue for the exhibition.[49] The same calculating model appeared in the 1983 IBM exhibition "A Calculator Chronicle" developed by I.B. Cohen.[50] The exhibition showcased "300 years of Counting and Reckoning Tools" which included print material, thirteenth-century tally sticks, an early abacus, Napier's Rods, slide rules, Pascal calculator, Leibniz calculating machine and the many increasingly complex calculating machines from the nineteenth-century into the electronic age.

[49] Charles Eames, Ray Eames, and International Business Machines Corporation, *A Computer Perspective* (Cambridge: Harvard University Press, 1973). Cohen's involvement with the IBM collection dates back to the late 1960s, see I. Bernard Cohen Papers, 1889–1987. Finding Aid. CBI 182 University of Minnesota Libraries. Charles Babbage Institute.

[50] I. B. M. Gallery of Science and Art, *A Calculator Chronicle: 300 Years of Counting and Reckoning Tools: The Mechanization of Arithmetic* (Armonk, N.Y.: IBM Co., 1983), p. 22. Report on the Exhibition "A Calculator Chronicle" at the IBM Gallery of Science and Art, Draft #2, I. Bernard Cohen, 1985. (Box 1). I. Bernard Cohen Papers, 1889–1987. Finding Aid. CBI 182 University of Minnesota Libraries. Charles Babbage Institute, accessed October 30, 2015.

Fig. 14.6 Replica of Babbage's "Trial Portion" of the Difference Engine from 1833. Made by Roberto Guatelli c. 1972. Photo: Canada Science and Technology Museum, Artifact Number 2011.0022

The replica was based on an original Babbage machine which ended up at the Science Museum in London.[51] Charles Babbage began developing the Difference Engine in 1821; he envisioned an automated machine that reduced human error while producing tables of mathematical functions using the method of finite differences.[52] He was inspired by the ever-present industrial machines of that period. He did not finish any of his mathematical conceptions, but the Science Museum holds the original "trial portion" of Difference Engine No. 1.[53] It is often referred to as

[51] Henry Prevost Babbage, *Babbage's Calculating Engines: A Collection of Papers* (Los Angeles: Tomash, 1982), 339–343. Science Museum Inventory No. 1862–89. David Baxandall and Jane Pugh, *Calculating Machines and Instruments* (London: Science Museum, 1975), 20.

[52] Bruce Collier and James H. MacLachlan, *Charles Babbage and the Engines of Perfection* (New York: Oxford University Press, 1998); Doron Swade, "Pre-Electronic Computing," in *Dependable and Historic Computing Essays Dedicated to Brian Randell on the Occasion of His 75th Birthday*, ed. Cliff Jones and John L. Lloyd, *Lecture Notes in Computer Science* (Springer, 2011), 58–83.

[53] Babbage, *Babbage's Calculating Engines: A Collection of Papers*, 339–343. Science Museum Inventory No. 1862–89; Baxandall and Pugh, *Calculating Machines and Instruments*, 20.

"the beautiful fragment."[54] There are also parts of his original Difference engines at Oxford, Harvard and Cambridge. These constitute a distributed physical memory of Babbage's original attempts to build a difference engine.[55]

It was from these models, well known to Cohen, that Guatelli and his technicians produced the replicas. The Computer History Museum in Mountain View, California owns a replica identical to the one now in Canada. Dated from c. 1972, it is a ¾ scale replica and has a plaque that reads "To T. Vincent Learson with appreciation and admiration from his associates in IBM, December 1972."[56] Learson was IBM's Chairman and CEO from 1971 to 1973.

The Babbage replica has deep roots in the 1960s, and IBM's investments in design, science education, arts and culture, and shaping their corporate identity. In 1961 at the California Museum of Science and Industry in Los Angeles, IBM sponsored a ground-breaking exhibition *Mathematica: A World of Numbers... and Beyond* in collaboration with the well-known California designers, Charles and Ray Eames.[57] The Eames Office used this opportunity, their first exhibition, to test their emerging ideas related to information and design.[58] They introduced highly influential notions of exhibit design and interactivity into the science museum world. "In doing an exhibition, as in *Mathematica*, one deliberately tries to let the fun out of the bag," Charles Eames wrote.[59] The show was an enormous success and went on to be duplicated at the New York World's Fair in 1964 and the Chicago Museum of Science and Industry. In making the exhibition, and several subsequent film and exhibition projects related to science, the Eames Office and IBM developed fruitful relationships with Guatelli, and Harvard historians of science I.B. Cohen and Owen Gingerich. The Eames also worked with UCLA mathematician Raymond Redheffer on what became an iconic timeline poster "Men of Modern Mathematics.".[60]

By the late 1960s, the *Mathematica* exhibit, films and poster projects inspired deeper interest by IBM in historic calculators and computers.[61] I.B. Cohen was instrumental in this effort, and Guatelli produced calculator reproductions for IBM's collection, corporate displays and public exhibitions.[62] Cohen's *Calculator Chronicle* had been an inaugural exhibit at the IBM Gallery of Art and Science at

[54] Henry Prevost Babbage, *Babbage's Calculating Engines Being a Collection of Papers Relating to Them, Their History and Construction*, Calculating Engines (London: E. and F.N. Spon, 1889), preface.

[55] D. Roegel, "Anecdotes: Prototype Fragments from Babbage's First Difference Engine," *IEEE Annals of the History of Computing* **31**, no. 2 (2009), 70–75.

[56] Computer History Museum, Catalogue No. 102711127

[57] International Business Machines Corporation. et al., "Mathematica: A World of Numbers... And Beyond, Presented by IBM," (New York: IBM, 1961).

[58] Pat Kirkham, *Charles and Ray Eames: Designers of the Twentieth Century* (Cambridge, Mass.: MIT Press, 1995), 263–308.

[59] Ibid., 297.

[60] Ibid., 263.

[61] Ibid., 300.

[62] I. Bernard Cohen Papers, 1889–1987. Finding Aid. CBI 182 University of Minnesota Libraries. Charles Babbage Institute.

their new tower at 590 Madison Ave.[63] The Gallery of Art and Science, which operated from 1983 to 1993, was a popular and influential destination in New York's cultural scene.[64] Before 1983, the "IBM Gallery" at the base of the IBM tower at Madison Ave and 57th Street had an impressive history of exhibits related to art, science, technology and American abstract art in the 1960s.[65] By the time of the *Mathematica* exhibitions in 1961, exhibitions and history had begun to serve prominent functions in the IBM corporate outlook.

14.1.3.1 IBM, Guatelli and da Vinci in Post-World War II America

Since the 1930s, IBM's founder, Thomas Watson Sr. had been collecting antiques and books related to the history of calculating and computing.[66] They were first exhibited at the 1939 World's Fair, and then in displays at IBM headquarters at Madison and 57th Street (built in 1938, demolished in 1977). In 1951, Watson channeled this interest toward Leonardo da Vinci and a travelling exhibition of his machines.[67] A few years earlier, Watson had seen an exhibit of replicas made by the Italian artisan and scholar Roberto Guatelli; he subsequently bought the collection and hired Guatelli. Thus began a long relationship between Guatelli and IBM.[68]

Roberto Guatelli (1904–1993), a science and engineering graduate from the University of Milan, developed his reputation by building and exhibiting interpretations based on the work of Leonardo da Vinci. In 1939, fitting the nationalist context of the time, the Italian government sponsored a major exhibition of Leonardo da Vinci models at the Palazzo dell'Arte Milan.[69] Several scholars and artisans became

[63] IBM Gallery of Science and Art, *A Calculator Chronicle: 300 Years of Counting and Reckoning Tools: The Mechanization of Arithmetic* (Armonk, N.Y.: IBM, 1983).

[64] Carol Vogel, "I.B.M. To Close Its Midtown Gallery," *New York Times*, March 23 1993.

[65] "American Abstract Artists 26th Annual Exhibition," IBM Gallery, New York, NY, Feb. 5–24.

[66] IBM Gallery of Science and Art, *A Calculator Chronicle: 300 Years of Counting and Reckoning Tools: The Mechanization of Arithmetic*, 29.

[67] Corporation International Business Machines et al., *Leonardo Da Vinci, an Exhibition of His Scientific Achievements: Collections of the Fine Arts Department, International Business Machines Corporation* (New York: IBM, 1951).

[68] In addition to the sources listed below, I have interviewed Guatelli's stepson, Joseph Mirabella who was Guatelli's stepson through marriage since 1947. Mirabella worked in Guatelli's workshop from 1964 to 1993. He continued running the business until 2005.

[69] Claudio Giorgione, "Leonardo Da Vinci: The Models Collection," ed. *Museo Nazionale della Scienza e della Tecnologia Leonardo da Vinci* (Milan: Museum Collections, 2009). The first effort to display models had been at the First National Exhibition of the History of Science in Florence. Ibid., 16–19. Claudio Giorgione, "The Birth of a Collection in Milan: From the Leonardo Exhibition of 1939 to the Opening of the National Museum of Science and Technology in 1953," *Science Museum Group Journal* **4**, no. 4 (March 22, 2016), doi:10.15180/150404; Matt Landrus, "Re-reading Heydenreich's positivist assessment of Leonardo's achievement," Romano Nanni and Maurizio Torrini (eds.), *Leonardo "1952" e la cultura dell'Europa nel dopoguerra*, Biblioteca Leonardiana (Firenze: L.S. Olschki, 2013), 321–338 on 323–325. Francis C. Moon, *The Machines of Leonardo Da Vinci and Franz Reuleaux* (New York: Springer, 2007), 201; *Time Magazine*, "Art Great Creator," **XXXIII**, no. 22, (May 29, 1939).

involved in interpreting and building the models based on studies of da Vinci's codices. The Italian ministry of culture then sponsored a travelling version for the United States. In 1940 Guatelli, emerging as one of the leading model makers, travelled to New York for an exhibition at the Museum of Science and Industry in the Rockefeller Center, at the base of what was then the RCA building.[70] As war loomed, Guatelli desperately tried to remain in New York, but his lawyer could not extend his stay and Guatelli was forced to flee to Japan with the models. In 1942, he exhibited the models in Tokyo, but following the surrender of Italy to the Allies, Japanese authorities sent him to a detention camp. Guatelli spent 22 months in captivity, and the models were eventually destroyed during the war.[71] Following the war, a curator at the Los Angeles Country Museum invited Guatelli to exhibit his models.[72] He built 56 new models and thus began another tour of the United States. It was through this tour that Guatelli came into contact with Thomas Watson Sr.[73]

Guatelli and his models were a travelling sensation in post-World War II America. He travelled as "Dr. Guatelli" with the models and built further replicas.[74] At times during the 1950s, the *New Yorker* posted announcements for the exhibits at the IBM building.[75] In 1952 and 1953 there was also a renewal of interest in Italy with major exhibitions in Milan, as well as the creation of a new museum in Vinci.[76]

---

[70] New York Museum of Science and Industry et al., *An Exhibition of the Scientific Achievements of Leonardo Da Vinci, Loaned by the Italian Ministry of Popular Culture* (New York: New York Museum of Science and Industry, 1940).

[71] Further research and documents are needed to verify these accounts and complete the biographical information about Guatelli. I interviewed Joseph Mirabella on March 4, 2015 to verify some of the main biographical claims mentioned in these sources: Popular Science, "Leonardo Comes to Life," *Popular Science* (October 1949), 164–65; Chicago Tribune, "Leonardo's Inventions," *Chicago Sunday Tribune Graphic Magazine* (March 19 1950), 7; Brendan Gill, "Early I.B.M. Man," *New Yorker* (December 13, 1951), 23–24; Erez Kaplan, "The Controversial Replica of Leonardo da Vinci's Adding Machine," *IEEE Annals of the History of Computing* **19**, no. 2 (1997), 62–63; Jim Strickland, "Who Was That Guy? Roberto Guatelli," *Computer History Museum Volunteer Information Exchange* **2**, no. 3 (2012); John P. Wiley Jr., "After Five Centuries, a Devoted Modeler Gives Shape to Genius," *Smithsonian* (1987), 90–95.

[72] Gill, "Early I.B.M. Man."; Los Angeles County Museum and Ludwig Heinrich Heydenreich, *Leonardo da Vinci; an exhibition of his scientific achievements* (Los Angeles, 1949), 40.

[73] Joseph Mirabella personal communication, March 4, 2015; Strickland, "Who Was That Guy? Roberto Guatelli."

[74] Popular Science, "Leonardo Comes to Life."; Chicago Tribune, "Leonardo's Inventions."; Gill, "Early I.B.M. Man."; "Making Modern Models from Designs Centuries Old, an Italian Scholar Is...Updating Da Vinci," *The Rotarian*(December 1952).

[75] *New Yorker* (Dec 13, 1958) and (Jan 10, 1959).

[76] The Milan models became the foundation of the newly formed National Museum of Science and Technology. Giorgione, "Leonardo Da Vinci: The Models Collection," 20–24. Giorgione, "The Birth of a Collection in Milan"; Jim Bennett, "The Science Museum and the Leonardo Da Vinci Quincentenary Exhibition of 1952," *Science Museum Group Journal* **4**, no. 4 (March 22, 2016), doi:10.15180/150403. In 1953 "IBM Italia" donated several of the models for the new Leonardo museum in Vinci. See, History of the Leonardiano Museo in Vinci, http://www.museoleonardiano.it/eng/museum/the-history accessed Oct. 28, 2015. Also see, Romano Nanni et al., *Leonardo and*

Guatelli eventually left IBM in 1961 and began his own business making replicas at a new workshop on Lafayette Street in New York, but most of his business continued to be with IBM.[77] In 1967 IBM exhibited several of his models at Brandeis University along with Bern Dibner's da Vinci collection.[78]

It was at this time that Guatelli became part of a historical controversy over one of his models. In 1968, he built a replica of "Leonardo's Calculator" after consulting copies of some newly discovered da Vinci documents related to the *Codex Madrid.* From those drawings, and others he had consulted previously, he built a Da Vinci calculator. IBM added this replica to their travelling exhibition, but some critics felt that the replica was a liberal interpretation from da Vinci's drawings. I.B Cohen, Bern Dibner and a small panel of experts were conscripted to judge the merits of the work. Due to uncertainty about Guatelli's calculator IBM removed the replica from exhibition.[79]

Guatelli also began making replicas of a wider variety of historic calculators such as the ones based on Babbage and Pascal's work. This period coincided with a more concerted effort on the part of IBM to build their collection and exhibits based on the history of calculating and display.

I.B. Cohen acted as IBM's consultant for the collection and research efforts.[80] He collaborated with Guatelli on the replicas needed for the travelling exhibits. In fact, Cohen and Bern Dibner were frequent visitors to the Lafayette workshop. Guatelli's stepson and partner, Joseph Mirabella remembered that Cohen was "always interested in making a cut-away to show and teach people how the machine worked."[81] The hinged door on the bottom of the Pascal replica, for example, was a suggestion from Cohen (Fig. 14.7). Once Guatelli and Mirabella had figured out how to build it, they produced several versions for IBM's displays worldwide and/or as corporate gifts. In this way, the Pascal calculator is as much an artifact of IBM's history as from Pascal's original history.

---

the Artes Mechanicae, First English language edition. ed. (Milano, Italy; New York, NY: Skira Editore Distributed in USA, Canada, Central & South America by Rizzoli International Publications, 2013); Nanni and Torrini, *Leonardo "1952."*

[77] Kaplan, "The Controversial Replica of Leonardo da Vinci's Adding Machine."

[78] University Brandeis et al., *Leonardo Da Vinci; an Exhibition of Scale Models from the IBM Collection and Books, Sketches and Drawings from the Bern Dibner Collection of Vinciana, May 29th through June 22, 1967. Being the First Exhibition in the New Samuel and Rieka Rapaporte Treasure Hall, Brandeis University, Waltham, Mass* (Waltham, Mass, 1957).

[79] Kaplan, "The Controversial Replica of Leonardo da Vinci's Adding Machine."

[80] I. Bernard Cohen Papers, 1889–1987. Finding Aid. CBI 182 University of Minnesota Libraries. Charles Babbage Institute.

[81] Joseph Mirabella personal communication, March 4, 2015.

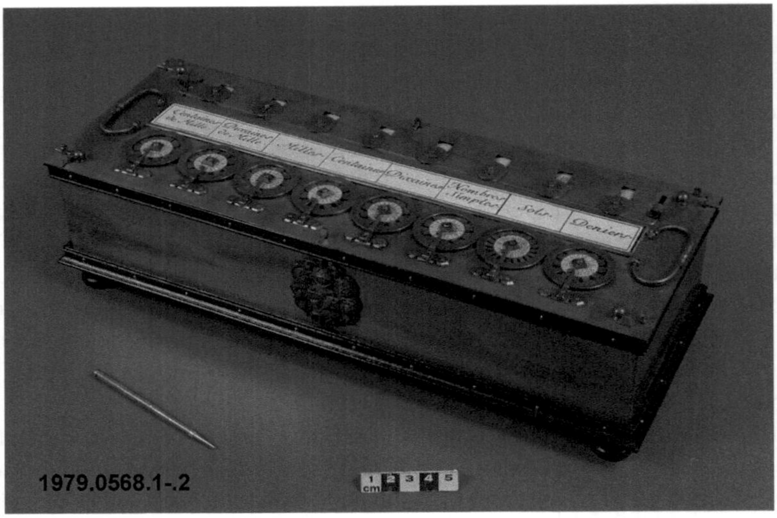

Fig. 14.7 Replica of Pascal Calculator made by Roberto Guatelli c. 1978. CSTM purchased this replica in 1978 directly from Guatelli's workshop on Lafayette St. in New York. It has a small, hinged door that resulted from a suggestion from I.B. Cohen. Only a few models have that door. Photo: Canada Science and Technology Museum, Artifact Number 1979.0568

14.2 Conclusions

In this essay, I have explored the role that collectors, displays and replicas have played in the construction of scientific heritage. In each case we go beyond the original function of the objects to understand their cultural impact in different contexts and periods.

The provenance of the Landmarks, and other scientific instruments in the Exploration and Surveying collection at the Canada Science and Technology Museum, provide a material, geographical and practical perspective on the history of surveying in Canada, but they are just as valuable as a window into the culture of the post-World War II Canadian surveying community. As the surveyors experienced radical technical and institutional change, they became collectors developing a keen sense of their own history and spending great efforts to preserve this heritage.

We gain a new perspective on the broader social and material history of mathematics by studying in detail the provenance and physical features of the Olivier models at the CSTM. We learn of their specific context of creation and use in a museum setting, which was different than a university classroom; we also learn of their changing cultural function in this setting. The deteriorated models remind us that their present condition is instructive about the culture (and value) of mathematics and science museums today.

The story behind the making of the Babbage replica takes us into the fascinating story of post-World War II industry in Europe and the United States. One could consider this object a Babbage replica, but just as much as an "authentic" Guatelli. In fact, one could produce an entire exhibition of Guatelli's work that would touch on the many historic themes that run through his career and more broadly, twentieth-century science—the growth of the Leonardo da Vinci industry; IBM's use of history for branding and identity; changes in science museums and education; the Eames exhibition influence on science communication; and a generation of historians of science that collaborated on these kinds of projects.

In the three case studies presented above, by following the succession of ownership and uses of scientific objects from creation to present context, we come to understand their role as cultural artifacts, beyond their initial scientific function. Provenance studies provide a powerful tool for critically evaluating our histories about science, and how we collect, preserve, study and share the material dimensions of it.

Chapter 15
Context, Connections and Culture: The History of Science in Canada as a Field of Study

Suzanne Zeller

Abstract This celebration of the broad range of Professor Trevor Levere's scholarship as a historian of science offers a welcome opportunity to highlight a few of his many contributions to our understanding of the Canadian historical experience of science, and to engage with some challenges that remain. Among the first to offer an overview of the field, Professor Levere earned the admiration – and influenced the work – of many who followed. Themes of context, connection, and culture permeates his influential contributions to the history of science in Canada, inspiring inquiries into ways of moving forward our historically (and historiographically) bound understanding. In a field that remains chronically under-researched, where do we go from here?

This celebration of the broad range of Professor Trevor Levere's scholarship as a historian of science offers a welcome opportunity to highlight a few of his many contributions to our understanding of the Canadian historical experience of science, and to engage with some challenges that remain. Among the first to offer an overview of the field, Professor Levere earned the admiration – and influenced the work – of many who followed. In so many ways a pioneer practitioner in the history of science in Canada, Professor Levere joined the faculty at the University of Toronto's Institute for the History and Philosophy of Science and Technology [IHPST] only a year after its founding in 1967. Having trained at Oxford University as a historian with an unusual specialization at the time, in the history of chemistry,[1] he also brought with him lifelong interests in natural history (especially birds) and the history of scientific instruments and their workings, both of which have recurred as major themes in his scholarly pursuits.

[1] Trevor H. Levere, *Affinity and Matter: Elements of Chemical Philosophy 1800–1865* (Oxford: Oxford University Press, 1971; reprint ed.: Reading, U.K.: Gordon and Breach Science Publishers, 1993).

S. Zeller (✉)
Wilfrid Laurier University, Waterloo, ON, Canada N2L 3C5
e-mail: szeller@wlu.ca

J. Buchwald, L. Stewart (eds.), *The Romance of Science: Essays in Honour of Trevor H. Levere*, Archimedes 52, DOI 10.1007/978-3-319-58436-2_15

Trevor Levere arrived in Canada during heady times for the country at large, in the afterglow of festivities that marked Confederation's centenary, on the one hand, while masking growing postwar concerns to defend – let alone define – a Canadian identity that seemed threatened, on the other hand, by foreign (read American) infiltration and control. As a result, according to F.R. Hayes (1904–1982), founding director of the Institute of Oceanography at Dalhousie University and chair of the Fisheries Research Board of Canada, Canadian policymakers kept Canadian science "continually on the couch pouring out its thoughts to one committee of social diagnosticians after another" during the 1960s and 70s, part of the process of formulating a national science policy. As Hayes observed in an important critique of this *Chaining of Prometheus* in 1973, "Probably the scientific endeavour of no other country has endured so much scrutiny."[2]

Where academic life was concerned, an influx, during those Viet Nam War years, of American scholars – welcomed *en masse* to staff Canada's expanding university system – also intensified demands for "Canadian content," in a backlash not only over popular culture, but also over educational curricula. In a further response, the Association of Universities and Colleges of Canada [AUCC] in 1972 appointed T.H.B. Symons (1929–), founding president of Trent University, to chair a national Commission on Canadian Studies (1972–1984), itself a remarkable move in a country where education famously falls under provincial jurisdiction. After conducting nationwide hearings, the Symons Commission issued the first two volumes of its influential report, *To Know Ourselves*, in 1975.[3]

Among many perceptive insights that have warranted every new Canadian generation's attention, the Symons Report highlighted an urgent need to embrace the history of Canadian science and technology in research and teaching. "Is there Canadian science?" the Commission wondered, "Are there, then, Canadian perspectives, Canadian applications, Canadian motivations or Canadian approaches to science that could be described accurately in Canadian Studies? The answer," it concluded, "is emphatically yes":

---

[2] Jeffrey Cormier, *The Canadianization Movement: Emergence, Survival, and Success* (Toronto: University of Toronto Press, 2004); Stephen Azzi, *Walter Gordon and the Rise of Canadian Nationalism* (Montreal and Kingston: McGill-Queen's University Press, 1999); Paul Litt, *The Muses, the Masses, and the Massey Commission* (Toronto: University of Toronto Press, 1992); F. Ronald Hayes, *The Chaining of Prometheus: Evolution of a Power Structure for Canadian Science* (Toronto: University of Toronto Press, 1973), xiii. See also J. Grant Glassco, *Report of the Royal Commission on Government Organization*, 5 vols. (Ottawa: Queen's Printer,1962–1963); George Grant, *Lament for a Nation: The Defeat of Canadian Nationalism*, [1965] (Montreal and Ithaca: McGill-Queen's University Press, 2005); Maurice Lamontagne, *Science Policy for Canada: Report of the Senate Special Committee on Science Policy*, 4 vols. (Ottawa: Special Committee on Science Policy, 1970); G. Bruce Doern, *Science and Politics in Canada* (Montreal and London: McGill-Queen's University Press, 1972).

[3] T.H.B. Symons, *To Know Ourselves: Report of the Commission on Canadian Studies*, 2 vols. (Ottawa: AUCC, 1975); Volume 3 appeared later as Thomas H.B. Symons and James E. Page, *Some Questions of Balance: Human Resources, Higher Education and Canadian Studies* (Ottawa: AUCC, 1982).

> To know ourselves we must have an understanding and appreciation of the enormously important role played by science in our lives and in the formulation of our values and viewpoints. Science inevitably plays a large part in shaping the character of any modern state.... [It] is a key ingredient in the cultural fabric of our society.[4]

Moreover, the Commission found, few Canadians "learn about such important facets of their country's history.... [T]hey learn virtually nothing about the accomplishments and impact of science in their own country." Among the obstacles to "historical studies in Canadian science," Symons noted, few university departments, let alone their senior administrations, "have been willing to recognize the significance of the history of Canadian science or to accept the academic validity of teaching and research in this field." The result was a regrettable "promotion among Canadians of the impression that this country contributes minimally to science, that Canadian science has no history, and, indeed, that ours is a second-rate scientific community."[5]

15.1 Context: What Science?

As part of those wide-ranging discussions of Canadian science past and present – and predating the publication of the Symons Report – Professor Levere collaborated with one of IHPST's first graduates, Richard Jarrell (1946–2013), to introduce their thoughtfully assembled collection of historical documents, *A Curious Field-Book: Science and Society in Canadian History,* in 1974. Levere followed up with "What Is Canadian About Science in Canadian History?" in 1978 at the first Kingston Conference, a preliminary meeting that led to the founding of the Canadian Science and Technology Historical Association [CSTHA] in 1980. A third essay, "The History of Science of Canada," addressed the question more broadly for an international audience in *The British Journal for the History of Science* in 1988.[6]

---

[4] Symons, *To Know Ourselves,* I: Chap. 4, esp. 141–5.

[5] *Ibid.,* 162–4; the earlier literature included Royal Society of Canada, *Fifty Years Retrospect* (Ottawa: Royal Society of Canada, 1932); H.M. Tory, ed., *A History of Science in Canada* (Toronto: Ryerson Press, 1939); and W.S. Wallace, ed., *The Royal Canadian Institute Centennial Volume* (Toronto: Royal Canadian Institute, 1949).

[6] Trevor H. Levere and Richard A. Jarrell, eds., *A Curious Field-Book: Science and Society in Canadian History* (Toronto: Oxford University Press, 1974); see also its companion volume, B. Sinclair, N.R. Ball, and J.O. Petersen, eds., *Let Us Be Honest and Modest: Technology and Society in Canadian History* (Toronto: Oxford University Press, 1974); Trevor Levere, "What is Canadian About Science in Canadian History?" in R.A. Jarrell and N.R. Ball, eds., *Science, Technology, and Canadian History/ Les sciences, la technologie et l'histoire canadienne* (Waterloo: Wilfrid Laurier University Press, 1980), 14–23; Trevor H. Levere, "The History of Science of Canada," *The British Journal for the History of Science* **21** (December, 1988), 419–25. Jarrell's doctoral dissertation concerned the early modern German astronomer Michael Mästlin (1550–1631); he went on to publish *The Cold Light of Dawn: A History of Canadian Astronomy* (Toronto: University of Toronto Press, 1988).

Concomitant organizational developments in the history of Canadian science and technology echoed important themes in the Symons Report. The CSTHA's future founders began publishing the *HSTC Bulletin: Journal of the History of Canadian Science, Technology and Medicine* under Richard Jarrell's editorial care in 1976, while working from within the Canadian Society for the History and Philosophy of Science (CSHPS) for recognition as a formal subgroup. Their activities included research workshops in collaboration with IHPST's counterpart, the Université de Montréal's Institut d'histoire et sociopolitique des sciences. After efforts to reach an agreement with the CSHPS fell through, they formed the CSTHA in 1980, upgrading the *HSTC Bulletin* to the peer-reviewed scholarly journal *Scientia Canadensis* in 1981. In an irony not unusual for the time, this separatist CSTHA leadership included among its most dedicated members several ex-patriot Americans, some of whom felt most acutely the CSHPS's "repudiation" of their adopted cause. Among the CSTHA founders more generally, the political scientist C.E.S. Franks of Queen's University observed that, in one way or another, "the persons interested in the study of the history of Canadian science and technology are mostly outsiders, at the periphery or beyond of their discipline or profession." Franks saw the first Kingston Conference as, "in fact, a conference of outsiders" struggling to build a source of mutual support in a "Canada-wide 'invisible university'."[7]

This struggle for scholarly visibility was intensified immeasurably by more than the longstanding perceptions of Canadian science and its history, for Professor Levere was not alone in 1988 when he described the CSHPS as "*the* national history of science society in Canada." From a broader perspective, the state of the historiography also posed challenges. In *A Curious Field Book*, Levere and Jarrell characterized Canadian science as "owing much" to American, British, and French "national styles of science" while "predominantly utilitarian" in its "unique" response to geography and environment. Under the French regime, they concluded, "Native-born Canadians were not much disposed to the study of science.... The most impressive intellects were ... French, not Canadian." Then under the British regime, Canadian scientists consisted mainly of "dedicated amateurs," government employees, and British military officers who "did not normally see [themselves] as part of a greater North American scientific community." Even later in the nineteenth century, a persistent "thrusting frontier mentality" in Canada remained "remote from the abstractions of mathematical physics."[8] In their companion volume on the

---

[7] By 1983 Franks recognized around him "a core of insiders focusing on the study of the History of Canadian Science and Technology": C.E.S. Franks, "Foreword," in Richard A. Jarrell and Arnold E. Roos, eds., *Critical Issues in the History of Canadian Science, Technology and Medicine* (Thornhill and Ottawa: HSTC Publications, 1983), vii; Richard Jarrell and Norman Ball, "The Study of the History of Canadian Science and Technology," in Jarrell and Ball, eds., *Science, Technology, and Canadian History*, 1–7; C.E.S. Franks, "The Kingston Conference and Beyond," in Jarrell and Ball, eds., *Science, Technology, and Canadian History*, 9–12; Levere, "History of Science of Canada," 422; see also Richard Jarrell, "The Infrastructure of an Emerging Field," *Scientia Canadensis* **11** (Spring/Summer, 1987), 37–45.

[8] Levere, "History of Science of Canada," 421 (author's italics); Levere and Jarrell, "General Introduction," in Levere and Jarrell, eds., *A Curious Field-Book*, esp. 2, 6,7, 9, 15, 20; cf. n. 44 below.

history of Canadian technology, Bruce Sinclair, Norman Ball, and J.O. Petersen sharpened these same points, urging a methodological transition from heroic to social history as more appropriate to their fledgling field of study:

> no one would expect to find any creative activity in Canada's past, especially any activity that required institutions, financial support, and a community of like-minded spirits. The land was sparsely populated and its people wholly engaged with a struggle for survival in a hostile environment. Against that background, the historian could expect to find only pale imitations of European culture, the stunted off-shoots of an undernourished soil. That view is a popular one and has its own history. Canadians of the nineteenth century also bemoaned the lack of genius in the arts or sciences, and excused it for the same reasons.[9]

These defensive characterizations of Canadian science and technology as colonial in both derivation and practice persisted into the 1980s and well beyond: "Our nineteenth century science was, at least in aspiration," Levere told his audience in 1978, "more useful than theoretically significant." Its practitioners were "amateurs who moulded themselves rapidly into professionals, forming a national style of science within the Empire, and depending heavily on the scientific education and entrepreneurial ideology of Scottish immigrants." This "emphasis upon practice" made nineteenth-century Canadian scientists, he concluded, "cautious, indeed conservative" in matters of scientific theory, including Darwinian evolution.[10]

The language of dependency so prevalent in works by this first generation of historians of Canadian science and technology should come as no surprise, in light of widely influential sociological models that sought to explain modern science in terms of its historic diffusion from its metropolitan centres. During the 1960s the American historians Donald Fleming and George Basalla, in particular, had schematized in a series of stages the crucial role of science as an imperialist instrument with powerful psychological repercussions. By definition, these largely anglocentric models marginalized colonial experiences of science as "incomplete," "truncated," even "failed" reflections of their metropolitan prototypes. With modern science recast, over the course of the nineteenth century, from the rational study of nature to a force of production, experimental research with its attendant laboratory

---

[9] Sinclair, Ball, and Petersen, "Introduction," in Sinclair, Ball, and Petersen, eds., *Let Us Be Honest and Modest*, 1; see also Bruce Sinclair, "Canadian Technology: British Traditions and American Influences," *Technology and Culture* **20** (January, 1979), 108–23.

[10] Levere, "What is Canadian," esp. 14, 19; R.A. Jarrell and A.E. Roos, "Preface," in Jarrell and Roos, eds., *Critical Issues*, viii–xi. Reviewers duly received the message: see Elizabeth C. Patterson in *American Scientist* **63** (January–February, 1975), 118–20; Morris Zaslow in *Histoire sociale*, **9** (1976), 436–38; Christopher C. Smart in *Isis* **67** (June, 1976), 292–294; Russell Moseley in *British Journal for the History of Science* **10** (March, 1977), 79–81. Cf. Suzanne Zeller, "Nature's Gullivers and Crusoes: The Scientific Exploration of British North America, 1800–1870," in *North American Exploration*, 3 vols., ed. John L. Allen, III: *A Continent Comprehended* (Lincoln: University of Nebraska Press, 1997), 190–243, 564–77; Suzanne Zeller, "Environment, Culture, and the Reception of Darwin in Canada, 1859–1909," in Ronald L. Numbers and John Stenhouse, eds., *Disseminating Darwinism: The Role of Place, Race, Religion, and Gender* (Cambridge: Cambridge University Press, 1999), 91–122.

settings came to signify escape from colonial backwardness: no laboratories, no science, ran the modernist syllogism; no science, no modern progress.[11]

For historical reasons, this positivist outlook weighed heavily in former British dominions such as Canada and Australia, rooting comparative histories of "national" styles of science inextricably in issues of cultural legitimacy.[12] While national comparisons offered useful starting points, their generalizations also indicated to critical observers that only more detailed primary research could test the validity of their assumptions. In Canada, the historian of science Raymond Duchesne offered a first telling critique based on unexpected revelations in his own research of the wide-ranging networks of nineteenth-century French-Canadian naturalists and their scientific correspondents. Duchesne emphasized the need to formulate new questions and approaches based on primary sources, rather than on traditional interpretive tropes in the writing of Canadian history. Others continued this critique of metropolitan-based assumptions over the years; in a volume of *Osiris* dedicated to a re-examination of the Basalla model 40 years on, the Canadian contribution explored the context in which Sir William Logan (1798–1875), founding director of the Geological Survey of Canada [GSC], resisted the British imperial authority of Sir Roderick Murchison (1792–1871), director of the Geological Survey of Great

[11] Richard Drayton, "Science, Medicine, and the British Empire," in Robin Winks and William Roger Louis, eds., *The Oxford History of the British Empire*, Vol. 5: *Historiography* (Oxford: Oxford University Press, 1999), 265–75. Donald Fleming, "Science in Australia, Canada, and the United States: Some Comparative Remarks," *Proceedings of the Tenth International Congress of the History of Science* **18** (1962),180–96; George Basalla, "The Spread of Western Science," *Science* **156** (5 May, 1967), 611–21; Lewis Pyenson, "The Incomplete Transmission of a European Image: Physics at Greater Buenos Aires and Montreal, 1890–1920," *American Philosophical Society Proceedings* **122** (April, 1978), 92–114; Yves Gingras, ""La Physique à McGill entre 1920 et 1940: la reception de la mécanique quantique par un communauté scientifique périphérique," *Scientia Canadensis* **5/** (January, 1981), 15–40; Richard A. Jarrell, "Colonialism and the Truncation of Science in Ireland and French Canada During the 19th c.," *HSTC Bulletin* **5** (May 1981), 140–57; Roy MacLeod, "On Visiting the 'Moving Metropolis': Reflections on the Architecture of Imperial Science," *Historical Records of Australian Science 5* (1982), 1–16; Clelia Pighetti, *Scienza e colonialismo nel Canada ottocentesco* (Florence: Leo S. Olschki Editore 1984); Nathan Reingold and Marc Rothenberg, eds., *Scientific Colonialism: A Cross-Cultural Comparison* (Washington: Smithsonian Institution Press, 1987), esp. Richard Jarrell, "Differential National Development and Science in the 19th c.: The Problems of Quebec and Ireland," 323–50; Yves Gingras, *Physics and the Rise of Scientific Research in Canada* (Montreal and Kingston: McGill-Queen's University Press, 1991). Cf. Marianne Gosztonyi Ainley, "Rowan vs. Tory: Conflicting Views of Scientific Research in Canada, 1920–1935," *Scientia Canadensis* **12** (Spring/ Summer 1988), 3–21.

[12] Drayton, "Science, Medicine, and the British Empire"; Roy MacLeod, "Introduction," in Roy MacLeod, ed., *The Commonwealth of Science: ANZAAS and the Scientific Enterprise in Australasia 1888–1988* (Oxford and Melbourne: Oxford University Press, 1988), 6–7; R.W. Home, "Introduction," in R.W. Home, ed., *Australian Science in the Making* (Cambridge: Cambridge University Press, 1988), vii–xxvii; Ian Inkster, "Scientific Enterprise and the Colonial 'Model': Observations on Australian Experience in Historical Context," *Social Studies of Science* **15** (1985), 677–704; Ian Inkster and Jack Morrell, eds., *Metropolis and Province: Science in British Culture, 1780–1850* (London and Melbourne: Hutchinson, 1983); Maurice Crosland, "Presidential Address: History of Science in a National Context," *The British Journal for the History of Science* **10** (July, 1977), 95–113.

Britain, over issues such as Logan's naming of Canada's Precambrian Laurentian and Huronian formations.[13]

Historians of Canadian science who continue to face a persistent colonialism in the treatment – whether current or historical – of their subject matter can take heart in a growing number of suggestions for moving forward. The Canadian political economist Harold Innis's (1894–1952) latest intellectual biographer reminds us that Innis thought about such matters firsthand already during the 1920s, valuing colonial hinterlands as "central to the continued cultural vitality of Western civilization." More recently, the British historical geographer David N. Livingstone's spatial perspective urged a reinstatement of colonized territories, including Canada, in the broader venues, cultures, and movements of scientific knowledge, acknowledging their active and meaningful participation in its production.[14]

Questions of periodization in the history of Canadian science may also warrant reconsideration, as the historian J.L. Heilbron has done so effectively for early modern science. Heilbron emphasizes the dynamic impact of the New World on Europe's classically-based knowledge system, in periods that distinguish an "Overloaded System" (1500–1600); "New Formulations and Responses" (1600–1660); "Attempted Reconstructions" (1640–1750); and, ultimately, its "Effective Dismemberment" (1720–1770). These and other recent approaches seek to "de-centre" science, to recover in its history the "regional modulations" and reciprocal movements that help to clarify in the production and transfer of knowledge a web of complex exchanges over time.[15]

For example, Pierre Boucher's (1622–1717) remarkable *Histoire véritable et naturelle des moeurs et productions du pays de la Nouvelle-France* (1664), a report solicited by the French Royal court, remains an undervalued text in the history of Canadian science and environment. Boucher had immigrated to New France as a

---

[13] Raymond Duchesne, "Science et société coloniale: les naturalistes du Canada français et leurs correspondants scientifiques (1860–1900)," *HSTC Bulletin* **5** (May, 1981), 140–57; Raymond Duchesne, "Historiographie des sciences et des techniques au Canada," *Revue d'histoire de l'Amérique française* **35** (September, 1981), 193–215; Jarrell and Ball, "The Study of the History of Canadian Science," 5–6; Suzanne Zeller, "The Colonial World as a Geological Metaphor: Strata(gems) of Empire in Victorian Canada," in Roy MacLeod, ed., *Nature and Empire: Science and the Colonial Enterprise*, *Osiris* 15 (2001), 85–107; see also n. 17 below.

[14] A.J. Watson, *Marginal Man: The Dark Visions of Harold Innis* (Toronto: University of Toronto Press, 2006), 121; David N. Livingstone, *Putting Science in Its Place: Geographies of Scientific Knowledge* (Chicago: University of Chicago Press, 2003); see also Diarmid. A. Finnegan, "The Spatial Turn: Geographical Approaches in the History of Science," *Journal of the History of Biology* **41** (2008), 369–388; David N. Livingstone, "Keeping Knowledge in Site," *History of Education* **39** (November, 2010), 779–85.

[15] J.L. Heilbron, "History of Science," in J.L. Heilbron, ed., *The Oxford Companion to the History of Modern Science* (Oxford: Oxford University Press, 2003) 370–74; Germaine Warkentin and Carolyn Podruchny, eds., *Decentring the Renaissance: Canada and Europe in Multidisciplinary Perspective 1500–1700* (Toronto: University of Toronto Press, 2001); Andrew Cunningham and Perry Williams, "De-centring the 'Big Picture': *The Origins of Modern Science* and the Modern Origins of Science," *The British Journal for the History of Science* **26** (1993), 429–30; see also David Wade Chambers, "Period and Process in Colonial and National Science," in Reingold and Rothenberg, eds., *Scientific Colonialism*, 297–321.

youth in 1635, assisted the Jesuit mission at Huronia, married an indigenous woman while moving up the ranks of the fur trade, and served as a soldier and interpreter before becoming governor of Trois-Rivières in 1653. His "naturalized" appreciation of New France differed from reports by mere sojourners, including the Jesuit Relations (1632–1672) and Louis Nicolas's *Codex Canadensis* (1700), in Boucher's direct association of nature with both the colony's destiny as "un Monde nouveau" still under construction, and his own developing New-World identity.[16]

Boucher's astute representations supported France's decision to elevate the colony of New France to provincial status. Within the larger scientific context of Europe's "attempted reconstructions" of natural knowledge, King Louis XIV's administration bureaucratized and centralized a "scientifico-colonial machine" to "mobilize, organize, centralize, and unify" France's natural and intellectual resources. In this vein, J.-B. Colbert (1619–1683), Minister of the Marine, ordered systematic floral and faunal collections in French territories by his *médecins du roi*. Nor was it only leading men who participated in European "reconstructions" of natural knowledge, as important contributions to botany and ethnology by the Ursuline teacher Marie de l'Incarnation (1599–1672) and the herbalist Catherine Jérémie (1664–1744) – both of whom consulted in these matters with their indigenous neighbours – attest. By the eighteenth century, Scholastic knowledge was in these ways deeply inflected by new information from northern North America and its peoples. The Jesuit P-F-X de Charlevoix's (1682–1761) *Histoire et description générale de la Nouvelle France* (3 vols., 1744), appended by the most complete list of plants for years to come, earned its author recognition in France as "the one unimpeachable historian of the New World," impressing Enlightened critics of the *ancien regime* as "the ultimate authority" on indigenous peoples, from whom, they believed, Europe had much to learn about both nature and humanity.[17]

---

[16] Pierre Boucher, *Histoire véritable et naturelle* (1664) reprint ed. (Boucherville: Societé Historique de Boucherville, 1964); Lynn Berry, "The Delights of Nature in the New World: A Seventeenth-Century Canadian View of the Environment," in Warkentin and Podruchny, eds., *Decentring the Renaissance*, 223–35; Jacques Rousseau, "Pierre Boucher, naturaliste et géographe," in Boucher, *Histoire véritable*, 264, 277–78; Auguste Vachon, "Louis Nicolas and the *Codex Canadensis*," *The Archivist* **12** (March-April 1985), 1–2; Raymond Douville, "Boucher, Pierre," in *DCB*, Vol. 2, accessed February 14, 2014, http://www.biographi.ca/en/bio/boucher_pierre_2E.html; Greer, ed., *Jesuit Relations*, chap. 5; François-Marc Gagnon with Nancy Senior and Réal Ouellet, *The Codex Canadensis and the Writings of Louis Nicolas: The Natural History of the New World* (Tulsa: Gilcrease Museum/Montreal and Kingston: McGill-Queen's University Press, 2011).

[17] J.E. McClellan III and François Regourd, "The Colonial Machine: French Science and Colonization in the Ancien Regime," in Roy MacLeod, ed., *Nature and Empire*, *Osiris 15* (2001), 31–50; K.A. Young, "Crown Agent-Canadian Correspondent: Michel Sarrazin and the Académie Royale des Sciences, 1697–1734," *French Historical Studies* **18**/(Fall, 1993), 416–33; Jacques Rousseau, "Jérémie, Lamontagne, Nicolas," in *DCB*, Vol. 2, accessed February 14, 2014, http://www.biographi.ca/en/bio/jeremie_nicolas_2E.html; Marie-Emmanuel Chabot OSU, "Guyart, Marie, Marie de l'Incarnation," in *DCB*, Vol. 1, accessed February 14, 2014, http://www.biographi.ca/en/bio/guyart_marie_1E.html; Natalie Zemon Davis, "Polarities, Hybridities: What Strategies for Decentring?" in Warkentin and Podruchny, eds., *Decentring the Renaissance*, 25; Denys Delâge, "L'Influence des amérindiens sur les canadiens et les français au temps de la Nouvelle-France," *Lekton*, **2** (Automne, 1992), 163–89; J.I. Israel, *Radical Enlightenment* (Oxford: Oxford University Press, 2001), 580–82.

Michel Foucault's (1926–1984) observation that "probably every culture, in every civilization" invents "counter-sites," mirror-image places where "all the other real sites that can be found within the culture, are simultaneously represented, contested, and inverted," included colonized territories among these so-called "heterotopias." Foucault's illumination of the power wielded by the imagined resemblances and presumed differences imposed from above highlights every colony as a counter-site to its respective empire, every hinterland as a counter-site to its respective metropolis, and every periphery as a counter-site to its respective centre, opening much of what we think we know about these presumed hierarchical relationships to more critical inquiry. The insight proves suggestive, for example, in studying the career of the Nova Scotian geologist David Honeyman (1817–1889), who found his fieldwork colonized – and devalued – not by his wide-ranging and supportive French, British, and American scientific contacts, but rather by the leading Canadian geological authorities, Sir J.W. Dawson (1820–1899), principal of McGill University, and Sir William Logan, who in turn felt their own work undervalued by imperial geological authorities in London and Edinburgh.[18] Overall, the resulting temporal, psychological, and other spaces help to situate a growing body of scholarly work relating the New World and its peoples, through their active engagement, to the changing knowledge systems of the Old.[19]

In her thought-provoking study of *Africa as a Living Laboratory*, the historian Helen Tilley argues persuasively that the term "colonial science" is in fact misleading. Scientific practices are too widely shared to be neatly confined to particular places; what really matters is "the dynamic interplay between scientific fieldwork and research across metropolitan and colonial contexts." This interplay in turn supported the development of "cross-disciplinary methods that would bring to light

[18] Michel Foucault and Jay Miskowiec, "Of Other Spaces," *Diacritics* **16** (Spring, 1986), 22–27; Michel Foucault, *The Order of Things* (New York: Vintage Books, 1973), 67–162; Suzanne Zeller, "Arctic Moment: The Nova Scotian Institute of Science and the Halifax-Hudson Bay Axis, 1870s–1890s," *Journal of the Royal Nova Scotia Historical Society* **17** (2014), 57–91.

[19] Julie Cruikshank, *Do Glaciers Listen?* (Vancouver: UBC Press, 2006); Allan Greer, "The Exchange of Medical Knowledge Between Natives and Jesuits in New France," in L.M. Figueroa and Domingo Ledezma, eds., *El Saber de los Jesuitas* (Frankfurt: Vervuert, 2005) 137; Deborah Doxtator, "Inclusive and Exclusive Perceptions of Difference: Native and Euro-Based Concepts of Time, History, and Change," in Warkentin and Podruchny, eds., *Decentring the Renaissance*, 33–47; Allan Greer, ed., *The Jesuit Relations* (Boston: Bedford/St. Martin's Press, 2000) 16–19; Sylvie Vincent, "Compatabilité apparente, incompatabilité réelle des versions autochtones et occidentales de l'histoire," *Recherches amérindiennes au Québec* **32** (2000), 96–106; Victoria Dickenson, *Drawn From Life: Science and Art in the Portrayal of the New World* (Toronto: University of Toronto Press, 1998); Ramsay Cook, "Donnacona Discovers Europe," in Ramsay Cook, ed., *The Voyages of Jacques Cartier* (Toronto: University of Toronto Press, 1993) ix-xli; Susan Rowley, "Inuit Oral History: The Voyages of Sir Martin Frobisher, 1576–78," in Stephen Alsford, ed., *The Meta Incognita Project* (Hull, Quebec: Canadian Museum of Civilization, 1993), 211–19; Sylvie Vincent, "L'arrivée des chercheurs de terres," *Recherches amérindiennes au Québec* **22** (1992),19–29; Malcolm Lewis, "Indicators of Unacknowledged Assimilations from the Amerindian Maps on Euro-American Maps," *Imago Mundi* **38** (1986), 9–34; Judith Hudson Beattie, "Indian Maps in the Hudson's Bay Company Archives," *Archivaria* **21** (Winter, 1985–86,: 166–75.

connections among diverse phenomena in nature." It makes sense to think about these developments, as Tilley suggests, as decentring Europe's modernizing role, in this case "Canadianizing" the science introduced there by Europeans. British North America's importance as a field for scientific investigations highlights, as Tilley reminds us, the "complexity and interdependence" of science, its partnership with empire representing by no means either a one-way or a monolithic colonizing process. Much as elsewhere, the need for extra-European networks in order to acquire site-specific knowledge of British North America held power to destabilize traditional assumptions and methods of science.[20]

15.2 Connections: What Boundaries?

Taking the lead in different directions was Professor Levere's *Science and the Canadian Arctic: A Century of Exploration, 1818–1918* (1993). Historians of Canada had to that time engaged a mythical North mainly intangibly, as a focal point for discussions of national identity; the centenary in 1980 of Britain's cession to Canada of its claims to the Arctic archipelago began turning Canadian interest northward in earnest. Levere's timely and well-researched study of Arctic exploratory programmes elaborated technical details, complex political and cultural connections, and tensions between exploratory ways and scientific means of proceeding. Having visited the place as part of his research, he helped to bring the history of Canadian Arctic science in out of the cold, from heroic tales of exploration into the scholarly mainstream, well before climate change and the International Polar Year (2007–2008) released an unprecedented torrent of interdisciplinary and collaborative Arctic research in humanities and social sciences.[21]

Science and the Canadian Arctic served as a crossroads for a wealth of scholarly insights and interconnections based on Trevor Levere's own wide-ranging interests. The interplay of its themes had enriched our seminar discussions with him several years before its publication, planting ideas that played out not only in his own later work, but also in that of his students. A sharp focus, for example, on the Icelandic-Canadian-born explorer Vilhjalmur Stefansson (1879–1962) raised issues of scientific legitimacy that resonated in Levere's later work on the Royal Society of Canada

[20] H. Tilley, *Africa as a Living Laboratory: Empire, Development, and the Problem of Scientific Knowledge, 1870–1950* (Chicago: University of Chicago Press, 2011). 12–13 and chap. 7.

[21] Trevor H. Levere, *Science and the Canadian Arctic: A Century of Exploration, 1818–1918* (Cambridge: Cambridge University Press, 1993). Morris Zaslow, *The Opening of the Canadian North, 1870–1914*, Canadian Centenary Series Vol. 16 (Toronto: McClelland and Stewart, 1971); Morris Zaslow, ed., *A Century of Canada's Arctic Islands, 1880–1980*, Royal Society of Canada, 23rd Symposium (Ottawa: Royal Society of Canada,1981; Bruce W. Hodgins and Shelagh Grant, "The Canadian North: Trends in Recent Historiography," *Acadiensis* **16** (Autumn, 1986), 173–86; Ronald E. Doel, Urban Wråkberg, and Suzanne Zeller, "Science, Environment, and the New Arctic," *Journal of Historical Geography* **43** (February, 2014). http://www.sciencedirect.com/science/article/pii/S0305748813001448.

(see Part III below). Moreover, in re-reading *Science and the Canadian Arctic* recently, I recognized its relevance to similar problems with which I had struggled in collaborative work with a Danish colleague, Christopher Jacob Ries, on two other rogue Arctic scientific explorers, the Canadian artist-naturalist Ernest Thompson Seton (1860–1946) and the Danish geologist of Greenland, Lauge Koch (1892–1964).

As "Wild Men In and Out of Science," Seton and Koch challenged established authority with their unorthodox outlooks and approaches, suffering professional (and very public) ostracism even while they modernized scientific practices in their respective ecological and geological fields. Stefansson appears to have succeeded Seton as an ambitious, influential, and well-connected propagandist for Canada's Arctic. Thus, in a pattern that has become all too familiar, between the time the federal government struck its Royal Commission to Investigate the Possibilities of the Reindeer and Musk-Ox Industries in the Arctic and Sub-Arctic Regions of Canada in 1919, and the publication of its report in 1922, Stefansson managed to wear out his welcome as one of the four commissioners. Levere's question of "explicitly and implicitly, what constitutes science in the eyes of scientists, explorers and governments," surely warrants further exploration, as does his conclusion that "Stefansson was neither civil servant nor twentieth-century scientist. Nor, certainly, was he 'a scientist of the Darwinian type'" – the latter a claim that Stefansson made for himself after parroting among his Inuit hosts his understanding of the ethnograhic procedures of his day. These aspects of Stefansson's career offer important entry points into early twentieth-century science, its practices, and the frameworks within which it continues to be understood.[22]

Among the most fruitful of the further insights that thread through *Science and the Canadian Arctic* was the intellectual influence of the Romantic sensibility, especially as it adopted analytical methods in the widely admired work of the Prussian scientific traveller Alexander von Humboldt (1769–1859). Humboldt's unfinished master work, *Cosmos* (4 vols., 1845–1862) attempted nothing less than a *Physical Description of the Universe* by investigating nature's underlying interconnectedness. He unified these analytical interconnections using powerful visual expression in iso-maps, tables, graphs, and charts, the forms in which science tended increasingly to be reported during the nineteenth century. Scientific knowledge accrued,

[22] Levere, *Science and the Canadian Arctic*, chapter 10; Trevor H. Levere, "Vilhjalmur Stefansson, the Continental Shelf, and a New Arctic Continent," *The British Journal for the History of Science* **21** (June, 1988), esp. 236, 247; Suzanne Zeller and Christopher Jacob Ries, "Wild Men In and Out of Science: Finding a Place in the Disciplinary Borderlands of Arctic Canada and Greenland," *Journal of Historical Geography* **43** (February, 2014). http://www.sciencedirect.com/science/article/pii/S0305748813001436; see also Ernest Thompson Seton, *The Arctic Prairies* (Toronto: William Briggs, 1911); Vilhjalmur Stefansson, *The Friendly Arctic* (New York: The Macmillan Company, 1921); *Report of the Royal Commission Appointed by Order-in-Council of Date May 20, 1919, to Investigate the Possibilities of the Reindeer and Musk-Ox Industries in the Arctic and Sub-Arctic Regions of Canada* (Ottawa: F.A. Acland, 1922); and Michael Bravo, "Measuring Danes and Eskimos," in Michael Bravo and Sverker Sörlin, eds., *Narrating the Arctic: A Cultural History of Nordic Scientific Practices* (Canton, MA: Science History Publications, 2002), 235–73.

in Humboldt's view, through cooperation among widely dispersed participants using standardized instruments in synchronized observations. In return, it promised the greatest possible accuracy in the search for patterns in nature that would, he believed, ultimately reveal natural laws.

Humboldt's creation of worldwide networks of scientific communication and exchange constituted a feat of incredible proportions long before the internet enabled later generations to take World-Wide Webs of knowledge for granted. Seen as the modern embodiment of Britain's proud Baconian tradition, he inspired the establishment of the British Association for the Advancement of Science (1831) along with the research projects of scientifically-minded British military officers – including Royal Navy officers who staffed Britain's Arctic voyages – and even fur traders in British North America who hoped to aid the growth of scientific knowledge by contributing observations from the far corners of the natural and physical world. The resulting chain of connections helped to shape the research agendas and methods of nineteenth-century British exploratory voyages, including those to the Arctic.[23]

Humboldt's influence in Canadian science has been manifested time and again in my own subsequent research, first and foremost in shaping scientific assessments of the Great Northwest in support of its annexation to Canada after the expiration of the Hudson's Bay Company's 200-year charter in 1869. It reappeared as the basis for a lengthy nationwide climatological project of seasonal observations undertaken in 1890 by the Nova Scotian educator and scientist A.H. Mackay (1848–1929) under the auspices of the Royal Society of Canada, and lasting well into the 1920s.[24]

[23] R.W. Home, "Humboldtian Science Revisited: An Australian Case Study," *History of Science* **33** (1995), 1–22; Trevor H. Levere, "Elements in the Structure of Victorian Science or Cannon Revisited," in J.D. North and J.J. Roche, eds., *The Light of Nature* (Dordrecht: Springer Verlag, 1985), 433–49; Trevor H. Levere, "Sabine, Sir Edward," in *Dictionary of Canadian Biography* [*DCB*], Vol. 11 (Toronto and Quebec: University of Toronto Press and les presses de l'Université Laval, 2003–) accessed February 16, 2014. http://www.biographi.ca/en/bio/sabine_edward_11E.html; Jack Morrell and Arnold Thackray, *Gentlemen of Science: Early Years of the British Association for the Advancement of Science* (Oxford: Oxford University Press, 1981); Susan Faye Cannon, "Humboldtian Science," in Susan Faye Cannon, *Science in Culture: The Early Victorian Period* (New York: Dawson and Science History Publications, 1978), 73–110; Suzanne Zeller, "Humboldt and the Habitability of Canada's Great Northwest," *The Geographical Review* **96** (July, 2006), 382–98; Suzanne Zeller, "Recalibrating Empire: Humboldtian Climatology in the Reports of the Palliser and Hind Expeditions to British North America's Great North West, 1857–58," in *Alexander von Humboldt and the Americas*, ed. Vera M. Kutzinski, Ottmar Ette, and Laura Dassow Walls (Berlin: Verlag Walter Frey, 2012), 70–116; Suzanne Zeller, "The Spirit of Bacon: Science and Self-Perception in the Hudson's Bay Company, 1830–1870," *Scientia Canadensis*, **8** (Fall/Winter, 1989), 79–101.

[24] Suzanne Zeller, "Reflections on Time and Place: The Nova Scotian Institute of Science in Its First 150 Years." *Proceedings of the Nova Scotian Institute of Science* **48** (2015), 5–61; Zeller, "Recalibrating Empire"; Suzanne Zeller, "Humboldt and the Habitability of Canada's Great Northwest," in Mathewson and Sluiter, eds., *Humboldt in the Americas*, 382–98; Suzanne Zeller, "Classical Codes: Biogeographical Assessments of Environment in Victorian Canada," *Journal of Historical Geography* **24** (1998), 20–35; Suzanne Zeller, *Inventing Canada: Early Victorian Science and the Idea of a Transcontinental Nation*, reprint ed. (Montreal and Kingston: McGill-Queen's University Press, 2009).

A growing body of sophisticated interdisciplinary literature has shown that, for all their fundamental geographical dimensions, Humboldtian modes of science also penetrated politics, including Humboldt's "unfulfilled project" of an "alternate modernity" that supported, among other things, nationhood and liberation for colonized and enslaved peoples, respectively. We know that Canadian politicians employed the language and ideas of Humboldtian science (not to mention maps of British North America criss-crossed by Humboldtian iso-lines) to frame their understanding of Canadian climatic patterns and settlement potential during the nineteenth century; these hints at a reciprocal relationship between science and politics in Canadian history too await further investigation.[25]

Still more unexpectedly, an important Humboldtian connection can be traced through the development of the first "scientific" Canadian census in 1871. In *The Politics of Population* (2001), the historical sociologist Bruce Curtis explained that, in the aftermath of the Rebellions of 1837, "Canadian statistical development and census making practice were firmly situated" in the international statistical network established by the Belgian scientist, social scientist, and administrator Adolphe Quetelet (1794–1874), himself a Humboldtian protégé. Humboldt's influence comes through in the leading role of J-C. Taché (1820–1894) as deputy Minister of Agriculture and Statistics for the United Province of Canada in 1864. Taché undertook a modern systematic (but not unproblematic) census that recognized as indispensable the achievement of a centralized administrative organization; educated and disciplined respondents; standardized forms, protocols, and procedures; trained field observers; and statistical experts to process the resulting data. Here was an important aspect of modern state formation, as Patrick Carroll explains, "crucially constituted by science." Most significantly, for their own political reasons Taché and his colleagues envisioned the project from an avowedly ultramontane perspective, as a "genealogical monument" to the *canadien* nationality of a rurally-based people tied historically to both their territory and the Catholic Church. Curtis here challenged fundamental assumptions in Canadian history by showing that Taché "belonged to the pro-science wing of the ultramontane movement, which insisted on the importance of empirical science for promoting social progress." He thus

[25] See esp. the essays by Laura Dassow Walls, Christiana Borchart Moreno and Segundo y Moreno Yánez, Christopher Iannini, Michael Dettelbach, and Ottmar Ette in Kutzinski, Ette, and Walls, eds., *Alexander von Humboldt and the Americas*; Kent Mathewson and Andrew Sluiter, eds., *Humboldt in the Americas*, special issue of *The Geographical Review* **96** (July, 2006); Ottmar Ette and Oliver Lubrich, "Die andere Reise durch das Universum," in Alexander von Humboldt, *Kosmos*, ed., Ottmar Ette and Oliver Lubrich (Frankfurt: Eichborn Verlag, 2004), 905–20; Ottmar Ette, *Weltbewusstsein: Alexander von Humboldt und das unvollendete Projekt einder anderen Moderne* (Weilerswist: Velbrück Wissenschaft, 2002); Ottmar Ette et al., eds., *Alexander von Humboldt: Aufbruch in die Moderne* (Berlin: Akademie Verlag, 2001); Michael Dettelbach, "The Face of Nature: Precise Measurement, Mapping, and Sensibility in the Works of Alexander von Humboldt," *Studies in the History and Philosophy of the Biological and Biomedical Sciences* **30** (1999), 473–504; Michael Dettelbach, "Global Physics and Aesthetic Empire: Humboldt's Physical Portraits of the Tropics," in David Philip Miller and Peter Hanns Reill, eds., *Visions of Empire: Voyages, Botany, and Representations of Nature* (Cambridge: Cambridge University Press, 1996), 258–92; Suzanne Zeller, "Recalibrating Empire."

extended "through particularizing scientific practices, the depth and reach of state power," through the same standardized instruments and protocols as those governing simultaneous geological, magnetic, botanical, climatic, and other inventory surveys.[26]

Science and the Canadian Arctic made a further point of explaining the importance of concurrent Scandinavian explorations with their lengthy Arctic experience. Trevor Levere's research has always demonstrated how deeply these kinds of interconnections and mutual cultural influences can run, for example, between the Romantic scientist Humphry Davy (1778–1829) and the Romantic poet Samuel Taylor Coleridge (1772–1834) in Levere's *Poetry Realized in Nature* (1981). Levere's work on the politically radical physician and scientist Dr. Thomas Beddoes (1760–1808) furthermore resonates directly with Canadian connections: Beddoes served as the educator and guardian of young John George Lambton (1792–1840) and his brother after the death of their father, the Earl of Durham, during the years when Humphry Davy, as Beddoes's laboratory assistant, also lived in the Beddoes household. Lambton the radical Reform politician, of course, went on as governor general of the British North American colonies to inquire – accompanied by an entourage of statisticians – into the causes of the Rebellions of 1837 in Upper and Lower Canada, using his analysis of the resulting information to chart the colonies' constitutional future.[27]

15.3 Culture: What Nation?

When Levere and Jarrell noted in *A Curious Field Book* that "The roots of twentieth-century science in Canada were established by the time of the Durham Report" in 1839, they turned this important connection in yet another direction. Durham's *Report on the Affairs of British North America* notoriously relegated

[26] Bruce Curtis, *The Politics of Population: State Formation, Statistics, and the Census of Canada, 1840–1875* (Toronto: University of Toronto Press, 2001) Introduction and Chapter 7, esp. 243; on Quetelet's connection to Humboldt, see Laura Dassow Walls, *The Passage to Cosmos: Alexander von Humboldt and the Shaping of America* (Chicago: University of Chicago Press, 2009), 108, 134; Frank H. Hankins, *Quetelet as Statistician* (New York: Columbia University, 1908), 20; Patrick Carroll, *Science, Culture, and Modern State Formation* (Berkeley: University of California Press, 2006), 166–67, 171; Jean-Guy Nadeau, "Taché, Joseph-Charles," *DCB*, Vol. 12, accessed February 20, 2014, http://www.biographi.ca/en/bio/tache_joseph_charles_12E.html; cf. Richard A. Jarrell, "L'ultramontanisme et la science au Canada français," in Fournier, Gingras, and Keel, *Sciences et médecine*, 41–68.

[27] On further Scandinavian connections, see Jennifer Hubbard, *A Science on the Scales: The Rise of Canadian Atlantic Fisheries Biology, 1898–1939* (Toronto: University of Toronto Press, 2006); Trevor H. Levere, *Poetry Realized in Nature: Samuel Taylor Coleridge and Early Nineteenth-Century Science*, reprint ed. (Cambridge: Cambridge University Press, 2002; f.p. 1981); Trevor H. Levere, "Dr. Thomas Beddoes: Chemistry, Medicine, and the Perils of Democracy," *Notes and Records of the Royal Society of London* **63** (20 September, 2009), 215–29; Trevor H. Levere, "Dr. Thomas Beddoes (1750–1808): Science and Medicine in Politics and Society," *The British Journal for the History of Science* **17** (July, 1984), 187–204; Trevor H. Levere, "Dr. Thomas Beddoes and the Establishment of His Pneumatic Institution: A Tale of Three Presidents," *Notes and records of the Royal Society of London* **32** (July, 1977), 41–49; Curtis, *Politics of Population*, 20–21.

French-Canadian culture to what we might now understand as a classic Foucauldian counter-site, for the apparently undifferentiated resistance of an entire society to the modern values associated with commercial and industrial progress. His far-reaching influence helped to legitimize a dominant theme in twentieth-century historical interpretations of Canada, as a country caught between "the France which the political revolution had destroyed and the England which the industrial revolution had created." As the leading edge of a great discursive divide in the writing of Canadian history,[28] these assumptions remain far removed from Heilbron's insistence on the need to accommodate "modalities of culture" as well as "regional modulations" in the history of science. Instead they normalized the use of partition and contrast as standard interpretive approaches to the Canadian historical experience.[29]

Professor Levere's *Research and Influence: A Century of Science in the Royal Society of Canada* (1993) reflected this ready acceptance of "two solitudes" in the history of Canadian science as derived not only from the country's history and historiography, but also from conceptual limitations imposed by connotations of science that only hardened over time. Founded in 1882 by Governor General Lord Lorne (1845–1914) as an elite national institution, the RSC emulated older academies such as the Royal Society of London (founded 1662) and the Académie française (founded 1635). Against the preferences of advisors who included Sir J.W. Dawson, Lorne rejected the model of the British Association for the Advancement of Science and its American counterpart (founded 1848), with their open memberships: he aimed thereby to ensure both the representation of arts as well as sciences, and the inclusion of French- as well as English Canadians.

[28] Heilbron, "History of Science," 373–74; Levere and Jarrell, "General Introduction," 2; Donald Creighton quoted in Carl Berger, *The Writing of Canadian History: Aspects of English-Canadian Historical Writing Since 1900*, 2nd ed. (Toronto: University of Toronto Press, 1988; f.p. 1976) 210–11; Serge Gagnon, *Quebec and Its Historians*, trans. Yves Brunelle and Jane Brierley, 2 vols. (Montreal: Harvest House, 1982, 1985); M. Brook Taylor, *Promoters, Patriots, and Partisans: Historiography in Nineteenth-Century English Canada* (Toronto: University of Toronto Press, 1989); by far the most controversial is Ronald Rudin, *Making History in Twentieth-Century Quebec: Historians and Their Society* (Toronto: University of Toronto Press, 1997): see also Ramsay Cook, review of Ronald Rudin, *Making History in Twentieth-Century Quebec*, *Histoire sociale/ Social History* **32** (1999), 120–22; Michael Behiels, "'Normalizing' the Writing of Quebec History," *Left History* **6** (1999), 91–99; Yves Gingras, "Making Up History," *Literary Review of Canada* (Summer, 1999), 19–22.

[29] Levere and Jarrell, "General Introduction"; the section "What is Canadian About the History of Canadian Science and Technology" in Jarrell and Ball, eds., *Science and Technology in Canadian History* consists of Levere, "What is Canadian," which restricts itself to nineteenth-century Upper Canada, and Raymond Duchesne, "Problèmes d'histoire des sciences au Canada français," 23–32; Yves Gingras, "Le développement du marché de la physique au Canada: 1879–1928," in Jarrell and Roos, eds., *Critical Issues*, 16–30; Richard A. Jarrell, "The Social Functions of the Scientific Society in Nineteenth-Century Canada," in Jarrell and Roos, eds., *Critical Issues*, 31–44; Carl Berger, *Science, God, and Nature in Victorian Canada* (Toronto: University of Toronto Press, 1983); Luc Chartrand, Raymond Duchesne, and Yves Gingras, *Histoire des sciences au Québec*, rev. ed. (Montreal: Les Editions du Boréal, 2008; f.p. 1987); Yves Gingras, "La réception des rayons X au Québec: radiographie des pratiques scientifiques," in Marcel Fournier, Yves Gingras, and Othmar Keel, eds., *Sciences et médécine au Québec: perspectives sociohistoriques* (Québec: Institute québecois de recherche sur la culture, 1987) 41–68.

The resulting RSC consisted of internally elected members in two literary and two (later three) science sections. The direct partition of anglophone and francophone literary cultures in the arts sections found indirect reinforcement in the science sections, where French Canadians remained heavily under-represented even long after they established their own Association canadienne-française pour l'avancement des sciences (ACFAS) in 1923. The RSC in these ways relinquished a golden opportunity to become the voice for science on the national level before the rival National Research Council was fully in place.[30]

Levere's account adopted sociological explanations that rejected an earlier generation's claims for a scientific "weft" in the fabric of Canadian history, at least where French Canada was concerned.[31] "A major reason for this divergence" among RSC science fellows, he agreed, "was the different development of institutions of scientific research in francophone and anglophone Canada." As Levere and Jarrell had noted in *A Curious Field Book*, the editor of the journal *Le naturaliste canadien*, Abbé V.-A. Huard (1859–1929) – who became Quebec's first provincial entomologist and a member of the RSC's Section IV (Biological and Geological Sciences) in 1913 – echoed earlier criticisms from the RSC's affiliated societies, pointing out in 1912 that the Society's own systemic rigidity exacerbated continuing inequities, with the voting majority in the science sections (in this case anglophone Protestants) naturally tending to replicate themselves. The RSC's "conservative strength of a self-renewing body" also drew complaints at the time from representatives of various sciences who jockeyed for position within their sections. Yet, as Huard urged, in a country with only one such "mark of honour" to bestow upon its "men of science,"

[30] The phrase gained common usage with the publication of Hugh MacLennan's classic novel, *Two Solitudes* (Toronto: Collins, 1945); Trevor Levere, *Research and Influence: A Century of Science in the Royal Society of Canada* (Ottawa: Royal Society of Canada, 1998), 1–13, 33, chapters 11–12, reprinted from "The Most Select and the Most Democratic: A Century of Science in the Royal Society of Canada," *Scientia Canadensis* **20** (1996), 3–99; Robert Daley and Paul Dufour, "Creating a 'Northern Minerva': John William Dawson and the Royal Society of Canada," *Scientia Canadensis* **5** (January, 1981), 3–14; the writing of the RSC's history was itself partitioned: see Andrée Desilets, *L'Académie des lettres et des sciences de la Société du Canada: un siècle d'histoire* (Ottawa: la Société Royale du Canada, 1997) and Carl Berger, *Honour and the Search for Influence: A History of the Royal Society of Canada* (Toronto: University of Toronto Press, 1996), esp. chapters 6–9; Yves Gingras, *Pour l'avancement des sciences: histoire de l'ACFAS, 1923–1993* (Montreal: Les editions de Boréal, 1994).

[31] Léon Lortie, "La trame scientifique de l'histoire du Canada," in G.F.G. Stanley, ed., *Pioneers of Canadian Science* (Toronto: University of Toronto Press, 1966), 3–35; Chartrand, Duchesne, and Gingras, *Histoire des Sciences au Québec*, 463; see also Fournier, Gingras, and Keel, "Introduction," in Fournier, Gingras, and Keel, eds., *Sciences et médecine*, 10–11; Léon Lortie, "Les sciences à Montréal et à Québec au XIXième siècle," *L'action universitaire* (February, 1936), 46–47, IX; Jacques Rousseau, "La Botanique canadienne á l'époque de Jacques Cartier," *Annales de l'ACFAS* **3** (1937), 151–236; Léon Lortie and Adrien Plouffe, eds., *Aux Sources du Présent* (Toronto: University of Toronto Press, 1960); Jacques Rousseau, "Des naturalistes à la découverte du Canada au XIXe siècle," *Cahiers des dix* **28** (1963), 179–208; Jacques Rousseau, "Pierre Boucher, naturaliste et géographe," in Pierre Boucher, *Histoire véritable et naturelle* (1664) reprint ed. (Boucherville: Societé Historique de Boucherville, 1964), 264, 277–78; Cyrias Ouellet, *The Sciences in French Canada* (Quebec: Department of Cultural Affairs, 1967).

a dearth of places for growing numbers of deserving anglophones could not outweigh the "practical injustice" that the few "*scientistes*" in French-Canadian society at the time "need not apply."[32]

It clarifies matters to temper this sociological overview – which insists that, "Prior to 1927," without the requisite scientific infrastructures in place, "there were no French-Canadian scientists in the scientific sections" of the RSC – with the historical details of individual experiences. On the one hand, the senior GSC officers A.R.C. Selwyn (1824–1902), Robert Bell (1841–1917), and G.M. Dawson (1849–1901) repeatedly sought out geological field research by their colleague, the naturalist Abbé J.-C.-K. Laflamme (1849–1910) of Laval University, after his appointment as a founding member of Section IV (Biological and Geological Sciences). Bell also strongly supported Laflamme's nomination of the naturalist Abbé Léon Provancher (1820–1892) as an RSC fellow in 1887: in third place after a close three-way split of the vote to fill two vacancies, Provancher won out over the McGill physiologist T. Wesley Mills (1847–1915) when the RSC Council urged Section IV to render a firm decision.[33]

On the other hand, when Laflamme and the science educator Abbé T.-E. Hamel (1830–1913), rector of Laval University, attended the inaugural RSC meetings of Sections IV (Biological and Geological Sciences) and III (Mathematical and Physical Sciences), respectively, they soon found themselves caught in the crosshairs of a critical Toronto press that accused the entire Quebec delegation of causing a "schism" by abandoning the proceedings "in a huff." One alleged reason was their objection to "heterodox" comments about evolution by their close friend and former colleague, the American geologist and chemist Thomas Sterry Hunt (1826–1892), in his opening address to Section III on "The Relations of the Natural Sciences." A first report announced that Hamel, Laflamme, and others had quit the RSC altogether; a follow-up modified this charge only slightly a few days later: "the fact remains that they withdrew from the meeting to consider their position, and did not return." The trumped-up story, which spread immediately to Montreal and Quebec, posed as much of a quandary for the RSC president, J.W. Dawson, as it did for Hamel, who had actually returned to Quebec with several colleagues to tend to their administrative responsibilities at the seminary. In response, Dawson and Hamel

[32] Levere, *Research and Influence*, 11, 32–33, 73 n. 18; Marcel Fournier, *L'entrée dans la modernité: science, culture et société au Québec* (Montreal: Les éditions Saint-Martin, 1986); Fournier, Gingras, and Keel, "Introduction," in Fournier, Gingras, and Keel, eds., *Sciences et médecine*, 9–18; Levere and Jarrell, "General Introduction," 24; V.-A. Huard, "Les sections scientifiques de la Société Royale du Canada et les canadiens-français," *Le naturaliste canadien*, **39** (August, 1912), 17–27; Mélanie Desmeules, "Huard, Victor-Alphonse," *DCB*, Vol. 15, accessed February 7, 2014, http://www.biographi.ca/en/bio/huard_victor_alphonse_15E.html; see also Suzanne Zeller, "Reflections on Time and Place."

[33] Levere, *Research and Influence*, 33–34. Archive du Séminaire de Québec [ASQ]: U62 no. 52, 2 June 1883; U60 no. 38, 20 June 1883; U60 no. 47, 29 July 1887 ff. Raymond Duchesne, "Laflamme, Joseph-Clovis-Kemner," *DCB*, Vol. 13, accessed February 9, 2014 http://www.biographi.ca/en/bio/laflamme_joseph_clovis_kemner_13E.html; Jean-Marie Perron, "Provancher, Léon," *DCB*, Vol. 12, accessed February 9, 2014, http://www.biographi.ca/en/bio/provancher_leon_12E.html; Royal Society of Canada *Proceedings and Transactions* **6** (1888), vi–viii, xxxvi.

contacted various newspaper editors with separate denials, rather than issuing a strong joint statement.[34]

One is reminded that a longstanding discomfort belied this new RSC proximity; and that, for its science fellows, at least ("les Dawson et les Lawson," to borrow Raymond Duchesne's delightful phrasing), RSC fellowship was not about boosting Canadian unity. Dawson the Presbyterian fundamentalist had long before committed himself to a "double-handed [linguistic and religious] fight" that precluded any alliance with these Quebec colleagues, despite their shared anti-Darwinian convictions. For Hamel's part, and despite some pleasant RSC experiences here and there over the years, in 1889 he described the meetings as "cahin-caha, comme d'habitude" [problematic, as usual]. The Literary and Historical Society of Quebec (founded 1824) suffered similar problems, in that its dwindling francophone membership felt uncomfortable at meetings and receptions that were only nominally bilingual. These tensions increased exponentially with the Great War and its aftermath, and may help to explain not only why the botanist Frère Marie-Victorin (1885–1944) allowed himself to be elected to the RSC's francophone literary Section I in 1924, but also why he and others then suggested that Section I create its own francophone subsection for science. It is difficult otherwise to comprehend these decisions, especially since the much-admired and influential Marie-Victorin, a founder of ACFAS, moved to Section V (Life Sciences, created in 1918) in 1927.[35]

It may, as a result, be time to draw finer lines of comparison, and to ask different sorts of questions. What factors, for example, fired the interest of Provancher, Laflamme, Hamel, Huard, Marie-Victorin, and their colleagues in science? One suspects that at least some of them shared Taché's understanding of its active partnership in statecraft. And while Yves Gingras's pioneering analysis of X-rays' introduction in Quebec contrasts the "production of knowledge" at McGill University with its "reproduction" at Laval University in 1895, McGill had shown itself in a starkly different light only 5 years earlier. In seeking a successor for Alexander Johnson (1830–1912) as Professor of Natural Philosophy in 1890, it rejected an application from the physicist James Gordon MacGregor (1852–1913), an active experimental researcher in thermodynamics at Dalhousie University who

[34] "From the Capital: Heterodoxy in the 'Royal Society': A 'Royal Society' Rumpus," Toronto *Daily Mail* (30 May, 1882); the Toronto *Globe* added a second explanation, about infighting within the French-Canadian delegation, 2 June. 1882; "Notes From the Capital: The Alleged Split in the 'Royal Society'," Toronto *Daily Mail* (3 June, 1882); Thomas Sterry Hunt, "The Relations of the Natural Sciences," RSC *Proceedings and Transactions* **1** (1882) Section III: 1–7; ASQ: MS 33, 1 June 1882, 155; U83 No. 33, n.d. J.W. Dawson quoted in Daley and Dufour, 7.

[35] Levere, *Research and Influence*, 33–34; Berger, *Honour and the Search for Influence*, 89; cf. Duchesne. "Problèmes," 30; and Peter J. Bowler, "The Early Development of Scientific Societies in Canada," in Alexandra Oleson and S.C. Brown, eds., *The Pursuit of Knowledge in the Early American Republic* (Baltimore: The Johns Hopkins University Press, 1976), 326–39; Zeller, "Reflections on Time and Place"; Frère Marie-Victorian, *Science, culture et nation*, ed. Yves Gingras (Montreal: Boréal, 1996); ASQ, Journal du Séminaire: IV, 18 May, 1895, 511 and V, 27 May, 1889, 153; Zeller, "Environment, Culture, and the Reception of Darwin," 93–94. Ginette Bernatchez, "La Société Littéraire et Historique de Québec (The Literary and Historical Society of Quebec) 1824–1890," *Revue d'Histoire de l'Amérique française* **35** (1981), 184–85.

went on to succeed the eminent mathematical physicist Peter Guthrie Tait (1831–1901) at the University of Edinburgh. Instead, McGill preferred the more traditionally Cambridge-educated John Cox (1851–1923), who lacked any research training whatsoever.[36]

In another example, one would be hard-pressed to imagine a contrast greater than that between two particular papers presented in the RSC's Section III (Mathematical and Physical Sciences): Hamel's philosophical "Essai sur la constitution atomique de la matière" (1884) and Ernest Rutherford's (1871–1937) experimental report on "The Existence of Bodies Smaller Than Atoms" (1902), the very notion of which Hamel had declared inconceivable and therefore physically impossible. Yet it also changes things to know that Section III's president, the chemist R.F. Ruttan (1856–1930) of McGill University, focused his own address that year on "[John] Dalton [1766–1844] and His Theory of Atoms." Rutherford – who became McGill's Macdonald Professor of Experimental Physics in 1898 and an RSC fellow in 1900 – inspired his audience at the RSC session to an unusually animated discussion. Ruttan's paper did not appear in the RSC *Transactions*. Nor did Hamel attend the RSC meetings that year, so that his response remains, regrettably, impossible to gauge.[37]

In this context, David Livingstone's spatial recognition in science of "an imaginary unity masking the disparate kinds of activity that trade under the label" has found a useful temporal counterpart in the historian John V. Pickstone's analysis of a sequence of *Ways of Knowing* in the history of science. "If we seek an explanation for the overall form" of a "scientific project," Pickstone argued, "then we must look to the contingencies of social life and to the politics of knowledge as favouring certain modes over others." Exploring a dynamic interplay among successive Natural Historical, Analytical, and Experimentalist Ways of Knowing in science, he emphasized historical context as much as socio-economic structure to explain their particular life-cycles. While each mode in its turn displaced – but emphatically did not replace – its predecessor as a dominant Way of Knowing, Pickstone pointed out, it continued even in decline to contribute to the production of scientific knowledge. By embedding in time and place the "changing boundaries and relations between different ways of knowing," which he depicted as nested one inside the other, Pickstone's insightful "displacement model" points up the need to elucidate in Canadian history just how, when, and why each Way of Knowing did – or did not – become dominant. As Levere demonstrated, the RSC with its elitist structure

[36] Carroll, *Science, Culture, and Modern State Formation,* 167; Gingras, "La réception des rayons X"; Yves Gingras, *Physics and the Rise of Scientific Research in Canada,* trans. Peter Keating (Montreal and Kingston: McGill University Press, 1991), 20; Zeller, "Reflections on Time and Place."

[37] T.-E. Hamel, "Essai sur la constitution atomique de la matière," RSC *Proceedings and Transactions* **2** (1884) Section III, 91–100; E. Rutherford, "The Existence of Bodies Smaller Than Atoms," RSC *Proceedings and Transactions*, ser.2, **8** (1902) Section III, 79–86; "Report of Section III," RSC *Proceedings and Transactions*, ser. 2, 8 (1902), XXXVI; see also Lawrence Badash, "The Influence of New Zealand on Rutherford's Scientific Development," in Reingold and Rothenberg, eds., *Scientific Colonialism*, 379–89.

assumed the role of gatekeeper for Experimentalist Ways of Knowing in science – whose accession to dominance coincided with the RSC's founding – a trend reinforced more widely during the twentieth century through the National Research Council, by many of the same people in both institutions.[38]

15.4 Conclusion

The history of science in Canada remains, in important ways, very much an under-researched field of study. Many of its fundamental questions remain unresolved, especially regarding the historical relationship between science and the state, with all of the complex cultural and contextual questions entailed therein. While the task remains daunting, the end of the twentieth century may have left us better positioned to historicize a persistent need to dichotomize amateur and professional, pure and applied, and – perhaps most important in the Canadian context – Ways of Knowing in science as valid or not. Among the perils of ignoring this history, we remain open not only to misunderstanding, but also to politically- and ideologically inspired misrepresentations, a situation made all the more serious by an increasingly present-minded public, government, and even elite cultural institutions that all too seldom factor history into their thinking, except for purely ideological reasons.[39]

There are, however, promising beginnings that historians of Canadian science can follow. Instead of assuming the traditional dualities and discontinuities that we have learned to take for granted, we could re-situate the evidence along a spectrum with room for nuances, gradations, and mixed positions. Abbé V.-A. Huard's sceptical response in 1912 to claims that "the grand scientific effort" of his day had "made the world a happier and better place" reminds us that other Canadians have, in various times and places, expressed similar doubts.[40] How much – and what kind of – science has seemed too much?

---

[38] Livingstone, 14–15; J.V. Pickstone, *Ways of Knowing: A New History of Science, Technology and Medicine* (Chicago: University of Chicago Press, 2000), 34–35, 157, and chapter 6; Marianne Ainley, *Restless Energy: A Biography of William Rowan 1891–1957* (Montreal: Véhicule Press, 1998); R.A. Jarrell and Yves Gingras, eds., *Building Canadian Science: The Role of the National Research Council* (Ottawa: CSTHA/AHSTC, 1991) and special issue of *Scientia Canadensis* **15** (1991), esp. Michel Girard, "The Commission of Conservation as a Forerunner to the National Research Council," 19–40. See also Suzanne Zeller, "Warp and Weft: The National Context(s) of Science in Canada," in David N. Livingstone, Ron Numbers, and Hugh Slotten, eds., *Science in National, International, and Global Context*, Cambridge History of Science Series, forthcoming.

[39] Simon Schaffer, "What Is Science?" in John Krige and Dominique Pestre, eds., *Companion to Science in the Twentieth Century* (London: Routledge, 2006), 27–44; Edward Grant, *A History of Natural Philosophy* (Cambridge: Cambridge University Press, 2007), 319–22; Donald J.C. Philippson, "The Steacie Myth and the Institutions of Industrial Research," *HSTC Bulletin 7* (September, 1983),117–34; Berger, *Honour and the Search for Influence*, 128.

[40] Huard, "Les sections scientifiques," 19 (author's translation); Berger, *Honour and the Search for Influence*, 24.

Indications for the nineteenth- and twentieth centuries point insistently to industrialization as key to these discussions. Pickstone's Analytical Ways of Knowing, which displaced traditional descriptive and classificatory Natural Historical *encounters* with nature in British North America after about 1815, sought to apply the information acquired over past centuries to *subdue* nature, to break it down into units (whether elements, atoms, cells, species, or energy) as components of larger systems or processes that could be standardized, controlled, and commodified. The goal of Analytical Ways of Knowing was "rationalized production," epitomized, as Pickstone elaborated, by industrialization as the modern path to economic, social, and even moral progress.[41] In Canada a generation of "revisionist" historians has finally shown French-Canadian society during the nineteenth- and twentieth centuries more open to these modernizing processes than those among their predecessors who took its formal ultramontane ideology at face value had been willing to acknowledge. As a new generation of "post-revisionist" historians fills in more of the cultural aspects, new light reveals J.-C. Taché's generation – many of whom had been committed Patriotes – resorting to an ultramontane spiritual revival in recoiling from the shattering experiences of Britain's heavy-handed responses to the Rebellions of 1837 and their aftermath. The Durham Report and its political consequences (including the reunion of Upper and Lower Canada as a more easily controlled economic unit, under a new constitution that also proscribed French as an official language) played out during that same generation's lifetime, with Taché a leader among those who faced down the enraged arsonists who torched the parliament building as well as Reform leaders' homes during the Montreal riots against the Rebellion Losses Bill in 1849.[42]

These events coincided crucially with the rise to dominance in Canada of Analytical Ways of Knowing that began in earnest, albeit at differential rates, to serve the forces of modernization and state formation that Durham had mapped out

[41] Pickstone, *Ways of Knowing*, chapters 3–4; Carroll, *Science, Culture, and Modern State Formation,* 167; see also Crosbie Smith, *The Science of Energy: A Cultural History of Energy Physics in Victorian Britain* (Chicago: University of Chicago Press, 1998).

[42] Curtis, *Politics of Population*, 239–40; Pickstone, *Ways of Knowing*, chapters 3–4. The term "revisionist" is Ronald Rudin's, in *Making History in Twentieth-Century Quebec* (Toronto: University of Toronto Press, 1997), and the resulting discussion of his broader argument was politically charged: cf. Michael D. Behiels, "Normalizing' the Writing of Quebec History," *Left History* **6** (1999), 91–99; Ramsay Cook, review of Ronald Rudin, *Making History in Twentieth-Century Quebec, Histoire sociale* **32** (1999), 120–23; Yves Gingras, "Making Up History," *Literary Review of Canada* (Summer, 1999), 19–22. Serge Courville, *Rêves d'empire: le Québec et le rêve colonial* (Ottawa: les presses de l'Université d'Ottawa, 2000); Serge Courville, Jean-Claude Robert, and Normand Séguin, *Atlas historique du Québec. Le pays laurentien au XIXè siècle: les morphologies de base* (Sainte-Foy: les presses de l'Université Laval, 1995); Serge Courville, *Entre ville et campagne: l'essor du village dans les seigneuries du Bas-Canada* (Sainte-Foy: les presses de l'Université Laval, 1990); Paul-André Linteau, René Durocher, and Jean-Claude Robert, *Quebec: A History 1867–1929*, translated by Robert Chodos (Toronto: James Lorimer and Company, 1979; Louis Rousseau, "Crise et réveil religieux dans le Québec du XIXe siècle, *Interface* **11** (1990), 24–31; Louis Rousseau, "À propos du 'réveil religieux' dans le Québec du XIXe siècle: où se loge le vrai débat?" *Revue d'histoire de l'Amérique française* **49** (1995), 223–45; cf. Jarrell, "L'ultramontanisme."

for British North America. In so doing, his Report aimed to eradicate what he discarded as mere cultural hindrances, to ease this transition by substituting a figuratively "flattened and emptied space." With the subsequent rise to dominance of industrial research and Experimentalist Ways of Knowing in the context of the Great War and the bitter Conscription Crisis, it is little wonder that, a century later in 1944, the pioneering Canadian ecologist and Quebec educator Pierre Dansereau (1911–2011) also thought in terms of dichotomies where science was concerned. Dansereau rejected, as the obvious alternative to Quebec's "humanistic" educational system, what he saw around him throughout North America as "a purely utilitarian conception of science" losing sight of the fact that "civilization … is to be measured not by numbers, but by spirit." It was not science that ultramontanes and their twentieth-century successors abhorred, but rather its aggressive identification with the homogenizing political, economic, social, and cultural demands of industrialization.[43]

The longstanding discursive partition in the history of Canadian science might best be seen to constitute a subset of the larger imperial/colonial divide, and there is room now to move beyond the restrictions of both approaches to explore some of the important themes suggested by Trevor Levere's explorations. The decision by Oxford University Press to publish *A Curious Field Book* as its introductory volume in a projected series of studies in Canadian social history constitutes a reminder of the many directions into which social history has since exploded as a field of study. While class, gender, race, and ethnicity were only beginning to influence the study of Canadian history in 1974, cultural history had yet to bring to bear its postmodern toolbox of critical theoretical insights, along with the vocabularies for discussing them. Not only has the field also been exceptionally well-served by research databases, but recent scholarship on the history of the book may be able finally to identify through booksellers' records the "many men of University education" who, the physical scientist J.H. Lefroy (1817–1890) had marvelled during the 1840s, were "scattered about on Farms in this country" while reading in isolation important works including Pierre Simon de Laplace's *Méchanique céleste* (5 vols., 1749–1825). Given as well the rise of scholarly enthusiasm for Canadian environmental history since then, there is all the more for historians of Canadian science to contribute from any and all of these various perspectives, clarifying science – including Darwinian theories of evolution – as a crucial basis for the historical understanding

[43] Curtis, *Politics of Population*, 253–54; Pickstone, *Ways of Knowing*, chapter 6; Carroll, *Science, Culture, and Modern State Formation,* 166; Berger, *Honour and the Search for Influence*, chapter 6 and p. 124; Pierre Dansereau, "Science in French Canada," *Scientific Monthly* **59** (September, 1944), 188–194 and **59** (October, 1944) 261–72, esp. 263–64; cf. Jarrell, "Differential National Development," 347.

of nature and environment, with its inextricable connections to issues of land usage and territoriality.[44]

Social and cultural history's welcome interdisciplinary approaches to the history of Canadian science underline a continuing need to decentre their subject matter by recognizing the (always) mutual inflections involved in the production and application of science as a form of knowledge. The historian Natalie Zemon Davis lays out some options and implications for the "seismic shift ... that jolts European perspectives even more than the relocation through considerations of class and gender." Where apparent differences suggest polarized outlooks, one could instead investigate "hybrid" examples that occupy spaces in between. Davis and others also recommend "symmetrical" comparisons that eschew universal models while respecting "strong and concrete cultural differences and the importance of divergent context." The ultimate aim of this "radical contemporaneity" is to "decolonize modernity" itself, acknowledging the historical relationships of "coproduction" between science, industry, and the state, on the one hand, while listening more attentively for variety in the experiences of this process, on the other hand. The imagined outcome, to paraphrase Davis, would resemble a "braided history" of Canadian science, retaining intact the individual strands that strengthen our understanding of the whole. Perhaps we should have listened to Léon Lortie after all, and looked more closely at the interwoven threads that he too had envisioned.[45] To Trevor Levere we owe thanks, for laying the groundwork by teaching us to pose the questions about how things in their own times and places were thought to work, and why.

---

[44] Levere and Jarrell, eds., *A Curious Field Book*, back cover; Carroll, *Science, Culture, and Modern State Formation*, 167; R.A. Richardson and B.H. MacDonald, *Science and Technology in Canadian History: A Bibliography of Primary Sources to 1914* (Thornhill, Ontario: HSTC Publications, 1987); B.H. MacDonald, *Science and Technology in Canadian History: A Bibliographic Database* (http://acsweb2.ucis.dal.ca/slis/main.htm); *Early Canadiana Online* (http://canadiana.org); *Dictionary of Canadian Biography Online* (http://www.biographi.ca); Lefroy quoted in Zeller, *Inventing Canada*, 150. See also Suzanne Zeller, "Exploration, Empire, and the Extractive Impulse: The Case of Victorian Geology," in Pierre Bélanger *et al.*, eds., *EXTRACTION* (Cambridge, MA: MIT Press, forthcoming).

[45] Davis, "Polarities, Hybridities," p. 19–32; Walter D. Mignolo, *The Darker Side of the Renaissance: Literacy, Territoriality, and Colonization* (Ann Arbor: The University of Michigan Press, 2003), esp. the Introduction; Carroll, *Science, Culture, and Modern State Formation*, 164–74; Jeff Horn, *The Path Not Taken: French Industrialization in the Age of Revolution 1750–1830* (Cambridge, Massachusetts: The MIT Press, 2006); Lortie, "La trame scientifique de l'histoire du Canada"; see also Jane Jacobs, *Systems of Survival: A Dialogue on the Moral Foundations of Commerce and Politics* (New York: Vintage Books, 1994). For a "hybrid" example and the power struggles involved, see Robert Gagnon, "Les discours sur l'enseignement pratique au Canada français: 1850–1900," in Fournier, Gingras, and Keel, eds., *Sciences et médecine*, 19–40; René Durocher and P.-A. Linteau, eds., *Le "retard" du Québec et l'infériorité économique des canadiens français* (Trois Rivières: Boréal, 1971) clarifies the context within which universal models had to be shaken off by "revisionist" historians.

Index

J. Buchwald, L. Stewart (eds.), *The Romance of Science: Essays in Honour of Trevor H. Levere*, Archimedes 52, DOI 10.1007/978-3-319-58436-2

S

T

U